Studies in Computational Intelligence

Volume 636

Series editor

Janusz Kacprzyk, Polish Academy of Sciences, Warsaw, Poland
e-mail: kacprzyk@ibspan.waw.pl

About this Series

The series "Studies in Computational Intelligence" (SCI) publishes new developments and advances in the various areas of computational intelligence—quickly and with a high quality. The intent is to cover the theory, applications, and design methods of computational intelligence, as embedded in the fields of engineering, computer science, physics and life sciences, as well as the methodologies behind them. The series contains monographs, lecture notes and edited volumes in computational intelligence spanning the areas of neural networks, connectionist systems, genetic algorithms, evolutionary computation, artificial intelligence, cellular automata, self-organizing systems, soft computing, fuzzy systems, and hybrid intelligent systems. Of particular value to both the contributors and the readership are the short publication timeframe and the worldwide distribution, which enable both wide and rapid dissemination of research output.

More information about this series at http://www.springer.com/series/7092

Sundarapandian Vaidyanathan
Christos Volos
Editors

Advances and Applications in Chaotic Systems

Springer

Editors
Sundarapandian Vaidyanathan
Research and Development Centre
Vel Tech University
Chennai
India

Christos Volos
Department of Physics
Aristotle University of Thessaloniki
Thessaloniki
Greece

ISSN 1860-949X ISSN 1860-9503 (electronic)
Studies in Computational Intelligence
ISBN 978-3-319-30278-2 ISBN 978-3-319-30279-9 (eBook)
DOI 10.1007/978-3-319-30279-9

Library of Congress Control Number: 2016932742

Printed on acid-free paper

This Springer imprint is published by Springer Nature
The registered company is Springer International Publishing AG Switzerland

Preface

About the Subject

Chaos theory is a field of study in mathematics with several applications in science and engineering. Chaotic systems are nonlinear dynamical systems and maps that are highly sensitive to initial conditions. The sensitivity to initial conditions is usually called the *butterfly effect* for dynamical systems and maps.

Chaotic systems can be observed in many natural systems such as weather and climate. Chaos theory has applications in several areas such as vibration control, electric circuits, chemical reactions, lasers, combustion engines, computers, cryptosystems, encryption, secure communications, biology, medicine, management, finance, etc. Chaotic behaviour of systems can be modelled by discrete-time or continuous-time mathematical models.

About the Book

The new Springer book, *Advances and Applications in Chaotic Systems,* consists of 25 contributed chapters by subject experts who are specialized in the various topics addressed in this book. The special chapters have been brought out in this book after a rigorous review process in the broad areas of modelling and application of chaotic systems. Special importance was given to chapters offering practical solutions and novel methods for the recent research problems in the modelling and application of chaotic systems.

This book discusses trends and applications of chaos modeling and chaotic systems in science and engineering.

Objectives of the Book

The objective of this book takes a modest attempt to cover the framework of advances and applications of chaotic systems in a single volume. The book is not only a valuable title on the publishing market, it is also a successful synthesis of control techniques applied to chaotic systems. Several multidisciplinary applications of chaotic systems in control, engineering and information technology are discussed in this book.

Organization of the Book

This well-structured book consists of 25 full chapters.

Book Features

- The chapters deal with the recent research problems in the areas of chaos theory, chaos modelling and applications.
- The chapters contain a good literature survey with a long list of references.
- The chapters are well written with a good exposition of the research problem, methodology and block diagrams.
- The chapters are lucidly illustrated with numerical examples and simulations.
- The chapters discuss details of engineering applications and future research areas.

Audience

The book is primarily meant for researchers from academia and industry, who are working in the research areas—chaos theory, control engineering, computer science and information technology. The book can also be used at the graduate or advanced undergraduate level as a textbook or major reference for courses such as nonlinear dynamical systems, control systems, mathematical modelling, computational science, numerical simulation and many others.

Acknowledgements

As the editors, we hope that the chapters in this well-structured book will stimulate further research in chaos theory, computational intelligence and control systems and utilize them in real-world applications.

We hope sincerely that this book, covering so many different topics, will be very useful for all readers.

We would like to thank all the reviewers for their diligence in reviewing the chapters.

Special thanks go to Springer, especially the book editorial team.

Sundarapandian Vaidyanathan
Christos Volos

Contents

Synchronization Phenomena in Coupled Hyperchaotic Oscillators with Hidden Attractors Using a Nonlinear Open Loop Controller

Ch.K. Volos, V.-T. Pham, S. Vaidyanathan, I.M. Kyprianidis and I.N. Stouboulos

Abstract In recent years the study of dynamical systems with hidden attractors, namely systems in which their basins of attraction do not intersect with small neighborhoods of equilibria, is a great challenge due to their application in many research fields such as in mechanics, secure communication and electronics. Especially, the investigation of hyperchaotic systems with hidden attractors plays a crucial role in this research approach. Motivated by the very complex dynamical behavior of hyperchaotic systems and the unusual features of hidden attractors, a bidirectionally and unidirectionally coupling scheme of systems of this family, by using a nonlinear open loop controller, is studied in this chapter. For this reason, a recently new proposed hyperchaotic system with hidden attractors, the four-dimensional modified Lorenz system, which is structurally the simplest hyperchaotic system with hidden attractors, is used. The simulation results show that the proposed scheme drives the coupled system either to complete synchronization or anti-synchronization depending on the choice of the signs of the error function's parameters. In addition, an electronic circuit emulating the control scheme of the coupled hyperchaotic systems with hidden attractors is also presented to verify the feasibility of the proposed model.

Ch.K. Volos (✉) · I.M. Kyprianidis · I.N. Stouboulos
Physics Department, Aristotle University of Thessaloniki, GR-54124 Thessaloniki, Greece
e-mail: volos@physics.auth.gr

I.M. Kyprianidis
e-mail: imkypr@auth.gr

I.N. Stouboulos
e-mail: stouboulos@physics.auth.gr

V.-T. Pham
School of Electronics and Telecommunications, Hanoi University of Science and Technology, 01 Dai Co Viet, Hanoi, Vietnam
e-mail: pvt3010@gmail.com

S. Vaidyanathan
Research and Development Centre, Vel Tech University, Avadi, Chennai 600062, Tamil Nadu, India
e-mail: sundarvtu@gmail.com

S. Vaidyanathan and C. Volos (eds.), *Advances and Applications in Chaotic Systems*, Studies in Computational Intelligence 636,
DOI 10.1007/978-3-319-30279-9_1

Keywords Chaos · Hidden oscillation · Complete synchronization · Anti-synchronization · Bidirectional coupling · Unidirectional coupling · Nonlinear open loop controller

1 Introduction

In the last three decades the phenomenon of synchronization between coupled chaotic systems has attracted the interest of the scientific community because it is a rich and multi-disciplinary phenomenon with broad range applications, such as in secure communications [19] and cryptography [14, 60], in broadband communications systems [7] and in a variety of complex physical, chemical, and biological systems [17, 37, 41, 51, 54, 57, 62]. In general, synchronization of chaos is a process, where two or more chaotic systems adjust a given property of their motion to a common behavior, such as equal trajectories or phase locking, due to coupling or forcing. Because of the exponential divergence of the nearby trajectories of a chaotic system, having two chaotic systems being synchronized, might be a surprise. However, today the synchronization of coupled chaotic oscillators is a phenomenon well established experimentally and reasonably well understood theoretically.

The history of chaotic synchronization's theory began with the study of the interaction between coupled chaotic systems in the 1980s and early 1990s by Fujisaka and Yamada [11], Pikovsky [49], Pecora and Carroll [48]. Since then, a wide range of research activity based on synchronization of nonlinear systems has risen and a variety of synchronization's forms depending on the nature of the interacting systems and of the coupling schemes has been presented. Complete or full chaotic synchronization [9, 24–26, 28, 39, 55, 63], phase synchronization [8, 45], lag synchronization [52, 56], generalized synchronization [53], anti-synchronization [22, 36], anti-phase synchronization [1, 5, 6, 27, 58, 64], projective synchronization [38], anticipating [61] and inverse lag synchronization [34] are the most interesting types of synchronization, that have been investigated numerically and experimentally by many research groups.

This work is referred to complete synchronization and to anti-synchronization. In the first case, which is the most studied type of synchronization, two identical coupled chaotic systems leads to a perfect coincidence of their chaotic trajectories i.e., $x_1(t) = x_2(t)$ as $t \rightarrow \infty$. In the anti-synchronization, on the other hand, which is also a very interesting type of synchronization, two systems x_1 and x_2, can be synchronized in amplitude, but with opposite sign, for initial conditions chosen from large regions in the phase space, that is $x_1(t) = -x_2(t)$ as $t \rightarrow \infty$.

As it is known, nonlinear systems and especially chaotic systems exhibit high sensitivity on initial conditions or system's parameters and thus, if they are identical and, possibly, starting from almost the same initial conditions, following trajectories which rapidly become uncorrelated. For this reason, many techniques have been set up to obtain the aim of chaotic synchronization. So, many of these techniques to couple two or more nonlinear chaotic systems can be mainly divided into two classes: bidirectional or mutual coupling and unidirectional coupling [13]. In the mutual coupling both the systems are connected and each system's dynamic behavior influences the dynamics of the other, while on the contrary in unidirectional coupling, only the first system drives the second one.

Recently, a great interest for dynamical systems with hidden attractors has been raised. The term "*hidden attractor*" is referred to the fact that in this class of systems the attractor is not associated with an unstable equilibrium and thus often remains undiscovered because it may occur in a small region of parameter space and with a small basin of attraction in the space of initial conditions [23, 31–33, 46, 47]. In 2010, for the first time, a chaotic hidden attractor was discovered in the most well-known nonlinear circuit, in Chua's circuit, which is described by a three-dimensional dynamical system [23, 31].

The problem of analyzing hidden oscillations arose for the first time in the second part of Hilbert's 16th problem (1900) for two-dimensional polynomial systems [16]. The first nontrivial results were obtained in Bautin's works [2, 3], which were devoted to constructing nested limit cycles in quadratic systems and showed the necessity of studying hidden oscillations for solving this problem.

Later, in the middle of the 20th century, Kapranov studied [21] the qualitative behavior of Phase-Locked Loop (PLL) systems, which are used in telecommunications and computer architectures, and estimated stability domains. In that work, Kapranov assumed that in PLL systems there were self-excited oscillations only. However, in 1961, Gubar' [15] revealed a gap in Kapranov's work and showed analytically the possibility of the existence of hidden oscillations in two-dimensional system of PLL, thus, from a computational point of view, the system considered was globally stable, but, in fact, there was only a bounded domain of attraction.

Also, in the same period, the investigations of widely known Markus–Yamabe [40] and Kalman [20] conjectures on absolute stability have led to the finding of hidden oscillations in automatic control systems with a unique stable stationary point and with a nonlinearity, which belongs to the sector of linear stability [4, 10, 30].

Furthermore, systems with hidden attractors have received attention due to their practical and theoretical importance in other scientific branches, such as in mechanics (unexpected responses to perturbations in a structure like a bridge or in an airplane wing) [18, 29]. So, the study of these systems is an interesting topic of a significant importance.

In this work a hyperchaotic four-dimensional modified Lorenz system with hidden attractors, is used for studying the bidirectional or unidirectional coupling by using the nonlinear open loop controller. The simulation results from system's numerical integration as well as the circuital implementation of the proposed system in SPICE, confirm the appearance of complete synchronization and anti-synchronization phenomena depending on the signs of the parameters of the error functions.

The chapter is organized as follows. In Sect. 2 the four-dimensional modified Lorenz system, which is used in this work, is presented. The scheme, by using the nonlinear open loop controller, in both coupling ways (bidirectional and unidirectional) as well as the simulation results are discussed in Sect. 3. Section 4 presents the circuital implementation of the unidirectional coupling system and the simulation results which are obtained by using SPICE. Finally, the conclusive remarks are drawn in the last section.

2 The Four-Dimensional Modified Lorenz System

In this work the simplest four-dimensional hyperchaotic Lorenz-type system, which has been proposed by Gao and Zhang [12], is used. This system is an extension of a modified Lorenz system, which was studied by Schrier and Maas as well as by Munmuangsaen and Srisuchinwong [42, 59]. The proposed system is described by the following set of differential equations.

$$\begin{cases} \dot{x} = y - x \\ \dot{y} = -xz + u \\ \dot{z} = xy - c \\ \dot{u} = -dy \end{cases} \tag{1}$$

It is structurally a very simple four-dimensional dynamical system having only two independent parameters (c, d). Also, as it is mentioned in [35], it has many interesting properties not found in other proposed systems, such as:

(i) It has very few terms, only seven with two quadratic nonlinearities, and two parameters.
(ii) All its attractors are hidden.
(iii) It exhibits hyperchaos over a large region of parameter space.
(iv) Its Jacobian matrix has rank less than four everywhere in the space of the parameters.
(v) It exhibits a quasi-periodic route to chaos with an attracting torus for some choice of parameter values.
(vi) It has regions in which the torus coexists with either a symmetric pair of strange attractors or a symmetric pair of limit cycles and other regions where three limit cycles coexist.
(vii) The basins of attraction have an intricate fractal structure.
(viii) There is a series of Arnold tongues [43] within the quasi-periodic region where the two fundamental oscillations mode-lock and form limit cycles of various periodicities.

All the afore-mentioned reasons make the dynamical system (1) an ideal candidate for the coupling scheme which is used in this work. Especially, the existence of hidden attractors and the hyperchaotic nature of a system like this have played a crucial role in our decision.

In this section the system's dynamic behavior is investigated numerically by employing a fourth order Runge–Kutta algorithm. For this reason, the bifurcation diagram, which is a very useful tool from nonlinear theory, is used. In Figs. 1, 2, 3, 4, 5, 6 and 7 the bifurcation diagrams of the variable y versus the parameter d, for

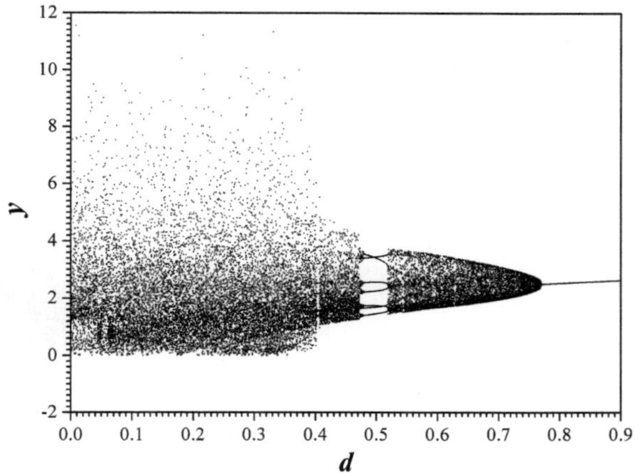

Fig. 1 Bifurcation diagram of y versus d for $c = 5$, with initial conditions $(x_0, y_0, z_0, u_0) = (0.55, -0.49, -0.08, 0.50)$

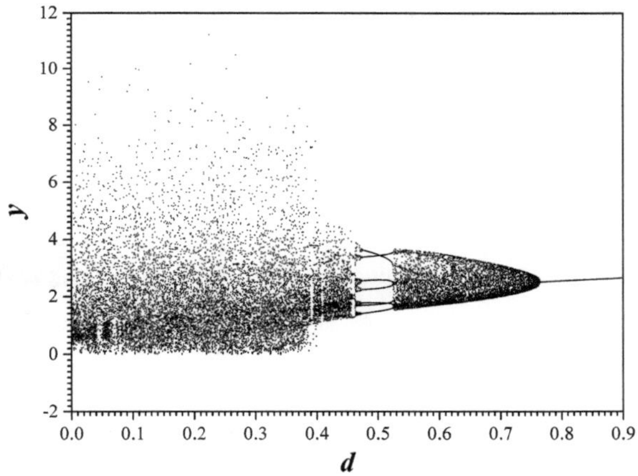

Fig. 2 Bifurcation diagram of y versus d for $c = 4$, with initial conditions $(x_0, y_0, z_0, u_0) = (0.55, -0.49, -0.08, 0.50)$

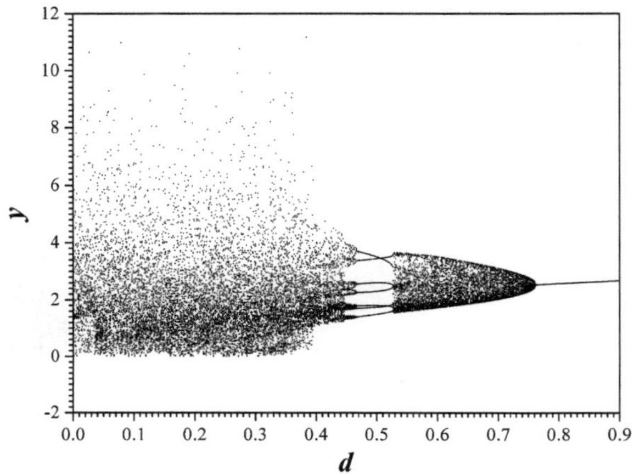

Fig. 3 Bifurcation diagram of y versus d for $c = 3.5$, with initial conditions $(x_0, y_0, z_0, u_0) = (0.55, -0.49, -0.08, 0.50)$

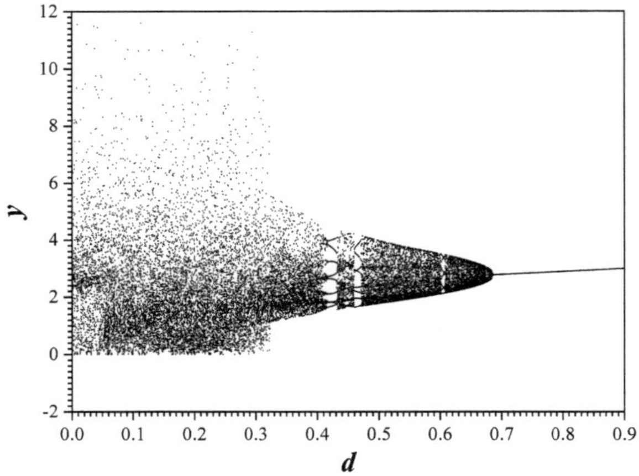

Fig. 4 Bifurcation diagram of y versus d for $c = 2.97$, with initial conditions (x_0, y_0, z_0, u_0) = (0.55, −0.49, −0.08, 0.50)

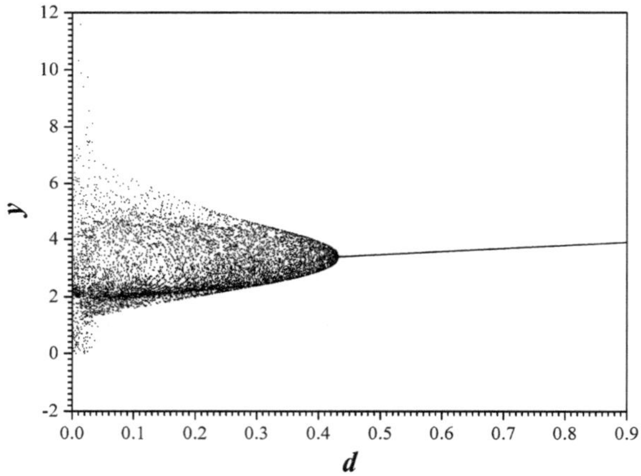

Fig. 5 Bifurcation diagram of y versus d for $c = 2.9$, with initial conditions (x_0, y_0, z_0, u_0) = (0.55, −0.49, −0.08, 0.50)

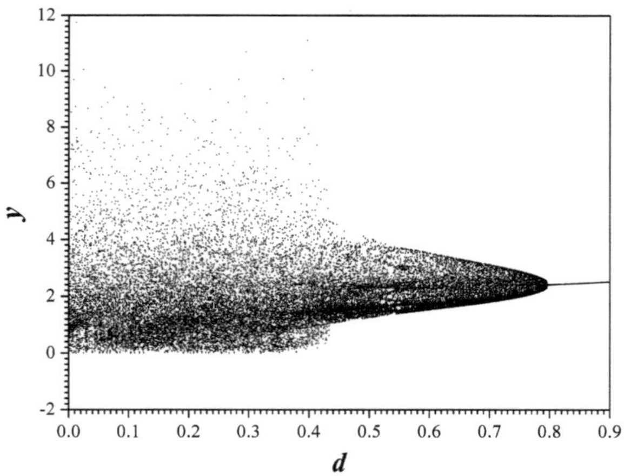

Fig. 6 Bifurcation diagram of y versus d, for $c = 2.7$, with initial conditions (x_0, y_0, z_0, u_0) = (0.55, −0.49, −0.08, 0.50)

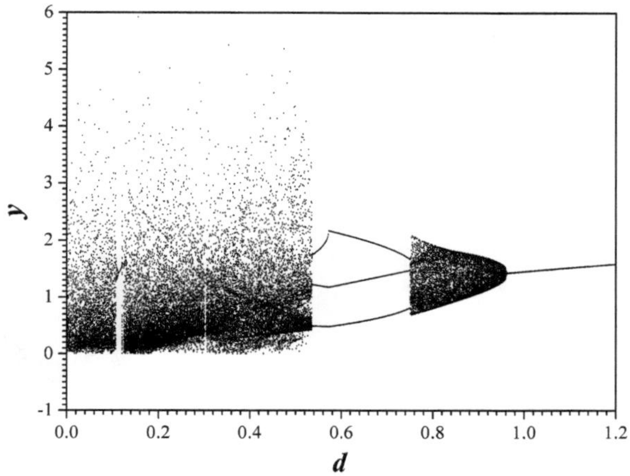

Fig. 7 Bifurcation diagram of y versus d, for $c = 1$, with initial conditions $(x_0, y_0, z_0, u_0) = (0.55, -0.49, -0.08, 0.50)$

various values of the parameter c, reveal the richness of system's dynamical behavior. Besides limit cycles, system (1) has quasi-periodicity, chaos, and hyperchaos, which can make the control of the system a difficult case in practical applications where a particular dynamic is desired. In more, details, as the value of d is decreased from $d = 0.9$ the system goes from a period-1 steady state (Fig. 8), through a quasi-periodic route (Figs. 9, 10, 11, 12 and 13), to a chaotic state, which is confirmed by the chaotic attractor in x–z plane, that is shown in Fig. 14. However, a very interesting feature of the specific system is the existence of hyperchaos for a range of parameters as it is shown in the phase portraits of Figs. 15, 16, 17, 18 and 19. Figure 20 shows

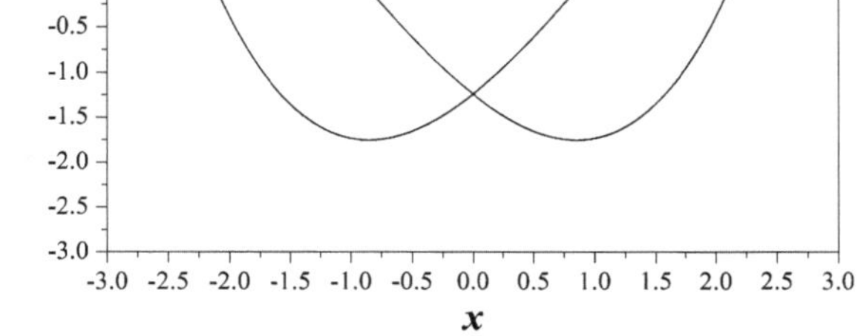

Fig. 8 Phase portrait of z versus x for $c = 2.7$ and $d = 0.9$ (period-1 state), with initial conditions $(x_0, y_0, z_0, u_0) = (0.55, -0.49, -0.08, 0.50)$

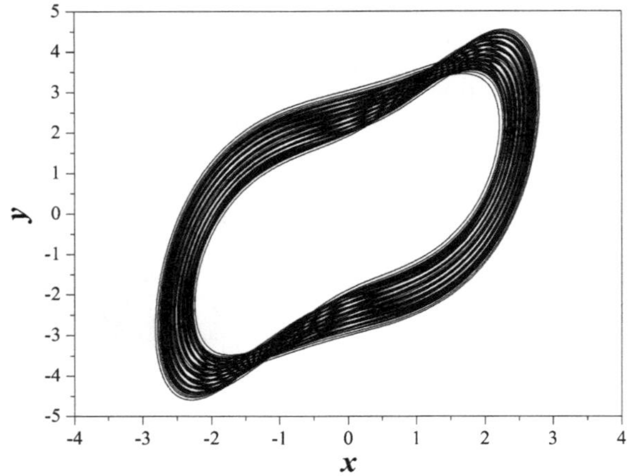

Fig. 9 Quasi-periodic attractor for $c = 2.7$ and $d = 0.75$, in x–y plane, with initial conditions $(x_0, y_0, z_0, u_0) = (0.55, -0.49, -0.08, 0.50)$

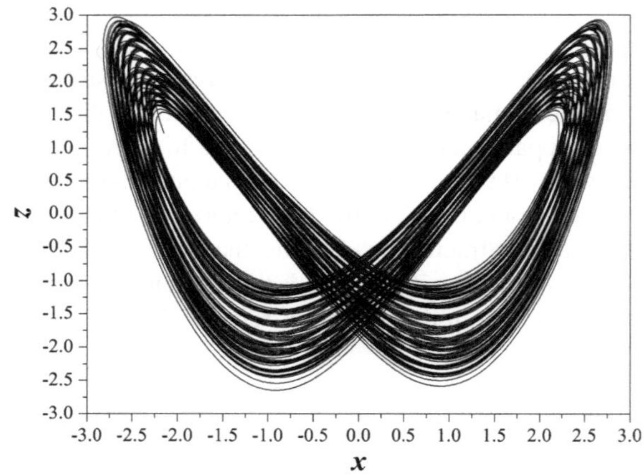

Fig. 10 Quasi-periodic attractor for $c = 2.7$ and $d = 0.75$, in x–z plane, with initial conditions $(x_0, y_0, z_0, u_0) = (0.55, -0.49, -0.08, 0.50)$

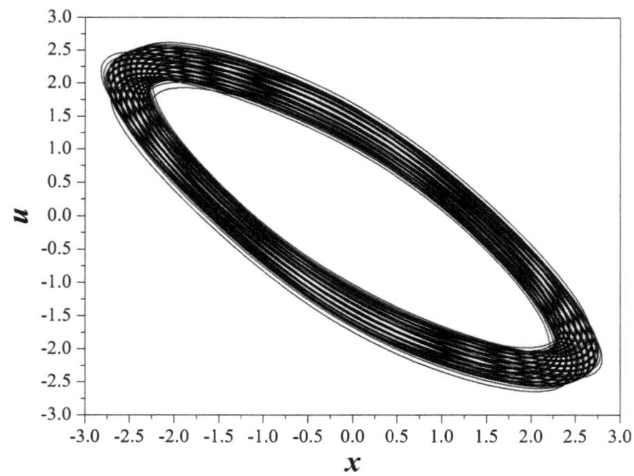

Fig. 11 Quasi-periodic attractor for $c = 2.7$ and $d = 0.75$, in x–u plane, with initial conditions $(x_0, y_0, z_0, u_0) = (0.55, -0.49, -0.08, 0.50)$

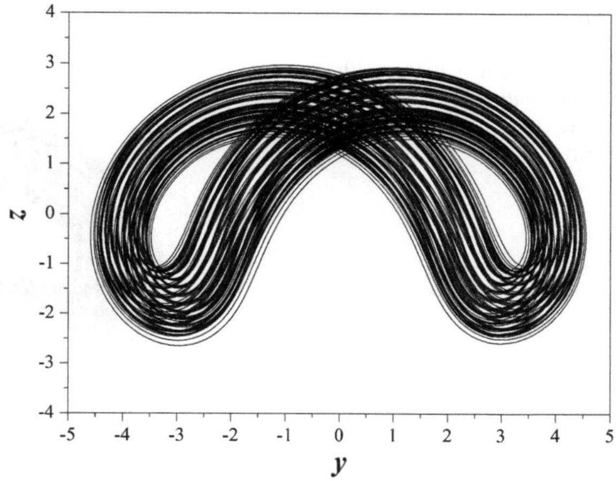

Fig. 12 Quasi-periodic attractor for $c = 2.7$ and $d = 0.75$, in y–z plane, with initial conditions (x_0, y_0, z_0, u_0) = (0.55, −0.49, −0.08, 0.50)

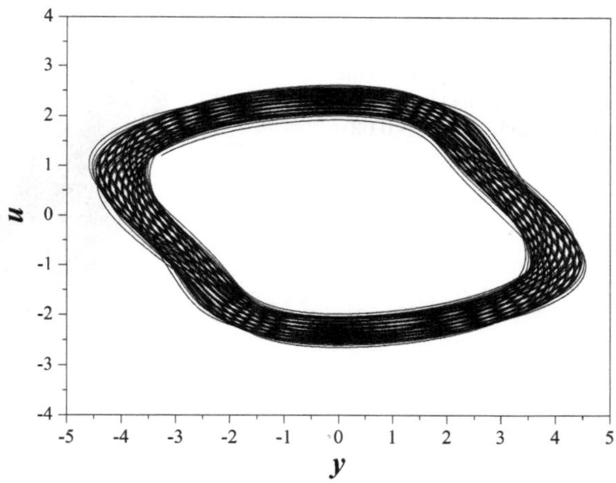

Fig. 13 Quasi-periodic attractor for $c = 2.7$ and $d = 0.75$, in y–u plane, with initial conditions (x_0, y_0, z_0, u_0) = (0.55, −0.49, −0.08, 0.50)

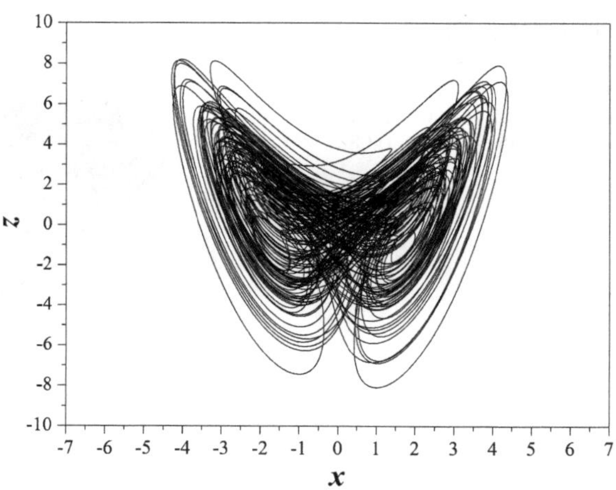

Fig. 14 Phase portrait of z versus x for $c = 2.7$ and $d = 0.2$ (chaotic state), with initial conditions (x_0, y_0, z_0, u_0) = (0.55, −0.49, −0.08, 0.50)

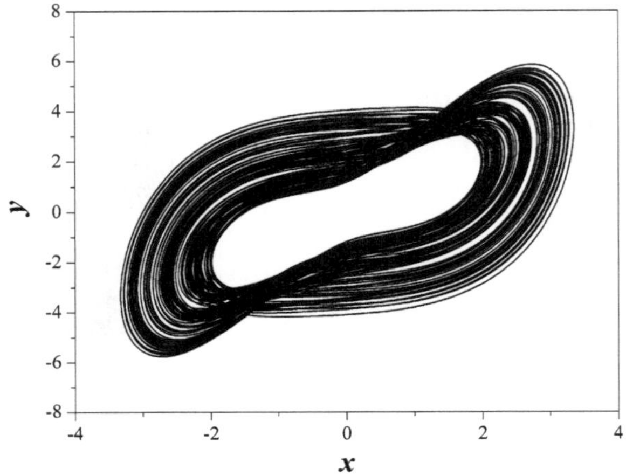

Fig. 15 Hyperchaotic attractor for $c = 2.7$ and $d = 0.44$, in x–y plane, with initial conditions $(x_0, y_0, z_0, u_0) = (0.55, -0.49, -0.08, 0.50)$

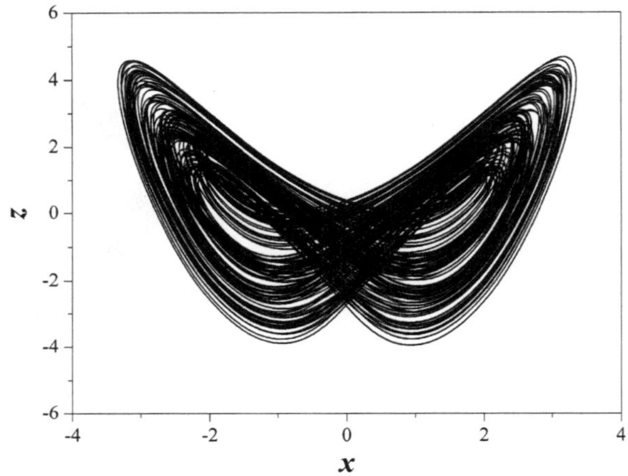

Fig. 16 Hyperchaotic attractor for $c = 2.7$ and $d = 0.44$, in x–z plane, with initial conditions $(x_0, y_0, z_0, u_0) = (0.55, -0.49, -0.08, 0.50)$

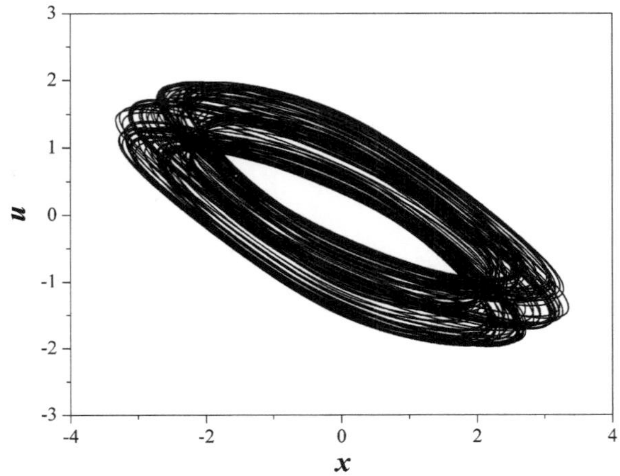

Fig. 17 Hyperchaotic attractor for $c = 2.7$ and $d = 0.44$, in x–u plane, with initial conditions $(x_0, y_0, z_0, u_0) = (0.55, -0.49, -0.08, 0.50)$

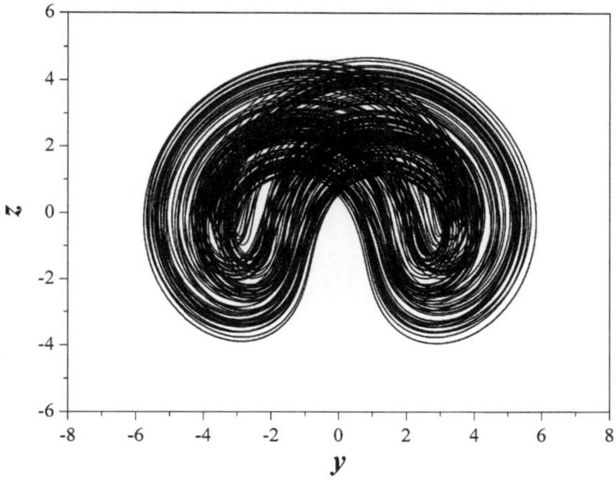

Fig. 18 Hyperchaotic attractor for $c = 2.7$ and $d = 0.44$, in y–z plane, with initial conditions $(x_0, y_0, z_0, u_0) = (0.55, -0.49, -0.08, 0.50)$

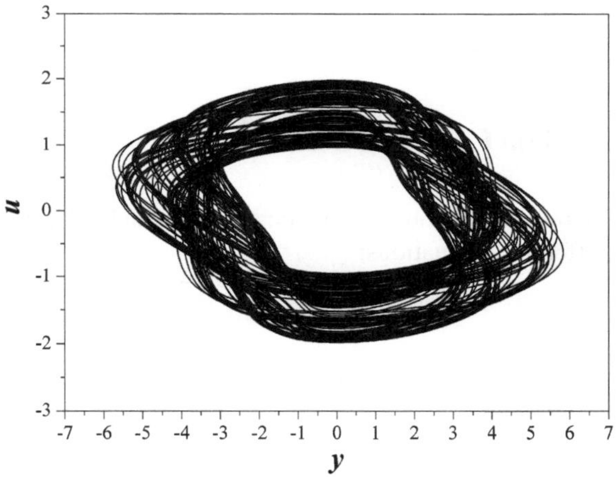

Fig. 19 Hyperchaotic attractor for $c = 2.7$ and $d = 0.44$, in y–u plane, with initial conditions $(x_0, y_0, z_0, u_0) = (0.55, -0.49, -0.08, 0.50)$

the Lyapunov exponents' spectra for chosen value of the parameter c ($c = 2.7$). The system's hyperchaotic behavior is found for $c = 2.7$ in the range of $d \in [0.388, 0.49]$ (Figs. 15, 16, 17, 18 and 19), where the system has two positive Lyapunov exponents, as it is shown in the embedded diagram in Fig. 20.

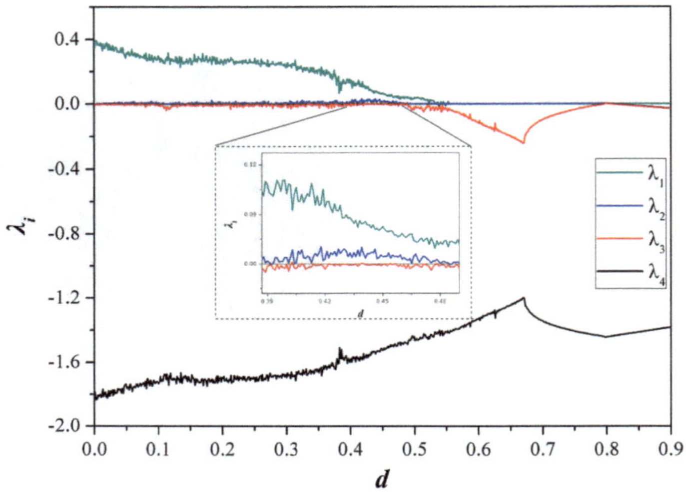

Fig. 20 The diagrams of Lyapunov exponents (λ_i) versus the parameter d, for $c = 2.7$

3 The Coupling Scheme

Two identical coupled chaotic systems can be described by the following system of differential equations:

$$\begin{cases} \dot{x} = f(x) + U_X \\ \dot{y} = f(y) + U_Y \end{cases} \tag{2}$$

where $(f(x), f(y)) \in R^n$ are the flows of the systems. The coupling of the systems is defined by the Nonlinear Open Loop Controllers (NOLCs), U_X and U_Y [44]. The error function is given by $e = \beta y - \alpha x$, where α and β are constants. If one applies the Lyapunov Function Stability (LFS) technique, a stable synchronization state will be realized when the error function of the coupled system follows the limit

$$\lim_{t \to \infty} ||e(t)|| \to 0 \tag{3}$$

so that $\alpha x = \beta y$.

As it is mentioned, the design process of the coupling scheme, is based on the Lyapunov function

$$V(e) = \frac{1}{2} e^T e \tag{4}$$

where T denotes transpose of a matrix and $V(e)$ is a positive definite function. For known system's parameters and with the appropriate choice of the controllers U_X and U_Y, the coupled system has $\dot{V}(e) < 0$. This ensures the asymptotic global stability of synchronization and thereby realizes any desired synchronization state.

By using the appropriate NOLCs functions U_X, U_Y and error function's parameters α, β, a unidirectional or bidirectional (mutual) coupling scheme can be implemented. In more details, for ($U_X = 0, \beta = 1$) or ($U_Y = 0, \alpha = 1$), a unidirectional coupling scheme is realized, while for $U_{X,Y} \neq 0$ and $\alpha, \beta \neq 0$, a bidirectional coupling scheme is realized, respectively. The signs of α, β play a crucial role to the type of synchronization (complete synchronization or anti-synchronization), which is observed in this work. On the other hand, the ratio of α over β decides the amplification or attenuation of one oscillator relative to another one.

Next, the results of the simulation process in the two coupling (bidirectional and unidirectional) schemes and for various values of parameters α and β are presented.

3.1 Bidirectional Coupling

Systems of chaotic oscillators bidirectionally (mutually) coupled are frequently found not only in the simulation environment or the laboratory but also in the natural world [41, 50]. This way of coupling, which is the simplest, is very interesting because it displays much of the phenomenology that is observed in more complex networks. Asymptotically stable synchronization between the coupled oscillators happens to be one of the basic phenomena that is observed.

As it is mentioned, the synchronization of coupled chaotic systems is a process where two or more systems adjust a given property of their motion to a common behavior, such as identical trajectories, due to coupling.

So, in the first case, the bidirectional coupling scheme of two coupled systems of Eq. (1), which is described by the following systems (5) and (6), is studied.

Coupled System-1:

$$\begin{cases} \dot{x}_1 = x_2 - x_1 + U_{X1} \\ \dot{x}_2 = -x_1 x_3 + x_4 + U_{X2} \\ \dot{x}_3 = x_1 x_2 - c + U_{X3} \\ \dot{x}_4 = -d x_2 + U_{X4} \end{cases} \tag{5}$$

Coupled System-2:

$$\begin{cases} \dot{y}_1 = y_2 - y_1 + U_{Y1} \\ \dot{y}_2 = -y_1 y_3 + y_4 + U_{Y2} \\ \dot{y}_3 = y_1 y_2 - c + U_{Y3} \\ \dot{y}_4 = -d y_2 + U_{Y4} \end{cases} \tag{6}$$

where $U_X = [U_{X1}, U_{X2}, U_{X3}, U_{X4}]^{\mathrm{T}}$ and $U_Y = [U_{Y1}, U_{Y2}, U_{Y3}, U_{Y4}]^{\mathrm{T}}$ are the NOLCs functions. The error function is defined by $\boldsymbol{e} = \beta \boldsymbol{y} - \alpha \boldsymbol{x}$, with $\boldsymbol{e} = [e_1, e_2, e_3, e_4]^{\mathrm{T}}$, $\boldsymbol{x} = [x_1, x_2, x_3, x_4]^{\mathrm{T}}$ and $\boldsymbol{y} = [y_1, y_2, y_3, y_4]^{\mathrm{T}}$. So, the errors dynamics, by taking the difference of Eqs. (5) and (6), are written as:

$$\begin{cases} \dot{e}_1 = e_2 - e_1 + \beta U_{Y1} - \alpha U_{X1} \\ \dot{e}_2 = \alpha x_1 x_3 - \beta y_1 y_3 + e_4 + \beta U_{Y2} - \alpha U_{X2} \\ \dot{e}_3 = -\alpha x_1 x_2 + \beta y_1 y_2 - c(\beta - \alpha) + \beta U_{Y3} - \alpha U_{X3} \\ \dot{e}_4 = -d e_2 + \beta U_{Y4} - \alpha U_{X4} \end{cases} \tag{7}$$

For stable synchronization $e \to 0$ as $t \to \infty$. By substituting the conditions in Eq. (7) and taking the time derivative of Lyapunov function

$$\begin{aligned} \dot{V}(e) &= e_1\dot{e}_1 + e_2\dot{e}_2 + e_3\dot{e}_3 + e_4\dot{e}_4 \\ &= e_1\left(e_2 - e_1 + \beta U_{Y1} - \alpha U_{X1}\right) \\ &\quad + e_2\left(\alpha x_1 x_3 - \beta y_1 y_3 + e_4 + \beta U_{Y2} - \alpha U_{X2}\right) \\ &\quad + e_3\left[-\alpha x_1 x_2 + \beta y_1 y_2 - c\left(\beta - \alpha\right) + \beta U_{Y1} - \alpha U_{X1}\right] \\ &\quad + e_4\left(-d e_2 + \beta U_{Y1} - \alpha U_{X1}\right) \end{aligned} \tag{8}$$

we consider the following NOLC controllers:

$$\begin{cases} U_{X1} = \frac{1}{2\alpha} e_2 \\ U_{X2} = \frac{1}{\alpha}\left(\alpha x_1 x_3 + e_2 + e_4\right) \\ U_{X3} = \frac{1}{\alpha}\left(-\alpha x_1 x_2 + e_3\right) \\ U_{X4} = \frac{1}{\alpha}\left(-\frac{d}{2} e_2 + e_4\right) \end{cases} \tag{9}$$

and

$$\begin{cases} U_{Y1} = -\frac{1}{2\beta} e_2 \\ U_{Y2} = \frac{1}{\beta}\left(\beta y_1 y_3\right) \\ U_{Y3} = \frac{1}{\beta}\left[-\beta y_1 y_2 + c(\beta - \alpha)\right] \\ U_{Y4} = \frac{1}{2\beta}\left(d e_2\right) \end{cases} \tag{10}$$

such that

$$\dot{V}(e) = -e_1^2 - e_2^2 - e_3^2 - e_4^2 < 0 \tag{11}$$

So, Eq. (11) ensures the asymptotic global stability of synchronization.

Next, the simulation results, in this coupling scheme, for three different cases of system's parameters (α, β), are presented.

3.1.1 The symmetric case ($\alpha = \beta$)

Firstly, the parameters α, β are chosen to be equal ($\alpha = \beta = 1$). This is the most studied type of mutual coupling and also the most interesting due to its applications in a variety of scientific fields. Also, by choosing, in this case, the systems' parameters as $c = 2.7$ and $d = 0.44$, each one of the coupled systems is in a hyperchaotic state. In this case of coupled identical systems with the proposed coupling scheme, only the complete synchronization is observed. This type of synchronization is confirmed

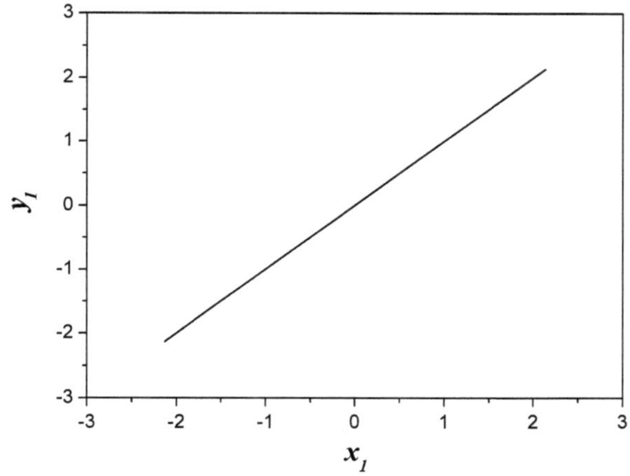

Fig. 21 The phase portrait of y_1 versus x_1, for $\alpha = \beta = 1, c = 2.7$ and $d = 0.44$

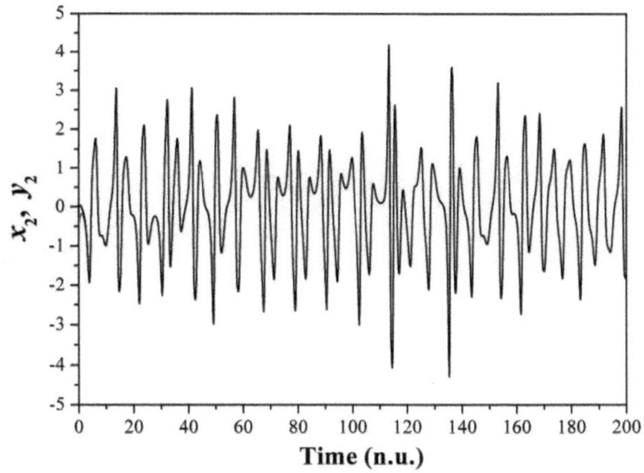

Fig. 22 The time-series of x_2, y_2, for $\alpha = \beta = 1$, $c = 2.7$ and $d = 0.44$

by the y_1 versus x_1 plot of Fig. 21. Furthermore, the time-series of the variables x_2, y_2 as well as the errors e_i ($i = 1, 2, 3, 4$) show the exponential convergence to zero which confirms the expected system's complete synchronization (Figs. 22 and 23).

3.1.2 The case $\alpha = 2, \beta = 1$

In this case, the parameters of the error functions are chosen to be $\alpha = 2$ and $\beta = 1$. By choosing again the systems' parameters as $c = 2.7$, $d = 0.44$ and for $\alpha = 2$ the hyperchaotic attractor of the second system is enlarged by two times, as it is shown with red color in Fig. 24, as well as by the time-series of signals y_1 and y_2 in regard to the signals x_1 and x_2 respectively (Figs. 26 and 27). The y_1 versus x_1 plot in Fig. 25 confirms that the coupled system is in complete synchronization state independently

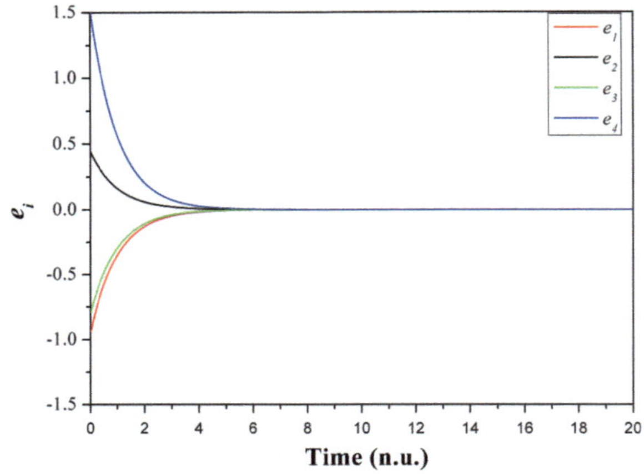

Fig. 23 The plot of errors $e_i(=\beta y_i - \alpha x_i)$, for $\alpha = \beta = 1$, $c = 2.7$ and $d = 0.44$

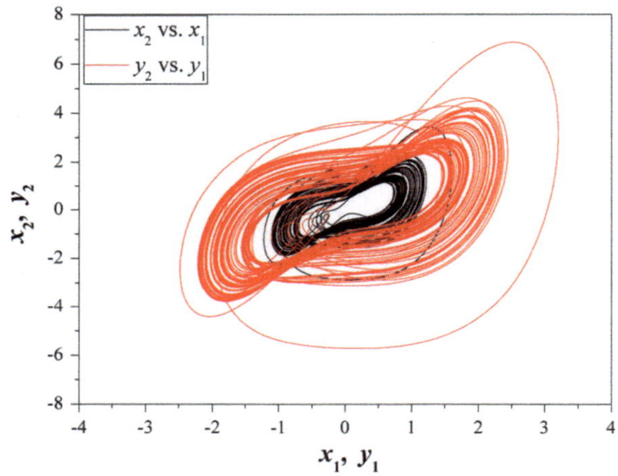

Fig. 24 The phase portraits of x_2 versus x_1 and y_2 versus y_1, for $\alpha = 2$, $\beta = 1$, $c = 2.7$ and $d = 0.44$

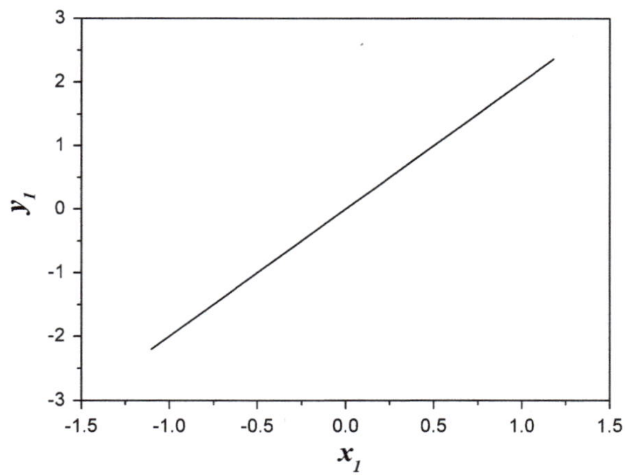

Fig. 25 The phase portrait of y_1 versus x_1, for $\alpha = 2$, $\beta = 1$, $c = 2.7$ and $d = 0.44$

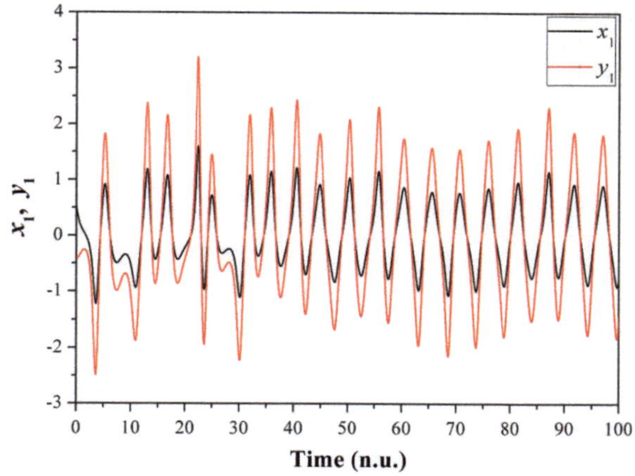

Fig. 26 The time-series of x_1, y_1, for $\alpha = 2, \beta = 1$, $c = 2.7$ and $d = 0.44$

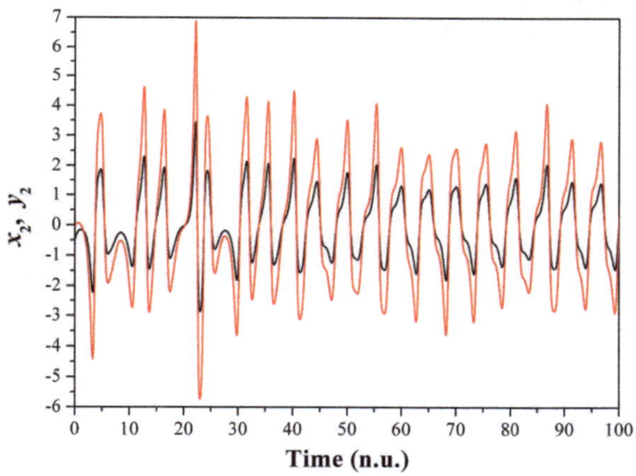

Fig. 27 The time-series of x_2, y_2, for $\alpha = 2, \beta = 1$, $c = 2.7$ and $d = 0.44$

of the values of the error's parameters α, β. The error plot $e_i = y_i - 2x_i$ $(i = 1, 2, 3, 4)$ in Fig. 28 shows the exponential convergence to zero that confirms the realization of system's complete synchronization state.

3.1.3 The Case $\alpha = -1, \beta = 2$

By choosing the parameters of the error functions as $\alpha = -1$ and $\beta = 2$, the attractor of the first coupled system has been enlarged by factor two, while the attractor of the second coupled system has been inverted in regard to the first one, as it is shown in Fig. 29. In this case the systems' parameters are chosen again as $c = 2.7$ and $d = 0.44$ so as both of the coupled systems are in hyperchaotic state. This process is shown more clearly in the plots of the time-series of x_1, y_1 and x_2, y_2 (Figs. 31 and 32). The phase portrait of y_1 versus x_1 in Fig. 30 indicates that the

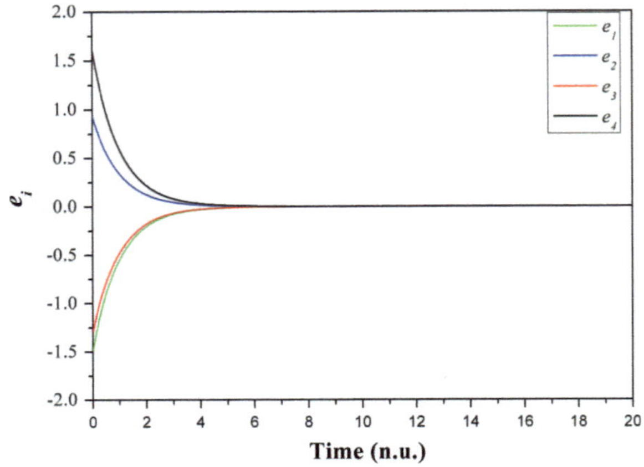

Fig. 28 The plot of errors $e_i(=\beta y_i - \alpha x_i)$, for $\alpha = 2$, $\beta = 1$, $c = 2.7$ and $d = 0.44$

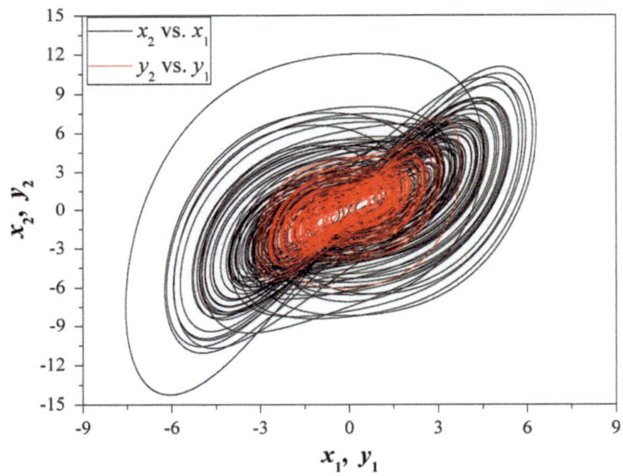

Fig. 29 The phase portraits of x_2 versus x_1 and y_2 versus y_1, for $\alpha = -1$, $\beta = 2$, $c = 2.7$ and $d = 0.44$

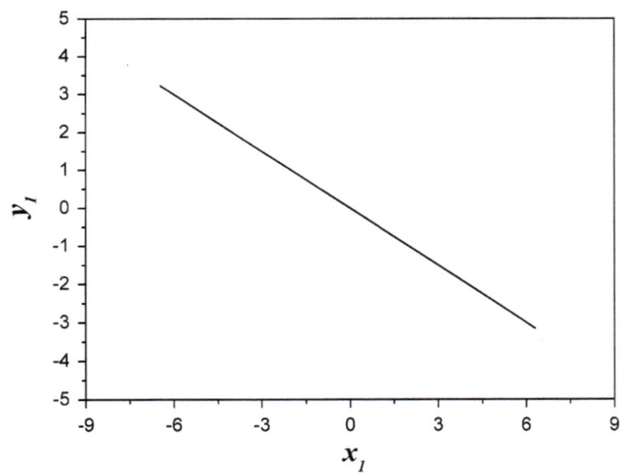

Fig. 30 The phase portrait of y_1 versus x_1, for $\alpha = -1$, $\beta = 2$, $c = 2.7$ and $d = 0.44$

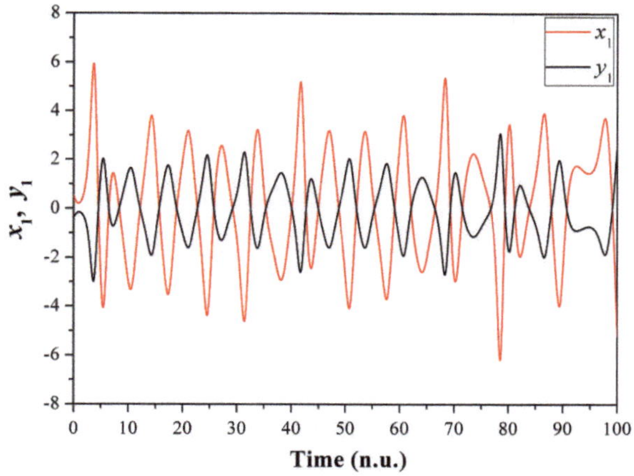

Fig. 31 The time-series of x_1, y_1, for $\alpha = -1$, $\beta = 2$, $c = 2.7$ and $d = 0.44$

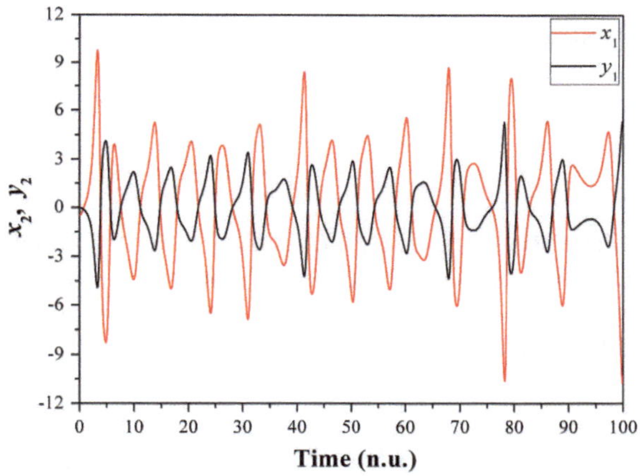

Fig. 32 The time-series of x_2, y_2, for $\alpha = -1$, $\beta = 2$, $c = 2.7$ and $d = 0.44$

coupled system is in anti-synchronization state, which is also confirmed by the error plot $e_i = 2y_1 + x_1$ $(i = 1, 2, 3, 4)$ in Fig. 33.

3.2 *Unidirectional Coupling*

In this section, the unidirectional coupling scheme, $U_X = 0$, for $\beta = 1$, given by Eq. (1), is presented.

Master System:

$$\begin{cases} \dot{x}_1 = x_2 - x_1 \\ \dot{x}_2 = -x_1 x_3 + x_4 \\ \dot{x}_3 = x_1 x_2 - c \\ \dot{x}_4 = -d x_2 \end{cases} \tag{12}$$

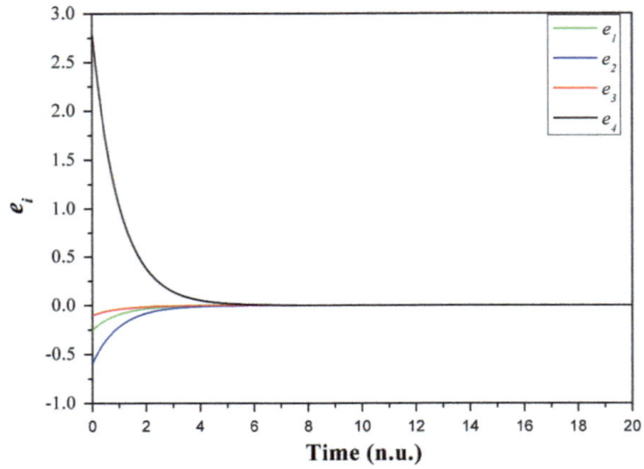

Fig. 33 The plot of errors $e_i(=\beta y_i - \alpha x_i)$, for $\alpha = -1$, $\beta = 2$, $c = 2.7$ and $d = 0.44$

Slave System:

$$\begin{cases} \dot{y}_1 = y_2 - y_1 + U_{Y1} \\ \dot{y}_2 = -y_1 y_3 + y_4 + U_{Y2} \\ \dot{y}_3 = y_1 y_2 - c + U_{Y3} \\ \dot{y}_4 = -d y_2 + U_{Y4} \end{cases} \tag{13}$$

where $\boldsymbol{U}_Y = [U_{Y1}, U_{Y2}, U_{Y3}, U_{Y4}]^{\mathrm{T}}$ are the Nonlinear Open Loop Controller (NOLC). The error function is defined by $\boldsymbol{e} = \beta \boldsymbol{y} - \alpha \boldsymbol{x}$, with $\boldsymbol{e} = [e_1, e_2, e_3, e_4]^{\mathrm{T}}$, $\boldsymbol{x} = [x_1, x_2, x_3, x_4]^{\mathrm{T}}$ and $\boldsymbol{y} = [y_1, y_2, y_3, y_4]^{\mathrm{T}}$. So, the error dynamics, by taking the difference of Eqs. (12) and (13), are written as:

$$\begin{cases} \dot{e}_1 = e_2 - e_1 + \beta U_{Y1} \\ \dot{e}_2 = \alpha x_1 x_3 - \beta y_1 y_3 + e_4 + \beta U_{Y2} \\ \dot{e}_3 = -\alpha x_1 x_2 + \beta y_1 y_2 + c(\alpha - \beta) + \beta U_{Y3} \\ \dot{e}_4 = -d e_2 + \beta U_{Y4} \end{cases} \tag{14}$$

For stable synchronization $e \to 0$ with $t \to \infty$. By substituting the conditions in Eq. (14) and taking the time derivative of Lyapunov function

$$\begin{aligned} \dot{V}(e) &= e_1 \dot{e}_1 + e_2 \dot{e}_2 + e_3 \dot{e}_3 + e_4 \dot{e}_4 \\ &= e_1 (e_2 - e_1 + \beta U_{Y1}) + e_2 (\alpha x_1 x_3 - \beta y_1 y_3 + e_4 + \beta U_{Y2}) \\ &\quad + e_3 [-\alpha x_1 x_2 + \beta y_1 y_2 + c (\alpha - \beta) + \beta U_{Y3}] + e_4 (-d e_2 + \beta U_{Y4}) \end{aligned} \tag{15}$$

we consider the following NOLC controllers

$$\begin{cases} U_{Y1} = -\frac{1}{\beta}e_2 \\ U_{Y2} = -\frac{1}{\beta}\left(e_2 + \alpha x_1 x_3 - \beta y_1 y_3 + e_4\right) \\ U_{Y3} = -\frac{1}{\beta}\left[e_3 - \alpha x_1 x_2 + \beta y_1 y_2 + c(\alpha - \beta)\right] \\ U_{Y4} = -\frac{1}{\beta}\left(e_4 - d e_2\right) \end{cases} \tag{16}$$

such that

$$\dot{V}(e) = -e_1^2 - e_2^2 - e_3^2 - e_4^2 < 0 \tag{17}$$

Equation (17) ensures the asymptotic global stability of synchronization.

3.2.1 The Case $\alpha = \beta = 1$

In this case, as it occurs in the mutual coupling, the phenomenon of complete synchronization is achieved for every value of $\alpha = \beta$. Especially, for $\alpha = \beta = 1$, the two coupled systems are in the same hyperchaotic state, due to the chosen values of system's parameters ($c = 2.7$ and $d = 0.44$). The goal of complete synchronization is achieved as it is shown from the plots of y_1 versus x_1, the time-series of x_2, y_2 and the errors e_i in Figs. 34, 35 and 36.

3.2.2 The Case for $\alpha = -\beta = -1$

By using opposing values for the parameters $\alpha = -\beta = -1$ the phenomenon of anti-synchronization is achieved, as it is shown in Fig. 37. Initially, the coupled systems are in different hyperchaotic states but the unidirectional coupling leads the slave system to an opposite hyperchaotic attractor in regard to the master system. This conclusion is derived from the phase portrait of y_1 versus x_1 (Fig. 37), as well as from the

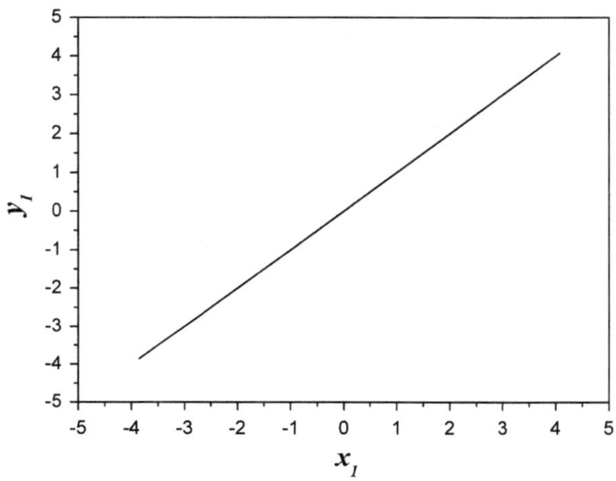

Fig. 34 The phase portrait of y_1 versus x_1, for $\alpha = \beta = 1$, $c = 2.7$ and $d = 0.44$

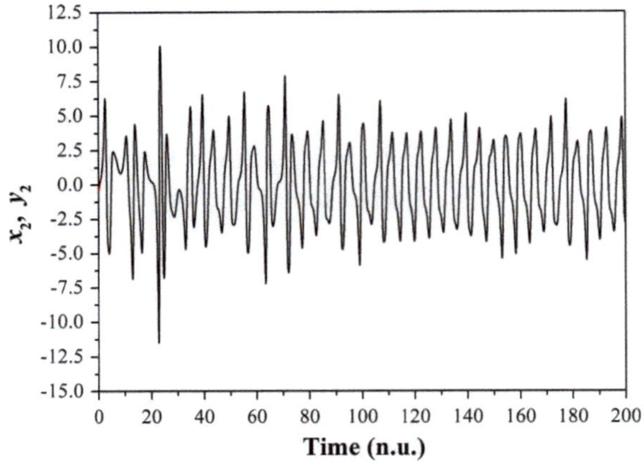

Fig. 35 The time-series of y_2, x_2, for $\alpha = \beta = 1$, $c = 2.7$ and $d = 0.44$

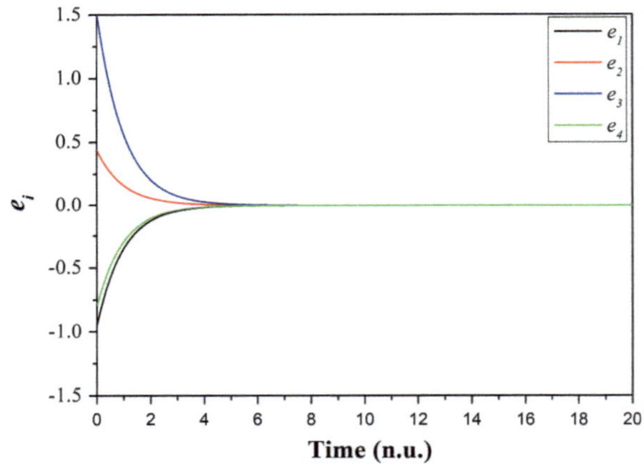

Fig. 36 The plot of errors $e_i(=\beta y_i - \alpha x_i)$, for $\alpha = \beta = 1$, $c = 2.7$ and $d = 0.44$

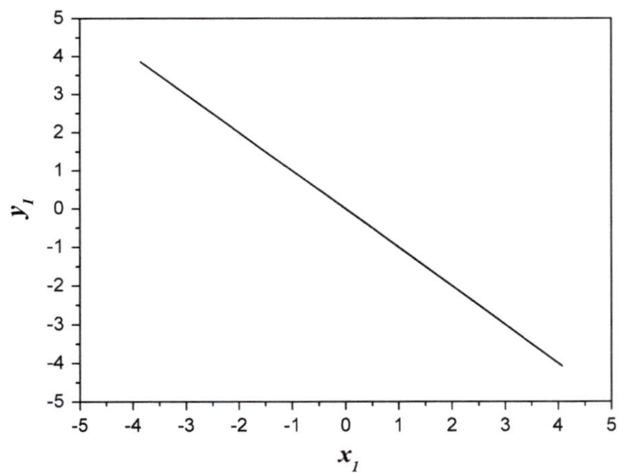

Fig. 37 The phase portrait of y_1 versus x_1, for $\alpha = -\beta = -1$, $c = 2.7$ and $d = 0.44$

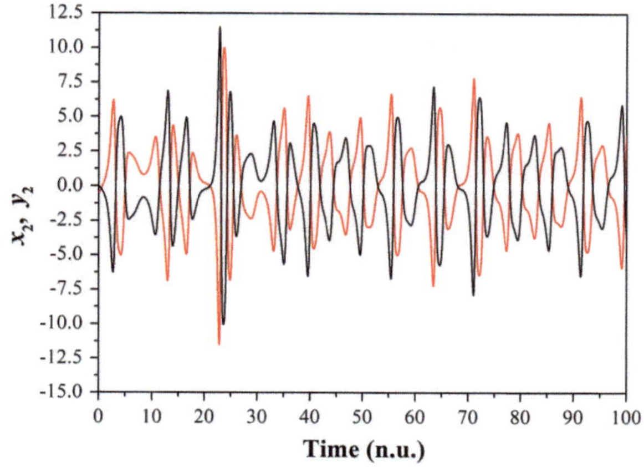

Fig. 38 The time-series of y_2, x_2, for $\alpha = -\beta = -1$, $c = 2.7$ and $d = 0.44$

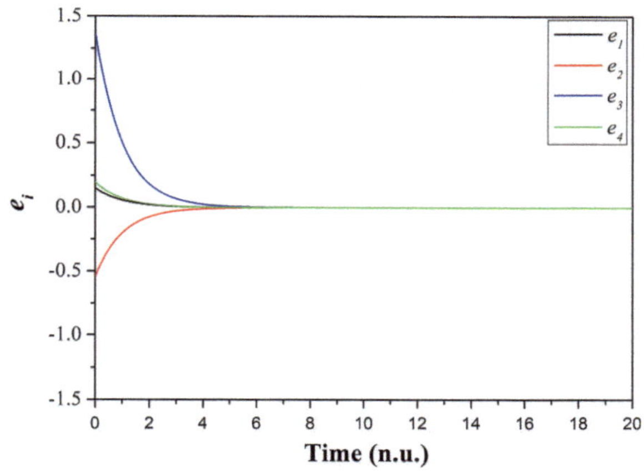

Fig. 39 The plot of errors $e_i (= \beta y_i - \alpha x_i)$, for $\alpha = -\beta = -1$, $c = 2.7$ and $d = 0.44$

time-series of x_2, y_2 (Fig. 38). Also, the plot of errors $e_i = y_i + x_i$ in Fig. 39 confirms the anti-synchronization of the coupled system.

3.2.3 The Case $\alpha = 2$, $\beta = 1$

In this case, the parameters of the error functions are chosen as $\alpha = 2$ and $\beta = 1$. By choosing the systems' parameters as $c = 2.7$, $d = 0.44$ and for $\alpha = 2$ the chaotic attractor of the second system is enlarged by two times, as it is shown with red color in Fig. 40, as well as by the time-series of signals y_1 and y_2 in regard to the signals x_1 and x_2 respectively (Figs. 42 and 43). The y_1 versus x_1 plot in Fig. 41 confirms that the coupled system is in complete synchronization state independently of the values of the error's parameters α, β. The error plot $e_i = y_1 - 2x_1$ $(i = 1, 2, 3, 4)$ in Fig. 44

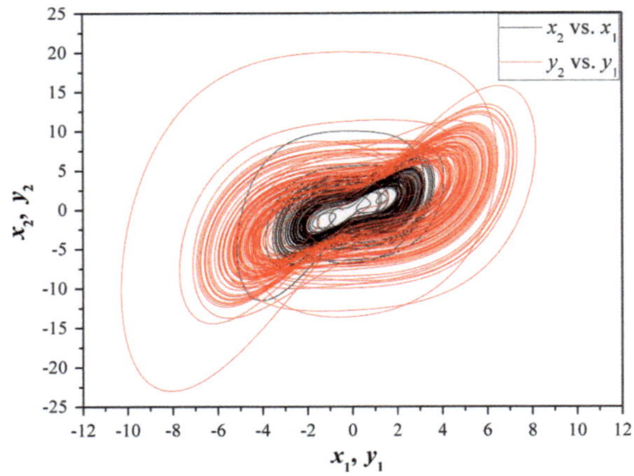

Fig. 40 The phase portraits of x_2 versus x_1 and y_2 versus y_1, for $\alpha = 2$, $\beta = 1$, $c = 2.7$ and $d = 0.44$

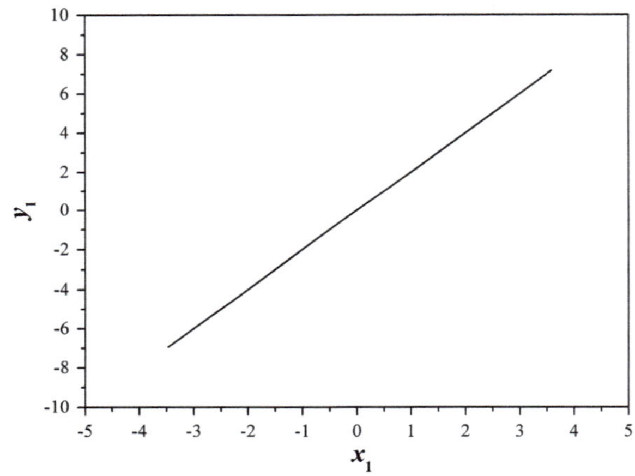

Fig. 41 The phase portrait of y_1 versus x_1, for $\alpha = 2$, $\beta = 1$, $c = 2.7$ and $d = 0.44$

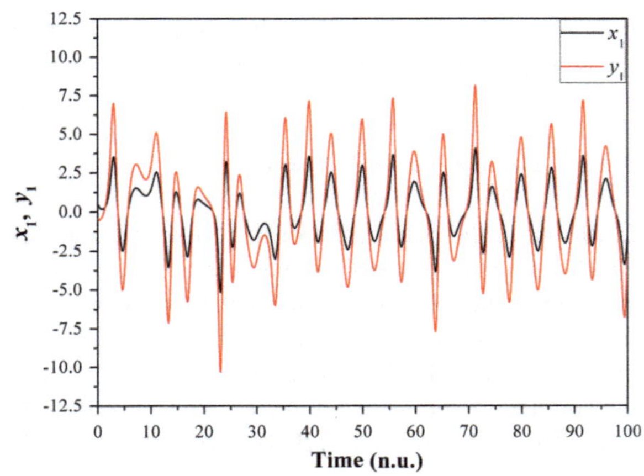

Fig. 42 The time-series of x_1, y_1, for $\alpha = 2$, $\beta = 1$, $c = 2.7$ and $d = 0.44$

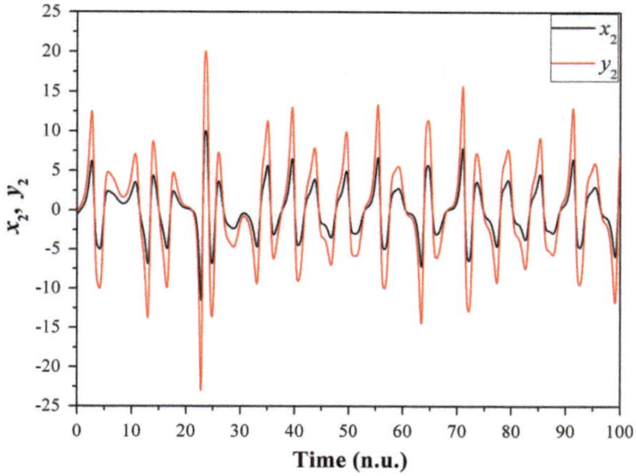

Fig. 43 The time-series of x_2, y_2, for $\alpha = 2$, $\beta = 1$, $c = 2.7$ and $d = 0.44$

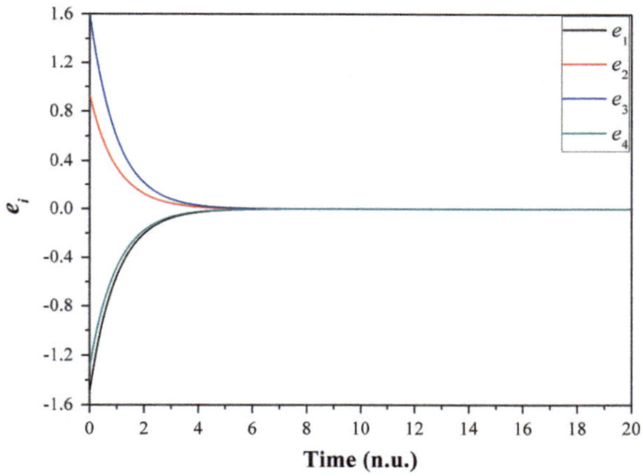

Fig. 44 The plot of errors $e_i(=\beta y_i - \alpha x_i)$, for $\alpha = 2$, $\beta = 1$, $c = 2.7$ and $d = 0.44$

shows the exponential convergence to zero that confirms the realization of system's complete synchronization state.

3.2.4 The Case $\alpha = -2$, $\beta = 1$

In the last case the parameters of the error function are chosen as $\alpha = -2$ and $\beta = 1$. So, the attractor of the first coupled system has been enlarged again by factor two, while the attractor of the second coupled system has been inverted in regard to the first one, as it is shown in Fig. 45. In this case the systems' parameters are chosen as $c = 2.7$ and $d = 0.44$ so as both of the coupled systems are in hyperchaotic state. This process is shown more clearly in the plots of the time series of x_1, y_1 and x_2, y_2 of Figs. 47 and 48. The phase portrait of y_1 versus x_1 in Fig. 46 indicates that the coupled system is in anti-synchronization state, which is also confirmed by the error plot $e_i = 2y_i + x_i$ $(i = 1, 2, 3, 4)$ in Fig. 49.

Fig. 45 The phase portrait of x_2 versus x_1, for $\alpha = -2$, $\beta = 1$, $c = 2.7$ and $d = 0.44$

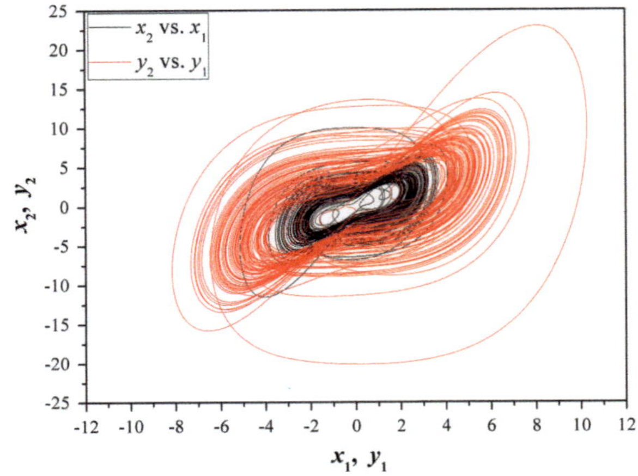

Fig. 46 The phase portrait of y_1 versus x_1, for $\alpha = -2$, $\beta = 1$, $c = 2.7$ and $d = 0.44$

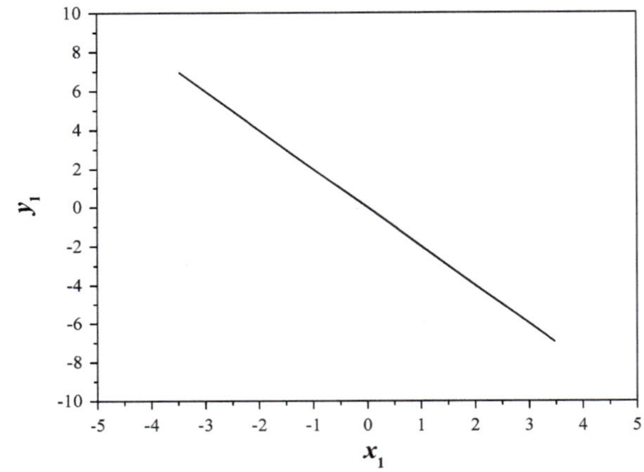

Fig. 47 The time-series of x_1, y_1, for $\alpha = -2$, $\beta = 1$, $c = 2.7$ and $d = 0.44$

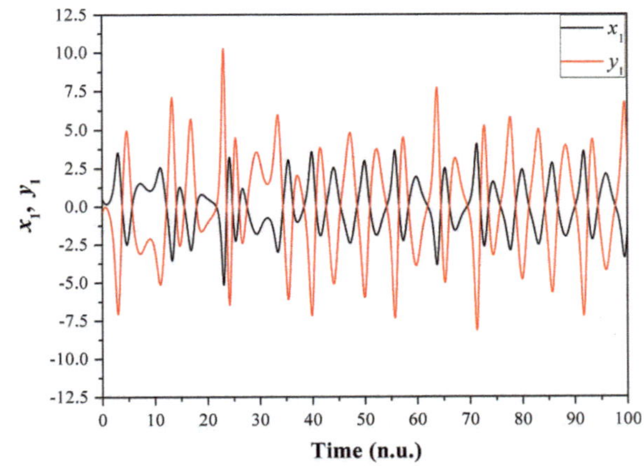

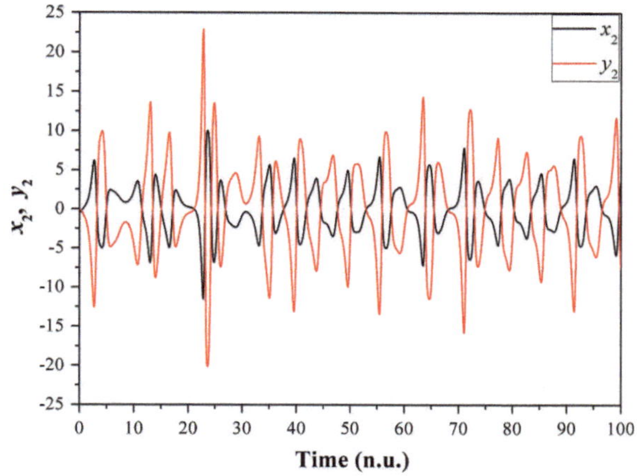

Fig. 48 The time-series of x_2, y_2, for $\alpha = -2$, $\beta = 1$, $c = 2.7$ and $d = 0.44$

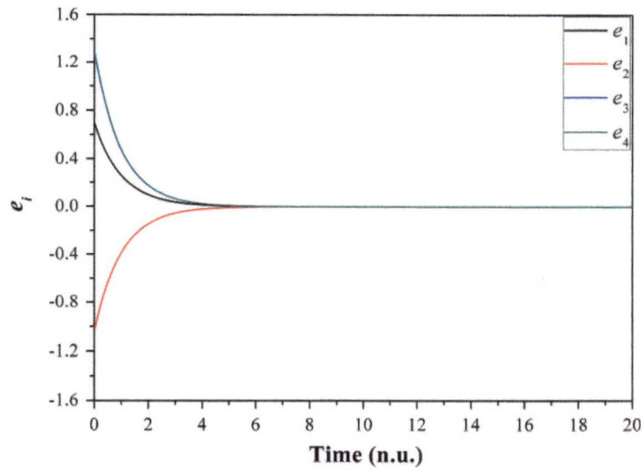

Fig. 49 The plot of errors $e_i (= \beta y_i - \alpha x_i)$, for $\alpha = -2$, $\beta = 1$, $c = 2.7$ and $d = 0.44$

4 Circuit's Implementation of the Proposed Scheme

In this section the circuit implementation of the proposed scheme, with the electronic simulation package Cadence OrCAD, in the case of unidirectional coupling systems with $a = \beta$ is presented, in order to prove the feasibility of this method. The coupling system is realized by common electronic components. The system's circuit consists of three sub-circuits, which are the master circuit, the slave circuit and the coupling circuit.

Figure 50 depicts the schematic of the master circuit. It has four integrators (U_1, U_2, U_3 and U_4) and two differential amplifiers (U_7, U_8), which are implemented with the TL084, as well as two signals multipliers (U_5, U_6) by using the AD633. By applying Kirchhoff's circuit laws, the corresponding circuital equations of designed master circuit can be written as:

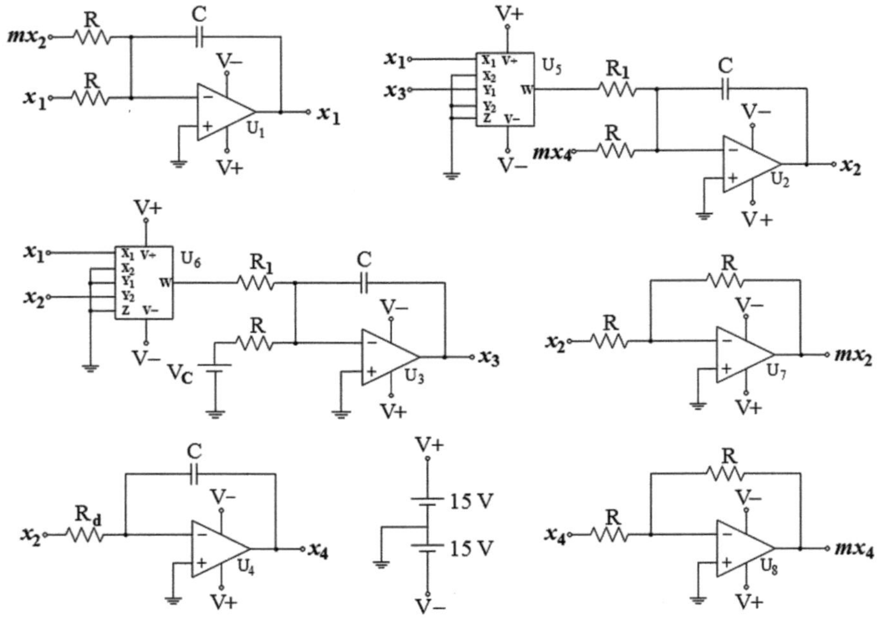

Fig. 50 The schematic of the master circuit

$$\begin{cases} \dot{x}_1 = \frac{1}{RC}(x_2 - x_1) \\ \dot{x}_2 = \frac{1}{RC}\left(-\frac{R}{R_1 10V}x_1 x_3 + x_4\right) \\ \dot{x}_3 = \frac{1}{RC}\left(\frac{R}{R_1 10V}x_1 x_2 - c\right) \\ \dot{x}_4 = \frac{1}{RC}\left(-\frac{R}{R_d}x_2\right) \end{cases} \tag{18}$$

where $x_i\,(i = 1, \ldots, 4)$ are the voltages in the outputs of the operational amplifiers U_1, U_2, U_3 and U_4. Normalizing the differential equations of system (18) by using $\tau = t/RC$ we could see that this system is equivalent to the system (12). The circuit components have been selected as: $R = 10\,k\Omega$, $R_1 = 1\,k\Omega$, $R_d = 22.727\,k\Omega$, $C = 10\,nF$, $V_C = 2.7\,V$, while the power supplies of all active devices are $\pm 15\,V_{DC}$. For the chosen set of components the master system's parameters are: $c = 2.7$ and $d = 0.44$. In Figs. 51, 52, 53, 54 and 55 the hyperchaotic attractors, which are obtained from Cadence OrCAD in various phase planes, are proved to be in a very good agreement with the respective phase portraits from system's simulation process (Figs. 15, 16, 17, 18 and 19). So, the proposed circuit emulates well the master system.

In Fig. 56 the schematic of the slave circuit, which is similar to the master circuit, is shown. The difference of this circuit in comparison to the previous one are the signals mu_2, mu_3 and mu_4, which are the opposites of the signals U_{Y2}, U_{Y3} and U_{Y4}, produced by the controllers of Eq. (16). Also, e_2 is the difference signal $(\beta y_2 - \alpha x_2)$.

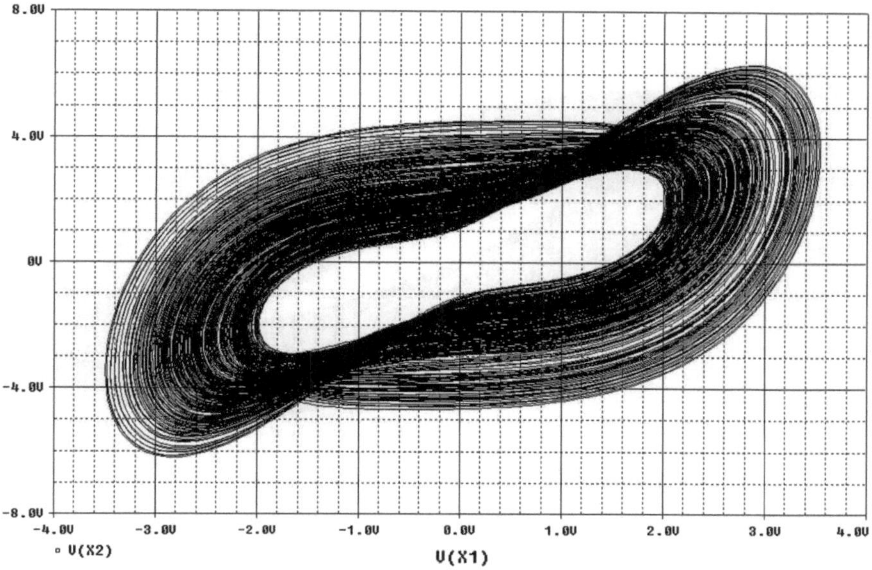

Fig. 51 Hyperchaotic attractor of the designed master circuit obtained from Cadence OrCAD in the (x_1, x_2) phase plane

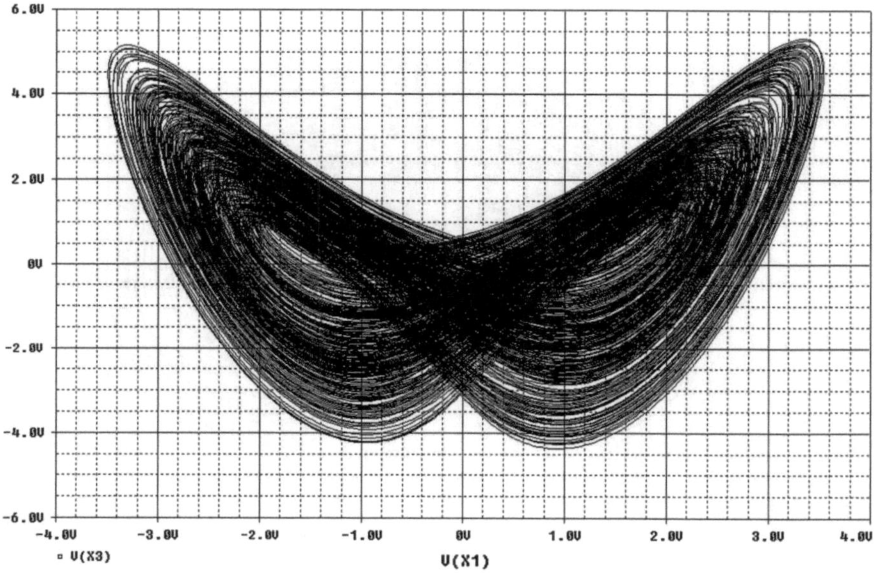

Fig. 52 Hyperchaotic attractor of the designed master circuit obtained from Cadence OrCAD in the (x_1, x_3) phase plane

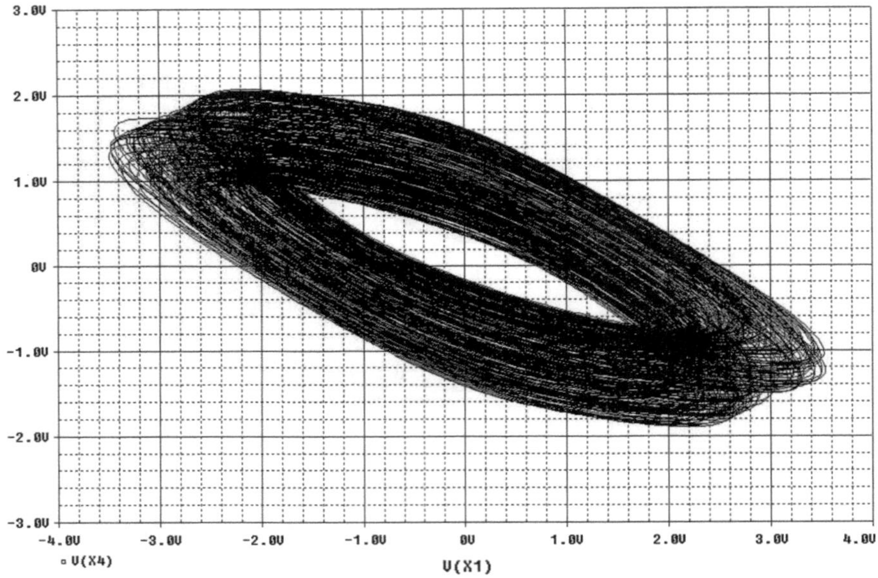

Fig. 53 Hyperchaotic attractor of the designed master circuit obtained from Cadence OrCAD in the (x_1, x_4) phase plane

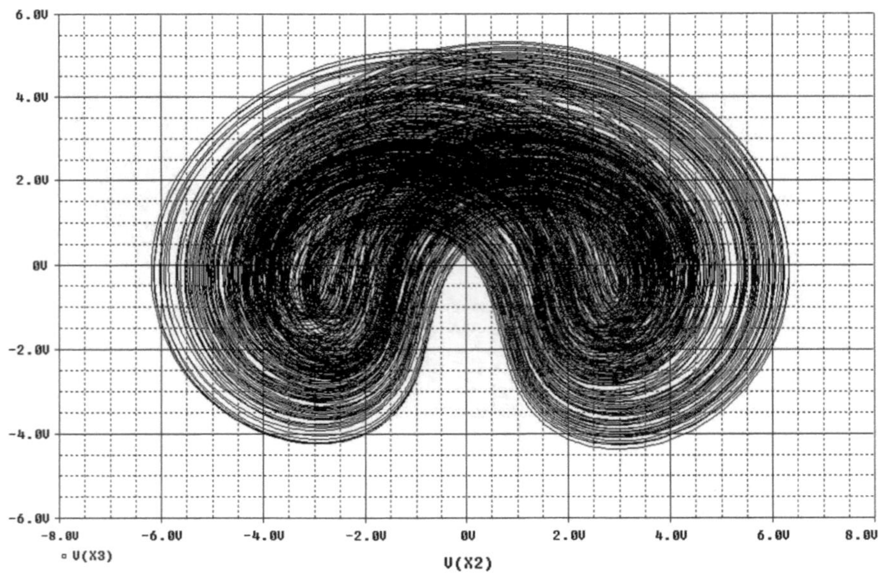

Fig. 54 Hyperchaotic attractor of the designed master circuit obtained from Cadence OrCAD in the (x_2, x_3) phase plane

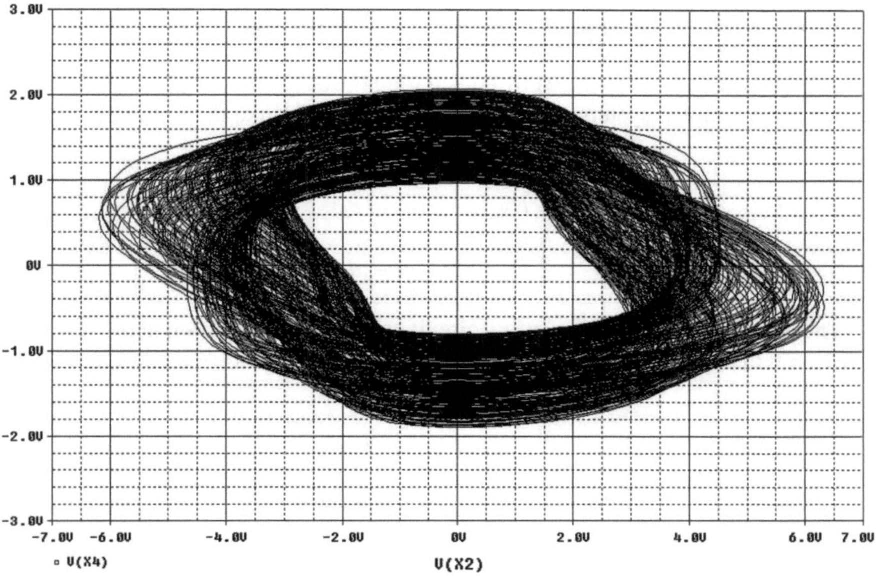

Fig. 55 Hyperchaotic attractor of the designed master circuit obtained from Cadence OrCAD in the (x_2, x_4) phase plane

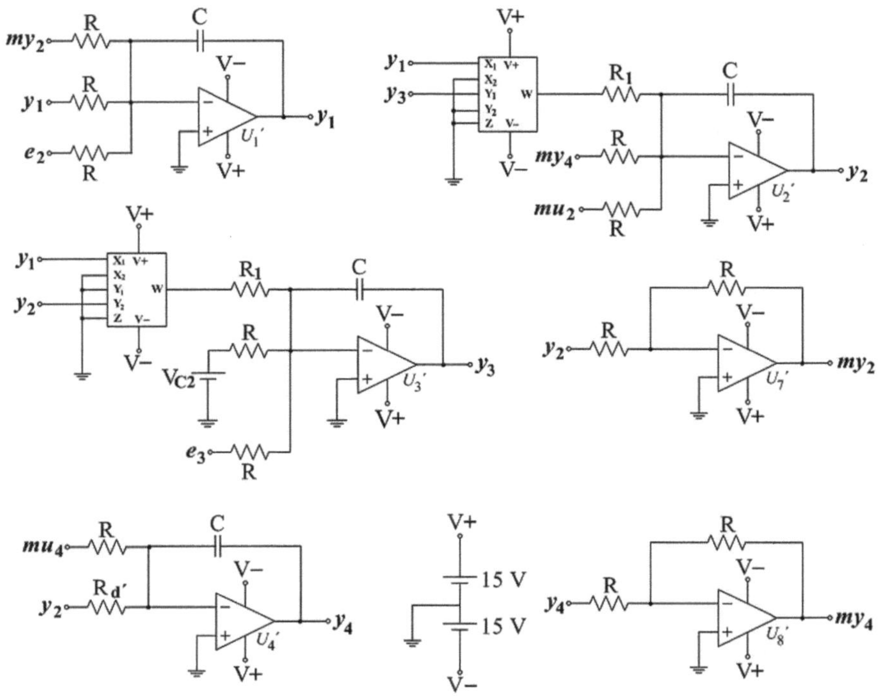

Fig. 56 The schematic of the slave circuit

Next, the design of the coupling circuit as well as the simulation results obtained from SPICE in the case of $\alpha = \beta$ is discussed in details.

In the case of $\alpha = \beta = 1$ and by considering the achievement of synchronization between the coupled systems (12) and (13), the NOLCs take the following forms.

$$\begin{cases} U_{Y1} = -e_2 \\ U_{Y2} = -\left(e_2 + e_4\right) \\ U_{Y3} = -e_3 \\ U_{Y4} = -\left(e_4 - de_2\right) \end{cases} \tag{19}$$

The units from which the coupling circuit is consisted, are shown in the schematic of Fig. 57. In this schematic u_2 and u_4 are the control signals U_{Y2} and U_{Y4} of Eq. (19)

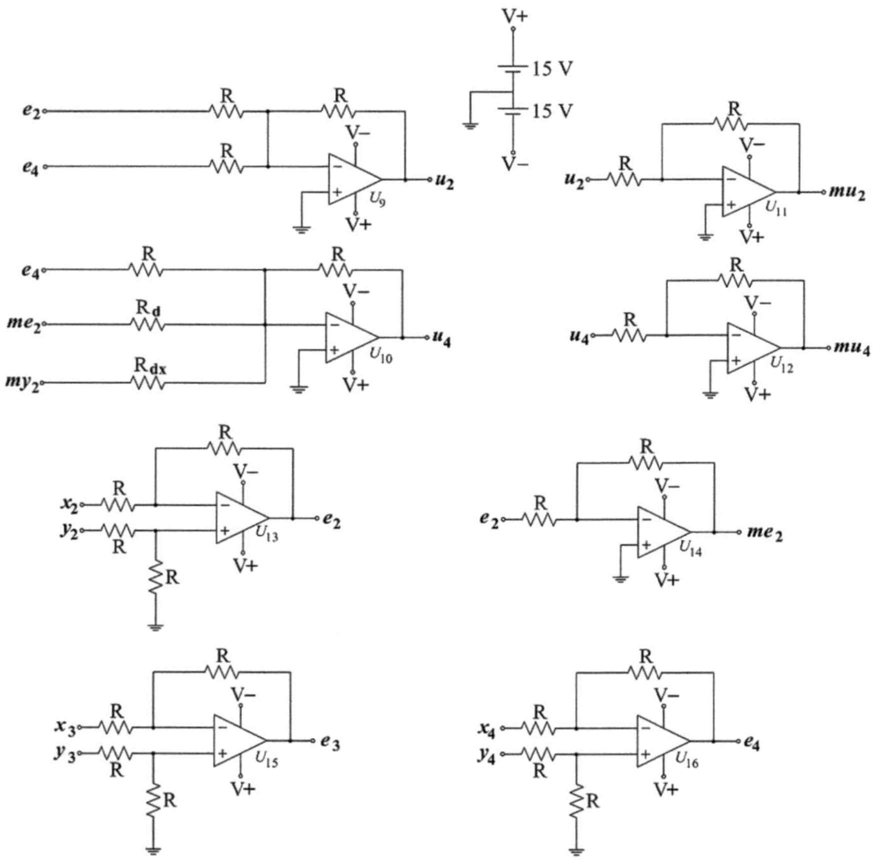

Fig. 57 The schematic of the coupling circuit

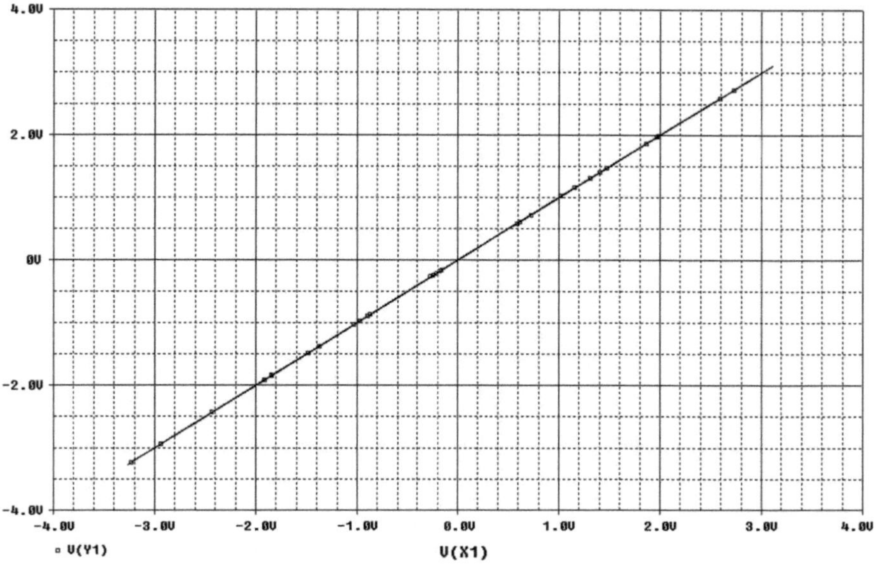

Fig. 58 The phase portrait in the (x_1, y_1) phase plane, for $\alpha = \beta = 1$, $c = 2.7$ and $d = 0.44$, obtained from Cadence OrCAD

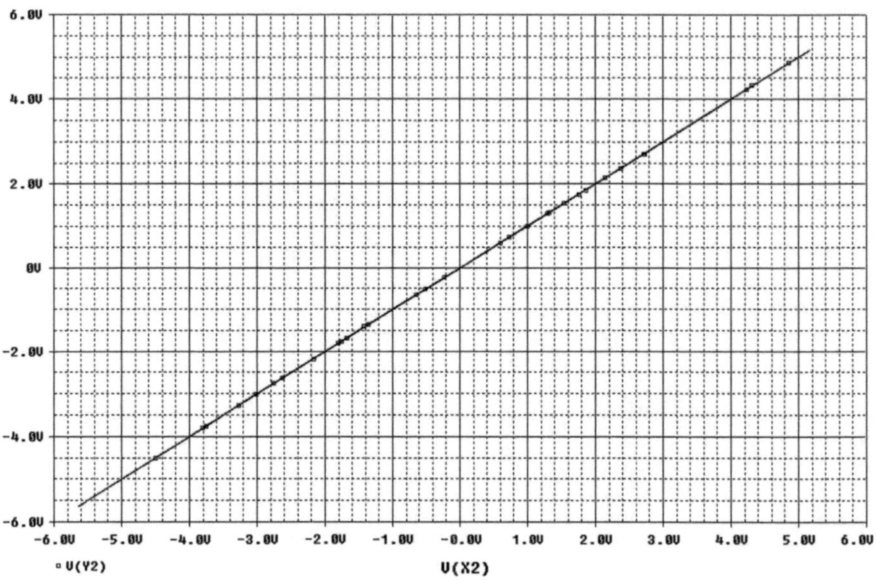

Fig. 59 The phase portrait in the (x_2, y_2) phase plane, for $\alpha = \beta = 1$, $c = 2.7$ and $d = 0.44$, obtained from Cadence OrCAD

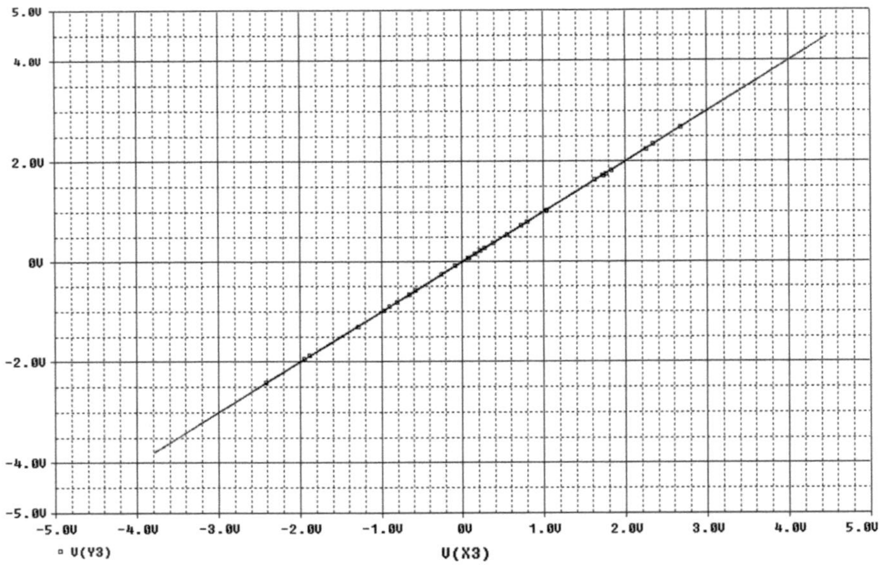

Fig. 60 The phase portrait in the (x_3, y_3) phase plane, for $\alpha = \beta = 1$, $c = 2.7$ and $d = 0.44$, obtained from Cadence OrCAD

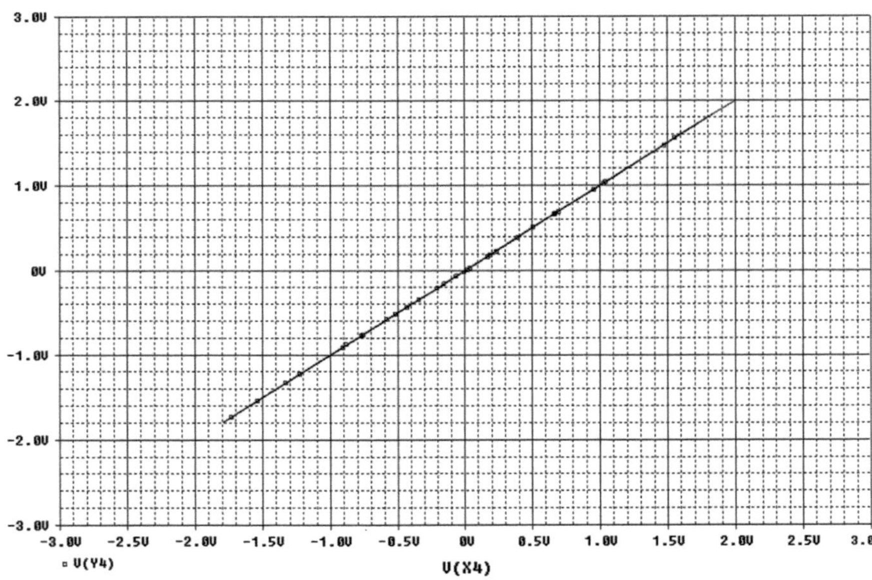

Fig. 61 The phase portrait in the (x_4, y_4) phase plane, for $\alpha = \beta = 1, c = 2.7$ and $d = 0.44$, obtained from Cadence OrCAD

respectively, while mu_2 and mu_4 are the opposite of these signals. Also, e_i, ($i = 2, 3, 4$) are the difference signals ($\beta y_i - \alpha x_i, i = 2, 3, 4$) and me_2 is the opposite of e_2.

Figures 58, 59, 60 and 61 depict the phase portraits in (x_i, y_i) phase plane, with $i = 1, \ldots, 4$, for $\alpha = \beta = 1$, $c = 2.7$ and $d = 0.44$, obtained from Cadence OrCAD. These figures confirm the achievement of complete synchronization in the case of unidirectionally coupled circuits with the proposed method.

5 Conclusion

In this chapter, the case of bidirectional and unidirectional coupling scheme of hyperchaotic dynamical systems with hidden attractors was studied. The proposed system is a four-dimensional modified Lorenz system, which is the simplest hyperchaotic system of this family. Furthermore, the coupling method was based on a recently new proposed scheme based on the nonlinear open loop controller.

According to the simulation results from system's numerical integration as well as the circuital implementation of the proposed system in SPICE, in the case of unidirectional coupling, the appearance of complete synchronization and anti-synchronization, depending on the signs of the parameters of the error functions, was investigated in various cases. So, by choosing an appropriate sign for the error functions one could drive the coupling system either in complete synchronization or anti-synchronization behavior.

As it is known, the complex behavior of hyperchaotic systems, like the aforementioned, makes the control difficult in practical applications where a particular dynamic is desired. So, this chapter presents an interesting research result for the family of hyperchaotic systems with hidden attractors, because this method could be very useful in many potential applications of these systems. As a next step in this direction is the application of the proposed method in non-identical coupling systems in order to satisfy the goal of control of systems, which are in totally different dynamical behaviors.

References

1. Astakhov V, Shabunin A, Anishchenko V (2000) Antiphase synchronization in symmetrically coupled self-oscillators. Int J Bifurc Chaos 10:849–857
2. Bautin NN (1939) On the number of limit cycles generated on varying the coefficients from a focus or centre type equilibrium state. Doklady Akademii Nauk SSSR 24:668–671
3. Bautin NN (1952) On the number of limit cycles appearing on varying the coefficients from a focus or centre type of equilibrium state. Mat Sb (N.S.) 30:181–196
4. Bernat J, Llibre J (1996) Counterexample to Kalman and Markus-Yamabe conjectures in dimension larger than 3. Dyn Contin Discret Impul Syst 2:337–379
5. Blazejczuk-Okolewska B, Brindley J, Czolczynski K, Kapitaniak T (2001) Antiphase synchronization of chaos by noncontinuous coupling: two impacting oscillators. Chaos Solitons Fract 12:1823–1826

6. Cao LY, Lai YC (1998) Antiphase synchronism in chaotic system. Phys Rev 58:382–386
7. Dimitriev AS, Kletsovi AV, Laktushkin AM, Panas AI, Starkov SO (2006) Ultrawide-band wireless communications based on dynamic chaos. J Commun Technol Electron 51: 1126–1140
8. Dykman GI, Landa PS, Neymark YI (1991) Synchronizing the chaotic oscillations by external force. Chaos Solitons Fract 1:339–353
9. Enjieu Kadji HG, Chabi Orou JB, Woafo P (2008) Synchronization dynamics in a ring of four mutually coupled biological systems. Commun Nonlinear Sci Numer Simul 13:1361–1372
10. Fitts RE (1966) Two counterexamples to Aizerman's conjecture. Trans IEEE AC-11:553–556
11. Fujisaka H, Yamada T (1983) Stability theory of synchronized motion in coupled-oscillator systems. Prog Theor Phys 69:32–47
12. Gao Z, Zhang C (2011) A novel hyperchaotic system. J Jishou Univ (Natural Science Edition) 32:65–68
13. Gonzalez-Miranda JM (2004) Synchronization and control of chaos. Imperial College Press, London
14. Grassi G, Mascolo S (1999) Synchronization of high-order oscillators by observer design with application to hyperchaos-based cryptography. Int J Circ Theor Appl 27:543–553
15. Gubar' NA (1961) Investigation of a piecewise linear dynamical system with three parameters. J Appl Math Mech 25:1011–1023
16. Hilbert D (1901) Mathematical problems. Bull Am Math Soc 8:437–479
17. Holstein-Rathlou NH, Yip KP, Sosnovtseva OV, Mosekilde E (2001) Synchronization phenomena in nephron-nephron interaction. Chaos 11:417–426
18. Jafari S, Sprott J (2013) Simple chaotic flows with a line equilibrium. Chaos Solit Fract 57:79–84
19. Jafari S, Haeri M, Tavazoei MS (2010) Experimental study of a chaos-based communication system in the presence of unknown transmission delay. Int J Circ Theor Appl 38:1013–1025
20. Kalman RE (1957) Physical and mathematical mechanisms of instability in nonlinear automatic control systems. Trans ASME 79:553–566
21. Kapranov M (1956) Locking band for phase-locked loop. Radiofizika 2:37–52
22. Kim CM, Rim S, Kye WH, Rye JW, Park YJ (2003) Anti-synchronization of chaotic oscillators. Phys Lett A 320:39–46
23. Kuznetsov NV, Leonov GA, Vagaitsev VI (2010) Analytical-numerical method for attractor localization of generalized Chua's system. IFAC Proc Vol (IFAC-PapersOnline) 4(1):29–33
24. Kyprianidis IM, Stouboulos IN (2003) Synchronization of two resistively coupled nonautonomous and hyperchaotic oscillators. Chaos Solitons Fract 17:317–325
25. Kyprianidis IM, Stouboulos IN (2003) Chaotic synchronization of three coupled oscillators with ring connection. Chaos Solitons Fract 17:327–336
26. Kyprianidis IM, Volos ChK, Stouboulos IN, Hadjidemetriou J (2006) Dynamics of two resistively coupled Duffing-type electrical oscillators. Int J Bifurc Chaos 16:1765–1775
27. Kyprianidis IM, Bogiatzi AN, Papadopoulou M, Stouboulos IN, Bogiatzis GN, Bountis T (2006) Synchronizing chaotic attractors of Chua's canonical circuit. The case of uncertainty in chaos synchronization. Int J Bifurc Chaos 16:1961–1976
28. Kyprianidis IM, Volos ChK, Stouboulos IN (2008) Experimental synchronization of two resistively coupled Duffing-type circuits. Nonlinear Phenom Complex Syst 11:187–192
29. Lauvdal T, Murray R, Fossen T (1997) Stabilization of integrator chains in the presence of magnitude and rate saturations: a gain scheduling approach. IEEE Control Decision Conf, pp 4004–4005
30. Leonov G, Kuznetsov NV (2013) Analytical-numerical methods for hidden attractors' localization: the 16th Hilbert problem, Aizerman and Kalman conjectures, and Chua circuits. Numerical methods for differential equations, optimization, and technological problems, computational methods in applied sciences. Springer 27:41–64
31. Leonov G, Kuznetsov N, Vagaitsev V (2011) Localization of hidden Chua's attractors. Phys Lett A 375:2230–2233

32. Leonov G, Kuznetsov N, Kuznetsova O, Seldedzhi S, Vagaitsev V (2011) Hidden oscillations in dynamical systems. Trans Syst Contr 6:54–67
33. Leonov G, Kuznetsov N, Vagaitsev V (2012) Hidden attractor in smooth Chua system. Physica D 241:1482–1486
34. Li GH (2009) Inverse lag synchronization in chaotic systems. Chaos Solitons Fract 40: 1076–1080
35. Li C, Sprott JC (2014) Coexisting hidden attractors in a 4-D simplified Lorenz system. Int J Bifurc Chaos 24(3):1450034
36. Liu W, Qian X, Yang J, Xiao J (2006) Antisynchronization in coupled chaotic oscillators. Phys Lett A 354:119–125
37. Liu X, Chen T (2010) Synchronization of identical neural networks and other systems with an adaptive coupling strength. Int J Circ Theor Appl 38:631–648
38. Mainieri R, Rehacek J (1999) Projective synchronization in three-dimensional chaotic system. Phys Rev Lett 82:3042–3045
39. Maritan A, Banavar J (1994) Chaos noise and synchronization. Phys Rev Lett 72:1451–1454
40. Markus L, Yamabe H (1960) Global stability criteria for differential systems. Osaka Math J 12:305–317
41. Mosekilde E, Maistrenko Y, Postnov D (2002) Chaotic synchronization: applications to living systems. World Scientific, Singapore
42. Munmuangsaen B, Srisuchinwong B (2009) A new five-term simple chaotic attractor. Phys Lett A 373:4038–4043
43. Paar V, Pavin N (1998) Intermingled fractal Arnold tongues. Phys Rev E 57:1544–1549
44. Padmanaban E, Hens C, Dana K (2011) Engineering synchronization of chaotic oscillator using controller based coupling design. Chaos 21:013110
45. Parlitz U, Junge L, Lauterborn W, Kocarev L (1996) Experimental observation of phase synchronization. Phys Rev E 54:2115–2217
46. Pham V-T, Volos ChK, Jafari S, Wang X, Vaidyanathan S (2014) Hidden hyperchaotic attractor in a novel simple memristive neural network. J Optoelectron Adv Mater Rapid Commun 8(11–12):1157–1163
47. Pham V-T, Jafari S, Volos ChK, Wang X, Golpayegani Mohammad Reza Hashemi S (2014) Is that really hidden? The presence of complex fixed-points in chaotic flows with no equilibria. Int J Bifurc Chaos 24(11):1450146
48. Pecora LM, Carroll TL (1990) Synchronization in chaotic systems. Phys Rev Lett 64:821–824
49. Pikovsky AS (1984) On the interaction of strange attractors. Z Phys B—Condensed Matter 55:149–154
50. Pikovsky AS, Maistrenko YL (2003) Synchronization: theory and application. Springer, Netherlands
51. Pikovsky AS, Rosenblum M, Kurths J (2003) Synchronization: a universal concept in nonlinear sciences. Cambridge University Press, Cambridge
52. Rosenblum MG, Pikovsky AS, Kurths J (1997) From phase to lag synchronization in coupled chaotic oscillators. Phys Rev Lett 78:4193–4196
53. Rulkov NF, Sushchik MM, Tsimring LS, Abarbanel HDI (1995) Generalized synchronization of chaos in directionally coupled chaotic systems. Phys Rev E 51:980–994
54. Szatmári I, Chua LO (2008) Awakening dynamics via passive coupling and synchronization mechanism in oscillatory cellular neural/nonlinear networks. Int J Circ Theor Appl 36:525–553
55. Tafo Wembe E, Yamapi R (2009) Chaos synchronization of resistively coupled Duffing systems: numerical and experimental investigations. Commun Nonlinear Sci Numer Simul 14: 1439–1453
56. Taherion S, Lai YC (1999) Observability of lag synchronization of coupled chaotic oscillators. Phys Rev E 59:R6247–R6250
57. Tognoli E, Kelso JAS (2009) Brain coordination dynamics: true and false faces of phase synchrony and metastability. Prog Neurobiol 87:31–40
58. Tsuji S, Ueta T, Kawakami H (2007) Bifurcation analysis of current coupled BVP oscillators. Int J Bifurc Chaos 17:837–850

59. Van der Schrier G, Maas LRM (2000) The diffusionless Lorenz equations: Shil'nikov bifurcations and reduction to an explicit map. Phys Nonlinear Phenom 141:19–36
60. Volos ChK, Kyprianidis IM, Stouboulos IN (2006) Experimental demonstration of a chaotic cryptographic scheme. WSEAS Trans Circ Syst 5:1654–1661
61. Voss HU (2000) Anticipating chaotic synchronization. Phys Rev E 61:5115–5119
62. Wang J, Che YQ, Zhou SS, Deng B (2009) Unidirectional synchronization of Hodgkin-Huxley neurons exposed to ELF electric field. Chaos Solitons Fract 39:1335–1345
63. Woafo P, Enjieu Kadji HG (2004) Synchronized states in a ring of mutually coupled self-sustained electrical oscillators. Phys Rev E 69:046206
64. Zhong GQ, Man KF, Ko KT (2001) Uncertainty in chaos synchronization. Int J Bifurc Chaos 11:1723–1735

A Chaotic Hyperjerk System Based on Memristive Device

Viet-Thanh Pham, Sundarapandian Vaidyanathan, Christos K. Volos, Sajad Jafari and Xiong Wang

Abstract From the mechanical system point of view, third-order derivatives of displacement or the time rate of change of acceleration is the jerk, while the fourth derivative has been known as a snap. As a result, a dynamical system which is presented by an nth order ordinary differential equation with $n > 3$ describing the time evolution of a single scalar variable is considered as a hyperjerk system. Hyperjerk system has received significant attention because of its elegant form. Motivated by reported attractive hyperjerk systems, a 4-D novel chaotic hyperjerk system has been introduced and studied in this work. Interestingly, this hyperjerk system displays an infinite number of equilibrium points because of the presence of a memristive device. In addition, an adaptive controller is proposed to achieve synchronization of such novel hyperjerk systems with two unknown parameters. In order to confirm the feasibility of the mathematical hyperjerk model, its electronic circuit is designed and implemented by using SPICE.

V.-T. Pham (✉)
School of Electronics and Telecommunications, Hanoi University of Science and Technology, Hanoi, Vietnam
e-mail: pvt3010@gmail.com

S. Vaidyanathan
Research and Development Centre, Vel Tech University, Tamil Nadu, India
e-mail: sundar@veltechuniv.edu.in

C.K. Volos
Physics Department, Aristotle University of Thessaloniki, Thessaloniki, Greece
e-mail: volos@physics.auth.gr

S. Jafari
Biomedical Engineering Faculty, Amirkabir University of Technology, Tehran, Iran
e-mail: sajadjafari@aut.ac.ir

X. Wang
Institute for Advanced Study, Shenzhen University, Guangdong, Shenzhen 518060, People's Republic of China
e-mail: wangxiong8686@szu.edu.cn

S. Vaidyanathan and C. Volos (eds.), *Advances and Applications in Chaotic Systems*, Studies in Computational Intelligence 636,
DOI 10.1007/978-3-319-30279-9_2

Keywords Chaos · Hidden attractor · Hyperjerk · Equilibrium · Memristive · Circuit · OrCAD

1 Introduction

Chaotic systems have applied in several fields of science and engineering [2, 3, 7, 9, 46, 50, 66] after the vital discovery of Lorenz's model for atmospheric convection [31]. There are well-known chaotic systems such as Rössler system [42], Arneodo system [1], Chen system [7], Lü system [32] etc. In addition, various new chaotic systems have been introduced recently [16, 20, 34, 37, 40, 57, 63].

There is significant interest in investigating novel jerk chaotic systems [47]. From the view point of mathematics, a jerk system is presented by an explicit third-order ordinary differential equation which describes the time evolution of a single scale variable, for example x. Therefore, a jerk system is given as

$$\frac{d^3x}{dt^3} = f\left(\frac{d^2x}{dt^2}, \frac{dx}{dt}, x\right) \tag{1}$$

From the view point of mechanics, system (1) is called jerk system because when the scalar x represents the position of a moving object at the time t, the third derivative indicates the jerk [44]. Interestingly, well-known chaotic systems, i.e. Lorenz and Rössler systems, can be represented in jerk forms [21, 28].

Different examples of jerk systems were reported in the literature. A piecewise exponential jerk system was investigated by Sun and Sprott [52]. Another simple chaotic jerk system with exponential nonlinearity was presented in Munmuangsaen et al. [35] while its elegant electronic circuital implementation, including six resistors, three capacitors, four operational amplifiers and a silicon diode only, was introduced in Sprott [48]. A six-term 3-D novel jerk chaotic system with two hyperbolic sinusoidal nonlinearities was proposed by Vaidyanathan et al. [59]. Multi-scroll chaotic attractors could be generated in the jerk mode [30] or jerk circuits [33, 67] while multi-scroll and hypercube attractors were also achieved from a general jerk circuit using Josephson junctions [65].

By generalizing the definition of a jerk system [45], a hyperjerk system can be considered as

$$\frac{d^{(n)}x}{dt^n} = f\left(\frac{d^{(n-1)}x}{dt^{n-1}}, \ldots, \frac{dx}{dt}, x\right), \tag{2}$$

with $n \geq 4$ [47]. Hyperjerk form can described all periodically forced oscillators and many of the coupled oscillators [29] while transformation of 4-D dynamical systems to hyperjerk form was reported in Elhadj and Sprott [12]. Chaotic hyperjerk system including fourth and fifth derivatives was introduced [8]. In addition, Chlouverakis and Sprott found hyperchaotic hyperjerk flows. More recently, Sundarapandian proposed a 4-D novel hyperchaotic hyperjerk system by adding a quadratic nonlinearity to the Chlouverakis–Sprott hyperjerk system [60].

It is easy to see that reported jerk/hyperjerk systems have a finite number of equilibrium points. It is very interesting to ask naturally whether there exists a chaotic jerk/hyperjerk system without equilibria or with an unlimited equilibrium set. Some authors have recently answered this attractive question. Wang and Chen [64] constructed a jerk system with no equilibrium point, but still generated a chaotic attractor. A chaotic memory system with infinitely many equilibria was designed by using the concept of memory element [4]. Studying such jerk/hyperjerk systems with special features is still an open research direction.

In this chapter, our work has concentrated on a hyperjerk system based on a memristive device which can exhibit chaotic attractors. Moreover, such hyperjerk system has an infinite number of equilibrium points. This research work is organized as follows. Section 2 gives a brief introduction to the memristive device. The memristive hyperjerk system is presented in Sect. 3 while its qualitative properties are analyzed in Sect. 4. In Sect. 5, we describe the adaptive synchronization design for achieving global chaos synchronization of the identical novel hyperjerk systems with two unknown parameters. Section 6 shows the circuital implementation of our memristive hyperjerk system. Finally, conclusions are drawn in the last section.

2 Model of Memristive Device

Memristor was proposed by L.O. Chua as the fourth basic circuit element beside the three conventional ones (the resistor, the inductor and the capacitor) [10]. Memristor presents the relationship between two fundamental circuit variables, the charge (q) and the flux (φ). Hence, there are two kinds of memristor: charge-controlled memristor and flux-controlled memristor [10, 54]. A charge-controlled memristor is described by

$$v_M = M(q)\, i_M, \tag{3}$$

where v_M is the voltage across the memristor and i_M is the current through the memristor. Here the memristance (M) is defined by

$$M(q) = \frac{d\varphi(q)}{dq}, \tag{4}$$

while the flux-controlled memristor is given by

$$i_M = W(\varphi)\, v_M, \tag{5}$$

where $W(\varphi)$ is the memductance, which is defined by

$$W(\varphi) = \frac{dq(\varphi)}{d\varphi}. \tag{6}$$

Moreover, by generalizing the original definition of a memristor [11, 54], a memristive system is given as:

$$\begin{cases} \dot{x} = F(x, u, t) \\ y = G(x, u, t)\, u, \end{cases} \tag{7}$$

where u, y, and x denote the input, output and state of the memristive system, respectively. The function F is a continuously differentiable, n-dimensional vector field and G is a continuous scalar function.

Based on the definition of memristive system [4, 11, 38, 54], a memristive device is introduced in this section and used in our whole chapter. The memristive device is described by the following form:

$$\begin{cases} \dot{x}_1 = x_2 \\ y = (1 - x_1)\, x_2. \end{cases} \tag{8}$$

Here x_2, y, and x_1 are the input, output and state of the memristive device, respectively.

An external bipolar periodic signal is applied across terminals of memristive device (8) to investigate its fingerprint [51, 54, 55]. The external sinusoidal stimulus is given by

$$x_2 = X_2 \sin(2\pi f t), \tag{9}$$

where X_2 is the amplitude and f is the frequency. From the first equation of (8), the state variable of the memristive device is obtained as

$$\begin{aligned} x_1(t) &= \int_{-\infty}^{t} x_2(\tau)\, d\tau = x_1(0) + \int_{0}^{t} X_2 \sin(2\pi f \tau)\, d\tau \\ &= x_1(0) + \frac{X_2}{2\pi f}\left(1 - \cos(2\pi f t)\right), \end{aligned} \tag{10}$$

with $x_1(0)$ is the initial condition of the internal state in the memristive device. Thus, the initial condition of the internal state variable is given by

$$x_1(0) = \int_{-\infty}^{0} x_2(\tau)\, d\tau. \tag{11}$$

Substituting (9) and (10) into (8), it is easy to derive the output of the memristive device

$$\begin{aligned} y(t) &= \left[1 - x_1(0) - \frac{X_2}{2\pi f}\left(1 - \cos(2\pi f t)\right)\right] X_2 \sin(2\pi f t) \\ &= \left(1 - x_1(0) - \frac{X_2}{2\pi f}\right) X_2 \sin(2\pi f t) + \frac{X_2^2}{4\pi f t} \sin(4\pi f t). \end{aligned} \tag{12}$$

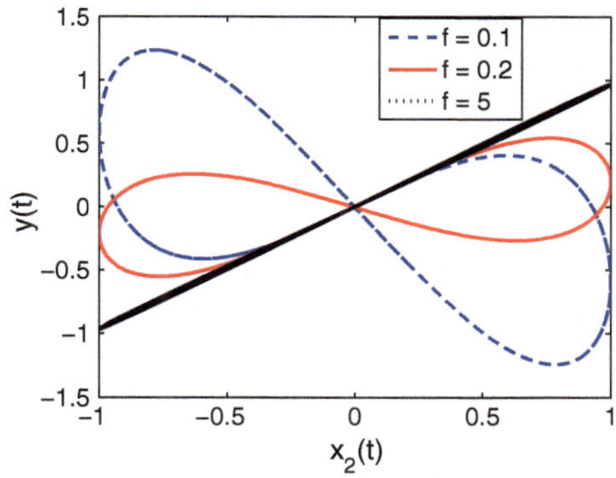

Fig. 1 Hysteresis loops of the proposed memristive device (8) driven by a sinusoidal stimulus (9) when changing the frequency f

From Eq. (12), it is easily seen that the output y depends on the frequency of the applied input stimulus. Hysteresis loop of the memristive device (8) when driven by a periodic signal (9) with different frequencies are shown in Fig. 1. Exhibited "pinched hysteresis loop" in the input–output plane indicates the vital fingerprint of memristive device (8).

3 A 4-D Novel Memristive Hyperjerk System

In this chapter, a novel 4-D memristive system is proposed by using the memristive device (8) and the reported approach in Bao et al. [4]. The novel memristive system is given in system form as

$$\begin{cases} \dot{x}_1 = x_2 \\ \dot{x}_2 = x_3 \\ \dot{x}_3 = x_4 \\ \dot{x}_4 = -x_3 - ax_4 - bx_3x_4 - y, \end{cases} \tag{13}$$

where a, b are positive parameters and $y = (1 - x_1)\,x_2$ is the output of memristive device (8).

The novel memristive system (13) can be rewritten by

$$\frac{d^4x_1}{dt^4} = f\left(\frac{d^3x_1}{dt^3}, \frac{d^2x_1}{dt^2}, \frac{dx_1}{dt}, x_1\right), \tag{14}$$

where

$$f = -\frac{d^2x_1}{dt^2} - a\frac{d^3x_1}{dt^3} - b\frac{d^2x_1}{dt^2}\frac{d^3x_1}{dt^3} - (1 - x_1)\,\frac{dx_1}{dt}. \tag{15}$$

Therefore, memristive system (13) is called a hyperjerk system because it involves time derivatives of a jerk function [45, 47]. In this chapter, the memristive system (13) is chaotic when the parameters a, and b take the values

$$a = 0.5, \quad b = 0.4. \tag{16}$$

For the selected parameter values in (16), the Lyapunov exponents of the novel memristive system (13) are obtained as

$$L_1 = 0.0730, \ \ L_2 = 0.0018, \ \ L_3 = 0, \ \ L_4 = -0.5755. \tag{17}$$

For numerical simulations, we take the initial conditions of the novel memristive system (13) as $x_1(0) = 0.06$, $x_2(0) = 10^{-6}$, $x_3(0) = 0$, and $x_4(0) = 0$. Here the initial condition of the input of the memristive device $x_2(0)$ should be tiny to guarantee an appropriate value of the internal state variable of the memristive device. Figures 2, 3 and 4 illustrate the 2-D projections and 3-D projections of the new memristive system (13).

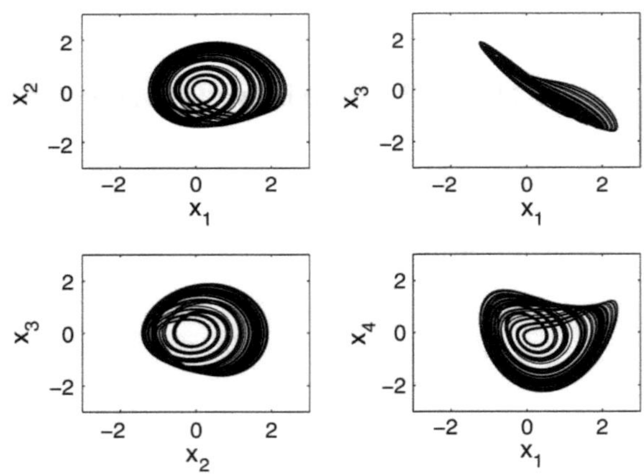

Fig. 2 2-D projections of the novel chaotic hyperjerk system (13) in (x_1, x_2)-plane, (x_1, x_3)-plane, (x_2, x_3)-plane, and (x_1, x_4)-plane

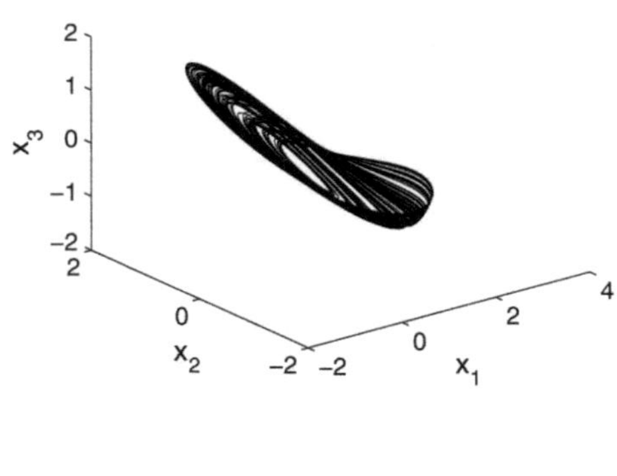

Fig. 3 Strange attractor of the novel chaotic hyperjerk system (13) in (x_1, x_2, x_3)-space

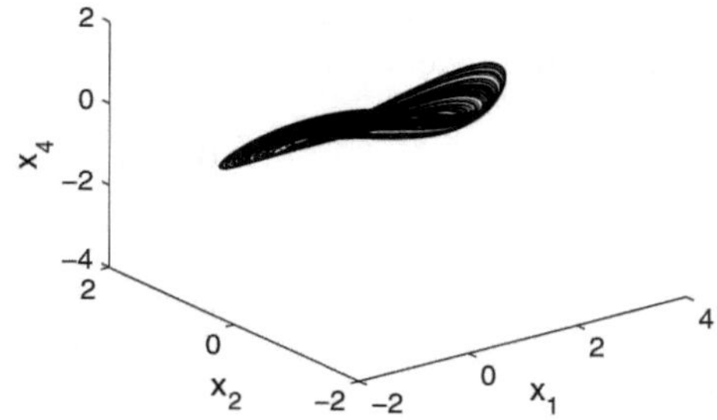

Fig. 4 Strange attractor of the novel chaotic hyperjerk system (13) in (x_1, x_2, x_4)-space

4 Analysis of the 4-D Novel Memristive Hyperjerk System

4.1 Equilibrium Points

The equilibrium points of the 4-D novel memristive hyperjerk system (13) are obtained by solving the equations

$$\begin{cases} f_1(x_1, x_2, x_3, x_4) = x_2 & = 0 \\ f_2(x_1, x_2, x_3, x_4) = x_3 & = 0 \\ f_3(x_1, x_2, x_3, x_4) = x_4 & = 0 \\ f_4(x_1, x_2, x_3, x_4) = -x_3 - ax_4 - bx_3x_4 - y & = 0 \end{cases}. \tag{18}$$

Thus, the equilibrium points of the system (13) are characterized by the equations

$$y = (1 - x_1)x_2 = 0, \quad x_2 = 0, \quad x_3 = 0, \quad x_4 = 0 \tag{19}$$

Solving the system (19), we get the equilibrium points of the hyperjerk system (13) as

$$E_c = \begin{bmatrix} c \\ 0 \\ 0 \\ 0 \end{bmatrix}, \tag{20}$$

where c is a real constant. Interestingly, the novel hyperjerk system (13) displays an infinite number of equilibrium points because of the presence of a memristive device (8). According to a new classification of chaotic dynamics [24–27], there are two kinds of attractors: self-excited attractors and hidden attractors. A self-excited attractor has a basin of attraction that is excited from unstable equilibria. In contrast, a hidden attractor cannot be discovered by using a numerical approach where a trajectory started from a point on the unstable manifold in the neighbourhood of an unstable equilibrium [15, 22, 23]. Therefore, hyperjerk system (13) can be considered as a chaotic memristive system with hidden attractor.

In order to discover the stability type of the equilibrium points E_c the Jacobian matrix of the novel memristive hyperjerk system (13) is calculated at any point $\mathbf{x}$ as

$$J(\mathbf{x}) = \begin{bmatrix} 0 & 1 & 0 & 0 \\ 0 & 0 & 1 & 0 \\ 0 & 0 & 0 & 1 \\ x_2 & x_1 - 1 & -1 - bx_4 & -a - bx_3 \end{bmatrix}, \quad (21)$$

It is noting that

$$J_0 \triangleq J(E_c) = \begin{bmatrix} 0 & 1 & 0 & 0 \\ 0 & 0 & 1 & 0 \\ 0 & 0 & 0 & 1 \\ 0 & c - 1 & -1 & -0.5 \end{bmatrix}, \quad (22)$$

which has the characteristic equation is

$$\lambda \left(\lambda^3 + 0.5\lambda^2 + \lambda + 1 - c\right) = 0. \quad (23)$$

When $c = 0.06$ the characteristic Eq. (23) has a zero eigenvalue and three nonzero eigenvalues

$$\lambda_1 = 0, \quad \lambda_2 = -0.7749, \quad \lambda_{3,4} = 0.1375 \pm 1.0928i \quad (24)$$

This shows that the equilibrium point E_c is an unstable saddle-focus point.

4.2 Lyapunov Exponents and Kaplan–Yorke Dimension

For the parameter values $a = 0.5, b = 0.4$ and $c = 0.06$, the Lyapunov exponents of the novel memristive hyperjerk system (13) are obtained using MATLAB as

$$L_1 = 0.0730, \quad L_2 = 0.0018, \quad L_3 = 0 \text{ and } L_4 = -0.5755 \quad (25)$$

There is one positive Lyapunov exponents in the LE spectrum (25), thus the novel memristive hyperjerk system (13) exhibits chaotic behavior.

In addition, since $L_1 + L_2 + L_3 + L_4 = -0.5007 < 0$, it indicates that the novel memristive system (13) is dissipative.

The Kaplan–Yorke fractional dimension, that presents the complexity of attractor [46, 50], is defined by

$$D_{KY} = j + \frac{1}{\left|L_{j+1}\right|} \sum_{i=1}^{j} L_i \quad (26)$$

where j is the largest interger satisfying $\sum_{i=1}^{j} L_i \geq 0$ and $\sum_{i=1}^{j+1} L_i < 0$. Therefore, the Kaplan–Yorke dimension of the novel memristive hyperjerk system (13) is calculated as

$$D_{KY} = 3 + \frac{L_1 + L_2 + L_3}{|L_4|} = 3.130, \tag{27}$$

which is fractional.

5 Adaptive Synchronization for the Hyperjerk Memristive System

One of the most important characteristics relating to chaotic systems and their applications is the possibility of synchronization of two chaotic ones [5, 13, 17, 36]. A wide range of research activities based on synchronization of nonlinear systems has been studied [6, 14, 18, 39, 49, 58]. For example, various synchronization phenomena in bidirectionally coupled double scroll circuits were reported in Volos et al. [61] or image encryption process based on chaotic synchronization phenomena was presented in [62]. Different synchronization schemes have proposed such as anti-synchronization [56], adaptive sycnchronization [59], or hybrid chaos synchronization [18], etc. Here we consider the adaptive synchronization of identical 4-D memristiive hyperjerk systems with two unknown parameters.

The master system is considered as the 4-D novel memristive hyperjerk system given by

$$\begin{cases} \dot{x}_1 = x_2 \\ \dot{x}_2 = x_3 \\ \dot{x}_3 = x_4 \\ \dot{x}_4 = -x_3 - ax_4 - bx_3x_4 - x_2 + x_1x_2 \end{cases} \tag{28}$$

where x_1, x_2, x_3, x_4 are the states of the system, and a, b are unknown constant parameters.

The slave system is considered as the 4-D novel memristive hyperjerk system given by

$$\begin{cases} \dot{y}_1 = y_2 \\ \dot{y}_2 = y_3 \\ \dot{y}_3 = y_4 \\ \dot{y}_4 = -y_3 - ay_4 - by_3y_4 - y_2 + y_1y_2 + u \end{cases} \tag{29}$$

where y_1, y_2, y_3, y_4 are the states of the system, and u is a backstepping control to be determined using estimates $\hat{a}(t)$ and $\hat{b}(t)$ for a and b, respectively.

The synchronization errors between the states of the master system (28) and the slave system (29) are defined as

$$\begin{cases} e_1 = y_1 - x_1 \\ e_2 = y_2 - x_2 \\ e_3 = y_3 - x_3 \\ e_4 = y_4 - x_4 \end{cases} \tag{30}$$

Thus, the error dynamics is easily obtained as follows

$$\begin{cases} \dot{e}_1 = e_2 \\ \dot{e}_2 = e_3 \\ \dot{e}_3 = e_4 \\ \dot{e}_4 = -e_3 - ae_4 - e_2 - b(y_3y_4 - x_3x_4) + y_1y_2 - x_1x_2 + u \end{cases} \tag{31}$$

The parameter estimation errors are defined as:

$$\begin{cases} e_a(t) = a - \hat{a}(t) \\ e_b(t) = b - \hat{b}(t) \end{cases} \tag{32}$$

Differentiating (32) with respect to t, we obtain the following equations:

$$\begin{cases} \dot{e}_a(t) = -\dot{\hat{a}}(t) \\ \dot{e}_b(t) = -\dot{\hat{b}}(t) \end{cases} \tag{33}$$

Next, the main result of this section will be presented and proved.

Theorem 5.1 *The identical 4-D novel memristive hyperjerk systems (28) and (29) with unknown parameters a and b are globally and exponentially synchronized by the adaptive control law*

$$\begin{cases} u(t) = -5e_1 - 9e_2 - 8e_3 - [4 - \hat{a}(t)]e_4 + \hat{b}(t)\,(y_3y_4 - x_3x_4) \\ \qquad\quad - (y_1y_2 - x_1x_2) - kz_4 \end{cases} \tag{34}$$

where $k > 0$ is a gain constant,

$$z_4 = 3e_1 + 5e_2 + 3e_3 + e_4, \tag{35}$$

and the update law for the parameter estimates $\hat{a}(t), \hat{b}(t), \hat{c}(t)$ is given by

$$\begin{cases} \dot{\hat{a}}(t) = -e_4z_4 \\ \dot{\hat{b}}(t) = -\,(y_3y_4 - x_3x_4)\,z_4 \end{cases} \tag{36}$$

Proof We prove this result via backstepping control method and Lyapunov stability theory.

First, we define a quadratic Lyapunov function

$$V_1(z_1) = \frac{1}{2}z_1^2 \tag{37}$$

where

$$z_1 = e_1 \tag{38}$$

Differentiating V_1 along the error dynamics (31), we get

$$\dot{V}_1 = z_1\dot{z}_1 = e_1e_2 = -z_1^2 + z_1(e_1 + e_2) \tag{39}$$

Here, we define

$$z_2 = e_1 + e_2 \tag{40}$$

Using (40), we can simplify the Eq. (39) as

$$\dot{V}_1 = -z_1^2 + z_1z_2 \tag{41}$$

Secondly, we define a quadratic Lyapunov function

$$V_2(z_1, z_2) = V_1(z_1) + \frac{1}{2}z_2^2 = \frac{1}{2}\left(z_1^2 + z_2^2\right) \tag{42}$$

Differentiating V_2 along the error dynamics (31), we get

$$\dot{V}_2 = -z_1^2 - z_2^2 + z_2(2e_1 + 2e_2 + e_3) \tag{43}$$

Now, we define

$$z_3 = 2e_1 + 2e_2 + e_3 \tag{44}$$

Using (44), we can simplify the Eq. (43) as

$$\dot{V}_2 = -z_1^2 - z_2^2 + z_2z_3 \tag{45}$$

Thirdly, we define a quadratic Lyapunov function

$$V_3(z_1, z_2, x_3) = V_2(z_1, z_2) + \frac{1}{2}z_3^2 = \frac{1}{2}\left(z_1^2 + z_2^2 + z_3^2\right) \tag{46}$$

Differentiating V_3 along the error dynamics (31), we get

$$\dot{V}_3 = -z_1^2 - z_2^2 - z_3^2 + z_3(3e_1 + 5e_2 + 3e_3 + e_4) \tag{47}$$

Now, we define

$$z_4 = 3e_1 + 5e_2 + 3e_3 + e_4 \tag{48}$$

Using (48), we can simplify the Eq. (47) as

$$\dot{V}_3 = -z_1^2 - z_2^2 - z_3^2 + z_3 z_4 \tag{49}$$

Finally, we define a quadratic Lyapunov function

$$V(z_1, z_2, z_3, z_4, e_a, e_b) = V_3(z_1, z_2, z_3) + \frac{1}{2}z_4^2 + \frac{1}{2}e_a^2 + \frac{1}{2}e_b^2 \tag{50}$$

which is a positive definite function on R^6.

Differentiating V along the error dynamics (31), we get

$$\dot{V} = -z_1^2 - z_2^2 - z_3^2 - z_4^2 + z_4(z_4 + z_3 + \dot{z}_4) - e_a\dot{\hat{a}} - e_b\dot{\hat{b}} \tag{51}$$

Equation (51) can be written compactly as

$$\dot{V} = -z_1^2 - z_2^2 - z_3^2 - z_4^2 + z_4 S - e_a\dot{\hat{a}} - e_b\dot{\hat{b}} \tag{52}$$

where

$$S = z_4 + z_3 + \dot{z}_4 = z_4 + z_3 + 3\dot{e}_1 + 5\dot{e}_2 + 3\dot{e}_3 + \dot{e}_4 \tag{53}$$

A simple calculation gives

$$S = 5e_1 + 9e_2 + 8e_3 + (4 - a)e_4 - b\,(y_3 y_4 - x_3 x_4) + (y_1 y_2 - x_1 x_2) + u \tag{54}$$

Substituting the adaptive control law (34) into (54), we obtain

$$S = -\left[a - \hat{a}(t)\right] e_4 - \left[b - \hat{b}(t)\right] (y_3 y_4 - x_3 x_4) - k z_4 \tag{55}$$

Using the definitions (33), we can simplify (55) as

$$S = -e_a e_4 - e_b\,(y_3 y_4 - x_3 x_4) - k z_4 \tag{56}$$

Substituting the value of S from (56) into (52), we obtain

$$\begin{cases} \dot{V} = -z_1^2 - z_2^2 - z_3^2 - (1 + k)z_4^2 + e_a(-e_4 z_4 - \dot{\hat{a}}) \\ \qquad + e_b\left[-\,(y_3 y_4 - x_3 x_4)\, z_4 - \dot{\hat{b}}\right] \end{cases} \tag{57}$$

Substituting the update law (36) into (57), we get

$$\dot{V} = -z_1^2 - z_2^2 - z_3^2 - (1+k)z_4^2, \tag{58}$$

which is a negative semi-definite function on R^6. Therefore, according to the Lyapunov stability theory [19, 43] we obtain $e_1(t) \to 0$, $e_2(t) \to 0$, $e_3(t) \to 0$, $e_4(t) \to 0$, $e_a(t) \to 0$, $e_b(t) \to 0$ exponentially when $t \to 0$ that is, synchronization between master and slave system.

In order to confirm and demonstrate the effectiveness of the proposed synchronization scheme, we consider a numerical example. In the numerical simulations, the fourth-order Runge–Kutta method is used to solve the systems. The parameters of the memristive hyperjerk systems are selected as $a = 0.5$, $b = 0.4$ and the positive gain constant as $k = 6$. The initial conditions of the master system (28) and the slave system (29) have been chosen as $x_1(0) = 0.06$, $x_2(0) = 10^{-6}$, $x_3(0) = 0$, $x_4(0) = 0$ and $y_1(0) = 0.02$, $y_2(0) = 10^{-4}$, $y_3(0) = 0$, $y_4(0) = 0$, respectively. We assumed that the initial values of the parameter estimates are $\hat{a}(0) = 0.46$ and $\hat{b}(0) = 0.01$.

When adaptive control law (34) and the update law for the parameter estimates (36) are applied, the master (28) and slave system (29) are synchronized completely

Fig. 5 Synchronization of the states $x_1(t)$ and $y_1(t)$

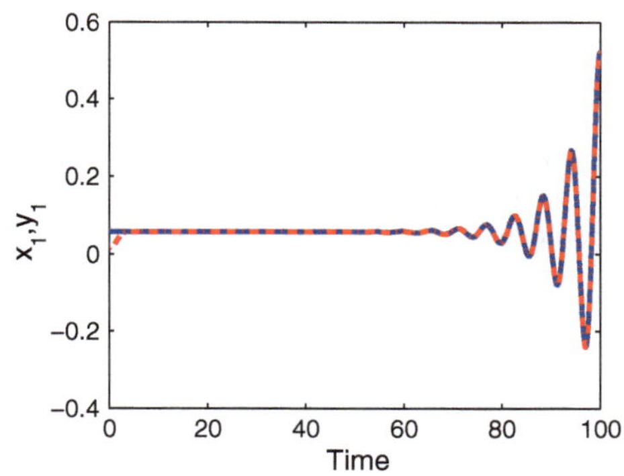

Fig. 6 Synchronization of the states $x_2(t)$ and $y_2(t)$

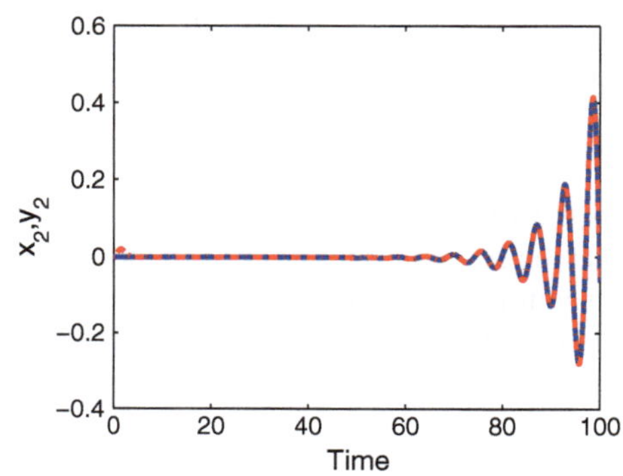

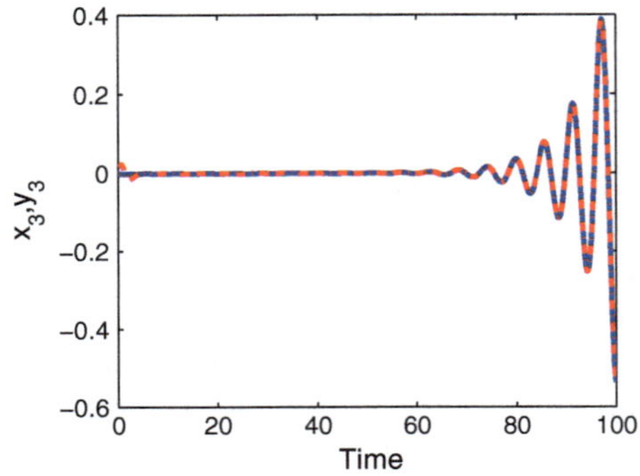

Fig. 7 Synchronization of the states $x_3(t)$ and $y_3(t)$

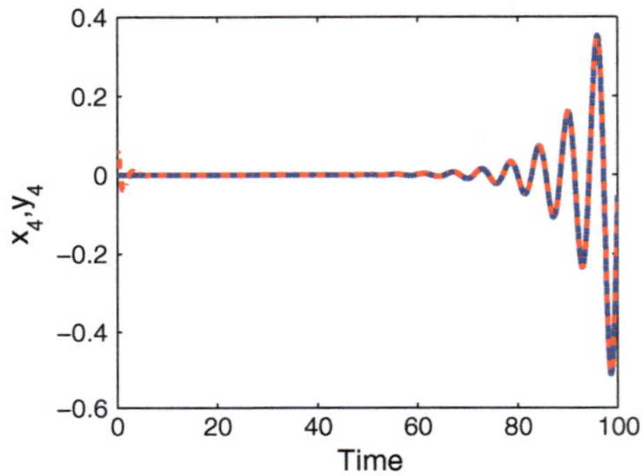

Fig. 8 Synchronization of the states $x_4(t)$ and $y_4(t)$

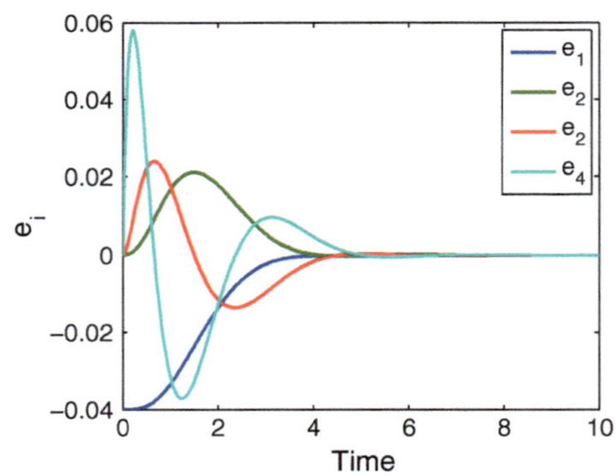

Fig. 9 Time series of the synchronization errors e_1, e_2, e_3, and e_4

as shown in Figs. 5, 6, 7 and 8. In such figures, time series of master states are denoted as blue solid lines while corresponding slave states are plotted as red dash-dot lines. In addition, the time-history of the complete synchronization errors e_1, e_2, e_3, and e_4 are reported in Fig. 9. The obtained results illustrate the correctness of used approach.

6 SPICE Implementation of the Memristive Hyperjerk System

In this section, an electronic circuit is proposed to implement memristive hyperjerk system (13). The circuit in Fig. 10 has been designed by applying the general approach with operational amplifiers [41, 53]. Thus, the variables x_1, x_2, x_3, x_4 of memristive system (13) are the voltages across the capacitor C1, C2, C3, and C4, respectively. As shown in Fig. 10 the memristive system is realized by common electronic components. Indeed the sub-circuit of memristive device in Fig. 10 only emulates the memristive device because there are not any commercial off-the-shelf memristive devices in the market yet. By applying Kirchhoff's circuit laws, the corresponding circuital equations of designed circuit can be written as:

$$\begin{cases} \frac{dv_{C_1}}{dt} = \frac{1}{R_1C_1}v_{C_2} \\ \frac{dv_{C_2}}{dt} = \frac{1}{R_2C_2}v_{C_3} \\ \frac{dv_{C_3}}{dt} = \frac{1}{R_3C_3}v_{C_4} \\ \frac{dv_{C_4}}{dt} = -\frac{1}{R_4C_4}v_{C_3} - \frac{1}{R_5C_4}v_{C_4} - \frac{1}{R_6C_4}v_{C_3}v_{C_4} - \frac{1}{R_7C_4}y, \end{cases} \tag{59}$$

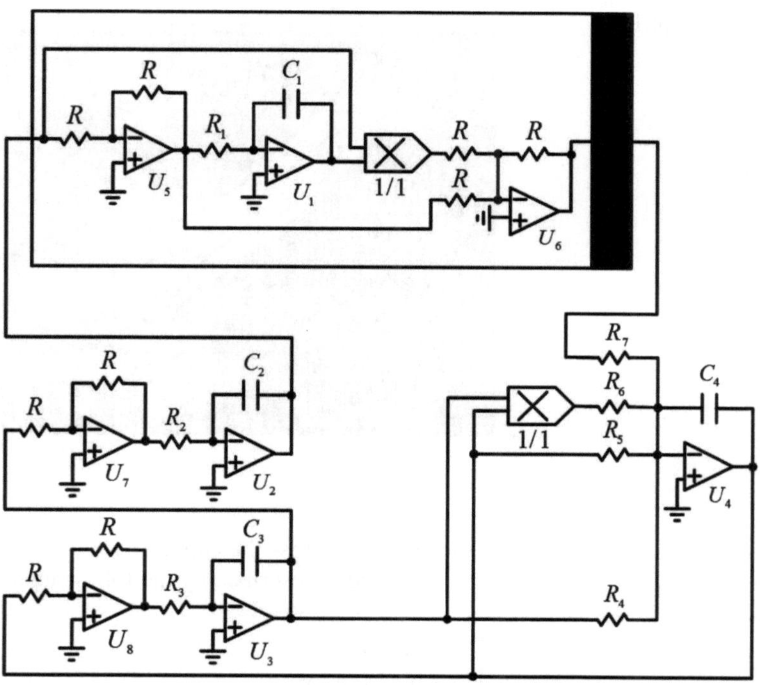

Fig. 10 Schematic of the circuit which emulating novel hyperjerk system (8) with the presence of the memristive device

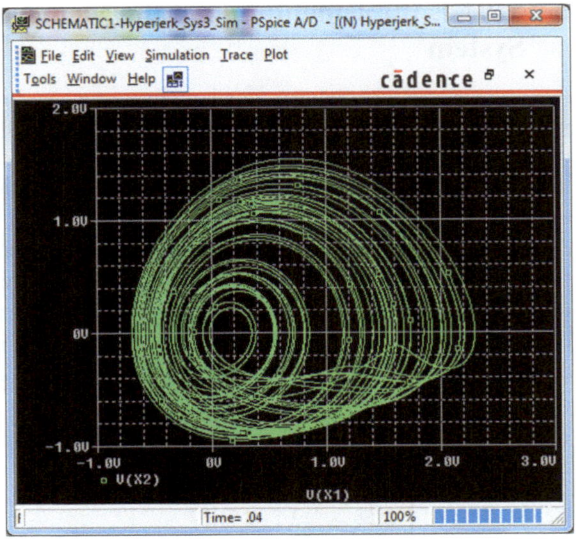

Fig. 11 Chaotic attractor of the designed circuit obtained from Cadence OrCAD in the (v_{C_1}, v_{C_2}) phase plane

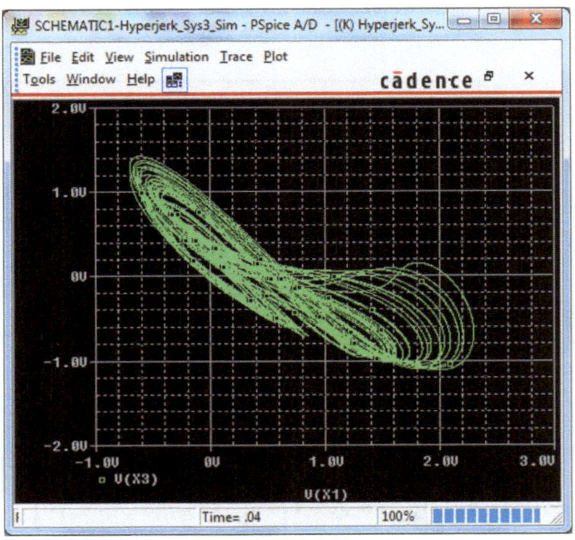

Fig. 12 Chaotic attractor of the designed circuit obtained from Cadence OrCAD in the (v_{C_1}, v_{C_3}) phase plane

where v_{C_1}, v_{C_2}, v_{C_3}, and v_{C_4} are the voltages across the capacitors C_1, C_2, C_3, and C_4, respectively. Here the memristive device is described by the following circuital equations:

$$\begin{cases} \frac{dv_{C_1}}{dt} = \frac{1}{R_1 C_1} v_{C_2} \\ y = v_{C_2} - v_{C_1} v_{C_2}. \end{cases} \tag{60}$$

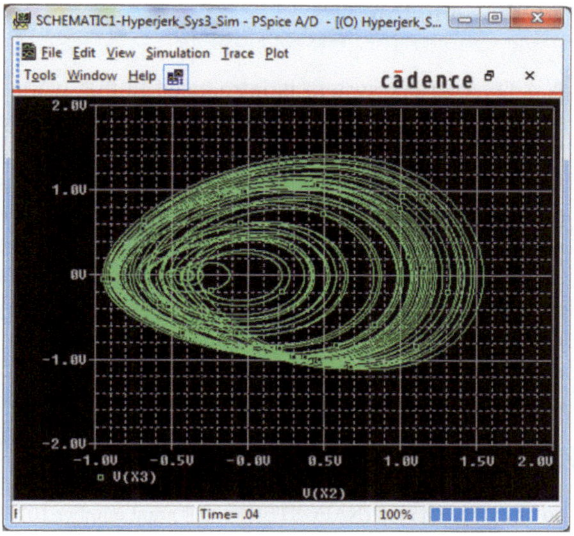

Fig. 13 Chaotic attractor of the designed circuit obtained from Cadence OrCAD in the (v_{C_2}, v_{C_3}) phase plane

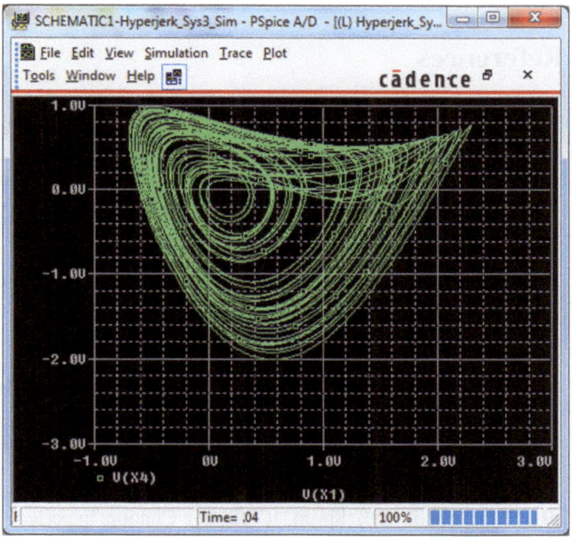

Fig. 14 Chaotic attractor of the designed circuit obtained from Cadence OrCAD in the (v_{C_1}, v_{C_4}) phase plane

The power supplies of all active devices are $\pm 15\,V_{DC}$ and the operational amplifiers TL084 are used in this work. The values of components are selected as follows: $R_1 = R_2 = R_3 = R_4 = R_7 = R = 100\,\text{k}\Omega$, $R_5 = 200\,\text{k}\Omega$, $R_6 = 250\,\text{k}\Omega$, and $C_1 = C_2 = C_3 = C_4 = 1\,\text{nF}$.

The designed circuit is implemented in the electronic simulation package Cadence OrCAD and the obtained results are reported in Figs. 11, 12, 13 and 14. Theoretical attractors (see Fig. 2) are similar with the circuital ones (see Figs. 11, 12, 13 and 14).

7 Conclusion

A 4-D hyperjerk system is introduced in this work. The hyperjerk system is constructed by using a memristive device which creates the special feature of such hyperjerk system, possessing an infinite number of equilibrium points. This special feature is rarely observed in other chaotic hyperjerk systems. Dynamical behaviors of the memristive hyperjerk system are investigated through equilibrium points, projections of chaotic attractors, Lyapunov exponents and Kaplan–Yorke dimension. In addition, the capacity of synchronization scheme of memristive hyperjerk systems is shown via backstepping control approach. To verify the feasibility of such hyperjerk system, we present its circuital implementation. Because the designed circuit modeling the hyperjerk system can generate chaos, it can applied into potential applications in various fields of chaos-based engineering, such as secure communications, random bit generation, liquid mixing or path planning for mobile robot, etc.

Acknowledgments This research is funded by Vietnam National Foundation for Science and Technology Development (NAFOSTED) under grant number 102.99–2013.06.

References

1. Arneodo A, Coullet P, Tresser C (1981) Possible new strange attractors with spiral structure. Comm Math Phys 79:573–579
2. Azar AT, Vaidyanathan S (2015) Chaos modeling and control systems design. Springer, New York
3. Azar AT, Vaidyanathan S (2015) Computational intelligence applications in modeling and control. Springer, New York
4. Bao B, Zou X, Liu Z, Hu F (2013) Generalized memory element and chaotic memory system. Int J Bif Chaos 23:1350135
5. Boccaletti S, Kurths J, Osipov G, Valladares DL, Zhou CS (2002) The synchronization of chaotic systems. Phys Rep 366:1–101
6. Buscarino A, Fortuna L, Frasca M (2009) Experimental robust synchronization of hyperchaotic circuits. Physica D 238:1917–1922
7. Chen GR (1999) Controlling chaos and bifurcations in engineering systems. CRC Press, Boca Raton
8. Chen G, Yu X (2003) Chaos control: theory and applications. Springer, Berlin
9. Chlouverakis KE, Sprott JC (2006) Chaotic hyperjerk systems. Chaos, Solitons Fractals 28:739–746
10. Chua LO (1971) Memristor–the missing circuit element. IEEE Trans Circuit Theory 18: 507–519
11. Chua LO, Kang SM (1976) Memristive devices and system. Proc IEEE 64:209–223
12. Elhadj Z, Sprott JC (2013) Transformation of 4-D dynamical systems to hyperjerk form. Palest J Maths 2:38–45
13. Fortuna L, Frasca M (2007) Experimental synchronization of single-transistor-based chaotic circuits. Chaos 17:043118-1–5
14. Gamez-Guzman L, Cruz-Hernandez C, Lopez-Gutierrez R, Garcia-Guerrero EE (2009) Synchronization of chua's circuits with multi-scroll attractors: application to communication. Commun Nonlinear Sci Numer Simul 14:2765–2775

15. Jafari S, Sprott JC (2013) Simple chaotic flows with a line equilibrium. Chaos, Solitons Fractals 57:79–84
16. Jafari S, Sprott JC, Golpayegani SMRH (2013) Elementary quadratic chaotic flows with no equilibria. Phys Lett A 377:699–702
17. Kapitaniak T (1994) Synchronization of chaos using continuous control. Phys Rev E 50: 1642–1644
18. Karthikeyan R, Vaidyanathan S (2014) Hybrid chaos synchronization of four-scroll systems via active control. J Electr Eng 65:97–103
19. Khalil H (2002) Nonlinear systems. Prentice Hall, New Jersey
20. Kingni ST, Jafari S, Simo H, Woafo P (2014) Three-dimensional chaotic autonomous system with only one stable equilibrium: analysis, circuit design, parameter estimation, control, synchronization and its fractional-order form. Eur Phys J Plus 129:76
21. Lainscsek C, Lettellier C, Gorodnitsky I (2003) Global modeling of the rössler system from the z-variable. Phys Lett A 314:409–427
22. Leonov GA, Kuznetsov NV (2011) Algorithms for searching for hidden oscillations in the Aizerman and Kalman problems. Dokl Math 84:475–481
23. Leonov GA, Kuznetsov NV (2013) Hidden attractors in dynamical systems: from hidden oscillation in Hilbert-Kolmogorov, Aizerman and Kalman problems to hidden chaotic attractor in Chua circuits. Int J Bifurc Chaos 23:1330002
24. Leonov GA, Kuznetsov NV, Kuznetsova OA, Seldedzhi SM, Vagaitsev VI (2011) Hidden oscillations in dynamical systems. Trans Syst Contr 6:54–67
25. Leonov GA, Kuznetsov NV, Vagaitsev VI (2011) Localization of hidden Chua's attractors. Phys Lett A 375:2230–2233
26. Leonov GA, Kuznetsov NV, Vagaitsev VI (2012) Hidden attractor in smooth Chua system. Physica D 241:1482–1486
27. Leonov GA, Kuznetsov NV, Kiseleva MA, Solovyeva EP, Zaretskiy AM (2014) Hidden oscillations in mathematical model of drilling system actuated by induction motor with a wound rotor. Nonlinear Dyn 77:277–288
28. Linz SJ (1997) Nonlinear dynamical models and jerky motion. Am J Phys 65:523–526
29. Linz SJ (2008) On hyperjerky systems. Chaos, Solitons Fractals 37:741–747
30. Liu C, Yi J, Xi X, An L, Fu Y (2012) Research on the multi-scroll chaos generation based on Jerk mode. Procedia Eng 29:957–961
31. Lorenz EN (1963) Deterministic non-periodic flow. J Atmos Sci 20:130–141
32. Lü J, Chen G (2002) A new chaotic attractor coined. Int J Bifurc Chaos 12:659–661
33. Ma J, Wu X, Chu R, Zhang L (2014) Selection of multi-scroll attractors in Jerk circuits and their verification using Pspice. Nonlinear Dyn 76:1951–1962
34. Molaei M, Jafari S, Sprott JC, Golpayegani S (2013) Simple chaotic flows with one stable equilibrium. Int J Bifurc Chaos 23:1350188
35. Munmuangsaen B, Srisuchinwong B, Sprott JC (2011) Generalization of the simplest autonomous chaotic system. Phys Lett A 375:1445–1450
36. Pecora LM, Carroll TL (1990) Synchronization in chaotic signals. Phys Rev A 64:821–824
37. Pehlivan I, Moroz I, Vaidyanathan S (2014) Analysis, synchronization and circuit design of a novel butterfly attractor. J Sound Vibr 333:5077–5096
38. Pershin YV, Fontaine SL, Ventra MD (2009) Memristive model of amoeba learning. Phys Rev E 80:021926
39. Pham VT, Rahma F, Frasca M, Fortuna L (2014) Dynamics and synchronization of a novel hyperchaotic system without equilibrium. Int J Bifurc Chaos 24:1450087
40. Pham V-T, Volos C, Jafari S, Wang X, Vaidyanathan S (2014) Hidden hyperchaotic attractor in a novel simple memristive neural network. Optoelectron. Adv Mater Rapid Comm 8:1157–1163
41. Pham VT, Volos CK, Jafari S, Wei Z, Wang X (2014) Constructing a novel no-equilibrium chaotic system. Int J Bifurc Chaos 24:1450073
42. Rössler OE (1976) An equation for continuous chaos. Phys Lett A 57:397–398
43. Sastry S (1999) Nonlinear systems: analysis, stability, and control. Springer, USA
44. Schot S (1978) Jerk: the time rate of change of acceleration. Am J Phys 46:1090–1094

45. Sprott JC (1997) Some simple chaotic jerk functions. Am J Phys 65:537–543
46. Sprott JC (2003) Chaos and times-series analysis. Oxford University Press, Oxford
47. Sprott JC (2010) Elegant chaos: algebraically simple chaotic flows. World Scientific, Singapore
48. Sprott JC (2011) A new chaotic jerk circuit. IEEE Trans Circuits Syst-II: Exp Briefs 58:240–243
49. Srinivasan K, Senthilkumar DV, Murali K, Lakshmanan M, Kurths J (2011) Synchronization transitions in coupled time-delay electronic circuits with a threshold nonlinearity. Chaos 21:023119
50. Strogatz SH (1994) Nonlinear dynamics and chaos: with applications to physics, biology, chemistry, and engineering. Perseus Books, Massachusetts
51. Strukov DB, Snider GS, Stewart DR, Williams RS (2008) The missing memristor found. Nature 453:80–83
52. Sun KH, Sprott JC (2009) A simple jerk system with piecewise exponential nonlinearity. Int J Nonlinear Sci Num Simu 10:1443–1450
53. Sundarapandian V, Pehlivan I (2012) Analysis, control, synchronization, and circuit design of a novel chaotic system. Math Comp Model 55:1904–1915
54. Tetzlaff R (2014) Memristor and memristive systems. Springer, New York
55. Tour JM, He T (2008) The fourth element. Nature 453:42–43
56. Vaidyanathan S (2012) Anti-synchronization of four-wing chaotic systems via sliding mode control. Int J Auto Compt 9:274–279
57. Vaidyanathan S (2013) A new six-term 3-D chaotic system with an exponential nonlineariry. Far East J Math Sci 79:135–143
58. Vaidyanathan S (2014) Analysis and adaptive synchronization of eight-term novel 3-D chaotic system with three quadratic nonlinearities. Eur Phys J Spec Topics 223:1519–1529
59. Vaidyanathan S, Volos C, Pham VT, Madhavan K, Idowo BA (2014) Adaptive backstepping control, synchronization and circuit simualtion of a 3-D novel jerk chaotic system with two hyperbolic sinusoidal nonlinearities. Arch Cont Sci 33:257–285
60. Vaidyanathan S, Volos C, Pham VT (2015) Analysis, adaptive control and synchronization of a novel 4-D hyperchaotic hyperjerk system and its SPICE implementation. Arch Cont Sci 25:257–285
61. Volos CK, Kyprianidis IM, Stouboulos IN (2011) Various synchronization phenomena in bidirectionally coupled double scroll circuits. Commun Nonlinear Sci Numer Simul 71:3356–3366
62. Volos CK, Kyprianidis IM, Stouboulos IN (2013) Image encryption process based on chaotic synchronization phenomena. Signal Process 93:1328–1340
63. Wang X, Chen G (2012) A chaotic system with only one stable equilibrium. Commun Nonlinear Sci Numer Simul 17:1264–1272
64. Wang X, Chen G (2013) Constructing a chaotic system with any number of equilibria. Nonlinear Dyn 71:429–436
65. Yalcin M (2007) Multi-scroll and hypercube attractors from a general Jerk circuit using Josephson junctions. Chaos, Solitons Fractals 34:1659–1666
66. Yalcin ME, Suykens JAK, Vandewalle J (2005) Cellular Neural Networks Multi-Scroll Chaos and Synchronization. World Scientific, Singapore
67. Yu S, Lü J, Leung H, Chen G (2005) Design and implementation of n-scroll chaotic attractors from a general Jerk circuit. IEEE Trans Circ Syst I 52:1459–1476

A Novel Hyperjerk System with Two Quadratic Nonlinearities and Its Adaptive Control

Sundarapandian Vaidyanathan

Abstract This work announces a novel 4-D hyperjerk system with two cubic nonlinearities. The proposed chaotic system is an eight-term polynomial system with two cubic nonlinearities. The phase portraits of the novel hyperjerk system are displayed and the qualitative properties of the system are discussed. The novel hyperjerk system has a unique equilibrium, which is unstable. The Lyapunov exponents of the novel hyperjerk system are obtained as $L_1 = 0.0622$, $L_2 = 0$, $L_3 = -0.4639$ and $L_4 = -0.5945$, which shows that the novel hyperjerk system is chaotic. The Kaplan–Yorke dimension of the novel hyperjerk system is obtained as $D_{KY} = 2.1341$. Next, an adaptive backstepping controller is designed to globally stabilize the novel hyperjerk system with unknown parameters. Moreover, an adaptive backstepping controller is also designed to achieve global chaos synchronization of the identical novel hyperjerk systems with unknown parameters. The main control results in this work are established using Lyapunov stability theory. MATLAB simulations have been shown to illustrate the phase portraits of the novel hyperjerk system and also the adaptive backstepping control results.

Keywords Chaos · Chaotic systems · Backstepping control · Adaptive control · Synchronization · Hyperjerk systems

1 Introduction

Chaos theory deals with the qualitative study of chaotic dynamical systems and their applications in science and engineering. A dynamical system is called *chaotic* if it satisfies the three properties: boundedness, infinite recurrence and sensitive dependence on initial conditions [3].

S. Vaidyanathan (✉)
Research and Development Centre, Vel Tech University,
Avadi, Chennai 600062, India
e-mail: sundarvtu@gmail.com

S. Vaidyanathan and C. Volos (eds.), *Advances and Applications in Chaotic Systems*, Studies in Computational Intelligence 636,
DOI 10.1007/978-3-319-30279-9_3

Some classical paradigms of 3-D chaotic systems in the literature are Lorenz system [15], Rössler system [25], ACT system [2], Sprott systems [32], Chen system [6], Lü system [16], Cai system [4], Tigan system [43], etc.

Many new chaotic systems have been discovered in the recent years such as Zhou system [117], Zhu system [118], Li system [13], Wei-Yang system [115], Sundarapandian systems [35, 40], Vaidyanathan systems [51, 52, 54–57, 60, 71, 72, 86, 89, 91, 100, 103, 105, 107, 109, 110], Pehlivan system [18], Sampath system [26], etc.

Chaos theory has many applications in science and engineering such as chemical systems [61, 65, 67, 69, 73, 77–79], biological systems [59, 62–64, 66, 68, 70, 74–76, 80–84], memristors [1, 19, 113], etc.

The study of control of a chaotic system investigates feedback control methods that globally or locally asymptotically stabilize or regulate the outputs of a chaotic system. Many methods have been designed for control and regulation of chaotic systems such as active control [33, 34, 45], adaptive control [101, 108, 111], backstepping control [14, 114], sliding mode control [48, 50], etc.

Synchronization of chaotic systems is a phenomenon that occurs when two or more chaotic systems are coupled or when a chaotic system drives another chaotic system. Because of the butterfly effect which causes exponential divergence of the trajectories of two identical chaotic systems started with nearly the same initial conditions, the synchronization of chaotic systems is a challenging research problem in the chaos literature [3].

Pecora and Carroll pioneered the research on synchronization of chaotic systems with their seminal papers [5, 17]. The active control method [11, 27, 28, 39, 44, 49, 92, 93, 96] is typically used when the system parameters are available for measurement. Adaptive control method [29–31, 36–38, 47, 53, 85, 90, 94, 95, 102, 106] is typically used when some or all the system parameters are not available for measurement and estimates for the uncertain parameters of the systems.

Sampled-data feedback control method [9, 116] and time-delay feedback control method [7, 10] are also used for synchronization of chaotic systems. Backstepping control method [20–24, 42, 97, 104, 112] is also used for the synchronization of chaotic systems. Backstepping control is a recursive method for stabilizing the origin of a control system in strict-feedback form [12]. In this research work, we apply backstepping control method for the adaptive control and synchronization of the novel hyperjerk system.

Sliding mode control method [41, 46, 58, 87, 88, 98, 99] is also a popular method for the synchronization of chaotic systems.

In the recent decades, there is some good interest in finding novel chaotic and hyperchaotic systems, which can be expressed by an explicit fourth order differential equation describing the time evolution of the single scalar variable x given by

$$\frac{d^4x}{dt} = j\left(x, \frac{dx}{dt}, \frac{d^2x}{dt^2}, \frac{d^3x}{dt^3}\right) \tag{1}$$

The differential equation (1) is called "hyperjerk system" because the fourth order time derivative in mechanical systems is called *hyperjerk* [8].

By defining phase variables

$$x_1 = x, \quad x_2 = \frac{dx}{dt}, \quad x_3 = \frac{d^2x}{dt^2}, \quad x_4 = \frac{d^3x}{dt^3}, \tag{2}$$

the hyperjerk differential equation (1) can be expressed as a 4-D system given by

$$\begin{cases} \dot{x}_1 = x_2 \\ \dot{x}_2 = x_3 \\ \dot{x}_3 = x_4 \\ \dot{x}_4 = j(x_1, x_2, x_3, x_4) \end{cases} \tag{3}$$

In this research work, we announce a 4-D novel hyperjerk system and discuss the qualitative properties of the novel hyperjerk system. We have designed adaptive backstepping controllers for stabilization and synchronization of the 4-D novel hyperjerk system.

This work is organized as follows. Section 2 describes the dynamic equations and phase portraits of the 4-D novel hyperjerk system. Section 3 details the qualitative properties of the novel hyperjerk system.

The Lyapunov exponents of the hyperjerk system are obtained as $L_1 = 0.0622$, $L_2 = 0$, $L_3 = -0.4639$ and $L_4 = -0.5945$. The Kaplan–Yorke dimension of the hyperjerk system is obtained as $D_{KY} = 2.1341$.

In Sect. 4, we design an adaptive backstepping controller to globally stabilize the novel hyperjerk system with unknown parameters. In Sect. 5, an adaptive backstepping controller is designed to achieve global chaos synchronization of the identical novel hyperjerk systems with unknown parameters. Section 6 contains a summary of the main results derived in this work.

2 A 4-D Novel Hyperjerk System

In this section, we describe a 4-D novel hyperjerk system with two quadratic nonlinearities, which is modeled by the 4-D dynamics

$$\begin{cases} \dot{x}_1 = x_2 \\ \dot{x}_2 = x_3 \\ \dot{x}_3 = x_4 \\ \dot{x}_4 = 1 - ax_1 - x_2^2 - x_3^2 - bx_3 - cx_4 \end{cases} \tag{4}$$

where x_1, x_2, x_3, x_4 are the states and a, b, c are constant positive parameters.

The system (4) is a nine-term polynomial system having two quadratic nonlinearities.

The equilibrium points of (4) are obtained by solving the system

$$x_2 = 0, \quad x_3 = 0, \quad x_4 = 0, \quad 1 - ax_1 - x_2^2 - x_3^2 - bx_3 - cx_4 = 0 \tag{5}$$

By solving the Eq. (5), we see that the system (4) has a unique equilibrium point given by

$$E_1 = \begin{bmatrix} 1/a \\ 0 \\ 0 \\ 0 \end{bmatrix} \tag{6}$$

The system (4) exhibits a *strange chaotic attractor* for the parameter values

$$a = 1, \quad b = 4, \quad c = 1 \tag{7}$$

For numerical simulations, we take the initial conditions as

$$x_1(0) = 0.5, \quad x_2(0) = 0.4, \quad x_3(0) = 0.3, \quad x_4(0) = 0.9 \tag{8}$$

Figures 1, 2, 3 and 4 show the 3-D projection of the novel hyperjerk system (4) on the (x_1, x_2, x_3), (x_1, x_2, x_4), (x_1, x_3, x_4) and (x_2, x_3, x_4) spaces, respectively.

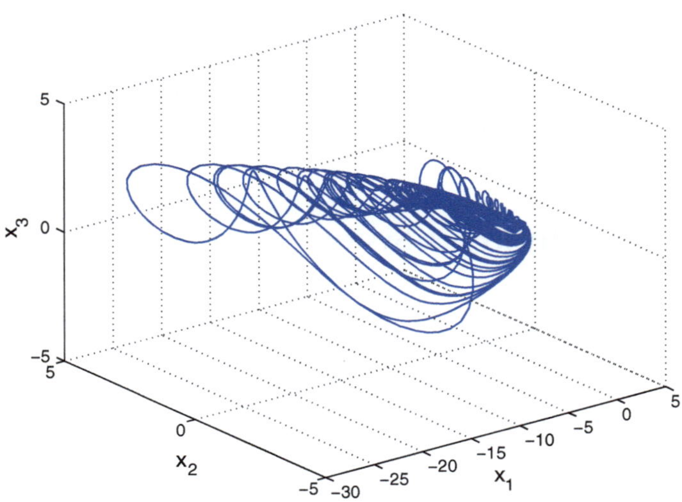

Fig. 1 3-D projection of the novel hyperjerk system on the (x_1, x_2, x_3) space

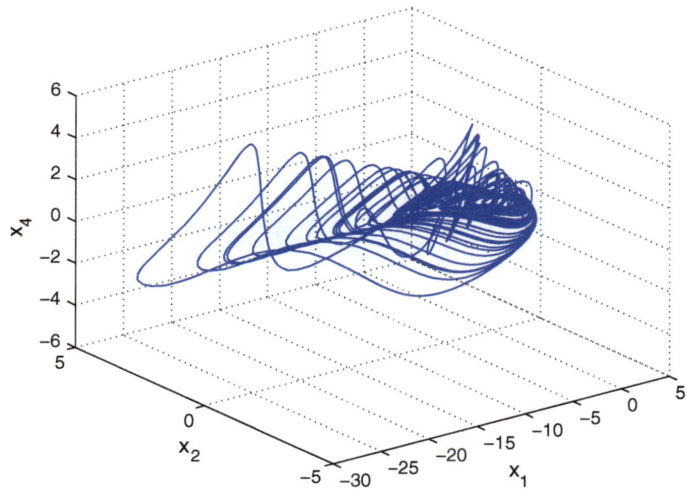

Fig. 2 3-D projection of the novel hyperjerk system on the (x_1, x_2, x_4) space

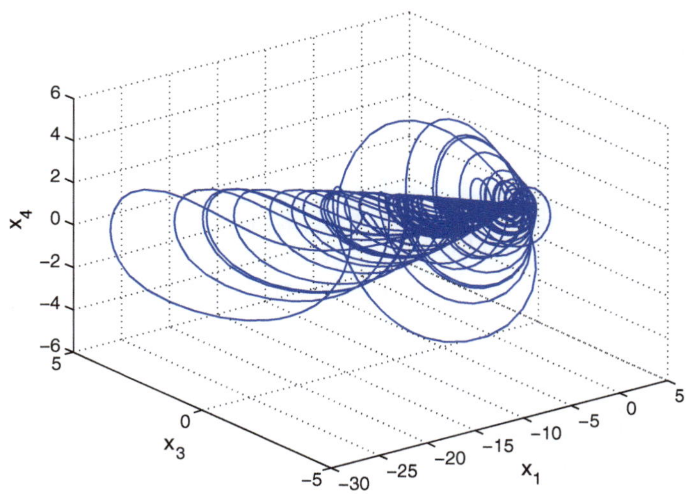

Fig. 3 3-D projection of the novel hyperjerk system on the (x_1, x_3, x_4) space

3 Analysis of the Novel Hyperjerk System

In this section, we give a dynamic analysis of the novel hyperjerk system (4).

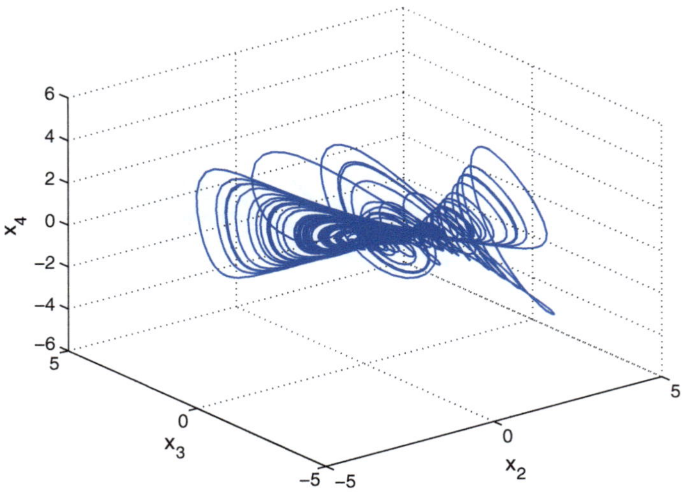

Fig. 4 3-D projection of the novel hyperjerk system on the (x_2, x_3, x_4) space

3.1 Dissipativity

In vector notation, the novel hyperjerk system (4) can be expressed as

$$\dot{\mathbf{x}} = f(\mathbf{x}) = \begin{bmatrix} f_1(x_1, x_2, x_3, x_4) \\ f_2(x_1, x_2, x_3, x_4) \\ f_3(x_1, x_2, x_3, x_4) \\ f_4(x_1, x_2, x_3, x_4) \end{bmatrix}, \tag{9}$$

where

$$\begin{cases} f_1(x_1, x_2, x_3, x_4) = x_2 \\ f_2(x_1, x_2, x_3, x_4) = x_3 \\ f_3(x_1, x_2, x_3, x_4) = x_4 \\ f_4(x_1, x_2, x_3, x_4) = 1 - ax_1 - x_2^2 - x_3^2 - bx_3 - cx_4 \end{cases} \tag{10}$$

Let Ω be any region in $\mathbf{R}^4$ with a smooth boundary and also, $\Omega(t) = \Phi_t(\Omega)$, where Φ_t is the flow of f. Furthermore, let $V(t)$ denote the hypervolume of $\Omega(t)$.

By Liouville's theorem, we know that

$$\dot{V}(t) = \int\limits_{\Omega(t)} (\nabla \cdot f)\, dx_1\, dx_2\, dx_3\, dx_4 \tag{11}$$

The divergence of the novel hyperjerk system (9) is found as:

$$\nabla \cdot f = \frac{\partial f_1}{\partial x_1} + \frac{\partial f_2}{\partial x_2} + \frac{\partial f_3}{\partial x_3} + \frac{\partial f_4}{\partial x_4} = -c < 0 \tag{12}$$

since c is a positive constant.

Inserting the value of $\nabla \cdot f$ from (12) into (11), we get

$$\dot{V}(t) = \int\limits_{\Omega(t)} (-1)\, dx_1\, dx_2\, dx_3\, dx_4 = -cV(t) \tag{13}$$

Integrating the first order linear differential equation (13), we get

$$V(t) = \exp(-ct)V(0) \tag{14}$$

From Eq. (14), it follows that $V(t) \to 0$ exponentially as $t \to \infty$. This shows that the hyperjerk system (4) is dissipative. Hence, the system limit sets are ultimately confined into a specific limit set of zero hypervolume, and the asymptotic motion of the hyperjerk system (4) settles onto a strange attractor of the system.

3.2 *Equilibrium Points*

We take the parameter values as in the chaotic case (7), i.e.

$$a = 1, \quad b = 4, \quad c = 1 \tag{15}$$

In Sect. 2, we showed that the novel hyperjerk system (4) has a unique equilibrium point given by

$$E_1 = \begin{bmatrix} 1/a \\ 0 \\ 0 \\ 0 \end{bmatrix} = \begin{bmatrix} 1 \\ 0 \\ 0 \\ 0 \end{bmatrix} \tag{16}$$

To test the stability type of the equilibrium point E_1, we calculate the Jacobian matrix of the novel hyperjerk system (4) at any point x:

$$J(x) = \begin{bmatrix} 0 & 1 & 0 & 0 \\ 0 & 0 & 1 & 0 \\ 0 & 0 & 0 & 1 \\ -a & -2x_2 & -2x_3 - b & -c \end{bmatrix} = \begin{bmatrix} 0 & 1 & 0 & 0 \\ 0 & 0 & 1 & 0 \\ 0 & 0 & 0 & 1 \\ -1 & -2x_2 & -2x_3 - 4 & -1 \end{bmatrix} \tag{17}$$

We find that

$$J_1 \triangleq J(E_1) = \begin{bmatrix} 0 & 1 & 0 & 0 \\ 0 & 0 & 1 & 0 \\ 0 & 0 & 0 & 1 \\ -1 & 0 & -4 & -1 \end{bmatrix} \tag{18}$$

The matrix J_1 has the eigenvalues

$$\lambda_{1,2} = -0.5368 \pm 1.8785i, \quad \lambda_{3,4} = 0.0368 \pm 0.5105i \tag{19}$$

This shows that the equilibrium point E_1 is a saddle-focus, which is unstable.

3.3 Lyapunov Exponents and Kaplan–Yorke Dimension

We take the parameter values of the novel hyperjerk system (4) as $a = 1, b = 4$ and $c = 1$. We take the initial state of the novel hyperjerk system (4) as given in (8).

Then the Lyapunov exponents of the novel hyperjerk system (4) are numerically obtained using MATLAB as

$$L_1 = 0.0622, \quad L_2 = 0, \quad L_3 = -0.4639, \quad L_4 = -0.5945 \tag{20}$$

Thus, the maximal Lyapunov exponent (MLE) of the novel hyperjerk system (4) is positive, which means that the system has a chaotic behavior.

Since $L_1 + L_2 + L_3 + L_4 = -0.9962 < 0$, it follows that the novel hyperjerk system (4) is dissipative.

Also, the Kaplan–Yorke dimension of the hyperjerk system (4) is obtained as

$$D_{KY} = 2 + \frac{L_1 + L_2}{|L_3|} = 2.1341 \tag{21}$$

which is fractional.

4 Adaptive Control of the Novel Hyperjerk System

In this section, we use backstepping control method to derive an adaptive feedback control law for globally stabilizing the 4-D novel hyperjerk system with unknown parameters.

Thus, we consider the 4-D novel hyperjerk system given by

$$\begin{cases} \dot{x}_1 = x_2 \\ \dot{x}_2 = x_3 \\ \dot{x}_3 = x_4 \\ \dot{x}_4 = 1 - ax_1 - x_2^2 - x_3^2 - bx_3 - cx_4 + u \end{cases} \tag{22}$$

where x_1, x_2, x_3, x_4 are the states, a, b, c are unknown constant parameters, and u is a backstepping control law to be determined using estimates $\hat{a}(t)$, $\hat{b}(t)$ and $\hat{c}(t)$ for a, b and c, respectively.

The parameter estimation errors are defined as:

$$\begin{cases} e_a(t) = a - \hat{a}(t) \\ e_b(t) = b - \hat{b}(t) \\ e_c(t) = c - \hat{c}(t) \end{cases} \tag{23}$$

Differentiating (23) with respect to t, we obtain the following equations:

$$\begin{cases} \dot{e}_a(t) = -\dot{\hat{a}}(t) \\ \dot{e}_b(t) = -\dot{\hat{b}}(t) \\ \dot{e}_c(t) = -\dot{\hat{c}}(t) \end{cases} \tag{24}$$

Next, we shall state and prove the main result of this section.

Theorem 1 *The 4-D novel hyperjerk system (22), with unknown parameters a, b and c, is globally and exponentially stabilized by the adaptive feedback control law,*

$$u(t) = -1 - [5 - \hat{a}(t)]\, x_1 - 10x_2 - \left[9 - \hat{b}(t)\right] x_3 + \hat{c}(t)x_4 + x_2^2 + x_3^2 - kz_4 \tag{25}$$

where $k > 0$ is a gain constant,

$$z_4 = 3x_1 + 5x_2 + 3x_3 + x_4 \tag{26}$$

and the update law for the parameter estimates $\hat{a}(t), \hat{b}(t), \hat{c}(t)$ is given by

$$\begin{cases} \dot{\hat{a}}(t) = -x_1 z_4 \\ \dot{\hat{b}}(t) = -x_3 z_4 \\ \dot{\hat{c}}(t) = -x_4 z_4 \end{cases} \tag{27}$$

Proof We prove this result via Lyapunov stability theory [12].

First, we define a quadratic Lyapunov function

$$V_1(z_1) = \frac{1}{2}\, z_1^2 \tag{28}$$

where

$$z_1 = x_1 \tag{29}$$

Differentiating V_1 along the dynamics (22), we get

$$\dot{V}_1 = z_1 \dot{z}_1 = x_1 x_2 = -z_1^2 + z_1(x_1 + x_2) \tag{30}$$

Now, we define

$$z_2 = x_1 + x_2 \tag{31}$$

Using (31), we can simplify the Eq. (30) as

$$\dot{V}_1 = -z_1^2 + z_1 z_2 \tag{32}$$

Secondly, we define a quadratic Lyapunov function

$$V_2(z_1, z_2) = V_1(z_1) + \frac{1}{2} z_2^2 = \frac{1}{2} \left(z_1^2 + z_2^2\right) \tag{33}$$

Differentiating V_2 along the dynamics (22), we get

$$\dot{V}_2 = -z_1^2 - z_2^2 + z_2(2x_1 + 2x_2 + x_3) \tag{34}$$

Now, we define

$$z_3 = 2x_1 + 2x_2 + x_3 \tag{35}$$

Using (35), we can simplify the Eq. (34) as

$$\dot{V}_2 = -z_1^2 - z_2^2 + z_2 z_3 \tag{36}$$

Thirdly, we define a quadratic Lyapunov function

$$V_3(z_1, z_2, x_3) = V_2(z_1, z_2) + \frac{1}{2} z_3^2 = \frac{1}{2} \left(z_1^2 + z_2^2 + z_3^2\right) \tag{37}$$

Differentiating V_3 along the dynamics (22), we get

$$\dot{V}_3 = -z_1^2 - z_2^2 - z_3^2 + z_3(3x_1 + 5x_2 + 3x_3 + x_4) \tag{38}$$

Now, we define

$$z_4 = 3x_1 + 5x_2 + 3x_3 + x_4 \tag{39}$$

Using (39), we can simplify the Eq. (38) as

$$\dot{V}_2 = -z_1^2 - z_2^2 - z_3^2 + z_3 z_4 \tag{40}$$

Finally, we define a quadratic Lyapunov function

$$V(z_1, z_2, z_3, z_4, e_a, e_b, e_c) = V_3(z_1, z_2, z_3) + \frac{1}{2}z_4^2 + \frac{1}{2}e_a^2 + \frac{1}{2}e_b^2 + \frac{1}{2}e_c^2 \quad (41)$$

which is a positive definite function on $\mathbf{R}^7$.

Differentiating V along the dynamics (22), we get

$$\dot{V} = -z_1^2 - z_2^2 - z_3^2 - z_4^2 + z_4(z_4 + z_3 + \dot{z}_4) - e_a\dot{\hat{a}} - e_b\dot{\hat{b}} - e_c\dot{\hat{c}} \quad (42)$$

Equation (42) can be written compactly as

$$\dot{V} = -z_1^2 - z_2^2 - z_3^2 - z_4^2 + z_4 S - e_a\dot{\hat{a}} - e_b\dot{\hat{b}} - e_c\dot{\hat{c}} \quad (43)$$

where

$$S = z_4 + z_3 + \dot{z}_4 = z_4 + z_3 + 3\dot{x}_1 + 5\dot{x}_2 + 3\dot{x}_3 + \dot{x}_4 \quad (44)$$

A simple calculation gives

$$S = 1 + (5 - a)x_1 + 10x_2 + (9 - b)x_3 - cx_4 - x_2^2 - x_3^2 + u \quad (45)$$

Substituting the adaptive control law (25) into (45), we obtain

$$S = -\left[a - \hat{a}(t)\right]x_1 - \left[b - \hat{b}(t)\right]x_3 - \left[c - \hat{c}(t)\right]x_4 - kz_4 \quad (46)$$

Using the definitions (24), we can simplify (46) as

$$S = -e_a x_1 - e_b x_3 - e_c x_4 - kz_4 \quad (47)$$

Substituting the value of S from (47) into (43), we obtain

$$\begin{aligned} \dot{V} = & -z_1^2 - z_2^2 - z_3^2 - (1 + k)z_4^2 + e_a\left[-x_1 z_4 - \dot{\hat{a}}\right] \\ & + e_b\left[-x_3 z_4 - \dot{\hat{b}}\right] + e_c\left[-x_4 z_4 - \dot{\hat{c}}\right] \end{aligned} \quad (48)$$

Substituting the update law (27) into (48), we get

$$\dot{V} = -z_1^2 - z_2^2 - z_3^2 - (1 + k)z_4^2, \quad (49)$$

which is a negative semi-definite function on $\mathbf{R}^7$.

From (49), it follows that the vector $\mathbf{z}(t) = (z_1(t), z_2(t), z_3(t), z_4(t))$ and the parameter estimation error $(e_a(t), e_b(t), e_c(t))$ are globally bounded, i.e.

$$\left[\, z_1(t)\ z_2(t)\ z_3(t)\ z_4(t)\ e_a(t)\ e_b(t)\ e_c(t) \,\right] \in \mathbf{L}_\infty \quad (50)$$

Also, it follows from (49) that

$$\dot{V} \le -z_1^2 - z_2^2 - z_3^2 - z_4^2 = -\|\mathbf{z}\|^2 \tag{51}$$

That is,

$$\|\mathbf{z}\|^2 \le -\dot{V} \tag{52}$$

Integrating the inequality (52) from 0 to t, we get

$$\int_0^t |\mathbf{z}(\tau)|^2 \, d\tau \le V(0) - V(t) \tag{53}$$

From (53), it follows that $\mathbf{z}(t) \in \mathbf{L}_2$.

From Eq. (22), it can be deduced that $\dot{\mathbf{z}}(t) \in \mathbf{L}_\infty$.

Thus, using Barbalat's lemma [12], we conclude that $\mathbf{z}(t) \to \mathbf{0}$ exponentially as $t \to \infty$ for all initial conditions $\mathbf{z}(0) \in \mathbf{R}^4$.

Hence, it is immediate that $\mathbf{x}(t) \to \mathbf{0}$ exponentially as $t \to \infty$ for all initial conditions $\mathbf{x}(0) \in \mathbf{R}^4$.

This completes the proof. □

For the numerical simulations, the classical fourth-order Runge–Kutta method with step size $h = 10^{-8}$ is used to solve the system of differential equations (22) and (27), when the adaptive control law (25) and the parameter update law (27) are applied.

The parameter values of the novel hyperjerk system (22) are taken as in the chaotic case (7), i.e.

$$a = 1, \quad b = 4, \quad c = 1 \tag{54}$$

We take the positive gain constant as

$$k = 8 \tag{55}$$

As initial conditions of the hyperjerk system (22), we take

$$x_1(0) = -2.3, \quad x_2(0) = 3.4, \quad x_3(0) = 4.7, \quad x_4(0) = -1.9 \tag{56}$$

Also, as initial conditions of the parameter estimates $\hat{a}(t)$ and $\hat{b}(t)$, we take

$$\hat{a}(0) = 5.2, \quad \hat{b}(0) = 1.4, \quad \hat{c}(0) = 8.5 \tag{57}$$

In Fig. 5, the exponential convergence of the controlled states $x_1(t)$, $x_2(t)$, $x_3(t)$, $x_4(t)$ is depicted, when the adaptive control law (25) and the parameter update law (27) are implemented.

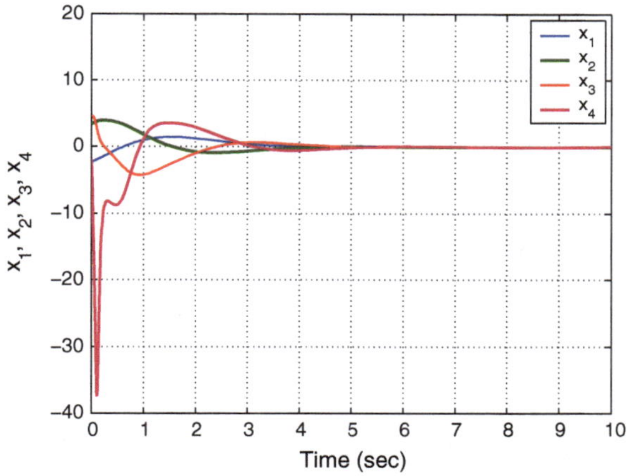

Fig. 5 Time-history of the controlled states x_1, x_2, x_3, x_4

5 Adaptive Synchronization of the Identical Novel Hyperjerk Systems

In this section, we use backstepping control to derive an adaptive control law for globally and exponentially synchronizing the identical novel hyperjerk systems with unknown parameters.

As the master system, we consider the 4-D novel hyperjerk system given by

$$\begin{cases} \dot{x}_1 = x_2 \\ \dot{x}_2 = x_3 \\ \dot{x}_3 = x_4 \\ \dot{x}_4 = 1 - ax_1 - x_2^2 - x_3^2 - bx_3 - cx_4 \end{cases} \tag{58}$$

where x_1, x_2, x_3, x_4 are the states of the system, and a, b, c are unknown constant parameters.

As the slave system, we consider the 4-D novel hyperjerk system given by

$$\begin{cases} \dot{y}_1 = y_2 \\ \dot{y}_2 = y_3 \\ \dot{y}_3 = y_4 \\ \dot{y}_4 = 1 - ay_1 - y_2^2 - y_3^2 - by_3 - cy_4 + u \end{cases} \tag{59}$$

where y_1, y_2, y_3, y_4 are the states of the system, and u is a backstepping control to be determined using estimates $\hat{a}(t)$, $\hat{b}(t)$ and $\hat{c}(t)$ for a, b and c, respectively.

We define the synchronization errors between the states of the master system (58) and the slave system (59) as

$$e_i = y_i - x_i, \quad (i = 1, 2, 3, 4) \tag{60}$$

Then the error dynamics is easily obtained as

$$\begin{cases} \dot{e}_1 = e_2 \\ \dot{e}_2 = e_3 \\ \dot{e}_3 = e_4 \\ \dot{e}_4 = -ae_1 - be_3 - ce_4 - y_2^2 + x_2^2 - y_3^2 + x_3^2 + u \end{cases} \tag{61}$$

The parameter estimation errors are defined as:

$$\begin{cases} e_a(t) = a - \hat{a}(t) \\ e_b(t) = b - \hat{b}(t) \\ e_c(t) = c - \hat{c}(t) \end{cases} \tag{62}$$

Differentiating (62) with respect to t, we obtain the following equations:

$$\begin{cases} \dot{e}_a(t) = -\dot{\hat{a}}(t) \\ \dot{e}_b(t) = -\dot{\hat{b}}(t) \\ \dot{e}_c(t) = -\dot{\hat{c}}(t) \end{cases} \tag{63}$$

Next, we shall state and prove the main result of this section.

Theorem 2 *The identical 4-D hyperjerk systems (58) and (59) with unknown parameters a, b and c are globally and exponentially synchronized by the adaptive control law*

$$\begin{aligned} u = &-\left[5 - \hat{a}(t)\right] e_1 - 10e_2 - \left[9 - \hat{b}(t)\right] e_3 - \left[4 - \hat{c}(t)\right] e_4 \\ &+ y_2^2 - x_2^2 + y_3^2 - x_3^2 - kz_4 \end{aligned} \tag{64}$$

where $k > 0$ is a gain constant,

$$z_4 = 3e_1 + 5e_2 + 3e_3 + e_4, \tag{65}$$

and the update law for the parameter estimates $\hat{a}(t)$, $\hat{b}(t)$, $\hat{c}(t)$ is given by

$$\begin{cases} \dot{\hat{a}}(t) = -e_1 z_4 \\ \dot{\hat{b}}(t) = -e_3 z_4 \\ \dot{\hat{c}}(t) = -e_4 z_4 \end{cases} \tag{66}$$

Proof We prove this result via backstepping control method and Lyapunov stability theory.

First, we define a quadratic Lyapunov function

$$V_1(z_1) = \frac{1}{2}\, z_1^2 \tag{67}$$

where

$$z_1 = e_1 \tag{68}$$

Differentiating V_1 along the error dynamics (61), we get

$$\dot{V}_1 = z_1 \dot{z}_1 = e_1 e_2 = -z_1^2 + z_1(e_1 + e_2) \tag{69}$$

Now, we define

$$z_2 = e_1 + e_2 \tag{70}$$

Using (70), we can simplify the Eq. (69) as

$$\dot{V}_1 = -z_1^2 + z_1 z_2 \tag{71}$$

Secondly, we define a quadratic Lyapunov function

$$V_2(z_1, z_2) = V_1(z_1) + \frac{1}{2} z_2^2 = \frac{1}{2}\left(z_1^2 + z_2^2\right) \tag{72}$$

Differentiating V_2 along the error dynamics (61), we get

$$\dot{V}_2 = -z_1^2 - z_2^2 + z_2(2e_1 + 2e_2 + e_3) \tag{73}$$

Now, we define

$$z_3 = 2e_1 + 2e_2 + e_3 \tag{74}$$

Using (74), we can simplify the Eq. (73) as

$$\dot{V}_2 = -z_1^2 - z_2^2 + z_2 z_3 \tag{75}$$

Thirdly, we define a quadratic Lyapunov function

$$V_3(z_1, z_2, x_3) = V_2(z_1, z_2) + \frac{1}{2} z_3^2 = \frac{1}{2}\left(z_1^2 + z_2^2 + z_3^2\right) \tag{76}$$

Differentiating V_3 along the error dynamics (61), we get

$$\dot{V}_3 = -z_1^2 - z_2^2 - z_3^2 + z_3(3e_1 + 5e_2 + 3e_3 + e_4) \tag{77}$$

Now, we define

$$z_4 = 3e_1 + 5e_2 + 3e_3 + e_4 \tag{78}$$

Using (78), we can simplify the Eq. (77) as

$$\dot{V}_2 = -z_1^2 - z_2^2 - z_3^2 + z_3 z_4 \tag{79}$$

Finally, we define a quadratic Lyapunov function

$$V(z_1, z_2, z_3, z_4, e_a, e_b, e_c) = V_3(z_1, z_2, z_3) + \frac{1}{2}z_4^2 + \frac{1}{2}e_a^2 + \frac{1}{2}e_b^2 + \frac{1}{2}e_c^2 \tag{80}$$

Differentiating V along the error dynamics (61), we get

$$\dot{V} = -z_1^2 - z_2^2 - z_3^2 - z_4^2 + z_4(z_4 + z_3 + \dot{z}_4) - e_a\dot{\hat{a}} - e_b\dot{\hat{b}} - e_c\dot{\hat{c}} \tag{81}$$

Equation (81) can be written compactly as

$$\dot{V} = -z_1^2 - z_2^2 - z_3^2 - z_4^2 + z_4 S - e_a\dot{\hat{a}} - e_b\dot{\hat{b}} - e_c\dot{\hat{c}} \tag{82}$$

where

$$S = z_4 + z_3 + \dot{z}_4 = z_4 + z_3 + 3\dot{e}_1 + 5\dot{e}_2 + 3\dot{e}_3 + \dot{e}_4 \tag{83}$$

A simple calculation gives

$$S = (5 - a)e_1 + 10e_2 + (9 - b)e_3 + (4 - c)e_4 - y_2^2 + x_2^2 - y_3^2 + x_3^2 + u \tag{84}$$

Substituting the adaptive control law (64) into (84), we obtain

$$S = -\left[a - \hat{a}(t)\right]e_1 - \left[b - \hat{b}(t)\right]e_3 - \left[c - \hat{c}(t)\right]e_4 - kz_4 \tag{85}$$

Using the definitions (63), we can simplify (85) as

$$S = -e_a e_1 - e_b e_3 - e_c e_4 - kz_4 \tag{86}$$

Substituting the value of S from (86) into (82), we obtain

$$\begin{cases} \dot{V} = -z_1^2 - z_2^2 - z_3^2 - (1 + k)z_4^2 + e_a\left[-e_1 z_4 - \dot{\hat{a}}\right] \\ \qquad +e_b\left[-e_3 z_4 - \dot{\hat{b}}\right] + e_c\left[-e_4 z_4 - \dot{\hat{c}}\right] \end{cases} \tag{87}$$

Substituting the update law (66) into (87), we get

$$\dot{V} = -z_1^2 - z_2^2 - z_3^2 - (1 + k)z_4^2, \tag{88}$$

which is a negative semi-definite function on $\mathbf{R}^7$.

From (88), it follows that the vector $\mathbf{z}(t) = (z_1(t), z_2(t), z_3(t), z_4(t))$ and the parameter estimation error $(e_a(t), e_b(t), e_c(t))$ are globally bounded, i.e.

$$\left[\, z_1(t)\ z_2(t)\ z_3(t)\ z_4(t)\ e_a(t)\ e_b(t)\ e_c(t) \,\right] \in \mathbf{L}_\infty \tag{89}$$

Also, it follows from (88) that

$$\dot{V} \leq -z_1^2 - z_2^2 - z_3^2 - z_4^2 = -\|\mathbf{z}\|^2 \tag{90}$$

That is,

$$\|\mathbf{z}\|^2 \leq -\dot{V} \tag{91}$$

Integrating the inequality (91) from 0 to t, we get

$$\int_0^t |\mathbf{z}(\tau)|^2 \, d\tau \leq V(0) - V(t) \tag{92}$$

From (92), it follows that $\mathbf{z}(t) \in \mathbf{L}_2$.

From Eq. (61), it can be deduced that $\dot{\mathbf{z}}(t) \in \mathbf{L}_\infty$.

Thus, using Barbalat's lemma [12], we conclude that $\mathbf{z}(t) \to \mathbf{0}$ exponentially as $t \to \infty$ for all initial conditions $\mathbf{z}(0) \in \mathbf{R}^4$.

Hence, it is immediate that $\mathbf{e}(t) \to \mathbf{0}$ exponentially as $t \to \infty$ for all initial conditions $\mathbf{e}(0) \in \mathbf{R}^4$. This completes the proof. □

For the numerical simulations, the classical fourth-order Runge–Kutta method with step size $h = 10^{-8}$ is used to solve the system of differential equations (58) and (59).

The parameter values of the novel hyperjerk system are taken as in the chaotic case, *viz.* $a = 1, b = 4$ and $c = 1$. The gain constant is taken as $k = 8$.

Also, as initial conditions of the master system (58), we take

$$x_1(0) = 1.8, \quad x_2(0) = -0.5, \quad x_3(0) = -2.7, \quad x_4(0) = 4.9 \tag{93}$$

Fig. 6 Synchronization of the states x_1 and y_1

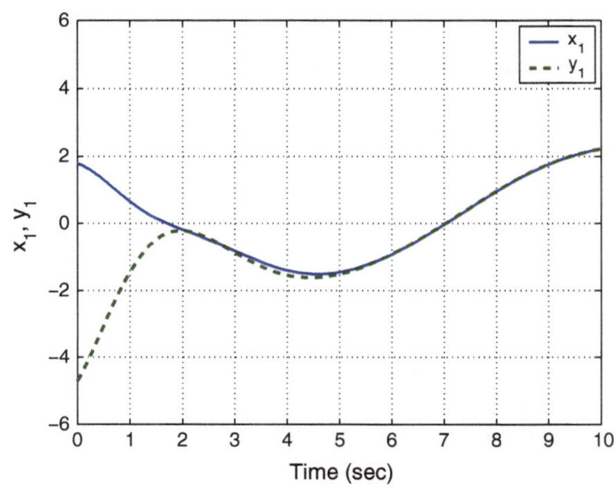

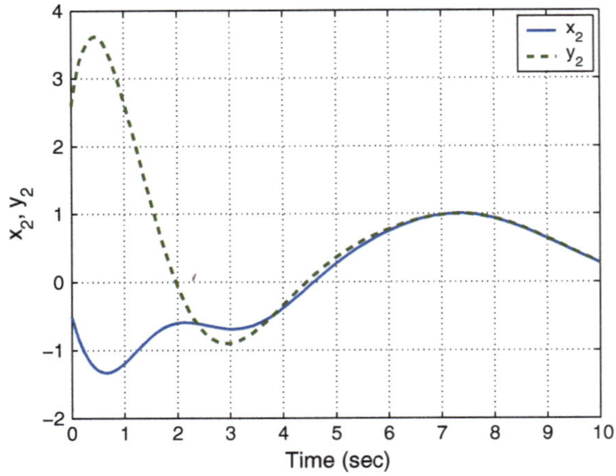

Fig. 7 Synchronization of the states x_2 and y_2

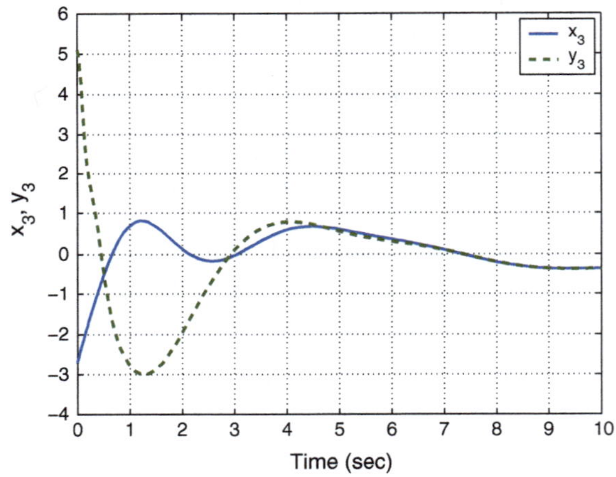

Fig. 8 Synchronization of the states x_3 and y_3

As initial conditions of the slave system (59), we take

$$y_1(0) = -4.7, \quad y_2(0) = 2.6, \quad y_3(0) = 5.1, \quad y_4(0) = -3.2 \tag{94}$$

Furthermore, as initial conditions of the parameter estimates $\hat{a}(t)$, $\hat{b}(t)$ and $\hat{c}(t)$, we take

$$\hat{a}(0) = 2.3, \quad \hat{b}(0) = 6.8, \quad \hat{c}(0) = 7.6 \tag{95}$$

In Figs. 6, 7, 8 and 9, the complete synchronization of the identical 4-D hyperjerk systems (58) and (59) is shown, when the adaptive control law and the parameter update law are implemented.

Also, in Fig. 10, the time-history of the complete synchronization errors is shown.

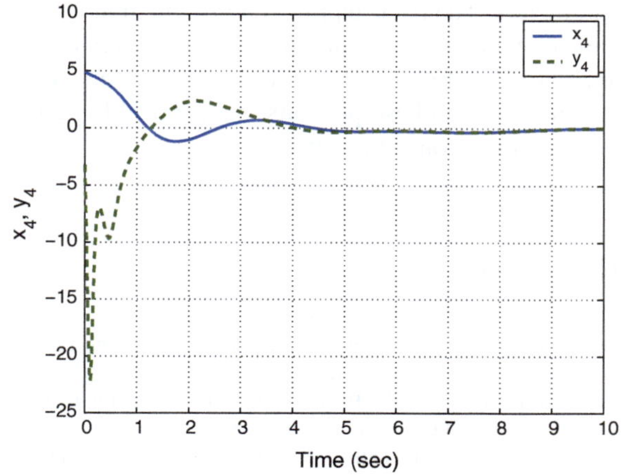

Fig. 9 Synchronization of the states x_4 and y_4

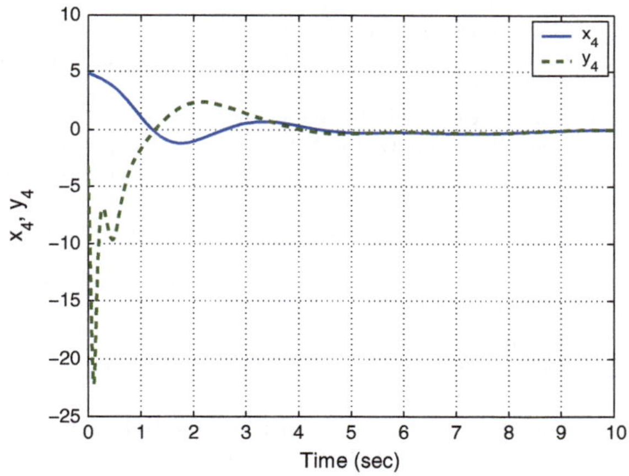

Fig. 10 Time-history of the synchronization errors e_1, e_2, e_3, e_4

6 Conclusions

This work announced a novel 4-D hyperjerk system with two cubic nonlinearities. The proposed chaotic system is an eight-term polynomial system with two cubic nonlinearities. The phase portraits of the novel hyperjerk system are displayed and the qualitative properties of the system are discussed. The novel hyperjerk system has a unique equilibrium, which is unstable. The Lyapunov exponents of the novel hyperjerk system have been obtained as $L_1 = 0.0622$, $L_2 = 0$, $L_3 = -0.4639$ and $L_4 = -0.5945$, while the Kaplan–Yorke dimension of the novel hyperjerk system has been found as $D_{KY} = 2.1341$. Next, an adaptive backstepping controller has been designed to globally stabilize the novel hyperjerk system with unknown parameters. Moreover, an adaptive backstepping controller has also been designed to achieve global chaos synchronization of the identical novel hyperjerk systems with unknown parameters. The backstepping control method is a recursive procedure that links the choice of a Lyapunov function with the design of a controller and guarantees global asymptotic stability of strict feedback systems. MATLAB simulations were shown to illustrate the phase portraits of the novel hyperjerk system and also the adaptive backstepping control results.

References

1. Abdurrahman A, Jiang H, Teng Z (2015) Finite-time synchronization for memristor-based neural networks with time-varying delays. Neural Netw 69:20–28
2. Arneodo A, Coullet P, Tresser C (1981) Possible new strange attractors with spiral structure. Commun Math Phys 79(4):573–576
3. Azar AT, Vaidyanathan S (2015) Chaos modeling and control systems design, vol 581. Springer, Germany
4. Cai G, Tan Z (2007) Chaos synchronization of a new chaotic system via nonlinear control. J Uncertain Syst. 1(3):235–240
5. Carroll TL, Pecora LM (1991) Synchronizing chaotic circuits. IEEE Trans Circuits Syst 38(4):453–456
6. Chen G, Ueta T (1999) Yet another chaotic attractor. Int J Bifurc Chaos 9(7):1465–1466
7. Chen WH, Wei D, Lu X (2014) Global exponential synchronization of nonlinear time-delay Lur'e systems via delayed impulsive control. Commun Nonlinear Sci Numer Simul 19(9):3298–3312
8. Chlouverakis KE, Sprott JC (2006) Chaotic hyperjerk systems. Chaos Solitons Fractals 28(3):739–746
9. Gan Q, Liang Y (2012) Synchronization of chaotic neural networks with time delay in the leakage term and parametric uncertainties based on sampled-data control. J Frankl Inst 349(6):1955–1971
10. Jiang GP, Zheng WX, Chen G (2004) Global chaos synchronization with channel time-delay. Chaos Solitons Fractals 20(2):267–275
11. Karthikeyan R, Sundarapandian V (2014) Hybrid chaos synchronization of four-scroll systems via active control. J Electr Eng 65(2):97–103
12. Khalil HK (2001) Nonlinear systems, 3rd edn. Prentice Hall, New Jersey
13. Li D (2008) A three-scroll chaotic attractor. Phys Lett A 372(4):387–393
14. Li GH, Zhou SP, Yang K (2007) Controlling chaos in Colpitts oscillator. Chaos Solitons Fractals 33:582–587
15. Lorenz EN (1963) Deterministic periodic flow. J Atmos Sci 20(2):130–141
16. Lü J, Chen G (2002) A new chaotic attractor coined. Int J Bifurc Chaos 12(3):659–661
17. Pecora LM, Carroll TL (1990) Synchronization in chaotic systems. Phys Rev Lett 64(8): 821–824
18. Pehlivan I, Moroz IM, Vaidyanathan S (2014) Analysis, synchronization and circuit design of a novel butterfly attractor. J Sound Vib 333(20):5077–5096
19. Pham VT, Volos CK, Vaidyanathan S, Le TP, Vu VY (2015) A memristor-based hyperchaotic system with hidden attractors: dynamics, synchronization and circuital emulating. J Eng Sci Technol Rev 8(2):205–214
20. Rasappan S, Vaidyanathan S (2012) Global chaos synchronization of WINDMI and Coullet chaotic systems by backstepping control. Far East J Math Sci 67(2):265–287
21. Rasappan S, Vaidyanathan S (2012) Hybrid synchronization of n-scroll Chua and Lur'e chaotic systems via backstepping control with novel feedback. Arch Control Sci 22(3):343–365
22. Rasappan S, Vaidyanathan S (2012) Synchronization of hyperchaotic Liu system via backstepping control with recursive feedback. Commun Comput Inf Sci 305:212–221

23. Rasappan S, Vaidyanathan S (2013) Hybrid synchronization of n-scroll chaotic Chua circuits using adaptive backstepping control design with recursive feedback. Malays J Math Sci 7(2):219–246
24. Rasappan S, Vaidyanathan S (2014) Global chaos synchronization of WINDMI and Coullet chaotic systems using adaptive backstepping control design. Kyungpook Math J 54(1): 293–320
25. Rössler OE (1976) An equation for continuous chaos. Phys Lett A 57(5):397–398
26. Sampath S, Vaidyanathan S, Volos CK, Pham VT (2015) An eight-term novel four-scroll chaotic system with cubic nonlinearity and its circuit simulation. J Eng Sci Technol Rev 8(2):1–6
27. Sarasu P, Sundarapandian V (2011) Active controller design for the generalized projective synchronization of four-scroll chaotic systems. Int J Syst Signal Control Eng Appl 4(2):26–33
28. Sarasu P, Sundarapandian V (2011) The generalized projective synchronization of hyperchaotic Lorenz and hyperchaotic Qi systems via active control. Int J Soft Comput 6(5): 216–223
29. Sarasu P, Sundarapandian V (2012) Adaptive controller design for the generalized projective synchronization of 4-scroll systems. Int J Syst Signal Control Eng Appl 5(2):21–30
30. Sarasu P, Sundarapandian V (2012) Generalized projective synchronization of three-scroll chaotic systems via adaptive control. Eur J Sci Res 72(4):504–522
31. Sarasu P, Sundarapandian V (2012) Generalized projective synchronization of two-scroll systems via adaptive control. Int J Soft Comput 7(4):146–156
32. Sprott JC (1994) Some simple chaotic flows. Phys Rev E 50(2):647–650
33. Sundarapandian V (2010) Output regulation of the Lorenz attractor. Far East J Math Sci 42(2):289–299
34. Sundarapandian V (2011) Output regulation of the Arneodo-Coullet chaotic system. Commun Comput Inf Sci 133:98–107
35. Sundarapandian V (2013) Analysis and anti-synchronization of a novel chaotic system via active and adaptive controllers. J Eng Sci Technol Rev 6(4):45–52
36. Sundarapandian V, Karthikeyan R (2011) Anti-synchronization of hyperchaotic Lorenz and hyperchaotic Chen systems by adaptive control. Int J Syst Signal Control Eng Appl 4(2):18–25
37. Sundarapandian V, Karthikeyan R (2011) Anti-synchronization of Lü and Pan chaotic systems by adaptive nonlinear control. Eur J Sci Res 64(1):94–106
38. Sundarapandian V, Karthikeyan R (2012) Adaptive anti-synchronization of uncertain Tigan and Li systems. J Eng Appl Sci 7(1):45–52
39. Sundarapandian V, Karthikeyan R (2012) Hybrid synchronization of hyperchaotic Lorenz and hyperchaotic Chen systems via active control. J Eng Appl Sci 7(3):254–264
40. Sundarapandian V, Pehlivan I (2012) Analysis, control, synchronization, and circuit design of a novel chaotic system. Math Comput Model 55(7–8):1904–1915
41. Sundarapandian V, Sivaperumal S (2011) Sliding controller design of hybrid synchronization of four-wing chaotic systems. Int J Soft Comput 6(5):224–231
42. Suresh R, Sundarapandian V (2013) Global chaos synchronization of a family of n-scroll hyperchaotic Chua circuits using backstepping control with recursive feedback. Far East J Math Sci 73(1):73–95
43. Tigan G, Opris D (2008) Analysis of a 3D chaotic system. Chaos Solitons Fractals 36: 1315–1319
44. Vaidyanathan S (2011) Hybrid chaos synchronization of Liu and Lü systems by active nonlinear control. Commun Comput Inf Sci 204:1–10
45. Vaidyanathan S (2011) Output regulation of the unified chaotic system. Commun Comput Inf Sci 204:84–93
46. Vaidyanathan S (2012) Analysis and synchronization of the hyperchaotic Yujun systems via sliding mode control. Adv Intell Syst Comput 176:329–337
47. Vaidyanathan S (2012) Anti-synchronization of Sprott-L and Sprott-M chaotic systems via adaptive control. Int J Control Theory Appl 5(1):41–59

48. Vaidyanathan S (2012) Global chaos control of hyperchaotic Liu system via sliding control method. Int J Control Theory Appl 5(2):117–123
49. Vaidyanathan S (2012) Output regulation of the Liu chaotic system. Appl Mech Mater 110–116:3982–3989
50. Vaidyanathan S (2012) Sliding mode control based global chaos control of Liu-Liu-Liu-Su chaotic system. Int J Control Theory Appl 5(1):15–20
51. Vaidyanathan S (2013) A new six-term 3-D chaotic system with an exponential nonlinearity. Far East J Math Sci 79(1):135–143
52. Vaidyanathan S (2013) Analysis and adaptive synchronization of two novel chaotic systems with hyperbolic sinusoidal and cosinusoidal nonlinearity and unknown parameters. J Eng Sci Technol Rev 6(4):53–65
53. Vaidyanathan S (2013) Analysis, control and synchronization of hyperchaotic Zhou system via adaptive control. Adv Intell Syst Comput 177:1–10
54. Vaidyanathan S (2014) A new eight-term 3-D polynomial chaotic system with three quadratic nonlinearities. Far East J Math Sci 84(2):219–226
55. Vaidyanathan S (2014) Analysis and adaptive synchronization of eight-term 3-D polynomial chaotic systems with three quadratic nonlinearities. Eur Phys J: Spec Top 223(8):1519–1529
56. Vaidyanathan S (2014) Analysis, control and synchronisation of a six-term novel chaotic system with three quadratic nonlinearities. Int J Model Identif Control 22(1):41–53
57. Vaidyanathan S (2014) Generalized projective synchronisation of novel 3-D chaotic systems with an exponential non-linearity via active and adaptive control. Int J Model Identif Control 22(3):207–217
58. Vaidyanathan S (2014) Global chaos synchronization of identical Li-Wu chaotic systems via sliding mode control. Int J Model Identif Control 22(2):170–177
59. Vaidyanathan S (2015) 3-cells Cellular Neural Network (CNN) attractor and its adaptive biological control. Int J PharmTech Res 8(4):632–640
60. Vaidyanathan S (2015) A 3-D novel highly chaotic system with four quadratic nonlinearities, its adaptive control and anti-synchronization with unknown parameters. J Eng Sci Technol Rev 8(2):106–115
61. Vaidyanathan S (2015) A novel chemical chaotic reactor system and its adaptive control. Int J ChemTech Res 8(7):146–158
62. Vaidyanathan S (2015) Adaptive backstepping control of enzymes-substrates system with ferroelectric behaviour in brain waves. Int J PharmTech Res 8(2):256–261
63. Vaidyanathan S (2015) Adaptive biological control of generalized Lotka-Volterra three-species biological system. Int J PharmTech Res 8(4):622–631
64. Vaidyanathan S (2015) Adaptive chaotic synchronization of enzymes-substrates system with ferroelectric behaviour in brain waves. Int J PharmTech Res 8(5):964–973
65. Vaidyanathan S (2015) Adaptive control of a chemical chaotic reactor. Int J PharmTech Res 8(3):377–382
66. Vaidyanathan S (2015) Adaptive control of the FitzHugh-Nagumo chaotic neuron model. Int J PharmTech Res 8(6):117–127
67. Vaidyanathan S (2015) Adaptive synchronization of chemical chaotic reactors. Int J ChemTech Res 8(2):612–621
68. Vaidyanathan S (2015) Adaptive synchronization of generalized Lotka-Volterra three-species biological systems. Int J PharmTech Res 8(5):928–937
69. Vaidyanathan S (2015) Adaptive synchronization of novel 3-D chemical chaotic reactor systems. Int J ChemTech Res 8(7):159–171
70. Vaidyanathan S (2015) Adaptive synchronization of the identical FitzHugh-Nagumo chaotic neuron models. Int J PharmTech Res 8(6):167–177
71. Vaidyanathan S (2015) Analysis, control and synchronization of a 3-D novel jerk chaotic system with two quadratic nonlinearities. Kyungpook Math J 55:563–586
72. Vaidyanathan S (2015) Analysis, properties and control of an eight-term 3-D chaotic system with an exponential nonlinearity. Int J Model Identif Control 23(2):164–172

73. Vaidyanathan S (2015) Anti-synchronization of brusselator chemical reaction systems via adaptive control. Int J ChemTech Res 8(6):759–768
74. Vaidyanathan S (2015) Chaos in neurons and adaptive control of Birkhoff-Shaw strange chaotic attractor. Int J PharmTech Res 8(5):956–963
75. Vaidyanathan S (2015) Chaos in neurons and synchronization of Birkhoff-Shaw strange chaotic attractors via adaptive control. Int J PharmTech Res 8(6):1–11
76. Vaidyanathan S (2015) Coleman-Gomatam logarithmic competitive biology models and their ecological monitoring. Int J PharmTech Res 8(6):94–105
77. Vaidyanathan S (2015) Dynamics and control of brusselator chemical reaction. Int J ChemTech Res 8(6):740–749
78. Vaidyanathan S (2015) Dynamics and control of tokamak system with symmetric and magnetically confined plasma. Int J ChemTech Res 8(6):795–803
79. Vaidyanathan S (2015) Global chaos synchronization of chemical chaotic reactors via novel sliding mode control method. Int J ChemTech Res 8(7):209–221
80. Vaidyanathan S (2015) Global chaos synchronization of the forced Van der Pol chaotic oscillators via adaptive control method. Int J PharmTech Res 8(6):156–166
81. Vaidyanathan S (2015) Global chaos synchronization of the Lotka-Volterra biological systems with four competitive species via active control. Int J PharmTech Res 8(6):206–217
82. Vaidyanathan S (2015) Lotka-Volterra population biology models with negative feedback and their ecological monitoring. Int J PharmTech Res 8(5):974–981
83. Vaidyanathan S (2015) Lotka-Volterra two species competitive biology models and their ecological monitoring. Int J PharmTech Res 8(6):32–44
84. Vaidyanathan S (2015) Output regulation of the forced Van der Pol chaotic oscillator via adaptive control method. Int J PharmTech Res 8(6):106–116
85. Vaidyanathan S, Azar AT (2015) Analysis and control of a 4-D novel hyperchaotic system. In: Azar AT, Vaidyanathan S (eds) Chaos modeling and control systems design, studies in computational intelligence, vol 581. Springer, Germany, pp 19–38
86. Vaidyanathan S, Azar AT (2015) Analysis, control and synchronization of a nine-term 3-D novel chaotic system. In: Azar AT, Vaidyanathan S (eds) Chaos modelling and control systems design, studies in computational intelligence, vol 581. Springer, Germany, pp 19–38
87. Vaidyanathan S, Azar AT (2015) Anti-synchronization of identical chaotic systems using sliding mode control and an application to Vaidhyanathan-Madhavan chaotic systems. Stud Comput Intell 576:527–547
88. Vaidyanathan S, Azar AT (2015) Hybrid synchronization of identical chaotic systems using sliding mode control and an application to Vaidhyanathan chaotic systems. Stud Comput Intell 576:549–569
89. Vaidyanathan S, Madhavan K (2013) Analysis, adaptive control and synchronization of a seven-term novel 3-D chaotic system. Int J Control Theory Appl 6(2):121–137
90. Vaidyanathan S, Pakiriswamy S (2013) Generalized projective synchronization of six-term Sundarapandian chaotic systems by adaptive control. Int J Control Theory Appl 6(2):153–163
91. Vaidyanathan S, Pakiriswamy S (2015) A 3-D novel conservative chaotic system and its generalized projective synchronization via adaptive control. J Eng Sci Technol Rev 8(2):52–60
92. Vaidyanathan S, Rajagopal K (2011) Anti-synchronization of Li and T chaotic systems by active nonlinear control. Commun Comput Inf Sci 198:175–184
93. Vaidyanathan S, Rajagopal K (2011) Global chaos synchronization of hyperchaotic Pang and Wang systems by active nonlinear control. Commun Comput Inf Sci 204:84–93
94. Vaidyanathan S, Rajagopal K (2011) Global chaos synchronization of Lü and Pan systems by adaptive nonlinear control. Commun Comput Inf Sci 205:193–202
95. Vaidyanathan S, Rajagopal K (2012) Global chaos synchronization of hyperchaotic Pang and hyperchaotic Wang systems via adaptive control. Int J Soft Comput 7(1):28–37
96. Vaidyanathan S, Rasappan S (2011) Global chaos synchronization of hyperchaotic Bao and Xu systems by active nonlinear control. Commun Comput Inf Sci 198:10–17

97. Vaidyanathan S, Rasappan S (2014) Global chaos synchronization of n-scroll Chua circuit and Lur'e system using backstepping control design with recursive feedback. Arab J Sci Eng 39(4):3351–3364
98. Vaidyanathan S, Sampath S (2011) Global chaos synchronization of hyperchaotic Lorenz systems by sliding mode control. Commun Comput Inf Sci 205:156–164
99. Vaidyanathan S, Sampath S (2012) Anti-synchronization of four-wing chaotic systems via sliding mode control. Int J Autom Comput 9(3):274–279
100. Vaidyanathan S, Volos C (2015) Analysis and adaptive control of a novel 3-D conservative no-equilibrium chaotic system. Arch Control Sci 25(3):333–353
101. Vaidyanathan S, Volos C, Pham VT (2014) Hyperchaos, adaptive control and synchronization of a novel 5-D hyperchaotic system with three positive Lyapunov exponents and its SPICE implementation. Arch Control Sci 24(4):409–446
102. Vaidyanathan S, Volos C, Pham VT (2014) Hyperchaos, adaptive control and synchronization of a novel 5-D hyperchaotic system with three positive Lyapunov exponents and its SPICE implementation. Arch Control Sci 24(4):409–446
103. Vaidyanathan S, Volos C, Pham VT, Madhavan K, Idowu BA (2014) Adaptive backstepping control, synchronization and circuit simulation of a 3-D novel jerk chaotic system with two hyperbolic sinusoidal nonlinearities. Arch Control Sci 24(3):375–403
104. Vaidyanathan S, Idowu BA, Azar AT (2015) Backstepping controller design for the global chaos synchronization of Sprott's jerk systems. Stud Comput Intell 581:39–58
105. Vaidyanathan S, Rajagopal K, Volos CK, Kyprianidis IM, Stouboulos IN (2015) Analysis, adaptive control and synchronization of a seven-term novel 3-D chaotic system with three quadratic nonlinearities and its digital implementation in LabVIEW. J Eng Sci Technol Rev 8(2):130–141
106. Vaidyanathan S, Volos C, Pham VT, Madhavan K (2015) Analysis, adaptive control and synchronization of a novel 4-D hyperchaotic hyperjerk system and its SPICE implementation. Arch Control Sci 25(1):5–28
107. Vaidyanathan S, Volos CK, Kyprianidis IM, Stouboulos IN, Pham VT (2015) Analysis, adaptive control and anti-synchronization of a six-term novel jerk chaotic system with two exponential nonlinearities and its circuit simulation. J Eng Sci Technol Rev 8(2):24–36
108. Vaidyanathan S, Volos CK, Madhavan K (2015) Analysis, control, synchronization and SPICE implementation of a novel 4-D hyperchaotic Rikitake dynamo System without equilibrium. J Eng Sci Technol Rev 8(2):232–244
109. Vaidyanathan S, Volos CK, Pham VT (2015) Analysis, adaptive control and adaptive synchronization of a nine-term novel 3-D chaotic system with four quadratic nonlinearities and its circuit simulation. J Eng Sci Technol Rev 8(2):181–191
110. Vaidyanathan S, Volos CK, Pham VT (2015) Global chaos control of a novel nine-term chaotic system via sliding mode control. In: Azar AT, Zhu Q (eds) Advances and applications in sliding mode control systems, studies in computational intelligence, vol 576. Springer, Germany, pp 571–590
111. Vaidyanathan S, Volos CK, Pham VT, Madhavan K (2015) Analysis, adaptive control and synchronization of a novel 4-D hyperchaotic hyperjerk system and its SPICE implementation. Arch Control Sci 25(1):135–158
112. Vaidyanathan S, Volos CK, Rajagopal K, Kyprianidis IM, Stouboulos IN (2015) Adaptive backstepping controller design for the anti-synchronization of identical WINDMI chaotic systems with unknown parameters and its SPICE implementation. J Eng Sci Technol Rev 8(2):74–82
113. Volos CK, Kyprianidis IM, Stouboulos IN, Tlelo-Cuautle E, Vaidyanathan S (2015) Memristor: a new concept in synchronization of coupled neuromorphic circuits. J Eng Sci Technol Rev 8(2):157–173
114. Wang X, Ge C (2008) Controlling and tracking of Newton-Leipnik system via backstepping design. Int J Nonlinear Sci 5(2):133–139
115. Wei Z, Yang Q (2010) Anti-control of Hopf bifurcation in the new chaotic system with two stable node-foci. Appl Math Comput 217(1):422–429

116. Xiao X, Zhou L, Zhang Z (2014) Synchronization of chaotic Lur'e systems with quantized sampled-data controller. Commun Nonlinear Sci Numer Simul 19(6):2039–2047
117. Zhou W, Xu Y, Lu H, Pan L (2008) On dynamics analysis of a new chaotic attractor. Phys Lett A 372(36):5773–5777
118. Zhu C, Liu Y, Guo Y (2010) Theoretic and numerical study of a new chaotic system. Intell Inf Manag 2:104–109

A Novel Conservative Jerk Chaotic System With Two Cubic Nonlinearities and Its Adaptive Backstepping Control

Sundarapandian Vaidyanathan and Christos K. Volos

Abstract First, this work announces a six-term novel 3-D conservative jerk chaotic system with two cubic nonlinearities. The conservative chaotic systems are characterized by the property that they are volume conserving. The phase portraits of the novel conservative jerk chaotic system are displayed and the qualitative properties of the novel system are discussed. The novel jerk chaotic system has three unstable equilibrium points. The Lyapunov exponents of the novel jerk chaotic system are obtained as $L_1 = 0.01562$, $L_2 = 0$ and $L_3 = -0.01562$. The Kaplan–Yorke dimension of the novel jerk chaotic system is obtained as $D_{KY} = 3$. Next, an adaptive backstepping controller is designed to globally stabilize the novel conservative chaotic system with unknown parameters. Moreover, an adaptive backstepping controller is also designed to achieve global chaos synchronization of the identical conservative jerk chaotic systems with unknown parameters. The backstepping control method is a recursive procedure that links the choice of a Lyapunov function with the design of a controller and guarantees global asymptotic stability of strict feedback systems. MATLAB simulations have been shown to illustrate the phase portraits of the novel conservative jerk chaotic system and also the adaptive backstepping control results.

Keywords Chaos · Chaotic systems · Conservative chaotic systems · Backstepping control · Adaptive control · Synchronization

S. Vaidyanathan (✉)
Research and Development Centre, Vel Tech University,
Avadi, Chennai 600062, Tamil Nadu, India
e-mail: sundarvtu@gmail.com

C.K. Volos
Physics Department, Aristotle University of Thessaloniki,
54124 Thessaloniki, Greece
e-mail: chvolos@gmail.com

S. Vaidyanathan and C. Volos (eds.), *Advances and Applications in Chaotic Systems*, Studies in Computational Intelligence 636,
DOI 10.1007/978-3-319-30279-9_4

1 Introduction

Chaos theory deals with the qualitative study of chaotic dynamical systems and their applications in science and engineering. A dynamical system is called *chaotic* if it satisfies the three properties: boundedness, infinite recurrence and sensitive dependence on initial conditions [3].

The Lyapunov exponent is a measure of the divergence of phase points that are initially very close and can be used to quantify chaotic systems. It is common to refer to the largest Lyapunov exponent as the *Maximal Lyapunov Exponent* (MLE).

Some classical paradigms of 3-D chaotic systems in the literature are Lorenz system [19], Rössler system [32], ACT system [2], Sprott systems [40], Chen system [7], Lü system [20], Cai system [5], Tigan system [53], etc.

Many new chaotic systems have been discovered in the recent years such as Zhou system [134], Zhu system [135], Li system [16], Wei-Yang system [129], Sundarapandian systems [45, 50], Vaidyanathan systems [63, 64, 66–69, 72, 83, 84, 98, 101, 103, 112, 115, 117, 119, 121, 122], Pehlivan system [23], Sampath system [33], etc.

Chaos theory has applications in several fields of science and engineering such as chemical reactors [73, 77, 79, 81, 85, 89–91], biological systems [71, 74–76, 78, 80, 82, 86–88, 92–96], memristors [1, 24, 127], lasers [4], oscillations [54], neural networks [11, 42], robotics [12, 126], electrical circuits [21, 125], cryptosystems [31, 55], secure communications [131, 132], etc.

The study of control of a chaotic system investigates feedback control methods that globally or locally asymptotically stabilize or regulate the outputs of a chaotic system. Many methods have been designed for control and regulation of chaotic systems such as active control [43, 44, 57], adaptive control [113, 120, 123], backstepping control [17, 128], sliding mode control [60, 62], etc.

Synchronization of chaotic systems is a phenomenon that occurs when two or more chaotic systems are coupled or when a chaotic system drives another chaotic system. Because of the butterfly effect which causes exponential divergence of the trajectories of two identical chaotic systems started with nearly the same initial conditions, the synchronization of chaotic systems is a challenging research problem in the chaos literature [3].

Major works on synchronization of chaotic systems deal with the complete synchronization of a pair of chaotic systems called the *master* and *slave* systems. The design goal of the complete synchronization problem is to apply the output of the master system to control the slave system so that the output of the slave system tracks the output of the master system asymptotically with time.

Pecora and Carroll pioneered the research on synchronization of chaotic systems with their seminal papers [6, 22]. The active control method [14, 34, 35, 49, 56, 61, 104, 105, 108] is typically used when the system parameters are available for measurement. Adaptive control method [36–38, 46–48, 59, 65, 97, 102, 106, 107,

114, 118] is typically used when some or all the system parameters are not available for measurement and estimates for the uncertain parameters of the systems.

Sampled-data feedback control method [9, 18, 130, 133] and time-delay feedback control method [8, 13, 39] are also used for synchronization of chaotic systems. Backstepping control method [26–30, 52, 109, 116, 124] is also used for the synchronization of chaotic systems. Backstepping control is a recursive method for stabilizing the origin of a control system in strict-feedback form [15]. Another popular method for the synchronization of chaotic systems is the sliding mode control method [51, 58, 70, 99, 100, 110, 111], which is a nonlinear control method that alters the dynamics of a nonlinear system by application of a discontinuous control signal that forces the system to "slide" along a cross-section of the system's normal behavior.

In the chaos literature, there is an active interest in the discovery of conservative chaotic systems [41], which have the special property that the volume of the flow is conserved. If the sum of the Lyapunov exponents of a chaotic system is zero, then the system is conservative. On the other hand, if the sum of the Lyapunov exponents of a chaotic system is negative, then the system is dissipative.

Classical examples of conservative chaotic systems are Nosé-Hoover system [25], Hénon-Heiles system [10], etc. Classical example of dissipative chaotic systems are Lorenz system [19], Rössler system [32], Chen system [7], etc.

In the chaos literature, numerous dissipative chaotic systems have been discovered, but only a very few conservative chaotic systems have been found.

In this research work, we announce a six-term novel 3-D conservative jerk chaotic system with two cubic nonlinearities. We have also designed adaptive backstepping controllers for stabilization and synchronization of the six-term novel 3-D conservative jerk chaotic system.

This work is organized as follows. Section 2 describes the dynamic equations and phase portraits of the novel 3-D conservative jerk chaotic system. Section 3 details the qualitative properties of the novel conservative jerk chaotic system. The novel jerk chaotic system has three unstable equilibrium points. Also, the Lyapunov exponents of the novel jerk chaotic system are obtained as $L_1 = 0.01562$, $L_2 = 0$ and $L_3 = -0.01562$, while the Kaplan–Yorke dimension of the novel jerk chaotic system is obtained as $D_{KY} = 3$.

In Sect. 4, we design an adaptive backstepping controller to globally stabilize the novel conservative chaotic system with unknown parameters. In Sect. 5, an adaptive backstepping controller is designed to achieve global chaos synchronization of the identical conservative jerk chaotic systems with unknown parameters. Section 6 provides a summary of the main results obtained in this work.

2 A 3-D Novel Conservative Jerk Chaotic System

In this section, we describe a six-term novel conservative chaotic system with two cubic nonlinearities.

Our novel 3-D conservative jerk chaotic system is modeled by the 3-D dynamics

$$\begin{cases} \dot{x}_1 = x_2 \\ \dot{x}_2 = x_3 \\ \dot{x}_3 = -ax_2 + x_1\left(x_1^2 + x_2^2 - b\right) \end{cases} \tag{1}$$

where x_1, x_2, x_3 are the states and a, b are constant positive parameters.

The system (1) exhibits *conservative chaotic behaviour* for the parameter values

$$a = 4, \quad b = 1 \tag{2}$$

For numerical simulations, we take the initial conditions as

$$x_1(0) = -0.5, \quad x_2(0) = 0.1, \quad x_3(0) = 0.4 \tag{3}$$

Figure 1 shows the 3-D phase portrait of the novel conservative jerk chaotic system (1). Figures 2, 3 and 4 show the 2-D projection of the novel conservative jerk chaotic system (1) on the (x_1, x_2), (x_2, x_3) and (x_1, x_3) planes, respectively.

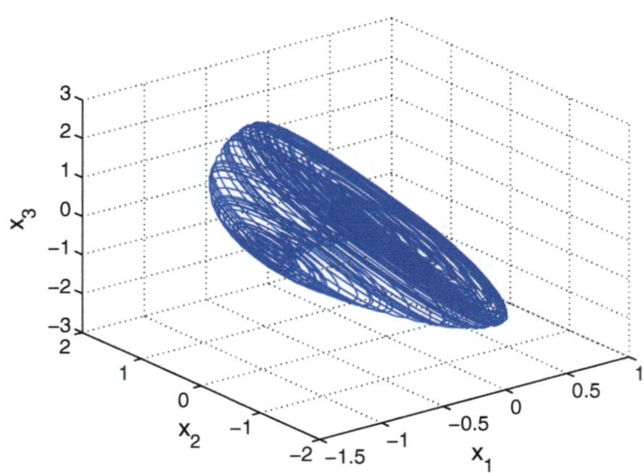

Fig. 1 3-D phase portrait of the novel conservative jerk chaotic system

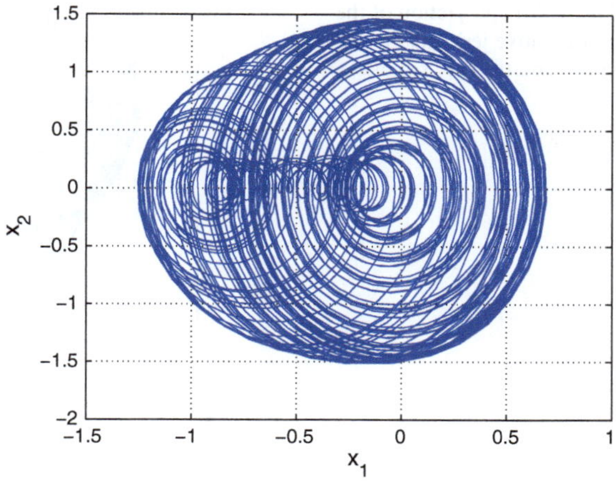

Fig. 2 2-D projection of the conservative jerk chaotic system on the (x_1, x_2) plane

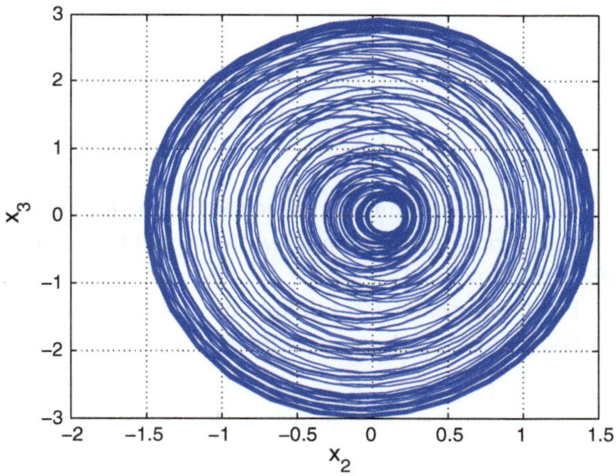

Fig. 3 2-D projection of the conservative jerk chaotic system on the (x_2, x_3) plane

3 Analysis of the 3-D Conservative Jerk Chaotic System

In this section, we give a dynamic analysis of the 3-D novel conservative jerk chaotic system (1).

3.1 Volume Conservation of the Flow

In vector notation, we may express the system (1) as

$$\dot{x} = f(x) = \begin{bmatrix} f_1(x_1, x_2, x_3) \\ f_2(x_1, x_2, x_3) \\ f_3(x_1, x_2, x_3) \end{bmatrix} \tag{4}$$

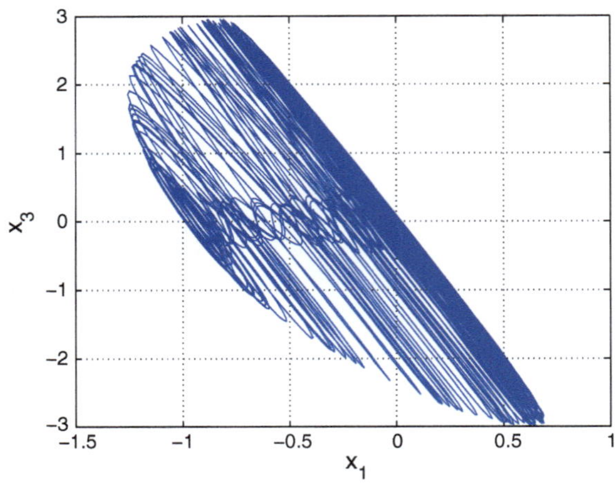

Fig. 4 2-D projection of the conservative jerk chaotic system on the (x_1, x_3) plane

where

$$\begin{cases} f_1(x_1, x_2, x_3) = x_2 \\ f_2(x_1, x_2, x_3) = x_3 \\ f_3(x_1, x_2, x_3) = -ax_2 + x_1\left(x_1^2 + x_2^2 - b\right) \end{cases} \tag{5}$$

We take the parameter values as in the chaotic case, viz. $a = 4$ and $b = 1$.

Let Ω be any region in $\mathbf{R}^3$ with smooth boundary and also $\Omega(t) = \Phi_t(\Omega)$, where Φ_t is the flow of f.

Furthermore, let $V(t)$ denote the volume of $\Omega(t)$.

By Liouville's theorem, we have

$$\dot{V} = \int\limits_{\Omega(t)} (\nabla \cdot f)\, dx_1\, dx_2\, dx_3 \tag{6}$$

The divergence of the novel chaotic system (1) is easily calculated as

$$\nabla \cdot f = \frac{\partial f_1}{\partial x_1} + \frac{\partial f_2}{\partial x_2} + \frac{\partial f_3}{\partial x_3} = 0 + 0 + 0 = 0 \tag{7}$$

Substituting (7) into (6), we get

$$\dot{V} = 0 \tag{8}$$

Integrating (8), we obtain the unique solution as

$$V(t) = V(0) \quad \text{for all } \ t \geq 0 \tag{9}$$

This shows that the 3-D novel chaotic system (1) is volume-conserving.

Hence, the system (1) is a conservative chaotic system.

3.2 Symmetry

It is easy to see that the system (1) is invariant under the coordinates transformation

$$(x_1, x_2, x_3) \mapsto (-x_1, -x_2, -x_3) \tag{10}$$

Thus, the novel conservative jerk chaotic system (1) has point reflection symmetry about the origin. Hence, it follows that any non-trivial trajectory of the novel conservative jerk chaotic system (1) must have a twin trajectory.

3.3 Equilibrium Points

The equilibrium points of the conservative jerk chaotic system (1) are obtained by solving the equations

$$\begin{cases} f_1(x_1, x_2, x_3) = x_2 & = 0 \\ f_2(x_1, x_2, x_3) = x_3 & = 0 \\ f_3(x_1, x_2, x_3) = -ax_2 + x_1\left(x_1^2 + x_2^2 - b\right) & = 0 \end{cases} \tag{11}$$

We take the parameter values as in the chaotic case, *viz.* $a = 4$ and $b = 1$.

Solving the Eq. (11), we get the equilibrium points of the jerk chaotic system (1) as

$$E_0 = \begin{bmatrix} 0 \\ 0 \\ 0 \end{bmatrix}, \quad E_1 = \begin{bmatrix} 1 \\ 0 \\ 0 \end{bmatrix} \text{ and } E_2 = \begin{bmatrix} -1 \\ 0 \\ 0 \end{bmatrix} \tag{12}$$

To test the stability type of the equilibrium points E_0, E_1 and E_2, we calculate the Jacobian matrix of the novel conservative jerk chaotic system (1) at any point x:

$$J(x) = \begin{bmatrix} 0 & 1 & 0 \\ 0 & 0 & 1 \\ 3x_1^2 + x_2^2 - 1 & -4 + 2x_1x_2 & 0 \end{bmatrix} \tag{13}$$

We find that

$$J_0 \triangleq J(E_0) = \begin{bmatrix} 0 & 1 & 0 \\ 0 & 0 & 1 \\ -1 & -4 & 0 \end{bmatrix} \tag{14}$$

The matrix J_0 has the eigenvalues

$$\lambda_1 = -0.2463, \quad \lambda_{2,3} = 0.1231 \pm 2.0113i \tag{15}$$

This shows that the equilibrium point E_0 is a saddle-focus.

Next, we find that

$$J_1 \triangleq J(E_1) = \begin{bmatrix} 0 & 1 & 0 \\ 0 & 0 & 1 \\ 2 & -4 & 0 \end{bmatrix} \tag{16}$$

The matrix J_1 has the eigenvalues

$$\lambda_1 = 0.4735, \quad \lambda_{2,3} = -0.2367 \pm 2.0416i \tag{17}$$

This shows that the equilibrium point E_1 is a saddle-focus.

Next, we find that

$$J_2 \triangleq J(E_2) = \begin{bmatrix} 0 & 1 & 0 \\ 0 & 0 & 1 \\ 2 & -4 & 0 \end{bmatrix} \tag{18}$$

The matrix J_2 has the eigenvalues

$$\lambda_1 = 0.4735, \quad \lambda_{2,3} = -0.2367 \pm 2.0416i \tag{19}$$

This shows that the equilibrium point E_2 is a saddle-focus.

Thus, the conservative jerk chaotic system (1) has three saddle-foci equilibria, which are all unstable.

3.4 Lyapunov Exponents and Kaplan–Yorke Dimension

We take the parameter values of the conservative jerk system (1) as $a = 4$ and $b = 1$. We take the initial state of the jerk system (1) as given in (3).

Then the Lyapunov exponents of the jerk system (1) are numerically obtained using MATLAB as

$$L_1 = 0.01562, \quad L_2 = 0, \quad L_3 = -0.01562 \tag{20}$$

It is noted that the sum of the Lyapunov exponents of the jerk system (1) is zero, which confirms the fact that the jerk system (1) is conservative.

Also, the Kaplan–Yorke dimension of the jerk system (1) is calculated as

$$D_{KL} = 2 + \frac{L_1 + L_2}{|L_3|} = 2 + 1 = 3 \tag{21}$$

4 Adaptive Control of the 3-D Conservative Jerk Chaotic System

In this section, we use backstepping control method to derive an adaptive feedback control law for globally stabilizing the 3-D novel conservative jerk chaotic system with unknown parameters.

Thus, we consider the 3-D novel conservative jerk chaotic system given by

$$\begin{cases} \dot{x}_1 = x_2 \\ \dot{x}_2 = x_3 \\ \dot{x}_3 = -ax_2 + x_1\left(x_1^2 + x_2^2 - b\right) + u \end{cases} \tag{22}$$

where a and b are unknown constant parameters, and u is a backstepping control law to be determined using estimates $\hat{a}(t)$ and $\hat{b}(t)$ for a and b, respectively.

The parameter estimation errors are defined as:

$$\begin{cases} e_a(t) = a - \hat{a}(t) \\ e_b(t) = b - \hat{b}(t) \end{cases} \tag{23}$$

Differentiating (23) with respect to t, we obtain the following equations:

$$\begin{cases} \dot{e}_a(t) = -\dot{\hat{a}}(t) \\ \dot{e}_b(t) = -\dot{\hat{b}}(t) \end{cases} \tag{24}$$

Next, we shall state and prove the main result of this section.

This theorem gives a backstepping-based adaptive control for globally stabilizing the 3-D novel conservative jerk chaotic system (22) with unknown parameters, and we establish theorem using Lyapunov stability theory [15].

Theorem 1 *The 3-D novel conservative jerk chaotic system (22), with unknown parameters a and b, is globally and exponentially stabilized by the adaptive feedback control law,*

$$u(t) = -(3 - \hat{b}(t))x_1 - (5 - \hat{a}(t))x_2 - 3x_3 - x_1\left(x_1^2 + x_2^2\right) - kz_3 \tag{25}$$

where $k > 0$ is a gain constant,

$$z_3 = 2x_1 + 2x_2 + x_3, \tag{26}$$

and the update law for the parameter estimates $\hat{a}(t)$, $\hat{b}(t)$ is given by

$$\begin{cases} \dot{\hat{a}}(t) = -x_2 z_3 \\ \dot{\hat{b}}(t) = -x_1 z_3 \end{cases} \tag{27}$$

Proof We prove this result via Lyapunov stability theory [15].
First, we define a quadratic Lyapunov function

$$V_1(z_1) = \frac{1}{2}\, z_1^2 \tag{28}$$

where

$$z_1 = x_1 \tag{29}$$

Differentiating V_1 along the dynamics (22), we get

$$\dot{V}_1 = z_1 \dot{z}_1 = x_1 x_2 = -z_1^2 + z_1(x_1 + x_2) \tag{30}$$

Now, we define

$$z_2 = x_1 + x_2 \tag{31}$$

Using (31), we can simplify the Eq. (30) as

$$\dot{V}_1 = -z_1^2 + z_1 z_2 \tag{32}$$

Secondly, we define a quadratic Lyapunov function

$$V_2(z_1, z_2) = V_1(z_1) + \frac{1}{2}\, z_2^2 = \frac{1}{2}\left(z_1^2 + z_2^2\right) \tag{33}$$

Differentiating V_2 along the dynamics (22), we get

$$\dot{V}_2 = -z_1^2 - z_2^2 + z_2(2x_1 + 2x_2 + x_3) \tag{34}$$

Now, we define

$$z_3 = 2x_1 + 2x_2 + x_3 \tag{35}$$

Using (35), we can simplify the Eq. (34) as

$$\dot{V}_2 = -z_1^2 - z_2^2 + z_2 z_3 \tag{36}$$

Finally, we define a quadratic Lyapunov function

$$V(z_1, z_2, z_3, e_a, e_b) = V_2(z_1, z_2) + \frac{1}{2} z_3^2 + \frac{1}{2} e_a^2 + \frac{1}{2} e_b^2 \tag{37}$$

which is a positive definite function on $\mathbf{R}^5$.
Differentiating V along the dynamics (22), we get

$$\dot{V} = -z_1^2 - z_2^2 - z_3^2 + z_3(z_3 + z_2 + \dot{z}_3) - e_a \dot{\hat{a}} - e_b \dot{\hat{b}} \tag{38}$$

Equation (38) can be written compactly as

$$\dot{V} = -z_1^2 - z_2^2 - z_3^2 + z_3 S - e_a \dot{\hat{a}} - e_b \dot{\hat{b}} \tag{39}$$

where

$$S = z_3 + z_2 + \dot{z}_3 = z_3 + z_2 + 2\dot{x}_1 + 2\dot{x}_2 + \dot{x}_3 \tag{40}$$

A simple calculation gives

$$S = (3 - b)x_1 + (5 - a)x_2 + 3x_3 + x_1 \left(x_1^2 + x_2^2\right) + u \tag{41}$$

Substituting the adaptive control law (25) into (41), we obtain

$$S = -\left(b - \hat{b}(t)\right) x_1 - \left(a - \hat{a}(t)\right) x_2 - k z_3 \tag{42}$$

Using the definitions (24), we can simplify (42) as

$$S = -e_b x_1 - e_a x_2 - k z_3 \tag{43}$$

Substituting the value of S from (43) into (39), we obtain

$$\dot{V} = -z_1^2 - z_2^2 - (1 + k)z_3^2 + e_a \left(-x_2 z_3 - \dot{\hat{a}}\right) + e_b \left(-x_1 z_3 - \dot{\hat{b}}\right) \tag{44}$$

Substituting the update law (27) into (44), we get

$$\dot{V} = -z_1^2 - z_2^2 - (1 + k)z_3^2, \tag{45}$$

which is a negative semi-definite function on $\mathbf{R}^5$.

From (45), it follows that the vector $\mathbf{z}(t) = (z_1(t), z_2(t), z_3(t))$ and the parameter estimation error $(e_a(t), e_b(t))$ are globally bounded, i.e.

$$\left[\, z_1(t)\ z_2(t)\ z_3(t)\ e_a(t)\ e_b(t) \,\right] \in \mathbf{L}_\infty \tag{46}$$

Also, it follows from (45) that

$$\dot{V} \le -z_1^2 - z_2^2 - z_3^2 = -\|\mathbf{z}\|^2 \tag{47}$$

That is,

$$\|\mathbf{z}\|^2 \le -\dot{V} \tag{48}$$

Integrating the inequality (48) from 0 to t, we get

$$\int_0^t |\mathbf{z}(\tau)|^2 \, d\tau \le V(0) - V(t) \tag{49}$$

From (49), it follows that $\mathbf{z}(t) \in \mathbf{L}_2$.

From Eq. (22), it can be deduced that $\dot{\mathbf{z}}(t) \in \mathbf{L}_\infty$.

Thus, using Barbalat's lemma [15], we conclude that $\mathbf{z}(t) \to \mathbf{0}$ exponentially as $t \to \infty$ for all initial conditions $\mathbf{z}(0) \in \mathbf{R}^3$.

Hence, it is immediate that $\mathbf{x}(t) \to \mathbf{0}$ exponentially as $t \to \infty$ for all initial conditions $\mathbf{x}(0) \in \mathbf{R}^3$.

This completes the proof. ∎

For the numerical simulations, the classical fourth-order Runge–Kutta method with step size $h = 10^{-8}$ is used to solve the system of differential equations (22) and (27), when the adaptive control law (25) is applied.

The parameter values of the novel conservative jerk chaotic system (22) are taken as

$$a = 4, \quad b = 1 \tag{50}$$

We take the positive gain constant as

$$k = 10 \tag{51}$$

Furthermore, as initial conditions of the novel conservative jerk chaotic system (22), we take

$$x_1(0) = 6.2, \quad x_2(0) = -8.3, \quad x_3(0) = 4.7 \tag{52}$$

Also, as initial conditions of the parameter estimates $\hat{a}(t)$ and $\hat{b}(t)$, we take

$$\hat{a}(0) = 8.2, \quad \hat{b}(0) = 9.5 \tag{53}$$

In Fig. 5, the exponential convergence of the controlled states $x_1(t), x_2(t), x_3(t)$ is depicted, when the adaptive control law (25) and (27) are implemented.

5 Adaptive Synchronization of the Identical 3-D Conservative Jerk Chaotic Systems

In this section, we use backstepping control method to derive an adaptive control law for globally and exponentially synchronizing the identical 3-D novel conservative jerk chaotic systems with unknown parameters.

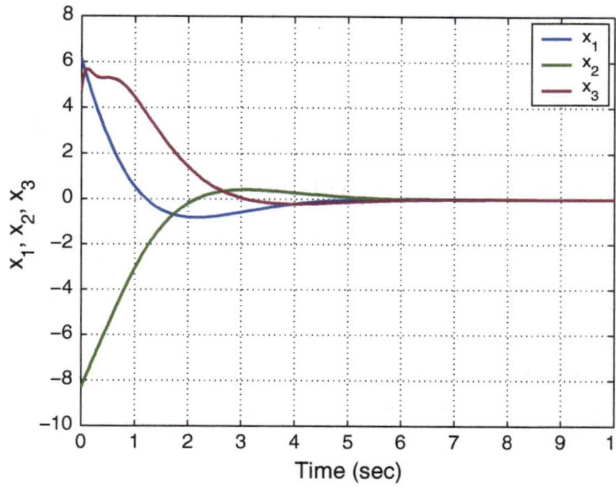

Fig. 5 Time-history of the controlled states x_1, x_2, x_3

As the master system, we consider the 3-D novel jerk chaotic system given by

$$\begin{cases} \dot{x}_1 = x_2 \\ \dot{x}_2 = x_3 \\ \dot{x}_3 = -ax_2 + x_1\left(x_1^2 + x_2^2 - b\right) \end{cases} \tag{54}$$

where x_1, x_2, x_3 are the states of the system, and a and b are unknown constant parameters.

As the slave system, we consider the 3-D novel jerk chaotic system given by

$$\begin{cases} \dot{y}_1 = y_2 \\ \dot{y}_2 = y_3 \\ \dot{y}_3 = -ay_2 + y_1\left(y_1^2 + y_2^2 - b\right) + u \end{cases} \tag{55}$$

where y_1, y_2, y_3 are the states of the system, and u is a backstepping control to be determined using estimates $\hat{a}(t)$ and $\hat{b}(t)$ for a and b, respectively.

We define the synchronization errors between the states of the master system (54) and the slave system (55) as

$$\begin{cases} e_1 = y_1 - x_1 \\ e_2 = y_2 - x_2 \\ e_3 = y_3 - x_3 \end{cases} \tag{56}$$

Then the error dynamics is easily obtained as

$$\begin{cases} \dot{e}_1 = e_2 \\ \dot{e}_2 = e_3 \\ \dot{e}_3 = -be_1 - ae_2 + y_1^3 - x_1^3 + y_1y_2^2 - x_1x_2^2 + u \end{cases} \tag{57}$$

The parameter estimation errors are defined as:

$$\begin{cases} e_a(t) = a - \hat{a}(t) \\ e_b(t) = b - \hat{b}(t) \end{cases} \tag{58}$$

Differentiating (58) with respect to t, we obtain the following equations:

$$\begin{cases} \dot{e}_a(t) = -\dot{\hat{a}}(t) \\ \dot{e}_b(t) = -\dot{\hat{b}}(t) \end{cases} \tag{59}$$

Theorem 2 *The identical 3-D novel jerk chaotic systems (54) and (55) with unknown parameters a and b are globally and exponentially synchronized by the adaptive control law*

$$u(t) = -\left[3 - \hat{b}(t)\right] e_1 - \left[5 - \hat{a}(t)\right] e_2 - 3e_3 - y_1^3 + x_1^3 - y_1 y_2^2 + x_1 x_2^2 - k z_3 \tag{60}$$

where $k > 0$ *is a gain constant,*

$$z_3 = 2e_1 + 2e_2 + e_3, \tag{61}$$

and the update law for the parameter estimates $\hat{a}(t), \hat{b}(t)$ *is given by*

$$\begin{cases} \dot{\hat{a}}(t) = -e_2 z_3 \\ \dot{\hat{b}}(t) = -e_1 z_3 \end{cases} \tag{62}$$

Proof First, we define a quadratic Lyapunov function

$$V_1(z_1) = \frac{1}{2} z_1^2 \tag{63}$$

where

$$z_1 = e_1 \tag{64}$$

Differentiating V_1 along the error dynamics (57), we get

$$\dot{V}_1 = z_1 \dot{z}_1 = e_1 e_2 = -z_1^2 + z_1(e_1 + e_2) \tag{65}$$

Now, we define

$$z_2 = e_1 + e_2 \tag{66}$$

Using (66), we can simplify the Eq. (65) as

$$\dot{V}_1 = -z_1^2 + z_1 z_2 \tag{67}$$

Secondly, we define a quadratic Lyapunov function

$$V_2(z_1, z_2) = V_1(z_1) + \frac{1}{2}z_2^2 = \frac{1}{2}\left(z_1^2 + z_2^2\right) \tag{68}$$

Differentiating V_2 along the error dynamics (57), we get

$$\dot{V}_2 = -z_1^2 - z_2^2 + z_2(2e_1 + 2e_2 + e_3) \tag{69}$$

Now, we define

$$z_3 = 2e_1 + 2e_2 + e_3 \tag{70}$$

Using (70), we can simplify the Eq. (69) as

$$\dot{V}_2 = -z_1^2 - z_2^2 + z_2 z_3 \tag{71}$$

Finally, we define a quadratic Lyapunov function

$$V(z_1, z_2, z_3, e_a, e_b) = V_2(z_1, z_2) + \frac{1}{2}z_3^2 + \frac{1}{2}e_a^2 + \frac{1}{2}e_b^2 \tag{72}$$

which is a positive definite function on $\mathbf{R}^5$.

Differentiating V along the error dynamics (57), we get

$$\dot{V} = -z_1^2 - z_2^2 - z_3^2 + z_3(z_3 + z_2 + \dot{z}_3) - e_a\dot{\hat{a}} - e_b\dot{\hat{b}} \tag{73}$$

Equation (73) can be written compactly as

$$\dot{V} = -z_1^2 - z_2^2 - z_3^2 + z_3 S - e_a\dot{\hat{a}} - e_b\dot{\hat{b}} \tag{74}$$

where

$$S = z_3 + z_2 + \dot{z}_3 = z_3 + z_2 + 2\dot{e}_1 + 2\dot{e}_2 + \dot{e}_3 \tag{75}$$

A simple calculation gives

$$S = (3 - b)e_1 + (5 - a)e_2 + 3e_3 + y_1^3 - x_1^3 + y_1 y_2^2 - x_1 x_2^2 + u \tag{76}$$

Substituting the adaptive control law (60) into (41), we obtain

$$S = -\left[b - \hat{b}(t)\right]e_1 - \left[a - \hat{a}(t)\right]e_2 - kz_3 \tag{77}$$

Using the definitions (59), we can simplify (77) as

$$S = -e_b e_1 - e_a e_2 - kz_3 \tag{78}$$

Substituting the value of S from (78) into (74), we obtain

$$\dot{V} = -z_1^2 - z_2^2 - (1+k)z_3^2 + e_a\left(-e_2 z_3 - \dot{\hat{a}}\right) + e_b\left(-e_1 z_3 - \dot{\hat{b}}\right) \tag{79}$$

Substituting the update law (62) into (79), we get

$$\dot{V} = -z_1^2 - z_2^2 - (1+k)z_3^2, \tag{80}$$

which is a negative semi-definite function on $\mathbf{R}^5$.

From (80), it follows that the vector $\mathbf{z}(t) = (z_1(t), z_2(t), z_3(t))$ and the parameter estimation error $(e_a(t), e_b(t))$ are globally bounded, i.e.

$$\left[\, z_1(t)\ z_2(t)\ z_3(t)\ e_a(t)\ e_b(t) \,\right] \in \mathbf{L}_\infty \tag{81}$$

Also, it follows from (80) that

$$\dot{V} \leq -z_1^2 - z_2^2 - z_3^2 = -\|\mathbf{z}\|^2 \tag{82}$$

That is,

$$\|\mathbf{z}\|^2 \leq -\dot{V} \tag{83}$$

Integrating the inequality (83) from 0 to t, we get

$$\int_0^t |\mathbf{z}(\tau)|^2\, d\tau \leq V(0) - V(t) \tag{84}$$

From (84), it follows that $\mathbf{z}(t) \in \mathbf{L}_2$.

From Eq. (57), it can be deduced that $\dot{\mathbf{z}}(t) \in \mathbf{L}_\infty$.

Thus, using Barbalat's lemma [15], we conclude that $\mathbf{z}(t) \to \mathbf{0}$ exponentially as $t \to \infty$ for all initial conditions $\mathbf{z}(0) \in \mathbf{R}^3$.

Hence, it is immediate that $\mathbf{e}(t) \to \mathbf{0}$ exponentially as $t \to \infty$ for all initial conditions $\mathbf{e}(0) \in \mathbf{R}^3$.

This completes the proof. ■

For the numerical simulations, the classical fourth-order Runge–Kutta method with step size $h = 10^{-8}$ is used to solve the system of differential equations (54) and (55), when the adaptive control law (60) and the parameter update law (62) are applied.

The parameter values of the novel conservative jerk chaotic systems are taken as in the chaotic case, i.e. $a = 4$ and $b = 1$.

The positive gain constant k is taken as $k = 10$.

Furthermore, as initial conditions of the master chaotic system (54), we take

$$x_1(0) = 0.1, \quad x_2(0) = -1.2, \quad x_3(0) = -1.6 \tag{85}$$

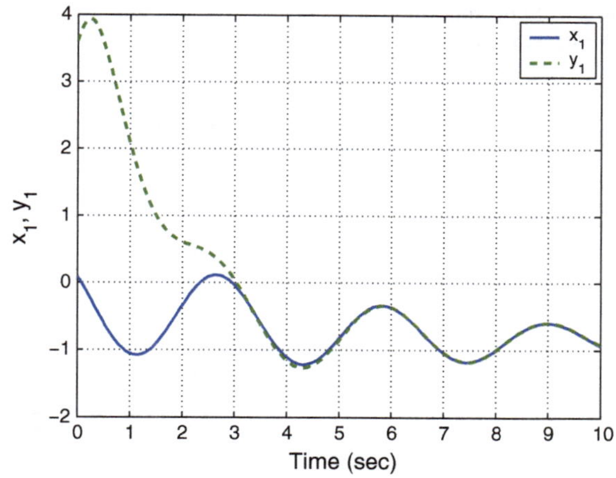

Fig. 6 Synchronization of the states x_1 and y_1

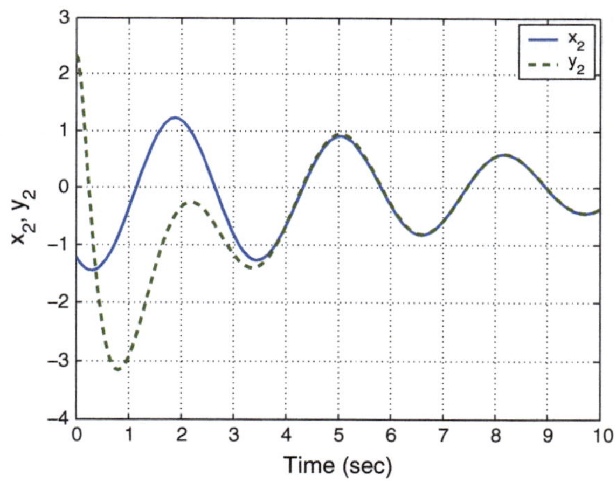

Fig. 7 Synchronization of the states x_2 and y_2

As initial conditions of the slave chaotic system (55), we take

$$y_1(0) = 3.6, \quad y_2(0) = 2.3, \quad y_3(0) = 1.2 \tag{86}$$

Also, as initial conditions of the parameter estimates $\hat{a}(t)$ and $\hat{b}(t)$, we take

$$\hat{a}(0) = 2.6, \quad \hat{b}(0) = 3.4 \tag{87}$$

In Figs. 6, 7 and 8, the complete synchronization of the identical 3-D conservative jerk chaotic systems (54) and (55) is shown.

Also, in Fig. 9, the time-history of the synchronization errors $e_1(t)$, $e_2(t)$, $e_3(t)$, is shown.

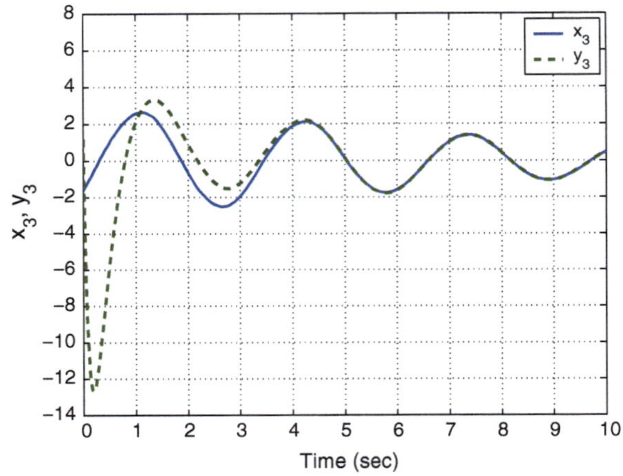

Fig. 8 Synchronization of the states x_3 and y_3

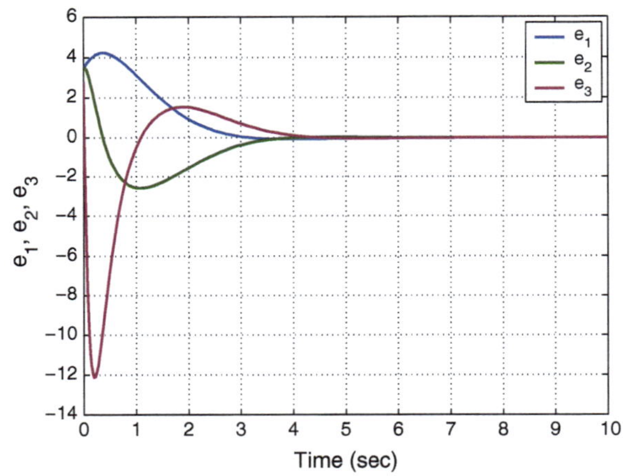

Fig. 9 Time-history of the synchronization errors e_1, e_2, e_3

6 Conclusions

In the chaos literature, numerous dissipative chaotic systems have been discovered, but only a very few conservative chaotic systems have been found. In this research work, we announced a six-term novel 3-D conservative jerk chaotic system with two cubic nonlinearities. The qualitative properties of the conservative jerk chaotic system were discussed in detail. We have also designed adaptive backstepping controllers for stabilization and synchronization of the six-term novel 3-D conservative jerk chaotic system.

References

1. Abdurrahman A, Jiang H, Teng Z (2015) Finite-time synchronization for memristor-based neural networks with time-varying delays. Neural Netw 69:20–28
2. Arneodo A, Coullet P, Tresser C (1981) Possible new strange attractors with spiral structure. Commun Math Phys 79(4):573–576
3. Azar AT, Vaidyanathan S (2015) Chaos modeling and control systems design, vol 581. Springer, Germany
4. Behnia S, Afrang S, Akhshani A, Mabhouti K (2013) A novel method for controlling chaos in external cavity semiconductor laser. Optik 124(8):757–764
5. Cai G, Tan Z (2007) Chaos synchronization of a new chaotic system via nonlinear control. J Uncertain Syst 1(3):235–240
6. Carroll TL, Pecora LM (1991) Synchronizing chaotic circuits. IEEE Trans Circuits Syst 38(4):453–456
7. Chen G, Ueta T (1999) Yet another chaotic attractor. Int J Bifurc Chaos 9(7):1465–1466
8. Chen WH, Wei D, Lu X (2014) Global exponential synchronization of nonlinear time-delay Lur'e systems via delayed impulsive control. Commun Nonlinear Sci Numer Simul 19(9):3298–3312
9. Gan Q, Liang Y (2012) Synchronization of chaotic neural networks with time delay in the leakage term and parametric uncertainties based on sampled-data control. J Frankl Inst 349(6):1955–1971
10. Hénon M, Heiles C (1964) The applicability of third integral of motion: some numerical experiments. Astron J 69(1):73–79
11. Huang WZ, Huang Y (2008) Chaos of a new class of Hopfield neural networks. Appl Math Comput 206(1):1–11
12. Islam MM, Murase K (2005) Chaotic dynamics of a behaviour-based miniature mobile robot: effects of environment and control structure. Neural Netw 18(2):123–144
13. Jiang GP, Zheng WX, Chen G (2004) Global chaos synchronization with channel time-delay. Chaos Solitons Fractals 20(2):267–275
14. Karthikeyan R, Sundarapandian V (2014) Hybrid chaos synchronization of four-scroll systems via active control. J Electr Eng 65(2):97–103
15. Khalil HK (2001) Nonlinear systems, 3rd edn. Prentice Hall, New Jersey
16. Li D (2008) A three-scroll chaotic attractor. Phys Lett A 372(4):387–393
17. Li GH, Zhou SP, Yang K (2007) Controlling chaos in Colpitts oscillator. Chaos Solitons Fractals 33:582–587
18. Li N, Zhang Y, Nie Z (2011) Synchronization for general complex dynamical networks with sampled-data. Neurocomputing 74(5):805–811
19. Lorenz EN (1963) Deterministic periodic flow. J Atmos Sci 20(2):130–141
20. Lü J, Chen G (2002) A new chaotic attractor coined. Int J Bifurc Chaos 12(3):659–661
21. Matouk AE (2011) Chaos, feedback control and synchronization of a fractional-order modified autonomous Van der Pol-Duffing circuit. Commun Nonlinear Sci Numer Simul 16(2):975–986
22. Pecora LM, Carroll TL (1990) Synchronization in chaotic systems. Phys Rev Lett 64(8):821–824
23. Pehlivan I, Moroz IM, Vaidyanathan S (2014) Analysis, synchronization and circuit design of a novel butterfly attractor. J Sound Vib 333(20):5077–5096
24. Pham VT, Volos CK, Vaidyanathan S, Le TP, Vu VY (2015) A memristor-based hyperchaotic system with hidden attractors: dynamics, synchronization and circuital emulating. J Eng Sci Technol Rev 8(2):205–214
25. Posch HA, Hoover WG, Vesely FJ (1986) Canonical dynamics of the Nosé oscillator: stability, order, and chaos. Phys Rev A 33(6):4253–4265
26. Rasappan S, Vaidyanathan S (2012) Global chaos synchronization of WINDMI and Coullet chaotic systems by backstepping control. Far East J Math Sci 67(2):265–287

27. Rasappan S, Vaidyanathan S (2012) Hybrid synchronization of n-scroll Chua and Lur'e chaotic systems via backstepping control with novel feedback. Arch Control Sci 22(3):343–365
28. Rasappan S, Vaidyanathan S (2012) Synchronization of hyperchaotic Liu system via backstepping control with recursive feedback. Commun Comput Inf Sci 305:212–221
29. Rasappan S, Vaidyanathan S (2013) Hybrid synchronization of n-scroll chaotic Chua circuits using adaptive backstepping control design with recursive feedback. Malays J Math Sci 7(2):219–246
30. Rasappan S, Vaidyanathan S (2014) Global chaos synchronization of WINDMI and Coullet chaotic systems using adaptive backstepping control design. Kyungpook Math J 54(1): 293–320
31. Rhouma R, Belghith S (2011) Cryptanalysis of a chaos-based cryptosystem. Commun Nonlinear Sci Numer Simul 16(2):876–884
32. Rössler OE (1976) An equation for continuous chaos. Phys Lett A 57(5):397–398
33. Sampath S, Vaidyanathan S, Volos CK, Pham VT (2015) An eight-term novel four-scroll chaotic system with cubic nonlinearity and its circuit simulation. J Eng Sci Technol Rev 8(2):1–6
34. Sarasu P, Sundarapandian V (2011) Active controller design for the generalized projective synchronization of four-scroll chaotic systems. Int J Syst Signal Control Eng Appl 4(2):26–33
35. Sarasu P, Sundarapandian V (2011) The generalized projective synchronization of hyperchaotic Lorenz and hyperchaotic Qi systems via active control. Int J Soft Comput 6(5): 216–223
36. Sarasu P, Sundarapandian V (2012) Adaptive controller design for the generalized projective synchronization of 4-scroll systems. Int J Syst Signal Control Eng Appl 5(2):21–30
37. Sarasu P, Sundarapandian V (2012) Generalized projective synchronization of three-scroll chaotic systems via adaptive control. Eur J Sci Res 72(4):504–522
38. Sarasu P, Sundarapandian V (2012) Generalized projective synchronization of two-scroll systems via adaptive control. Int J Soft Comput 7(4):146–156
39. Shahverdiev EM, Shore KA (2009) Impact of modulated multiple optical feedback time delays on laser diode chaos synchronization. Opt Commun 282(17):3568–3572
40. Sprott JC (1994) Some simple chaotic flows. Phys Rev E 50(2):647–650
41. Sprott JC (2010) Elegant chaos. World Scientific, Singapore
42. Sun Y, Babovic V, Chan ES (2010) Multi-step-ahead model error prediction using time-delay neural networks combined with chaos theory. J Hydrol 395:109–116
43. Sundarapandian V (2010) Output regulation of the Lorenz attractor. Far East J Math Sci 42(2):289–299
44. Sundarapandian V (2011) Output regulation of the Arneodo-Coullet chaotic system. Commun Comput Inf Sci 133:98–107
45. Sundarapandian V (2013) Analysis and anti-synchronization of a novel chaotic system via active and adaptive controllers. J Eng Sci Technol Rev 6(4):45–52
46. Sundarapandian V, Karthikeyan R (2011) Anti-synchronization of hyperchaotic Lorenz and hyperchaotic Chen systems by adaptive control. Int J Syst Signal Control Eng Appl 4(2):18–25
47. Sundarapandian V, Karthikeyan R (2011) Anti-synchronization of Lü and Pan chaotic systems by adaptive nonlinear control. Eur J Sci Res 64(1):94–106
48. Sundarapandian V, Karthikeyan R (2012) Adaptive anti-synchronization of uncertain Tigan and Li systems. J Eng Appl Sci 7(1):45–52
49. Sundarapandian V, Karthikeyan R (2012) Hybrid synchronization of hyperchaotic Lorenz and hyperchaotic Chen systems via active control. J Eng Appl Sci 7(3):254–264
50. Sundarapandian V, Pehlivan I (2012) Analysis, control, synchronization, and circuit design of a novel chaotic system. Math Comput Model 55(7–8):1904–1915
51. Sundarapandian V, Sivaperumal S (2011) Sliding controller design of hybrid synchronization of Four-Wing Chaotic systems. Int J Soft Comput 6(5):224–231
52. Suresh R, Sundarapandian V (2013) Global chaos synchronization of a family of n-scroll hyperchaotic Chua circuits using backstepping control with recursive feedback. Far East J Math Sci 73(1):73–95

53. Tigan G, Opris D (2008) Analysis of a 3D chaotic system. Chaos Solitons Fractals 36:1315–1319
54. Tuwankotta JM (2006) Chaos in a coupled oscillators system with widely spaced frequencies and energy-preserving non-linearity. Int J Nonlinear Mech 41(2):180–191
55. Usama M, Khan MK, Alghathbar K, Lee C (2010) Chaos-based secure satellite imagery cryptosystem. Comput Math Appl 60(2):326–337
56. Vaidyanathan S (2011) Hybrid chaos synchronization of Liu and Lü systems by active non-linear control. Commun Comput Inf Sci 204:1–10
57. Vaidyanathan S (2011) Output regulation of the unified chaotic system. Commun Comput Inf Sci 204:84–93
58. Vaidyanathan S (2012) Analysis and synchronization of the hyperchaotic Yujun systems via sliding mode control. Adv Intell Syst Comput 176:329–337
59. Vaidyanathan S (2012) Anti-synchronization of Sprott-L and Sprott-M chaotic systems via adaptive control. Int J Control Theory Appl 5(1):41–59
60. Vaidyanathan S (2012) Global chaos control of hyperchaotic Liu system via sliding control method. Int J Control Theory Appl 5(2):117–123
61. Vaidyanathan S (2012) Output regulation of the Liu chaotic system. Appl Mech Mater 110–116:3982–3989
62. Vaidyanathan S (2012) Sliding mode control based global chaos control of Liu-Liu-Liu-Su chaotic system. Int J Control Theory Appl 5(1):15–20
63. Vaidyanathan S (2013) A new six-term 3-D chaotic system with an exponential nonlinearity. Far East J Math Sci 79(1):135–143
64. Vaidyanathan S (2013) Analysis and adaptive synchronization of two novel chaotic systems with hyperbolic sinusoidal and cosinusoidal nonlinearity and unknown parameters. J Eng Sci Technol Rev 6(4):53–65
65. Vaidyanathan S (2013) Analysis, control and synchronization of hyperchaotic Zhou system via adaptive control. Adv Intell Syst Comput 177:1–10
66. Vaidyanathan S (2014) A new eight-term 3-D polynomial chaotic system with three quadratic nonlinearities. Far East J Math Sci 84(2):219–226
67. Vaidyanathan S (2014) Analysis and adaptive synchronization of eight-term 3-D polynomial chaotic systems with three quadratic nonlinearities. Eur Phys J: Spec Top 223(8):1519–1529
68. Vaidyanathan S (2014) Analysis, control and synchronisation of a six-term novel chaotic system with three quadratic nonlinearities. Int J Model Identif Control 22(1):41–53
69. Vaidyanathan S (2014) Generalized projective synchronisation of novel 3-D chaotic systems with an exponential non-linearity via active and adaptive control. Int J Model Identif Control 22(3):207–217
70. Vaidyanathan S (2014) Global chaos synchronization of identical Li-Wu chaotic systems via sliding mode control. Int J Model Identif Control 22(2):170–177
71. Vaidyanathan S (2015) 3-cells Cellular Neural Network (CNN) attractor and its adaptive biological control. Int J PharmTech Res 8(4):632–640
72. Vaidyanathan S (2015) A 3-D novel highly chaotic system with four quadratic nonlinearities, its adaptive control and anti-synchronization with unknown parameters. J Eng Sci Technol Rev 8(2):106–115
73. Vaidyanathan S (2015) A novel chemical chaotic reactor system and its adaptive control. Int J ChemTech Res 8(7):146–158
74. Vaidyanathan S (2015) Adaptive backstepping control of enzymes-substrates system with ferroelectric behaviour in brain waves. Int J PharmTech Res 8(2):256–261
75. Vaidyanathan S (2015) Adaptive biological control of generalized Lotka-Volterra three-species biological system. Int J PharmTech Res 8(4):622–631
76. Vaidyanathan S (2015) Adaptive chaotic synchronization of enzymes-substrates system with ferroelectric behaviour in brain waves. Int J PharmTech Res 8(5):964–973
77. Vaidyanathan S (2015) Adaptive control of a chemical chaotic reactor. Int J PharmTech Res 8(3):377–382

78. Vaidyanathan S (2015) Adaptive control of the FitzHugh-Nagumo chaotic neuron model. Int J PharmTech Res 8(6):117–127
79. Vaidyanathan S (2015) Adaptive synchronization of chemical chaotic reactors. Int J ChemTech Res 8(2):612–621
80. Vaidyanathan S (2015) Adaptive synchronization of generalized Lotka-Volterra three-species biological systems. Int J PharmTech Res 8(5):928–937
81. Vaidyanathan S (2015) Adaptive synchronization of novel 3-D chemical chaotic reactor systems. Int J ChemTech Res 8(7):159–171
82. Vaidyanathan S (2015) Adaptive synchronization of the identical FitzHugh-Nagumo chaotic neuron models. Int J PharmTech Res 8(6):167–177
83. Vaidyanathan S (2015) Analysis, control and synchronization of a 3-D novel jerk chaotic system with two quadratic nonlinearities. Kyungpook Math J 55:563–586
84. Vaidyanathan S (2015) Analysis, properties and control of an eight-term 3-D chaotic system with an exponential nonlinearity. Int J Model Identif Control 23(2):164–172
85. Vaidyanathan S (2015) Anti-synchronization of brusselator chemical reaction systems via adaptive control. Int J ChemTech Res 8(6):759–768
86. Vaidyanathan S (2015) Chaos in neurons and adaptive control of Birkhoff-Shaw strange chaotic attractor. Int J PharmTech Res 8(5):956–963
87. Vaidyanathan S (2015) Chaos in neurons and synchronization of Birkhoff-Shaw strange chaotic attractors via adaptive control. Int J PharmTech Res 8(6):1–11
88. Vaidyanathan S (2015) Coleman-Gomatam logarithmic competitive biology models and their ecological monitoring. Int J PharmTech Res 8(6):94–105
89. Vaidyanathan S (2015) Dynamics and control of brusselator chemical reaction. Int J ChemTech Res 8(6):740–749
90. Vaidyanathan S (2015) Dynamics and control of tokamak system with symmetric and magnetically confined plasma. Int J ChemTech Res 8(6):795–803
91. Vaidyanathan S (2015) Global chaos synchronization of chemical chaotic reactors via novel sliding mode control method. Int J ChemTech Res 8(7):209–221
92. Vaidyanathan S (2015) Global chaos synchronization of the forced Van der Pol chaotic oscillators via adaptive control method. Int J PharmTech Res 8(6):156–166
93. Vaidyanathan S (2015) Global chaos synchronization of the Lotka-Volterra biological systems with four competitive species via active control. Int J PharmTech Res 8(6):206–217
94. Vaidyanathan S (2015) Lotka-Volterra population biology models with negative feedback and their ecological monitoring. Int J PharmTech Res 8(5):974–981
95. Vaidyanathan S (2015) Lotka-Volterra two species competitive biology models and their ecological monitoring. Int J PharmTech Res 8(6):32–44
96. Vaidyanathan S (2015) Output regulation of the forced Van der Pol chaotic oscillator via adaptive control method. Int J PharmTech Res 8(6):106–116
97. Vaidyanathan S, Azar AT (2015) Analysis and control of a 4-D novel hyperchaotic system. In: Azar AT, Vaidyanathan S (eds) Chaos modeling and control systems design, studies in computational intelligence, vol 581. Springer, Germany, pp 19–38
98. Vaidyanathan S, Azar AT (2015) Analysis, control and synchronization of a nine-term 3-D novel chaotic system. In: Azar AT, Vaidyanathan S (eds) Chaos modelling and control systems design, studies in computational intelligence, vol 581. Springer, Germany, pp 19–38
99. Vaidyanathan S, Azar AT (2015) Anti-synchronization of identical chaotic systems using sliding mode control and an application to Vaidhyanathan-Madhavan chaotic systems. Stud Comput Intell 576:527–547
100. Vaidyanathan S, Azar AT (2015) Hybrid synchronization of identical chaotic systems using sliding mode control and an application to Vaidhyanathan chaotic systems. Stud Comput Intell 576:549–569
101. Vaidyanathan S, Madhavan K (2013) Analysis, adaptive control and synchronization of a seven-term novel 3-D chaotic system. Int J Control Theory Appl 6(2):121–137
102. Vaidyanathan S, Pakiriswamy S (2013) Generalized projective synchronization of six-term Sundarapandian chaotic systems by adaptive control. Int J Control Theory Appl 6(2):153–163

103. Vaidyanathan S, Pakiriswamy S (2015) A 3-D novel conservative chaotic system and its generalized projective synchronization via adaptive control. J Eng Sci Technol Rev 8(2):52–60
104. Vaidyanathan S, Rajagopal K (2011) Anti-synchronization of Li and T chaotic systems by active nonlinear control. Commun Comput Inf Sci 198:175–184
105. Vaidyanathan S, Rajagopal K (2011) Global chaos synchronization of hyperchaotic Pang and Wang systems by active nonlinear control. Commun Comput Inf Science 204:84–93
106. Vaidyanathan S, Rajagopal K (2011) Global chaos synchronization of Lü and Pan systems by adaptive nonlinear control. Commun Comput Inf Sci 205:193–202
107. Vaidyanathan S, Rajagopal K (2012) Global chaos synchronization of hyperchaotic Pang and hyperchaotic Wang systems via adaptive control. Int J Soft Comput 7(1):28–37
108. Vaidyanathan S, Rasappan S (2011) Global chaos synchronization of hyperchaotic Bao and Xu systems by active nonlinear control. Commun Comput Inf Sci 198:10–17
109. Vaidyanathan S, Rasappan S (2014) Global chaos synchronization of n-scroll Chua circuit and Lur'e system using backstepping control design with recursive feedback. Arab J Sci Eng 39(4):3351–3364
110. Vaidyanathan S, Sampath S (2011) Global chaos synchronization of hyperchaotic Lorenz systems by sliding mode control. Commun Comput Inf Sci 205:156–164
111. Vaidyanathan S, Sampath S (2012) Anti-synchronization of four-wing chaotic systems via sliding mode control. Int J Autom Comput 9(3):274–279
112. Vaidyanathan S, Volos C (2015) Analysis and adaptive control of a novel 3-D conservative no-equilibrium chaotic system. Arch Control Sci 25(3):333–353
113. Vaidyanathan S, Volos C, Pham VT (2014) Hyperchaos, adaptive control and synchronization of a novel 5-D hyperchaotic system with three positive Lyapunov exponents and its SPICE implementation. Arch Control Sci 24(4):409–446
114. Vaidyanathan S, Volos C, Pham VT (2014) Hyperchaos, adaptive control and synchronization of a novel 5-D hyperchaotic system with three positive Lyapunov exponents and its SPICE implementation. Arch Control Sci 24(4):409–446
115. Vaidyanathan S, Volos C, Pham VT, Madhavan K, Idowu BA (2014) Adaptive backstepping control, synchronization and circuit simulation of a 3-D novel jerk chaotic system with two hyperbolic sinusoidal nonlinearities. Arch Control Sci 24(3):375–403
116. Vaidyanathan S, Idowu BA, Azar AT (2015) Backstepping controller design for the global chaos synchronization of Sprott's jerk systems. Stud Comput Intell 581:39–58
117. Vaidyanathan S, Rajagopal K, Volos CK, Kyprianidis IM, Stouboulos IN (2015) Analysis, adaptive control and synchronization of a seven-term novel 3-D chaotic system with three quadratic nonlinearities and its digital implementation in LabVIEW. J Eng Sci Technol Rev 8(2):130–141
118. Vaidyanathan S, Volos C, Pham VT, Madhavan K (2015) Analysis, adaptive control and synchronization of a novel 4-D hyperchaotic hyperjerk system and its SPICE implementation. Arch Control Sci 25(1):5–28
119. Vaidyanathan S, Volos CK, Kyprianidis IM, Stouboulos IN, Pham VT (2015) Analysis, adaptive control and anti-synchronization of a six-term novel jerk chaotic system with two exponential nonlinearities and its circuit simulation. J Eng Sci Technol Rev 8(2):24–36
120. Vaidyanathan S, Volos CK, Madhavan K (2015) Analysis, control, synchronization and SPICE implementation of a novel 4-D hyperchaotic Rikitake dynamo System without equilibrium. J Eng Sci Technol Rev 8(2):232–244
121. Vaidyanathan S, Volos CK, Pham VT (2015) Analysis, adaptive control and adaptive synchronization of a nine-term novel 3-D chaotic system with four quadratic nonlinearities and its circuit simulation. J Eng Sci Technol Rev 8(2):181–191
122. Vaidyanathan S, Volos CK, Pham VT (2015) Global chaos control of a novel nine-term chaotic system via sliding mode control. In: Azar AT, Zhu Q (eds) Advances and applications in sliding mode control systems, studies in computational intelligence, vol 576. Springer, Germany, pp 571–590

123. Vaidyanathan S, Volos CK, Pham VT, Madhavan K (2015) Analysis, adaptive control and synchronization of a novel 4-D hyperchaotic hyperjerk system and its SPICE implementation. Arch Control Sci 25(1):135–158
124. Vaidyanathan S, Volos CK, Rajagopal K, Kyprianidis IM, Stouboulos IN (2015) Adaptive backstepping controller design for the anti-synchronization of identical WINDMI chaotic systems with unknown parameters and its SPICE implementation. J Eng Sci Technol Rev 8(2):74–82
125. Volos CK, Kyprianidis IM, Stouboulos IN, Anagnostopoulos AN (2009) Experimental study of the dynamic behavior of a double scroll circuit. J Appl Funct Anal 4:703–711
126. Volos CK, Kyprianidis IM, Stouboulos IN (2013) Experimental investigation on coverage performance of a chaotic autonomous mobile robot. Robot Auton Syst 61(12):1314–1322
127. Volos CK, Kyprianidis IM, Stouboulos IN, Tlelo-Cuautle E, Vaidyanathan S (2015) Memristor: a new concept in synchronization of coupled neuromorphic circuits. J Eng Sci Technol Rev 8(2):157–173
128. Wang X, Ge C (2008) Controlling and tracking of Newton-Leipnik system via backstepping design. Int Jo Nonlinear Sci 5(2):133–139
129. Wei Z, Yang Q (2010) Anti-control of Hopf bifurcation in the new chaotic system with two stable node-foci. Appl Math Comput 217(1):422–429
130. Xiao X, Zhou L, Zhang Z (2014) Synchronization of chaotic Lur'e systems with quantized sampled-data controller. Commun Nonlinear Sci Numer Simul 19(6):2039–2047
131. Yang J, Zhu F (2013) Synchronization for chaotic systems and chaos-based secure communications via both reduced-order and step-by-step sliding mode observers. Commun Nonlinear Sci Numer Simul 18(4):926–937
132. Yang J, Chen Y, Zhu F (2014) Singular reduced-order observer-based synchronization for uncertain chaotic systems subject to channel disturbance and chaos-based secure communication. Appl Math Comput 229:227–238
133. Zhang H, Zhou J (2012) Synchronization of sampled-data coupled harmonic oscillators with control inputs missing. Syst Control Lett 61(12):1277–1285
134. Zhou W, Xu Y, Lu H, Pan L (2008) On dynamics analysis of a new chaotic attractor. Phys Lett A 372(36):5773–5777
135. Zhu C, Liu Y, Guo Y (2010) Theoretic and numerical study of a new chaotic system. Intell Inf Manag 2:104–109

Adaptive Backstepping Control, Synchronization and Circuit Simulation of a Novel Jerk Chaotic System with a Quartic Nonlinearity

Sundarapandian Vaidyanathan, Viet-Thanh Pham and Christos K. Volos

Abstract This work describes a six-term novel 3-D jerk chaotic system with a quartic nonlinearity. The phase portraits of the novel jerk chaotic system are displayed and the qualitative properties of the novel jerk system are discussed. The novel jerk chaotic system has exactly one equilibrium point, which is saddle-focus. The Lyapunov exponents of the novel jerk chaotic system are obtained as $L_1 = 0.1443$, $L_2 = 0$ and $L_3 = -2.8439$. The Kaplan–Yorke dimension of the novel jerk chaotic system is obtained as $D_{KY} = 2.0507$. Next, an adaptive backstepping controller is designed to globally stabilize the novel jerk chaotic system with unknown parameters. Moreover, an adaptive backstepping controller is also designed to achieve global chaos synchronization of the identical jerk chaotic systems with unknown parameters. The backstepping control method is a recursive procedure that links the choice of a Lyapunov function with the design of a controller and guarantees global asymptotic stability of strict feedback systems. MATLAB simulations have been shown to illustrate the phase portraits of the novel jerk chaotic system and also the adaptive backstepping control results. Finally, an electronic circuit realization of the novel jerk chaotic system using Spice is presented in detail to confirm the feasibility of the theoretical model.

Keywords Chaos · Chaotic systems · Backstepping control · Adaptive control · Synchronization · Circuit simulation

---

S. Vaidyanathan (✉)
Research and Development Centre, Vel Tech University, Avadi, Chennai 600062, Tamil Nadu, India
e-mail: sundarvtu@gmail.com

V.-T. Pham
School of Electronics and Telecommunications, Hanoi University of Science and Technology, 01 Dai Co Viet, Hanoi, Vietnam
e-mail: pvt3010@gmail.com

C.K. Volos
Physics Department, Aristotle University of Thessaloniki, 54124 Thessaloniki, Greece
e-mail: chvolos@gmail.com

S. Vaidyanathan and C. Volos (eds.), *Advances and Applications in Chaotic Systems*, Studies in Computational Intelligence 636,
DOI 10.1007/978-3-319-30279-9_5

1 Introduction

Chaos theory deals with the qualitative study of chaotic dynamical systems and their applications in science and engineering. A dynamical system is called *chaotic* if it satisfies the three properties: boundedness, infinite recurrence and sensitive dependence on initial conditions [3].

Some classical paradigms of 3-D chaotic systems in the literature are Lorenz system [17], Rössler system [29], ACT system [2], Sprott systems [37], Chen system [7], Lü system [18], Cai system [5], Tigan system [48], etc.

Many new chaotic systems have been discovered in the recent years such as Zhou system [129], Zhu system [130], Li system [14], Wei-Yang system [124], Sundarapandian systems [40, 45], Vaidyanathan systems [58, 59, 61–64, 67, 78, 79, 93, 96, 98, 107, 110, 112, 114, 116, 117], Pehlivan system [21], Sampath system [30], etc.

Chaos theory has applications in several fields of science and engineering such as chemical reactors [68, 72, 74, 76, 80, 84–86], biological systems [66, 69–71, 73, 75, 77, 81–83, 87–91], memristors [1, 22, 122], lasers [4], oscillations [49], robotics [10, 121], electrical circuits [19, 120], cryptosystems [28, 50], secure communications [126, 127], etc.

The study of control of a chaotic system investigates feedback control methods that globally or locally asymptotically stabilize or regulate the outputs of a chaotic system. Many methods have been designed for control and regulation of chaotic systems such as active control [38, 39, 52], adaptive control [108, 115, 118], backstepping control [15, 123], sliding mode control [55, 57], etc.

Synchronization of chaotic systems is a phenomenon that occurs when two or more chaotic systems are coupled or when a chaotic system drives another chaotic system. Because of the butterfly effect which causes exponential divergence of the trajectories of two identical chaotic systems started with nearly the same initial conditions, the synchronization of chaotic systems is a challenging research problem in the chaos literature [3].

Major works on synchronization of chaotic systems deal with the complete synchronization of a pair of chaotic systems called the *master* and *slave* systems. The design goal of the complete synchronization problem is to apply the output of the master system to control the slave system so that the output of the slave system tracks the output of the master system asymptotically with time.

Pecora and Carroll pioneered the research on synchronization of chaotic systems with their seminal papers [6, 20]. The active control method [12, 31, 32, 44, 51, 56, 99, 100, 103] is typically used when the system parameters are available for measurement. Adaptive control method [33–35, 41–43, 54, 60, 92, 97, 101, 102, 109, 113] is typically used when some or all the system parameters are not available for measurement and estimates for the uncertain parameters of the systems.

Sampled-data feedback control method [9, 16, 125, 128] and time-delay feedback control method [8, 11, 36] are also used for synchronization of chaotic systems. Backstepping control method [23–27, 47, 104, 111, 119] is also used for the synchronization of chaotic systems. Backstepping control is a recursive method for stabilizing the origin of a control system in strict-feedback form [13]. Another popular method for the synchronization of chaotic systems is the sliding mode control method [46, 53, 65, 94, 95, 105, 106], which is a nonlinear control method that alters the dynamics of a nonlinear system by application of a discontinuous control signal that forces the system to "slide" along a cross-section of the system's normal behavior.

In the recent decades, there is some good interest in finding novel chaotic systems, which can be expressed by an explicit third order differential equation describing the time evolution of the single scalar variable x given by

$$\dddot{x} = j(x, \dot{x}, \ddot{x}) \tag{1}$$

The differential equation (1) is called "jerk system" because the third order time derivative in mechanical systems is called *jerk*.

By defining phase variables $x_1 = x$, $x_2 = \dot{x}$ and $x_3 = \ddot{x}$, the jerk differential equation (1) can be expressed as a 3-D system given by

$$\begin{cases} \dot{x}_1 = x_2 \\ \dot{x}_2 = x_3 \\ \dot{x}_3 = j(x_1, x_2, x_3) \end{cases} \tag{2}$$

In this research work, we announce a six-term novel 3-D jerk chaotic system with a quartic nonlinearity. We have also designed adaptive backstepping controllers for stabilization and synchronization of the six-term novel 3-D jerk chaotic system.

This work is organized as follows. Section 2 describes the dynamic equations and phase portraits of the novel 3-D jerk chaotic system. Section 3 details the qualitative properties of the novel jerk chaotic system. The novel jerk chaotic system has exactly one equilibrium point, which is a saddle-focus. Thus, the system has an unstable equilibrium point. The Lyapunov exponents of the novel jerk chaotic system are obtained as $L_1 = 0.1443$, $L_2 = 0$ and $L_3 = -2.8439$. Since the sum of the Lyapunov exponents is negative, the novel jerk chaotic system is dissipative. Thus, the system limit sets are ultimately confined into a specific limit set of zero volume, and the asymptotic motion of the novel jerk chaotic system settles onto a strange attractor of the system. The Kaplan–Yorke dimension of the novel jerk chaotic system is obtained as $D_{KY} = 2.0507$.

In Sect. 4, we design an adaptive backstepping controller to globally stabilize the novel jerk chaotic system with unknown parameters. In Sect. 5, an adaptive backstepping controller is designed to achieve global chaos synchronization of the identical novel jerk chaotic systems with unknown parameters. In Sect. 6, an electronic circuit realization of the novel jerk chaotic system using Spice is presented to confirm the feasibility of the theoretical model. Section 7 contains the conclusions of this work.

2 A 3-D Novel Jerk Chaotic System

In this section, we describe a six-term novel jerk chaotic system with a quartic nonlinearity.

Our novel 3-D jerk chaotic system is modeled by the 3-D dynamics

$$\begin{cases} \dot{x}_1 = x_2 \\ \dot{x}_2 = x_3 \\ \dot{x}_3 = -ax_1 - bx_3 + cx_1^3 x_2 - 1 \end{cases} \tag{3}$$

where x_1, x_2, x_3 are the states and a, b, c are constant positive parameters.

The system (3) exhibits a *strange chaotic attractor* for the parameter values

$$a = 4, \quad b = 2.7, \quad c = 0.6 \tag{4}$$

For numerical simulations, we take the initial conditions as

$$x_1(0) = 0.6, \quad x_2(0) = 0.2, \quad x_3(0) = 0.4 \tag{5}$$

Figure 1 shows the 3-D phase portrait of the novel system (3). Figures 2, 3 and 4 show the 2-D projection of the novel jerk chaotic system (3) on the (x_1, x_2), (x_2, x_3) and (x_1, x_3) planes, respectively.

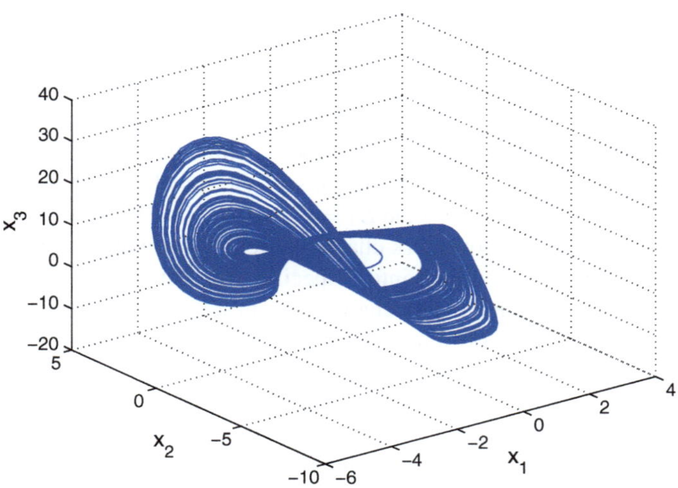

Fig. 1 3-D phase portrait of the novel jerk chaotic system

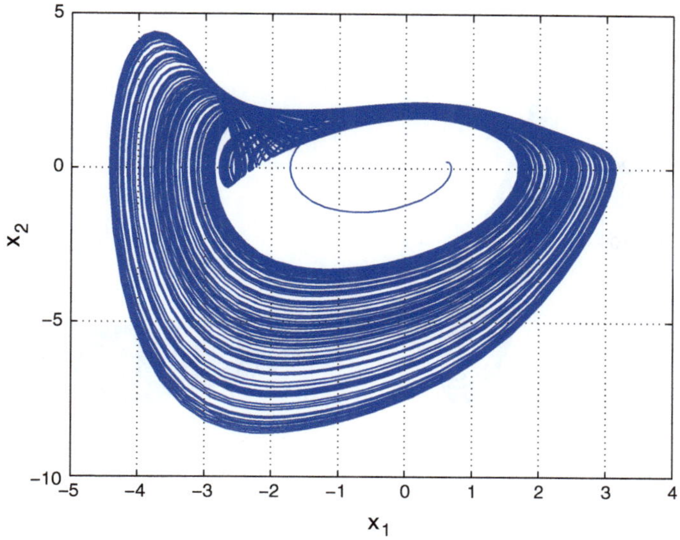

Fig. 2 2-D projection of the jerk chaotic system on the (x_1, x_2) plane

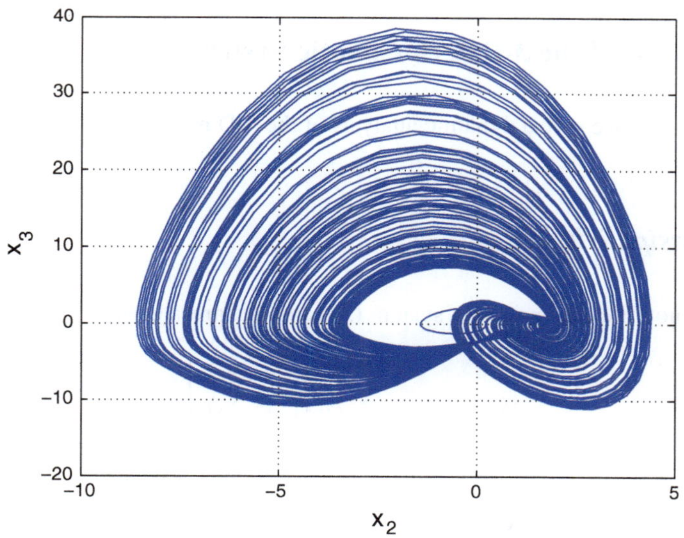

Fig. 3 2-D projection of the jerk chaotic system on the (x_2, x_3) plane

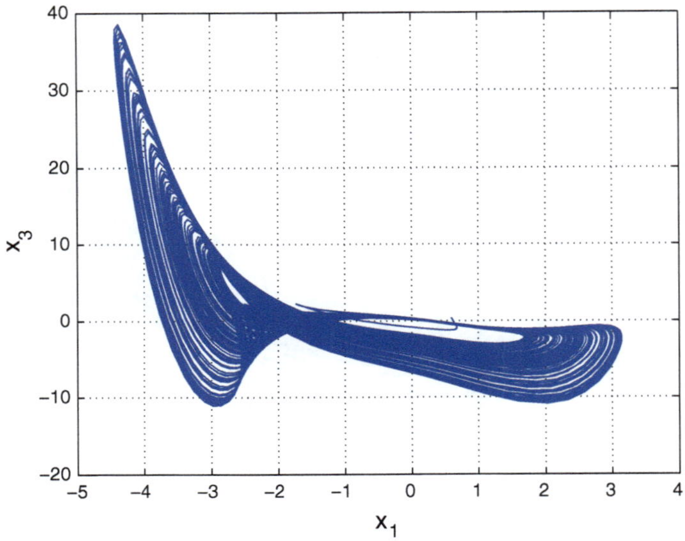

Fig. 4 2-D projection of the jerk chaotic system on the (x_1, x_3) plane

3 Analysis of the 3-D Jerk Chaotic System

In this section, we give a dynamic analysis of the 3-D novel jerk chaotic system (3).

3.1 *Dissipativity*

In vector notation, the new jerk system (3) can be expressed as

$$\dot{\mathbf{x}} = f(\mathbf{x}) = \begin{bmatrix} f_1(x_1, x_2, x_3) \\ f_2(x_1, x_2, x_3) \\ f_3(x_1, x_2, x_3) \end{bmatrix}, \tag{6}$$

where

$$\begin{cases} f_1(x_1, x_2, x_3) = x_2 \\ f_2(x_1, x_2, x_3) = x_3 \\ f_3(x_1, x_2, x_3) = -ax_1 - bx_3 + cx_1^3 x_2 - 1 \end{cases} \tag{7}$$

Let Ω be any region in $\mathbf{R}^3$ with a smooth boundary and also, $\Omega(t) = \Phi_t(\Omega)$, where Φ_t is the flow of f. Furthermore, let $V(t)$ denote the volume of $\Omega(t)$.

By Liouville's theorem, we know that

$$\dot{V}(t) = \int\limits_{\Omega(t)} (\nabla \cdot f)\, dx_1\, dx_2\, dx_3 \tag{8}$$

The divergence of the novel jerk system (6) is found as:

$$\nabla \cdot f = \frac{\partial f_1}{\partial x_1} + \frac{\partial f_2}{\partial x_2} + \frac{\partial f_3}{\partial x_3} = -b < 0 \tag{9}$$

since b is a positive parameter.

Inserting the value of $\nabla \cdot f$ from (9) into (8), we get

$$\dot{V}(t) = \int\limits_{\Omega(t)} (-b)\, dx_1\, dx_2\, dx_3 = -bV(t) \tag{10}$$

Integrating the first order linear differential equation (10), we get

$$V(t) = \exp(-bt)V(0) \tag{11}$$

Since $b > 0$, it follows from Eq. (11) that $V(t) \to 0$ exponentially as $t \to \infty$. This shows that the novel 3-D jerk chaotic system (3) is dissipative. Hence, the system limit sets are ultimately confined into a specific limit set of zero volume, and the asymptotic motion of the novel jerk chaotic system (3) settles onto a strange attractor of the system.

3.2 Equilibrium Points

The equilibrium points of the jerk chaotic system (3) are obtained by solving the equations

$$\begin{cases} f_1(x_1, x_2, x_3) = x_2 & = 0 \\ f_2(x_1, x_2, x_3) = x_3 & = 0 \\ f_3(x_1, x_2, x_3) = -ax_1 - bx_3 + cx_1^3 x_2 - 1 & = 0 \end{cases} \tag{12}$$

We take the parameter values as in the chaotic case, viz. $a = 4$, $b = 2.7$ and $c = 0.6$.

Solving the Eq. (12), we get a unique equilibrium point of the jerk chaotic system (3) as

$$E_1 = \begin{bmatrix} -1/a \\ 0 \\ 0 \end{bmatrix} = \begin{bmatrix} -0.25 \\ 0 \\ 0 \end{bmatrix} \tag{13}$$

To test the stability type of the equilibrium point E_1, we calculate the Jacobian matrix of the novel jerk chaotic system (3) at any point x:

$$J(x) = \begin{bmatrix} 0 & 1 & 0 \\ 0 & 0 & 1 \\ -a + 3cx_1^2 x_2 & cx_1^3 & -b \end{bmatrix} \tag{14}$$

We find that

$$J_1 \triangleq J(E_1) = \begin{bmatrix} 0 & 1 & 0 \\ 0 & 0 & 1 \\ -4 & -0.0094 & -2.7 \end{bmatrix} \tag{15}$$

The matrix J_1 has the eigenvalues

$$\lambda_1 = -3.1104, \quad \lambda_{2,3} = 0.2052 \pm 1.1153i \tag{16}$$

This shows that the equilibrium point E_1 is a saddle-focus, which is unstable.

3.3 Lyapunov Exponents and Kaplan–Yorke Dimension

We take the parameter values of the novel jerk system (3) as $a = 4$, $b = 2.7$ and $c = 0.6$. We take the initial state of the jerk system (3) as given in (5).

Then the Lyapunov exponents of the jerk system (3) are numerically obtained using MATLAB as

$$L_1 = 0.1443, \quad L_2 = 0, \quad L_3 = -2.8439 \tag{17}$$

Thus, the maximal Lyapunov exponent (MLE) of the novel jerk system (3) is positive, which means that the system has a chaotic behavior.

Since $L_1 + L_2 + L_3 = -0.082 < 0$, it follows that the novel jerk chaotic system (3) is dissipative.

Also, the Kaplan–Yorke dimension of the novel jerk chaotic system (3) is obtained as

$$D_{KY} = 2 + \frac{L_1 + L_2}{|L_3|} = 2.0507 \tag{18}$$

which is fractional.

4 Adaptive Control of the 3-D Novel Jerk Chaotic System

In this section, we use backstepping control method to derive an adaptive feedback control law for globally stabilizing the 3-D novel novel jerk chaotic system with unknown parameters.

Thus, we consider the 3-D novel jerk chaotic system given by

$$\begin{cases} \dot{x}_1 = x_2 \\ \dot{x}_2 = x_3 \\ \dot{x}_3 = -ax_1 - bx_3 + cx_1^3x_2 - 1 + u \end{cases} \tag{19}$$

where a, b and c are unknown constant parameters, and u is a backstepping control law to be determined using estimates $\hat{a}(t)$, $\hat{b}(t)$ and $\hat{c}(t)$ for a, b and c, respectively.

The parameter estimation errors are defined as:

$$\begin{cases} e_a(t) = a - \hat{a}(t) \\ e_b(t) = b - \hat{b}(t) \\ e_c(t) = c - \hat{c}(t) \end{cases} \tag{20}$$

Differentiating (20) with respect to t, we obtain the following equations:

$$\begin{cases} \dot{e}_a(t) = -\dot{\hat{a}}(t) \\ \dot{e}_b(t) = -\dot{\hat{b}}(t) \\ \dot{e}_c(t) = -\dot{\hat{c}}(t) \end{cases} \tag{21}$$

Next, we shall state and prove the main result of this section.

This theorem gives a backstepping-based adaptive control for globally stabilizing the 3-D novel jerk chaotic system (19) with unknown parameters, and we establish theorem using Lyapunov stability theory [13].

Theorem 1 *The 3-D novel jerk chaotic system (19), with unknown parameters a, b and c, is globally and exponentially stabilized by the adaptive feedback control law,*

$$u(t) = -(3 - \hat{a}(t))x_1 - 5x_2 - (3 - \hat{b}(t))x_3 - \hat{c}(t)x_1^3x_2 + 1 - kz_3 \tag{22}$$

where $k > 0$ is a gain constant,

$$z_3 = 2x_1 + 2x_2 + x_3, \tag{23}$$

and the update law for the parameter estimates $\hat{a}(t)$, $\hat{b}(t)$, $\hat{c}(t)$ is given by

$$\begin{cases} \dot{\hat{a}}(t) = -x_1z_3 \\ \dot{\hat{b}}(t) = -x_3z_3 \\ \dot{\hat{c}}(t) = x_1^3x_2z_3 \end{cases} \tag{24}$$

Proof We prove this result via Lyapunov stability theory [13].

First, we define a quadratic Lyapunov function

$$V_1(z_1) = \frac{1}{2}z_1^2 \tag{25}$$

where

$$z_1 = x_1 \tag{26}$$

Differentiating V_1 along the dynamics (19), we get

$$\dot{V}_1 = z_1\dot{z}_1 = x_1x_2 = -z_1^2 + z_1(x_1 + x_2) \tag{27}$$

Now, we define

$$z_2 = x_1 + x_2 \tag{28}$$

Using (28), we can simplify the Eq. (27) as

$$\dot{V}_1 = -z_1^2 + z_1z_2 \tag{29}$$

Secondly, we define a quadratic Lyapunov function

$$V_2(z_1, z_2) = V_1(z_1) + \frac{1}{2}\,z_2^2 = \frac{1}{2}\left(z_1^2 + z_2^2\right) \tag{30}$$

Differentiating V_2 along the dynamics (19), we get

$$\dot{V}_2 = -z_1^2 - z_2^2 + z_2(2x_1 + 2x_2 + x_3) \tag{31}$$

Now, we define

$$z_3 = 2x_1 + 2x_2 + x_3 \tag{32}$$

Using (32), we can simplify the Eq. (31) as

$$\dot{V}_2 = -z_1^2 - z_2^2 + z_2z_3 \tag{33}$$

Finally, we define a quadratic Lyapunov function

$$V(z_1, z_2, z_3, e_a, e_b, e_c) = V_2(z_1, z_2) + \frac{1}{2}z_3^2 + \frac{1}{2}e_a^2 + \frac{1}{2}e_b^2 + \frac{1}{2}e_c^2 \tag{34}$$

which is a positive definite function on $\mathbf{R}^6$.

Differentiating V along the dynamics (19), we get

$$\dot{V} = -z_1^2 - z_2^2 - z_3^2 + z_3(z_3 + z_2 + \dot{z}_3) - e_a\dot{\hat{a}} - e_b\dot{\hat{b}} - e_c\dot{\hat{c}} \tag{35}$$

Equation (35) can be written compactly as

$$\dot{V} = -z_1^2 - z_2^2 - z_3^2 + z_3S - e_a\dot{\hat{a}} - e_b\dot{\hat{b}} - e_b\dot{\hat{b}} - e_c\dot{\hat{c}} \tag{36}$$

where

$$S = z_3 + z_2 + \dot{z}_3 = z_3 + z_2 + 2\dot{x}_1 + 2\dot{x}_2 + \dot{x}_3 \quad (37)$$

A simple calculation gives

$$S = (3 - a)x_1 + 5x_2 + (3 - b)x_3 + cx_1^3 x_2 - 1 + u \quad (38)$$

Substituting the adaptive control law (22) into (38), we obtain

$$S = -\left[a - \hat{a}(t)\right]x_1 - \left[b - \hat{b}(t)\right]x_3 + \left[c - \hat{c}(t)\right]x_1^3 x_2 - kz_3 \quad (39)$$

Using the definitions (21), we can simplify (39) as

$$S = -e_a x_1 - e_b x_3 + e_c x_1^3 x_2 - kz_3 \quad (40)$$

Substituting the value of S from (40) into (36), we obtain

$$\dot{V} = -z_1^2 - z_2^2 - (1 + k)z_3^2 + e_a\left(-x_1 z_3 - \dot{\hat{a}}\right) + e_b\left(-x_3 z_3 - \dot{\hat{b}}\right) + e_c\left(x_1^3 x_2 z_3 - \dot{\hat{c}}\right) \quad (41)$$

Substituting the update law (24) into (41), we get

$$\dot{V} = -z_1^2 - z_2^2 - (1 + k)z_3^2, \quad (42)$$

which is a negative semi-definite function on $\mathbf{R}^6$.

From (42), it follows that the vector $\mathbf{z}(t) = (z_1(t), z_2(t), z_3(t))$ and the parameter estimation error $(e_a(t), e_b(t), e_c(t))$ are globally bounded, i.e.

$$\left[\, z_1(t)\ z_2(t)\ z_3(t)\ e_a(t)\ e_b(t)\ e_c(t) \,\right] \in \mathbf{L}_\infty \quad (43)$$

Also, it follows from (42) that

$$\dot{V} \le -z_1^2 - z_2^2 - z_3^2 = -\|\mathbf{z}\|^2 \quad (44)$$

That is,

$$\|\mathbf{z}\|^2 \le -\dot{V} \quad (45)$$

Integrating the inequality (45) from 0 to t, we get

$$\int_0^t |\mathbf{z}(\tau)|^2\, d\tau \le V(0) - V(t) \quad (46)$$

From (46), it follows that $\mathbf{z}(t) \in \mathbf{L}_2$.

From Eq. (19), it can be deduced that $\dot{\mathbf{z}}(t) \in \mathbf{L}_\infty$.

Thus, using Barbalat's lemma [13], we conclude that $\mathbf{z}(t) \to \mathbf{0}$ exponentially as $t \to \infty$ for all initial conditions $\mathbf{z}(0) \in \mathbf{R}^3$.

Hence, it is immediate that $\mathbf{x}(t) \to \mathbf{0}$ exponentially as $t \to \infty$ for all initial conditions $\mathbf{x}(0) \in \mathbf{R}^3$.

This completes the proof. ■

For the numerical simulations, the classical fourth-order Runge–Kutta method with step size $h = 10^{-8}$ is used to solve the system of differential equations (19) and (24), when the adaptive control law (22) is applied.

The parameter values of the novel jerk chaotic system (19) are taken as

$$a = 4, \quad b = 2.7, \quad c = 0.6 \tag{47}$$

We take the positive gain constant as

$$k = 8 \tag{48}$$

Furthermore, as initial conditions of the novel jerk chaotic system (19), we take

$$x_1(0) = 5.4, \quad x_2(0) = 3.2, \quad x_3(0) = 2.7 \tag{49}$$

Also, as initial conditions of the parameter estimates $\hat{a}(t)$, $\hat{b}(t)$ and $\hat{c}(t)$, we take

$$\hat{a}(0) = 3.2, \quad \hat{b}(0) = 5.4, \quad \hat{c}(0) = 10.4 \tag{50}$$

In Fig. 5, the exponential convergence of the controlled states $x_1(t), x_2(t), x_3(t)$ is depicted, when the adaptive control law (22) and (24) are implemented.

5 Adaptive Synchronization of the Identical 3-D Jerk Chaotic Systems

In this section, we use backstepping control method to derive an adaptive control law for globally and exponentially synchronizing the identical 3-D novel jerk chaotic systems with unknown parameters.

As the master system, we consider the 3-D novel jerk chaotic system given by

$$\begin{cases} \dot{x}_1 = x_2 \\ \dot{x}_2 = x_3 \\ \dot{x}_3 = -ax_1 - bx_3 + cx_1^3 x_2 - 1 \end{cases} \tag{51}$$

where x_1, x_2, x_3 are the states of the system, and a, b and c are unknown constant parameters.

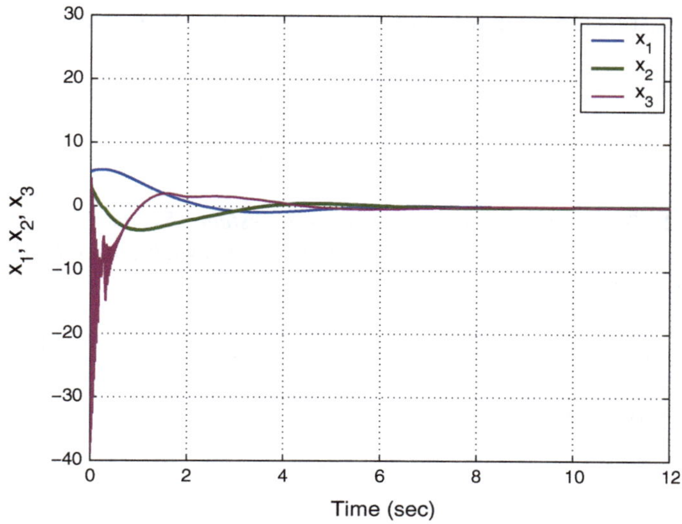

Fig. 5 Time-history of the controlled states x_1, x_2, x_3

As the slave system, we consider the 3-D novel jerk chaotic system given by

$$\begin{cases} \dot{y}_1 = y_2 \\ \dot{y}_2 = y_3 \\ \dot{y}_3 = -ay_1 - by_3 + cy_1^3 y_2 - 1 + u \end{cases} \tag{52}$$

where y_1, y_2, y_3 are the states of the system, and u is a backstepping control to be determined using estimates $\hat{a}(t)$, $\hat{b}(t)$ and $\hat{c}(t)$ for a, b and c, respectively.

We define the synchronization error between the states of the master system (51) and the slave system (52) as

$$\begin{cases} e_1 = y_1 - x_1 \\ e_2 = y_2 - x_2 \\ e_3 = y_3 - x_3 \end{cases} \tag{53}$$

Then the error dynamics is easily obtained as

$$\begin{cases} \dot{e}_1 = e_2 \\ \dot{e}_2 = e_3 \\ \dot{e}_3 = -ae_1 - be_3 + c\left(y_1^3 y_2 - x_1^3 x_2\right) + u \end{cases} \tag{54}$$

The parameter estimation errors are defined as:

$$\begin{cases} e_a(t) = a - \hat{a}(t) \\ e_b(t) = b - \hat{b}(t) \\ e_c(t) = c - \hat{c}(t) \end{cases} \tag{55}$$

Differentiating (55) with respect to t, we obtain the following equations:

$$\begin{cases} \dot{e}_a(t) = -\dot{\hat{a}}(t) \\ \dot{e}_b(t) = -\dot{\hat{b}}(t) \\ \dot{e}_c(t) = -\dot{\hat{c}}(t) \end{cases} \tag{56}$$

Theorem 2 *The identical 3-D novel jerk chaotic systems (51) and (52) with unknown parameters a, b and c are globally and exponentially synchronized by the adaptive control law*

$$u(t) = -\left[3 - \hat{a}(t)\right]e_1 - 5e_2 - \left[3 - \hat{b}(t)\right]e_3 - \hat{c}(t)\left[y_1^3 y_2 - x_1^3 x_2\right] - kz_3 \tag{57}$$

where $k > 0$ is a gain constant,

$$z_3 = 2e_1 + 2e_2 + e_3, \tag{58}$$

and the update law for the parameter estimates $\hat{a}(t)$, $\hat{b}(t)$, $\hat{c}(t)$ is given by

$$\begin{cases} \dot{\hat{a}}(t) = -e_1 z_3 \\ \dot{\hat{b}}(t) = -e_3 z_3 \\ \dot{\hat{c}}(t) = z_3\left(y_1^3 y_2 - x_1^3 x_2\right) \end{cases} \tag{59}$$

Proof First, we define a quadratic Lyapunov function

$$V_1(z_1) = \frac{1}{2}\, z_1^2 \tag{60}$$

where

$$z_1 = e_1 \tag{61}$$

Differentiating V_1 along the error dynamics (54), we get

$$\dot{V}_1 = z_1 \dot{z}_1 = e_1 e_2 = -z_1^2 + z_1(e_1 + e_2) \tag{62}$$

Now, we define

$$z_2 = e_1 + e_2 \tag{63}$$

Using (63), we can simplify the Eq. (62) as

$$\dot{V}_1 = -z_1^2 + z_1 z_2 \tag{64}$$

Secondly, we define a quadratic Lyapunov function

$$V_2(z_1, z_2) = V_1(z_1) + \frac{1}{2} z_2^2 = \frac{1}{2} \left(z_1^2 + z_2^2\right) \tag{65}$$

Differentiating V_2 along the error dynamics (54), we get

$$\dot{V}_2 = -z_1^2 - z_2^2 + z_2(2e_1 + 2e_2 + e_3) \tag{66}$$

Now, we define

$$z_3 = 2e_1 + 2e_2 + e_3 \tag{67}$$

Using (67), we can simplify the Eq. (66) as

$$\dot{V}_2 = -z_1^2 - z_2^2 + z_2 z_3 \tag{68}$$

Finally, we define a quadratic Lyapunov function

$$V(z_1, z_2, z_3, e_a, e_b, e_c) = V_2(z_1, z_2) + \frac{1}{2} z_3^2 + \frac{1}{2} e_a^2 + \frac{1}{2} e_b^2 + \frac{1}{2} e_c^2 \tag{69}$$

Differentiating V along the error dynamics (54), we get

$$\dot{V} = -z_1^2 - z_2^2 - z_3^2 + z_3(z_3 + z_2 + \dot{z}_3) - e_a \dot{\hat{a}} - e_b \dot{\hat{b}} \tag{70}$$

Equation (70) can be written compactly as

$$\dot{V} = -z_1^2 - z_2^2 - z_3^2 + z_3 S - e_a \dot{\hat{a}} - e_b \dot{\hat{b}} \tag{71}$$

where

$$S = z_3 + z_2 + \dot{z}_3 = z_3 + z_2 + 2\dot{e}_1 + 2\dot{e}_2 + \dot{e}_3 \tag{72}$$

A simple calculation gives

$$S = (3 - a)e_1 + 5e_2 + (3 - b)e_3 + c\left(y_1^3 y_2 - x_1^3 x_2\right) + u \tag{73}$$

Substituting the adaptive control law (57) into (73), we obtain

$$S = -\left[a - \hat{a}(t)\right] e_1 - \left[b - \hat{b}(t)\right] e_3 + \left[c - \hat{c}(t)\right] \left(y_1^3 y_2 - x_1^3 x_2\right) - k z_3 \tag{74}$$

Using the definitions (56), we can simplify (74) as

$$S = -e_a e_1 - e_b e_3 + e_c \left(y_1^3 y_2 - x_1^3 x_2\right) - k z_3 \tag{75}$$

Substituting the value of S from (75) into (71), we obtain

$$\begin{aligned}\dot{V} &= -z_1^2 - z_2^2 - (1+k)z_3^2 + e_a\left(-e_1z_3 - \dot{\hat{a}}\right) + e_b\left(-e_3z_3 - \dot{\hat{b}}\right) \\ &\quad + e_c\left[z_3\left(y_1^3y_2 - x_1^3x_2\right) - \dot{\hat{c}}\right]\end{aligned} \tag{76}$$

Substituting the update law (59) into (76), we get

$$\dot{V} = -z_1^2 - z_2^2 - (1+k)z_3^2, \tag{77}$$

which is a negative semi-definite function on $\mathbf{R}^6$.

From (77), it follows that the vector $\mathbf{z}(t) = (z_1(t), z_2(t), z_3(t))$ and the parameter estimation error $(e_a(t), e_b(t), e_c(t))$ are globally bounded, i.e.

$$\left[\, z_1(t)\ z_2(t)\ z_3(t)\ e_a(t)\ e_b(t)\ e_c(t) \,\right] \in \mathbf{L}_\infty \tag{78}$$

Also, it follows from (77) that

$$\dot{V} \le -z_1^2 - z_2^2 - z_3^2 = -\|\mathbf{z}\|^2 \tag{79}$$

That is,

$$\|\mathbf{z}\|^2 \le -\dot{V} \tag{80}$$

Integrating the inequality (80) from 0 to t, we get

$$\int_0^t |\mathbf{z}(\tau)|^2\, d\tau \le V(0) - V(t) \tag{81}$$

From (81), it follows that $\mathbf{z}(t) \in \mathbf{L}_2$.

From Eq. (54), it can be deduced that $\dot{\mathbf{z}}(t) \in \mathbf{L}_\infty$.

Thus, using Barbalat's lemma [13], we conclude that $\mathbf{z}(t) \to \mathbf{0}$ exponentially as $t \to \infty$ for all initial conditions $\mathbf{z}(0) \in \mathbf{R}^3$.

Hence, it is immediate that $\mathbf{e}(t) \to \mathbf{0}$ exponentially as $t \to \infty$ for all initial conditions $\mathbf{e}(0) \in \mathbf{R}^3$.

This completes the proof. ■

For the numerical simulations, the classical fourth-order Runge–Kutta method with step size $h = 10^{-8}$ is used to solve the system of novel jerk chaotic systems, which are taken as the master and slave systems.

The parameter values of the novel jerk chaotic systems are taken as in the chaotic case, i.e. $a = 4$, $b = 2.7$ and $c = 0.6$.

We take the positive gain constant as $k = 10$.

Furthermore, as initial conditions of the master system (51), we take

$$x_1(0) = 0.5, \quad x_2(0) = 0.3, \quad x_3(0) = 0.6 \tag{82}$$

As initial conditions of the slave system (52), we take

$$y_1(0) = -0.4, \quad y_2(0) = -0.2, \quad y_3(0) = 0.2 \tag{83}$$

Also, as initial conditions of the parameter estimates $\hat{a}(t)$, $\hat{b}(t)$ and $\hat{c}(t)$, we take

$$\hat{a}(0) = 0.6, \quad \hat{b}(0) = 1.4, \quad \hat{c}(0) = 0.5 \tag{84}$$

In Figs. 6, 7, and 8, the complete synchronization of the identical 3-D jerk chaotic systems (51) and (52) is depicted.

Also, in Fig. 9, the time-history of the synchronization errors $e_1(t)$, $e_2(t)$, $e_3(t)$, is depicted.

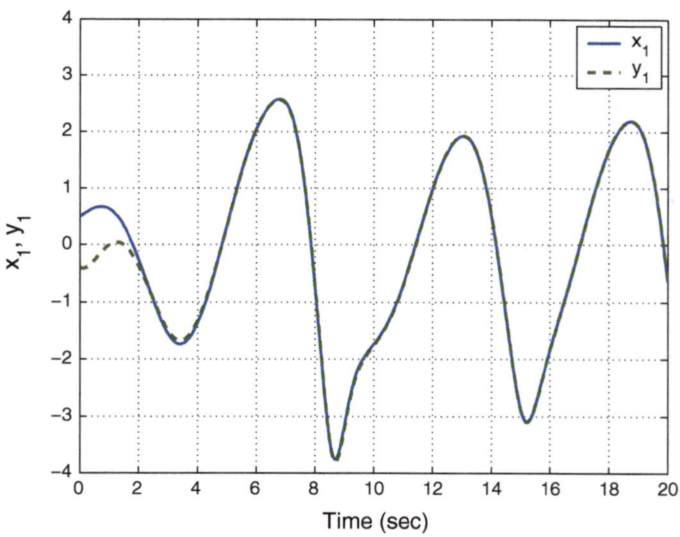

Fig. 6 Synchronization of the states x_1 and y_1

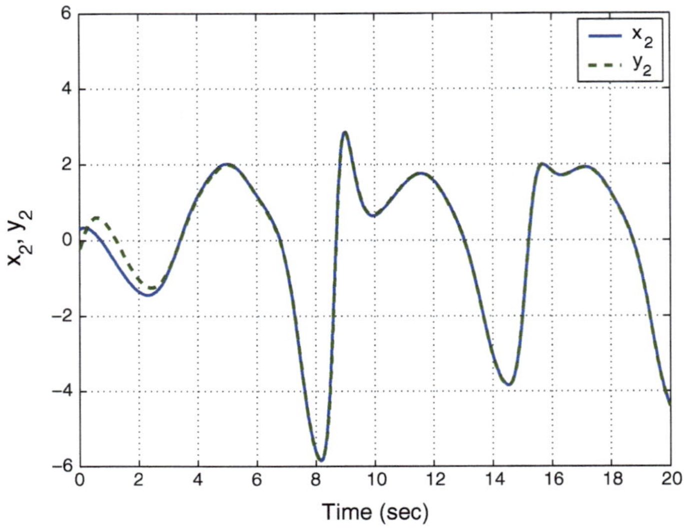

Fig. 7 Synchronization of the states x_2 and y_2

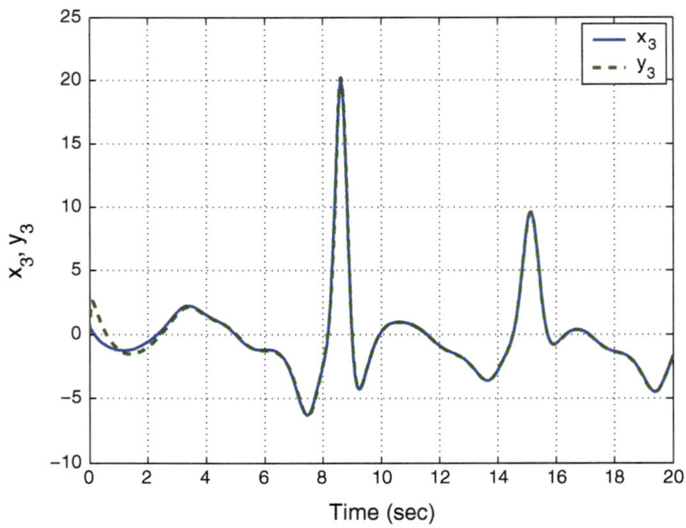

Fig. 8 Synchronization of the states x_3 and y_3

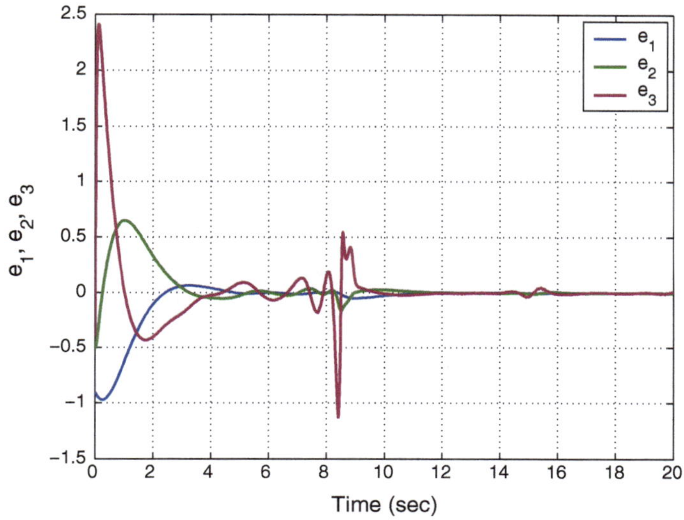

Fig. 9 Time-history of the synchronization errors e_1, e_2, e_3

6 Circuit Realization of the Novel Jerk System

In this section, circuit realization of the novel jerk system is reported. The state variable x_3 of jerk system (3) is scaled down. As a result, the novel jerk system (3) has been changed to

$$\begin{cases} \dot{X}_1 = X_2 \\ \dot{X}_2 = 4X_3 \\ \dot{X}_3 = -\frac{a}{4}X_1 - bX_3 + \frac{c}{4}X_1^3 X_2 - \frac{1}{4}, \end{cases} \tag{85}$$

where $X_1 = x_1$, $X_2 = x_2$, and $X_3 = \frac{1}{4}x_3$.

The electronic circuit realizing the system (85) is designed by using off-the-shelf components and shown in Fig. 10. It is easy to obtained the following circuital equations

$$\begin{cases} \frac{dv_{C_1}}{dt} = \frac{1}{R_1C_1}v_{C_2} \\ \frac{dv_{C_2}}{dt} = \frac{1}{R_2C_2}v_{C_3} \\ \frac{dv_{C_3}}{dt} = -\frac{1}{R_3C_3}v_{C_1} - \frac{1}{R_4C_3}v_{C_3} + \frac{1}{1000R_5C_3}v_{C_1}^3 v_{C_2} - \frac{1}{R_6C_3}V_e \end{cases} \tag{86}$$

where v_{C_1}, v_{C_2}, and v_{C_3} are the voltages across the capacitors C_1, C_2, and C_3, respectively. Here the design approach based on the operational amplifiers [98, 110] is applied. Therefore, each state variable of system (85), i.e. X_1, X_2, X_3 is implemented as the voltage across the corresponding capacitors C_1, C_2, and C_3, respectively.

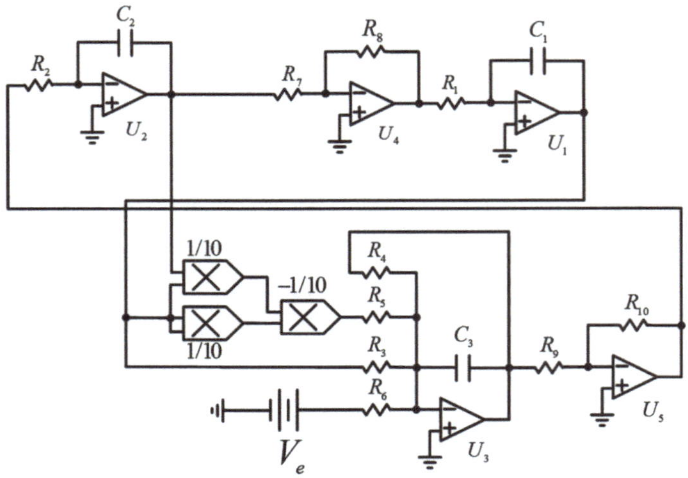

Fig. 10 The designed electronic circuit schematic of the novel jerk chaotic system

The power supplies of all active devices are $\pm 15 V_{DC}$ and the TL084 operational amplifiers are used in this work. The values of components in Fig. 10 are chosen to match the parameters of system (85) as follows: $R_1 = R_3 = R_7 = R_8 = R_9 = R_{10} = 400\,\text{k}\Omega$, $R_2 = 100\,\text{k}\Omega$, $R_4 = 148.148\,\text{k}\Omega$, $R_5 = 2.666\,\text{k}\Omega$, $R_6 = 1.6\,\text{M}\Omega$, $V_e = 1\,V_{DC}$, and $C_1 = C_2 = C_3 = 1\,\text{nF}$.

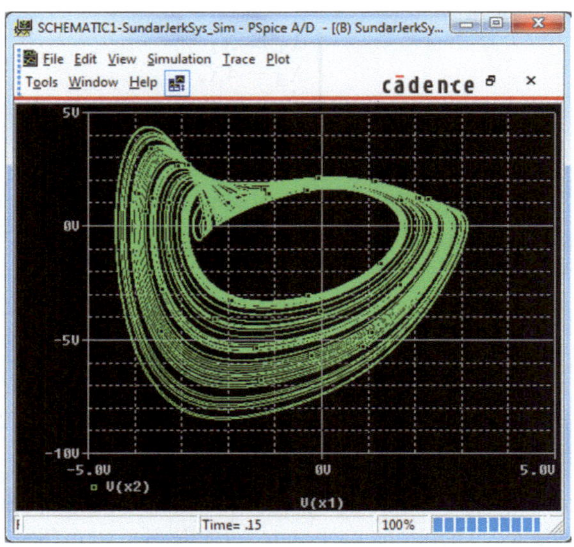

Fig. 11 Phase portrait result of the designed electronic circuit obtained from OrCAD in v_{C_1}–v_{C_2} plane

The designed circuit is implemented in the electronic simulation package Cadence OrCAD. The obtained results are displayed in Figs. 11, 12 and 13, which show the chaotic attractors in v_{C_1}–v_{C_2}, v_{C_2}–v_{C_3}, and v_{C_1}–v_{C_3} planes. Thus the feasibility of the proposed chaotic jerk system is confirmed.

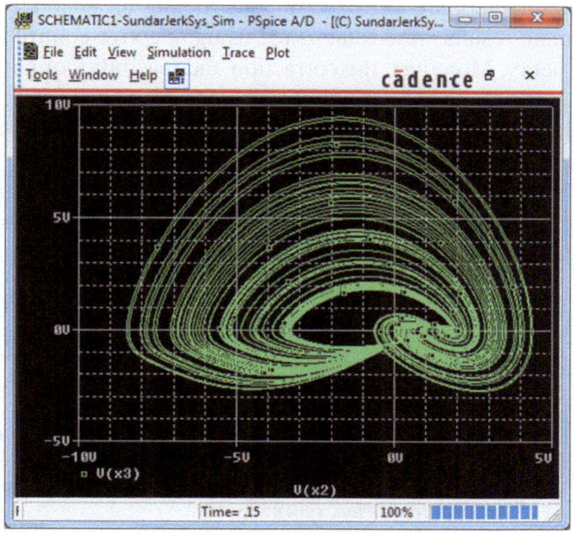

Fig. 12 Phase portrait result of the designed electronic circuit obtained from OrCAD in v_{C_2}–v_{C_3} plane

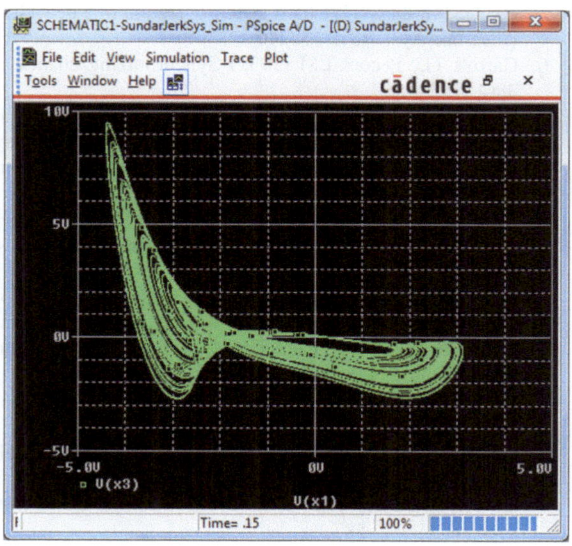

Fig. 13 Phase portrait result of the designed electronic circuit obtained from OrCAD in v_{C_1}–v_{C_3} plane

7 Conclusions

In this paper, we proposed a novel six-term jerk chaotic system with a quartic nonlinearity. Dynamic characteristics of new system has been discovered. It is worth noting that the possibilities of control and synchronization of such system with unknown parameters are verified by constructing an adaptive backstepping controller. The main results were established using adaptive control theory and Lyapunov stability theory. Moreover, the correction and feasibility of novel theoretical system are confirmed through Spice results which are obtained from the designed electronic circuit. It is possible to use the new jerk system in potential chaos-based applications such as secure communications, random generation, or path planning for autonomous mobile robots. It is believed that the unknown dynamical behaviors of such strange chaotic jerk systems should be further investigated in the future research on chaos theory.

References

1. Abdurrahman A, Jiang H, Teng Z (2015) Finite-time synchronization for memristor-based neural networks with time-varying delays. Neural Netw 69:20–28
2. Arneodo A, Coullet P, Tresser C (1981) Possible new strange attractors with spiral structure. Commun Math Phys 79(4):573–576
3. Azar AT, Vaidyanathan S (2015) Chaos modeling and control systems design, vol 581. Springer, Germany
4. Behnia S, Afrang S, Akhshani A, Mabhouti K (2013) A novel method for controlling chaos in external cavity semiconductor laser. Optik 124(8):757–764
5. Cai G, Tan Z (2007) Chaos synchronization of a new chaotic system via nonlinear control. J Uncertain Syst 1(3):235–240
6. Carroll TL, Pecora LM (1991) Synchronizing chaotic circuits. IEEE Trans Circuits Syst 38(4):453–456
7. Chen G, Ueta T (1999) Yet another chaotic attractor. Int J Bifurc Chaos 9(7):1465–1466
8. Chen WH, Wei D, Lu X (2014) Global exponential synchronization of nonlinear time-delay Lur'e systems via delayed impulsive control. Commun Nonlinear Sci Numer Simul 19(9):3298–3312
9. Gan Q, Liang Y (2012) Synchronization of chaotic neural networks with time delay in the leakage term and parametric uncertainties based on sampled-data control. J Frankl Inst 349(6):1955–1971
10. Islam MM, Murase K (2005) Chaotic dynamics of a behaviour-based miniature mobile robot: effects of environment and control structure. Neural Netw 18(2):123–144
11. Jiang GP, Zheng WX, Chen G (2004) Global chaos synchronization with channel time-delay. Chaos Solitons Fractals 20(2):267–275
12. Karthikeyan R, Sundarapandian V (2014) Hybrid chaos synchronization of four-scroll systems via active control. J Electr Eng 65(2):97–103
13. Khalil HK (2001) Nonlinear systems, 3rd edn. Prentice Hall, New Jersey
14. Li D (2008) A three-scroll chaotic attractor. Phys Lett A 372(4):387–393
15. Li GH, Zhou SP, Yang K (2007) Controlling chaos in Colpitts oscillator. Chaos Solitons Fractals 33:582–587
16. Li N, Zhang Y, Nie Z (2011) Synchronization for general complex dynamical networks with sampled-data. Neurocomputing 74(5):805–811

17. Lorenz EN (1963) Deterministic periodic flow. J Atmos Sci 20(2):130–141
18. Lü J, Chen G (2002) A new chaotic attractor coined. Int J Bifurc Chaos 12(3):659–661
19. Matouk AE (2011) Chaos, feedback control and synchronization of a fractional-order modified autonomous Van der Pol-Duffing circuit. Commun Nonlinear Sci Numer Simul 16(2):975–986
20. Pecora LM, Carroll TL (1990) Synchronization in chaotic systems. Phys Rev Lett 64(8):821–824
21. Pehlivan I, Moroz IM, Vaidyanathan S (2014) Analysis, synchronization and circuit design of a novel butterfly attractor. J Sound Vib 333(20):5077–5096
22. Pham VT, Volos CK, Vaidyanathan S, Le TP, Vu VY (2015) A memristor-based hyperchaotic system with hidden attractors: dynamics, synchronization and circuital emulating. J Eng Sci Technol Rev 8(2):205–214
23. Rasappan S, Vaidyanathan S (2012) Global chaos synchronization of WINDMI and Coullet chaotic systems by backstepping control. Far East J Math Sci 67(2):265–287
24. Rasappan S, Vaidyanathan S (2012) Hybrid synchronization of n-scroll Chua and Lur'e chaotic systems via backstepping control with novel feedback. Arch Control Sci 22(3):343–365
25. Rasappan S, Vaidyanathan S (2012) Synchronization of hyperchaotic Liu system via backstepping control with recursive feedback. Commun Comput Inf Sci 305:212–221
26. Rasappan S, Vaidyanathan S (2013) Hybrid synchronization of n-scroll chaotic Chua circuits using adaptive backstepping control design with recursive feedback. Malays J Math Sci 7(2):219–246
27. Rasappan S, Vaidyanathan S (2014) Global chaos synchronization of WINDMI and Coullet chaotic systems using adaptive backstepping control design. Kyungpook Math J 54(1): 293–320
28. Rhouma R, Belghith S (2011) Cryptanalysis of a chaos-based cryptosystem. Commun Nonlinear Sci Numer Simul 16(2):876–884
29. Rössler OE (1976) An equation for continuous chaos. Phys Lett A 57(5):397–398
30. Sampath S, Vaidyanathan S, Volos CK, Pham VT (2015) An eight-term novel four-scroll chaotic system with cubic nonlinearity and its circuit simulation. J Eng Sci Technol Rev 8(2):1–6
31. Sarasu P, Sundarapandian V (2011) Active controller design for the generalized projective synchronization of four-scroll chaotic systems. Int J Syst Signal Control Eng Appl 4(2):26–33
32. Sarasu P, Sundarapandian V (2011) The generalized projective synchronization of hyperchaotic Lorenz and hyperchaotic Qi systems via active control. Int J Soft Comput 6(5): 216–223
33. Sarasu P, Sundarapandian V (2012) Adaptive controller design for the generalized projective synchronization of 4-scroll systems. Int J Syst Signal Control Eng Appl 5(2):21–30
34. Sarasu P, Sundarapandian V (2012) Generalized projective synchronization of three-scroll chaotic systems via adaptive control. Eur J Sci Res 72(4):504–522
35. Sarasu P, Sundarapandian V (2012) Generalized projective synchronization of two-scroll systems via adaptive control. Int J Soft Comput 7(4):146–156
36. Shahverdiev EM, Shore KA (2009) Impact of modulated multiple optical feedback time delays on laser diode chaos synchronization. Opt Commun 282(17):3568–3572
37. Sprott JC (1994) Some simple chaotic flows. Phys Rev E 50(2):647–650
38. Sundarapandian V (2010) Output regulation of the Lorenz attractor. Far East J Math Sci 42(2):289–299
39. Sundarapandian V (2011) Output regulation of the Arneodo-Coullet chaotic system. Commun Comput Inf Sci 133:98–107
40. Sundarapandian V (2013) Analysis and anti-synchronization of a novel chaotic system via active and adaptive controllers. J Eng Sci Technol Rev 6(4):45–52
41. Sundarapandian V, Karthikeyan R (2011) Anti-synchronization of hyperchaotic Lorenz and hyperchaotic Chen systems by adaptive control. Int J Syst Signal Control Eng Appl 4(2):18–25
42. Sundarapandian V, Karthikeyan R (2011) Anti-synchronization of Lü and Pan chaotic systems by adaptive nonlinear control. Eur J Sci Res 64(1):94–106

43. Sundarapandian V, Karthikeyan R (2012) Adaptive anti-synchronization of uncertain Tigan and Li systems. J Eng Appl Sci 7(1):45–52
44. Sundarapandian V, Karthikeyan R (2012) Hybrid synchronization of hyperchaotic Lorenz and hyperchaotic Chen systems via active control. J Eng Appl Sci 7(3):254–264
45. Sundarapandian V, Pehlivan I (2012) Analysis, control, synchronization, and circuit design of a novel chaotic system. Math Comput Model 55(7–8):1904–1915
46. Sundarapandian V, Sivaperumal S (2011) Sliding controller design of hybrid synchronization of four-wing chaotic systems. Int J Soft Comput 6(5):224–231
47. Suresh R, Sundarapandian V (2013) Global chaos synchronization of a family of n-scroll hyperchaotic Chua circuits using backstepping control with recursive feedback. Far East J Math Sci 73(1):73–95
48. Tigan G, Opris D (2008) Analysis of a 3D chaotic system. Chaos Solitons Fractals 36: 1315–1319
49. Tuwankotta JM (2006) Chaos in a coupled oscillators system with widely spaced frequencies and energy-preserving non-linearity. Int J Non-Linear Mech 41(2):180–191
50. Usama M, Khan MK, Alghathbar K, Lee C (2010) Chaos-based secure satellite imagery cryptosystem. Comput Math Appl 60(2):326–337
51. Vaidyanathan S (2011) Hybrid chaos synchronization of Liu and Lü systems by active nonlinear control. Commun Comput Inf Sci 204:1–10
52. Vaidyanathan S (2011) Output regulation of the unified chaotic system. Commun Comput Inf Sci 204:84–93
53. Vaidyanathan S (2012) Analysis and synchronization of the hyperchaotic Yujun systems via sliding mode control. Adv Intell Syst Comput 176:329–337
54. Vaidyanathan S (2012) Anti-synchronization of Sprott-L and Sprott-M chaotic systems via adaptive control. Int J Control Theory Appl 5(1):41–59
55. Vaidyanathan S (2012) Global chaos control of hyperchaotic Liu system via sliding control method. Int J Control Theory Appl 5(2):117–123
56. Vaidyanathan S (2012) Output regulation of the Liu chaotic system. Appl Mech Mater 110–116:3982–3989
57. Vaidyanathan S (2012) Sliding mode control based global chaos control of Liu-Liu-Liu-Su chaotic system. Int J Control Theory Appl 5(1):15–20
58. Vaidyanathan S (2013) A new six-term 3-D chaotic system with an exponential nonlinearity. Far East J Math Sci 79(1):135–143
59. Vaidyanathan S (2013) Analysis and adaptive synchronization of two novel chaotic systems with hyperbolic sinusoidal and cosinusoidal nonlinearity and unknown parameters. J Eng Sci Technol Rev 6(4):53–65
60. Vaidyanathan S (2013) Analysis, control and synchronization of hyperchaotic Zhou system via adaptive control. Adv Intell Syst Comput 177:1–10
61. Vaidyanathan S (2014) A new eight-term 3-D polynomial chaotic system with three quadratic nonlinearities. Far East J Math Sci 84(2):219–226
62. Vaidyanathan S (2014) Analysis and adaptive synchronization of eight-term 3-D polynomial chaotic systems with three quadratic nonlinearities. Eur Phys J Spec Top 223(8):1519–1529
63. Vaidyanathan S (2014) Analysis, control and synchronisation of a six-term novel chaotic system with three quadratic nonlinearities. Int J Model Identif Control 22(1):41–53
64. Vaidyanathan S (2014) Generalized projective synchronisation of novel 3-D chaotic systems with an exponential non-linearity via active and adaptive control. Int J Model Identif Control 22(3):207–217
65. Vaidyanathan S (2014) Global chaos synchronization of identical Li-Wu chaotic systems via sliding mode control. Int J Model Identif Control 22(2):170–177
66. Vaidyanathan S (2015) 3-cells cellular neural network (CNN) attractor and its adaptive biological control. Int J PharmTech Res 8(4):632–640
67. Vaidyanathan S (2015) A 3-D novel highly chaotic system with four quadratic nonlinearities, its adaptive control and anti-synchronization with unknown parameters. J Eng Sci Technol Rev 8(2):106–115

68. Vaidyanathan S (2015) A novel chemical chaotic reactor system and its adaptive control. Int J ChemTech Res 8(7):146–158
69. Vaidyanathan S (2015) Adaptive backstepping control of enzymes-substrates system with ferroelectric behaviour in brain waves. Int J PharmTech Res 8(2):256–261
70. Vaidyanathan S (2015) Adaptive biological control of generalized Lotka-Volterra three-species biological system. Int J PharmTech Res 8(4):622–631
71. Vaidyanathan S (2015) Adaptive chaotic synchronization of enzymes-substrates system with ferroelectric behaviour in brain waves. Int J PharmTech Res 8(5):964–973
72. Vaidyanathan S (2015) Adaptive control of a chemical chaotic reactor. Int J PharmTech Res 8(3):377–382
73. Vaidyanathan S (2015) Adaptive control of the FitzHugh-Nagumo chaotic neuron model. Int J PharmTech Res 8(6):117–127
74. Vaidyanathan S (2015) Adaptive synchronization of chemical chaotic reactors. Int J ChemTech Res 8(2):612–621
75. Vaidyanathan S (2015) Adaptive synchronization of generalized Lotka-Volterra three-species biological systems. Int J PharmTech Res 8(5):928–937
76. Vaidyanathan S (2015) Adaptive synchronization of novel 3-D chemical chaotic reactor systems. Int J ChemTech Res 8(7):159–171
77. Vaidyanathan S (2015) Adaptive synchronization of the identical FitzHugh-Nagumo chaotic neuron models. Int J PharmTech Res 8(6):167–177
78. Vaidyanathan S (2015) Analysis, control and synchronization of a 3-D novel jerk chaotic system with two quadratic nonlinearities. Kyungpook Math J 55:563–586
79. Vaidyanathan S (2015) Analysis, properties and control of an eight-term 3-D chaotic system with an exponential nonlinearity. Int J Model Identif Control 23(2):164–172
80. Vaidyanathan S (2015) Anti-synchronization of Brusselator chemical reaction systems via adaptive control. Int J ChemTech Res 8(6):759–768
81. Vaidyanathan S (2015) Chaos in neurons and adaptive control of Birkhoff-Shaw strange chaotic attractor. Int J PharmTech Res 8(5):956–963
82. Vaidyanathan S (2015) Chaos in neurons and synchronization of Birkhoff-Shaw strange chaotic attractors via adaptive control. Int J PharmTech Res 8(6):1–11
83. Vaidyanathan S (2015) Coleman-Gomatam logarithmic competitive biology models and their ecological monitoring. Int J PharmTech Res 8(6):94–105
84. Vaidyanathan S (2015) Dynamics and control of Brusselator chemical reaction. Int J ChemTech Res 8(6):740–749
85. Vaidyanathan S (2015) Dynamics and control of tokamak system with symmetric and magnetically confined plasma. Int J ChemTech Res 8(6):795–803
86. Vaidyanathan S (2015) Global chaos synchronization of chemical chaotic reactors via novel sliding mode control method. Int J ChemTech Res 8(7):209–221
87. Vaidyanathan S (2015) Global chaos synchronization of the forced Van der Pol chaotic oscillators via adaptive control method. Int J PharmTech Res 8(6):156–166
88. Vaidyanathan S (2015) Global chaos synchronization of the Lotka-Volterra biological systems with four competitive species via active control. Int J PharmTech Res 8(6):206–217
89. Vaidyanathan S (2015) Lotka-Volterra population biology models with negative feedback and their ecological monitoring. Int J PharmTech Res 8(5):974–981
90. Vaidyanathan S (2015) Lotka-Volterra two species competitive biology models and their ecological monitoring. Int J PharmTech Res 8(6):32–44
91. Vaidyanathan S (2015) Output regulation of the forced Van der Pol chaotic oscillator via adaptive control method. Int J PharmTech Res 8(6):106–116
92. Vaidyanathan S, Azar AT (2015) Analysis and control of a 4-D novel hyperchaotic system. In: Azar AT, Vaidyanathan S (eds) Chaos modeling and control systems design, Studies in computational intelligence, vol 581. Springer, Germany, pp 19–38
93. Vaidyanathan S, Azar AT (2015) Analysis, control and synchronization of a nine-term 3-D novel chaotic system. In: Azar AT, Vaidyanathan S (eds) Chaos modelling and control systems design, Studies in computational intelligence, vol 581. Springer, Germany, pp 19–38

94. Vaidyanathan S, Azar AT (2015) Anti-synchronization of identical chaotic systems using sliding mode control and an application to Vaidhyanathan-Madhavan chaotic systems. Stud Comput Intell 576:527–547
95. Vaidyanathan S, Azar AT (2015) Hybrid synchronization of identical chaotic systems using sliding mode control and an application to Vaidhyanathan chaotic systems. Stud Comput Intell 576:549–569
96. Vaidyanathan S, Madhavan K (2013) Analysis, adaptive control and synchronization of a seven-term novel 3-D chaotic system. Int J Control Theory Appl 6(2):121–137
97. Vaidyanathan S, Pakiriswamy S (2013) Generalized projective synchronization of six-term Sundarapandian chaotic systems by adaptive control. Int J Control Theory Appl 6(2):153–163
98. Vaidyanathan S, Pakiriswamy S (2015) A 3-D novel conservative chaotic system and its generalized projective synchronization via adaptive control. J Eng Sci Technol Rev 8(2):52–60
99. Vaidyanathan S, Rajagopal K (2011) Anti-synchronization of Li and T chaotic systems by active nonlinear control. Commun Comput Inf Sci 198:175–184
100. Vaidyanathan S, Rajagopal K (2011) Global chaos synchronization of hyperchaotic Pang and Wang systems by active nonlinear control. Commun Comput Inf Sci 204:84–93
101. Vaidyanathan S, Rajagopal K (2011) Global chaos synchronization of Lü and Pan systems by adaptive nonlinear control. Commun Comput Inf Sci 205:193–202
102. Vaidyanathan S, Rajagopal K (2012) Global chaos synchronization of hyperchaotic Pang and hyperchaotic Wang systems via adaptive control. Int J Soft Comput 7(1):28–37
103. Vaidyanathan S, Rasappan S (2011) Global chaos synchronization of hyperchaotic Bao and Xu systems by active nonlinear control. Commun Comput Inf Sci 198:10–17
104. Vaidyanathan S, Rasappan S (2014) Global chaos synchronization of n-scroll Chua circuit and Lur'e system using backstepping control design with recursive feedback. Arab J Sci Eng 39(4):3351–3364
105. Vaidyanathan S, Sampath S (2011) Global chaos synchronization of hyperchaotic Lorenz systems by sliding mode control. Commun Comput Inf Sci 205:156–164
106. Vaidyanathan S, Sampath S (2012) Anti-synchronization of four-wing chaotic systems via sliding mode control. Int J Autom Comput 9(3):274–279
107. Vaidyanathan S, Volos C (2015) Analysis and adaptive control of a novel 3-D conservative no-equilibrium chaotic system. Arch Control Sci 25(3):333–353
108. Vaidyanathan S, Volos C, Pham VT (2014) Hyperchaos, adaptive control and synchronization of a novel 5-D hyperchaotic system with three positive Lyapunov exponents and its SPICE implementation. Arch Control Sci 24(4):409–446
109. Vaidyanathan S, Volos C, Pham VT (2014) Hyperchaos, adaptive control and synchronization of a novel 5-D hyperchaotic system with three positive Lyapunov exponents and its SPICE implementation. Arch Control Sci 24(4):409–446
110. Vaidyanathan S, Volos C, Pham VT, Madhavan K, Idowu BA (2014) Adaptive backstepping control, synchronization and circuit simulation of a 3-D novel jerk chaotic system with two hyperbolic sinusoidal nonlinearities. Arch Control Sci 24(3):375–403
111. Vaidyanathan S, Idowu BA, Azar AT (2015) Backstepping controller design for the global chaos synchronization of Sprott's jerk systems. Stud Comput Intell 581:39–58
112. Vaidyanathan S, Rajagopal K, Volos CK, Kyprianidis IM, Stouboulos IN (2015) Analysis, adaptive control and synchronization of a seven-term novel 3-D chaotic system with three quadratic nonlinearities and its digital implementation in LabVIEW. J Eng Sci Technol Rev 8(2):130–141
113. Vaidyanathan S, Volos C, Pham VT, Madhavan K (2015) Analysis, adaptive control and synchronization of a novel 4-D hyperchaotic hyperjerk system and its SPICE implementation. Arch Control Sci 25(1):5–28
114. Vaidyanathan S, Volos CK, Kyprianidis IM, Stouboulos IN, Pham VT (2015) Analysis, adaptive control and anti-synchronization of a six-term novel jerk chaotic system with two exponential nonlinearities and its circuit simulation. J Eng Sci Technol Rev 8(2):24–36

115. Vaidyanathan S, Volos CK, Madhavan K (2015) Analysis, control, synchronization and SPICE implementation of a novel 4-D hyperchaotic Rikitake dynamo system without equilibrium. J Eng Sci Technol Rev 8(2):232–244
116. Vaidyanathan S, Volos CK, Pham VT (2015) Analysis, adaptive control and adaptive synchronization of a nine-term novel 3-D chaotic system with four quadratic nonlinearities and its circuit simulation. J Eng Sci Technol Rev 8(2):181–191
117. Vaidyanathan S, Volos CK, Pham VT (2015) Global chaos control of a novel nine-term chaotic system via sliding mode control. In: Azar AT, Zhu Q (eds) Advances and applications in sliding mode control systems, Studies in computational intelligence, vol 576. Springer, Germany, pp 571–590
118. Vaidyanathan S, Volos CK, Pham VT, Madhavan K (2015) Analysis, adaptive control and synchronization of a novel 4-D hyperchaotic hyperjerk system and its SPICE implementation. Arch Control Sci 25(1):135–158
119. Vaidyanathan S, Volos CK, Rajagopal K, Kyprianidis IM, Stouboulos IN (2015) Adaptive backstepping controller design for the anti-synchronization of identical WINDMI chaotic systems with unknown parameters and its SPICE implementation. J Eng Sci Technol Rev 8(2):74–82
120. Volos CK, Kyprianidis IM, Stouboulos IN, Anagnostopoulos AN (2009) Experimental study of the dynamic behavior of a double scroll circuit. J Appl Funct Anal 4:703–711
121. Volos CK, Kyprianidis IM, Stouboulos IN (2013) Experimental investigation on coverage performance of a chaotic autonomous mobile robot. Robot Auton Syst 61(12):1314–1322
122. Volos CK, Kyprianidis IM, Stouboulos IN, Tlelo-Cuautle E, Vaidyanathan S (2015) Memristor: a new concept in synchronization of coupled neuromorphic circuits. J Eng Sci Technol Rev 8(2):157–173
123. Wang X, Ge C (2008) Controlling and tracking of Newton-Leipnik system via backstepping design. Int J Nonlinear Sci 5(2):133–139
124. Wei Z, Yang Q (2010) Anti-control of Hopf bifurcation in the new chaotic system with two stable node-foci. Appl Math Comput 217(1):422–429
125. Xiao X, Zhou L, Zhang Z (2014) Synchronization of chaotic Lur'e systems with quantized sampled-data controller. Commun Nonlinear Sci Numer Simul 19(6):2039–2047
126. Yang J, Zhu F (2013) Synchronization for chaotic systems and chaos-based secure communications via both reduced-order and step-by-step sliding mode observers. Commun Nonlinear Sci Numer Simul 18(4):926–937
127. Yang J, Chen Y, Zhu F (2014) Singular reduced-order observer-based synchronization for uncertain chaotic systems subject to channel disturbance and chaos-based secure communication. Appl Math Comput 229:227–238
128. Zhang H, Zhou J (2012) Synchronization of sampled-data coupled harmonic oscillators with control inputs missing. Syst Control Lett 61(12):1277–1285
129. Zhou W, Xu Y, Lu H, Pan L (2008) On dynamics analysis of a new chaotic attractor. Phys Lett A 372(36):5773–5777
130. Zhu C, Liu Y, Guo Y (2010) Theoretic and numerical study of a new chaotic system. Intell Inf Manag 2:104–109

A Seven-Term Novel Jerk Chaotic System and Its Adaptive Control

Sundarapandian Vaidyanathan

Abstract In this work, we describe a seven-term novel 3-D jerk chaotic system with two nonlinearities (quadratic and cubic). The phase portraits of the novel jerk chaotic system are displayed and the dynamic properties of the novel jerk chaotic system are discussed. The novel jerk chaotic system has three saddle-foci equilibrium points, which are unstable. The Lyapunov exponents of the novel jerk chaotic system are obtained as $L_1 = 0.5565$, $L_2 = 0$ and $L_3 = -1.5566$. The Kaplan–Yorke dimension of the novel jerk chaotic system is obtained as $D_{KY} = 2.3575$. Next, an adaptive backstepping controller is designed to globally stabilize the novel jerk chaotic system with unknown parameters. Moreover, an adaptive backstepping controller is also designed to achieve global chaos synchronization of the identical jerk chaotic systems with unknown parameters. The backstepping control method is a recursive procedure that links the choice of a Lyapunov function with the design of a controller and guarantees global asymptotic stability of strict feedback systems. MATLAB simulations have been shown to illustrate all the main results derived in this work.

Keywords Chaos · Chaotic systems · Jerk systems · Backstepping control · Adaptive control · Synchronization

1 Introduction

Chaos theory deals with the qualitative study of chaotic dynamical systems and their applications in science and engineering. A dynamical system is called *chaotic* if it satisfies the three properties: boundedness, infinite recurrence and sensitive dependence on initial conditions [3].

S. Vaidyanathan (✉)
Research and Development Centre, Vel Tech University,
Avadi, Chennai 600062, Tamil Nadu, India
e-mail: sundarvtu@gmail.com

S. Vaidyanathan and C. Volos (eds.), *Advances and Applications in Chaotic Systems*, Studies in Computational Intelligence 636,
DOI 10.1007/978-3-319-30279-9_6

Some classical paradigms of 3-D chaotic systems in the literature are Lorenz system [17], Rössler system [29], ACT system [2], Sprott systems [37], Chen system [7], Lü system [18], Cai system [5], Tigan system [48], etc.

Many new chaotic systems have been discovered in the recent years such as Zhou system [128], Zhu system [129], Li system [14], Wei-Yang system [123], Sundarapandian systems [40, 45], Vaidyanathan systems [58, 59, 61–64, 67, 78, 79, 93, 96, 98, 107, 109, 111, 113, 115, 116], Pehlivan system [21], Sampath system [30], etc.

Chaos theory has applications in several fields of science and engineering such as chemical reactors [68, 72, 74, 76, 80, 84–86], biological systems [66, 69–71, 73, 75, 77, 81–83, 87–91], memristors [1, 22, 121], lasers [4], oscillations [49], robotics [10, 120], electrical circuits [19, 119], cryptosystems [28, 50], secure communications [125, 126], etc.

Many methods have been designed for control and regulation of chaotic systems such as active control [38, 39, 52], adaptive control [108, 114, 117], backstepping control [15, 122], sliding mode control [55, 57], etc.

Synchronization of chaotic systems is a phenomenon that occurs when two or more chaotic systems are coupled or when a chaotic system drives another chaotic system. Because of the butterfly effect which causes exponential divergence of the trajectories of two identical chaotic systems started with nearly the same initial conditions, the synchronization of chaotic systems is a challenging research problem in the chaos literature [3].

Pecora and Carroll pioneered the research on synchronization of chaotic systems with their seminal papers [6, 20]. The active control method [12, 31, 32, 44, 51, 56, 99, 100, 103] is typically used when the system parameters are available for measurement. Adaptive control method [33–35, 41–43, 54, 60, 92, 97, 101, 102, 108, 112] is typically used when some or all the system parameters are not available for measurement and estimates for the uncertain parameters of the systems.

Sampled-data feedback control method [9, 16, 124, 127] and time-delay feedback control method [8, 11, 36] are also used for synchronization of chaotic systems. Backstepping control method [23–27, 47, 104, 110, 118] is also used for the synchronization of chaotic systems. Backstepping control is a recursive method for stabilizing the origin of a control system in strict-feedback form [13]. Another popular method for the synchronization of chaotic systems is the sliding mode control method [46, 53, 65, 94, 95, 105, 106], which is a nonlinear control method that alters the dynamics of a nonlinear system by application of a discontinuous control signal that forces the system to "slide" along a cross-section of the system's normal behavior.

In the recent decades, there is some good interest in finding jerk chaotic systems, which are described by the third-order ODE

$$\dddot{x} = j(x, \dot{x}, \ddot{x}) \tag{1}$$

The differential equation (1) is called "jerk system" because the third order time derivative in mechanical systems is called *jerk*.

By defining phase variables $x_1 = x, x_2 = \dot{x}$ and $x_3 = \ddot{x}$, the jerk differential equation (1) can be expressed as a 3-D system given by

$$\begin{cases} \dot{x}_1 = x_2 \\ \dot{x}_2 = x_3 \\ \dot{x}_3 = j(x_1, x_2, x_3) \end{cases} \tag{2}$$

In this research work, we announce a six-term novel 3-D jerk chaotic system with a quartic nonlinearity. We have also designed adaptive backstepping controllers for stabilization and synchronization of the six-term novel 3-D jerk chaotic system.

This work is organized as follows. Section 2 describes the dynamic equations and phase portraits of the novel 3-D jerk chaotic system. Section 3 details the qualitative properties of the novel jerk chaotic system. The novel jerk chaotic system has three unstable equilibrium points. The Lyapunov exponents of the novel jerk chaotic system are obtained as $L_1 = 0.5565$, $L_2 = 0$ and $L_3 = -1.5566$, while the Kaplan–Yorke dimension of the novel jerk chaotic system is obtained as $D_{KY} = 2.3575$.

In Sect. 4, we design an adaptive backstepping controller to globally stabilize the novel jerk chaotic system with unknown parameters. In Sect. 5, an adaptive backstepping controller is designed to achieve global chaos synchronization of the identical novel jerk chaotic systems with unknown parameters. Section 6 contains the conclusions of this work.

2 A 3-D Novel Jerk Chaotic System

In this section, we describe a seven-term novel 3-D jerk chaotic system with two nonlinearities (quadratic and cubic) described by

$$\begin{cases} \dot{x}_1 = x_2 \\ \dot{x}_2 = x_3 \\ \dot{x}_3 = ax_1 - x_1^2 - x_1^3 - bx_2 - x_3 \end{cases} \tag{3}$$

where x_1, x_2, x_3 are the states and a, b are constant, positive parameters.

The system (3) exhibits a *strange chaotic attractor* for the parameter values

$$a = 134, \quad b = 50 \tag{4}$$

For numerical simulations, we take the initial conditions as

$$x_1(0) = 0.2, \quad x_2(0) = 0, \quad x_3(0) = 0.4 \tag{5}$$

Figure 1 shows the 3-D phase portrait of the novel system (3). Figures 2, 3 and 4 show the 2-D projection of the novel jerk chaotic system (3) on the (x_1, x_2), (x_2, x_3) and (x_1, x_3) planes, respectively.

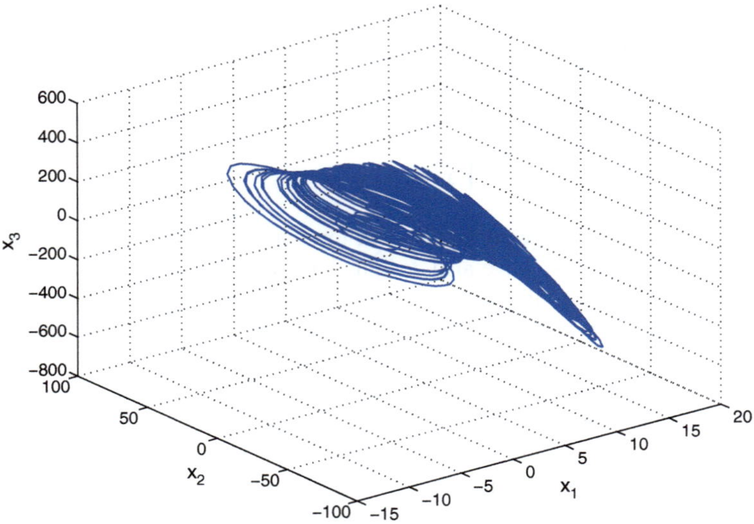

Fig. 1 3-D phase portrait of the novel jerk chaotic system

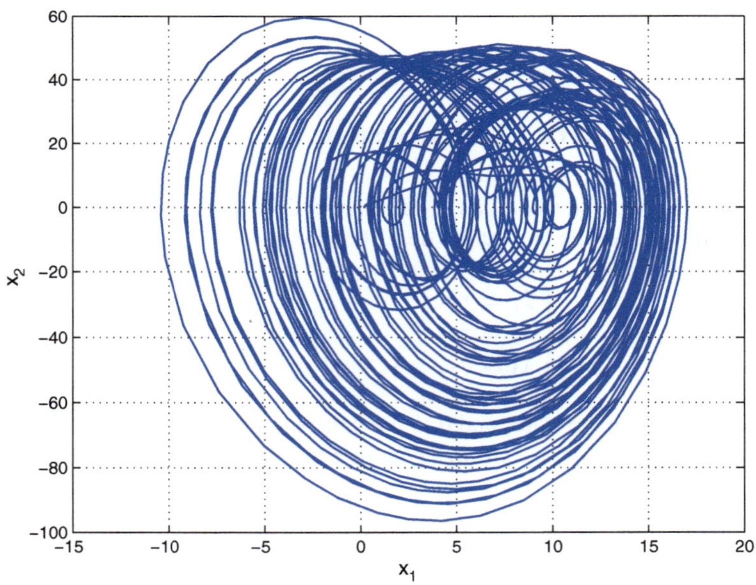

Fig. 2 2-D projection of the jerk chaotic system on the (x_1, x_2) plane

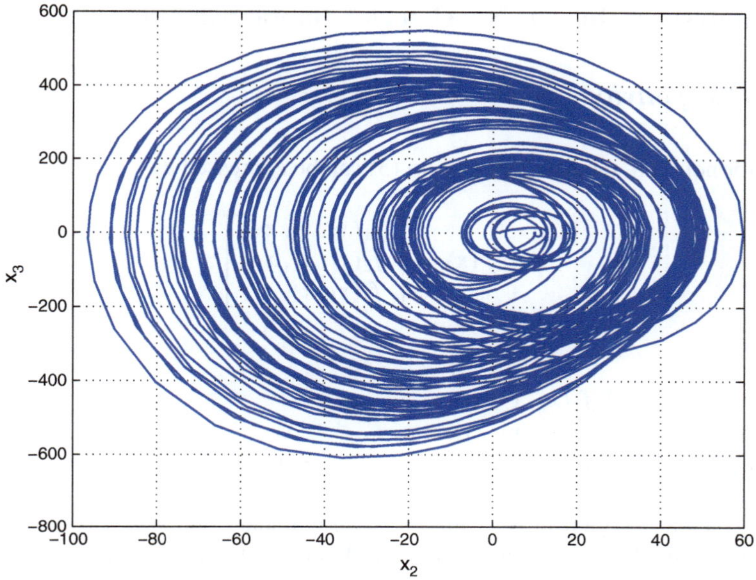

Fig. 3 2-D projection of the jerk chaotic system on the (x_2, x_3) plane

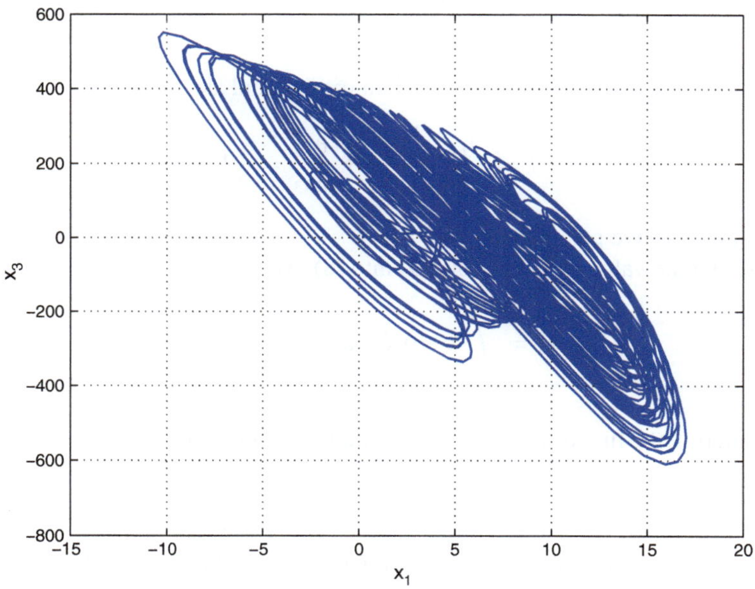

Fig. 4 2-D projection of the jerk chaotic system on the (x_1, x_3) plane

3 Analysis of the 3-D Novel Jerk Chaotic System

3.1 Dissipativity

In vector notation, the new jerk system (3) can be expressed as

$$\dot{\mathbf{x}} = f(\mathbf{x}) = \begin{bmatrix} f_1(x_1, x_2, x_3) \\ f_2(x_1, x_2, x_3) \\ f_3(x_1, x_2, x_3) \end{bmatrix}, \tag{6}$$

where

$$\begin{cases} f_1(x_1, x_2, x_3) = x_2 \\ f_2(x_1, x_2, x_3) = x_3 \\ f_3(x_1, x_2, x_3) = ax_1 - x_1^2 - x_1^3 - bx_2 - x_3 \end{cases} \tag{7}$$

Let Ω be any region in $\mathbf{R}^3$ with a smooth boundary and also, $\Omega(t) = \Phi_t(\Omega)$, where Φ_t is the flow of f. Furthermore, let $V(t)$ denote the volume of $\Omega(t)$.

By Liouville's theorem, we know that

$$\dot{V}(t) = \int\limits_{\Omega(t)} (\nabla \cdot f)\, dx_1\, dx_2\, dx_3 \tag{8}$$

The divergence of the novel jerk system (6) is found as:

$$\nabla \cdot f = \frac{\partial f_1}{\partial x_1} + \frac{\partial f_2}{\partial x_2} + \frac{\partial f_3}{\partial x_3} = -1 < 0 \tag{9}$$

since b is a positive parameter.

Inserting the value of $\nabla \cdot f$ from (9) into (8), we get

$$\dot{V}(t) = \int\limits_{\Omega(t)} (-b)\, dx_1\, dx_2\, dx_3 = -V(t) \tag{10}$$

Integrating the first order linear differential equation (10), we get

$$V(t) = \exp(-t)V(0) \tag{11}$$

Since $b > 0$, it follows from Eq. (11) that $V(t) \to 0$ exponentially as $t \to \infty$. This shows that the novel 3-D jerk chaotic system (3) is dissipative. Hence, the system limit sets are ultimately confined into a specific limit set of zero volume, and the asymptotic motion of the novel jerk chaotic system (3) settles onto a strange attractor of the system.

3.2 Equilibrium Points

The equilibrium points of the novel jerk chaotic system (3) are obtained by solving the equations

$$\begin{cases} f_1(x_1, x_2, x_3) = x_2 & = 0 \\ f_2(x_1, x_2, x_3) = x_3 & = 0 \\ f_3(x_1, x_2, x_3) = ax_1 - x_1^2 - x_1^3 - bx_2 - x_3 & = 0 \end{cases} \tag{12}$$

We take the parameter values as in the chaotic case, *viz.* $a = 134$ and $b = 50$.

Solving the Eq. (12), we get three equilibrium points of the novel jerk chaotic system (3) as

$$E_0 = \begin{bmatrix} 0 \\ 0 \\ 0 \end{bmatrix}, \quad E_1 = \begin{bmatrix} 11.0866 \\ 0 \\ 0 \end{bmatrix}, \quad E_2 = \begin{bmatrix} -12.0866 \\ 0 \\ 0 \end{bmatrix} \tag{13}$$

To test the stability type of the equilibrium points, we calculate the Jacobian matrix of the novel jerk chaotic system (3) at any point x:

$$J(x) = \begin{bmatrix} 0 & 1 & 0 \\ 0 & 0 & 1 \\ a - 2x_1 - 3x_1^2 & -b & -1 \end{bmatrix} \tag{14}$$

We find that

$$J_0 \triangleq J(E_0) = \begin{bmatrix} 0 & 1 & 0 \\ 0 & 0 & 1 \\ 134 & -50 & -1 \end{bmatrix} \tag{15}$$

The matrix J_0 has the eigenvalues

$$\lambda_1 = 2.3218, \quad \lambda_{2,3} = -1.6609 \pm 7.4131\, i \tag{16}$$

This shows that the equilibrium point E_0 is a saddle-focus, which is unstable.
Next, we find that

$$J_1 \triangleq J(E_0) = \begin{bmatrix} 0 & 1 & 0 \\ 0 & 0 & 1 \\ -256.9113 & -50 & -1 \end{bmatrix} \tag{17}$$

The matrix J_1 has the eigenvalues

$$\lambda_1 = -4.0978, \quad \lambda_{2,3} = 1.5489 \pm 7.7650\, i \tag{18}$$

This shows that the equilibrium point E_1 is a saddle-focus, which is unstable.

We also find that

$$J_2 \triangleq J(E_0) = \begin{bmatrix} 0 & 1 & 0 \\ 0 & 0 & 1 \\ -280.0845 & -50 & -1 \end{bmatrix} \tag{19}$$

The matrix J_1 has the eigenvalues

$$\lambda_1 = -4.3418, \quad \lambda_{2,3} = 1.6709 \pm 7.8560\, i \tag{20}$$

This shows that the equilibrium point E_2 is a saddle-focus, which is unstable. Thus, the novel jerk chaotic system (3) has three unstable equilibrium points.

3.3 Lyapunov Exponents and Kaplan–Yorke Dimension

We take the parameter values of the novel jerk system (3) as $a = 134$ and $b = 50$. We take the initial state of the jerk system (3) as given in (5).

Then the Lyapunov exponents of the novel jerk system (3) are numerically obtained using MATLAB as

$$L_1 = 0.5565, \quad L_2 = 0, \quad L_3 = -1.5566 \tag{21}$$

Thus, the maximal Lyapunov exponent (MLE) of the novel jerk system (3) is positive, which means that the system has a chaotic behavior.

Since $L_1 + L_2 + L_3 = -1.001 < 0$, it follows that the novel jerk chaotic system (3) is dissipative.

Also, the Kaplan–Yorke dimension of the novel jerk chaotic system (3) is obtained as

$$D_{KY} = 2 + \frac{L_1 + L_2}{|L_3|} = 2.3575 \tag{22}$$

which is fractional.

4 Adaptive Control of the 3-D Novel Jerk Chaotic System

In this section, we use backstepping control method to derive an adaptive feedback control law for globally stabilizing the 3-D novel jerk chaotic system with unknown parameters.

Thus, we consider the 3-D novel jerk chaotic system given by

$$\begin{cases} \dot{x}_1 = x_2 \\ \dot{x}_2 = x_3 \\ \dot{x}_3 = ax_1 - x_1^2 - x_1^3 - bx_2 - x_3 + u \end{cases} \tag{23}$$

In (23), x_1, x_2, x_3 are the states, a, b are unknown constant parameters, and u is a backstepping control law to be determined using estimates $\hat{a}(t)$ and $\hat{b}(t)$ for the unknown parameters a and b, respectively.

The parameter estimation errors are defined as:

$$\begin{cases} e_a(t) = a - \hat{a}(t) \\ e_b(t) = b - \hat{b}(t) \end{cases} \tag{24}$$

Differentiating (24) with respect to t, we obtain the following equations:

$$\begin{cases} \dot{e}_a(t) = -\dot{\hat{a}}(t) \\ \dot{e}_b(t) = -\dot{\hat{b}}(t) \end{cases} \tag{25}$$

Next, we shall state and prove the main result of this section.

Theorem 1 *The 3-D novel jerk chaotic system (23), with unknown parameters a and b, is globally and exponentially stabilized by the adaptive feedback control law,*

$$u(t) = -\left[3 + \hat{a}(t)\right] x_1 - \left[5 - \hat{b}(t)\right] x_2 - 2x_3 + x_1^2 + x_1^3 - kz_3 \tag{26}$$

where $k > 0$ is a gain constant,

$$z_3 = 2x_1 + 2x_2 + x_3, \tag{27}$$

and the update law for the parameter estimates $\hat{a}(t), \hat{b}(t)$ is given by

$$\begin{cases} \dot{\hat{a}}(t) = x_1 z_3 \\ \dot{\hat{b}}(t) = -x_2 z_3 \end{cases} \tag{28}$$

Proof We prove this result via Lyapunov stability theory [13].

First, we define a quadratic Lyapunov function

$$V_1(z_1) = \frac{1}{2} z_1^2 \tag{29}$$

where

$$z_1 = x_1 \tag{30}$$

Differentiating V_1 along the dynamics (23), we get

$$\dot{V}_1 = z_1\dot{z}_1 = x_1x_2 = -z_1^2 + z_1(x_1 + x_2) \tag{31}$$

Now, we define

$$z_2 = x_1 + x_2 \tag{32}$$

Using (32), we can simplify the Eq. (31) as

$$\dot{V}_1 = -z_1^2 + z_1z_2 \tag{33}$$

Secondly, we define a quadratic Lyapunov function

$$V_2(z_1, z_2) = V_1(z_1) + \frac{1}{2}\, z_2^2 = \frac{1}{2}\left(z_1^2 + z_2^2\right) \tag{34}$$

Differentiating V_2 along the dynamics (23), we get

$$\dot{V}_2 = -z_1^2 - z_2^2 + z_2(2x_1 + 2x_2 + x_3) \tag{35}$$

Now, we define

$$z_3 = 2x_1 + 2x_2 + x_3 \tag{36}$$

Using (36), we can simplify the Eq. (35) as

$$\dot{V}_2 = -z_1^2 - z_2^2 + z_2z_3 \tag{37}$$

Finally, we define a quadratic Lyapunov function

$$V(z_1, z_2, z_3, e_a, e_b) = V_2(z_1, z_2) + \frac{1}{2}z_3^2 + \frac{1}{2}e_a^2 + \frac{1}{2}e_b^2 \tag{38}$$

Differentiating V along the dynamics (23), we get

$$\dot{V} = -z_1^2 - z_2^2 - z_3^2 + z_3(z_3 + z_2 + \dot{z}_3) - e_a\dot{\hat{a}} - e_b\dot{\hat{b}} \tag{39}$$

Equation (39) can be written compactly as

$$\dot{V} = -z_1^2 - z_2^2 - z_3^2 + z_3S - e_a\dot{\hat{a}} - e_b\dot{\hat{b}} \tag{40}$$

where

$$S = z_3 + z_2 + \dot{z}_3 = z_3 + z_2 + 2\dot{x}_1 + 2\dot{x}_2 + \dot{x}_3 \tag{41}$$

A simple calculation gives

$$S = (3 + a)x_1 + (5 - b)x_2 + 2x_3 - x_1^2 - x_1^3 + u \tag{42}$$

Substituting the adaptive control law (26) into (42), we obtain

$$S = \left[a - \hat{a}(t)\right] x_1 - \left[b - \hat{b}(t)\right] x_2 - kz_3 \tag{43}$$

Using the definitions (25), we can simplify (43) as

$$S = e_a x_1 - e_b x_2 - kz_3 \tag{44}$$

Substituting the value of S from (44) into (40), we obtain

$$\dot{V} = -z_1^2 - z_2^2 - (1 + k)z_3^2 + e_a \left(x_1 z_3 - \dot{\hat{a}}\right) + e_b \left(-x_2 z_3 - \dot{\hat{b}}\right) \tag{45}$$

Substituting the update law (28) into (45), we get

$$\dot{V} = -z_1^2 - z_2^2 - (1 + k)z_3^2, \tag{46}$$

which is a negative semi-definite function on $\mathbf{R}^5$.

From (46), it follows that the vector $\mathbf{z}(t) = (z_1(t), z_2(t), z_3(t))$ and the parameter estimation error $(e_a(t), e_b(t))$ are globally bounded, i.e.

$$\left[\, z_1(t)\ z_2(t)\ z_3(t)\ e_a(t)\ e_b(t) \,\right] \in \mathbf{L}_\infty \tag{47}$$

Also, it follows from (46) that

$$\dot{V} \le -z_1^2 - z_2^2 - z_3^2 = -\|\mathbf{z}\|^2 \tag{48}$$

That is,

$$\|\mathbf{z}\|^2 \le -\dot{V} \tag{49}$$

Integrating the inequality (49) from 0 to t, we get

$$\int_0^t |\mathbf{z}(\tau)|^2 \, d\tau \le V(0) - V(t) \tag{50}$$

From (50), it follows that $\mathbf{z}(t) \in \mathbf{L}_2$.

From Eq. (23), it can be deduced that $\dot{\mathbf{z}}(t) \in \mathbf{L}_\infty$.

Thus, using Barbalat's lemma [13], we conclude that $\mathbf{z}(t) \to \mathbf{0}$ exponentially as $t \to \infty$ for all initial conditions $\mathbf{z}(0) \in \mathbf{R}^3$.

Hence, it is immediate that $\mathbf{x}(t) \to \mathbf{0}$ exponentially as $t \to \infty$ for all initial conditions $\mathbf{x}(0) \in \mathbf{R}^3$.

This completes the proof. □

For the numerical simulations, the classical fourth-order Runge–Kutta method with step size $h = 10^{-8}$ is used to solve the system of differential equations (23) and (28), when the adaptive control law (26) is applied.

The parameter values of the novel jerk chaotic system (23) are taken as

$$a = 134, \quad b = 50 \tag{51}$$

We take the positive gain constant as

$$k = 8 \tag{52}$$

Furthermore, as initial conditions of the novel jerk chaotic system (23), we take

$$x_1(0) = -5.4, \quad x_2(0) = 6.2, \quad x_3(0) = 4.7 \tag{53}$$

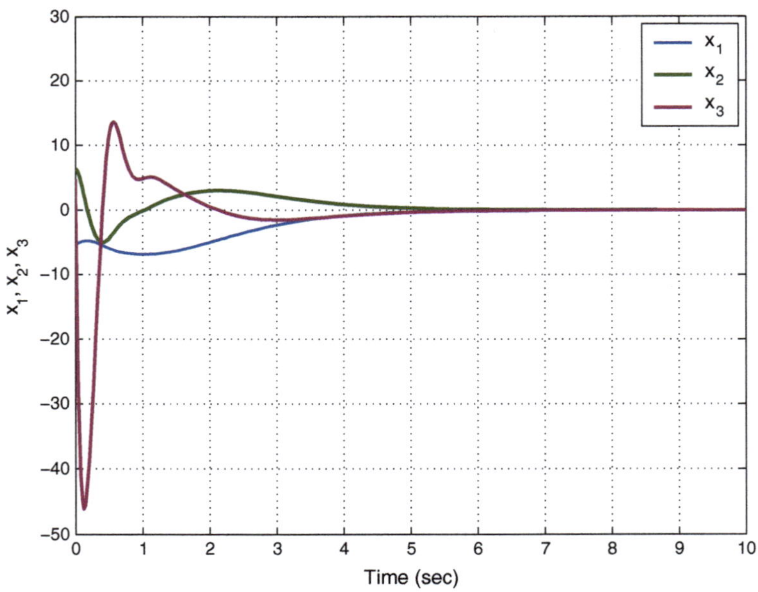

Fig. 5 Time-history of the controlled states x_1, x_2, x_3

Also, as initial conditions of the parameter estimates $\hat{a}(t)$ and $\hat{b}(t)$, we take

$$\hat{a}(0) = 10.2, \quad \hat{b}(0) = 12.4 \tag{54}$$

In Fig. 5, the exponential convergence of the controlled states $x_1(t), x_2(t), x_3(t)$ is depicted, when the adaptive control law (26) and (28) are implemented.

5 Adaptive Synchronization of the Identical 3-D Jerk Chaotic Systems

In this section, we use backstepping control method to derive an adaptive control law for globally and exponentially synchronizing the identical 3-D novel jerk chaotic systems with unknown parameters.

As the master system, we consider the 3-D novel jerk chaotic system given by

$$\begin{cases} \dot{x}_1 = x_2 \\ \dot{x}_2 = x_3 \\ \dot{x}_3 = ax_1 - x_1^2 - x_1^3 - bx_2 - x_3 \end{cases} \tag{55}$$

where x_1, x_2, x_3 are the states of the system, and a, b are unknown, constant parameters.

As the slave system, we consider the 3-D novel jerk chaotic system given by

$$\begin{cases} \dot{y}_1 = y_2 \\ \dot{y}_2 = y_3 \\ \dot{y}_3 = ay_1 - y_1^2 - y_1^3 - by_2 - y_3 + u \end{cases} \tag{56}$$

where y_1, y_2, y_3 are the states of the system, and u is a backstepping control to be determined using estimates $\hat{a}(t)$ and $\hat{b}(t)$ for the unknown parameters a and b, respectively.

We define the synchronization error between the states of the master system (55) and the slave system (56) as

$$\begin{cases} e_1 = y_1 - x_1 \\ e_2 = y_2 - x_2 \\ e_3 = y_3 - x_3 \end{cases} \tag{57}$$

Then the error dynamics is easily obtained as

$$\begin{cases} \dot{e}_1 = e_2 \\ \dot{e}_2 = e_3 \\ \dot{e}_3 = ae_1 - be_2 - e_3 - y_1^2 + x_1^2 - y_1^3 + x_1^3 + u \end{cases} \tag{58}$$

The parameter estimation errors are defined as:

$$\begin{cases} e_a(t) = a - \hat{a}(t) \\ e_b(t) = b - \hat{b}(t) \end{cases} \tag{59}$$

Differentiating (59) with respect to t, we obtain

$$\begin{cases} \dot{e}_a(t) = -\dot{\hat{a}}(t) \\ \dot{e}_b(t) = -\dot{\hat{b}}(t) \end{cases} \tag{60}$$

Theorem 2 *The identical 3-D novel jerk chaotic systems (55) and (56) with unknown parameters a and b are globally and exponentially synchronized by the adaptive control law*

$$u = -\left[3 + \hat{a}(t)\right] e_1 - \left[5 - \hat{b}(t)\right] e_2 - 2e_3 + y_1^2 - x_1^2 + y_1^3 - x_1^3 - kz_3 \tag{61}$$

where $k > 0$ is a gain constant,

$$z_3 = 2e_1 + 2e_2 + e_3, \tag{62}$$

and the update law for the parameter estimates $\hat{a}(t)$, $\hat{b}(t)$ is given by

$$\begin{cases} \dot{\hat{a}}(t) = e_1 z_3 \\ \dot{\hat{b}}(t) = -e_2 z_3 \end{cases} \tag{63}$$

Proof First, we define a quadratic Lyapunov function

$$V_1(z_1) = \frac{1}{2} z_1^2 \tag{64}$$

where

$$z_1 = e_1 \tag{65}$$

Differentiating V_1 along the error dynamics (58), we get

$$\dot{V}_1 = z_1 \dot{z}_1 = e_1 e_2 = -z_1^2 + z_1(e_1 + e_2) \tag{66}$$

Now, we define

$$z_2 = e_1 + e_2 \tag{67}$$

Using (67), we can simplify the Eq. (66) as

$$\dot{V}_1 = -z_1^2 + z_1 z_2 \tag{68}$$

Secondly, we define a quadratic Lyapunov function

$$V_2(z_1, z_2) = V_1(z_1) + \frac{1}{2}\, z_2^2 = \frac{1}{2}\left(z_1^2 + z_2^2\right) \tag{69}$$

Differentiating V_2 along the error dynamics (58), we get

$$\dot{V}_2 = -z_1^2 - z_2^2 + z_2(2e_1 + 2e_2 + e_3) \tag{70}$$

Now, we define

$$z_3 = 2e_1 + 2e_2 + e_3 \tag{71}$$

Using (71), we can simplify the Eq. (70) as

$$\dot{V}_2 = -z_1^2 - z_2^2 + z_2 z_3 \tag{72}$$

Finally, we define a quadratic Lyapunov function

$$V(z_1, z_2, z_3, e_a, e_b) = V_2(z_1, z_2) + \frac{1}{2}z_3^2 + \frac{1}{2}e_a^2 + \frac{1}{2}e_b^2 \tag{73}$$

Differentiating V along the error dynamics (58), we get

$$\dot{V} = -z_1^2 - z_2^2 - z_3^2 + z_3(z_3 + z_2 + \dot{z}_3) - e_a\dot{\hat{a}} - e_b\dot{\hat{b}} \tag{74}$$

Equation (74) can be written compactly as

$$\dot{V} = -z_1^2 - z_2^2 - z_3^2 + z_3 S - e_a\dot{\hat{a}} - e_b\dot{\hat{b}} \tag{75}$$

where

$$S = z_3 + z_2 + \dot{z}_3 = z_3 + z_2 + 2\dot{e}_1 + 2\dot{e}_2 + \dot{e}_3 \tag{76}$$

A simple calculation gives

$$S = (3 + a)e_1 + (5 - b)e_2 + 2e_3 - y_1^2 + x_1^2 - y_1^3 + x_1^3 + u \tag{77}$$

Substituting the adaptive control law (61) into (77), we obtain

$$S = \left[a - \hat{a}(t)\right] e_1 - \left[b - \hat{b}(t)\right] e_2 - kz_3 \tag{78}$$

Using the definitions (60), we can simplify (78) as

$$S = e_a e_1 - e_b e_2 - kz_3 \tag{79}$$

Substituting the value of S from (79) into (75), we obtain

$$\dot{V} = -z_1^2 - z_2^2 - (1+k)z_3^2 + e_a\left[e_1 z_3 - \dot{\hat{a}}\right] + e_b\left[-e_2 z_3 - \dot{\hat{b}}\right] \tag{80}$$

Substituting the update law (63) into (80), we get

$$\dot{V} = -z_1^2 - z_2^2 - (1+k)z_3^2, \tag{81}$$

which is a negative semi-definite function on $\mathbf{R}^5$.

From (81), it follows that the vector $\mathbf{z}(t) = (z_1(t), z_2(t), z_3(t))$ and the parameter estimation error $(e_a(t), e_b(t))$ are globally bounded, i.e.

$$\left[\, z_1(t)\ z_2(t)\ z_3(t)\ e_a(t)\ e_b(t) \,\right] \in \mathbf{L}_\infty \tag{82}$$

Also, it follows from (81) that

$$\dot{V} \le -z_1^2 - z_2^2 - z_3^2 = -\|\mathbf{z}\|^2 \tag{83}$$

That is,

$$\|\mathbf{z}\|^2 \le -\dot{V} \tag{84}$$

Integrating the inequality (84) from 0 to t, we get

$$\int_0^t |\mathbf{z}(\tau)|^2\, d\tau \le V(0) - V(t) \tag{85}$$

From (85), it follows that $\mathbf{z}(t) \in \mathbf{L}_2$.

From Eq. (58), it can be deduced that $\dot{\mathbf{z}}(t) \in \mathbf{L}_\infty$.

Thus, using Barbalat's lemma [13], we conclude that $\mathbf{z}(t) \to \mathbf{0}$ exponentially as $t \to \infty$ for all initial conditions $\mathbf{z}(0) \in \mathbf{R}^3$.

Hence, it is immediate that $\mathbf{e}(t) \to \mathbf{0}$ exponentially as $t \to \infty$ for all initial conditions $\mathbf{e}(0) \in \mathbf{R}^3$.

This completes the proof. □

For the numerical simulations, the classical fourth-order Runge–Kutta method with step size $h = 10^{-8}$ is used to solve the system of novel jerk chaotic systems, which are taken as the master and slave systems.

The parameter values of the novel jerk chaotic systems are taken as in the chaotic case, i.e. $a = 134$ and $b = 50$.

We take the positive gain constant as $k = 8$.

Furthermore, as initial conditions of the master system (55), we take

$$x_1(0) = 2.5, \quad x_2(0) = 1.8, \quad x_3(0) = -1.7 \tag{86}$$

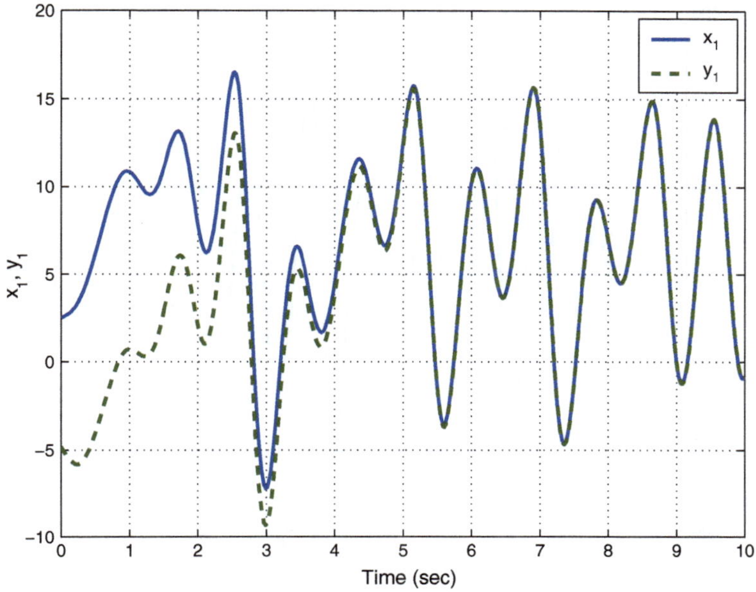

Fig. 6 Synchronization of the states x_1 and y_1

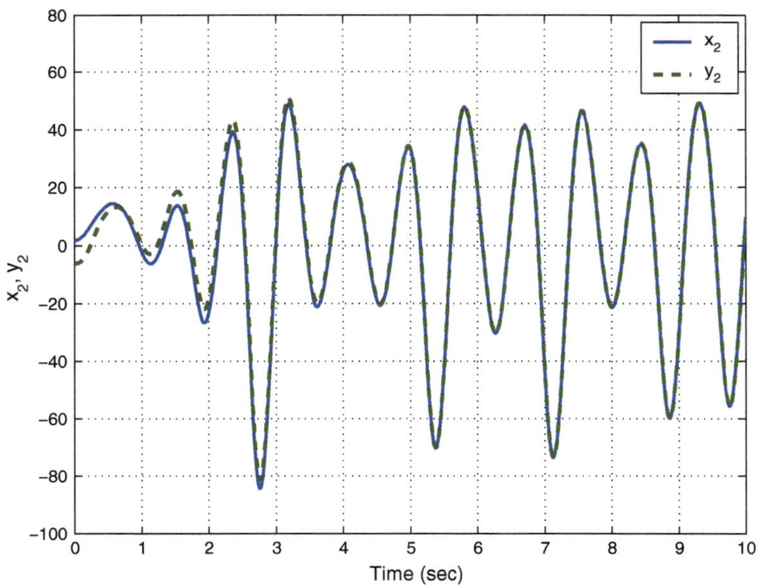

Fig. 7 Synchronization of the states x_2 and y_2

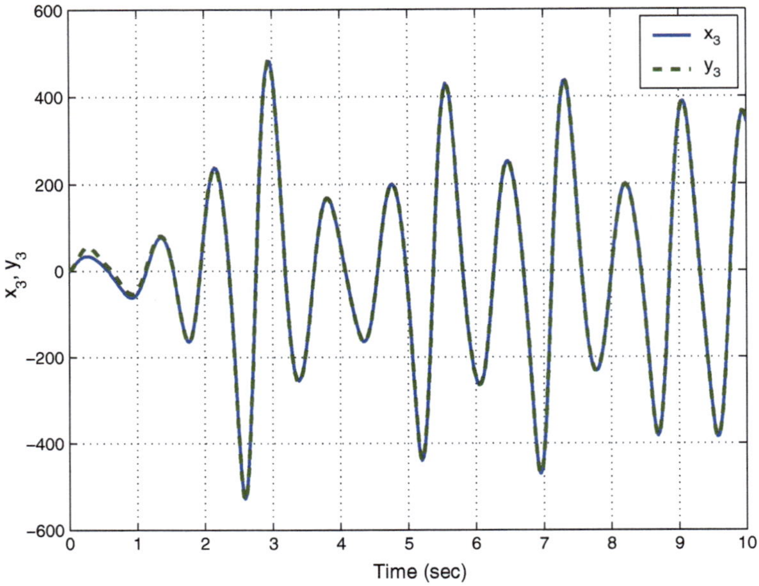

Fig. 8 Synchronization of the states x_3 and y_3

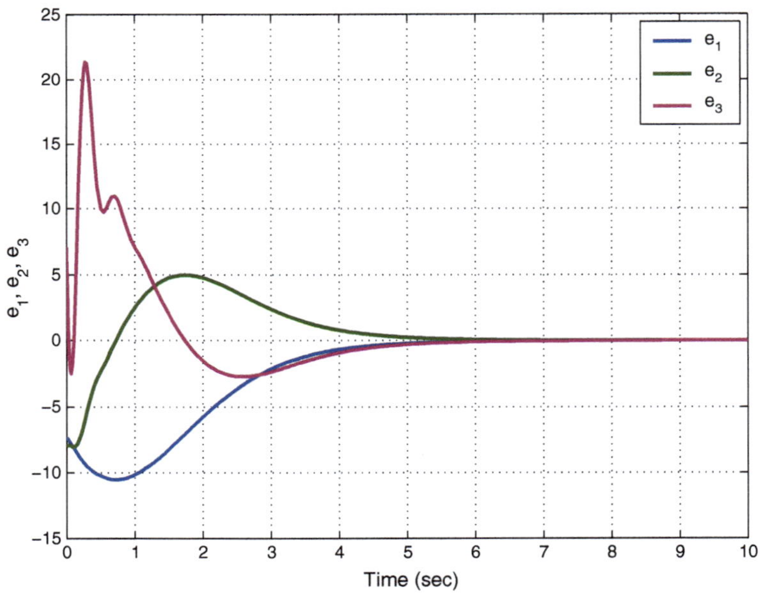

Fig. 9 Time-history of the synchronization errors e_1, e_2, e_3

As initial conditions of the slave system (56), we take

$$y_1(0) = -4.8, \quad y_2(0) = -6.2, \quad y_3(0) = 5.4 \tag{87}$$

Also, as initial conditions of the parameter estimates $\hat{a}(t)$ and $\hat{b}(t)$, we take

$$\hat{a}(0) = 16.3, \quad \hat{b}(0) = 8.5 \tag{88}$$

In Figs. 6, 7 and 8, the complete synchronization of the identical 3-D jerk chaotic systems (55) and (56) is depicted.

Also, in Fig. 9, the time-history of the synchronization errors $e_1(t)$, $e_2(t)$, $e_3(t)$, is depicted.

6 Conclusions

This work announced a seven-term novel 3-D jerk chaotic system with two nonlinearities (quadratic and cubic). The novel jerk chaotic system has three saddle-foci equilibrium points, which are unstable. The Lyapunov exponents of the novel jerk chaotic system have been obtained as $L_1 = 0.5565$, $L_2 = 0$ and $L_3 = -1.5566$, while the Kaplan–Yorke dimension of the novel jerk chaotic system is obtained as $D_{KY} = 2.3575$. In this work, adaptive backstepping controllers have been designed for the global stabilization and global chaos synchronization for the identical novel jerk chaotic systems. The main results were proved using Lyapunov stability theory. Numerical simulations using MATLAB have been depicted to illustrate all the main results derived in this work.

References

1. Abdurrahman A, Jiang H, Teng Z (2015) Finite-time synchronization for memristor-based neural networks with time-varying delays. Neural Netw 69:20–28
2. Arneodo A, Coullet P, Tresser C (1981) Possible new strange attractors with spiral structure. Commun Math Phys 79(4):573–576
3. Azar AT, Vaidyanathan S (2015) Chaos modeling and control systems design, vol 581. Springer, Germany
4. Behnia S, Afrang S, Akhshani A, Mabhouti K (2013) A novel method for controlling chaos in external cavity semiconductor laser. Optik 124(8):757–764
5. Cai G, Tan Z (2007) Chaos synchronization of a new chaotic system via nonlinear control. J Uncertain Syst 1(3):235–240
6. Carroll TL, Pecora LM (1991) Synchronizing chaotic circuits. IEEE Trans Circuits Syst 38(4):453–456
7. Chen G, Ueta T (1999) Yet another chaotic attractor. Int J Bifurc Chaos 9(7):1465–1466

8. Chen WH, Wei D, Lu X (2014) Global exponential synchronization of nonlinear time-delay Lur'e systems via delayed impulsive control. Commun Nonlinear Sci Numer Simul 19(9):3298–3312
9. Gan Q, Liang Y (2012) Synchronization of chaotic neural networks with time delay in the leakage term and parametric uncertainties based on sampled-data control. J Frankl Inst 349(6):1955–1971
10. Islam MM, Murase K (2005) Chaotic dynamics of a behaviour-based miniature mobile robot: effects of environment and control structure. Neural Netw 18(2):123–144
11. Jiang GP, Zheng WX, Chen G (2004) Global chaos synchronization with channel time-delay. Chaos, Solitons and Fractals 20(2):267–275
12. Karthikeyan R, Sundarapandian V (2014) Hybrid chaos synchronization of four-scroll systems via active control. J Electr Eng 65(2):97–103
13. Khalil HK (2001) Nonlinear Syst, 3rd edn. Prentice Hall, New Jersey
14. Li D (2008) A three-scroll chaotic attractor. Phys Lett A 372(4):387–393
15. Li GH, Zhou SP, Yang K (2007) Controlling chaos in Colpitts oscillator. Chaos, Solitons and Fractals 33:582–587
16. Li N, Zhang Y, Nie Z (2011) Synchronization for general complex dynamical networks with sampled-data. Neurocomputing 74(5):805–811
17. Lorenz EN (1963) Deterministic periodic flow. J Atmos Sci 20(2):130–141
18. Lü J, Chen G (2002) A new chaotic attractor coined. Int J Bifurc Chaos 12(3):659–661
19. Matouk AE (2011) Chaos, feedback control and synchronization of a fractional-order modified autonomous Van der Pol-Duffing circuit. Commun Nonlinear Sci Numer Simul 16(2): 975–986
20. Pecora LM, Carroll TL (1990) Synchronization in chaotic systems. Phys Rev Lett 64(8):821–824
21. Pehlivan I, Moroz IM, Vaidyanathan S (2014) Analysis, synchronization and circuit design of a novel butterfly attractor. J Sound Vib 333(20):5077–5096
22. Pham VT, Volos CK, Vaidyanathan S, Le TP, Vu VY (2015) A memristor-based hyperchaotic system with hidden attractors: dynamics, synchronization and circuital emulating. J Eng Sci Technol Rev 8(2):205–214
23. Rasappan S, Vaidyanathan S (2012) Global chaos synchronization of WINDMI and Coullet chaotic systems by backstepping control. Far East J Math Sci 67(2):265–287
24. Rasappan S, Vaidyanathan S (2012) Hybrid synchronization of n-scroll Chua and Lur'e chaotic systems via backstepping control with novel feedback. Arch Control Sci 22(3):343–365
25. Rasappan S, Vaidyanathan S (2012) Synchronization of hyperchaotic Liu system via backstepping control with recursive feedback. Commun Comput Inf Sci 305:212–221
26. Rasappan S, Vaidyanathan S (2013) Hybrid synchronization of n-scroll chaotic Chua circuits using adaptive backstepping control design with recursive feedback. Malaysian J Math Sci 7(2):219–246
27. Rasappan S, Vaidyanathan S (2014) Global chaos synchronization of WINDMI and Coullet chaotic systems using adaptive backstepping control design. Kyungpook Math J 54(1):293–320
28. Rhouma R, Belghith S (2011) Cryptanalysis of a chaos-based cryptosystem. Commun Nonlinear Sci Numer Simul 16(2):876–884
29. Rössler OE (1976) An equation for continuous chaos. Phys Lett A 57(5):397–398
30. Sampath S, Vaidyanathan S, Volos CK, Pham VT (2015) An eight-term novel four-scroll chaotic system with cubic nonlinearity and its circuit simulation. J Eng Sci Technol Rev 8(2):1–6
31. Sarasu P, Sundarapandian V (2011) Active controller design for the generalized projective synchronization of four-scroll chaotic systems. Int J Syst Signal Control Eng Appl 4(2):26–33
32. Sarasu P, Sundarapandian V (2011) The generalized projective synchronization of hyperchaotic Lorenz and hyperchaotic Qi systems via active control. Int J Soft Comput 6(5):216–223
33. Sarasu P, Sundarapandian V (2012) Adaptive controller design for the generalized projective synchronization of 4-scroll systems. Int J Syst Signal Control Eng Appl 5(2):21–30

34. Sarasu P, Sundarapandian V (2012) Generalized projective synchronization of three-scroll chaotic systems via adaptive control. Eur J Sci Res 72(4):504–522
35. Sarasu P, Sundarapandian V (2012) Generalized projective synchronization of two-scroll systems via adaptive control. Int J Soft Comput 7(4):146–156
36. Shahverdiev EM, Shore KA (2009) Impact of modulated multiple optical feedback time delays on laser diode chaos synchronization. Opt Commun 282(17):2572–3568
37. Sprott JC (1994) Some simple chaotic flows. Phys Rev E 50(2):647–650
38. Sundarapandian V (2010) Output regulation of the Lorenz attractor. Far East J Math Sci 42(2):289–299
39. Sundarapandian V (2011) Output regulation of the Arneodo-Coullet chaotic system. Commun Comput Inf Sci 133:98–107
40. Sundarapandian V (2013) Analysis and anti-synchronization of a novel chaotic system via active and adaptive controllers. J Eng Sci Technol Rev 6(4):45–52
41. Sundarapandian V, Karthikeyan R (2011) Anti-synchronization of hyperchaotic Lorenz and hyperchaotic Chen systems by adaptive control. Int J Syst Signal Control Eng Appl 4(2):18–25
42. Sundarapandian V, Karthikeyan R (2011) Anti-synchronization of Lü and Pan chaotic systems by adaptive nonlinear control. Eur J Sci Res 64(1):94–106
43. Sundarapandian V, Karthikeyan R (2012) Adaptive anti-synchronization of uncertain Tigan and Li systems. J Eng Appl Sci 7(1):45–52
44. Sundarapandian V, Karthikeyan R (2012) Hybrid synchronization of hyperchaotic Lorenz and hyperchaotic Chen systems via active control. J Eng Appl Sci 7(3):254–264
45. Sundarapandian V, Pehlivan I (2012) Analysis, control, synchronization, and circuit design of a novel chaotic system. Math Comput Model 55(7–8):1904–1915
46. Sundarapandian V, Sivaperumal S (2011) Sliding controller design of hybrid synchronization of four-wing chaotic systems. Int J Soft Comput 6(5):224–231
47. Suresh R, Sundarapandian V (2013) Global chaos synchronization of a family of n-scroll hyperchaotic Chua circuits using backstepping control with recursive feedback. Far East J Math Sci 73(1):73–95
48. Tigan G, Opris D (2008) Analysis of a 3D chaotic system. Chaos, Solitons and Fractals 36:1315–1319
49. Tuwankotta JM (2006) Chaos in a coupled oscillators system with widely spaced frequencies and energy-preserving non-linearity. Int J Non-Linear Mech 41(2):180–191
50. Usama M, Khan MK, Alghathbar K, Lee C (2010) Chaos-based secure satellite imagery cryptosystem. Comput Math Appl 60(2):326–337
51. Vaidyanathan S (2011) Hybrid chaos synchronization of Liu and Lü systems by active nonlinear control. Communications in computer and information science 204
52. Vaidyanathan S (2011) Output regulation of the unified chaotic system. Commun Comput Inf Sci 204:84–93
53. Vaidyanathan S (2012) Analysis and synchronization of the hyperchaotic Yujun systems via sliding mode control. Adv Intell Syst Comput 176:329–337
54. Vaidyanathan S (2012) Anti-synchronization of Sprott-L and Sprott-M chaotic systems via adaptive control. Int J Control Theory Appl 5(1):41–59
55. Vaidyanathan S (2012) Global chaos control of hyperchaotic Liu system via sliding control method. Int J Control Theory Appl 5(2):117–123
56. Vaidyanathan S (2012) Output regulation of the Liu chaotic system. Appl Mech Mat 110–116:3982–3989
57. Vaidyanathan S (2012) Sliding mode control based global chaos control of Liu-Liu-Liu-Su chaotic system. Int J Control Theory Appl 5(1):15–20
58. Vaidyanathan S (2013) A new six-term 3-D chaotic system with an exponential nonlinearity. Far East J Math Sci 79(1):135–143
59. Vaidyanathan S (2013) Analysis and adaptive synchronization of two novel chaotic systems with hyperbolic sinusoidal and cosinusoidal nonlinearity and unknown parameters. J Eng Sci Technol Rev 6(4):53–65

60. Vaidyanathan S (2013) Analysis, control and synchronization of hyperchaotic Zhou system via adaptive control. Adv Intell Syst Comput 177:1–10
61. Vaidyanathan S (2014) A new eight-term 3-D polynomial chaotic system with three quadratic nonlinearities. Far East J Math Sci 84(2):219–226
62. Vaidyanathan S (2014) Analysis and adaptive synchronization of eight-term 3-D polynomial chaotic systems with three quadratic nonlinearities. Eur Phys J Spec Top 223(8):1519–1529
63. Vaidyanathan S (2014) Analysis, control and synchronisation of a six-term novel chaotic system with three quadratic nonlinearities. Int J Model Identif Control 22(1):41–53
64. Vaidyanathan S (2014) Generalized projective synchronisation of novel 3-D chaotic systems with an exponential non-linearity via active and adaptive control. Int J Model Identif Control 22(3):207–217
65. Vaidyanathan S (2014) Global chaos synchronization of identical Li-Wu chaotic systems via sliding mode control. Int J Model Identif Control 22(2):170–177
66. Vaidyanathan S (2015) 3-cells cellular neural network (CNN) attractor and its adaptive biological control. Int J Pharmtech Res 8(4):632–640
67. Vaidyanathan S (2015) A 3-D novel highly chaotic system with four quadratic nonlinearities, its adaptive control and anti-synchronization with unknown parameters. J Eng Sci Technol Rev 8(2):106–115
68. Vaidyanathan S (2015) A novel chemical chaotic reactor system and its adaptive control. Int J ChemTech Res 8(7):146–158
69. Vaidyanathan S (2015) Adaptive backstepping control of enzymes-substrates system with ferroelectric behaviour in brain waves. Int J PharmTech Res 8(2):256–261
70. Vaidyanathan S (2015) Adaptive biological control of generalized Lotka-Volterra three-species biological system. Int J PharmTech Res 8(4):622–631
71. Vaidyanathan S (2015) Adaptive chaotic synchronization of enzymes-substrates system with ferroelectric behaviour in brain waves. Int J PharmTech Res 8(5):964–973
72. Vaidyanathan S (2015) Adaptive control of a chemical chaotic reactor. Int J PharmTech Res 8(3):377–382
73. Vaidyanathan S (2015) Adaptive control of the FitzHugh-Nagumo chaotic neuron model. Int J PharmTech Res 8(6):117–127
74. Vaidyanathan S (2015) Adaptive synchronization of chemical chaotic reactors. Int J ChemTech Res 8(2):612–621
75. Vaidyanathan S (2015) Adaptive synchronization of generalized Lotka-Volterra three-species biological systems. Int J PharmTech Res 8(5):928–937
76. Vaidyanathan S (2015) Adaptive synchronization of novel 3-D chemical chaotic reactor systems. Int J ChemTech Res 8(7):159–171
77. Vaidyanathan S (2015) Adaptive synchronization of the identical FitzHugh-Nagumo chaotic neuron models. Int J PharmTech Res 8(6):167–177
78. Vaidyanathan S (2015) Analysis, control and synchronization of a 3-D novel jerk chaotic system with two quadratic nonlinearities. Kyungpook Math J 55:563–586
79. Vaidyanathan S (2015) Analysis, properties and control of an eight-term 3-D chaotic system with an exponential nonlinearity. Int J Model Identif Control 23(2):164–172
80. Vaidyanathan S (2015) Anti-synchronization of brusselator chemical reaction systems via adaptive control. Int J ChemTech Res 8(6):759–768
81. Vaidyanathan S (2015) Chaos in neurons and adaptive control of Birkhoff-Shaw strange chaotic attractor. Int J PharmTech Res 8(5):956–963
82. Vaidyanathan S (2015) Chaos in neurons and synchronization of Birkhoff-Shaw strange chaotic attractors via adaptive control. Int J PharmTech Res 8(6):1–11
83. Vaidyanathan S (2015) Coleman–Gomatam logarithmic competitive biology models and their ecological monitoring. Int J PharmTech Res 8(6):94–105
84. Vaidyanathan S (2015) Dynamics and control of brusselator chemical reaction. Int J ChemTech Res 8(6):740–749
85. Vaidyanathan S (2015) Dynamics and control of tokamak system with symmetric and magnetically confined plasma. Int J ChemTech Res 8(6):795–803

86. Vaidyanathan S (2015) Global chaos synchronization of chemical chaotic reactors via novel sliding mode control method. Int J ChemTech Res 8(7):209–221
87. Vaidyanathan S (2015) Global chaos synchronization of the forced Van der Pol chaotic oscillators via adaptive control method. Int J PharmTech Res 8(6):156–166
88. Vaidyanathan S (2015) Global chaos synchronization of the Lotka-Volterra biological systems with four competitive species via active control. Int J PharmTech Res 8(6):206–217
89. Vaidyanathan S (2015) Lotka-Volterra population biology models with negative feedback and their ecological monitoring. Int J PharmTech Res 8(5):974–981
90. Vaidyanathan S (2015) Lotka-Volterra two species competitive biology models and their ecological monitoring. Int J PharmTech Res 8(6):32–44
91. Vaidyanathan S (2015) Output regulation of the forced Van der Pol chaotic oscillator via adaptive control method. Int J PharmTech Res 8(6):106–116
92. Vaidyanathan S, Azar AT (2015) Analysis and control of a 4-D novel hyperchaotic system. In: Azar AT, Vaidyanathan S (eds) Chaos modeling and control systems design, studies in computational intelligence, vol 581. Springer, Germany, pp 19–38
93. Vaidyanathan S, Azar AT (2015) Analysis, control and synchronization of a nine-term 3-D novel chaotic system. In: Azar AT, Vaidyanathan S (eds) Chaos modelling and control systems design, studies in computational intelligence, vol 581. Springer, Germany, pp 19–38
94. Vaidyanathan S, Azar AT (2015) Anti-synchronization of identical chaotic systems using sliding mode control and an application to Vaidhyanathan-Madhavan chaotic systems. Stud Comput Intell 576:527–547
95. Vaidyanathan S, Azar AT (2015) Hybrid synchronization of identical chaotic systems using sliding mode control and an application to Vaidhyanathan chaotic systems. Stud Comput Intell 576:549–569
96. Vaidyanathan S, Madhavan K (2013) Analysis, adaptive control and synchronization of a seven-term novel 3-D chaotic system. Int J Control Theory Appl 6(2):121–137
97. Vaidyanathan S, Pakiriswamy S (2013) Generalized projective synchronization of six-term Sundarapandian chaotic systems by adaptive control. Int J Control Theory Appl 6(2):153–163
98. Vaidyanathan S, Pakiriswamy S (2015) A 3-D novel conservative chaotic system and its generalized projective synchronization via adaptive control. J Eng Sci Technol Rev 8(2):52–60
99. Vaidyanathan S, Rajagopal K (2011) Anti-synchronization of Li and T chaotic systems by active nonlinear control. Commun Comput Inf Sci 198:175–184
100. Vaidyanathan S, Rajagopal K (2011) Global chaos synchronization of hyperchaotic Pang and Wang systems by active nonlinear control. Commun Comput Inf Sci 204:84–93
101. Vaidyanathan S, Rajagopal K (2011) Global chaos synchronization of Lü and Pan systems by adaptive nonlinear control. Commun Comput Inf Sci 205:193–202
102. Vaidyanathan S, Rajagopal K (2012) Global chaos synchronization of hyperchaotic Pang and hyperchaotic Wang systems via adaptive control. Int J Soft Comput 7(1):28–37
103. Vaidyanathan S, Rasappan S (2011) Global chaos synchronization of hyperchaotic Bao and Xu systems by active nonlinear control. Commun Comput Inf Sci 198:10–17
104. Vaidyanathan S, Rasappan S (2014) Global chaos synchronization of n-scroll Chua circuit and Lur'e system using backstepping control design with recursive feedback. Arabian J Sci Eng 39(4):3351–3364
105. Vaidyanathan S, Sampath S (2011) Global chaos synchronization of hyperchaotic Lorenz systems by sliding mode control. Commun Comput Inf Sci 205:156–164
106. Vaidyanathan S, Sampath S (2012) Anti-synchronization of four-wing chaotic systems via sliding mode control. Int J Autom Comput 9(3):274–279
107. Vaidyanathan S, Volos C (2015) Analysis and adaptive control of a novel 3-D conservative no-equilibrium chaotic system. Arch Control Sci 25(3):333–353
108. Vaidyanathan S, Volos C, Pham VT (2014) Hyperchaos, adaptive control and synchronization of a novel 5-D hyperchaotic system with three positive Lyapunov exponents and its SPICE implementation. Arch Control Sci 24(4):409–446

109. Vaidyanathan S, Volos C, Pham VT, Madhavan K, Idowu BA (2014) Adaptive backstepping control, synchronization and circuit simulation of a 3-D novel jerk chaotic system with two hyperbolic sinusoidal nonlinearities. Arch Control Sci 24(3):375–403
110. Vaidyanathan S, Idowu BA, Azar AT (2015) Backstepping controller design for the global chaos synchronization of Sprott's jerk systems. Stud Comput Intell 581:39–58
111. Vaidyanathan S, Rajagopal K, Volos CK, Kyprianidis IM, Stouboulos IN (2015) Analysis, adaptive control and synchronization of a seven-term novel 3-D chaotic system with three quadratic nonlinearities and its digital implementation in LabVIEW. J Eng Sci Technol Rev 8(2):130–141
112. Vaidyanathan S, Volos C, Pham VT, Madhavan K (2015) Analysis, adaptive control and synchronization of a novel 4-D hyperchaotic hyperjerk system and its SPICE implementation. Arch Control Sci 25(1):5–28
113. Vaidyanathan S, Volos CK, Kyprianidis IM, Stouboulos IN, Pham VT (2015) Analysis, adaptive control and anti-synchronization of a six-term novel jerk chaotic system with two exponential nonlinearities and its circuit simulation. J Eng Sci Technol Rev 8(2):24–36
114. Vaidyanathan S, Volos CK, Madhavan K (2015) Analysis, control, synchronization and SPICE implementation of a novel 4-D hyperchaotic Rikitake dynamo System without equilibrium. J Eng Sci Technol Rev 8(2):232–244
115. Vaidyanathan S, Volos CK, Pham VT (2015) Analysis, adaptive control and adaptive synchronization of a nine-term novel 3-D chaotic system with four quadratic nonlinearities and its circuit simulation. J Eng Sci Technol Rev 8(2):181–191
116. Vaidyanathan S, Volos CK, Pham VT (2015) Global chaos control of a novel nine-term chaotic system via sliding mode control. In: Azar AT, Zhu Q (eds) Advances and applications in sliding mode control systems, studies in computational intelligence, vol 576. Springer, Germany, pp 571–590
117. Vaidyanathan S, Volos CK, Pham VT, Madhavan K (2015) Analysis, adaptive control and synchronization of a novel 4-D hyperchaotic hyperjerk system and its SPICE implementation. Arch Control Sci 25(1):135–158
118. Vaidyanathan S, Volos CK, Rajagopal K, Kyprianidis IM, Stouboulos IN (2015) Adaptive backstepping controller design for the anti-synchronization of identical WINDMI chaotic systems with unknown parameters and its SPICE implementation. J Eng Sci Technol Rev 8(2):74–82
119. Volos CK, Kyprianidis IM, Stouboulos IN, Anagnostopoulos AN (2009) Experimental study of the dynamic behavior of a double scroll circuit. J Appl Funct Anal 4:703–711
120. Volos CK, Kyprianidis IM, Stouboulos IN (2013) Experimental investigation on coverage performance of a chaotic autonomous mobile robot. Robot Auton Syst 61(12):1314–1322
121. Volos CK, Kyprianidis IM, Stouboulos IN, Tlelo-Cuautle E, Vaidyanathan S (2015) Memristor: a new concept in synchronization of coupled neuromorphic circuits. J Eng Sci Technol Rev 8(2):157–173
122. Wang X, Ge C (2008) Controlling and tracking of Newton-Leipnik system via backstepping design. Int J Nonlinear Sci 5(2):133–139
123. Wei Z, Yang Q (2010) Anti-control of Hopf bifurcation in the new chaotic system with two stable node-foci. Appl Math Comput 217(1):422–429
124. Xiao X, Zhou L, Zhang Z (2014) Synchronization of chaotic Lur'e systems with quantized sampled-data controller. Commun Nonlinear Sci Numer Simul 19(6):2039–2047
125. Yang J, Zhu F (2013) Synchronization for chaotic systems and chaos-based secure communications via both reduced-order and step-by-step sliding mode observers. Commun Nonlinear Sci Numer Simul 18(4):926–937
126. Yang J, Chen Y, Zhu F (2014) Singular reduced-order observer-based synchronization for uncertain chaotic systems subject to channel disturbance and chaos-based secure communication. Appl Math Comput 229:227–238
127. Zhang H, Zhou J (2012) Synchronization of sampled-data coupled harmonic oscillators with control inputs missing. Syst Control Lett 61(12):1277–1285

128. Zhou W, Xu Y, Lu H, Pan L (2008) On dynamics analysis of a new chaotic attractor. Phys Lett A 372(36):5773–5777
129. Zhu C, Liu Y, Guo Y (2010) Theoretic and numerical study of a new chaotic system. Intell Inf Manag 2:104–109

Adaptive Control and Circuit Simulation of a Novel 4-D Hyperchaotic System with Two Quadratic Nonlinearities

Sundarapandian Vaidyanathan, Christos K. Volos and Viet-Thanh Pham

Abstract This work describes a ten-term novel 4-D hyperchaotic system with two quadratic nonlinearities. The phase portraits of the novel hyperchaotic system are depicted and the qualitative properties of the novel hyperchaotic system are discussed. The novel hyperchaotic system has a unique equilibrium at the origin, which is a saddle point. The Lyapunov exponents of the novel hyperchaotic system are obtained as $L_1 = 1.0784$, $L_2 = 0.1114$, $L_3 = 0$ and $L_4 = -18.1714$, while the Kaplan–Yorke dimension of the novel hyperchaotic system is obtained as $D_{KY} = 3.0655$. Since the sum of the Lyapunov exponents is negative, the novel hyperchaotic system is dissipative. Next, an adaptive controller is designed to globally stabilize the novel hyperchaotic system with unknown parameters. Moreover, an adaptive controller is also designed to achieve global chaos synchronization of the identical hyperchaotic systems with unknown parameters. MATLAB simulations are depicted to illustrate all the main results derived in this work. Finally, an electronic circuit realization of the novel hyperchaotic system using Spice is presented in detail to confirm the feasibility of the theoretical model.

Keywords Chaos · Chaotic systems · Hyperchaos · Hyperchaotic systems · Adaptive control · Synchronization · Circuit simulation

S. Vaidyanathan (✉)
Research and Development Centre, Vel Tech University, Avadi, Chennai 600062, Tamil Nadu, India
e-mail: sundarvtu@gmail.com

C.K. Volos
Physics Department, Aristotle University of Thessaloniki, 54124 Thessaloniki, Greece
e-mail: chvolos@gmail.com

V.-T. Pham
School of Electronics and Telecommunications, Hanoi University of Science and Technology, 01 Dai Co Viet, Hanoi, Vietnam
e-mail: pvt3010@gmail.com

S. Vaidyanathan and C. Volos (eds.), *Advances and Applications in Chaotic Systems*, Studies in Computational Intelligence 636,
DOI 10.1007/978-3-319-30279-9_7

1 Introduction

Chaos theory deals with the qualitative study of chaotic dynamical systems and their applications in science and engineering. A dynamical system is called *chaotic* if it satisfies the three properties: boundedness, infinite recurrence and sensitive dependence on initial conditions [3].

Some classical paradigms of 3-D chaotic systems in the literature are Lorenz system [14], Rössler system [22], ACT system [2], Sprott systems [27], Chen system [7], Lü system [15], Cai system [5], Tigan system [31], etc.

Many new chaotic systems have been discovered in the recent years such as Zhou system [93], Zhu system [94], Li system [12], Wei-Yang system [88], Sundarapandian systems [29, 30], Vaidyanathan systems [35, 37–41, 44, 55, 56, 70, 71, 73, 74, 76, 77, 79, 80, 82], Pehlivan system [17], Sampath system [24], etc.

Chaos theory has applications in several fields of science and engineering such as chemical reactors [45, 49, 51, 53, 57, 61–63], biological systems [43, 46–48, 50, 52, 54, 58–60, 64–68], memristors [1, 19, 85], lasers [4], oscillations [32], robotics [9, 84], electrical circuits [16, 83], cryptosystems [21, 33], secure communications [89, 90], etc.

A hyperchaotic system is defined as a chaotic system with at least two positive Lyapunov exponents [3]. Thus, the dynamics of a hyperchaotic system can expand in several different directions simultaneously. Thus, the hyperchaotic systems have more complex dynamical behaviour and they have miscellaneous applications in engineering such as secure communications [89, 90], cryptosystems [8, 20], fuzzy logic [26, 92], electrical circuits [87, 91], etc.

The minimum dimension of an autonomous, continuous-time, hyperchaotic system is four. The first 4-D hyperchaotic system was found by Rössler [23]. Many hyperchaotic systems have been reported in the chaos literature such as hyperchaotic Lorenz system [10], hyperchaotic Lü system [6], hyperchaotic Chen system [13], hyperchaotic Wang system [86], hyperchaotic Vaidyanathan systems [36, 42, 69, 75, 78, 81], hyperchaotic Pham system [18], etc.

In this research work, we announce a ten-term novel 4-D hyperchaotic system with two quadratic nonlinearities. We have also designed adaptive controllers for stabilization and synchronization of the novel hyperchaotic systems. Adaptive control method [25, 28, 34, 72] is a popular method used in the control literature for the synchronization of nonlinear systems when the system parameters are unknown.

This work is organized as follows. Section 2 describes the dynamic equations and phase portraits of the novel 4-D hyperchaotic system. Section 3 details the qualitative properties of the novel hyperchaotic chaotic system. The novel hyperchaotic system has a unique equilibrium point at the origin, which is a saddle point. Thus, the system has an unstable equilibrium point. The Lyapunov exponents of the novel hyperchaotic system are obtained as $L_1 = 1.0784$, $L_2 = 0.1114$, $L_3 = 0$ and $L_4 = -18.1714$, while the Kaplan–Yorke dimension of the novel hyperchaotic system is obtained as $D_{KY} = 3.0655$. Since the sum of the Lyapunov exponents is negative, the novel hyperchaotic system is dissipative.

In Sect. 4, we design an adaptive controller to globally stabilize the novel hyperchaotic system with unknown parameters. In Sect. 5, an adaptive controller is designed to achieve global chaos synchronization of the identical novel hyperchaotic systems with unknown parameters. In Sect. 6, an electronic circuit realization of the novel hyperchaotic system using Spice is presented to confirm the feasibility of the theoretical model. Section 7 contains the conclusions of this work.

2 A Novel 4-D Hyperchaotic System

In this section, we describe a ten-term novel hyperchaotic system, which is given by the 4-D dynamics

$$\begin{cases} \dot{x}_1 = a(x_2 - x_1) + x_4 \\ \dot{x}_2 = bx_1 - px_1x_3 + x_4 \\ \dot{x}_3 = x_1x_2 - cx_3 \\ \dot{x}_4 = -x_1 - x_2 \end{cases} \tag{1}$$

where x_1, x_2, x_3, x_4 are the states and a, b, c, p are constant positive parameters.

The novel 4-D system (1) is a ten-term polynomial system with two quadratic nonlinearities.

The system (1) exhibits a *strange chaotic attractor* for the parameter values

$$a = 12, \quad b = 36, \quad c = 5, \quad p = 12 \tag{2}$$

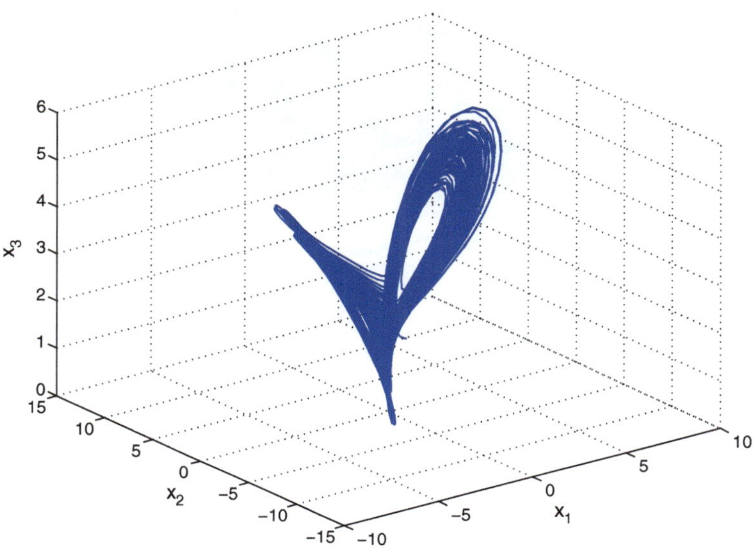

Fig. 1 3-D projection of the novel hyperchaotic system on the (x_1, x_2, x_3) space

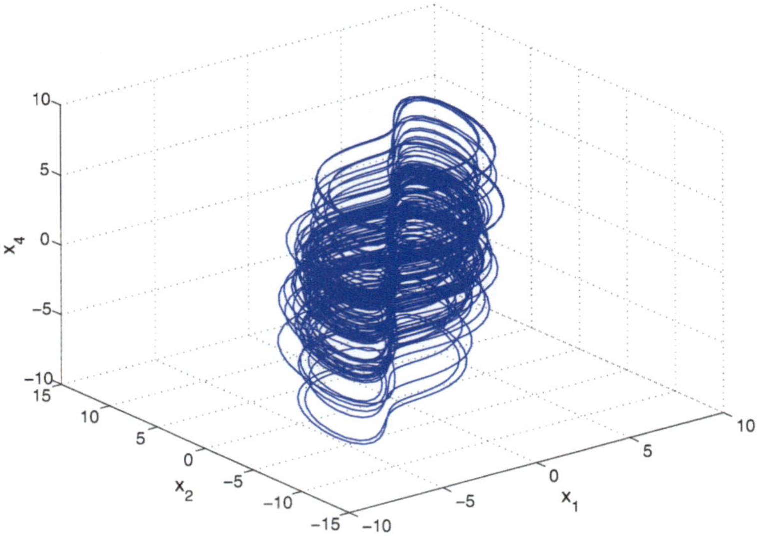

Fig. 2 3-D projection of the novel hyperchaotic system on the (x_1, x_2, x_4) space

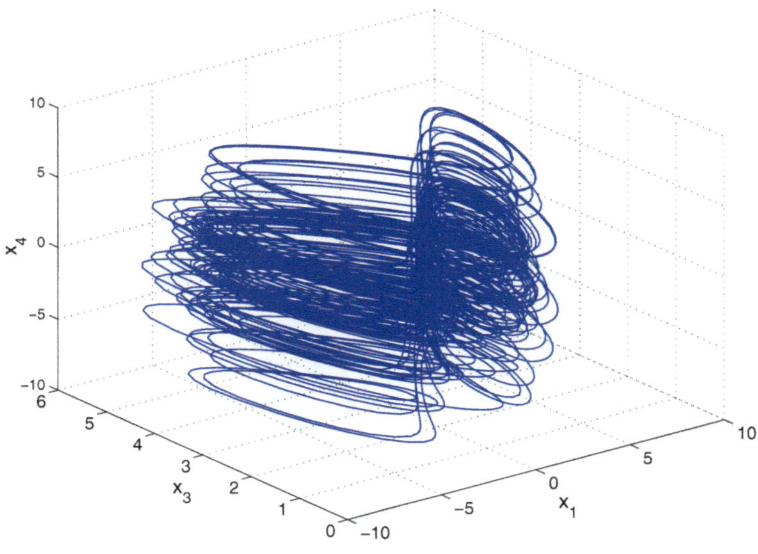

Fig. 3 3-D projection of the novel hyperchaotic system on the (x_1, x_3, x_4) space

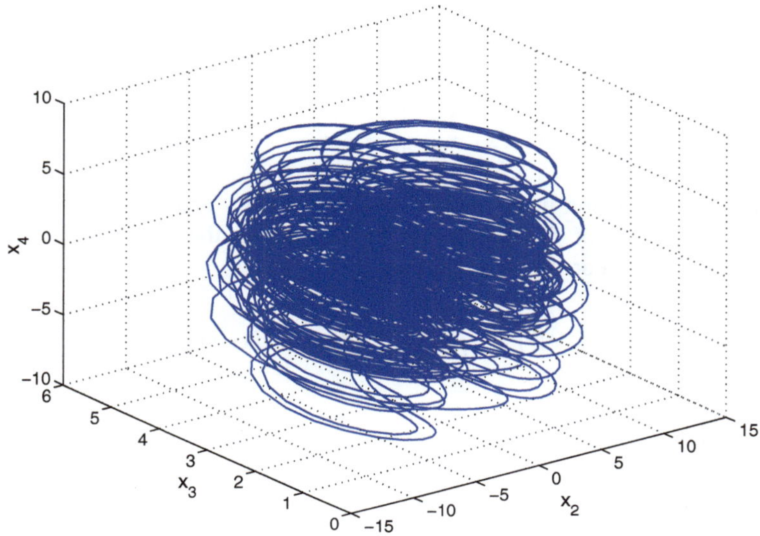

Fig. 4 3-D projection of the novel hyperchaotic system on the (x_2, x_3, x_4) space

For numerical simulations, we take the initial conditions as

$$x_1(0) = 1.8, \quad x_2(0) = 1.6, \quad x_3(0) = 1.2, \quad x_4(0) = 1.4 \tag{3}$$

Figures 1, 2, 3 and 4 the 3-D projection of the novel hyperchaotic system (1) on the (x_1, x_2, x_3), (x_1, x_2, x_4), (x_1, x_3, x_4) and (x_2, x_3, x_4) spaces, respectively.

3 Analysis of the Novel Hyperchaotic System

In this section, we give a dynamic analysis of the 4-D novel hyperchaotic system (1). We take the parameter values as in the hyperchaotic case (2).

3.1 Dissipativity

In vector notation, the novel hyperchaotic system (1) can be expressed as

$$\dot{\mathbf{x}} = f(\mathbf{x}) = \begin{bmatrix} f_1(x_1, x_2, x_3, x_4) \\ f_2(x_1, x_2, x_3, x_4) \\ f_3(x_1, x_2, x_3, x_4) \\ f_4(x_1, x_2, x_3, x_4) \end{bmatrix}, \tag{4}$$

where

$$\begin{cases} f_1(x_1, x_2, x_3, x_4) = a(x_2 - x_1) + x_4 \\ f_2(x_1, x_2, x_3, x_4) = bx_1 - px_1x_3 + x_4 \\ f_3(x_1, x_2, x_3, x_4) = x_1x_2 - cx_3 \\ f_4(x_1, x_2, x_3, x_4) = -x_1 - x_2 \end{cases} \tag{5}$$

Let Ω be any region in $\mathbf{R}^4$ with a smooth boundary and also, $\Omega(t) = \Phi_t(\Omega)$, where Φ_t is the flow of f. Furthermore, let $V(t)$ denote the hypervolume of $\Omega(t)$.

By Liouville's theorem, we know that

$$\dot{V}(t) = \int\limits_{\Omega(t)} (\nabla \cdot f)\, dx_1\, dx_2\, dx_3\, dx_4 \tag{6}$$

The divergence of the novel hyperchaotic system (4) is found as:

$$\nabla \cdot f = \frac{\partial f_1}{\partial x_1} + \frac{\partial f_2}{\partial x_2} + \frac{\partial f_3}{\partial x_3} + \frac{\partial f_4}{\partial x_4} = -(a + c) = -\mu < 0 \tag{7}$$

Inserting the value of $\nabla \cdot f$ from (7) into (6), we get

$$\dot{V}(t) = \int\limits_{\Omega(t)} (-\mu)\, dx_1\, dx_2\, dx_3\, dx_4 = -\mu V(t) \tag{8}$$

Integrating the first order linear differential equation (8), we get

$$V(t) = \exp(-\mu t) V(0) \tag{9}$$

Since $\mu > 0$, it follows from Eq. (9) that $V(t) \to 0$ exponentially as $t \to \infty$. This shows that the novel hyperchaotic system (1) is dissipative.

Hence, the system limit sets are ultimately confined into a specific limit set of zero hypervolume, and the asymptotic motion of the novel jerk chaotic system (1) settles onto a strange attractor of the system.

3.2 Equilibrium Points

We take the parameter values as in the hyperchaotic case (2).

It is easy to see that the system (1) has a unique equilibrium at the origin.

To test the stability type of the equilibrium point $E_0 = \mathbf{0}$, we calculate the Jacobian matrix of the novel hyperchaotic system (1) at $\mathbf{x} = \mathbf{0}$:

$$J \triangleq J(E_0) = \begin{bmatrix} -12 & 12 & 0 & 1 \\ 36 & 0 & 0 & 1 \\ 0 & 0 & -5 & 0 \\ -1 & -1 & 0 & 0 \end{bmatrix} \tag{10}$$

The matrix J has the eigenvalues

$$\lambda_1 = -5, \quad \lambda_2 = -27.6373, \quad \lambda_3 = 0.1401, \quad \lambda_4 = 15.4972 \tag{11}$$

This shows that the equilibrium point $E_0 = \mathbf{0}$ is a saddle-point, which is unstable.

3.3 Rotation Symmetry About the x_3-axis

We find that the novel 4-D hyperchaotic system (1) is invariant under the change of coordinates

$$(x_1, x_2, x_3, x_4) \mapsto (-x_1, -x_2, x_3, -x_4) \tag{12}$$

Since the transformation (12) persists for all values of the system parameters, it follows that the novel 4-D hyperchaotic system (1) has rotation symmetry about the x_3-axis and that any non-trivial trajectory must have a twin trajectory.

3.4 Invariance

We find that the x_3-axis is invariant under the flow of the novel 4-D hyperchaotic system (1). The invariant motion along the x_3-axis is characterized by the scalar dynamics

$$\dot{x}_3 = -cx_3, \quad (c > 0) \tag{13}$$

which is globally exponentially stable.

3.5 Lyapunov Exponents and Kaplan–Yorke Dimension

We take the parameter values of the novel system (1) as in the hyperchaotic case (2), i.e. $a = 12$, $b = 36$, $c = 5$ and $p = 12$.

We take the initial state of the novel system (1) as given in (3).

Then the Lyapunov exponents of the system (1) are numerically obtained using MATLAB as

$$L_1 = 1.0784, \quad L_2 = 0.1114, \quad L_3 = 0, \quad L_4 = -18.1714 \tag{14}$$

Since there are two positive Lyapunov exponents in (14), the novel system (1) exhibits *hyperchaotic* behavior.

Since $L_1 + L_2 + L_3 + L_4 = -16.9816 < 0$, it follows that the novel hyperchaotic system (1) is dissipative.

Also, the Kaplan–Yorke dimension of the novel hyperchaotic system (1) is calculated as

$$D_{KY} = 3 + \frac{L_1 + L_2 + L_3}{|L_4|} = 3.0655, \tag{15}$$

which is fractional.

4 Adaptive Control of the Novel Hyperchaotic System

In this section, we use adaptive control method to derive an adaptive feedback control law for globally stabilizing the novel 4-D hyperchaotic system with unknown parameters.

Thus, we consider the novel 4-D hyperchaotic system given by

$$\begin{cases} \dot{x}_1 = a(x_2 - x_1) + x_4 + u_1 \\ \dot{x}_2 = bx_1 - px_1x_3 + x_4 + u_2 \\ \dot{x}_3 = x_1x_2 - cx_3 + u_3 \\ \dot{x}_4 = -x_1 - x_2 + u_4 \end{cases} \tag{16}$$

In (16), x_1, x_2, x_3, x_4 are the states and u_1, u_2, u_3, u_4 are the adaptive controls to be determined using estimates of the unknown parameters.

We consider the adaptive feedback control law

$$\begin{cases} u_1 = -\hat{a}(t)(x_2 - x_1) - x_4 - k_1x_1 \\ u_2 = -\hat{b}(t)x_1 + \hat{p}(t)x_1x_3 - x_4 - k_2x_2 \\ u_3 = -x_1x_2 + \hat{c}(t)x_3 - k_3x_3 \\ u_4 = x_1 + x_2 - k_4x_4 \end{cases} \tag{17}$$

where k_1, k_2, k_3, k_4 are positive gain constants.

Substituting (17) into (16), we get the closed-loop plant dynamics as

$$\begin{cases} \dot{x}_1 = [a - \hat{a}(t)](x_2 - x_1) - k_1x_1 \\ \dot{x}_2 = [b - \hat{b}(t)]x_1 - [p - \hat{p}(t)]x_1x_3 - k_2x_2 \\ \dot{x}_3 = -[c - \hat{c}(t)]x_3 - k_3x_3 \\ \dot{x}_4 = -k_4x_4 \end{cases} \tag{18}$$

The parameter estimation errors are defined as

$$\begin{cases} e_a(t) = a - \hat{a}(t) \\ e_b(t) = b - \hat{b}(t) \\ e_c(t) = c - \hat{c}(t) \\ e_p(t) = p - \hat{p}(t) \end{cases} \tag{19}$$

In view of (19), we can simplify the plant dynamics (18) as

$$\begin{cases} \dot{x}_1 = e_a(x_2 - x_1) - k_1 x_1 \\ \dot{x}_2 = e_b x_1 - e_p x_1 x_3 - k_2 x_2 \\ \dot{x}_3 = -e_c x_3 - k_3 x_3 \\ \dot{x}_4 = -k_4 x_4 \end{cases} \tag{20}$$

Differentiating (19) with respect to t, we obtain

$$\begin{cases} \dot{e}_a(t) = -\dot{\hat{a}}(t) \\ \dot{e}_b(t) = -\dot{\hat{b}}(t) \\ \dot{e}_c(t) = -\dot{\hat{c}}(t) \\ \dot{e}_p(t) = -\dot{\hat{p}}(t) \end{cases} \tag{21}$$

We consider the quadratic candidate Lyapunov function defined by

$$V(\mathbf{x}, e_a, e_b, e_c, e_p) = \frac{1}{2}\left(x_1^2 + x_2^2 + x_3^2 + x_4^2\right) + \frac{1}{2}\left(e_a^2 + e_b^2 + e_c^2 + e_p^2\right) \tag{22}$$

Differentiating V along the trajectories of (20) and (21), we obtain

$$\begin{aligned} \dot{V} &= -k_1 x_1^2 - k_2 x_2^2 - k_3 x_3^2 - k_4 x_4^2 + e_a\left[x_1(x_2 - x_1) - \dot{\hat{a}}\right] \\ &\quad + e_b\left[x_1 x_2 - \dot{\hat{b}}\right] + e_c\left[-x_3^2 - \dot{\hat{c}}\right] + e_p\left[-x_1 x_2 x_3 - \dot{\hat{p}}\right] \end{aligned} \tag{23}$$

In view of (23), we take the parameter update law as

$$\begin{cases} \dot{\hat{a}}(t) = x_1(x_2 - x_1) \\ \dot{\hat{b}}(t) = x_1 x_2 \\ \dot{\hat{c}}(t) = -x_3^2 \\ \dot{\hat{p}}(t) = -x_1 x_2 x_3 \end{cases} \tag{24}$$

Next, we state and prove the main result of this section.

Theorem 1 *The novel 4-D hyperchaotic system (16) with unknown system parameters is globally and exponentially stabilized for all initial conditions by the adaptive control law (17) and the parameter update law (24), where k_1, k_2, k_3, k_4 are positive gain constants.*

Proof We prove this result by applying Lyapunov stability theory [11].

We consider the quadratic Lyapunov function defined by (22), which is clearly a positive definite function on $\mathbf{R}^8$.

By substituting the parameter update law (24) into (23), we obtain the time-derivative of V as

$$\dot{V} = -k_1 x_1^2 - k_2 x_2^2 - k_3 x_3^2 - k_4 x_4^2 \tag{25}$$

From (25), it is clear that $\dot{V}$ is a negative semi-definite function on $\mathbf{R}^8$.

Thus, we can conclude that the state vector $\mathbf{x}(t)$ and the parameter estimation error are globally bounded, i.e.

$$\left[x_1(t)\ x_2(t)\ x_3(t)\ x_4(t)\ e_a(t)\ e_b(t)\ e_c(t)\ e_p(t)\right]^T \in \mathbf{L}_\infty.$$

We define $k = \min\{k_1, k_2, k_3, k_4\}$.

Then it follows from (25) that

$$\dot{V} \leq -k\|\mathbf{x}(t)\|^2 \tag{26}$$

Thus, we have

$$k\|\mathbf{x}(t)\|^2 \leq -\dot{V} \tag{27}$$

Integrating the inequality (27) from 0 to t, we get

$$k \int_0^t \|\mathbf{x}(\tau)\|^2 \, d\tau \ \leq \ V(0) - V(t) \tag{28}$$

From (28), it follows that $\mathbf{x} \in \mathbf{L}_2$.

Using (20), we can conclude that $\dot{\mathbf{x}} \in \mathbf{L}_\infty$.

Using Barbalat's lemma [11], we conclude that $\mathbf{x}(t) \to 0$ exponentially as $t \to \infty$ for all initial conditions $\mathbf{x}(0) \in \mathbf{R}^4$.

This completes the proof. □

For the numerical simulations, the classical fourth-order Runge–Kutta method with step size $h = 10^{-8}$ is used to solve the systems (16) and (24), when the adaptive control law (17) is applied.

The parameter values of the novel 4-D hyperchaotic system (16) are taken as in the hyperchaotic case (2), i.e.

$$a = 12, \quad b = 36, \quad c = 5, \quad p = 12 \tag{29}$$

We take the positive gain constants as $k_i = 6$ for $i = 1, \ldots, 4$.

Furthermore, as initial conditions of the novel 4-D hyperchaotic system (16), we take

$$x_1(0) = 6.2, \quad x_2(0) = 3.8, \quad x_3(0) = -7.2, \quad x_4(0) = -5.6 \tag{30}$$

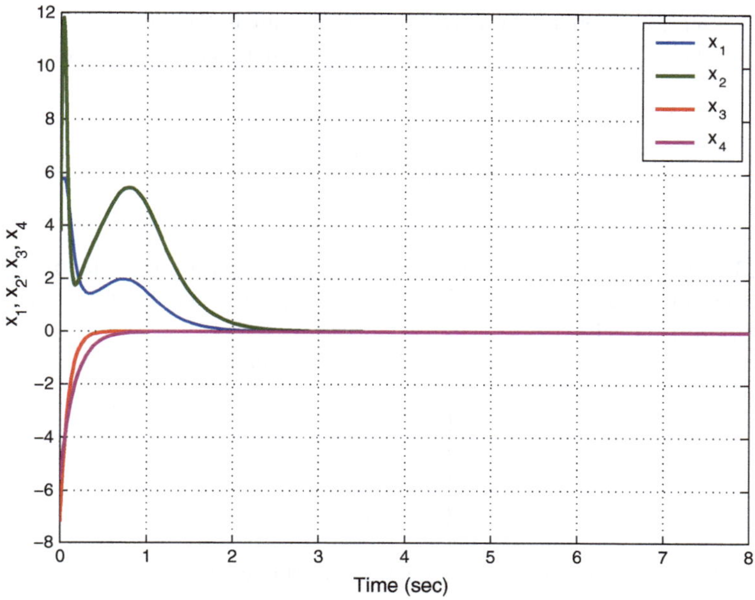

Fig. 5 Time-history of the controlled states x_1, x_2, x_3, x_4

Also, as initial conditions of the parameter estimates, we take

$$\hat{a}(0) = 5.1, \quad \hat{b}(0) = 9.3, \quad \hat{c}(0) = 1.9, \quad \hat{p}(0) = 3.4 \tag{31}$$

In Fig. 5, the exponential convergence of the controlled states of the novel 4-D hyperchaotic system (16) is depicted.

5 Adaptive Synchronization of the Identical Novel Hyperchaotic Systems

In this section, we use adaptive control method to derive an adaptive feedback control law for globally synchronizing identical novel 4-D hyperchaotic systems with unknown parameters.

As the master system, we consider the novel 4-D hyperchaotic system given by

$$\begin{cases} \dot{x}_1 = a(x_2 - x_1) + x_4 \\ \dot{x}_2 = bx_1 - px_1x_3 + x_4 \\ \dot{x}_3 = x_1x_2 - cx_3 \\ \dot{x}_4 = -x_1 - x_2 \end{cases} \tag{32}$$

In (32), x_1, x_2, x_3, x_4 are the states and a, b, c, p are unknown system parameters.

As the slave system, we consider the novel 4-D hyperchaotic system given by

$$\begin{cases} \dot{y}_1 = a(y_2 - y_1) + y_4 + u_1 \\ \dot{y}_2 = by_1 - py_1 y_3 + y_4 + u_2 \\ \dot{y}_3 = y_1 y_2 - cy_3 + u_3 \\ \dot{y}_4 = -y_1 - y_2 + u_4 \end{cases} \tag{33}$$

In (33), y_1, y_2, y_3, y_4 are the states and u_1, u_2, u_3, u_4 are the adaptive controls to be determined using estimates $\hat{a}(t), \hat{b}(t), \hat{c}(t), \hat{p}(t)$ for the unknown parameters a, b, c, p, respectively.

The synchronization error between the novel hyperchaotic systems (32) and (33) is defined by

$$\begin{cases} e_1 = y_1 - x_1 \\ e_2 = y_2 - x_2 \\ e_3 = y_3 - x_3 \\ e_4 = y_4 - x_4 \end{cases} \tag{34}$$

Then the synchronization error dynamics is obtained as

$$\begin{cases} \dot{e}_1 = a(e_2 - e_1) + e_4 + u_1 \\ \dot{e}_2 = be_1 + e_4 - p(y_1 y_3 - x_1 x_3) + u_2 \\ \dot{e}_3 = -ce_3 + y_1 y_2 - x_1 x_2 + u_3 \\ \dot{e}_4 = -e_1 - e_2 + u_4 \end{cases} \tag{35}$$

We consider the adaptive feedback control law

$$\begin{cases} u_1 = -\hat{a}(t)(e_2 - e_1) - e_4 - k_1 e_1 \\ u_2 = -\hat{b}(t)e_1 - e_4 + \hat{p}(t)(y_1 y_3 - x_1 x_3) - k_2 e_2 \\ u_3 = \hat{c}(t)e_3 - y_1 y_2 + x_1 x_2 - k_3 e_3 \\ u_4 = e_1 + e_2 - k_4 e_4 \end{cases} \tag{36}$$

where k_1, k_2, k_3, k_4 are positive gain constants.

Substituting (36) into (35), we get the closed-loop error dynamics as

$$\begin{cases} \dot{e}_1 = \left[a - \hat{a}(t)\right](e_2 - e_1) - k_1 e_1 \\ \dot{e}_2 = \left[b - \hat{b}(t)\right] e_1 - \left[p - \hat{p}(t)\right](y_1 y_3 - x_1 x_3) - k_2 e_2 \\ \dot{e}_3 = -\left[c - \hat{c}(t)\right] e_3 - k_3 e_3 \\ \dot{e}_4 = -k_4 e_4 \end{cases} \tag{37}$$

The parameter estimation errors are defined as

$$\begin{cases} e_a(t) = a - \hat{a}(t) \\ e_b(t) = b - \hat{b}(t) \\ e_c(t) = c - \hat{c}(t) \\ e_p(t) = p - \hat{p}(t) \end{cases} \tag{38}$$

In view of (38), we can simplify the error dynamics (37) as

$$\begin{cases} \dot{e}_1 = e_a(e_2 - e_1) - k_1 e_1 \\ \dot{e}_2 = e_b e_1 - e_p(y_1 y_3 - x_1 x_3) - k_2 e_2 \\ \dot{e}_3 = -e_c e_3 - k_3 e_3 \\ \dot{e}_4 = -k_4 e_4 \end{cases} \tag{39}$$

Differentiating (38) with respect to t, we obtain

$$\begin{cases} \dot{e}_a(t) = -\dot{\hat{a}}(t) \\ \dot{e}_b(t) = -\dot{\hat{b}}(t) \\ \dot{e}_c(t) = -\dot{\hat{c}}(t) \\ \dot{e}_p(t) = -\dot{\hat{p}}(t) \end{cases} \tag{40}$$

We use adaptive control theory to find an update law for the parameter estimates. We consider the quadratic candidate Lyapunov function defined by

$$V(\mathbf{e}, e_a, e_b, e_c, e_p) = \frac{1}{2}\left(e_1^2 + e_2^2 + e_3^2 + e_4^2\right) + \frac{1}{2}\left(e_a^2 + e_b^2 + e_c^2 + e_p^2\right) \tag{41}$$

Differentiating V along the trajectories of (39) and (40), we obtain

$$\begin{aligned} \dot{V} = & -k_1 e_1^2 - k_2 e_2^2 - k_3 e_3^2 - k_4 e_4^2 + e_a\left[e_1(e_2 - e_1) - \dot{\hat{a}}\right] \\ & + e_b\left[e_1 e_2 - \dot{\hat{b}}\right] + e_c\left[-e_3^2 - \dot{\hat{c}}\right] + e_p\left[-e_2(y_1 y_3 - x_1 x_3) - \dot{\hat{p}}\right] \end{aligned} \tag{42}$$

In view of (42), we take the parameter update law as

$$\begin{cases} \dot{\hat{a}}(t) = e_1(e_2 - e_1) \\ \dot{\hat{b}}(t) = e_1 e_2 \\ \dot{\hat{c}}(t) = -e_3^2 \\ \dot{\hat{p}}(t) = -e_2(y_1 y_3 - x_1 x_3) \end{cases} \tag{43}$$

Next, we state and prove the main result of this section.

Theorem 2 *The novel hyperchaotic systems (32) and (33) with unknown system parameters are globally and exponentially synchronized for all initial conditions by* *the adaptive control law (36) and the parameter update law (43), where k_1, k_2, k_3, k_4 are positive gain constants.*

Proof We prove this result by applying Lyapunov stability theory [11].

We consider the quadratic Lyapunov function defined by (41), which is clearly a positive definite function on $\mathbf{R}^8$.

By substituting the parameter update law (43) into (42), we obtain

$$\dot{V} = -k_1 e_1^2 - k_2 e_2^2 - k_3 e_3^2 - k_4 e_4^2 \tag{44}$$

From (44), it is clear that $\dot{V}$ is a negative semi-definite function on $\mathbf{R}^8$.

Thus, we can conclude that the error vector $\mathbf{e}(t)$ and the parameter estimation error are globally bounded, i.e.

$$\left[e_1(t)\ e_2(t)\ e_3(t)\ e_4(t)\ e_a(t)\ e_b(t)\ e_c(t)\ e_p(t) \right]^T \in \mathbf{L}_\infty. \tag{45}$$

We define $k = \min\{k_1, k_2, k_3, k_4\}$.

Then it follows from (44) that

$$\dot{V} \le -k \|\mathbf{e}(t)\|^2 \tag{46}$$

Thus, we have

$$k \|\mathbf{e}(t)\|^2 \le -\dot{V} \tag{47}$$

Integrating the inequality (47) from 0 to t, we get

$$k \int_0^t \|\mathbf{e}(\tau)\|^2 \, d\tau \le V(0) - V(t) \tag{48}$$

From (48), it follows that $\mathbf{e} \in \mathbf{L}_2$.

Using (39), we can conclude that $\dot{\mathbf{e}} \in \mathbf{L}_\infty$.

Using Barbalat's lemma [11], we conclude that $\mathbf{e}(t) \to 0$ exponentially as $t \to \infty$ for all initial conditions $\mathbf{e}(0) \in \mathbf{R}^4$.

This completes the proof. □

For the numerical simulations, the classical fourth-order Runge–Kutta method with step size $h = 10^{-8}$ is used to solve the systems (32), (33) and (43), when the adaptive control law (36) is applied.

The parameter values of the novel hyperchaotic systems are taken as in the hyperchaotic case (2), i.e. $a = 12$, $b = 36$, $c = 5$ and $p = 12$.

We take the positive gain constants as $k_i = 6$ for $i = 1, \ldots, 4$.

Furthermore, as initial conditions of the master system (32), we take

$$x_1(0) = 4.2, \quad x_2(0) = -5.8, \quad x_3(0) = 7.3, \quad x_4(0) = 9.1 \tag{49}$$

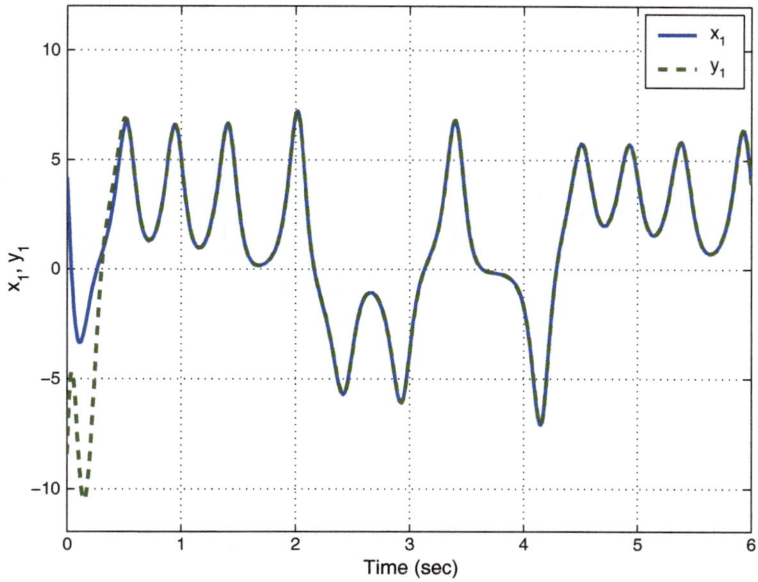

Fig. 6 Synchronization of the states x_1 and y_1

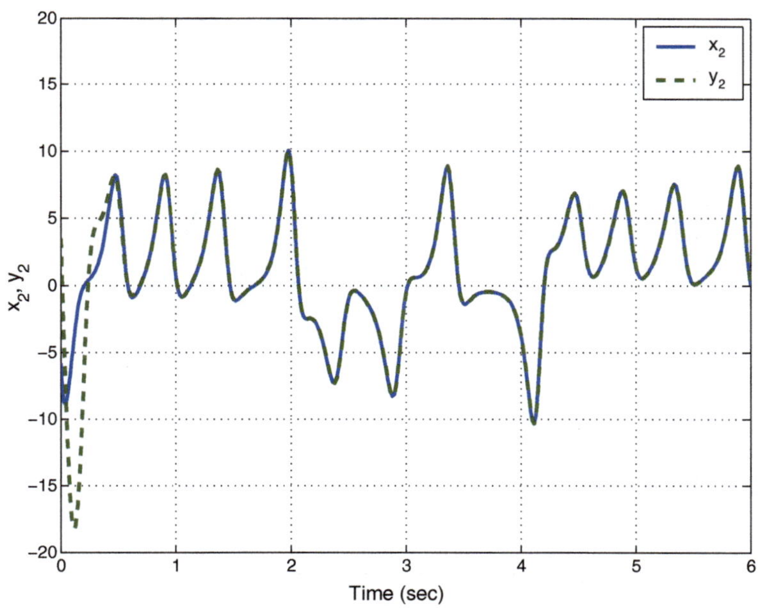

Fig. 7 Synchronization of the states x_2 and y_2

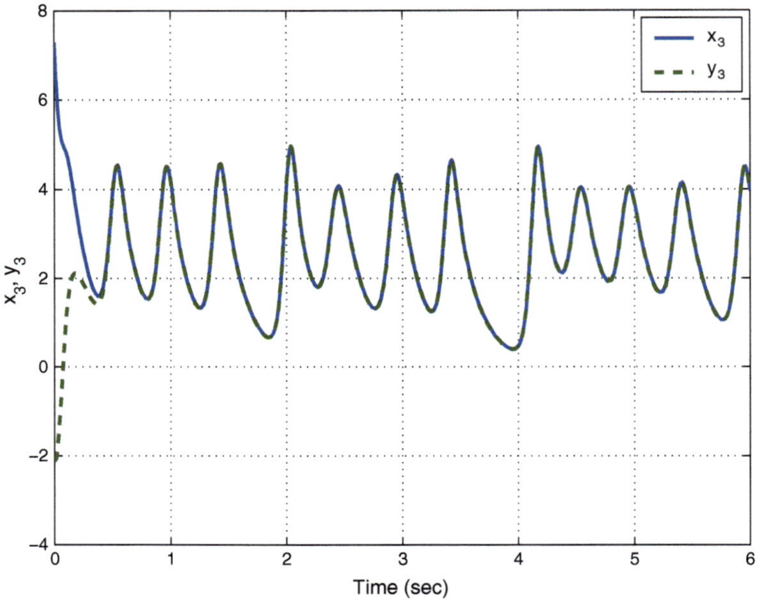

Fig. 8 Synchronization of the states x_3 and y_3

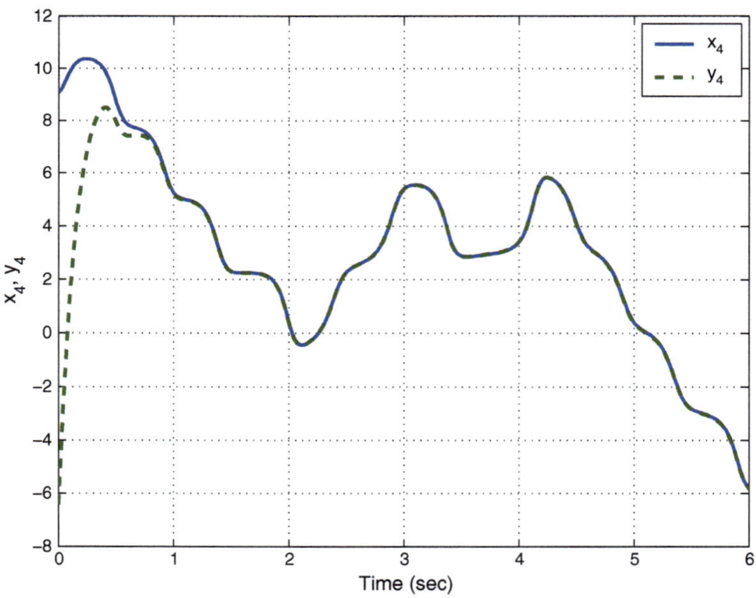

Fig. 9 Synchronization of the states x_4 and y_4

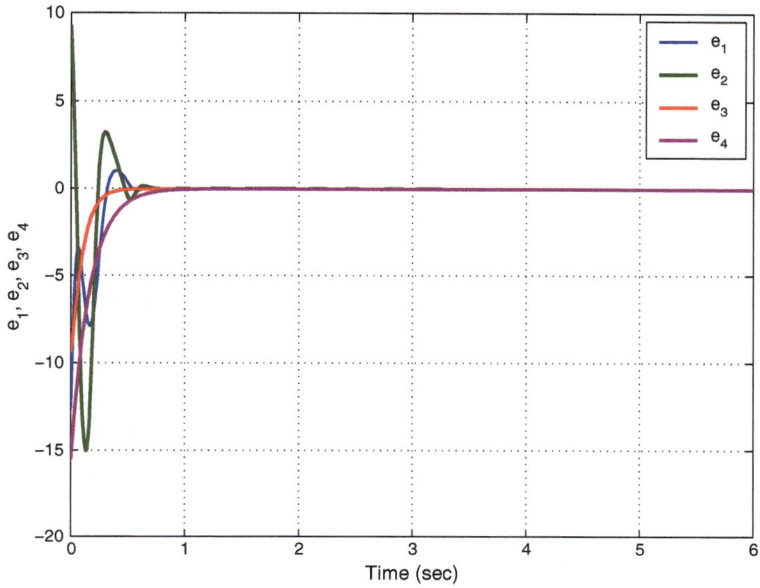

Fig. 10 Time-history of the synchronization errors e_1, e_2, e_3, e_4

As initial conditions of the slave system (33), we take

$$y_1(0) = -8.4, \quad y_2(0) = 3.5, \quad y_3(0) = -2.1, \quad y_4(0) = -6.4 \tag{50}$$

Also, as initial conditions of the parameter estimates, we take

$$\hat{a}(0) = 3.1, \quad \hat{b}(0) = 12.4, \quad \hat{c}(0) = 4.7, \quad \hat{p}(0) = -5.8 \tag{51}$$

Figures 6, 7, 8 and 9 describe the complete synchronization of the novel hyperchaotic systems (32) and (33), while Fig. 10 describes the time-history of the synchronization errors e_1, e_2, e_3, e_4.

6 Circuit Realization of the Novel 4-D Hyperchaotic System

The electronic circuit modelling the hyperchaotic system (1) is realized by using off-the-shelf components such as resistors, capacitors, operational amplifiers and multipliers. Using the design approach based on the operational amplifiers [17, 73, 76], we have the circuit as shown in Fig. 11 where each state variable of system (1), i.e. x_1, x_2, x_3, x_4 is implemented as the voltage across the corresponding capacitors C_1, C_2, C_3, and C_4, respectively.

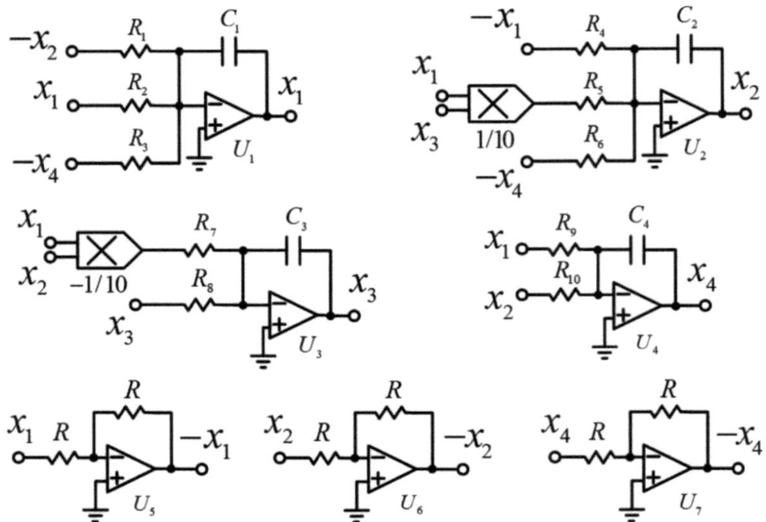

Fig. 11 The designed electronic circuit schematic of the ten-term novel 4-D hyperchaotic system with two quadratic nonlinearities

The circuital equations of the circuit in Fig. 11 are obtained as

$$\begin{cases} \dfrac{dv_{C_1}}{dt} = \dfrac{1}{R_1C_1}v_{C_2} - \dfrac{1}{R_2C_1}v_{C_1} + \dfrac{1}{R_3C_1}v_{C_4} \\ \dfrac{dv_{C_2}}{dt} = \dfrac{1}{R_4C_2}v_{C_1} - \dfrac{1}{10R_5C_2}v_{C_1}v_{C_3} + \dfrac{1}{R_6C_2}v_{C_4} \\ \dfrac{dv_{C_3}}{dt} = \dfrac{1}{10R_7C_3}v_{C_1}v_{C_2} - \dfrac{1}{R_8C_3}v_{C_3} \\ \dfrac{dv_{C_4}}{dt} = -\dfrac{1}{R_9C_4}v_{C_1} - \dfrac{1}{R_{10}C_4}v_{C_3} \end{cases} \tag{52}$$

where v_{C_1}, v_{C_2}, v_{C_3}, and v_{C_4} are the voltages across the capacitors C_1, C_2, C_3, and C_4, respectively.

The power supplies of all active devices are $\pm 15V_{DC}$. The TL084 operational amplifiers are used in this work. The values of components in Fig. 11 are seleceted to match the parameters of system (1) as follows: $R_1 = R_2 = 30\,\text{k}\Omega$, $R_3 = R_6 = R_9 = R_{10} = R = 360\,\text{k}\Omega$, $R_4 = 10\,\text{k}\Omega$, $R_5 = 3\,\text{k}\Omega$, $R_7 = 36\,\text{k}\Omega$, $R_8 = 72\,\text{k}\Omega$, and $C_1 = C_2 = C_3 = C_4 = 1nF$.

The designed circuit is implemented in the electronic simulation package Cadence OrCAD. The obtained results are summarized in Figs. 12, 13, 14 and 15, which indicate the hyperchaotic attractors in $v_{C_1} - v_{C_2}$, $v_{C_1} - v_{C_3}$, $v_{C_1} - v_{C_4}$ and $v_{C_2} - v_{C_3}$ planes.

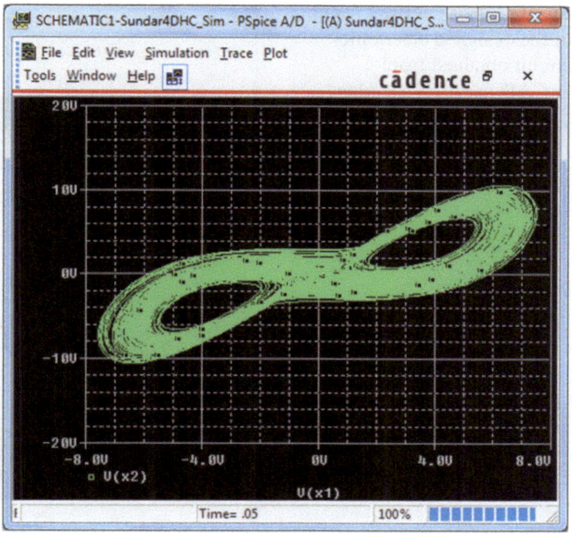

Fig. 12 Phase portrait result of the designed electronic circuit obtained from OrCAD in $v_{C_1} - v_{C_2}$ plane

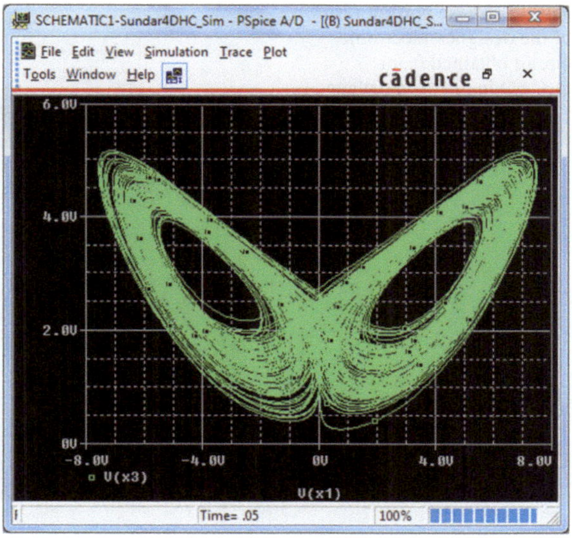

Fig. 13 Phase portrait result of the designed electronic circuit obtained from OrCAD in $v_{C_1} - v_{C_3}$ plane

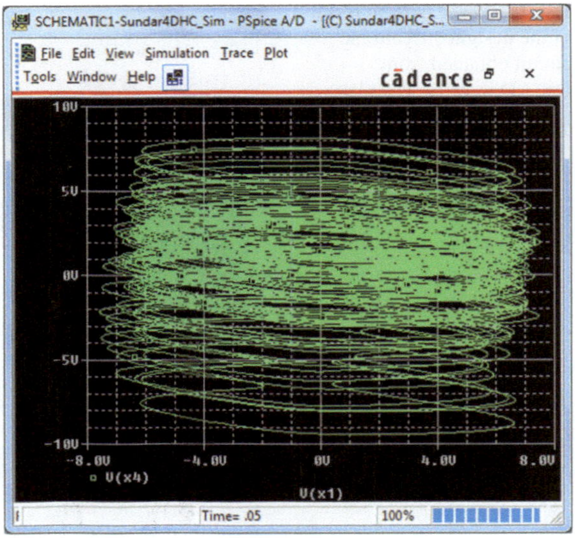

Fig. 14 Phase portrait result of the designed electronic circuit obtained from OrCAD in $v_{C_1} - v_{C_4}$ plane

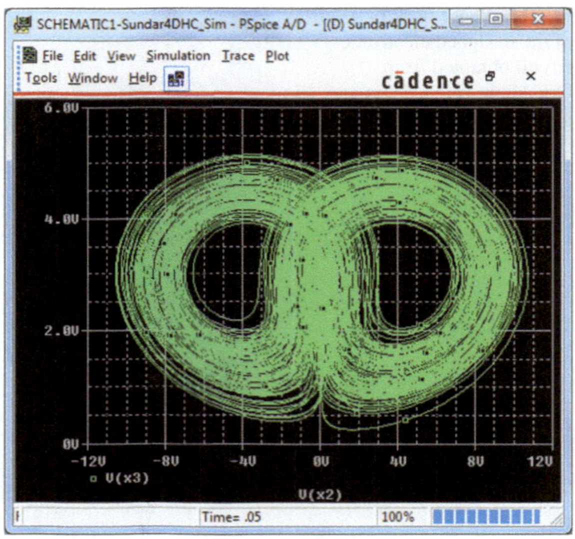

Fig. 15 Phase portrait result of the designed electronic circuit obtained from OrCAD in $v_{C_2} - v_{C_3}$ plane

7 Conclusions

In this work, a ten-term novel 4-D hyperchaotic system with two quadratic nonlinearities was presented. Fundamental dynamics of the new system were investigated through dissipativity, equilibria, rotation symmetry, Lyapunov exponents and Kaplan–Yorke dimension. In addition, an adaptive controller was designed not only to stabilize the novel hyperchaotic system with unknown parameters but also to achieve global chaos synchronization of two identical such systems with unknown system parameters. MATLAB simulations were depicted to illustrate all the main results derived in this work. Furthermore, an electronic circuit realization of the novel hyperchaotic system using the electronic simulation package Cadence OrCAD confirmed the feasibility of the theoretical model. Hence, it is believed that the new system due to its hyperchaotic nature, can be used in diverse chaos-based applications, such as in secure communication schemes. So, the complex dynamical behavior of this system and its applications in various scientific fields will be further explored in future research.

References

1. Abdurrahman A, Jiang H, Teng Z (2015) Finite-time synchronization for memristor-based neural networks with time-varying delays. Neural Netw 69:20–28
2. Arneodo A, Coullet P, Tresser C (1981) Possible new strange attractors with spiral structure. Commun Math Phys 79(4):573–576
3. Azar AT, Vaidyanathan S (2015) Chaos modeling and control systems design, vol 581. Springer, Germany
4. Behnia S, Afrang S, Akhshani A, Mabhouti K (2013) A novel method for controlling chaos in external cavity semiconductor laser. Optik 124(8):757–764
5. Cai G, Tan Z (2007) Chaos synchronization of a new chaotic system via nonlinear control. J Uncertain Syst 1(3):235–240
6. Chen A, Lu J, Lü J, Yu S (2006) Generating hyperchaotic Lü attractor via state feedback control. Physica A 364:103–110
7. Chen G, Ueta T (1999) Yet another chaotic attractor. Int J Bifurc Chaos 9(7):1465–1466
8. Hammami S (2015) State feedback-based secure image cryptosystem using hyperchaotic synchronization. ISA Trans 54:52–59
9. Islam MM, Murase K (2005) Chaotic dynamics of a behaviour-based miniature mobile robot: effects of environment and control structure. Neural Netw 18(2):123–144
10. Jia Q (2007) Hyperchaos generated from the Lorenz chaotic system and its control. Phys Lett A 366:217–222
11. Khalil HK (2001) Nonlinear Syst, 3rd edn. Prentice Hall, New Jersey
12. Li D (2008) A three-scroll chaotic attractor. Phys Lett A 372(4):387–393
13. Li X (2009) Modified projective synchronization of a new hyperchaotic system via nonlinear control. Commun Theor Phys 52:274–278
14. Lorenz EN (1963) Deterministic periodic flow. J Atmos Sci 20(2):130–141
15. Lü J, Chen G (2002) A new chaotic attractor coined. Int Bifurc Chaos 12(3):659–661
16. Matouk AE (2011) Chaos, feedback control and synchronization of a fractional-order modified autonomous Van der Pol-Duffing circuit. Commun Nonlinear Sci Numer Simul 16(2):975–986
17. Pehlivan I, Moroz IM, Vaidyanathan S (2014) Analysis, synchronization and circuit design of a novel butterfly attractor. J Sound Vib 333(20):5077–5096

18. Pham VT, Volos C, Jafari S, Wang X, Vaidyanathan S (2014) Hidden hyperchaotic attractor in a novel simple memristive neural network. Optoelectron Adv Mat Rapid Commun 8(11–12):1157–1163
19. Pham VT, Volos CK, Vaidyanathan S, Le TP, Vu VY (2015) A memristor-based hyperchaotic system with hidden attractors: dynamics, synchronization and circuital emulating. J Eng Sci Technol Rev 8(2):205–214
20. Rhouma R, Belghith S (2008) Cryptanalysis of a new image encryption algorithm based on hyper-chaos. Phys Lett A 372(38):5973–5978
21. Rhouma R, Belghith S (2011) Cryptanalysis of a chaos-based cryptosystem. Commun Nonlinear Sci Numer Simul 16(2):876–884
22. Rössler OE (1976) An equation for continuous chaos. Phys Lett A 57(5):397–398
23. Rössler OE (1979) An equation for hyperchaos. Phys Lett A 71:155–157
24. Sampath S, Vaidyanathan S, Volos CK, Pham VT (2015) An eight-term novel four-scroll chaotic System with cubic nonlinearity and its circuit simulation. J Eng Sci Technol Rev 8(2):1–6
25. Sarasu P, Sundarapandian V (2012) Generalized projective synchronization of two-scroll systems via adaptive control. Int J Soft Comput 7(4):146–156
26. Senouci A, Boukabou A (2014) Predictive control and synchronization of chaotic and hyperchaotic systems based on a $T - S$ fuzzy model. Math Comput Simul 105:62–78
27. Sprott JC (1994) Some simple chaotic flows. Phys Rev E 50(2):647–650
28. Sundarapandian V (2013) Adaptive control and synchronization design for the Lu-Xiao chaotic system. Lect Notes Electr Eng 131:319–327
29. Sundarapandian V (2013) Analysis and anti-synchronization of a novel chaotic system via active and adaptive controllers. J Eng Sci Technol Rev 6(4):45–52
30. Sundarapandian V, Pehlivan I (2012) Analysis, control, synchronization, and circuit design of a novel chaotic system. Math Comput Model 55(7–8):1904–1915
31. Tigan G, Opris D (2008) Analysis of a 3D chaotic system. Chaos, Solitons and Fractals 36:1315–1319
32. Tuwankotta JM (2006) Chaos in a coupled oscillators system with widely spaced frequencies and energy-preserving non-linearity. Int J Non-Linear Mech 41(2):180–191
33. Usama M, Khan MK, Alghathbar K, Lee C (2010) Chaos-based secure satellite imagery cryptosystem. Comput Math Appl 60(2):326–337
34. Vaidyanathan S (2012) Anti-synchronization of Sprott-L and Sprott-M chaotic systems via adaptive control. Int J Control Theory Appl 5(1):41–59
35. Vaidyanathan S (2013) A new six-term 3-D chaotic system with an exponential nonlinearity. Far East J Math Sci 79(1):135–143
36. Vaidyanathan S (2013) A ten-term novel 4-D hyperchaotic system with three quadratic nonlinearities and its control. Int J Control Theory Appl 6(2):97–109
37. Vaidyanathan S (2013) Analysis and adaptive synchronization of two novel chaotic systems with hyperbolic sinusoidal and cosinusoidal nonlinearity and unknown parameters. J Eng Sci Technol Rev 6(4):53–65
38. Vaidyanathan S (2014) A new eight-term 3-D polynomial chaotic system with three quadratic nonlinearities. Far East J Math Sci 84(2):219–226
39. Vaidyanathan S (2014) Analysis and adaptive synchronization of eight-term 3-D polynomial chaotic systems with three quadratic nonlinearities. Eur Phys J Spec Top 223(8):1519–1529
40. Vaidyanathan S (2014) Analysis, control and synchronisation of a six-term novel chaotic system with three quadratic nonlinearities. Int J Model Identif Control 22(1):41–53
41. Vaidyanathan S (2014) Generalized projective synchronisation of novel 3-D chaotic systems with an exponential non-linearity via active and adaptive control. Int J Model Identif Control 22(3):207–217
42. Vaidyanathan S (2014) Qualitative analysis and control of an eleven-term novel 4-D hyperchaotic system with two quadratic nonlinearities. Int J Control Theory Appl 7:35–47
43. Vaidyanathan S (2015) 3-cells cellular neural network (CNN) attractor and its adaptive biological control. Int J PharmTech Res 8(4):632–640

44. Vaidyanathan S (2015) A 3-D novel highly chaotic system with four quadratic nonlinearities, its adaptive control and anti-synchronization with unknown parameters. J Eng Sci Technol Rev 8(2):106–115
45. Vaidyanathan S (2015) A novel chemical chaotic reactor system and its adaptive control. Int J ChemTech Res 8(7):146–158
46. Vaidyanathan S (2015) Adaptive backstepping control of enzymes-substrates system with ferroelectric behaviour in brain waves. Int J PharmTech Res 8(2):256–261
47. Vaidyanathan S (2015) Adaptive biological control of generalized Lotka-Volterra three-species biological system. Int J PharmTech Res 8(4):622–631
48. Vaidyanathan S (2015) Adaptive chaotic synchronization of enzymes-substrates system with ferroelectric behaviour in brain waves. Int J PharmTech Res 8(5):964–973
49. Vaidyanathan S (2015) Adaptive control of a chemical chaotic reactor. Int J PharmTech Res 8(3):377–382
50. Vaidyanathan S (2015) Adaptive control of the FitzHugh-Nagumo chaotic neuron model. Int J PharmTech Res 8(6):117–127
51. Vaidyanathan S (2015) Adaptive synchronization of chemical chaotic reactors. Int J ChemTech Res 8(2):612–621
52. Vaidyanathan S (2015) Adaptive synchronization of generalized Lotka-Volterra three-species biological systems. Int J PharmTech Res 8(5):928–937
53. Vaidyanathan S (2015) Adaptive synchronization of novel 3-D chemical chaotic reactor systems. Int J ChemTech Res 8(7):159–171
54. Vaidyanathan S (2015) Adaptive synchronization of the identical FitzHugh-Nagumo chaotic neuron models. Int J PharmTech Res 8(6):167–177
55. Vaidyanathan S (2015) Analysis, control and synchronization of a 3-D novel jerk chaotic system with two quadratic nonlinearities. Kyungpook Math J 55:563–586
56. Vaidyanathan S (2015) Analysis, properties and control of an eight-term 3-D chaotic system with an exponential nonlinearity. Int J Model Identif Control 23(2):164–172
57. Vaidyanathan S (2015) Anti-synchronization of brusselator chemical reaction systems via adaptive control. Int J ChemTech Res 8(6):759–768
58. Vaidyanathan S (2015) Chaos in neurons and adaptive control of Birkhoff-Shaw strange chaotic attractor. Int J PharmTech Res 8(5):956–963
59. Vaidyanathan S (2015) Chaos in neurons and synchronization of Birkhoff-Shaw strange chaotic attractors via adaptive control. Int J PharmTech Res 8(6):1–11
60. Vaidyanathan S (2015) Coleman-Gomatam logarithmic competitive biology models and their ecological monitoring. Int J PharmTech Res 8(6):94–105
61. Vaidyanathan S (2015) Dynamics and control of brusselator chemical reaction. Int J ChemTech Res 8(6):740–749
62. Vaidyanathan S (2015) Dynamics and control of tokamak system with symmetric and magnetically confined plasma. Int J ChemTech Res 8(6):795–803
63. Vaidyanathan S (2015) Global chaos synchronization of chemical chaotic reactors via novel sliding mode control method. Int J ChemTech Res 8(7):209–221
64. Vaidyanathan S (2015) Global chaos synchronization of the forced Van der Pol chaotic oscillators via adaptive control method. Int J PharmTech Res 8(6):156–166
65. Vaidyanathan S (2015) Global chaos synchronization of the Lotka-Volterra biological systems with four competitive species via active control. Int J PharmTech Res 8(6):206–217
66. Vaidyanathan S (2015) Lotka–Volterra population biology models with negative feedback and their ecological monitoring. Int J PharmTech Res 8(5):974–981
67. Vaidyanathan S (2015) Lotka-Volterra two species competitive biology models and their ecological monitoring. Int J PharmTech Res 8(6):32–44
68. Vaidyanathan S (2015) Output regulation of the forced Van der Pol chaotic oscillator via adaptive control method. Int J PharmTech Res 8(6):106–116
69. Vaidyanathan S, Azar AT (2015) Analysis and control of a 4-D novel hyperchaotic system. Stud Comput Intell 581:3–17

70. Vaidyanathan S, Azar AT (2015) Analysis, control and synchronization of a nine-term 3-D novel chaotic system. In: Azar AT, Vaidyanathan S (eds) Chaos modelling and control systems design, studies in computational intelligence, vol 581. Springer, Germany, pp 19–38
71. Vaidyanathan S, Madhavan K (2013) Analysis, adaptive control and synchronization of a seven-term novel 3-D chaotic system. Int J Control Theory Appl 6(2):121–137
72. Vaidyanathan S, Pakiriswamy S (2013) Generalized projective synchronization of six-term Sundarapandian chaotic systems by adaptive control. Int J Control Theory Appl 6(2):153–163
73. Vaidyanathan S, Pakiriswamy S (2015) A 3-D novel conservative chaotic system and its generalized projective synchronization via adaptive control. J Eng Sci Technol Rev 8(2):52–60
74. Vaidyanathan S, Volos C (2015) Analysis and adaptive control of a novel 3-D conservative no-equilibrium chaotic system. Arch Control Sci 25(3):333–353
75. Vaidyanathan S, Volos C, Pham VT (2014) Hyperchaos, adaptive control and synchronization of a novel 5-D hyperchaotic system with three positive Lyapunov exponents and its SPICE implementation. Arch Control Sci 24(4):409–446
76. Vaidyanathan S, Volos C, Pham VT, Madhavan K, Idowu BA (2014) Adaptive backstepping control, synchronization and circuit simulation of a 3-D novel jerk chaotic system with two hyperbolic sinusoidal nonlinearities. Arch Control Sci 24(3):375–403
77. Vaidyanathan S, Rajagopal K, Volos CK, Kyprianidis IM, Stouboulos IN (2015) Analysis, adaptive control and synchronization of a seven-term novel 3-D chaotic system with three quadratic nonlinearities and its digital implementation in LabVIEW. J Eng Sci Technol Rev 8(2):130–141
78. Vaidyanathan S, Volos C, Pham VT, Madhavan K (2015) Analysis, adaptive control and synchronization of a novel 4-D hyperchaotic hyperjerk system and its SPICE implementation. Nonlinear Dyn 25(1):135–158
79. Vaidyanathan S, Volos CK, Kyprianidis IM, Stouboulos IN, Pham VT (2015) Analysis, adaptive control and anti-synchronization of a six-term novel jerk chaotic system with two exponential nonlinearities and its circuit simulation. J Eng Sci Technol Rev 8(2):24–36
80. Vaidyanathan S, Volos CK, Pham VT (2015) Analysis, adaptive control and adaptive synchronization of a nine-term novel 3-D chaotic system with four quadratic nonlinearities and its circuit simulation. J Eng Sci Technol Rev 8(2):181–191
81. Vaidyanathan S, Volos CK, Pham VT (2015) Analysis, control, synchronization and SPICE implementation of a novel 4-D hyperchaotic Rikitake dynamo system without equilibrium. J Eng Sci Technol Rev 8(2):232–244
82. Vaidyanathan S, Volos CK, Pham VT (2015) Global chaos control of a novel nine-term chaotic system via sliding mode control. In: Azar AT, Zhu Q (eds) Advances and applications in sliding mode control systems, studies in computational intelligence, vol 576. Springer, Germany, pp 571–590
83. Volos CK, Kyprianidis IM, Stouboulos IN, Anagnostopoulos AN (2009) Experimental study of the dynamic behavior of a double scroll circuit. J Appl Funct Anal 4:703–711
84. Volos CK, Kyprianidis IM, Stouboulos IN (2013) Experimental investigation on coverage performance of a chaotic autonomous mobile robot. Robot Autonom Syst 61(12):1314–1322
85. Volos CK, Kyprianidis IM, Stouboulos IN, Tlelo-Cuautle E, Vaidyanathan S (2015) Memristor: a new concept in synchronization of coupled neuromorphic circuits. J Eng Sci Technol Rev 8(2):157–173
86. Wang J, Chen Z (2008) A novel hyperchaotic system and its complex dynamics. Int J Bifurc Chaos 18:3309–3324
87. Wei X, Yunfei F, Qiang L (2012) A novel four-wing hyper-chaotic system and its circuit implementation. Procedia Eng 29:1264–1269
88. Wei Z, Yang Q (2010) Anti-control of Hopf bifurcation in the new chaotic system with two stable node-foci. Appl Math Comput 217(1):422–429
89. Yang J, Zhu F (2013) Synchronization for chaotic systems and chaos-based secure communications via both reduced-order and step-by-step sliding mode observers. Commun Nonlinear Sci Numer Simul 18(4):926–937

90. Yang J, Chen Y, Zhu F (2014) Singular reduced-order observer-based synchronization for uncertain chaotic systems subject to channel disturbance and chaos-based secure communication. Appl Math Comput 229:227–238
91. Yujun N, Xingyuan W, Mingjun W, Huaguang Z (2010) A new hyperchaotic system and its circuit implementation. Commun Nonlinear Sci Numer Simul 15(11):3518–3524
92. Zhang H, Liao X, Yu J (2005) Fuzzy modeling and synchronization of hyperchaotic systems. Chaos, Solitons and Fractals 26(3):835–843
93. Zhou W, Xu Y, Lu H, Pan L (2008) On dynamics analysis of a new chaotic attractor. Phys Lett A 372(36):5773–5777
94. Zhu C, Liu Y, Guo Y (2010) Theoretic and numerical study of a new chaotic system. Intell Inf Manag 2:104–109

Analysis, Adaptive Control and Synchronization of a Novel 3-D Highly Chaotic System

Sundarapandian Vaidyanathan

Abstract In this work, we describe an eight-term novel highly chaotic system with three quadratic nonlinearities. The phase portraits of the novel highly chaotic system are illustrated and the dynamic properties of the highly chaotic system are discussed. The novel highly chaotic system has three unstable equilibrium points. We show that the equilibrium point at the origin is a saddle point, while the other two equilibrium points are a saddle-focus and a critical point. The novel highly chaotic system has rotation symmetry about the x_3 axis. The Lyapunov exponents of the novel highly chaotic system are obtained as $L_1 = 7.6557$, $L_2 = 0$ and $L_3 = -24.6796$, while the Kaplan–Yorke dimension of the novel chaotic system is obtained as $D_{KY} = 2.3102$. Since the maximal Lyapunov exponent of the novel chaotic system has a high value, viz. $L_1 = 7.6557$, the novel chaotic system is highly chaotic. Since the sum of the Lyapunov exponents is negative, the novel chaotic system is dissipative. Next, we derive new results for the global chaos control of the novel highly chaotic system with unknown parameters using adaptive control method. We also derive new results for the global chaos synchronization of the identical novel highly chaotic systems with unknown parameters using adaptive control method. The main control results are established using Lyapunov stability theory. MATLAB simulations are depicted to illustrate the phase portraits of the novel highly chaotic system and also the adaptive control results derived in this work.

Keywords Chaos · Chaotic systems · Highly chaotic system · Chaos control · Adaptive control · Synchronization

---

S. Vaidyanathan (✉)
Research and Development Centre, Vel Tech University,
Avadi, Chennai 600062, Tamil Nadu, India
e-mail: sundarvtu@gmail.com

S. Vaidyanathan and C. Volos (eds.), *Advances and Applications in Chaotic Systems*, Studies in Computational Intelligence 636,
DOI 10.1007/978-3-319-30279-9_8

1 Introduction

In the last few decades, Chaos theory has become a very important and active research field, employing many applications in different disciplines like physics, chemistry, biology, ecology, engineering and economics, among others [3].

Some classical paradigms of 3-D chaotic systems in the literature are Lorenz system [11], Rössler system [20], ACT system [2], Sprott systems [25], Chen system [6], Lü system [12], Cai system [5], Tigan system [35], etc.

Many new chaotic systems have been discovered in the recent years such as Zhou system [108], Zhu system [109], Li system [10], Sundarapandian systems [28, 32], Vaidyanathan systems [44, 46, 48–51, 55, 66, 67, 81, 82, 84, 90, 92, 95, 98, 99, 101], Pehlivan system [15], Sampath system [21], etc.

Chaos theory has applications in several fields of science and engineering such as chemical reactors [56, 60, 62, 64, 68, 72–74], biological systems [54, 57–59, 61, 63, 65, 69–71, 75–79], memristors [1, 16, 105], lasers [4], oscillations [36], robotics [7, 104], electrical circuits [13, 103], cryptosystems [19, 37], secure communications [106, 107], etc.

The control of a chaotic system aims to stabilize or regulate the system with the help of a feedback control. There are many methods available for controlling a chaotic system such as active control [26, 38, 39], adaptive control [27, 40, 45, 47, 53, 80, 91, 97, 100], sliding mode control [42, 43], backstepping control [14, 94, 102], etc.

There are many methods available for chaos synchronization such as active control [8, 22, 23, 85, 87, 93], adaptive control [24, 29–31, 41, 83, 86], sliding mode control [33, 52, 89, 96], backstepping control [17, 18, 34, 88], etc.

In this research work, we announce an eight-term novel highly chaotic system with three quadratic nonlinearities. Using adaptive control method, we have also derived new results for the global chaos control of the novel highly chaotic system and global chaos synchronization of the identical novel highly chaotic systems when the system parameters are unknown.

This work is organized as follows. Section 2 describes the dynamic equations and phase portraits of the eight-term novel highly chaotic system. Section 3 details the dynamic analysis and properties of the novel highly chaotic system. The Lyapunov exponents of the novel chaotic system are obtained as $L_1 = 7.6557$, $L_2 = 0$ and $L_3 = -24.6796$, while the Kaplan–Yorke dimension of the novel chaotic system is obtained as $D_{KY} = 2.3102$. Since the maximal Lyapunov exponent of the novel chaotic system has a high value, viz. $L_1 = 7.6557$, the novel chaotic system is highly chaotic.

In Sect. 4, we derive new results for the global chaos control of the novel highly chaotic system with unknown parameters. In Sect. 5, we derive new results for the global chaos synchronization of the identical novel highly chaotic systems with unknown parameters. Section 6 contains a summary of the main results derived in this work.

2 A Novel 3-D Highly Chaotic System

In this section, we describe an eight-term novel chaotic system, which is given by the 3-D dynamics

$$\begin{cases} \dot{x}_1 = a(x_2 - x_1) + c x_2 x_3 \\ \dot{x}_2 = p x_1 - b x_2 - x_1 x_3 \\ \dot{x}_3 = x_1^2 - x_3 \end{cases} \tag{1}$$

where x_1, x_2, x_3 are the states and a, b, c, p are constant, positive parameters.

The novel 3-D system (1) is an eight-term polynomial system with three quadratic nonlinearities.

The system (1) exhibits a *highly chaotic attractor* for the parameter values

$$a = 15, \quad b = 1.1, \quad c = 12, \quad p = 90 \tag{2}$$

For numerical simulations, we take the initial conditions as

$$x_1(0) = 0.2, \quad x_2(0) = 0, \quad x_3(0) = 0.4 \tag{3}$$

Figure 1 depicts the 3-D phase portrait of the novel highly chaotic system (1), while Figs. 2, 3 and 4 depict the 2-D projection of the novel highly chaotic system (1) on the (x_1, x_2), (x_2, x_3) and (x_1, x_3) planes, respectively.

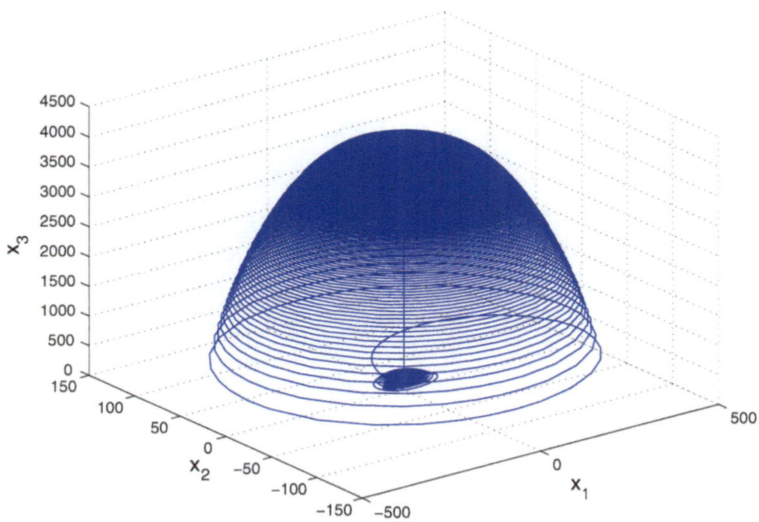

Fig. 1 3-D phase portrait of the novel highly chaotic system

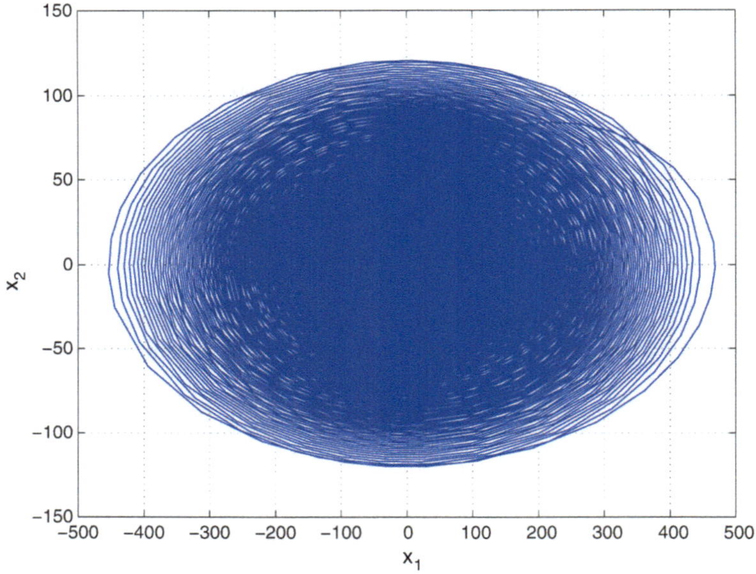

Fig. 2 2-D projection of the novel highly chaotic system on the (x_1, x_2) plane

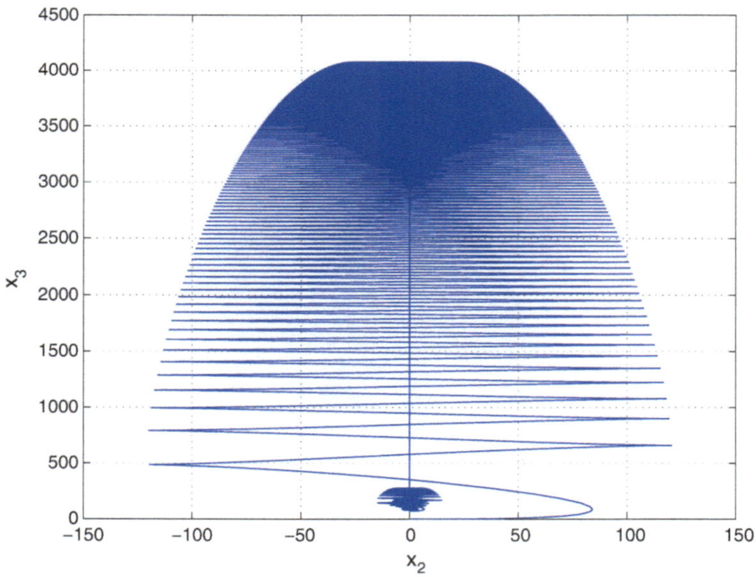

Fig. 3 2-D projection of the novel highly chaotic system on the (x_2, x_3) plane

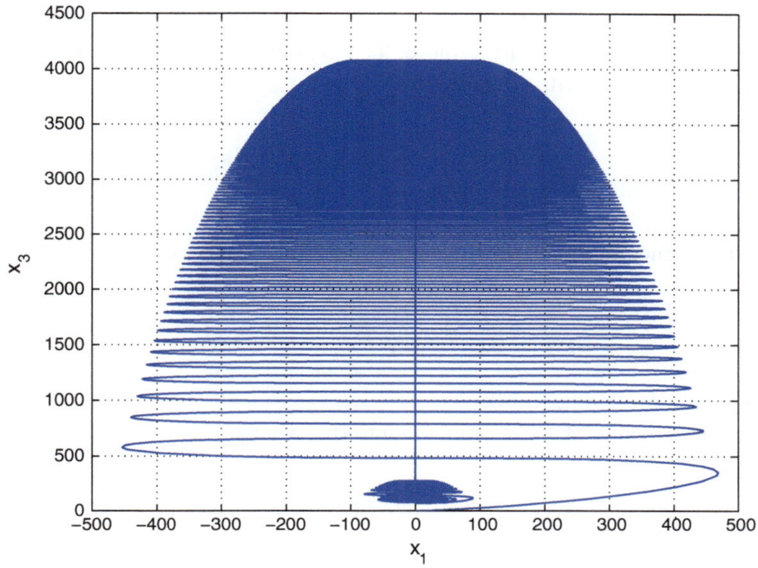

Fig. 4 2-D projection of the novel highly chaotic system on the (x_1, x_3) plane

3 Analysis of the Novel 3-D Highly Chaotic System

In this section, we give a dynamic analysis of the 3-D novel highly chaotic system (1). We take the parameter values as in the chaotic case (2), viz. $a = 15$, $b = 1.1$, $c = 12$ and $p = 90$.

3.1 Dissipativity

In vector notation, the novel chaotic system (1) can be expressed as

$$\dot{\mathbf{x}} = f(\mathbf{x}) = \begin{bmatrix} f_1(x_1, x_2, x_3) \\ f_2(x_1, x_2, x_3) \\ f_3(x_1, x_2, x_3) \end{bmatrix}, \tag{4}$$

where

$$\begin{cases} f_1(x_1, x_2, x_3) = a(x_2 - x_1) + cx_2x_3 \\ f_2(x_1, x_2, x_3) = px_1 - bx_2 - x_1x_3 \\ f_3(x_1, x_2, x_3) = x_1^2 - x_3 \end{cases} \tag{5}$$

Let Ω be any region in $\mathbf{R}^3$ with a smooth boundary and also, $\Omega(t) = \Phi_t(\Omega)$, where Φ_t is the flow of f. Furthermore, let $V(t)$ denote the volume of $\Omega(t)$.

By Liouville's theorem, we know that

$$\dot{V}(t) = \int\limits_{\Omega(t)} (\nabla \cdot f)\, dx_1\, dx_2\, dx_3 \tag{6}$$

The divergence of the novel chaotic system (4) is found as

$$\nabla \cdot f = \frac{\partial f_1}{\partial x_1} + \frac{\partial f_2}{\partial x_2} + \frac{\partial f_3}{\partial x_3} = -(a + b + 1) = -\mu < 0 \tag{7}$$

since $\mu = a - b + c = 21 > 0$.

Inserting the value of $\nabla \cdot f$ from (7) into (6), we get

$$\dot{V}(t) = \int\limits_{\Omega(t)} (-\mu)\, dx_1\, dx_2\, dx_3 = -\mu V(t) \tag{8}$$

Integrating the first order linear differential equation (8), we get

$$V(t) = \exp(-\mu t) V(0) \tag{9}$$

Since $\mu > 0$, it follows from Eq. (9) that $V(t) \to 0$ exponentially as $t \to \infty$. This shows that the novel chaotic system (1) is dissipative.

Hence, the system limit sets are ultimately confined into a specific limit set of zero volume, and the asymptotic motion of the novel chaotic system (1) settles onto a strange attractor of the system.

3.2 Equilibrium Points

We take the parameter values as in the chaotic case (2), viz. $a = 15, b = 1.1, c = 12$ and $p = 90$.

It is easy to see that the system (1) has three equilibrium points, viz.

$$E_0 = \begin{bmatrix} 0 \\ 0 \\ 0 \end{bmatrix}, \quad E_1 = \begin{bmatrix} 9.4860 \\ 0.1300 \\ 89.9849 \end{bmatrix}, \quad E_2 = \begin{bmatrix} -9.4860 \\ 0.1300 \\ -89.9849 \end{bmatrix} \tag{10}$$

The Jacobian of the system (1) at any point $\mathbf{x} \in \mathbf{R}^3$ is calculated as

$$J(\mathbf{x}) = \begin{bmatrix} -a & a + cx_3 & cx_2 \\ p - x_3 & -b & -x_1 \\ 2x_2 & 0 & -1 \end{bmatrix} = \begin{bmatrix} -15 & 15 + 12x_3 & 12x_2 \\ 90 - x_3 & -1.1 & -x_1 \\ 2x_2 & 0 & -1 \end{bmatrix} \tag{11}$$

We find that the matrix $J_0 = J(E_0)$ has the eigenvalues

$$\lambda_1 = -1, \;\; \lambda_2 = -45.4439, \;\; \lambda_3 = 29.3439 \tag{12}$$

This shows that the equilibrium point E_0 is a saddle-point, which is unstable.
We find that the matrix $J_1 = J(E_1)$ has the eigenvalues

$$\lambda_1 = -21.9770, \;\; \lambda_{2,3} = 2.4385 \pm 10.8119i \tag{13}$$

This shows that the equilibrium point E_1 is a saddle-focus, which is unstable.
We find that the matrix $J_2 = J(E_2)$ has the eigenvalues

$$\lambda_1 = 0, \;\; \lambda_{2,3} = -8.55 \pm 437.77i \tag{14}$$

Thus, E_2 is a critical point, which can be easily seen as unstable.

3.3 *Symmetry and Invariance*

It is easy to see that the system (1) is invariant under the change of coordinates

$$(x_1, x_2, x_3) \mapsto (-x_1, -x_2, x_3) \tag{15}$$

Thus, it follows that the 3-D novel chaotic system (1) has rotation symmetry about the x_3-axis and that any non-trivial trajectory must have a twin trajectory.

Next, it is easy to see that the x_3-axis is invariant under the flow of the 3-D novel chaotic system (1). The invariant motion along the x_3-axis is characterized by

$$\dot{x}_3 = -x_3, \tag{16}$$

which is globally exponentially stable.

3.4 *Lyapunov Exponents and Kaplan–Yorke Dimension*

We take the parameter values of the novel system (1) as in the chaotic case (2). We take the initial state of the novel system (1) as given in (3).

Then the Lyapunov exponents of the system (1) are numerically obtained as

$$L_1 = 7.6557, \quad L_2 = 0, \quad L_3 = -24.6796 \tag{17}$$

Also, the Kaplan–Yorke dimension of the novel chaotic system (1) is found as

$$D_{KY} = 2 + \frac{L_1 + L_2}{|L_3|} = 2.3102 \tag{18}$$

4 Adaptive Control of the Novel Highly Chaotic System

In this section, we use adaptive control method to derive an adaptive feedback control law for globally stabilizing the novel 3-D highly chaotic system with unknown parameters.

Thus, we consider the novel highly chaotic system given by

$$\begin{cases} \dot{x}_1 = a(x_2 - x_1) + c x_2 x_3 + u_1 \\ \dot{x}_2 = p x_1 - b x_2 - x_1 x_3 + u_2 \\ \dot{x}_3 = x_1^2 - x_3 + u_3 \end{cases} \tag{19}$$

In (19), x_1, x_2, x_3 are the states and u_1, u_2, u_3 are the adaptive controls to be determined using estimates $\hat{a}(t)$, $\hat{b}(t)$, $\hat{c}(t)$, $\hat{p}(t)$ for the unknown parameters a, b, c, p, respectively.

We consider the adaptive feedback control law

$$\begin{cases} u_1 = -\hat{a}(t)(x_2 - x_1) - \hat{c}(t) x_2 x_3 - k_1 x_1 \\ u_2 = -\hat{p}(t) x_1 + \hat{b}(t) x_2 + x_1 x_3 - k_2 x_2 \\ u_3 = -x_1^2 + x_3 - k_3 x_3 \end{cases} \tag{20}$$

where k_1, k_2, k_3 are positive gain constants.

Substituting (20) into (19), we get the closed-loop plant dynamics as

$$\begin{cases} \dot{x}_1 = \left[a - \hat{a}(t)\right](x_2 - x_1) + \left[c - \hat{c}(t)\right] x_2 x_3 - k_1 x_1 \\ \dot{x}_2 = \left[p - \hat{p}(t)\right] x_1 - \left[b - \hat{b}(t)\right] x_2 - k_2 x_2 \\ \dot{x}_3 = -k_3 x_3 \end{cases} \tag{21}$$

The parameter estimation errors are defined as

$$\begin{cases} e_a(t) = a - \hat{a}(t) \\ e_b(t) = b - \hat{b}(t) \\ e_c(t) = c - \hat{c}(t) \\ e_p(t) = p - \hat{p}(t) \end{cases} \tag{22}$$

In view of (22), we can simplify the plant dynamics (21) as

$$\begin{cases} \dot{x}_1 = e_a(x_2 - x_1) + e_c x_2 x_3 - k_1 x_1 \\ \dot{x}_2 = e_p x_1 - e_b x_2 - k_2 x_2 \\ \dot{x}_3 = -k_3 x_3 \end{cases} \tag{23}$$

Differentiating (22) with respect to t, we obtain

$$\begin{cases} \dot{e}_a(t) = -\dot{\hat{a}}(t) \\ \dot{e}_b(t) = -\dot{\hat{b}}(t) \\ \dot{e}_c(t) = -\dot{\hat{c}}(t) \\ \dot{e}_p(t) = -\dot{\hat{p}}(t) \end{cases} \tag{24}$$

We consider the quadratic candidate Lyapunov function defined by

$$V(\mathbf{x}, e_a, e_b, e_c, e_p) = \frac{1}{2}\left(x_1^2 + x_2^2 + x_3^2\right) + \frac{1}{2}\left(e_a^2 + e_b^2 + e_c^2 + e_p^2\right) \tag{25}$$

Differentiating V along the trajectories of (23) and (24), we obtain

$$\begin{aligned} \dot{V} = & -k_1 x_1^2 - k_2 x_2^2 - k_3 x_3^2 + e_a\left[x_1(x_2 - x_1) - \dot{\hat{a}}\right] \\ & + e_b\left[-x_2^2 - \dot{\hat{b}}\right] + e_c\left[x_1 x_2 x_3 - \dot{\hat{c}}\right] + e_p\left[x_1 x_2 - \dot{\hat{p}}\right] \end{aligned} \tag{26}$$

In view of (26), we take the parameter update law as

$$\begin{cases} \dot{\hat{a}}(t) = x_1(x_2 - x_1) \\ \dot{\hat{b}}(t) = -x_2^2 \\ \dot{\hat{c}}(t) = x_1 x_2 x_3 \\ \dot{\hat{p}}(t) = x_1 x_2 \end{cases} \tag{27}$$

Next, we state and prove the main result of this section.

Theorem 1 *The novel 3-D highly chaotic system (19) with unknown system parameters is globally and exponentially stabilized for all initial conditions by the adaptive control law (20) and the parameter update law (27), where k_1, k_2, k_3 are positive gain constants.*

Proof We prove this result by applying Lyapunov stability theory [9].

We consider the quadratic Lyapunov function defined by (25), which is clearly a positive definite function on $\mathbf{R}^7$.

By substituting the parameter update law (27) into (26), we obtain the time-derivative of V as

$$\dot{V} = -k_1 x_1^2 - k_2 x_2^2 - k_3 x_3^2 \tag{28}$$

From (28), it is clear that $\dot{V}$ is a negative semi-definite function on $\mathbf{R}^7$.

Thus, we can conclude that the state vector $\mathbf{x}(t)$ and the parameter estimation error are globally bounded i.e.

$$\left[x_1(t)\ x_2(t)\ x_3(t)\ e_a(t)\ e_b(t)\ e_c(t)\ e_p(t) \right]^T \in \mathbf{L}_\infty.$$

We define $k = \min\{k_1, k_2, k_3\}$.

Then it follows from (28) that

$$\dot{V} \leq -k\|\mathbf{x}(t)\|^2 \tag{29}$$

Thus, we have

$$k\|\mathbf{x}(t)\|^2 \leq -\dot{V} \tag{30}$$

Integrating the inequality (30) from 0 to t, we get

$$k \int_0^t \|\mathbf{x}(\tau)\|^2 \, d\tau \leq V(0) - V(t) \tag{31}$$

From (31), it follows that $\mathbf{x} \in \mathbf{L}_2$.

Using (23), we can conclude that $\dot{\mathbf{x}} \in \mathbf{L}_\infty$.

Using Barbalat's lemma [9], we conclude that $\mathbf{x}(t) \to 0$ exponentially as $t \to \infty$ for all initial conditions $\mathbf{x}(0) \in \mathbf{R}^3$.

Hence, the novel highly chaotic system (19) with unknown system parameters is globally and exponentially stabilized for all initial conditions by the adaptive control law (20) and the parameter update law (27).

This completes the proof. □

For the numerical simulations, the classical fourth-order Runge–Kutta method with step size $h = 10^{-8}$ is used to solve the systems (19) and (27), when the adaptive control law (20) is applied.

The parameter values of the novel 3-D highly chaotic system (19) are taken as in the chaotic case (2), i.e.

$$a = 15, \quad b = 1.1, \quad c = 12, \quad p = 90 \tag{32}$$

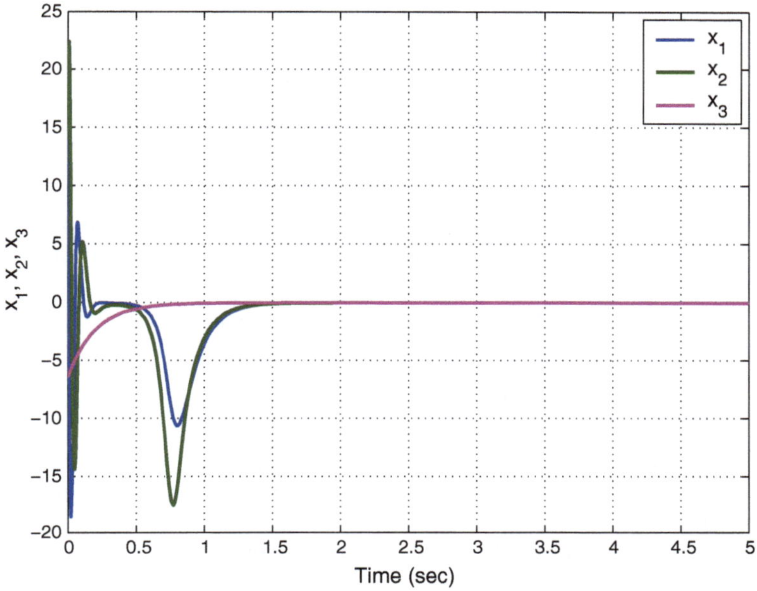

Fig. 5 Time-history of the controlled states x_1, x_2, x_3

We take the positive gain constants as

$$k_1 = 5, \quad k_2 = 5, \quad k_3 = 5 \tag{33}$$

Furthermore, as initial conditions of the novel highly chaotic system (19), we take

$$x_1(0) = 22.3, \quad x_2(0) = 14.7, \quad x_3(0) = -6.4 \tag{34}$$

Also, as initial conditions of the parameter estimates, we take

$$\hat{a}(0) = 3.7, \quad \hat{b}(0) = 4.3, \quad \hat{c}(0) = 1.5, \quad \hat{p}(0) = 18.1 \tag{35}$$

In Fig. 5, the exponential convergence of the controlled states of the 3-D novel highly chaotic system (19) is depicted.

5 Adaptive Synchronization of the Identical Novel Highly Chaotic Systems

In this section, we use adaptive control method to derive an adaptive feedback control law for globally synchronizing identical 3-D novel highly chaotic systems with unknown parameters. The main result is established using Lyapunov stability theory.

As the master system, we consider the novel 3-D chaotic system given by

$$\begin{cases} \dot{x}_1 = a(x_2 - x_1) + cx_2x_3 \\ \dot{x}_2 = px_1 - bx_2 - x_1x_3 \\ \dot{x}_3 = x_1^2 - x_3 \end{cases} \tag{36}$$

In (36), x_1, x_2, x_3 are the states and a, b, c, p are unknown system parameters.

As the slave system, we consider the novel 3-D chaotic system given by

$$\begin{cases} \dot{y}_1 = a(y_2 - y_1) + cy_2y_3 + u_1 \\ \dot{y}_2 = py_1 - by_2 - y_1y_3 + u_2 \\ \dot{y}_3 = y_1^2 - y_3 + u_3 \end{cases} \tag{37}$$

In (37), y_1, y_2, y_3 are the states and u_1, u_2, u_3 are the adaptive controls to be determined using estimates $\hat{a}(t), \hat{b}(t), \hat{c}(t), \hat{p}(t)$ for the unknown parameters a, b, c, p, respectively.

The synchronization error between the novel chaotic systems is defined by

$$\begin{cases} e_1 = y_1 - x_1 \\ e_2 = y_2 - x_2 \\ e_3 = y_3 - x_3 \end{cases} \tag{38}$$

Then the error dynamics is obtained as

$$\begin{cases} \dot{e}_1 = a(e_2 - e_1) + c(y_2y_3 - x_2x_3) + u_1 \\ \dot{e}_2 = pe_1 - be_2 - y_1y_3 + x_1x_3 + u_2 \\ \dot{e}_3 = -e_3 + y_1^2 - x_1^2 + u_3 \end{cases} \tag{39}$$

We consider the adaptive feedback control law

$$\begin{cases} u_1 = -\hat{a}(t)(e_2 - e_1) - \hat{c}(t)(y_2y_3 - x_2x_3) - k_1e_1 \\ u_2 = -\hat{p}(t)e_1 + \hat{b}(t)e_2 + y_1y_3 - x_1x_3 - k_2e_2 \\ u_3 = e_3 - y_1^2 + x_1^2 - k_3e_3 \end{cases} \tag{40}$$

where k_1, k_2, k_3 are positive gain constants.

Substituting (40) into (39), we get the closed-loop error dynamics as

$$\begin{cases} \dot{e}_1 = \left[a - \hat{a}(t)\right](e_2 - e_1) + \left[c - \hat{c}(t)\right](y_2y_3 - x_2x_3) - k_1e_1 \\ \dot{e}_2 = \left[p - \hat{p}(t)\right]e_1 - \left[b - \hat{b}(t)\right]e_2 - k_2e_2 \\ \dot{e}_3 = -k_3e_3 \end{cases} \tag{41}$$

The parameter estimation errors are defined as

$$\begin{cases} e_a(t) = a - \hat{a}(t) \\ e_b(t) = b - \hat{b}(t) \\ e_c(t) = c - \hat{c}(t) \\ e_p(t) = p - \hat{p}(t) \end{cases} \tag{42}$$

In view of (42), we can simplify the error dynamics (41) as

$$\begin{cases} \dot{e}_1 = e_a(e_2 - e_1) + e_c(y_2y_3 - x_2x_3) - k_1e_1 \\ \dot{e}_2 = e_pe_1 - e_be_2 - k_2e_2 \\ \dot{e}_3 = -k_3e_3 \end{cases} \tag{43}$$

Differentiating (42) with respect to t, we obtain

$$\begin{cases} \dot{e}_a(t) = -\dot{\hat{a}}(t) \\ \dot{e}_b(t) = -\dot{\hat{b}}(t) \\ \dot{e}_c(t) = -\dot{\hat{c}}(t) \\ \dot{e}_p(t) = -\dot{\hat{p}}(t) \end{cases} \tag{44}$$

We consider the quadratic candidate Lyapunov function defined by

$$V(\mathbf{e}, e_a, e_b, e_c, e_p) = \frac{1}{2}\left(e_1^2 + e_2^2 + e_3^2\right) + \frac{1}{2}\left(e_a^2 + e_b^2 + e_c^2 + e_p^2\right) \tag{45}$$

Differentiating V along the trajectories of (43) and (44), we obtain

$$\begin{aligned} \dot{V} = & -k_1e_1^2 - k_2e_2^2 - k_3e_3^2 + e_a\left[e_1(e_2 - e_1) - \dot{\hat{a}}\right] + e_b\left[-e_2^2 - \dot{\hat{b}}\right] \\ & + e_c\left[e_1(y_2y_3 - x_2x_3) - \dot{\hat{c}}\right] + e_p\left[e_1e_2 - \dot{\hat{p}}\right] \end{aligned} \tag{46}$$

In view of (46), we take the parameter update law as

$$\begin{cases} \dot{\hat{a}}(t) = e_1(e_2 - e_1) \\ \dot{\hat{b}}(t) = -e_2^2 \\ \dot{\hat{c}}(t) = e_1(y_2y_3 - x_2x_3) \\ \dot{\hat{p}}(t) = e_1e_2 \end{cases} \tag{47}$$

Next, we state and prove the main result of this section.

Theorem 2 *The novel highly chaotic systems (36) and (37) with unknown system parameters are globally and exponentially synchronized for all initial conditions by the adaptive control law (40) and the parameter update law (47), where k_1, k_2, k_3 are positive gain constants.*

Proof We prove this result by applying Lyapunov stability theory [9].

We consider the quadratic Lyapunov function defined by (45), which is clearly a positive definite function on $\mathbf{R}^7$.

By substituting the parameter update law (47) into (46), we obtain

$$\dot{V} = -k_1 e_1^2 - k_2 e_2^2 - k_3 e_3^2 \tag{48}$$

From (48), it is clear that $\dot{V}$ is a negative semi-definite function on $\mathbf{R}^7$.

Thus, we can conclude that the error vector $\mathbf{e}(t)$ and the parameter estimation error are globally bounded, i.e.

$$\left[\, e_1(t)\ e_2(t)\ e_3(t)\ e_a(t)\ e_b(t)\ e_c(t)\ e_p(t)\,\right]^T \in \mathbf{L}_\infty. \tag{49}$$

We define $k = \min\{k_1, k_2, k_3\}$.

Then it follows from (48) that

$$\dot{V} \le -k\|\mathbf{e}(t)\|^2 \tag{50}$$

Thus, we have

$$k\|\mathbf{e}(t)\|^2 \le -\dot{V} \tag{51}$$

Integrating the inequality (51) from 0 to t, we get

$$k\int_0^t \|\mathbf{e}(\tau)\|^2\, d\tau \;\le\; V(0) - V(t) \tag{52}$$

From (52), it follows that $\mathbf{e} \in \mathbf{L}_2$.

Using (43), we can conclude that $\dot{\mathbf{e}} \in \mathbf{L}_\infty$.

Using Barbalat's lemma [9], we conclude that $\mathbf{e}(t) \to 0$ exponentially as $t \to \infty$ for all initial conditions $\mathbf{e}(0) \in \mathbf{R}^3$.

Hence, we have proved that novel highly chaotic systems (36) and (37) with unknown system parameters are globally and exponentially synchronized for all initial conditions.

This completes the proof. □

For the numerical simulations, the classical fourth-order Runge–Kutta method with step size $h = 10^{-8}$ is used to solve the systems (36), (37) and (47), when the adaptive control law (40) is applied.

The parameter values of the novel chaotic systems are taken as in the chaotic case (2), i.e.

$$a = 15, \quad b = 1.1, \quad c = 12, \quad p = 90 \tag{53}$$

We take the positive gain constants as $k_i = 5$ for $i = 1, 2, 3$.
Furthermore, as initial conditions of the master system (36), we take

$$x_1(0) = 2.4, \quad x_2(0) = 3.7, \quad x_3(0) = 6.3 \tag{54}$$

As initial conditions of the slave system (37), we take

$$y_1(0) = 3.6, \quad y_2(0) = 1.5, \quad y_3(0) = 9.2 \tag{55}$$

Also, as initial conditions of the parameter estimates, we take

$$\hat{a}(0) = 12.1, \quad \hat{b}(0) = 10.3, \quad \hat{c}(0) = 15.4, \quad \hat{p}(0) = 20.9 \tag{56}$$

Figures 6, 7 and 8 describe the complete synchronization of the novel highly chaotic systems (36) and (37), while Fig. 9 describes the time-history of the synchronization errors e_1, e_2, e_3.

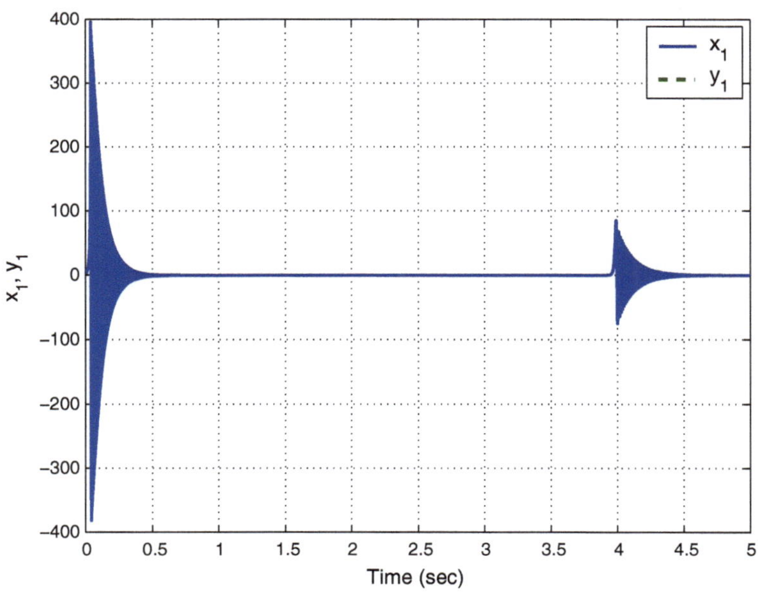

Fig. 6 Synchronization of the states x_1 and y_1

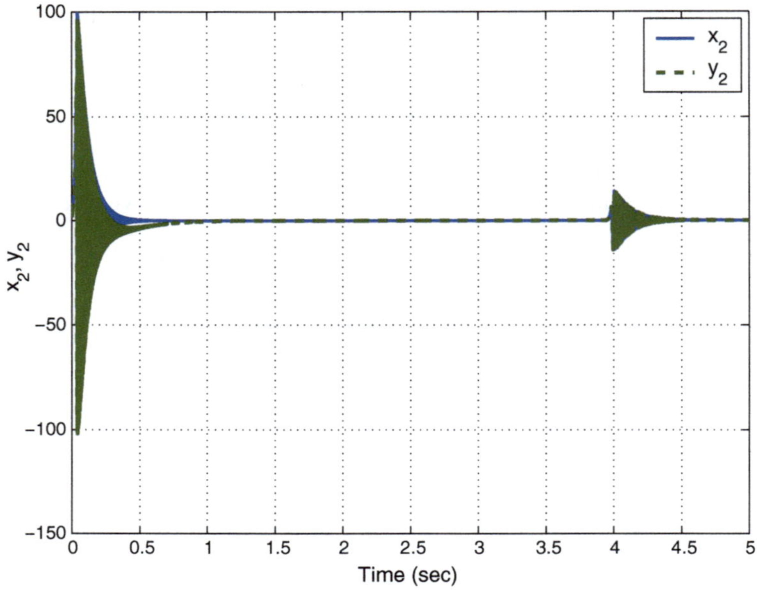

Fig. 7 Synchronization of the states x_2 and y_2

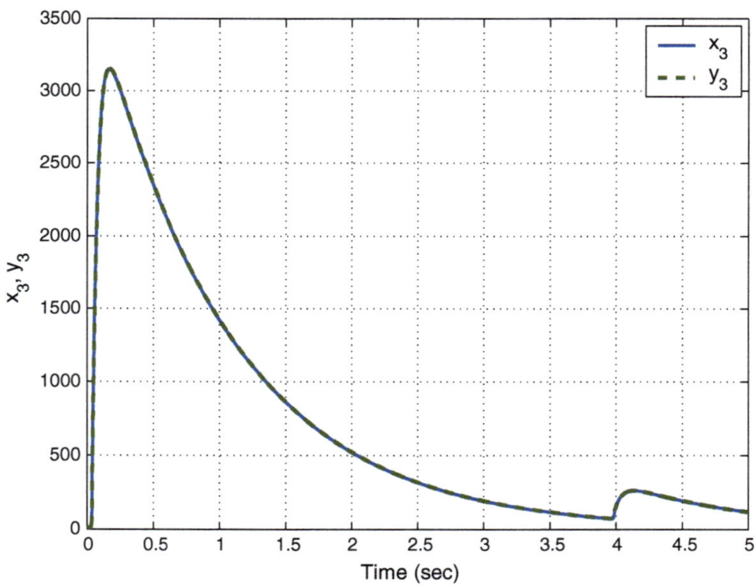

Fig. 8 Synchronization of the states x_3 and y_3

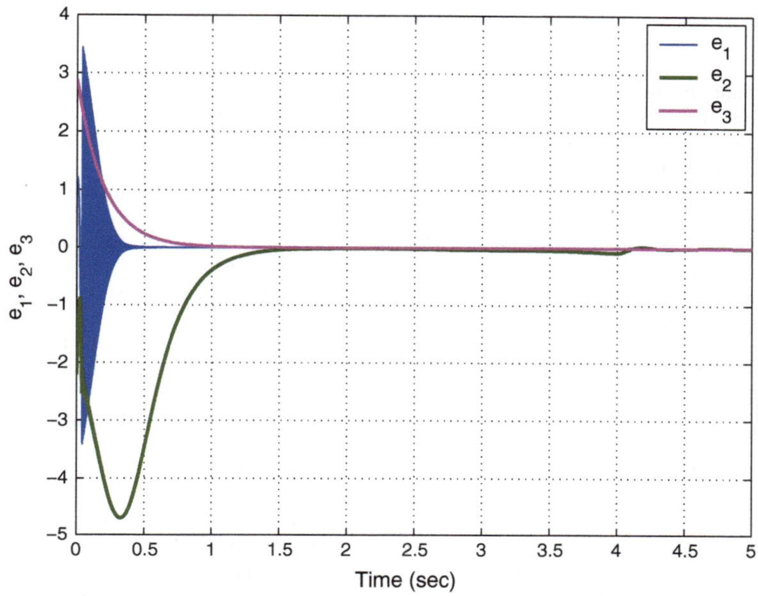

Fig. 9 Time-history of the synchronization errors e_1, e_2, e_3

6 Conclusions

This work announced an eight-term novel highly chaotic system with three quadratic nonlinearities. First, the qualitative properties of the highly chaotic system are discussed. We showed that the novel highly chaotic system has three unstable equilibrium points. The novel highly chaotic system has rotation symmetry about the x_3 axis. The Lyapunov exponents of the novel highly chaotic system have been obtained as $L_1 = 7.6557$, $L_2 = 0$ and $L_3 = -24.6796$, while the Kaplan–Yorke dimension of the novel chaotic system has been derived as $D_{KY} = 2.3102$. Since the maximal Lyapunov exponent of the novel chaotic system has a high value, viz. $L_1 = 7.6557$, the novel chaotic system is highly chaotic. Since the sum of the Lyapunov exponents is negative, the novel chaotic system is dissipative. Then we derived new results for the global chaos control of the novel highly chaotic system with unknown parameters using adaptive control method. We also derived new results for the global chaos synchronization of the identical novel highly chaotic systems with unknown parameters using adaptive control method. The main control results have been proved using Lyapunov stability theory. MATLAB simulations were shown to illustrate all the results derived in this work.

References

1. Abdurrahman A, Jiang H, Teng Z (2015) Finite-time synchronization for memristor-based neural networks with time-varying delays. Neural Netw 69:20–28
2. Arneodo A, Coullet P, Tresser C (1981) Possible new strange attractors with spiral structure. Commun Math Phys 79(4):573–576
3. Azar AT, Vaidyanathan S (2015) Chaos modeling and control systems design, vol 581. Springer, Cham
4. Behnia S, Afrang S, Akhshani A, Mabhouti K (2013) A novel method for controlling chaos in external cavity semiconductor laser. Optik 124(8):757–764
5. Cai G, Tan Z (2007) Chaos synchronization of a new chaotic system via nonlinear control. J Uncertain Syst 1(3):235–240
6. Chen G, Ueta T (1999) Yet another chaotic attractor. Int J Bifur Chaos 9(7):1465–1466
7. Islam MM, Murase K (2005) Chaotic dynamics of a behaviour-based miniature mobile robot: effects of environment and control structure. Neural Netw 18(2):123–144
8. Karthikeyan R, Sundarapandian V (2014) Hybrid chaos synchronization of four-scroll systems via active control. J Electr Eng 65(2):97–103
9. Khalil HK (2001) Nonlinear Syst, 3rd edn. Prentice Hall, New Jersey
10. Li D (2008) A three-scroll chaotic attractor. Phys Lett A 372(4):387–393
11. Lorenz EN (1963) Deterministic periodic flow. J Atmos Sci 20(2):130–141
12. Lü J, Chen G (2002) A new chaotic attractor coined. Int J Bifur Chaos 12(3):659–661
13. Matouk AE (2011) Chaos, feedback control and synchronization of a fractional-order modified autonomous Van der Pol-Duffing circuit. Commun Nonlinear Sci Numer Simul 16(2): 975–986
14. Njah AN, Sunday OD (2009) Generalization on the chaos control of 4-D chaotic systems using recursive backstepping nonlinear controller. Chaos, Solitons Fractals 41(5):2371–2376
15. Pehlivan I, Moroz IM, Vaidyanathan S (2014) Analysis, synchronization and circuit design of a novel butterfly attractor. J Sound Vib 333(20):5077–5096
16. Pham VT, Volos CK, Vaidyanathan S, Le TP, Vu VY (2015) A memristor-based hyperchaotic system with hidden attractors: dynamics, synchronization and circuital emulating. J Eng Sci Technol Rev 8(2):205–214
17. Rasappan S, Vaidyanathan S (2013) Hybrid synchronization of n-scroll Chua circuits using adaptive backstepping control design with recursive feedback. Malays J Math Sci 73(1):73–95
18. Rasappan S, Vaidyanathan S (2014) Global chaos synchronization of WINDMI and Coullet chaotic systems using adaptive backstepping control design. Kyungpook Math J 54(1): 293–320
19. Rhouma R, Belghith S (2011) Cryptanalysis of a chaos-based cryptosystem. Commun Nonlinear Sci Numer Simul 16(2):876–884
20. Rössler OE (1976) An equation for continuous chaos. Phys Lett A 57(5):397–398
21. Sampath S, Vaidyanathan S, Volos CK, Pham VT (2015) An eight-term novel four-scroll chaotic system with cubic nonlinearity and its circuit simulation. J Eng Sci Technol Rev 8(2):1–6
22. Sarasu P, Sundarapandian V (2011) Active controller design for generalized projective synchronization of four-scroll chaotic systems. Int J Syst Signal Control Eng Appl 4(2):26–33
23. Sarasu P, Sundarapandian V (2011) The generalized projective synchronization of hyperchaotic Lorenz and hyperchaotic Qi systems via active control. Int J Soft Comput 6(5):216–223
24. Sarasu P, Sundarapandian V (2012) Generalized projective synchronization of two-scroll systems via adaptive control. Int J Soft Comput 7(4):146–156
25. Sprott JC (1994) Some simple chaotic flows. Phys Rev E 50(2):647–650
26. Sundarapandian V (2010) Output regulation of the Lorenz attractor. Far East J Math Sci 42(2):289–299
27. Sundarapandian V (2013) Adaptive control and synchronization design for the Lu-Xiao chaotic system. Lect Notes Electr Eng 131:319–327

28. Sundarapandian V (2013) Analysis and anti-synchronization of a novel chaotic system via active and adaptive controllers. J Eng Sci Technol Rev 6(4):45–52
29. Sundarapandian V, Karthikeyan R (2011) Anti-synchronization of hyperchaotic Lorenz and hyperchaotic Chen systems by adaptive control. Int J Syst Signal Control Eng Appl 4(2):18–25
30. Sundarapandian V, Karthikeyan R (2011) Anti-synchronization of Lü and Pan chaotic systems by adaptive nonlinear control. Eur J Sci Res 64(1):94–106
31. Sundarapandian V, Karthikeyan R (2012) Adaptive anti-synchronization of uncertain Tigan and Li systems. J Eng Appl Sci 7(1):45–52
32. Sundarapandian V, Pehlivan I (2012) Analysis, control, synchronization, and circuit design of a novel chaotic system. Math Comput Model 55(7–8):1904–1915
33. Sundarapandian V, Sivaperumal S (2011) Sliding controller design of hybrid synchronization of four-wing chaotic systems. Int J Soft Comput 6(5):224–231
34. Suresh R, Sundarapandian V (2013) Global chaos synchronization of a family of n-scroll hyperchaotic Chua circuits using backstepping control with recursive feedback. Far East J Math Sci 7(2):219–246
35. Tigan G, Opris D (2008) Analysis of a 3D chaotic system. Chaos, Solitons Fractals 36: 1315–1319
36. Tuwankotta JM (2006) Chaos in a coupled oscillators system with widely spaced frequencies and energy-preserving non-linearity. Int J Non-Linear Mech 41(2):180–191
37. Usama M, Khan MK, Alghathbar K, Lee C (2010) Chaos-based secure satellite imagery cryptosystem. Comput Math Appl 60(2):326–337
38. Vaidyanathan S (2011) Output regulation of Arneodo-Coullet chaotic system. Commun Comput Inf Sci 133:98–107
39. Vaidyanathan S (2011) Output regulation of the unified chaotic system. Commun Comput Inf Sci 198:1–9
40. Vaidyanathan S (2012) Adaptive controller and syncrhonizer design for the Qi-Chen chaotic system. Advances in computer science and information technology. Computer science and engineering, vol 84, Lecture notes of the institute for computer sciences, social-informatics and telecommunications engineeringSpringer, Berlin, pp 73–82
41. Vaidyanathan S (2012) Anti-synchronization of Sprott-L and Sprott-M chaotic systems via adaptive control. Int J Control Theory Appl 5(1):41–59
42. Vaidyanathan S (2012) Global chaos control of hyperchaotic Liu system via sliding control method. Int J Control Theory Appl 5(2):117–123
43. Vaidyanathan S (2012) Sliding mode control based global chaos control of Liu-Liu-Liu-Su chaotic system. Int J Control Theory Appl 5(1):15–20
44. Vaidyanathan S (2013) A new six-term 3-D chaotic system with an exponential nonlinearity. Far East J Math Sci 79(1):135–143
45. Vaidyanathan S (2013) A ten-term novel 4-D hyperchaotic system with three quadratic nonlinearities and its control. Int J Control Theory Appl 6(2):97–109
46. Vaidyanathan S (2013) Analysis and adaptive synchronization of two novel chaotic systems with hyperbolic sinusoidal and cosinusoidal nonlinearity and unknown parameters. J Eng Sci Technol Rev 6(4):53–65
47. Vaidyanathan S (2013) Analysis, control and synchronization of hyperchaotic Zhou system via adaptive control. Adv Intell Syst Comput 177:1–10
48. Vaidyanathan S (2014) A new eight-term 3-D polynomial chaotic system with three quadratic nonlinearities. Far East J Math Sci 84(2):219–226
49. Vaidyanathan S (2014) Analysis and adaptive synchronization of eight-term 3-D polynomial chaotic systems with three quadratic nonlinearities. Eur Phys J: Special Top 223(8): 1519–1529
50. Vaidyanathan S (2014) Analysis, control and synchronisation of a six-term novel chaotic system with three quadratic nonlinearities. Int J Model, Ident Control 22(1):41–53
51. Vaidyanathan S (2014) Generalised projective synchronisation of novel 3-D chaotic systems with an exponential non-linearity via active and adaptive control. Int J Model, Identif Control 22(3):207–217

52. Vaidyanathan S (2014) Global chaos synchronisation of identical Li-Wu chaotic systems via sliding mode control. Int J Model, Identif Control 22(2):170–177
53. Vaidyanathan S (2014) Qualitative analysis and control of an eleven-term novel 4-D hyperchaotic system with two quadratic nonlinearities. Int J Control Theory Appl 7:35–47
54. Vaidyanathan S (2015) 3-cells cellular neural network (CNN) attractor and its adaptive biological control. Int J PharmTech Res 8(4):632–640
55. Vaidyanathan S (2015) A 3-D novel highly chaotic system with four quadratic nonlinearities, its adaptive control and anti-synchronization with unknown parameters. J Eng Sci Technol Rev 8(2):106–115
56. Vaidyanathan S (2015) A novel chemical chaotic reactor system and its adaptive control. Int J ChemTech Res 8(7):146–158
57. Vaidyanathan S (2015) Adaptive backstepping control of enzymes-substrates system with ferroelectric behaviour in brain waves. Int J PharmTech Res 8(2):256–261
58. Vaidyanathan S (2015) Adaptive biological control of generalized Lotka-Volterra three-species biological system. Int J PharmTech Res 8(4):622–631
59. Vaidyanathan S (2015) Adaptive chaotic synchronization of enzymes-substrates system with ferroelectric behaviour in brain waves. Int J PharmTech Res 8(5):964–973
60. Vaidyanathan S (2015) Adaptive control of a chemical chaotic reactor. Int J PharmTech Res 8(3):377–382
61. Vaidyanathan S (2015) Adaptive control of the FitzHugh-Nagumo chaotic neuron model. Int PharmTech Res 8(6):117–127
62. Vaidyanathan S (2015) Adaptive synchronization of chemical chaotic reactors. Int J ChemTech Res 8(2):612–621
63. Vaidyanathan S (2015) Adaptive synchronization of generalized Lotka-Volterra three-species biological systems. Int J PharmTech Res 8(5):928–937
64. Vaidyanathan S (2015) Adaptive synchronization of novel 3-D chemical chaotic reactor systems. Int J ChemTech Res 8(7):159–171
65. Vaidyanathan S (2015) Adaptive synchronization of the identical FitzHugh-Nagumo chaotic neuron models. Int J PharmTech Res 8(6):167–177
66. Vaidyanathan S (2015) Analysis, control and synchronization of a 3-D novel jerk chaotic system with two quadratic nonlinearities. Kyungpook Math J 55:563–586
67. Vaidyanathan S (2015) Analysis, properties and control of an eight-term 3-D chaotic system with an exponential nonlinearity. Int J Model, Identif Control 23(2):164–172
68. Vaidyanathan S (2015) Anti-synchronization of brusselator chemical reaction systems via adaptive control. Int J ChemTech Res 8(6):759–768
69. Vaidyanathan S (2015) Chaos in neurons and adaptive control of Birkhoff-Shaw strange chaotic attractor. Int J PharmTech Res 8(5):956–963
70. Vaidyanathan S (2015) Chaos in neurons and synchronization of Birkhoff-Shaw strange chaotic attractors via adaptive control. Int J PharmTech Res 8(6):1–11
71. Vaidyanathan S (2015) Coleman-Gomatam logarithmic competitive biology models and their ecological monitoring. Int J PharmTech Res 8(6):94–105
72. Vaidyanathan S (2015) Dynamics and control of brusselator chemical reaction. Int J ChemTech Res 8(6):740–749
73. Vaidyanathan S (2015) Dynamics and control of tokamak system with symmetric and magnetically confined plasma. Int J ChemTech Res 8(6):795–803
74. Vaidyanathan S (2015) Global chaos synchronization of chemical chaotic reactors via novel sliding mode control method. Int J ChemTech Res 8(7):209–221
75. Vaidyanathan S (2015) Global chaos synchronization of the forced Van der Pol chaotic oscillators via adaptive control method. Inte J PharmTech Res 8(6):156–166
76. Vaidyanathan S (2015) Global chaos synchronization of the Lotka-Volterra biological systems with four competitive species via active control. Int J PharmTech Res 8(6):206–217
77. Vaidyanathan S (2015) Lotka-Volterra population biology models with negative feedback and their ecological monitoring. Int J PharmTech Res 8(5):974–981

78. Vaidyanathan S (2015) Lotka-Volterra two species competitive biology models and their ecological monitoring. Int J PharmTech Res 8(6):32–44
79. Vaidyanathan S (2015) Output regulation of the forced Van der Pol chaotic oscillator via adaptive control method. Int J PharmTech Res 8(6):106–116
80. Vaidyanathan S, Azar AT (2015) Analysis and control of a 4-D novel hyperchaotic system. Stud Comput Intell 581:3–17
81. Vaidyanathan S, Azar AT (2015) Analysis, control and synchronization of a nine-term 3-D novel chaotic system. In: Azar AT, Vaidyanathan S (eds) Chaos Modelling and Control Systems Design, Studies in Computational Intelligence, vol 581. Springer, Cham, pp 19–38
82. Vaidyanathan S, Madhavan K (2013) Analysis, adaptive control and synchronization of a seven-term novel 3-D chaotic system. Int J Control Theory Appl 6(2):121–137
83. Vaidyanathan S, Pakiriswamy S (2013) Generalized projective synchronization of six-term Sundarapandian chaotic systems by adaptive control. Int J Control Theory Appl 6(2):153–163
84. Vaidyanathan S, Pakiriswamy S (2015) A 3-D novel conservative chaotic system and its generalized projective synchronization via adaptive control. J Eng Sci Technol Rev 8(2):52–60
85. Vaidyanathan S, Rajagopal K (2011) Hybrid synchronization of hyperchaotic Wang-Chen and hyperchaotic Lorenz systems by active non-linear control. Int J Syst Signal Control Eng Appl 4(3):55–61
86. Vaidyanathan S, Rajagopal K (2012) Global chaos synchronization of hyperchaotic Pang and hyperchaotic Wang systems via adaptive control. Int J Soft Comput 7(1):28–37
87. Vaidyanathan S, Rasappan S (2011) Global chaos synchronization of hyperchaotic Bao and Xu systems by active nonlinear control. Commun Comput Inf Sci 198:10–17
88. Vaidyanathan S, Rasappan S (2014) Global chaos synchronization of n-scroll Chua circuit and Lur'e system using backstepping control design with recursive feedback. Arabian J Sci Eng 39(4):3351–3364
89. Vaidyanathan S, Sampath S (2012) Anti-synchronization of four-wing chaotic systems via sliding mode control. Int J Autom Comput 9(3):274–279
90. Vaidyanathan S, Volos C (2015) Analysis and adaptive control of a novel 3-D conservative no-equilibrium chaotic system. Arch Control Sci 25(3):333–353
91. Vaidyanathan S, Volos C, Pham VT (2014) Hyperchaos, adaptive control and synchronization of a novel 5-D hyperchaotic system with three positive Lyapunov exponents and its SPICE implementation. Arch Control Sci 24(4):409–446
92. Vaidyanathan S, Volos C, Pham VT, Madhavan K, Idowu BA (2014) Adaptive backstepping control, synchronization and circuit simulation of a 3-D novel jerk chaotic system with two hyperbolic sinusoidal nonlinearities. Arch Control Sci 24(3):375–403
93. Vaidyanathan S, Azar AT, Rajagopal K, Alexander P (2015) Design and SPICE implementation of a 12-term novel hyperchaotic system and its synchronisation via active control. Int J Model, Identif Control 23(3):267–277
94. Vaidyanathan S, Idowu BA, Azar AT (2015) Backstepping controller design for the global chaos synchronization of Sprott's jerk systems. Stud Comput Intell 581:39–58
95. Vaidyanathan S, Rajagopal K, Volos CK, Kyprianidis IM, Stouboulos IN (2015) Analysis, adaptive control and synchronization of a seven-term novel 3-D chaotic system with three quadratic nonlinearities and its digital implementation in LabVIEW. J Eng Sci Technol Rev 8(2):130–141
96. Vaidyanathan S, Sampath S, Azar AT (2015) Global chaos synchronisation of identical chaotic systems via novel sliding mode control method and its application to Zhu system. Int J Model, Identif Control 23(1):92–100
97. Vaidyanathan S, Volos C, Pham VT, Madhavan K (2015) Analysis, adaptive control and synchronization of a novel 4-D hyperchaotic hyperjerk system and its SPICE implementation. Nonlinear Dyn 25(1):135–158
98. Vaidyanathan S, Volos CK, Kyprianidis IM, Stouboulos IN, Pham VT (2015) Analysis, adaptive control and anti-synchronization of a six-term novel jerk chaotic system with two exponential nonlinearities and its circuit simulation. J Eng Sci Technol Rev 8(2):24–36

99. Vaidyanathan S, Volos CK, Pham VT (2015) Analysis, adaptive control and adaptive synchronization of a nine-term novel 3-D chaotic system with four quadratic nonlinearities and its circuit simulation. J Eng Sci Technol Rev 8(2):181–191
100. Vaidyanathan S, Volos CK, Pham VT (2015) Analysis, control, synchronization and SPICE implementation of a novel 4-D hyperchaotic Rikitake dynamo system without equilibrium. J Eng Sci Technol Rev 8(2):232–244
101. Vaidyanathan S, Volos CK, Pham VT (2015) Global chaos control of a novel nine-term chaotic system via sliding mode control. In: Azar AT, Zhu Q (eds) Advances and Applications in Sliding Mode Control Systems, Studies in Computational Intelligence, vol 576. Springer, Cham, pp 571–590
102. Vincent UE, Njah AN, Laoye JA (2007) Controlling chaos and deterministic directed transport in inertia ratchets using backstepping control. Phys D 231(2):130–136
103. Volos CK, Kyprianidis IM, Stouboulos IN, Anagnostopoulos AN (2009) Experimental study of the dynamic behavior of a double scroll circuit. J Appl Funct Anal 4:703–711
104. Volos CK, Kyprianidis IM, Stouboulos IN (2013) Experimental investigation on coverage performance of a chaotic autonomous mobile robot. Robot Autonom Syst 61(12):1314–1322
105. Volos CK, Kyprianidis IM, Stouboulos IN, Tlelo-Cuautle E, Vaidyanathan S (2015) Memristor: a new concept in synchronization of coupled neuromorphic circuits. J Eng Sci Technol Rev 8(2):157–173
106. Yang J, Zhu F (2013) Synchronization for chaotic systems and chaos-based secure communications via both reduced-order and step-by-step sliding mode observers. Commun Nonlinear Sci Numer Simul 18(4):926–937
107. Yang J, Chen Y, Zhu F (2014) Singular reduced-order observer-based synchronization for uncertain chaotic systems subject to channel disturbance and chaos-based secure communication. Appl Math Comput 229:227–238
108. Zhou W, Xu Y, Lu H, Pan L (2008) On dynamics analysis of a new chaotic attractor. Phys Lett A 372(36):5773–5777
109. Zhu C, Liu Y, Guo Y (2010) Theoretic and numerical study of a new chaotic system. Intell Inf Manage 2:104–109

Qualitative Analysis and Adaptive Control of a Novel 4-D Hyperchaotic System

Sundarapandian Vaidyanathan

Abstract In this work, we announce a nine-term novel 4-D hyperchaotic system with two quadratic nonlinearities. The phase portraits of the nine-term novel hyperchaotic system are depicted and the qualitative properties of the novel hyperchaotic system are discussed. The novel hyperchaotic system has a unique equilibrium at the origin, which is a saddle point. The Lyapunov exponents of the novel hyperchaotic system are obtained as $L_1 = 5.3131, L_2 = 0.1122, L_3 = 0$ and $L_4 = -38.3607$. Since the maximal Lyapunov exponent of the novel hyperchaotic system has a high value, viz. $L_1 = 5.3131$, the system shows highly hyperchaotic behavior. Also, the Kaplan–Yorke dimension of the novel hyperchaotic system is obtained as $D_{KY} = 3.1414$. Since the sum of the Lyapunov exponents is negative, the novel hyperchaotic system is dissipative. Next, an adaptive controller is designed to globally stabilize the novel hyperchaotic system with unknown parameters. Finally, an adaptive controller is also designed to achieve global chaos synchronization of the identical hyperchaotic systems with unknown parameters. MATLAB simulations are depicted to illustrate all the main results derived in this work.

Keywords Chaos · Chaotic systems · Hyperchaos · Hyperchaotic systems · Adaptive control · Synchronization

1 Introduction

In the last few decades, Chaos theory has become a very important and active research field, employing many applications in different disciplines like physics, chemistry, biology, ecology, engineering and economics, among others [3].

S. Vaidyanathan (✉)
Research and Development Centre, Vel Tech University,
Avadi, Chennai 600062, Tamil Nadu, India
e-mail: sundarvtu@gmail.com

S. Vaidyanathan and C. Volos (eds.), *Advances and Applications in Chaotic Systems*, Studies in Computational Intelligence 636,
DOI 10.1007/978-3-319-30279-9_9

Some classical paradigms of 3-D chaotic systems in the literature are Lorenz system [15], Rössler system [23], ACT system [2], Sprott systems [30], Chen system [5], Lü system [16], Cai system [4], Tigan system [40], etc.

Many new chaotic systems have been discovered in the recent years such as Zhou system [111], Zhu system [113], Li system [12], Wei-Yang system [106], Sundarapandian systems [33, 37], Vaidyanathan systems [47, 49, 51–54, 58, 69, 70, 84, 85, 87, 93, 95, 97, 100, 101, 103], Pehlivan system [17], Sampath system [25], etc.

Chaos theory has applications in several fields of science and engineering such as chemical reactors [59, 63, 65, 67, 71, 75–77], biological systems [57, 60–62, 64, 66, 68, 72–74, 78–82], memristors [1, 19, 104], etc.

A hyperchaotic system is defined as a chaotic system with at least two positive Lyapunov exponents [3]. Thus, the dynamics of a hyperchaotic system can expand in several different directions simultaneously. Thus, the hyperchaotic systems have miscellaneous applications in engineering such as secure communications [7, 14, 108], cryptosystems [8, 22, 112], fuzzy logic [29, 110], electrical circuits [107, 109], etc.

The first 4-D hyperchaotic system was found by Rössler [24]. Many hyperchaotic systems have been reported in the chaos literature such as hyperchaotic Lorenz system [9], hyperchaotic Lü system [6], hyperchaotic Chen system [13], hyperchaotic Wang system [105], hyperchaotic Vaidyanathan systems [48, 56, 83, 94, 99, 102], hyperchaotic Pham system [18], etc.

The control of a chaotic or hyperchaotic system aims to stabilize or regulate the system with the help of a feedback control. There are many methods available for controlling a chaotic system such as active control [31, 41, 42], adaptive control [32, 43, 50], sliding mode control [45, 46], backstepping control [96], etc.

The synchronization of chaotic systems aims to synchronize the states of master and slave systems asymptotically with time. There are many methods available for chaos synchronization such as active control [10, 26, 27, 88, 90], adaptive control [28, 34–36, 44, 86, 89], sliding mode control [38, 55, 92, 98], backstepping control [20, 21, 39, 91], etc.

In this research work, we announce a nine-term novel 4-D hyperchaotic system with two quadratic nonlinearities. We have also designed adaptive controllers for stabilization and synchronization of the novel hyperchaotic systems when the system parameters are unknown.

This work is organized as follows. Section 2 describes the dynamics of the novel 4-D hyperchaotic system. Section 3 details the qualitative properties of the novel hyperchaotic system. The Lyapunov exponents of the novel hyperchaotic system are obtained as $L_1 = 5.3131$, $L_2 = 0.1122$, $L_3 = 0$ and $L_4 = -38.3607$, while the Kaplan–Yorke dimension of the novel hyperchaotic system is obtained as $D_{KY} = 3.1414$.

In Sect. 4, we design an adaptive controller to globally stabilize the novel hyperchaotic system with unknown parameters. In Sect. 5, an adaptive controller is designed to achieve global chaos synchronization of the identical novel hyperchaotic systems with unknown parameters. Section 6 summarizes the main results derived in this work.

2 A Novel 4-D Hyperchaotic System

In this section, we describe a nine-term novel hyperchaotic system, which is given by the 4-D dynamics

$$\begin{cases} \dot{x}_1 = a(x_2 - x_1) + x_4 \\ \dot{x}_2 = bx_1 - x_1x_3 \\ \dot{x}_3 = x_1x_2 - cx_3 \\ \dot{x}_4 = -p(x_1 + x_2) \end{cases} \tag{1}$$

where x_1, x_2, x_3, x_4 are the states and a, b, c, p are constant positive parameters.

The novel 4-D system (1) is a nine-term polynomial system with two quadratic nonlinearities.

The system (1) exhibits a *strange hyperchaotic attractor* for the parameter values

$$a = 30, \quad b = 340, \quad c = 3, \quad p = 6 \tag{2}$$

For numerical simulations, we take the initial conditions as

$$x_1(0) = 2.8, \quad x_2(0) = 2.4, \quad x_3(0) = 2.5, \quad x_4(0) = 2.6 \tag{3}$$

Figures 1, 2, 3 and 4 show the 3-D projection of the novel hyperchaotic system (1) on the (x_1, x_2, x_3), (x_1, x_2, x_4), (x_1, x_3, x_4) and (x_2, x_3, x_4) spaces, respectively.

3 Analysis of the Novel 4-D Hyperchaotic System

In this section, we give a dynamic analysis of the 4-D novel hyperchaotic system (1). We take the parameter values as in the hyperchaotic case (2).

3.1 Dissipativity

In vector notation, the novel hyperchaotic system (1) can be expressed as

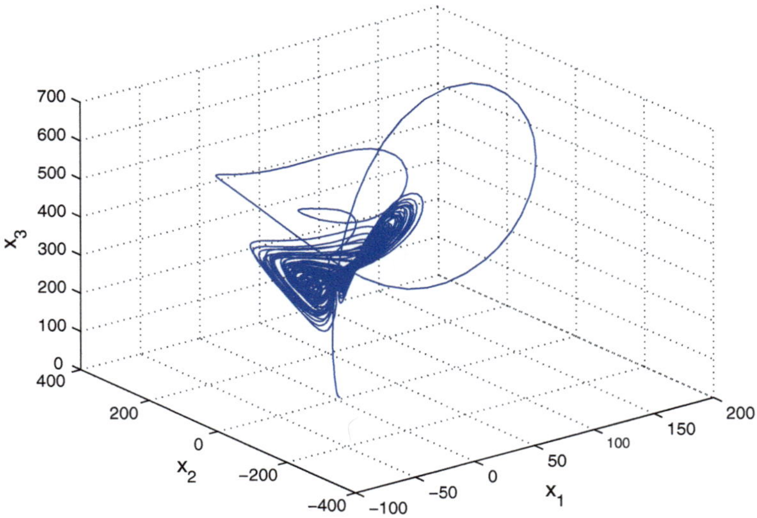

Fig. 1 3-D projection of the novel hyperchaotic system on the (x_1, x_2, x_3) space

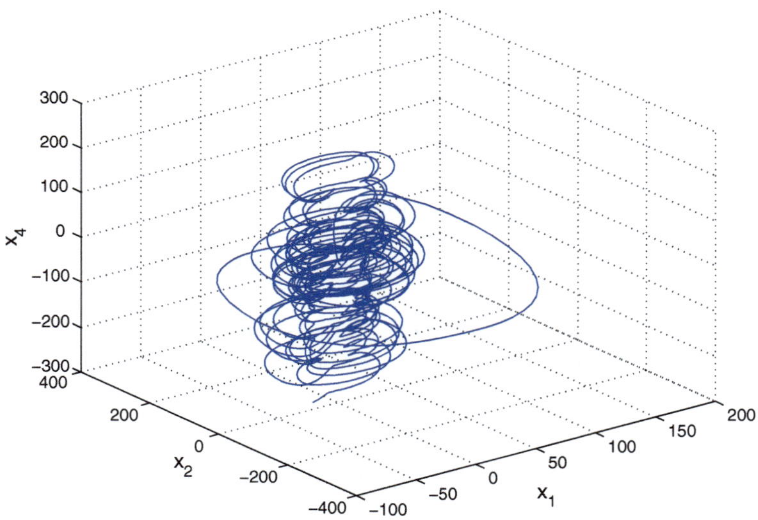

Fig. 2 3-D projection of the novel hyperchaotic system on the (x_1, x_2, x_4) space

$$\dot{\mathbf{x}} = f(\mathbf{x}) = \begin{bmatrix} f_1(x_1, x_2, x_3, x_4) \\ f_2(x_1, x_2, x_3, x_4) \\ f_3(x_1, x_2, x_3, x_4) \\ f_4(x_1, x_2, x_3, x_4) \end{bmatrix}, \tag{4}$$

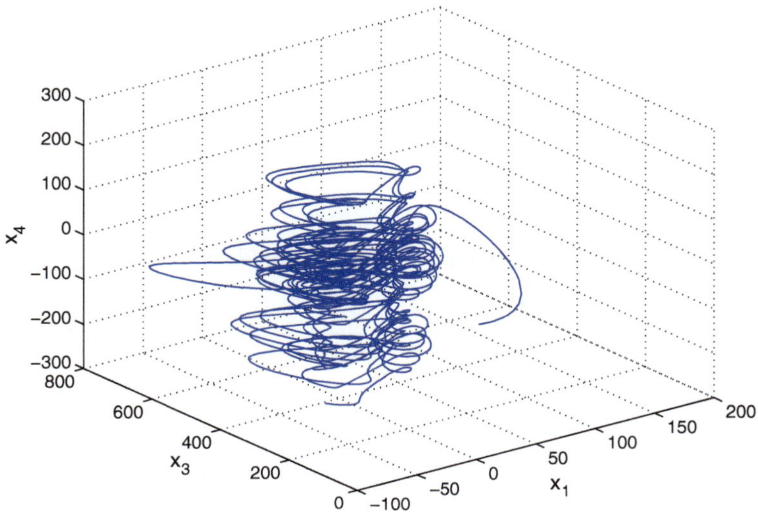

Fig. 3 3-D projection of the novel hyperchaotic system on the (x_1, x_3, x_4) space

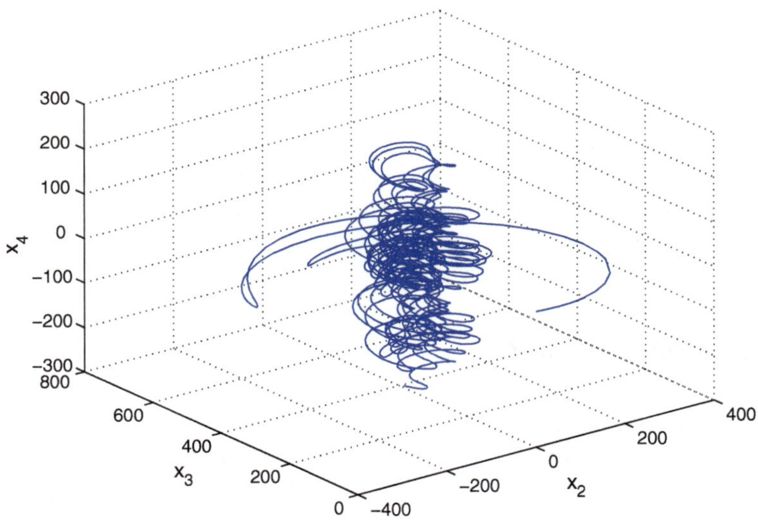

Fig. 4 3-D projection of the novel hyperchaotic system on the (x_2, x_3, x_4) space

where

$$\begin{cases} f_1(x_1, x_2, x_3, x_4) = a(x_2 - x_1) + x_4 \\ f_2(x_1, x_2, x_3, x_4) = bx_1 - x_1x_3 \\ f_3(x_1, x_2, x_3, x_4) = x_1x_2 - cx_3 \\ f_4(x_1, x_2, x_3, x_4) = -p(x_1 + x_2) \end{cases} \tag{5}$$

Let Ω be any region in $\mathbf{R}^4$ with a smooth boundary and also, $\Omega(t) = \Phi_t(\Omega)$, where Φ_t is the flow of f. Furthermore, let $V(t)$ denote the hypervolume of $\Omega(t)$.

By Liouville's theorem, we know that

$$\dot{V}(t) = \int\limits_{\Omega(t)} (\nabla \cdot f)\, dx_1\, dx_2\, dx_3\, dx_4 \tag{6}$$

The divergence of the novel hyperchaotic system (4) is found as:

$$\nabla \cdot f = \frac{\partial f_1}{\partial x_1} + \frac{\partial f_2}{\partial x_2} + \frac{\partial f_3}{\partial x_3} + \frac{\partial f_4}{\partial x_4} = -(a + c) = -\mu < 0 \tag{7}$$

Inserting the value of $\nabla \cdot f$ from (7) into (6), we get

$$\dot{V}(t) = \int\limits_{\Omega(t)} (-\mu)\, dx_1\, dx_2\, dx_3\, dx_4 = -\mu V(t) \tag{8}$$

Integrating the first order linear differential equation (8), we get

$$V(t) = \exp(-\mu t) V(0) \tag{9}$$

Since $\mu > 0$, it follows from Eq. (9) that $V(t) \to 0$ exponentially as $t \to \infty$. This shows that the novel hyperchaotic system (1) is dissipative. Hence, the system limit sets are ultimately confined into a specific limit set of zero hypervolume, and the asymptotic motion of the novel hyperchaotic system (1) settles onto a strange attractor of the system.

3.2 Equilibrium Points

We take the parameter values as in the hyperchaotic case (2).

It is easy to see that the system (1) has a unique equilibrium at the origin.

To test the stability type of the equilibrium point $E_0 = \mathbf{0}$, we calculate the Jacobian matrix of the novel hyperchaotic system (1) at $\mathbf{x} = \mathbf{0}$:

We find that

$$J \triangleq J(E_0) = \begin{bmatrix} -30 & 30 & 0 & 1 \\ 340 & 0 & 0 & 0 \\ 0 & 0 & -3 & 0 \\ -6 & -6 & 0 & 0 \end{bmatrix} \tag{10}$$

The matrix J has the eigenvalues

$$\lambda_1 = -117.1588, \quad \lambda_2 = -3, \quad \lambda_3 = 0.2002, \quad \lambda_4 = 86.9585 \tag{11}$$

This shows that the equilibrium point $E_0 = \mathbf{0}$ is a saddle-point, which is unstable.

3.3 Rotation Symmetry About the x_3-Axis

It is easy to see that the novel 4-D hyperchaotic system (1) is invariant under the change of coordinates

$$(x_1, x_2, x_3, x_4) \mapsto (-x_1, -x_2, x_3, -x_4) \tag{12}$$

Since the transformation (12) persists for all values of the system parameters, it follows that the novel 4-D hyperchaotic system (1) has rotation symmetry about the x_3-axis and that any non-trivial trajectory must have a twin trajectory.

3.4 Invariance

It is easy to see that the x_3-axis is invariant under the flow of the 4-D novel hyperchaotic system (1). The invariant motion along the x_3-axis is characterized by the scalar dynamics

$$\dot{x}_3 = -cx_3, \quad (c > 0) \tag{13}$$

which is globally exponentially stable.

3.5 Lyapunov Exponents and Kaplan–Yorke Dimension

We take the parameter values of the novel system (1) as in the hyperchaotic case (2), i.e. $a = 30$, $b = 340$, $c = 3$ and $p = 6$.

We take the initial state of the novel system (1) as given in (3).

Then the Lyapunov exponents of the system (1) are numerically obtained using MATLAB as

$$L_1 = 5.3131, \quad L_2 = 0.1122, \quad L_3 = 0, \quad L_4 = -38.3607 \tag{14}$$

Since there are two positive Lyapunov exponents in (14), the novel system (1) exhibits *hyperchaotic* behavior.

Since the maximal Lyapunov exponent of the system (1) has a high value, viz. $L_1 = 5.3131$, the system is highly hyperchaotic.

Since $L_1 + L_2 + L_3 + L_4 = -32.9354 < 0$, it follows that the novel hyperchaotic system (1) is dissipative.

Also, the Kaplan–Yorke dimension of the novel hyperchaotic system (1) is calculated as

$$D_{KY} = 3 + \frac{L_1 + L_2 + L_3}{|L_4|} = 3.1414, \tag{15}$$

which is fractional.

4 Adaptive Control of the Novel Hyperchaotic System

In this section, we use adaptive control method to derive an adaptive feedback control law for globally stabilizing the novel 4-D hyperchaotic system with unknown parameters.

Thus, we consider the novel 4-D hyperchaotic system given by

$$\begin{cases} \dot{x}_1 = a(x_2 - x_1) + x_4 + u_1 \\ \dot{x}_2 = bx_1 - x_1x_3 + u_2 \\ \dot{x}_3 = x_1x_2 - cx_3 + u_3 \\ \dot{x}_4 = -p(x_1 + x_2) + u_4 \end{cases} \tag{16}$$

In (16), x_1, x_2, x_3, x_4 are the states and u_1, u_2, u_3, u_4 are the adaptive controls to be determined using estimates $\hat{a}(t), \hat{b}(t), \hat{c}(t), \hat{p}(t)$ for the unknown parameters a, b, c, p, respectively.

We consider the adaptive feedback control law

$$\begin{cases} u_1 = -\hat{a}(t)(x_2 - x_1) - x_4 - k_1x_1 \\ u_2 = -\hat{b}(t)x_1 + x_1x_3 - k_2x_2 \\ u_3 = -x_1x_2 + \hat{c}(t)x_3 - k_3x_3 \\ u_4 = \hat{p}(t)(x_1 + x_2) - k_4x_4 \end{cases} \tag{17}$$

where k_1, k_2, k_3, k_4 are positive gain constants.

Substituting (17) into (16), we get the closed-loop plant dynamics as

$$\begin{cases} \dot{x}_1 = [a - \hat{a}(t)](x_2 - x_1) - k_1x_1 \\ \dot{x}_2 = [b - \hat{b}(t)]x_1 - k_2x_2 \\ \dot{x}_3 = -[c - \hat{c}(t)]x_3 - k_3x_3 \\ \dot{x}_4 = -[p - \hat{p}(t)](x_1 + x_2) - k_4x_4 \end{cases} \tag{18}$$

The parameter estimation errors are defined as

$$\begin{cases} e_a(t) = a - \hat{a}(t) \\ e_b(t) = b - \hat{b}(t) \\ e_c(t) = c - \hat{c}(t) \\ e_p(t) = p - \hat{p}(t) \end{cases} \tag{19}$$

In view of (19), we can simplify the plant dynamics (18) as

$$\begin{cases} \dot{x}_1 = e_a(x_2 - x_1) - k_1 x_1 \\ \dot{x}_2 = e_b x_1 - k_2 x_2 \\ \dot{x}_3 = -e_c x_3 - k_3 x_3 \\ \dot{x}_4 = -e_p(x_1 + x_2) - k_4 x_4 \end{cases} \tag{20}$$

Differentiating (19) with respect to t, we obtain

$$\begin{cases} \dot{e}_a(t) = -\dot{\hat{a}}(t) \\ \dot{e}_b(t) = -\dot{\hat{b}}(t) \\ \dot{e}_c(t) = -\dot{\hat{c}}(t) \\ \dot{e}_p(t) = -\dot{\hat{p}}(t) \end{cases} \tag{21}$$

We consider the quadratic candidate Lyapunov function defined by

$$V(\mathbf{x}, e_a, e_b, e_c, e_p) = \frac{1}{2}\left(x_1^2 + x_2^2 + x_3^2 + x_4^2\right) + \frac{1}{2}\left(e_a^2 + e_b^2 + e_c^2 + e_p^2\right) \tag{22}$$

Differentiating V along the trajectories of (20) and (21), we obtain

$$\begin{aligned} \dot{V} = & -k_1 x_1^2 - k_2 x_2^2 - k_3 x_3^2 - k_4 x_4^2 + e_a\left[x_1(x_2 - x_1) - \dot{\hat{a}}\right] \\ & + e_b\left[x_1 x_2 - \dot{\hat{b}}\right] + e_c\left[-x_3^2 - \dot{\hat{c}}\right] + e_p\left[-x_4(x_1 + x_2) - \dot{\hat{p}}\right] \end{aligned} \tag{23}$$

In view of (23), we take the parameter update law as

$$\begin{cases} \dot{\hat{a}}(t) = x_1(x_2 - x_1) \\ \dot{\hat{b}}(t) = x_1 x_2 \\ \dot{\hat{c}}(t) = -x_3^2 \\ \dot{\hat{p}}(t) = -x_4(x_1 + x_2) \end{cases} \tag{24}$$

Next, we state and prove the main result of this section.

Theorem 1 *The novel 4-D hyperchaotic system (16) with unknown system parameters is globally and exponentially stabilized for all initial conditions by the adaptive control law (17) and the parameter update law (24), where k_1, k_2, k_3, k_4 are positive gain constants.*

Proof We prove this result by applying Lyapunov stability theory [11].

We consider the quadratic Lyapunov function defined by (22), which is clearly a positive definite function on $\mathbf{R}^8$.

By substituting the parameter update law (24) into (23), we obtain the time-derivative of V as

$$\dot{V} = -k_1 x_1^2 - k_2 x_2^2 - k_3 x_3^2 - k_4 x_4^2 \tag{25}$$

From (25), it is clear that $\dot{V}$ is a negative semi-definite function on $\mathbf{R}^8$.

Thus, we can conclude that the state vector $\mathbf{x}(t)$ and the parameter estimation error are globally bounded, i.e.

$$\left[x_1(t)\ x_2(t)\ x_3(t)\ x_4(t)\ e_a(t)\ e_b(t)\ e_c(t)\ e_p(t)\right]^T \in \mathbf{L}_\infty.$$

We define $k = \min\{k_1, k_2, k_3, k_4\}$.

Then it follows from (25) that

$$\dot{V} \leq -k\|\mathbf{x}(t)\|^2 \tag{26}$$

Thus, we have

$$k\|\mathbf{x}(t)\|^2 \leq -\dot{V} \tag{27}$$

Integrating the inequality (27) from 0 to t, we get

$$k\int_0^t \|\mathbf{x}(\tau)\|^2\, d\tau \ \leq\ V(0) - V(t) \tag{28}$$

From (28), it follows that $\mathbf{x} \in \mathbf{L}_2$.

Using (20), we can conclude that $\dot{\mathbf{x}} \in \mathbf{L}_\infty$.

Using Barbalat's lemma [11], we conclude that $\mathbf{x}(t) \to 0$ exponentially as $t \to \infty$ for all initial conditions $\mathbf{x}(0) \in \mathbf{R}^4$.

This completes the proof. ■

For the numerical simulations, the classical fourth-order Runge–Kutta method with step size $h = 10^{-8}$ is used to solve the systems (16) and (24), when the adaptive control law (17) is applied.

The parameter values of the novel 4-D hyperchaotic system (16) are taken as in the hyperchaotic case (2), i.e.

$$a = 30, \quad b = 340, \quad c = 3, \quad p = 6 \tag{29}$$

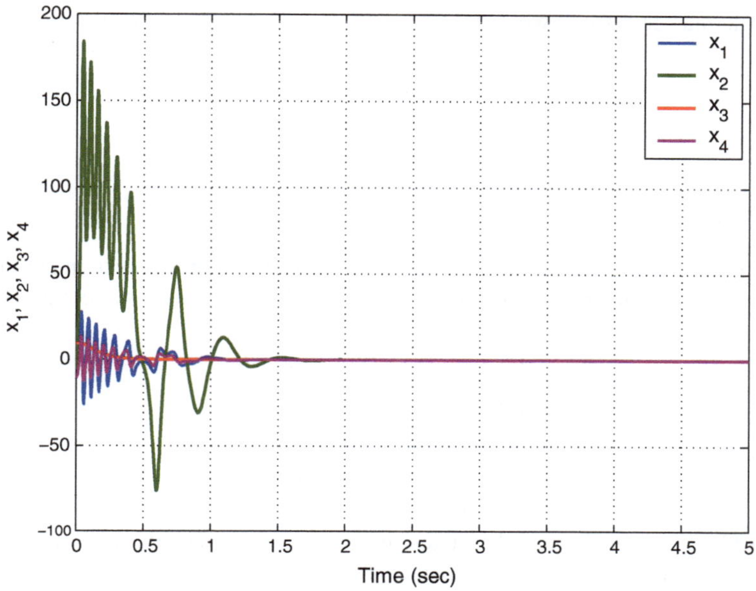

Fig. 5 Time-history of the controlled states x_1, x_2, x_3, x_4

We take the positive gain constants as $k_i = 6$ for $i = 1, \ldots, 4$.

Furthermore, as initial conditions of the novel 4-D hyperchaotic system (16), we take

$$x_1(0) = 7.5, \quad x_2(0) = -6.2, \quad x_3(0) = 9.3, \quad x_4(0) = -10.8 \tag{30}$$

Also, as initial conditions of the parameter estimates, we take

$$\hat{a}(0) = 1.4, \quad \hat{b}(0) = 20.3, \quad \hat{c}(0) = 10.5, \quad \hat{p}(0) = 15.7 \tag{31}$$

In Fig. 5, the exponential convergence of the controlled states of the novel 4-D hyperchaotic system (16) is shown.

5 Adaptive Synchronization of the Identical Novel Hyperchaotic Systems

In this section, we use adaptive control method to derive an adaptive feedback control law for globally synchronizing identical novel 4-D hyperchaotic systems with unknown parameters.

As the master system, we consider the novel 4-D hyperchaotic system given by

$$\begin{cases} \dot{x}_1 = a(x_2 - x_1) + x_4 \\ \dot{x}_2 = bx_1 - x_1x_3 \\ \dot{x}_3 = x_1x_2 - cx_3 \\ \dot{x}_4 = -p(x_1 + x_2) \end{cases} \tag{32}$$

In (32), x_1, x_2, x_3, x_4 are the states and a, b, c, p are unknown system parameters. As the slave system, we consider the 4-D novel hyperchaotic system given by

$$\begin{cases} \dot{y}_1 = a(y_2 - y_1) + y_4 + u_1 \\ \dot{y}_2 = by_1 - y_1y_3 + u_2 \\ \dot{y}_3 = y_1y_2 - cy_3 + u_3 \\ \dot{y}_4 = -p(y_1 + y_2) + u_4 \end{cases} \tag{33}$$

In (33), y_1, y_2, y_3, y_4 are the states and u_1, u_2, u_3, u_4 are the adaptive controls to be determined using estimates $\hat{a}(t), \hat{b}(t), \hat{c}(t), \hat{p}(t)$ for the unknown parameters a, b, c, p, respectively.

The synchronization error between the novel hyperchaotic systems (32) and (33) is defined by

$$\begin{cases} e_1 = y_1 - x_1 \\ e_2 = y_2 - x_2 \\ e_3 = y_3 - x_3 \\ e_4 = y_4 - x_4 \end{cases} \tag{34}$$

Then the synchronization error dynamics is obtained as

$$\begin{cases} \dot{e}_1 = a(e_2 - e_1) + e_4 + u_1 \\ \dot{e}_2 = be_1 - y_1y_3 + x_1x_3 + u_2 \\ \dot{e}_3 = -ce_3 + y_1y_2 - x_1x_2 + u_3 \\ \dot{e}_4 = -p(e_1 + e_2) + u_4 \end{cases} \tag{35}$$

We consider the adaptive feedback control law

$$\begin{cases} u_1 = -\hat{a}(t)(e_2 - e_1) - e_4 - k_1e_1 \\ u_2 = -\hat{b}(t)e_1 + y_1y_3 - x_1x_3 - k_2e_2 \\ u_3 = \hat{c}(t)e_3 - y_1y_2 + x_1x_2 - k_3e_3 \\ u_4 = \hat{p}(t)(e_1 + e_2) - k_4e_4 \end{cases} \tag{36}$$

where k_1, k_2, k_3, k_4 are positive gain constants.

Substituting (36) into (35), we get the closed-loop error dynamics as

$$\begin{cases} \dot{e}_1 = \left[a - \hat{a}(t)\right](e_2 - e_1) - k_1 e_1 \\ \dot{e}_2 = \left[b - \hat{b}(t)\right] e_1 - k_2 e_2 \\ \dot{e}_3 = -\left[c - \hat{c}(t)\right] e_3 - k_3 e_3 \\ \dot{e}_4 = -\left[p - \hat{p}(t)\right](e_1 + e_2) - k_4 e_4 \end{cases} \tag{37}$$

The parameter estimation errors are defined as

$$\begin{cases} e_a(t) = a - \hat{a}(t) \\ e_b(t) = b - \hat{b}(t) \\ e_c(t) = c - \hat{c}(t) \\ e_p(t) = p - \hat{p}(t) \end{cases} \tag{38}$$

In view of (38), we can simplify the error dynamics (37) as

$$\begin{cases} \dot{e}_1 = e_a(e_2 - e_1) - k_1 e_1 \\ \dot{e}_2 = e_b e_1 - k_2 e_2 \\ \dot{e}_3 = -e_c e_3 - k_3 e_3 \\ \dot{e}_4 = -e_p(e_1 + e_2) - k_4 e_4 \end{cases} \tag{39}$$

Differentiating (38) with respect to t, we obtain

$$\begin{cases} \dot{e}_a(t) = -\dot{\hat{a}}(t) \\ \dot{e}_b(t) = -\dot{\hat{b}}(t) \\ \dot{e}_c(t) = -\dot{\hat{c}}(t) \\ \dot{e}_p(t) = -\dot{\hat{p}}(t) \end{cases} \tag{40}$$

We use adaptive control theory to find an update law for the parameter estimates. We consider the quadratic candidate Lyapunov function defined by

$$V(\mathbf{e}, e_a, e_b, e_c, e_p) = \frac{1}{2}\left(e_1^2 + e_2^2 + e_3^2 + e_4^2\right) + \frac{1}{2}\left(e_a^2 + e_b^2 + e_c^2 + e_p^2\right) \tag{41}$$

Differentiating V along the trajectories of (39) and (40), we obtain

$$\begin{aligned} \dot{V} = & -k_1 e_1^2 - k_2 e_2^2 - k_3 e_3^2 - k_4 e_4^2 + e_a\left[e_1(e_2 - e_1) - \dot{\hat{a}}\right] \\ & + e_b\left[e_1 e_2 - \dot{\hat{b}}\right] + e_c\left[-e_3^2 - \dot{\hat{c}}\right] + e_p\left[-e_4(e_1 + e_2) - \dot{\hat{p}}\right] \end{aligned} \tag{42}$$

In view of (42), we take the parameter update law as

$$\begin{cases} \dot{\hat{a}}(t) = e_1(e_2 - e_1) \\ \dot{\hat{b}}(t) = e_1 e_2 \\ \dot{\hat{c}}(t) = -e_3^2 \\ \dot{\hat{p}}(t) = -e_4(e_1 + e_2) \end{cases} \tag{43}$$

Next, we state and prove the main result of this section.

Theorem 2 *The novel hyperchaotic systems (32) and (33) with unknown system parameters are globally and exponentially synchronized for all initial conditions by the adaptive control law (36) and the parameter update law (43), where* k_1, k_2, k_3, k_4 *are positive gain constants.*

Proof We prove this result by applying Lyapunov stability theory [11].

We consider the quadratic Lyapunov function defined by (41), which is clearly a positive definite function on $\mathbf{R}^8$.

By substituting the parameter update law (43) into (42), we obtain

$$\dot{V} = -k_1 e_1^2 - k_2 e_2^2 - k_3 e_3^2 - k_4 e_4^2 \tag{44}$$

From (44), it is clear that $\dot{V}$ is a negative semi-definite function on $\mathbf{R}^8$.

Thus, we can conclude that the error vector $\mathbf{e}(t)$ and the parameter estimation error are globally bounded, i.e.

$$\left[e_1(t)\ e_2(t)\ e_3(t)\ e_4(t)\ e_a(t)\ e_b(t)\ e_c(t)\ e_p(t)\right]^T \in \mathbf{L}_\infty. \tag{45}$$

We define $k = \min\{k_1, k_2, k_3, k_4\}$.

Then it follows from (44) that

$$\dot{V} \leq -k\|\mathbf{e}(t)\|^2 \tag{46}$$

Thus, we have

$$k\|\mathbf{e}(t)\|^2 \leq -\dot{V} \tag{47}$$

Integrating the inequality (47) from 0 to t, we get

$$k\int_0^t \|\mathbf{e}(\tau)\|^2\, d\tau \leq V(0) - V(t) \tag{48}$$

From (48), it follows that $\mathbf{e} \in \mathbf{L}_2$.

Using (39), we can conclude that $\dot{\mathbf{e}} \in \mathbf{L}_\infty$.

Using Barbalat's lemma [11], we conclude that $\mathbf{e}(t) \to 0$ exponentially as $t \to \infty$ for all initial conditions $\mathbf{e}(0) \in \mathbf{R}^4$.

This completes the proof. ∎

For the numerical simulations, the classical fourth-order Runge–Kutta method with step size $h = 10^{-8}$ is used to solve the systems (32), (33) and (43), when the adaptive control law (36) is applied.

The parameter values of the novel hyperchaotic systems are taken as in the hyperchaotic case (2), i.e. $a = 30$, $b = 340$, $c = 3$ and $p = 6$.

We take the positive gain constants as $k_i = 6$ for $i = 1, \ldots, 4$.

Furthermore, as initial conditions of the master system (32), we take

$$x_1(0) = -15.3, \quad x_2(0) = 2.8, \quad x_3(0) = -6.7, \quad x_4(0) = 12.9 \tag{49}$$

As initial conditions of the slave system (33), we take

$$y_1(0) = 25.4, \quad y_2(0) = -3.5, \quad y_3(0) = 4.2, \quad y_4(0) = -9.4 \tag{50}$$

Also, as initial conditions of the parameter estimates, we take

$$\hat{a}(0) = 12.1, \quad \hat{b}(0) = 10.4, \quad \hat{c}(0) = 14.7, \quad \hat{p}(0) = 25.8 \tag{51}$$

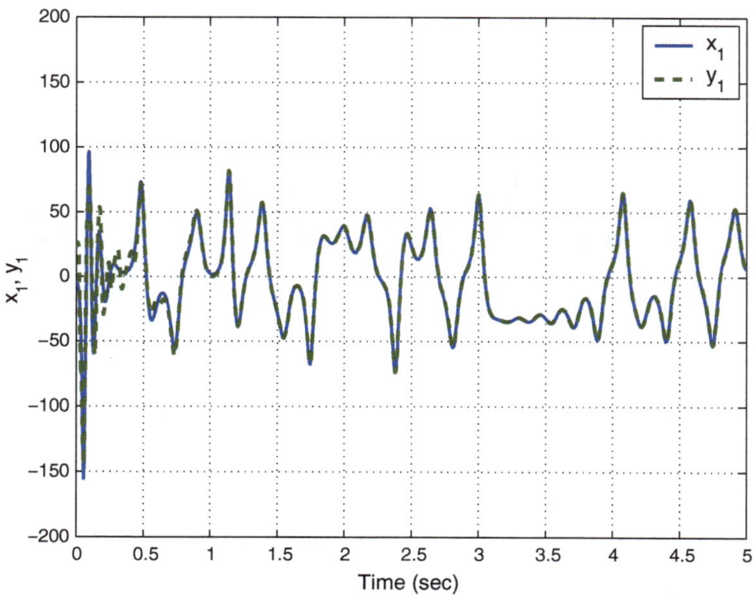

Fig. 6 Synchronization of the states x_1 and y_1

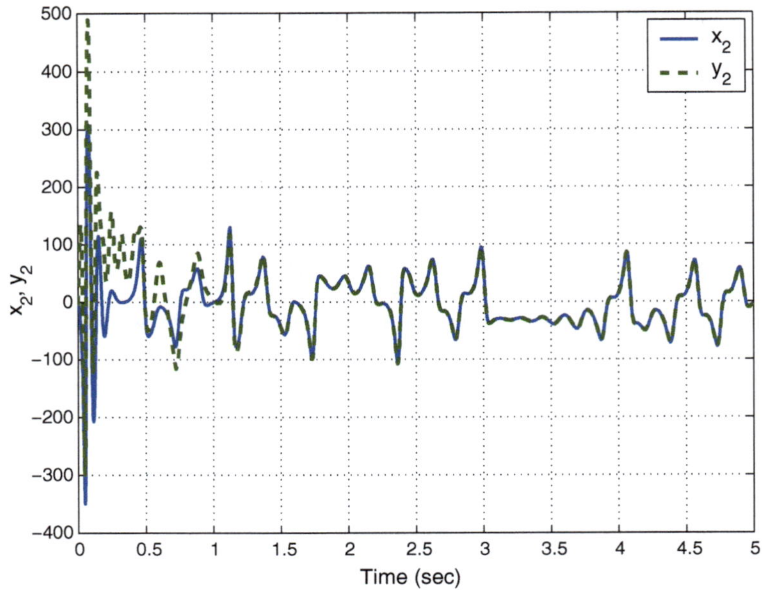

Fig. 7 Synchronization of the states x_2 and y_2

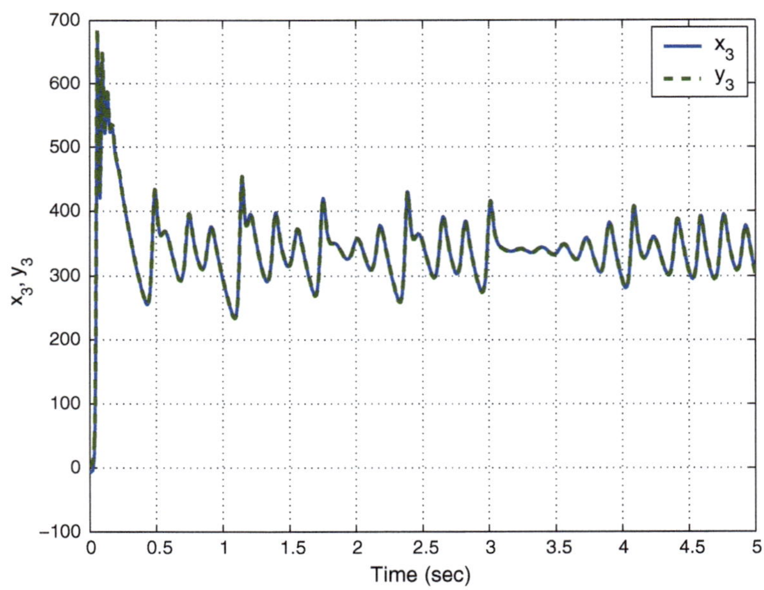

Fig. 8 Synchronization of the states x_3 and y_3

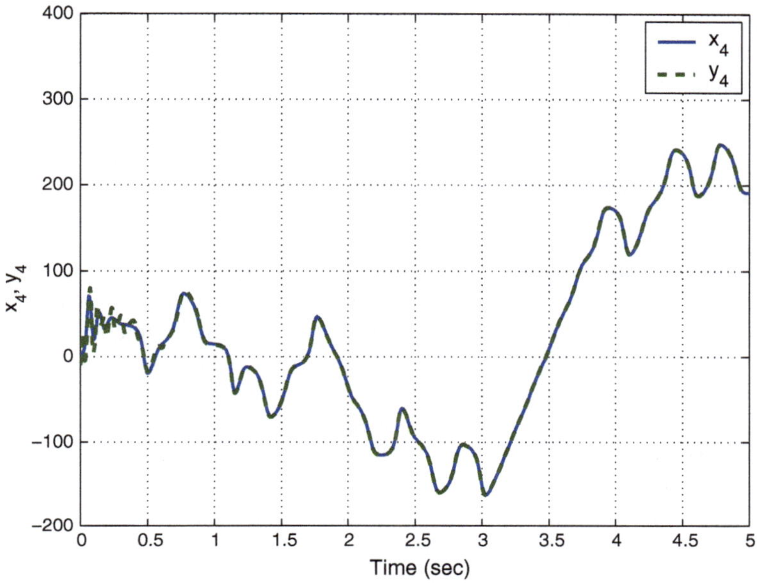

Fig. 9 Synchronization of the states x_4 and y_4

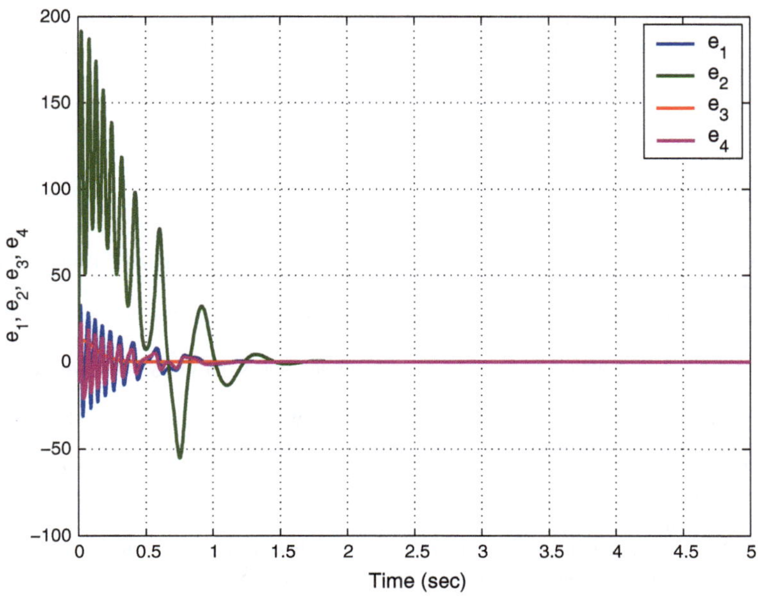

Fig. 10 Time-history of the synchronization errors e_1, e_2, e_3, e_4

Figures 6, 7, 8 and 9 describe the complete synchronization of the novel hyperchaotic systems (32) and (33), while Fig. 10 describes the time-history of the synchronization errors e_1, e_2, e_3, e_4.

6 Conclusions

In this work, we described a nine-term novel 4-D hyperchaotic system with two quadratic nonlinearities. We discussed the qualitative properties of the novel hyperchaotic system in detail. The novel hyperchaotic system has a unique equilibrium at the origin, which is a saddle-point and unstable. The novel hyperchaotic system has a rotation symmetry about the x_3-axis. The x_3 axis is an invariant manifold for the novel hyperchaotic system. The Lyapunov exponents of the novel hyperchaotic system have been obtained as $L_1 = 5.3131$, $L_2 = 0.1122$, $L_3 = 0$ and $L_4 = -38.3607$. Since the maximal Lyapunov exponent of the novel hyperchaotic system has a high value, viz. $L_1 = 5.3131$, the system shows highly hyperchaotic behavior. Also, the Kaplan–Yorke dimension of the novel hyperchaotic system is obtained as $D_{KY} = 3.1414$. Since the sum of the Lyapunov exponents is negative, the novel hyperchaotic system is dissipative. Next, an adaptive controller was designed to globally stabilize the novel hyperchaotic system with unknown parameters. Finally, an adaptive controller was also designed to achieve global chaos synchronization of the identical hyperchaotic systems with unknown parameters. MATLAB simulations were shown to depict all the main results derived in this work.

References

1. Abdurrahman A, Jiang H, Teng Z (2015) Finite-time synchronization for memristor-based neural networks with time-varying delays. Neural Netw 69:20–28
2. Arneodo A, Coullet P, Tresser C (1981) Possible new strange attractors with spiral structure. Commun Math Phys 79(4):573–576
3. Azar AT, Vaidyanathan S (2015) Chaos Modeling and Control Systems Design, vol 581. Springer, Germany
4. Cai G, Tan Z (2007) Chaos synchronization of a new chaotic system via nonlinear control. J Uncertain Syst 1(3):235–240
5. Chen G, Ueta T (1999) Yet another chaotic attractor. Int J Bifurc Chaos 9(7):1465–1466
6. Chen A, Lu J, Lü J, Yu S (2006) Generating hyperchaotic Lü attractor via state feedback control. Phys A 364:103–110
7. Filali RL, Benrejeb M, Borne P (2014) On observer-based secure communication design using discrete-time hyperchaotic systems. Commun Nonlinear Sci Numer Simul 19(5):1424–1432
8. Hammami S (2015) State feedback-based secure image cryptosystem using hyperchaotic synchronization. ISA Trans 54:52–59
9. Jia Q (2007) Hyperchaos generated from the Lorenz chaotic system and its control. Phys Lett A 366:217–222
10. Karthikeyan R, Sundarapandian V (2014) Hybrid chaos synchronization of four-scroll systems via active control. J Electr Eng 65(2):97–103

11. Khalil HK (2001) Nonlinear Syst, 3rd edn. Prentice Hall, New Jersey
12. Li D (2008) A three-scroll chaotic attractor. Phys Lett A 372(4):387–393
13. Li X (2009) Modified projective synchronization of a new hyperchaotic system via nonlinear control. Commun Theor Phys 52:274–278
14. Li C, Liao X, Wong KW (2005) Lag synchronization of hyperchaos with application to secure communications. Chaos, Solitons Fractals 23(1):183–193
15. Lorenz EN (1963) Deterministic periodic flow. J Atmos Sci 20(2):130–141
16. Lü J, Chen G (2002) A new chaotic attractor coined. Int J Bifurc and Chaos 12(3):659–661
17. Pehlivan I, Moroz IM, Vaidyanathan S (2014) Analysis, synchronization and circuit design of a novel butterfly attractor. J Sound Vib 333(20):5077–5096
18. Pham VT, Volos C, Jafari S, Wang X, Vaidyanathan S (2014) Hidden hyperchaotic attractor in a novel simple memristive neural network. Optoelectron Adv Mater Rapid Commun 8(11–12):1157–1163
19. Pham VT, Volos CK, Vaidyanathan S, Le TP, Vu VY (2015) A memristor-based hyperchaotic system with hidden attractors: dynamics, synchronization and circuital emulating. J Eng Sci Technol Rev 8(2):205–214
20. Rasappan S, Vaidyanathan S (2013) Hybrid synchronization of n-scroll Chua circuits using adaptive backstepping control design with recursive feedback. Malays J Math Sci 73(1):73–95
21. Rasappan S, Vaidyanathan S (2014) Global chaos synchronization of WINDMI and Coullet chaotic systems using adaptive backstepping control design. Kyungpook Math J 54(1):293–320
22. Rhouma R, Belghith S (2008) Cryptanalysis of a new image encryption algorithm based on hyper-chaos. Phys Lett A 372(38):5973–5978
23. Rössler OE (1976) An equation for continuous chaos. Phys Lett A 57(5):397–398
24. Rössler OE (1979) An equation for hyperchaos. Phys Lett A 71:155–157
25. Sampath S, Vaidyanathan S, Volos CK, Pham VT (2015) An eight-term novel four-scroll chaotic System with cubic nonlinearity and its circuit simulation. J Eng Sci Technol Rev 8(2):1–6
26. Sarasu P, Sundarapandian V (2011) Active controller design for generalized projective synchronization of four-scroll chaotic systems. Int J Syst Signal Control Eng Appl 4(2):26–33
27. Sarasu P, Sundarapandian V (2011) The generalized projective synchronization of hyperchaotic Lorenz and hyperchaotic Qi systems via active control. Int J Soft Comput 6(5):216–223
28. Sarasu P, Sundarapandian V (2012) Generalized projective synchronization of two-scroll systems via adaptive control. Int J Soft Comput 7(4):146–156
29. Senouci A, Boukabou A (2014) Predictive control and synchronization of chaotic and hyperchaotic systems based on a $T-S$ fuzzy model. Math Comput Simul 105:62–78
30. Sprott JC (1994) Some simple chaotic flows. Phys Rev E 50(2):647–650
31. Sundarapandian V (2010) Output regulation of the Lorenz attractor. Far East J Math Sci 42(2):289–299
32. Sundarapandian V (2013) Adaptive control and synchronization design for the Lu-Xiao chaotic system. Lect Notes Electr Eng 131:319–327
33. Sundarapandian V (2013) Analysis and anti-synchronization of a novel chaotic system via active and adaptive controllers. J Eng Sci Technol Rev 6(4):45–52
34. Sundarapandian V, Karthikeyan R (2011) Anti-synchronization of hyperchaotic Lorenz and hyperchaotic Chen systems by adaptive control. Int J Syst Signal Control Eng Appl 4(2):18–25
35. Sundarapandian V, Karthikeyan R (2011) Anti-synchronization of Lü and Pan chaotic systems by adaptive nonlinear control. Eur J Sci Res 64(1):94–106
36. Sundarapandian V, Karthikeyan R (2012) Adaptive anti-synchronization of uncertain Tigan and Li systems. J Eng Appl Sci 7(1):45–52
37. Sundarapandian V, Pehlivan I (2012) Analysis, control, synchronization, and circuit design of a novel chaotic system. Math Comput Model 55(7–8):1904–1915
38. Sundarapandian V, Sivaperumal S (2011) Sliding controller design of hybrid synchronization of four-wing chaotic systems. Int J Soft Comput 6(5):224–231

39. Suresh R, Sundarapandian V (2013) Global chaos synchronization of a family of n-scroll hyperchaotic Chua circuits using backstepping control with recursive feedback. Far East J Math Sci 7(2):219–246
40. Tigan G, Opris D (2008) Analysis of a 3D chaotic system. Chaos, Solitons Fractals 36:1315–1319
41. Vaidyanathan S (2011) Output regulation of Arneodo-Coullet chaotic system. Commun Comput Inform Sci 133:98–107
42. Vaidyanathan S (2011) Output regulation of the unified chaotic system. Commun Comput Inform Sci 198:1–9
43. Vaidyanathan S (2012) Adaptive controller and syncrhonizer design for the Qi-Chen chaotic system. Lect Notes Inst Comput Sci Social-Inform Telecommun Eng 84:73–82
44. Vaidyanathan S (2012) Anti-synchronization of Sprott-L and Sprott-M chaotic systems via adaptive control. Int J Control Theory Appl 5(1):41–59
45. Vaidyanathan S (2012) Global chaos control of hyperchaotic Liu system via sliding control method. Int J Control Theory Appl 5(2):117–123
46. Vaidyanathan S (2012) Sliding mode control based global chaos control of Liu-Liu-Liu-Su chaotic system. Int J Control Theory Appl 5(1):15–20
47. Vaidyanathan S (2013) A new six-term 3-D chaotic system with an exponential nonlinearity. Far East J Math Sci 79(1):135–143
48. Vaidyanathan S (2013) A ten-term novel 4-D hyperchaotic system with three quadratic nonlinearities and its control. Int J Control Theory Appl 6(2):97–109
49. Vaidyanathan S (2013) Analysis and adaptive synchronization of two novel chaotic systems with hyperbolic sinusoidal and cosinusoidal nonlinearity and unknown parameters. J Eng Sci Technol Rev 6(4):53–65
50. Vaidyanathan S (2013) Analysis, control and synchronization of hyperchaotic Zhou system via adaptive control. Adv Intell Syst Comput 177:1–10
51. Vaidyanathan S (2014) A new eight-term 3-D polynomial chaotic system with three quadratic nonlinearities. Far East J Math Sci 84(2):219–226
52. Vaidyanathan S (2014) Analysis and adaptive synchronization of eight-term 3-D polynomial chaotic systems with three quadratic nonlinearities. Eur Phys J: Spec Top 223(8):1519–1529
53. Vaidyanathan S (2014) Analysis, control and synchronisation of a six-term novel chaotic system with three quadratic nonlinearities. Int J Model Identif Control 22(1):41–53
54. Vaidyanathan S (2014) Generalized projective synchronisation of novel 3-D chaotic systems with an exponential non-linearity via active and adaptive control. Int J Model Identif Control 22(3):207–217
55. Vaidyanathan S (2014) Global chaos synchronisation of identical Li-Wu chaotic systems via sliding mode control. Int J Model Identif Control 22(2):170–177
56. Vaidyanathan S (2014) Qualitative analysis and control of an eleven-term novel 4-D hyperchaotic system with two quadratic nonlinearities. Int J Control Theory Appl 7:35–47
57. Vaidyanathan S (2015) 3-cells Cellular Neural Network (CNN) attractor and its adaptive biological control. Int J PharmTech Res 8(4):632–640
58. Vaidyanathan S (2015) A 3-D novel highly chaotic system with four quadratic nonlinearities, its adaptive control and anti-synchronization with unknown parameters. J Eng Sci Technol Rev 8(2):106–115
59. Vaidyanathan S (2015) A novel chemical chaotic reactor system and its adaptive control. Int J ChemTech Res 8(7):146–158
60. Vaidyanathan S (2015) Adaptive backstepping control of enzymes-substrates system with ferroelectric behaviour in brain waves. Int J PharmTech Res 8(2):256–261
61. Vaidyanathan S (2015) Adaptive biological control of generalized Lotka-Volterra three-species biological system. Int J PharmTech Res 8(4):622–631
62. Vaidyanathan S (2015) Adaptive chaotic synchronization of enzymes-substrates system with ferroelectric behaviour in brain waves. Int J PharmTech Res 8(5):964–973
63. Vaidyanathan S (2015) Adaptive control of a chemical chaotic reactor. Int J PharmTech Res 8(3):377–382

64. Vaidyanathan S (2015) Adaptive control of the FitzHugh–Nagumo chaotic neuron model. Int J PharmTech Res 8(6):117–127
65. Vaidyanathan S (2015) Adaptive synchronization of chemical chaotic reactors. Int J ChemTech Res 8(2):612–621
66. Vaidyanathan S (2015) Adaptive synchronization of generalized Lotka–Volterra three-species biological systems. Int J PharmTech Res 8(5):928–937
67. Vaidyanathan S (2015) Adaptive synchronization of novel 3-D chemical chaotic reactor systems. Int J ChemTech Res 8(7):159–171
68. Vaidyanathan S (2015) Adaptive synchronization of the identical FitzHugh–Nagumo chaotic neuron models. Int J PharmTech Res 8(6):167–177
69. Vaidyanathan S (2015) Analysis, control and synchronization of a 3-D novel jerk chaotic system with two quadratic nonlinearities. Kyungpook Math J 55:563–586
70. Vaidyanathan S (2015) Analysis, properties and control of an eight-term 3-D chaotic system with an exponential nonlinearity. Int J Model Identif Control 23(2):164–172
71. Vaidyanathan S (2015) Anti-synchronization of brusselator chemical reaction systems via adaptive control. Int J ChemTech Res 8(6):759–768
72. Vaidyanathan S (2015) Chaos in neurons and adaptive control of Birkhoff-Shaw strange chaotic attractor. Int J PharmTech Res 8(5):956–963
73. Vaidyanathan S (2015) Chaos in neurons and synchronization of Birkhoff-Shaw strange chaotic attractors via adaptive control. Int J PharmTech Res 8(6):1–11
74. Vaidyanathan S (2015) Coleman-Gomatam logarithmic competitive biology models and their ecological monitoring. Int J PharmTech Res 8(6):94–105
75. Vaidyanathan S (2015) Dynamics and control of brusselator chemical reaction. Int J ChemTech Res 8(6):740–749
76. Vaidyanathan S (2015) Dynamics and control of tokamak system with symmetric and magnetically confined plasma. Int J ChemTech Res 8(6):795–803
77. Vaidyanathan S (2015) Global chaos synchronization of chemical chaotic reactors via novel sliding mode control method. Int J ChemTech Res 8(7):209–221
78. Vaidyanathan S (2015) Global chaos synchronization of the forced Van der Pol chaotic oscillators via adaptive control method. Int J PharmTech Res 8(6):156–166
79. Vaidyanathan S (2015) Global chaos synchronization of the Lotka-Volterra biological systems with four competitive species via active control. Int J PharmTech Res 8(6):206–217
80. Vaidyanathan S (2015) Lotka-Volterra population biology models with negative feedback and their ecological monitoring. Int J PharmTech Res 8(5):974–981
81. Vaidyanathan S (2015) Lotka-Volterra two species competitive biology models and their ecological monitoring. Int J PharmTech Res 8(6):32–44
82. Vaidyanathan S (2015) Output regulation of the forced Van der Pol chaotic oscillator via adaptive control method. Int J PharmTech Res 8(6):106–116
83. Vaidyanathan S, Azar AT (2015) Analysis and control of a 4-D novel hyperchaotic system. Stud Comput Intell 581:3–17
84. Vaidyanathan S, Azar AT (2015) Analysis, control and synchronization of a nine-term 3-D novel chaotic system. In: Azar AT, Vaidyanathan S (eds) Chaos Modelling and Control Systems Design, Studies in Computational Intelligence, vol 581. Springer, Germany, pp 19–38
85. Vaidyanathan S, Madhavan K (2013) Analysis, adaptive control and synchronization of a seven-term novel 3-D chaotic system. Int J Control Theory Appl 6(2):121–137
86. Vaidyanathan S, Pakiriswamy S (2013) Generalized projective synchronization of six-term Sundarapandian chaotic systems by adaptive control. Int J Control Theory Appl 6(2):153–163
87. Vaidyanathan S, Pakiriswamy S (2015) A 3-D novel conservative chaotic System and its generalized projective synchronization via adaptive control. J Eng Sci Technol Rev 8(2):52–60
88. Vaidyanathan S, Rajagopal K (2011) Hybrid synchronization of hyperchaotic Wang-Chen and hyperchaotic Lorenz systems by active non-linear control. Int J Syst Signal Control Eng Appl 4(3):55–61

89. Vaidyanathan S, Rajagopal K (2012) Global chaos synchronization of hyperchaotic Pang and hyperchaotic Wang systems via adaptive control. Int J Soft Comput 7(1):28–37
90. Vaidyanathan S, Rasappan S (2011) Global chaos synchronization of hyperchaotic Bao and Xu systems by active nonlinear control. Commun Comput Inform Sci 198:10–17
91. Vaidyanathan S, Rasappan S (2014) Global chaos synchronization of n-scroll Chua circuit and Lur'e system using backstepping control design with recursive feedback. Arab J Sci Eng 39(4):3351–3364
92. Vaidyanathan S, Sampath S (2012) Anti-synchronization of four-wing chaotic systems via sliding mode control. Int J Autom Comput 9(3):274–279
93. Vaidyanathan S, Volos C (2015) Analysis and adaptive control of a novel 3-D conservative no-equilibrium chaotic system. Arch Control Sci 25(3):333–353
94. Vaidyanathan S, Volos C, Pham VT (2014) Hyperchaos, adaptive control and synchronization of a novel 5-D hyperchaotic system with three positive Lyapunov exponents and its SPICE implementation. Arch Control Sci 24(4):409–446
95. Vaidyanathan S, Volos C, Pham VT, Madhavan K, Idowu BA (2014) Adaptive backstepping control, synchronization and circuit simulation of a 3-D novel jerk chaotic system with two hyperbolic sinusoidal nonlinearities. Arch Control Sci 24(3):375–403
96. Vaidyanathan S, Idowu BA, Azar AT (2015) Backstepping controller design for the global chaos synchronization of Sprott's jerk systems. Stud Comput Intell 581:39–58
97. Vaidyanathan S, Rajagopal K, Volos CK, Kyprianidis IM, Stouboulos IN (2015) Analysis, adaptive control and synchronization of a seven-term novel 3-D chaotic system with three quadratic nonlinearities and its digital implementation in LabVIEW. J Eng Sci Technol Rev 8(2):130–141
98. Vaidyanathan S, Sampath S, Azar AT (2015) Global chaos synchronisation of identical chaotic systems via novel sliding mode control method and its application to Zhu system. Int J Model Identif Control 23(1):92–100
99. Vaidyanathan S, Volos C, Pham VT, Madhavan K (2015) Analysis, adaptive control and synchronization of a novel 4-D hyperchaotic hyperjerk system and its SPICE implementation. Nonlinear Dyn 25(1):135–158
100. Vaidyanathan S, Volos CK, Kyprianidis IM, Stouboulos IN, Pham VT (2015) Analysis, adaptive control and anti-synchronization of a six-term novel jerk chaotic system with two exponential nonlinearities and its circuit simulation. J Eng Sci Technol Rev 8(2):24–36
101. Vaidyanathan S, Volos CK, Pham VT (2015) Analysis, adaptive control and adaptive synchronization of a nine-term novel 3-D chaotic system with four quadratic nonlinearities and its circuit simulation. J Eng Sci Technol Rev 8(2):181–191
102. Vaidyanathan S, Volos CK, Pham VT (2015) Analysis, control, synchronization and SPICE implementation of a novel 4-D hyperchaotic Rikitake dynamo system without equilibrium. J Eng Sci Technol Rev 8(2):232–244
103. Vaidyanathan S, Volos CK, Pham VT (2015) Global chaos control of a novel nine-term chaotic system via sliding mode control. In: Azar AT, Zhu Q (eds) Advances and Applications in Sliding Mode Control Systems, Studies in Computational Intelligence, vol 576. Springer, Germany, pp 571–590
104. Volos CK, Kyprianidis IM, Stouboulos IN, Tlelo-Cuautle E, Vaidyanathan S (2015) Memristor: a new concept in synchronization of coupled neuromorphic circuits. J Eng Sci Technol Rev 8(2):157–173
105. Wang J, Chen Z (2008) A novel hyperchaotic system and its complex dynamics. Int J Bifurc Chaos 18:3309–3324
106. Wei Z, Yang Q (2010) Anti-control of Hopf bifurcation in the new chaotic system with two stable node-foci. Appl Math Comput 217(1):422–429
107. Wei X, Yunfei F, Qiang L (2012) A novel four-wing hyper-chaotic system and its circuit implementation. Proced Eng 29:1264–1269
108. Wu X, Zhu C, Kan H (2015) An improved secure communication scheme based passive synchronization of hyperchaotic complex nonlinear system. Appl Math Comput 252:201–214

109. Yujun N, Xingyuan W, Mingjun W, Huaguang Z (2010) A new hyperchaotic system and its circuit implementation. Commun Nonlinear Sci Numer Simul 15(11):3518–3524
110. Zhang H, Liao X, Yu J (2005) Fuzzy modeling and synchronization of hyperchaotic systems. Chaos, Solitons Fractals 26(3):835–843
111. Zhou W, Xu Y, Lu H, Pan L (2008) On dynamics analysis of a new chaotic attractor. Phys Lett A 372(36):5773–5777
112. Zhu C (2012) A novel image encryption scheme based on improved hyperchaotic sequences. Opt Commun 285(1):29–37
113. Zhu C, Liu Y, Guo Y (2010) Theoretic and numerical study of a new chaotic system. Intell Inform Manag 2:104–109

Global Chaos Control and Synchronization of a Novel Two-Scroll Chaotic System with Three Quadratic Nonlinearities

Sundarapandian Vaidyanathan

Abstract In this work, we describe a seven-term two-scroll novel chaotic system with three quadratic nonlinearities. The phase portraits of the two-scroll novel chaotic system are illustrated and the dynamic properties of the novel chaotic system are discussed. The novel chaotic system has two unstable equilibrium points. We show that the equilibrium point at the origin is a saddle point, while the other equilibrium point is a saddle-focus. The novel chaotic system has rotation symmetry about the x_3 axis. The Lyapunov exponents of the novel chaotic system are obtained as $L_1 = 3.1464$, $L_2 = 0$ and $L_3 = -24.0635$, while the Kaplan–Yorke dimension of the novel chaotic system is obtained as $D_{KY} = 2.1308$. Since the sum of the Lyapunov exponents is negative, the novel chaotic system is dissipative. Next, we derive new results for the global chaos control of the novel two-scroll chaotic system with unknown parameters using adaptive control method. We also derive new results for the global chaos synchronization of the identical novel two-scroll chaotic systems using adaptive control method. The main control results are established using Lyapunov stability theory. MATLAB simulations are shown to illustrate the phase portraits of the novel two-scroll chaotic system and also the adaptive control results derived in this work.

Keywords Chaos · Chaotic systems · Chaos control · Adaptive control · Synchronization

1 Introduction

In the last few decades, Chaos theory has become a very important and active research field, employing many applications in different disciplines like physics, chemistry, biology, ecology, engineering and economics, among others [2].

S. Vaidyanathan (✉)
Research and Development Centre, Vel Tech University, Avadi, Chennai 600062, Tamil Nadu, India
e-mail: sundarvtu@gmail.com

S. Vaidyanathan and C. Volos (eds.), *Advances and Applications in Chaotic Systems*, Studies in Computational Intelligence 636,
DOI 10.1007/978-3-319-30279-9_10

Some classical paradigms of 3-D chaotic systems in the literature are Lorenz system [9], Rössler system [18], ACT system [1], Sprott systems [23], Chen system [4], Lü system [10], Cai system [3], Tigan system [33], etc.

Many new chaotic systems have been discovered in the recent years such as Zhou system [105], Zhu system [106], Li system [8], Sundarapandian systems [26, 30], Vaidyanathan systems [41, 43, 45–48, 52, 63, 64, 78, 79, 81, 87, 89, 92, 95, 96, 98], Pehlivan system [13], Sampath system [19], etc.

Chaos theory has applications in several fields of science and engineering such as chemical reactors [53, 57, 59, 61, 65, 69–71], biological systems [51, 54–56, 58, 60, 62, 66–68, 72–76], memristors [14, 102], robotics [5, 101], electrical circuits [11, 100], cryptosystems [17, 34], secure communications [103, 104], etc.

The control of a chaotic system aims to stabilize or regulate the system with the help of a feedback control. There are many methods available for controlling a chaotic system such as active control [24, 35, 36], adaptive control [25, 37, 42, 44, 50, 77, 88, 94, 97], sliding mode control [39, 40], backstepping control [12, 91, 99], etc.

Major works on synchronization of chaotic systems deal with the complete synchronization (CS) which has the design goal of using the output of the master system to control the slave system so that the output of the slave system tracks the output of the master system asymptotically with time. Thus, if $\mathbf{x}(t)$ and $\mathbf{y}(t)$ denote the state of the master and slave systems, then the design goal of complete synchronization (CS) is to satisfy the condition

$$\lim_{t \to \infty} \|\mathbf{x}(t) - \mathbf{y}(t)\| = 0 \text{ for all } \mathbf{x}(0), \mathbf{y}(0) \in \mathbf{R}^n \tag{1}$$

There are many methods available for chaos synchronization such as active control [6, 20, 21, 82, 84, 90], adaptive control [22, 27–29, 38, 80, 83], sliding mode control [31, 49, 86, 93], backstepping control [15, 16, 32, 85], etc.

In this research work, we announce a seven-term novel two-scroll chaotic system with three quadratic nonlinearities. Using adaptive control method, we have also derived new results for the global chaos control of the novel two-scroll chaotic system and global chaos synchronization of the identical novel two-scroll chaotic systems when the system parameters are unknown.

This work is organized as follows. Section 2 describes the dynamic equations and phase portraits of the seven-term novel two-scroll chaotic system. Section 3 details the dynamic analysis and properties of the novel two-scroll chaotic system. The Lyapunov exponents of the novel chaotic system are obtained as $L_1 = 3.1464$, $L_2 = 0$ and $L_3 = -24.0635$, while the Kaplan–Yorke dimension of the novel chaotic system is obtained as $D_{KY} = 2.1308$.

In Sect. 4, we derive new results for the global chaos control of the novel two-scroll chaotic system with unknown parameters. In Sect. 5, we derive new results for the global chaos synchronization of the identical novel two-scroll chaotic systems with unknown parameters. Section 6 contains a summary of the main results derived in this work.

2 A Novel 3-D Two-Scroll Chaotic System

In this section, we describe a seven-term novel chaotic system, which is given by the 3-D dynamics

$$\begin{cases} \dot{x}_1 = a(x_2 - x_1) + x_2x_3 \\ \dot{x}_2 = bx_2 - x_1x_3 \\ \dot{x}_3 = x_1x_2 - cx_3 \end{cases} \tag{2}$$

where x_1, x_2, x_3 are the states and a, b, c are constant positive parameters.

The novel 3-D system (2) is a seven-term polynomial system with three quadratic nonlinearities.

The system (2) exhibits a *two-scroll chaotic attractor* for the parameter values

$$a = 36, \quad b = 20, \quad c = 5 \tag{3}$$

For numerical simulations, we take the initial conditions as

$$x_1(0) = 0.2, \quad x_2(0) = 0, \quad x_3(0) = 0.4 \tag{4}$$

Figure 1 depicts the 3-D phase portrait of the novel two-scroll chaotic system (2), while Figs. 2, 3 and 4 depict the 2-D projection of the novel two-scroll chaotic system (2) on the (x_1, x_2), (x_2, x_3) and (x_1, x_3) planes, respectively.

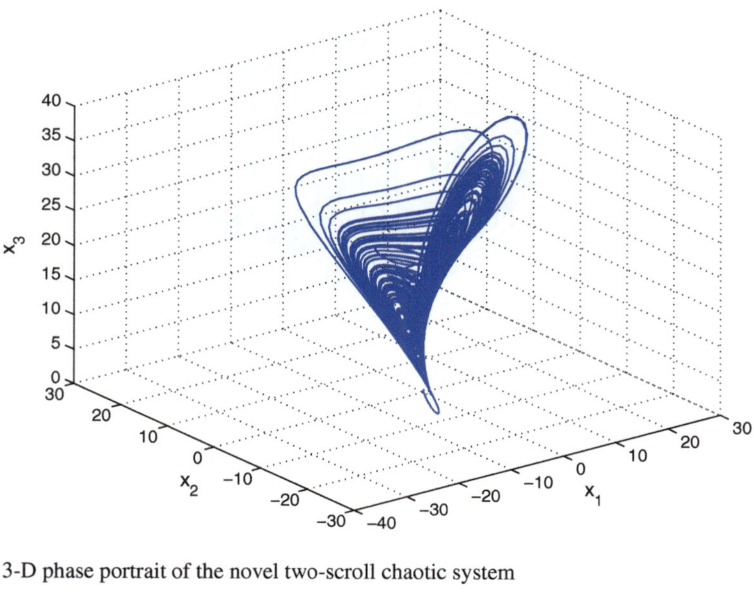

Fig. 1 3-D phase portrait of the novel two-scroll chaotic system

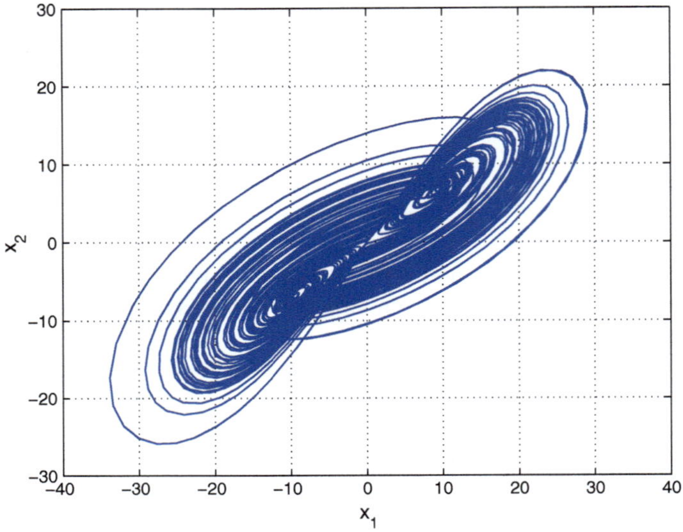

Fig. 2 2-D projection of the novel two-scroll chaotic system on the (x_1, x_2) plane

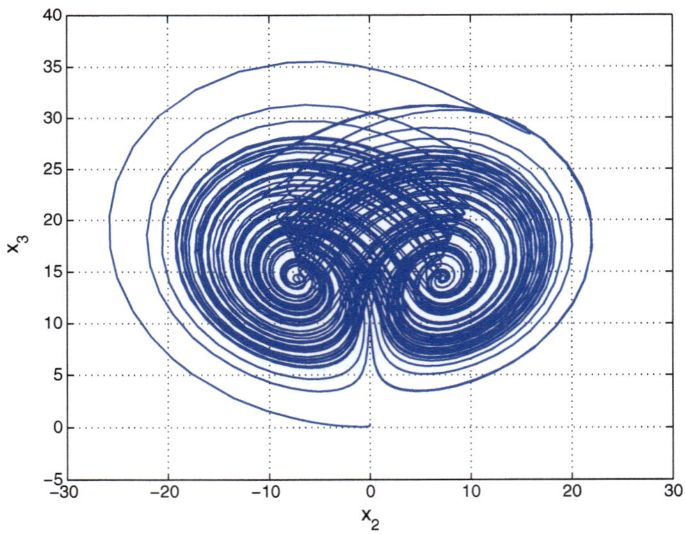

Fig. 3 2-D projection of the novel two-scroll chaotic system on the (x_2, x_3) plane

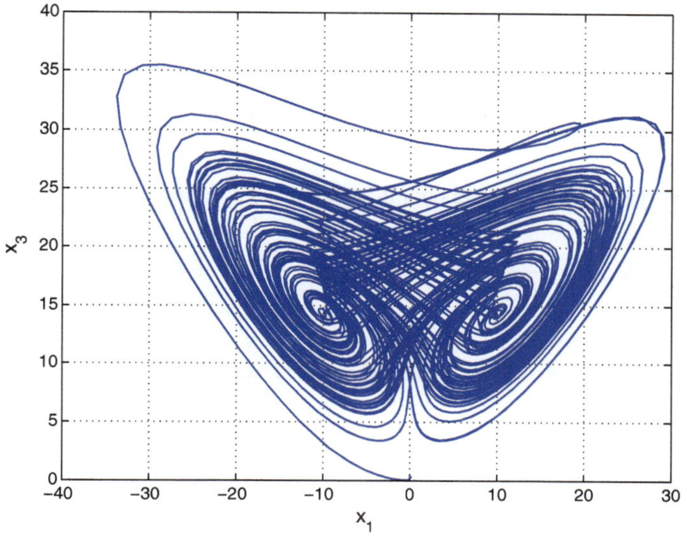

Fig. 4 2-D projection of the novel two-scroll chaotic system on the (x_1, x_3) plane

3 Analysis of the Novel 3-D Two-Scroll Chaotic System

In this section, we give a dynamic analysis of the 3-D novel chaotic system (2). We take the parameter values as in the chaotic case (3), viz. $a = 36$, $b = 20$ and $c = 5$.

3.1 Dissipativity

In vector notation, the novel chaotic system (2) can be expressed as

$$\dot{\mathbf{x}} = f(\mathbf{x}) = \begin{bmatrix} f_1(x_1, x_2, x_3) \\ f_2(x_1, x_2, x_3) \\ f_3(x_1, x_2, x_3) \end{bmatrix}, \tag{5}$$

where

$$\begin{cases} f_1(x_1, x_2, x_3) = a(x_2 - x_1) + x_2 x_3 \\ f_2(x_1, x_2, x_3) = b x_2 - x_1 x_3 \\ f_3(x_1, x_2, x_3) = x_1 x_2 - c x_3 \end{cases} \tag{6}$$

Let Ω be any region in $\mathbf{R}^3$ with a smooth boundary and also, $\Omega(t) = \Phi_t(\Omega)$, where Φ_t is the flow of f. Furthermore, let $V(t)$ denote the volume of $\Omega(t)$.

By Liouville's theorem, we know that

$$\dot{V}(t) = \int\limits_{\Omega(t)} (\nabla \cdot f)\, dx_1\, dx_2\, dx_3 \tag{7}$$

The divergence of the novel chaotic system (5) is found as

$$\nabla \cdot f = \frac{\partial f_1}{\partial x_1} + \frac{\partial f_2}{\partial x_2} + \frac{\partial f_3}{\partial x_3} = -(a - b + c) = -\mu < 0 \tag{8}$$

since $\mu = a - b + c = 21 > 0$.

Inserting the value of $\nabla \cdot f$ from (8) into (7), we get

$$\dot{V}(t) = \int\limits_{\Omega(t)} (-\mu)\, dx_1\, dx_2\, dx_3 = -\mu V(t) \tag{9}$$

Integrating the first order linear differential equation (9), we get

$$V(t) = \exp(-\mu t) V(0) \tag{10}$$

Since $\mu > 0$, it follows from Eq. (10) that $V(t) \to 0$ exponentially as $t \to \infty$. This shows that the novel chaotic system (2) is dissipative.

Hence, the system limit sets are ultimately confined into a specific limit set of zero volume, and the asymptotic motion of the novel chaotic system (2) settles onto a strange attractor of the system.

3.2 Equilibrium Points

We take the parameter values as in the chaotic case (3).

It is easy to see that the system (2) has two equilibrium points, viz.

$$E_0 = \begin{bmatrix} 0 \\ 0 \\ 0 \end{bmatrix} \quad \text{and} \quad E_1 = \begin{bmatrix} 10.0000 \\ 7.1555 \\ 14.3110 \end{bmatrix} \tag{11}$$

The Jacobian of the system (2) at any point $\mathbf{x} \in \mathbf{R}^3$ is calculated as

$$J(\mathbf{x}) = \begin{bmatrix} -a & a + x_3 & x_2 \\ -x_3 & b & -x_1 \\ x_2 & x_1 & -c \end{bmatrix} = \begin{bmatrix} -36 & 36 + x_3 & x_2 \\ -x_3 & 20 & -x_1 \\ x_2 & x_1 & -5 \end{bmatrix} \tag{12}$$

We find that the matrix $J_0 = J(E_0)$ has the eigenvalues

$$\lambda_1 = -5, \quad \lambda_2 = -36, \quad \lambda_3 = 20 \tag{13}$$

This shows that the equilibrium point E_0 is a saddle-point, which is unstable.
We also find that the matrix $J_1 = J(E_1)$ has the eigenvalues

$$\lambda_1 = -28.1171, \quad \lambda_{2,3} = 3.5586 \pm 17.7834i \tag{14}$$

This shows that the equilibrium point E_1 is a saddle-focus, which is unstable.

3.3 Rotation Symmetry About the x_3-Axis

It is easy to see that the system (2) is invariant under the change of coordinates

$$(x_1, x_2, x_3) \mapsto (-x_1, -x_2, x_3) \tag{15}$$

Thus, the 3-D novel chaotic system (2) has rotation symmetry about the x_3-axis and that any non-trivial trajectory must have a twin trajectory.

3.4 Invariance

It is easy to see that the x_3-axis is invariant under the flow of the 3-D novel chaotic system (2). The invariant motion along the x_3-axis is characterized by

$$\dot{x}_3 = -cx_3, \quad (c > 0) \tag{16}$$

which is globally exponentially stable.

3.5 Lyapunov Exponents and Kaplan–Yorke Dimension

We take the parameter values of the novel system (2) as in the chaotic case (3). We take the initial state of the novel system (2) as given in (4).

Then the Lyapunov exponents of the system (2) are numerically obtained as

$$L_1 = 3.1464, \quad L_2 = 0, \quad L_3 = -24.0635 \tag{17}$$

From (17), we note that the Maximal Lyapunov Exponent (MLE) of the novel chaotic system (2) is given by $L_1 = 3.1464$. Since the sum of the Lyapunov exponents in (17) is negative, we conclude that the novel chaotic system (2) is dissipative.

Also, the Kaplan–Yorke dimension of the novel chaotic system (2) is found as

$$D_{KY} = 2 + \frac{L_1 + L_2}{|L_3|} = 2.1308 \tag{18}$$

4 Adaptive Control of the Novel 3-D Two-Scroll Chaotic System

In this section, we use adaptive control method to derive an adaptive feedback control law for globally stabilizing the novel 3-D two-scroll chaotic system with unknown parameters.

Thus, we consider the novel 3-D chaotic system given by

$$\begin{cases} \dot{x}_1 = a(x_2 - x_1) + x_2 x_3 + u_1 \\ \dot{x}_2 = b x_2 - x_1 x_3 + u_2 \\ \dot{x}_3 = x_1 x_2 - c x_3 + u_3 \end{cases} \tag{19}$$

In (19), x_1, x_2, x_3 are the states and u_1, u_2, u_3 are the adaptive controls to be determined using estimates $\hat{a}(t)$, $\hat{b}(t)$, $\hat{c}(t)$ for the unknown parameters a, b, c, respectively.

We consider the adaptive feedback control law

$$\begin{cases} u_1 = -\hat{a}(t)(x_2 - x_1) - x_2 x_3 - k_1 x_1 \\ u_2 = -\hat{b}(t) x_2 + x_1 x_3 - k_2 x_2 \\ u_3 = -x_1 x_2 + \hat{c}(t) x_3 - k_3 x_3 \end{cases} \tag{20}$$

where k_1, k_2, k_3 are positive gain constants.

Substituting (20) into (19), we get the closed-loop plant dynamics as

$$\begin{cases} \dot{x}_1 = [a - \hat{a}(t)](x_2 - x_1) - k_1 x_1 \\ \dot{x}_2 = [b - \hat{b}(t)] x_2 - k_2 x_2 \\ \dot{x}_3 = -[c - \hat{c}(t)] x_3 - k_3 x_3 \end{cases} \tag{21}$$

The parameter estimation errors are defined as

$$\begin{cases} e_a(t) = a - \hat{a}(t) \\ e_b(t) = b - \hat{b}(t) \\ e_c(t) = c - \hat{c}(t) \end{cases} \tag{22}$$

In view of (22), we can simplify the plant dynamics (21) as

$$\begin{cases} \dot{x}_1 = e_a(x_2 - x_1) - k_1 x_1 \\ \dot{x}_2 = e_b x_2 - k_2 x_2 \\ \dot{x}_3 = -e_c x_3 - k_3 x_3 \end{cases} \tag{23}$$

Differentiating (22) with respect to t, we obtain

$$\begin{cases} \dot{e}_a(t) = -\dot{\hat{a}}(t) \\ \dot{e}_b(t) = -\dot{\hat{b}}(t) \\ \dot{e}_c(t) = -\dot{\hat{c}}(t) \end{cases} \tag{24}$$

We consider the quadratic candidate Lyapunov function defined by

$$V(\mathbf{x}, e_a, e_b, e_c) = \frac{1}{2}\left(x_1^2 + x_2^2 + x_3^2\right) + \frac{1}{2}\left(e_a^2 + e_b^2 + e_c^2\right) \tag{25}$$

Differentiating V along the trajectories of (23) and (24), we obtain

$$\begin{aligned} \dot{V} = & -k_1 x_1^2 - k_2 x_2^2 - k_3 x_3^2 + e_a\left[x_1(x_2 - x_1) - \dot{\hat{a}}\right] \\ & + e_b\left[x_2^2 - \dot{\hat{b}}\right] + e_c\left[-x_3^2 - \dot{\hat{c}}\right] \end{aligned} \tag{26}$$

In view of (26), we take the parameter update law as

$$\begin{cases} \dot{\hat{a}}(t) = x_1(x_2 - x_1) \\ \dot{\hat{b}}(t) = x_2^2 \\ \dot{\hat{c}}(t) = -x_3^2 \end{cases} \tag{27}$$

Next, we state and prove the main result of this section.

Theorem 1 *The novel 3-D two-scroll chaotic system (19) with unknown system parameters is globally and exponentially stabilized for all initial conditions by the adaptive control law (20) and the parameter update law (27), where k_1, k_2, k_3 are positive gain constants.*

Proof We prove this result by applying Lyapunov stability theory [7].

We consider the quadratic Lyapunov function defined by (25), which is clearly a positive definite function on $\mathbf{R}^6$.

By substituting the parameter update law (27) into (26), we obtain the time-derivative of V as

$$\dot{V} = -k_1 x_1^2 - k_2 x_2^2 - k_3 x_3^2 \tag{28}$$

From (28), it is clear that $\dot{V}$ is a negative semi-definite function on $\mathbf{R}^6$.

Thus, we can conclude that the state vector $\mathbf{x}(t)$ and the parameter estimation error are globally bounded, i.e.

$$\left[x_1(t)\ x_2(t)\ x_3(t)\ e_a(t)\ e_b(t)\ e_c(t) \right]^T \in \mathbf{L}_\infty.$$

We define $k = \min\{k_1, k_2, k_3\}$.

Then it follows from (28) that

$$\dot{V} \leq -k\|\mathbf{x}(t)\|^2 \tag{29}$$

Thus, we have

$$k\|\mathbf{x}(t)\|^2 \leq -\dot{V} \tag{30}$$

Integrating the inequality (30) from 0 to t, we get

$$k \int_0^t \|\mathbf{x}(\tau)\|^2 \, d\tau \ \leq \ V(0) - V(t) \tag{31}$$

From (31), it follows that $\mathbf{x} \in \mathbf{L}_2$.

Using (23), we can conclude that $\dot{\mathbf{x}} \in \mathbf{L}_\infty$.

Using Barbalat's lemma [7], we conclude that $\mathbf{x}(t) \to 0$ exponentially as $t \to \infty$ for all initial conditions $\mathbf{x}(0) \in \mathbf{R}^3$.

This completes the proof. ■

For the numerical simulations, the classical fourth-order Runge–Kutta method with step size $h = 10^{-8}$ is used to solve the systems (19) and (27), when the adaptive control law (20) is applied.

The parameter values of the novel 3-D two-scroll chaotic system (19) are taken as in the chaotic case (3), i.e.

$$a = 36, \quad b = 20, \quad c = 5 \tag{32}$$

We take the positive gain constants as

$$k_1 = 5, \quad k_2 = 5, \quad k_3 = 5 \tag{33}$$

Furthermore, as initial conditions of the novel 3-D chaotic system (19), we take

$$x_1(0) = 12.5, \quad x_2(0) = 32.7, \quad x_3(0) = -15.4 \tag{34}$$

Also, as initial conditions of the parameter estimates, we take

$$\hat{a}(0) = 2.7, \quad \hat{b}(0) = 10.3, \quad \hat{c}(0) = 7.4 \tag{35}$$

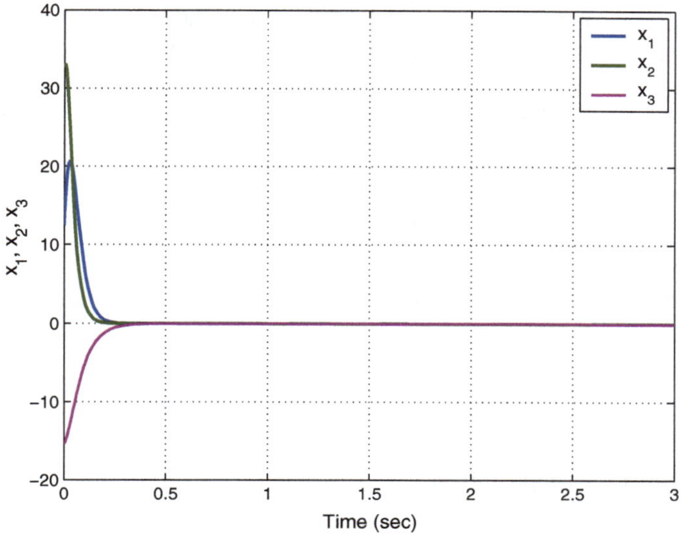

Fig. 5 Time-history of the controlled states x_1, x_2, x_3

In Fig. 5, the exponential convergence of the controlled states of the 3-D novel two-scroll chaotic system (19) is depicted.

5 Adaptive Synchronization of the Identical Novel Two-Scroll Chaotic Systems

In this section, we use adaptive control method to derive an adaptive feedback control law for globally synchronizing identical 3-D novel two-scroll chaotic systems with unknown parameters.

As the master system, we consider the novel 3-D chaotic system given by

$$\begin{cases} \dot{x}_1 = a(x_2 - x_1) + x_2 x_3 \\ \dot{x}_2 = b x_2 - x_1 x_3 \\ \dot{x}_3 = x_1 x_2 - c x_3 \end{cases} \tag{36}$$

In (36), x_1, x_2, x_3 are the states and a, b, c are unknown system parameters.

As the slave system, we consider the novel 3-D chaotic system given by

$$\begin{cases} \dot{y}_1 = a(y_2 - y_1) + y_2 y_3 + u_1 \\ \dot{y}_2 = b y_2 - y_1 y_3 + u_2 \\ \dot{y}_3 = y_1 y_2 - c y_3 + u_3 \end{cases} \tag{37}$$

In (37), y_1, y_2, y_3 are the states and u_1, u_2, u_3 are the adaptive controls to be determined using estimates $\hat{a}(t), \hat{b}(t), \hat{c}(t)$ for the unknown parameters a, b, c, respectively.

The synchronization error between the novel chaotic systems is defined by

$$\begin{cases} e_1 = y_1 - x_1 \\ e_2 = y_2 - x_2 \\ e_3 = y_3 - x_3 \end{cases} \tag{38}$$

Then the error dynamics is obtained as

$$\begin{cases} \dot{e}_1 = a(e_2 - e_1) + y_2 y_3 - x_2 x_3 + u_1 \\ \dot{e}_2 = b e_2 - y_1 y_3 + x_1 x_3 + u_2 \\ \dot{e}_3 = -c e_3 + y_1 y_2 - x_1 x_2 + u_3 \end{cases} \tag{39}$$

We consider the adaptive feedback control law

$$\begin{cases} u_1 = -\hat{a}(t)(e_2 - e_1) - y_2 y_3 + x_2 x_3 - k_1 e_1 \\ u_2 = -\hat{b}(t) e_2 + y_1 y_3 - x_1 x_3 - k_2 e_2 \\ u_3 = \hat{c}(t) e_3 - y_1 y_2 + x_1 x_2 - k_3 e_3 \end{cases} \tag{40}$$

where k_1, k_2, k_3 are positive gain constants.

Substituting (40) into (39), we get the closed-loop error dynamics as

$$\begin{cases} \dot{e}_1 = \left[a - \hat{a}(t)\right](e_2 - e_1) - k_1 e_1 \\ \dot{e}_2 = \left[b - \hat{b}(t)\right] e_2 - k_2 e_2 \\ \dot{e}_3 = -\left[c - \hat{c}(t)\right] e_3 - k_3 e_3 \end{cases} \tag{41}$$

The parameter estimation errors are defined as

$$\begin{cases} e_a(t) = a - \hat{a}(t) \\ e_b(t) = b - \hat{b}(t) \\ e_c(t) = c - \hat{c}(t) \end{cases} \tag{42}$$

In view of (42), we can simplify the error dynamics (41) as

$$\begin{cases} \dot{e}_1 = e_a(e_2 - e_1) - k_1 e_1 \\ \dot{e}_2 = e_b e_2 - k_2 e_2 \\ \dot{e}_3 = -e_c e_3 - k_3 e_3 \end{cases} \tag{43}$$

Differentiating (42) with respect to t, we obtain

$$\begin{cases} \dot{e}_a(t) = -\dot{\hat{a}}(t) \\ \dot{e}_b(t) = -\dot{\hat{b}}(t) \\ \dot{e}_c(t) = -\dot{\hat{c}}(t) \end{cases} \tag{44}$$

We consider the quadratic candidate Lyapunov function defined by

$$V(\mathbf{e}, e_a, e_b, e_c) = \frac{1}{2}\left(e_1^2 + e_2^2 + e_3^2\right) + \frac{1}{2}\left(e_a^2 + e_b^2 + e_c^2\right) \tag{45}$$

Differentiating V along the trajectories of (43) and (44), we obtain

$$\begin{aligned} \dot{V} &= -k_1 e_1^2 - k_2 e_2^2 - k_3 e_3^2 + e_a\left[e_1(e_2 - e_1) - \dot{\hat{a}}\right] \\ &\quad + e_b\left[e_2^2 - \dot{\hat{b}}\right] + e_c\left[-e_3^2 - \dot{\hat{c}}\right] \end{aligned} \tag{46}$$

In view of (46), we take the parameter update law as

$$\begin{cases} \dot{\hat{a}}(t) = e_1(e_2 - e_1) \\ \dot{\hat{b}}(t) = e_2^2 \\ \dot{\hat{c}}(t) = -e_3^2 \end{cases} \tag{47}$$

Next, we state and prove the main result of this section.

Theorem 2 *The novel two-scroll chaotic systems (36) and (37) with unknown system parameters are globally and exponentially synchronized for all initial conditions by the adaptive control law (40) and the parameter update law (47), where k_1, k_2, k_3 are positive gain constants.*

Proof We prove this result by applying Lyapunov stability theory [7].

We consider the quadratic Lyapunov function defined by (45), which is clearly a positive definite function on $\mathbf{R}^6$.

By substituting the parameter update law (47) into (46), we obtain

$$\dot{V} = -k_1 e_1^2 - k_2 e_2^2 - k_3 e_3^2 \tag{48}$$

From (48), it is clear that $\dot{V}$ is a negative semi-definite function on $\mathbf{R}^6$.

Thus, we can conclude that the error vector $\mathbf{e}(t)$ and the parameter estimation error are globally bounded, i.e.

$$\left[\, e_1(t)\ e_2(t)\ e_3(t)\ e_a(t)\ e_b(t)\ e_c(t) \,\right]^T \in \mathbf{L}_\infty. \tag{49}$$

We define $k = \min\{k_1, k_2, k_3\}$.

Then it follows from (48) that

$$\dot{V} \le -k\|\mathbf{e}(t)\|^2 \tag{50}$$

Thus, we have

$$k\|\mathbf{e}(t)\|^2 \leq -\dot{V} \tag{51}$$

Integrating the inequality (51) from 0 to t, we get

$$k\int_0^t \|\mathbf{e}(\tau)\|^2 \, d\tau \;\leq\; V(0) - V(t) \tag{52}$$

From (52), it follows that $\mathbf{e} \in \mathbf{L}_2$.

Using (43), we can conclude that $\dot{\mathbf{e}} \in \mathbf{L}_\infty$.

Using Barbalat's lemma [7], we conclude that $\mathbf{e}(t) \to 0$ exponentially as $t \to \infty$ for all initial conditions $\mathbf{e}(0) \in \mathbf{R}^3$.

Hence, we have proved that novel two-scroll chaotic systems (36) and (37) with unknown system parameters are globally and exponentially synchronized for all initial conditions.

This completes the proof. ■

For the numerical simulations, the classical fourth-order Runge–Kutta method with step size $h = 10^{-8}$ is used to solve the systems (36), (37) and (47), when the adaptive control law (40) is applied.

The parameter values of the novel chaotic systems are taken as in the chaotic case (3), i.e.

$$a = 36, \quad b = 20, \quad c = 5 \tag{53}$$

We take the positive gain constants as $k_i = 5$ for $i = 1, 2, 3$.

Furthermore, as initial conditions of the master system (36), we take

$$x_1(0) = 32.4, \quad x_2(0) = -15.8, \quad x_3(0) = -26.7 \tag{54}$$

As initial conditions of the slave system (37), we take

$$y_1(0) = 12.7, \quad y_2(0) = 3.5, \quad y_3(0) = 14.2 \tag{55}$$

Also, as initial conditions of the parameter estimates, we take

$$\hat{a}(0) = 2.1, \quad \hat{b}(0) = 20.4, \quad \hat{c}(0) = 25.3 \tag{56}$$

Figures 6, 7 and 8 describe the complete synchronization of the novel chaotic systems (36) and (37), while Fig. 9 describes the time-history of the synchronization errors e_1, e_2, e_3.

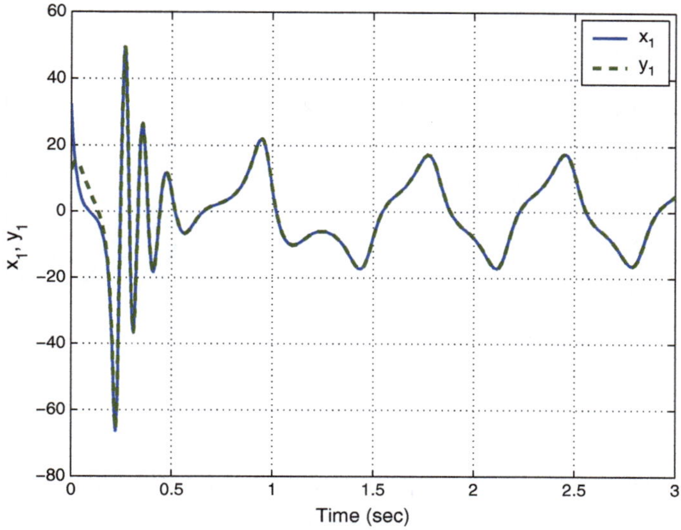

Fig. 6 Synchronization of the states x_1 and y_1

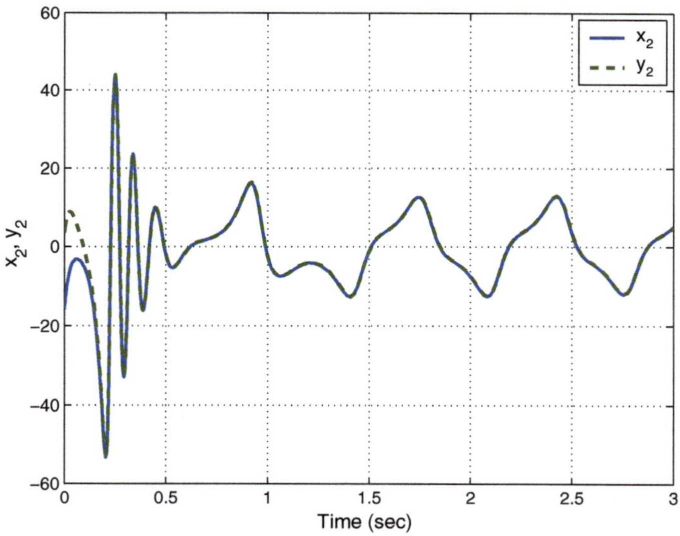

Fig. 7 Synchronization of the states x_2 and y_2

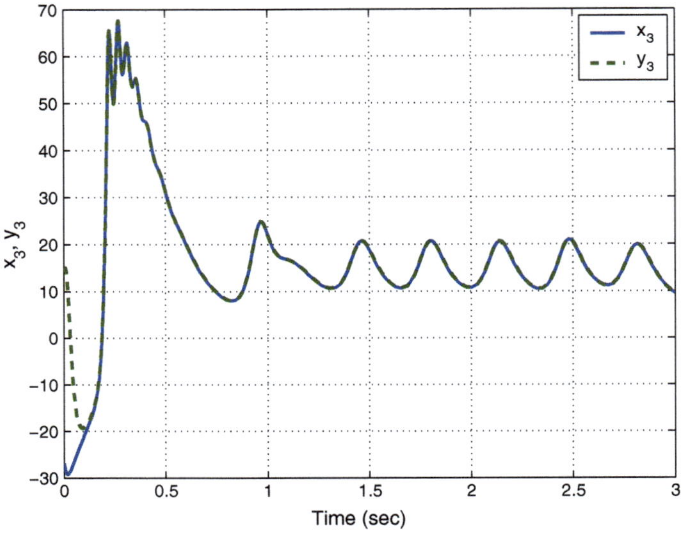

Fig. 8 Synchronization of the states x_3 and y_3

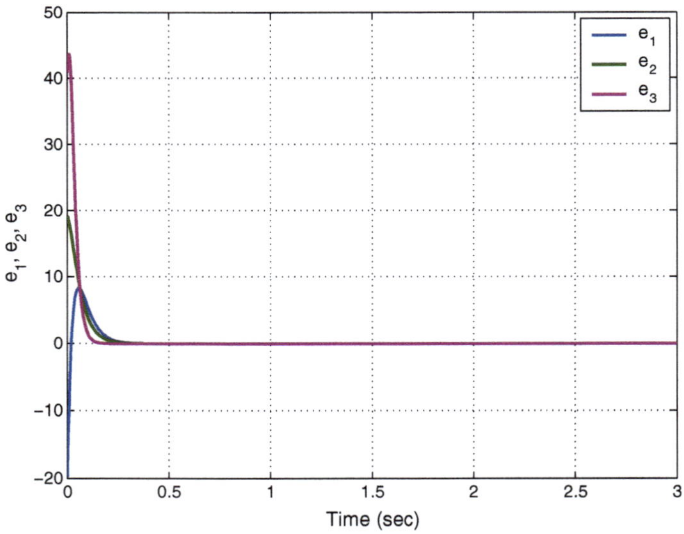

Fig. 9 Time-history of the synchronization errors e_1, e_2, e_3

6 Conclusions

In this work, we described a seven-term two-scroll novel chaotic system with three quadratic nonlinearities and discussed the qualitative properties of the system. We derived new results for the global chaos control of the novel two-scroll chaotic system with unknown parameters using adaptive control method. We also derived new results for the global chaos synchronization of the identical two-scroll chaotic systems using adaptive control method. MATLAB simulations were shown to illustrate all the main results derived in this work.

References

1. Arneodo A, Coullet P, Tresser C (1981) Possible new strange attractors with spiral structure. Commun Math Phys 79(4):573–576
2. Azar AT, Vaidyanathan S (2015) Chaos modeling and control systems design, vol 581. Springer, Germany
3. Cai G, Tan Z (2007) Chaos synchronization of a new chaotic system via nonlinear control. J Uncertain Syst 1(3):235–240
4. Chen G, Ueta T (1999) Yet another chaotic attractor. Intern J Bifurc Chaos 9(7):1465–1466
5. Islam MM, Murase K (2005) Chaotic dynamics of a behaviour-based miniature mobile robot: effects of environment and control structure. Neural Netw 18(2):123–144
6. Karthikeyan R, Sundarapandian V (2014) Hybrid chaos synchronization of four-scroll systems via active control. J Electr Eng 65(2):97–103
7. Khalil HK (2001) Nonlinear systems, 3rd edn. Prentice Hall, New Jersey
8. Li D (2008) A three-scroll chaotic attractor. Phys Lett A 372(4):387–393
9. Lorenz EN (1963) Deterministic periodic flow. J Atmos Sci 20(2):130–141
10. Lü J, Chen G (2002) A new chaotic attractor coined. Int J Bifurc Chaos 12(3):659–661
11. Matouk AE (2011) Chaos, feedback control and synchronization of a fractional-order modified autonomous Van der Pol-Duffing circuit. Commun Nonlinear Sci Numer Simul 16(2): 975–986
12. Njah AN, Sunday OD (2009) Generalization on the chaos control of 4-D chaotic systems using recursive backstepping nonlinear controller. Chaos, Solitons Fractals 41(5):2371–2376
13. Pehlivan I, Moroz IM, Vaidyanathan S (2014) Analysis, synchronization and circuit design of a novel butterfly attractor. J Sound Vib 333(20):5077–5096
14. Pham VT, Volos CK, Vaidyanathan S, Le TP, Vu VY (2015) A memristor-based hyperchaotic system with hidden attractors: dynamics, synchronization and circuital emulating. J Eng Sci Technol Rev 8(2):205–214
15. Rasappan S, Vaidyanathan S (2013) Hybrid synchronization of n-scroll Chua circuits using adaptive backstepping control design with recursive feedback. Malaysian J Math Sci 73(1): 73–95
16. Rasappan S, Vaidyanathan S (2014) Global chaos synchronization of WINDMI and Coullet chaotic systems using adaptive backstepping control design. Kyungpook Math J 54(1): 293–320
17. Rhouma R, Belghith S (2011) Cryptanalysis of a chaos-based cryptosystem. Commun Nonlinear Sci Numer Simul 16(2):876–884
18. Rössler OE (1976) An equation for continuous chaos. Phys Lett A 57(5):397–398
19. Sampath S, Vaidyanathan S, Volos CK, Pham VT (2015) An eight-term novel four-scroll chaotic system with cubic nonlinearity and its circuit simulation. J Eng Sci Technol Rev 8(2):1–6

20. Sarasu P, Sundarapandian V (2011) Active controller design for generalized projective synchronization of four-scroll chaotic systems. Int J Syst Signal Control Eng Appl 4(2):26–33
21. Sarasu P, Sundarapandian V (2011) The generalized projective synchronization of hyperchaotic Lorenz and hyperchaotic Qi systems via active control. Int J Soft Comput 6(5):216–223
22. Sarasu P, Sundarapandian V (2012) Generalized projective synchronization of two-scroll systems via adaptive control. Int J Soft Comput 7(4):146–156
23. Sprott JC (1994) Some simple chaotic flows. Phys Rev E 50(2):647–650
24. Sundarapandian V (2010) Output regulation of the Lorenz attractor. Far East J Math Sci 42(2):289–299
25. Sundarapandian V (2013) Adaptive control and synchronization design for the Lu-Xiao chaotic system. Lect Notes Electr Eng 131:319–327
26. Sundarapandian V (2013) Analysis and anti-synchronization of a novel chaotic system via active and adaptive controllers. J Eng Sci Technol Rev 6(4):45–52
27. Sundarapandian V, Karthikeyan R (2011) Anti-synchronization of hyperchaotic Lorenz and hyperchaotic Chen systems by adaptive control. Int J Syst Signal Control Eng Appl 4(2):18–25
28. Sundarapandian V, Karthikeyan R (2011) Anti-synchronization of Lü and Pan chaotic systems by adaptive nonlinear control. Eur J Sci Res 64(1):94–106
29. Sundarapandian V, Karthikeyan R (2012) Adaptive anti-synchronization of uncertain Tigan and Li systems. J Eng Appl Sci 7(1):45–52
30. Sundarapandian V, Pehlivan I (2012) Analysis, control, synchronization, and circuit design of a novel chaotic system. Math Comput Model 55(7–8):1904–1915
31. Sundarapandian V, Sivaperumal S (2011) Sliding controller design of hybrid synchronization of four-wing chaotic systems. Int J Soft Comput 6(5):224–231
32. Suresh R, Sundarapandian V (2013) Global chaos synchronization of a family of n-scroll hyperchaotic Chua circuits using backstepping control with recursive feedback. Far East J Math Sci 7(2):219–246
33. Tigan G, Opris D (2008) Analysis of a 3D chaotic system. Chaos Solitons Fractals 36:1315–1319
34. Usama M, Khan MK, Alghathbar K, Lee C (2010) Chaos-based secure satellite imagery cryptosystem. Comput Math Appl 60(2):326–337
35. Vaidyanathan S (2011) Output regulation of Arneodo-Coullet chaotic system. Commun Comput Inf Sci 133:98–107
36. Vaidyanathan S (2011) Output regulation of the unified chaotic system. Commun Comput Inf Sci 198:1–9
37. Vaidyanathan S (2012) Adaptive controller and syncrhonizer design for the Qi-Chen chaotic system. Lect Notes Inst Comput Sci Soc-Inf Telecommun Eng 84:73–82
38. Vaidyanathan S (2012) Anti-synchronization of Sprott-L and Sprott-M chaotic systems via adaptive control. Int J Control Theory Appl 5(1):41–59
39. Vaidyanathan S (2012) Global chaos control of hyperchaotic Liu system via sliding control method. Int J Control Theory Appl 5(2):117–123
40. Vaidyanathan S (2012) Sliding mode control based global chaos control of Liu-Liu-Liu-Su chaotic system. Int J Control Theory Appl 5(1):15–20
41. Vaidyanathan S (2013) A new six-term 3-D chaotic system with an exponential nonlinearity. Far East J Math Sci 79(1):135–143
42. Vaidyanathan S (2013) A ten-term novel 4-D hyperchaotic system with three quadratic nonlinearities and its control. Int J Control Theory Appl 6(2):97–109
43. Vaidyanathan S (2013) Analysis and adaptive synchronization of two novel chaotic systems with hyperbolic sinusoidal and cosinusoidal nonlinearity and unknown parameters. J Eng Sci Technol Rev 6(4):53–65
44. Vaidyanathan S (2013) Analysis, control and synchronization of hyperchaotic Zhou system via adaptive control. Adv Intell Syst Comput 177:1–10
45. Vaidyanathan S (2014) A new eight-term 3-D polynomial chaotic system with three quadratic nonlinearities. Far East J Math Sci 84(2):219–226

46. Vaidyanathan S (2014) Analysis and adaptive synchronization of eight-term 3-D polynomial chaotic systems with three quadratic nonlinearities. Eur Phys J: Spec Top 223(8):1519–1529
47. Vaidyanathan S (2014) Analysis, control and synchronisation of a six-term novel chaotic system with three quadratic nonlinearities. Int J Model Identif Control 22(1):41–53
48. Vaidyanathan S (2014) Generalised projective synchronisation of novel 3-D chaotic systems with an exponential non-linearity via active and adaptive control. Int J Model Identif Control 22(3):207–217
49. Vaidyanathan S (2014) Global chaos synchronisation of identical Li-Wu chaotic systems via sliding mode control. Int J Model Identif Control 22(2):170–177
50. Vaidyanathan S (2014) Qualitative analysis and control of an eleven-term novel 4-D hyperchaotic system with two quadratic nonlinearities. Int J Control Theory Appl 7:35–47
51. Vaidyanathan S (2015) 3-cells Cellular Neural Network (CNN) attractor and its adaptive biological control. Int J PharmTech Res 8(4):632–640
52. Vaidyanathan S (2015) A 3-D novel highly chaotic system with four quadratic nonlinearities, its adaptive control and anti-synchronization with unknown parameters. J Eng Sci Technol Rev 8(2):106–115
53. Vaidyanathan S (2015) A novel chemical chaotic reactor system and its adaptive control. Int J ChemTech Res 8(7):146–158
54. Vaidyanathan S (2015) Adaptive backstepping control of enzymes-substrates system with ferroelectric behaviour in brain waves. Int J PharmTech Res 8(2):256–261
55. Vaidyanathan S (2015) Adaptive biological control of generalized Lotka-Volterra three-species biological system. Int J PharmTech Res 8(4):622–631
56. Vaidyanathan S (2015) Adaptive chaotic synchronization of enzymes-substrates system with ferroelectric behaviour in brain waves. Int J PharmTech Res 8(5):964–973
57. Vaidyanathan S (2015) Adaptive control of a chemical chaotic reactor. Int J PharmTech Res 8(3):377–382
58. Vaidyanathan S (2015) Adaptive control of the FitzHugh-Nagumo chaotic neuron model. Int J PharmTech Res 8(6):117–127
59. Vaidyanathan S (2015) Adaptive synchronization of chemical chaotic reactors. Int J ChemTech Res 8(2):612–621
60. Vaidyanathan S (2015) Adaptive synchronization of generalized Lotka-Volterra three-species biological systems. Int J PharmTech Res 8(5):928–937
61. Vaidyanathan S (2015) Adaptive synchronization of novel 3-D chemical chaotic reactor systems. Int J ChemTech Res 8(7):159–171
62. Vaidyanathan S (2015) Adaptive synchronization of the identical FitzHugh-Nagumo chaotic neuron models. Int J PharmTech Res 8(6):167–177
63. Vaidyanathan S (2015) Analysis, control and synchronization of a 3-D novel jerk chaotic system with two quadratic nonlinearities. Kyungpook Math J 55:563–586
64. Vaidyanathan S (2015) Analysis, properties and control of an eight-term 3-D chaotic system with an exponential nonlinearity. Int J Model Identif Control 23(2):164–172
65. Vaidyanathan S (2015) Anti-synchronization of brusselator chemical reaction systems via adaptive control. Int J ChemTech Res 8(6):759–768
66. Vaidyanathan S (2015) Chaos in neurons and adaptive control of Birkhoff-Shaw strange chaotic attractor. Int J PharmTech Res 8(5):956–963
67. Vaidyanathan S (2015) Chaos in neurons and synchronization of Birkhoff-Shaw strange chaotic attractors via adaptive control. Int J PharmTech Res 8(6):1–11
68. Vaidyanathan S (2015) Coleman-Gomatam logarithmic competitive biology models and their ecological monitoring. Int J PharmTech Res 8(6):94–105
69. Vaidyanathan S (2015) Dynamics and control of brusselator chemical reaction. Int J ChemTech Res 8(6):740–749
70. Vaidyanathan S (2015) Dynamics and control of tokamak system with symmetric and magnetically confined plasma. Int J ChemTech Res 8(6):795–803
71. Vaidyanathan S (2015) Global chaos synchronization of chemical chaotic reactors via novel sliding mode control method. Int J ChemTech Res 8(7):209–221

72. Vaidyanathan S (2015) Global chaos synchronization of the forced Van der Pol chaotic oscillators via adaptive control method. Int J PharmTech Res 8(6):156–166
73. Vaidyanathan S (2015) Global chaos synchronization of the Lotka-Volterra biological systems with four competitive species via active control. Int J PharmTech Res 8(6):206–217
74. Vaidyanathan S (2015) Lotka-Volterra population biology models with negative feedback and their ecological monitoring. Int J PharmTech Res 8(5):974–981
75. Vaidyanathan S (2015) Lotka-Volterra two species competitive biology models and their ecological monitoring. Int J PharmTech Res 8(6):32–44
76. Vaidyanathan S (2015) Output regulation of the forced Van der Pol chaotic oscillator via adaptive control method. Int J PharmTech Res 8(6):106–116
77. Vaidyanathan S, Azar AT (2015) Analysis and control of a 4-D novel hyperchaotic system. Stud Comput Intell 581:3–17
78. Vaidyanathan S, Azar AT (2015) Analysis, control and synchronization of a nine-term 3-d novel chaotic system. In: Azar AT, Vaidyanathan S (eds) Chaos modelling and control systems design, studies in computational intelligence, vol 581. Springer, cham, pp 19–38
79. Vaidyanathan S, Madhavan K (2013) Analysis, adaptive control and synchronization of a seven-term novel 3-D chaotic system. Int J Control Theory Appl 6(2):121–137
80. Vaidyanathan S, Pakiriswamy S (2013) Generalized projective synchronization of six-term Sundarapandian chaotic systems by adaptive control. Int J Control Theory Appl 6(2):153–163
81. Vaidyanathan S, Pakiriswamy S (2015) A 3-D novel conservative chaotic system and its generalized projective synchronization via adaptive control. J Eng Sci Technol Rev 8(2):52–60
82. Vaidyanathan S, Rajagopal K (2011) Hybrid synchronization of hyperchaotic Wang-Chen and hyperchaotic Lorenz systems by active non-linear control. Int J Syst Signal Control Eng Appl 4(3):55–61
83. Vaidyanathan S, Rajagopal K (2012) Global chaos synchronization of hyperchaotic Pang and hyperchaotic Wang systems via adaptive control. Int J Soft Comput 7(1):28–37
84. Vaidyanathan S, Rasappan S (2011) Global chaos synchronization of hyperchaotic Bao and Xu systems by active nonlinear control. Commun Comput Inf Sci 198:10–17
85. Vaidyanathan S, Rasappan S (2014) Global chaos synchronization of n-scroll Chua circuit and Lur'e system using backstepping control design with recursive feedback. Arab J Sci Eng 39(4):3351–3364
86. Vaidyanathan S, Sampath S (2012) Anti-synchronization of four-wing chaotic systems via sliding mode control. Int J Autom Comput 9(3):274–279
87. Vaidyanathan S, Volos C (2015) Analysis and adaptive control of a novel 3-D conservative no-equilibrium chaotic system. Arch Control Sci 25(3):333–353
88. Vaidyanathan S, Volos C, Pham VT (2014) Hyperchaos, adaptive control and synchronization of a novel 5-D hyperchaotic system with three positive Lyapunov exponents and its SPICE implementation. Arch Control Sci 24(4):409–446
89. Vaidyanathan S, Volos C, Pham VT, Madhavan K, Idowu BA (2014) Adaptive backstepping control, synchronization and circuit simulation of a 3-D novel jerk chaotic system with two hyperbolic sinusoidal nonlinearities. Arch Control Sci 24(3):375–403
90. Vaidyanathan S, Azar AT, Rajagopal K, Alexander P (2015) Design and SPICE implementation of a 12-term novel hyperchaotic system and its synchronisation via active control. Int J Model Identif Control 23(3):267–277
91. Vaidyanathan S, Idowu BA, Azar AT (2015) Backstepping controller design for the global chaos synchronization of Sprott's jerk systems. Stud Comput Intell 581:39–58
92. Vaidyanathan S, Rajagopal K, Volos CK, Kyprianidis IM, Stouboulos IN (2015) Analysis, adaptive control and synchronization of a seven-term novel 3-D chaotic system with three quadratic nonlinearities and its digital implementation in LabVIEW. J Eng Sci Technol Rev 8(2):130–141
93. Vaidyanathan S, Sampath S, Azar AT (2015) Global chaos synchronisation of identical chaotic systems via novel sliding mode control method and its application to Zhu system. Int J Model Identif Control 23(1):92–100

94. Vaidyanathan S, Volos C, Pham VT, Madhavan K (2015) Analysis, adaptive control and synchronization of a novel 4-D hyperchaotic hyperjerk system and its SPICE implementation. Nonlinear Dyn 25(1):135–158
95. Vaidyanathan S, Volos CK, Kyprianidis IM, Stouboulos IN, Pham VT (2015) Analysis, adaptive control and anti-synchronization of a six-term novel jerk chaotic system with two exponential nonlinearities and its circuit simulation. J Eng Sci Technol Rev 8(2):24–36
96. Vaidyanathan S, Volos CK, Pham VT (2015) Analysis, adaptive control and adaptive synchronization of a nine-term novel 3-D chaotic system with four quadratic nonlinearities and its circuit simulation. J Eng Sci Technol Rev 8(2):181–191
97. Vaidyanathan S, Volos CK, Pham VT (2015) Analysis, control, synchronization and SPICE implementation of a novel 4-D hyperchaotic Rikitake dynamo system without equilibrium. J Eng Sci Technol Rev 8(2):232–244
98. Vaidyanathan S, Volos CK, Pham VT (2015) Global chaos control of a novel nine-term chaotic system via sliding mode control. In: Azar AT, Zhu Q (eds) Advances and applications in sliding mode control systems, studies in computational intelligence, vol 576. Springer, Germany, pp 571–590
99. Vincent UE, Njah AN, Laoye JA (2007) Controlling chaos and deterministic directed transport in inertia ratchets using backstepping control. Phys D 231(2):130–136
100. Volos CK, Kyprianidis IM, Stouboulos IN, Anagnostopoulos AN (2009) Experimental study of the dynamic behavior of a double scroll circuit. J Appl Funct Anal 4:703–711
101. Volos CK, Kyprianidis IM, Stouboulos IN (2013) Experimental investigation on coverage performance of a chaotic autonomous mobile robot. Robot Auton Syst 61(12):1314–1322
102. Volos CK, Kyprianidis IM, Stouboulos IN, Tlelo-Cuautle E, Vaidyanathan S (2015) Memristor: A new concept in synchronization of coupled neuromorphic circuits. J Eng Sci Technol Rev 8(2):157–173
103. Yang J, Zhu F (2013) Synchronization for chaotic systems and chaos-based secure communications via both reduced-order and step-by-step sliding mode observers. Commun Nonlinear Sci Numer Simul 18(4):926–937
104. Yang J, Chen Y, Zhu F (2014) Singular reduced-order observer-based synchronization for uncertain chaotic systems subject to channel disturbance and chaos-based secure communication. Appl Math Comput 229:227–238
105. Zhou W, Xu Y, Lu H, Pan L (2008) On dynamics analysis of a new chaotic attractor. Phys Lett A 372(36):5773–5777
106. Zhu C, Liu Y, Guo Y (2010) Theoretic and numerical study of a new chaotic system. Intell Inf Manag 2:104–109

A Novel 3-D Circulant Chaotic System with Labyrinth Chaos and Its Adaptive Control

Sundarapandian Vaidyanathan

Abstract In this work, we describe a novel 3-D dissipative circulant chaotic system with labyrinth chaos. The novel chaotic system is a nine-term polynomial system with six sinusoidal nonlinearities. The phase portraits of the novel circulant chaotic system are illustrated and the dynamic properties of the novel circulant chaotic system are discussed. The novel circulant chaotic system has infinitely many equilibrium points and it exhibits labyrinth chaos. We show that all the equilibrium points of the novel circulant chaotic system are saddle-foci and hence they are unstable. The Lyapunov exponents of the novel circulant chaotic system are obtained as $L_1 = 2.1714$, $L_2 = 0$ and $L_3 = -2.2373$, while the Kaplan–Yorke dimension of the novel circulant chaotic system is obtained as $D_{KY} = 2.9705$. Since the Kaplan–Yorke dimension of the the novel circulant chaotic system has a very large value and close to three, the novel circulant chaotic system with labyrinth chaos exhibits highly complex behaviour. Since the sum of the Lyapunov exponents is negative, the novel chaotic system is dissipative. Next, we derive new results for the global chaos control of the novel circulant chaotic system with unknown parameters using adaptive control method. We also derive new results for the global chaos synchronization of the identical novel circulant chaotic systems with unknown parameters using adaptive control method. The main control results are established using Lyapunov stability theory. MATLAB simulations are depicted to illustrate the phase portraits of the novel circulant chaotic system and also the adaptive control results derived in this work.

Keywords Chaos · Chaotic systems · Circulant chaotic system · Chaos control · Adaptive control · Synchronization

---

S. Vaidyanathan (✉)
Research and Development Centre, Vel Tech University, Avadi, Chennai 600062, Tamil Nadu, India
e-mail: sundarvtu@gmail.com

S. Vaidyanathan and C. Volos (eds.), *Advances and Applications in Chaotic Systems*, Studies in Computational Intelligence 636,
DOI 10.1007/978-3-319-30279-9_11

1 Introduction

In the last few decades, chaos theory has become a very important and active research field, employing many applications in different disciplines like physics, chemistry, biology, ecology, engineering and economics, among others [3].

Some classical paradigms of 3-D chaotic systems in the literature are Lorenz system [11], Rössler system [20], ACT system [2], Sprott systems [25], Chen system [6], Lü system [12], Cai system [5], Tigan system [37], etc.

Many new chaotic systems have been discovered in the recent years such as Zhou system [110], Zhu system [111], Li system [10], Sundarapandian systems [29, 33], Vaidyanathan systems [46, 48, 50–53, 57, 68, 69, 83, 84, 86, 92, 94, 97, 100, 101, 103], Pehlivan system [15], Sampath system [21], etc.

Chaos theory has applications in several fields of science and engineering such as chemical reactors [58, 62, 64, 66, 70, 74–76], biological systems [56, 59–61, 63, 65, 67, 71–73, 77–81], memristors [1, 16, 107], lasers [4], oscillations [38], robotics [7, 106], electrical circuits [13, 105], cryptosystems [19, 39], secure communications [108, 109], etc.

The control of a chaotic system aims to stabilize or regulate the system with the help of a feedback control. There are many methods available for controlling a chaotic system such as active control [27, 40, 41], adaptive control [28, 42, 47, 49, 55, 82, 93, 99, 102], sliding mode control [44, 45], backstepping control [14, 96, 104], etc.

There are many methods available for chaos synchronization such as active control [8, 22, 23, 87, 89, 95], adaptive control [24, 30–32, 43, 85, 88], sliding mode control [34, 54, 91, 98], backstepping control [17, 18, 35, 90], etc.

In this research work, we announce a novel circulant chaotic system with Labyrinth chaos. Using adaptive control method, we have also derived new results for the global chaos control of the novel circulant chaotic system and global chaos synchronization of the identical novel highly chaotic systems when the system parameters are unknown.

This work is organized as follows. Section 2 describes the dynamic equations and phase portraits of the novel circulant chaotic system. Section 3 details the qualitative properties of the novel highly chaotic system. The novel circulant chaotic system has infinitely many equilibrium points and it exhibits labyrinth chaos. The Lyapunov exponents of the novel chaotic system are obtained as $L_1 = 2.1714$, $L_2 = 0$ and $L_3 = -2.2373$, while the Kaplan–Yorke dimension of the novel chaotic system is obtained as $D_{KY} = 2.9705$. The large value of D_{KY} indicates the high complexity of the novel circulant chaotic system.

In Sect. 4, we derive new results for the global chaos control of the novel circulant chaotic system with unknown parameters. In Sect. 5, we derive new results for the global chaos synchronization of the identical novel circulant chaotic systems with unknown parameters. Section 6 contains the conclusions of this work.

2 A Novel 3-D Circulant System with Labyrinth Chaos

A particularly elegant chaos system is one in which the variables are cyclically symmetric [26]. Thus, a 3-D circulant system has the form

$$\begin{cases} \dot{x}_1 = \varphi(x_1, x_2, x_3) \\ \dot{x}_2 = \varphi(x_2, x_3, x_1) \\ \dot{x}_3 = \varphi(x_3, x_1, x_2) \end{cases} \tag{1}$$

where all the functions are the same except the state variables which are rotated.

A famous circulant chaotic system is the Thomas system [36], which can be expressed as

$$\begin{cases} \dot{x}_1 = \sin x_2 - bx_1 \\ \dot{x}_2 = \sin x_3 - bx_2 \\ \dot{x}_3 = \sin x_1 - bx_3 \end{cases} \tag{2}$$

where b is a constant that corresponds to how *dissipative* the system is, and acts as a bifurcation parameter. The Thomas system (2) is found chaotic when

$$b = 0.2082 \tag{3}$$

For numerical simulations, we take the initial state of the Thomas system (2) as

$$x_1(0) = 0.4, \quad x_2(0) = 0, \quad x_3(0) = 0 \tag{4}$$

The Lyapunov exponents of the Thomas circulant system (2) for the initial state (4) and the parameter value (3) are numerically found as

$$L_1 = 0.0179, \quad L_2 = 0, \quad L_3 = -0.6376 \tag{5}$$

Thus, the Kaplan–Yorke dimension of the Thomas circulant system (2) is derived as

$$D_{KY} = 2 + \frac{L_1 + L_2}{|L_3|} = 2.0281 \tag{6}$$

It is also easy to see that the Thomas circulant system (2) has infinitely many equilibrium points given by

$$E_\theta = \begin{bmatrix} \theta \\ \theta \\ \theta \end{bmatrix}, \tag{7}$$

where θ is a root of the transcendental equation

$$\sin\theta = b\theta, \quad (b = 0.2082) \tag{8}$$

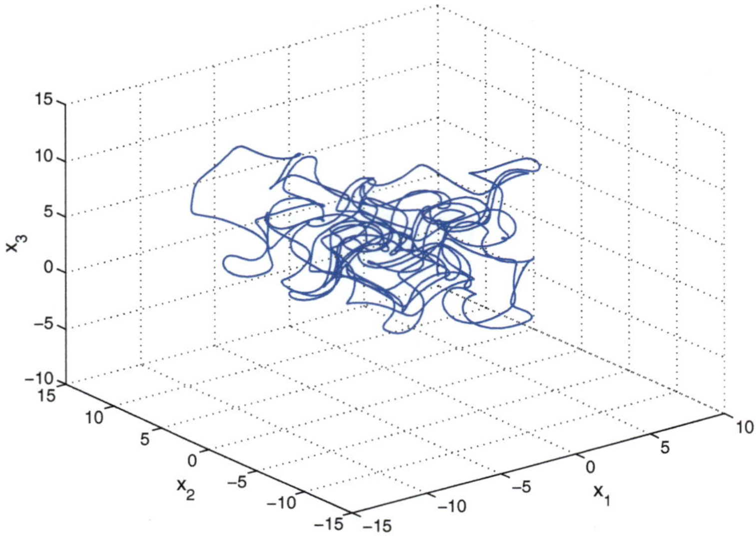

Fig. 1 3-D phase portrait of the Thomas circulant chaotic system with labyrinth chaos

Since the Eq. (8) has infinitely many roots θ, it follows that the Thomas circulant system (2) has infinitely many equilibrium points E_θ given by (7).

Figure 1 depicts the 3-D phase portrait of the Thomas circulant chaotic system (2) with labyrinth chaos.

In this research work, we announce a novel circulant chaotic system with labyrinth chaos, which is described by

$$\begin{cases} \dot{x}_1 = a(\sin x_2 - \cos x_2) - bx_1 \\ \dot{x}_2 = a(\sin x_3 - \cos x_3) - bx_2 \\ \dot{x}_3 = a(\sin x_1 - \cos x_1) - bx_3 \end{cases} \tag{9}$$

where a and b are constant, positive, parameters.

The novel circulant system (9) is chaotic when we take the parameter values as

$$a = 14, \quad b = 0.02 \tag{10}$$

For numerical simulations, we take the initial state of the circulant system (9) as

$$x_1(0) = 2.5, \quad x_2(0) = 2.8, \quad x_3(0) = 2.5 \tag{11}$$

The Lyapunov exponents of the novel circulant system (9) for the parameter values (10) and the initial state (11) are numerically found as

$$L_1 = 2.1714, \quad L_2 = 0, \quad L_3 = -2.2373 \tag{12}$$

Thus, the Kaplan–Yorke dimension of the novel circulant system (9) is derived as

$$D_{KY} = 2 + \frac{L_1 + L_2}{|L_3|} = 2.9705 \tag{13}$$

The maximal Lyapunov exponent of the novel circulant chaotic system (9) is $L_1 = 2.1714$, which is much higher than the maximal Lyapunov exponent of the Thomas circulant chaotic system (2), viz. $L_1 = 0.0179$.

Also, the Kaplan–Yorke dimension of the novel circulant chaotic system (9) is $D_{KY} = 2.9705$, which is much higher than the Kaplan–Yorke dimension of the Thomas circulant chaotic system (2), viz. $D_{KY} = 2.0281$.

This shows that the novel circulant chaotic system (9) exhibits more chaotic behaviour than the Thomas circulant system (2). Also, the large value of D_{KY}, which is close to three, indicates that it exhibits high complexity.

It is also easy to see that the novel circulant circulant system (9) has infinitely many equilibrium points given by

$$E_\theta = \begin{bmatrix} \theta \\ \theta \\ \theta \end{bmatrix}, \tag{14}$$

where θ is a root of the transcendental equation

$$\sin\theta - \cos\theta = \frac{b}{a}\,\theta, \quad (a = 14, \quad b = 0.02) \tag{15}$$

If we define

$$\mu = \frac{b}{a}, \tag{16}$$

then we can express (15) equivalently as

$$\sin\theta - \cos\theta = \mu\,\theta, \quad (\mu = 0.0014) \tag{17}$$

Since the Eq. (17) has infinitely many roots θ, it follows that the novel chaotic circulant system (9) has infinitely many equilibrium points E_θ given by (14).

In this work, we shall show that all the equilibrium points E_θ, $(\theta \in \mathbf{R})$ are saddle-focus points, which are unstable.

Figure 2 depicts the 3-D phase portrait of the novel circulant chaotic system (9) with labyrinth chaos.

Figures 3, 4 and 5 depict the 2-D projection of the novel circulant chaotic system (9) on the (x_1, x_2), (x_2, x_3) and (x_1, x_3) planes, respectively.

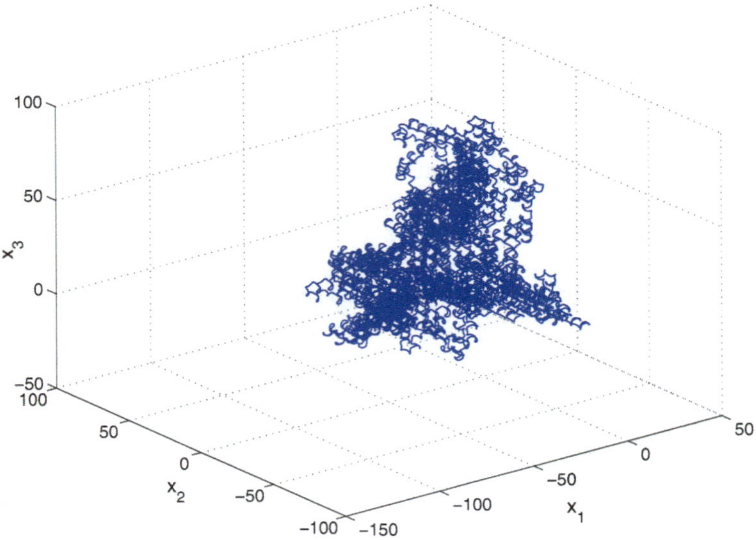

Fig. 2 3-D phase portrait of the novel circulant chaotic system with labyrinth chaos

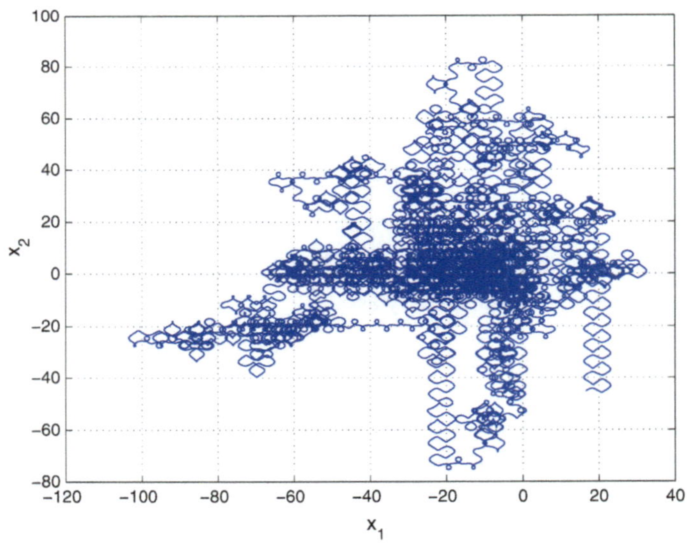

Fig. 3 2-D projection of the novel circulant chaotic system on the (x_1, x_2) *plane*

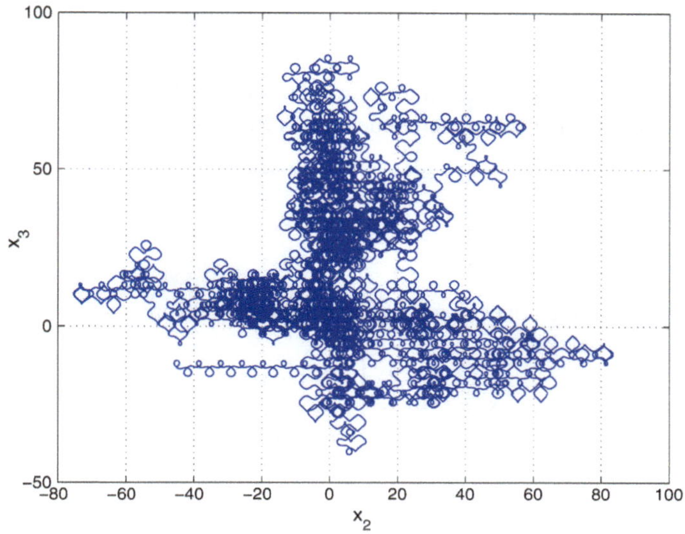

Fig. 4 2-D projection of the novel circulant chaotic system on the (x_2, x_3) plane

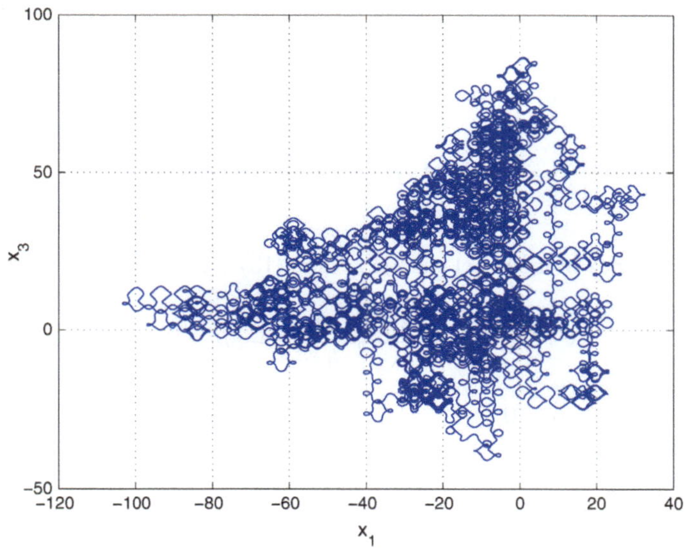

Fig. 5 2-D projection of the novel circulant chaotic system on the (x_1, x_3) plane

3 Analysis of the Novel 3-D Circulant Chaotic System

In this section, we study the qualitative properties of the 3-D novel circulant chaotic system (9). We take the parameter values as in (10), viz. $a = 14$ and $b = 0.02$.

3.1 Dissipativity

In vector notation, the novel chaotic system (9) can be expressed as

$$\dot{\mathbf{x}} = f(\mathbf{x}) = \begin{bmatrix} f_1(x_1, x_2, x_3) \\ f_2(x_1, x_2, x_3) \\ f_3(x_1, x_2, x_3) \end{bmatrix}, \tag{18}$$

where

$$\begin{cases} f_1(x_1, x_2, x_3) = a(\sin x_2 - \cos x_2) - bx_1 \\ f_2(x_1, x_2, x_3) = a(\sin x_3 - \cos x_3) - bx_2 \\ f_3(x_1, x_2, x_3) = a(\sin x_1 - \cos x_1) - bx_3 \end{cases} \tag{19}$$

Let Ω be any region in $\mathbf{R}^3$ with a smooth boundary and also, $\Omega(t) = \Phi_t(\Omega)$, where Φ_t is the flow of f. Furthermore, let $V(t)$ denote the volume of $\Omega(t)$.

By Liouville's theorem, we know that

$$\dot{V}(t) = \int\limits_{\Omega(t)} (\nabla \cdot f) \, dx_1 \, dx_2 \, dx_3 \tag{20}$$

The divergence of the novel chaotic system (18) is found as

$$\nabla \cdot f = \frac{\partial f_1}{\partial x_1} + \frac{\partial f_2}{\partial x_2} + \frac{\partial f_3}{\partial x_3} = -3b = -\varepsilon < 0 \tag{21}$$

where $\varepsilon = 3b = 0.06 > 0$.

Inserting the value of $\nabla \cdot f$ from (21) into (20), we get

$$\dot{V}(t) = \int\limits_{\Omega(t)} (-\varepsilon) \, dx_1 \, dx_2 \, dx_3 = -\varepsilon V(t) \tag{22}$$

Integrating the first order linear differential equation (22), we get

$$V(t) = \exp(-\varepsilon t) V(0) \tag{23}$$

Since $\varepsilon > 0$, it follows from Eq. (23) that $V(t) \to 0$ exponentially as $t \to \infty$. This shows that the novel chaotic system (9) is dissipative.

Hence, the system limit sets are ultimately confined into a specific limit set of zero volume, and the asymptotic motion of the novel chaotic system (9) settles onto a strange attractor of the system.

3.2 Equilibrium Points

We take the parameter values as in the chaotic case (10), viz. $a = 14$ and $b = 0.02$.

It is easy to see that the novel circulant circulant system (9) has infinitely many equilibrium points given by

$$E_\theta = \begin{bmatrix} \theta \\ \theta \\ \theta \end{bmatrix}, \tag{24}$$

where θ is a root of the transcendental equation

$$\sin\theta - \cos\theta = \frac{b}{a}\,\theta, \quad (a = 14, \quad b = 0.02) \tag{25}$$

If we define

$$\mu = \frac{b}{a}, \tag{26}$$

then we can express (25) equivalently as

$$\sin\theta - \cos\theta = \mu\,\theta, \quad (\mu = 0.0014) \tag{27}$$

Since the Eq. (27) has infinitely many roots θ, the novel chaotic circulant system (9) has infinitely many equilibrium points E_θ given by (24).

Using MATLAB, some equilibrium points of the novel chaotic circulant system (9) can be listed as follows:

$$\ldots, \begin{bmatrix} -2.3538 \\ -2.3538 \\ -2.3538 \end{bmatrix}, \begin{bmatrix} 0.7862 \\ 0.7862 \\ 0.7862 \end{bmatrix}, \begin{bmatrix} 3.9230 \\ 3.9230 \\ 3.9230 \end{bmatrix}, \begin{bmatrix} 7.0757 \\ 7.0757 \\ 7.0757 \end{bmatrix}, \ldots \tag{28}$$

The Jacobian matrix of the novel circulant chaotic system (9) at any point $\mathbf{x} \in \mathbf{R}^3$ is obtained as

$$J(\mathbf{x}) = \begin{bmatrix} -b & a(\cos x_2 + \sin x_2) & 0 \\ 0 & -b & a(\cos x_3 + \sin x_3) \\ a(\cos x_1 + \sin x_1) & 0 & -b \end{bmatrix} \tag{29}$$

For all equilibrium points E_θ, the matrix $J(E_\theta)$ has the same eigenvalues viz.

$$\lambda_1 = -19.8188, \quad \lambda_{2,3} = 9.8794 \pm 17.1463\, i \tag{30}$$

Thus, all the equilibrium points E_θ of the novel circulant chaotic system (9) are saddle-focus points, which are unstable.

3.3 Lyapunov Exponents and Kaplan–Yorke Dimension

We take the parameter values of the novel system (9) as in the chaotic case (10), viz. $a = 14$ and $b = 0.02$. We take the initial state of the novel system (9) as given in (11).

Then the Lyapunov exponents of the system (9) are numerically obtained as

$$L_1 = 2.1714, \quad L_2 = 0, \quad L_3 = -2.2373 \tag{31}$$

Thus, the maximal Lyapunov exponent of the novel circulant chaotic system (9) is $L_1 = 2.1714 > 0$.

Since the sum of the Lyapunov exponents of the novel circulant chaotic system (9) is negative, the system is dissipative.

Also, the Kaplan–Yorke dimension of the novel circulant chaotic system (9) is found as

$$D_{KY} = 2 + \frac{L_1 + L_2}{|L_3|} = 2.9705 \tag{32}$$

which is a very high value as it is close to three. This shows the high complexity of the novel circulant chaotic system (9). Hence, it is suitable for many engineering applications such as cryptosystems, secure communications, etc.

4 Adaptive Control of the Novel Circulant Chaotic System

In this section, we use adaptive control method to derive an adaptive feedback control law for globally stabilizing the novel 3-D circulant chaotic system with unknown parameters.

Thus, we consider the novel circulant chaotic system given by

$$\begin{cases} \dot{x}_1 = a(\sin x_2 - \cos x_2) - bx_1 + u_1 \\ \dot{x}_2 = a(\sin x_3 - \cos x_3) - bx_2 + u_2 \\ \dot{x}_3 = a(\sin x_1 - \cos x_1) - bx_3 + u_3 \end{cases} \tag{33}$$

In (33), x_1, x_2, x_3 are the states and u_1, u_2, u_3 are the adaptive controls to be determined using estimates $\hat{a}(t)$, $\hat{b}(t)$ for the unknown parameters a, b, respectively.

To simplify the notation, we define

$$F(\alpha) = \sin\alpha - \cos\alpha \tag{34}$$

Using (34), we can represent (33) in a simple form as

$$\begin{cases} \dot{x}_1 = aF(x_2) - bx_1 + u_1 \\ \dot{x}_2 = aF(x_3) - bx_2 + u_2 \\ \dot{x}_3 = aF(x_1) - bx_3 + u_3 \end{cases} \tag{35}$$

We consider the adaptive feedback control law

$$\begin{cases} u_1 = -\hat{a}(t)F(x_2) + \hat{b}(t)x_1 - k_1x_1 \\ u_2 = -\hat{a}(t)F(x_3) + \hat{b}(t)x_2 - k_2x_2 \\ u_3 = -\hat{a}(t)F(x_1) + \hat{b}(t)x_3 - k_3x_3 \end{cases} \tag{36}$$

where k_1, k_2, k_3 are positive gain constants.

Substituting (36) into (35), we get the closed-loop plant dynamics as

$$\begin{cases} \dot{x}_1 = \left[a - \hat{a}(t)\right] F(x_2) - \left[b - \hat{b}(t)\right] x_1 - k_1x_1 \\ \dot{x}_2 = \left[a - \hat{a}(t)\right] F(x_3) - \left[b - \hat{b}(t)\right] x_2 - k_2x_2 \\ \dot{x}_3 = \left[a - \hat{a}(t)\right] F(x_1) - \left[b - \hat{b}(t)\right] x_3 - k_2x_3 \end{cases} \tag{37}$$

The parameter estimation errors are defined as

$$\begin{cases} e_a(t) = a - \hat{a}(t) \\ e_b(t) = b - \hat{b}(t) \end{cases} \tag{38}$$

In view of (38), we can simplify the plant dynamics (37) as

$$\begin{cases} \dot{x}_1 = e_aF(x_2) - e_bx_1 - k_1x_1 \\ \dot{x}_2 = e_aF(x_3) - e_bx_2 - k_2x_2 \\ \dot{x}_3 = e_aF(x_1) - e_bx_3 - k_2x_3 \end{cases} \tag{39}$$

Differentiating (38) with respect to t, we obtain

$$\begin{cases} \dot{e}_a(t) = -\dot{\hat{a}}(t) \\ \dot{e}_b(t) = -\dot{\hat{b}}(t) \end{cases} \tag{40}$$

We consider the quadratic candidate Lyapunov function defined by

$$V(\mathbf{x}, e_a, e_b) = \frac{1}{2}\left(x_1^2 + x_2^2 + x_3^2\right) + \frac{1}{2}\left(e_a^2 + e_b^2\right) \tag{41}$$

Differentiating V along the trajectories of (39) and (40), we obtain

$$\begin{aligned} \dot{V} &= -k_1 x_1^2 - k_2 x_2^2 - k_3 x_3^2 + e_a \left[x_1 F(x_2) + x_2 F(x_3) + x_3 F(x_1) - \dot{\hat{a}} \right] \\ &\quad + e_b \left[-x_1^2 - x_2^2 - x_3^2 - \dot{\hat{b}} \right] \end{aligned} \tag{42}$$

In view of (42), we take the parameter update law as

$$\begin{cases} \dot{\hat{a}}(t) = x_1 F(x_2) + x_2 F(x_3) + x_3 F(x_1) \\ \dot{\hat{b}}(t) = -x_1^2 - x_2^2 - x_3^2 \end{cases} \tag{43}$$

Next, we state and prove the main result of this section.

Theorem 1 *The novel 3-D circulant chaotic system (35) with unknown system parameters is globally and exponentially stabilized for all initial conditions by the adaptive control law (36) and the parameter update law (43), where k_1, k_2, k_3 are positive gain constants and $F(\alpha)$ is defined by (34).*

Proof We prove this result by applying Lyapunov stability theory [9].

We consider the quadratic Lyapunov function defined by (41), which is clearly a positive definite function on $\mathbf{R}^5$.

By substituting the parameter update law (43) into (42), we obtain the time-derivative of V as

$$\dot{V} = -k_1 x_1^2 - k_2 x_2^2 - k_3 x_3^2 \tag{44}$$

From (44), it is clear that $\dot{V}$ is a negative semi-definite function on $\mathbf{R}^5$.

Thus, we can conclude that the state vector $\mathbf{x}(t)$ and the parameter estimation error are globally bounded i.e.

$$\left[x_1(t)\ x_2(t)\ x_3(t)\ e_a(t)\ e_b(t) \right]^T \in \mathbf{L}_\infty.$$

We define $k = \min\{k_1, k_2, k_3\}$.

Then it follows from (44) that

$$\dot{V} \leq -k \|\mathbf{x}(t)\|^2 \tag{45}$$

Thus, we have

$$k \|\mathbf{x}(t)\|^2 \leq -\dot{V} \tag{46}$$

Integrating the inequality (46) from 0 to t, we get

$$k \int_0^t \|\mathbf{x}(\tau)\|^2 \, d\tau \leq V(0) - V(t) \tag{47}$$

From (47), it follows that $\mathbf{x} \in \mathbf{L}_2$.

Using (39), we can conclude that $\dot{\mathbf{x}} \in \mathbf{L}_\infty$.

Using Barbalat's lemma [9], we conclude that $\mathbf{x}(t) \to 0$ exponentially as $t \to \infty$ for all initial conditions $\mathbf{x}(0) \in \mathbf{R}^3$.

This completes the proof. □

For the numerical simulations, the classical fourth-order Runge–Kutta method with step size $h = 10^{-8}$ is used to solve the systems (35) and (43), when the adaptive control law (36) is applied.

The parameter values of the novel 3-D circulant chaotic system (35) are taken as in the chaotic case (10), i.e. $a = 14$ and $b = 0.02$.

We take the positive gain constants as $k_i = 5$ for $i = 1, 2, 3$.

Furthermore, as initial conditions of the novel highly chaotic system (35), we take

$$x_1(0) = 15.3, \quad x_2(0) = 12.7, \quad x_3(0) = -16.9 \tag{48}$$

Also, as initial conditions of the parameter estimates, we take

$$\hat{a}(0) = 25.3, \quad \hat{b}(0) = 14.8 \tag{49}$$

In Fig. 6, the exponential convergence of the controlled states of the 3-D novel highly chaotic system (35) is depicted.

5 Adaptive Synchronization of the Identical Novel Circulant Chaotic Systems

In this section, we apply adaptive control method to derive an adaptive feedback control law for globally synchronizing identical 3-D novel circulant chaotic systems with unknown parameters.

To simplify the notation, we define

$$F(\alpha) = \sin\alpha - \cos\alpha \tag{50}$$

As the master system, we consider the novel circulant chaotic system given by

$$\begin{cases} \dot{x}_1 = aF(x_2) - bx_1 \\ \dot{x}_2 = aF(x_3) - bx_2 \\ \dot{x}_3 = aF(x_1) - bx_3 \end{cases} \tag{51}$$

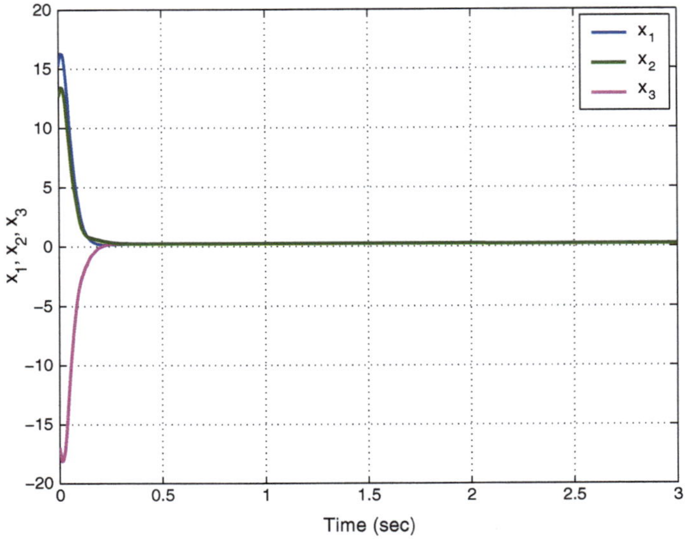

Fig. 6 Time-history of the controlled states x_1, x_2, x_3

In (51), x_1, x_2, x_3 are the states and a, b are unknown system parameters.

As the slave system, we consider the novel circulant chaotic system given by

$$\begin{cases} \dot{y}_1 = aF(y_2) - by_1 + u_1 \\ \dot{y}_2 = aF(y_3) - by_2 + u_2 \\ \dot{y}_3 = aF(y_1) - by_3 + u_3 \end{cases} \tag{52}$$

In (52), y_1, y_2, y_3 are the states and u_1, u_2, u_3 are the adaptive controls to be determined using estimates of the unknown system parameters.

The synchronization error between the novel chaotic systems is defined by

$$\begin{cases} e_1 = y_1 - x_1 \\ e_2 = y_2 - x_2 \\ e_3 = y_3 - x_3 \end{cases} \tag{53}$$

Then the error dynamics is obtained as

$$\begin{cases} \dot{e}_1 = a[F(y_2) - F(x_2)] - be_1 + u_1 \\ \dot{e}_2 = a[F(y_3) - F(x_3)] - be_2 + u_2 \\ \dot{e}_3 = a[F(y_1) - F(x_1)] - be_3 + u_3 \end{cases} \tag{54}$$

To simplify the notation, we define

$$G(\alpha, \beta) = F(\beta) - F(\alpha) \tag{55}$$

Then the error dynamics (54) can be simplified as

$$\begin{cases} \dot{e}_1 = a\,G(x_2, y_2) - be_1 + u_1 \\ \dot{e}_2 = a\,G(x_3, y_3) - be_2 + u_2 \\ \dot{e}_3 = a\,G(x_1, y_1) - be_3 + u_3 \end{cases} \tag{56}$$

We consider the adaptive feedback control law

$$\begin{cases} u_1 = -\hat{a}(t)\,G(x_2, y_2) + \hat{b}(t)e_1 - k_1 e_1 \\ u_2 = -\hat{a}(t)\,G(x_3, y_3) + \hat{b}(t)e_2 - k_2 e_2 \\ u_3 = -\hat{a}(t)\,G(x_1, y_1) + \hat{b}(t)e_3 - k_3 e_3 \end{cases} \tag{57}$$

where k_1, k_2, k_3 are positive gain constants.

Substituting (57) into (56), we get the closed-loop error dynamics as

$$\begin{cases} \dot{e}_1 = \left[a - \hat{a}(t)\right] G(x_2, y_2) - \left[b - \hat{b}(t)\right] e_1 - k_1 e_1 \\ \dot{e}_2 = \left[a - \hat{a}(t)\right] G(x_3, y_3) - \left[b - \hat{b}(t)\right] e_2 - k_2 e_2 \\ \dot{e}_3 = \left[a - \hat{a}(t)\right] G(x_1, y_1) - \left[b - \hat{b}(t)\right] e_3 - k_3 e_3 \end{cases} \tag{58}$$

The parameter estimation errors are defined as

$$\begin{cases} e_a(t) = a - \hat{a}(t) \\ e_b(t) = b - \hat{b}(t) \end{cases} \tag{59}$$

In view of (59), we can simplify the error dynamics (58) as

$$\begin{cases} \dot{e}_1 = e_a G(x_2, y_2) - e_b e_1 - k_1 e_1 \\ \dot{e}_2 = e_a G(x_3, y_3) - e_b e_2 - k_2 e_2 \\ \dot{e}_3 = e_a G(x_1, y_1) - e_b e_3 - k_3 e_3 \end{cases} \tag{60}$$

Differentiating (59) with respect to t, we obtain

$$\begin{cases} \dot{e}_a(t) = -\dot{\hat{a}}(t) \\ \dot{e}_b(t) = -\dot{\hat{b}}(t) \end{cases} \tag{61}$$

We consider the quadratic candidate Lyapunov function defined by

$$V(\mathbf{e}, e_a, e_b, e_c) = \frac{1}{2}\left(e_1^2 + e_2^2 + e_3^2\right) + \frac{1}{2}\left(e_a^2 + e_b^2\right) \tag{62}$$

Differentiating V along the trajectories of (60) and (61), we obtain

$$\begin{aligned} \dot{V} &= -k_1 e_1^2 - k_2 e_2^2 - k_3 e_3^2 \\ &\quad + e_a \left[e_1 G(x_2, y_2) + e_2 G(x_3, y_3) + e_3 G(x_1, y_1) - \dot{\hat{a}} \right] \\ &\quad + e_b \left[-e_1^2 - e_2^2 - e_3^2 - \dot{\hat{b}} \right] \end{aligned} \tag{63}$$

In view of (63), we take the parameter update law as

$$\begin{cases} \dot{\hat{a}}(t) = e_1 G(x_2, y_2) + e_2 G(x_3, y_3) + e_3 G(x_1, y_1) \\ \dot{\hat{b}}(t) = -e_1^2 - e_2^2 - e_3^2 \end{cases} \tag{64}$$

Next, we state and prove the main result of this section.

This result is proved by applying adaptive control theory and Lyapunov stability theory.

Theorem 2 *The novel circulant chaotic systems (51) and (52) with unknown system parameters are globally and exponentially synchronized for all initial conditions by the adaptive control law (57) and the parameter update law (64), where k_1, k_2, k_3 are positive gain constants and F, G are defined by the Eqs. (50) and (55), respectively.*

Proof We prove this result by applying Lyapunov stability theory [9].

We consider the quadratic Lyapunov function defined by (62), which is clearly a positive definite function on $\mathbf{R}^5$.

By substituting the parameter update law (64) into (63), we obtain

$$\dot{V} = -k_1 e_1^2 - k_2 e_2^2 - k_3 e_3^2 \tag{65}$$

From (65), it is clear that $\dot{V}$ is a negative semi-definite function on $\mathbf{R}^5$.

Thus, we can conclude that the error vector $\mathbf{e}(t)$ and the parameter estimation error are globally bounded, i.e.

$$\left[e_1(t) \; e_2(t) \; e_3(t) \; e_a(t) \; e_b(t) \right]^T \in \mathbf{L}_\infty. \tag{66}$$

We define $k = \min\{k_1, k_2, k_3\}$.

Then it follows from (65) that

$$\dot{V} \le -k \|\mathbf{e}(t)\|^2 \tag{67}$$

Thus, we have

$$k \|\mathbf{e}(t)\|^2 \le -\dot{V} \tag{68}$$

Integrating the inequality (68) from 0 to t, we get

$$k \int_0^t \|\mathbf{e}(\tau)\|^2 \, d\tau \leq V(0) - V(t) \tag{69}$$

From (69), it follows that $\mathbf{e} \in \mathbf{L}_2$.

Using (60), we can conclude that $\dot{\mathbf{e}} \in \mathbf{L}_\infty$.

Using Barbalat's lemma [9], we conclude that $\mathbf{e}(t) \to 0$ exponentially as $t \to \infty$ for all initial conditions $\mathbf{e}(0) \in \mathbf{R}^3$.

Hence, we have proved that novel circulant chaotic systems (51) and (52) with unknown system parameters are globally and exponentially synchronized for all initial conditions.

This completes the proof. □

For the numerical simulations, the classical fourth-order Runge–Kutta method with step size $h = 10^{-8}$ is used to solve the systems (51), (52) and (64), when the adaptive control law (57) is applied.

The parameter values of the novel chaotic systems are taken as in the chaotic case (10), i.e.

$$a = 14, \quad b = 0.02 \tag{70}$$

We take the positive gain constants as

$$k_1 = 5, \quad k_2 = 5, \quad k_3 = 5 \tag{71}$$

Furthermore, as initial conditions of the master system (51), we take

$$x_1(0) = 5.4, \quad x_2(0) = 12.7, \quad x_3(0) = 4.9 \tag{72}$$

As initial conditions of the slave system (52), we take

$$y_1(0) = 14.5, \quad y_2(0) = 3.5, \quad y_3(0) = 10.2 \tag{73}$$

Also, as initial conditions of the parameter estimates, we take

$$\hat{a}(0) = 6.1, \quad \hat{b}(0) = 24.8 \tag{74}$$

Figures 7, 8 and 9 describe the complete synchronization of the novel circulant chaotic systems (51) and (52), while Fig. 10 describes the time-history of the synchronization errors e_1, e_2, e_3.

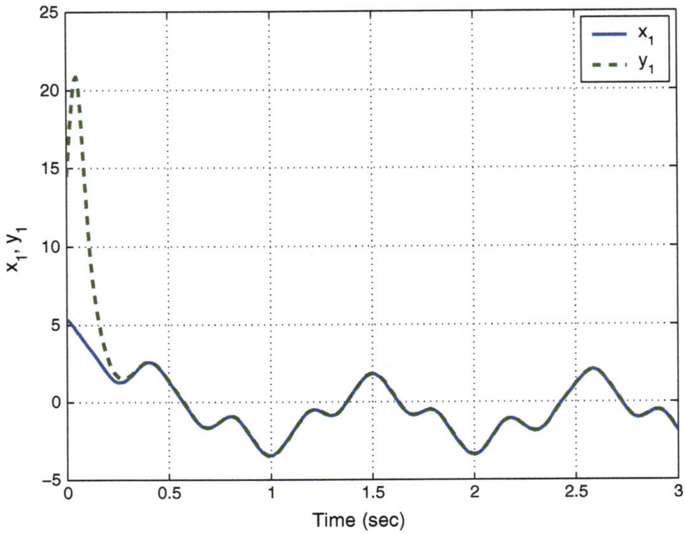

Fig. 7 Synchronization of the states x_1 and y_1

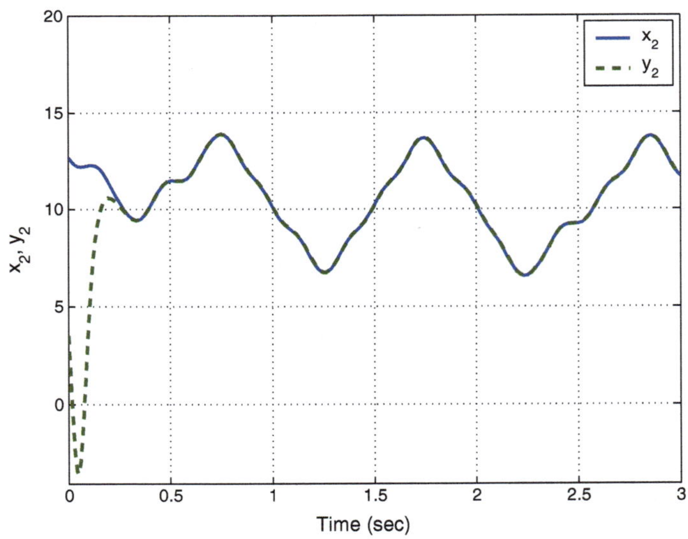

Fig. 8 Synchronization of the states x_2 and y_2

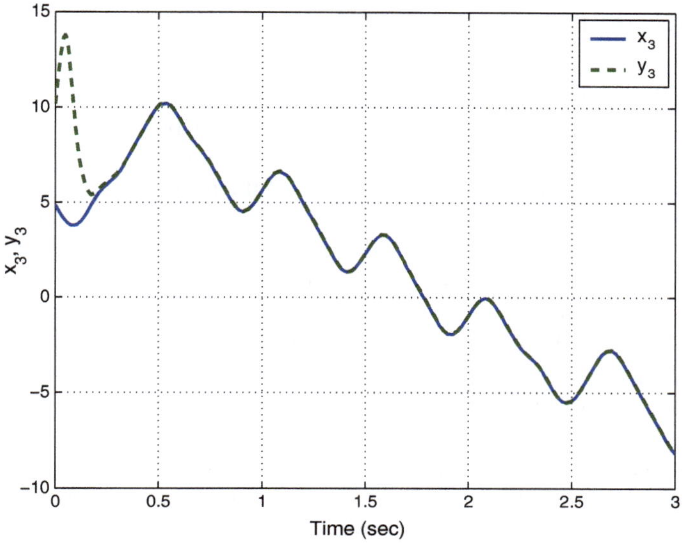

Fig. 9 Synchronization of the states x_3 and y_3

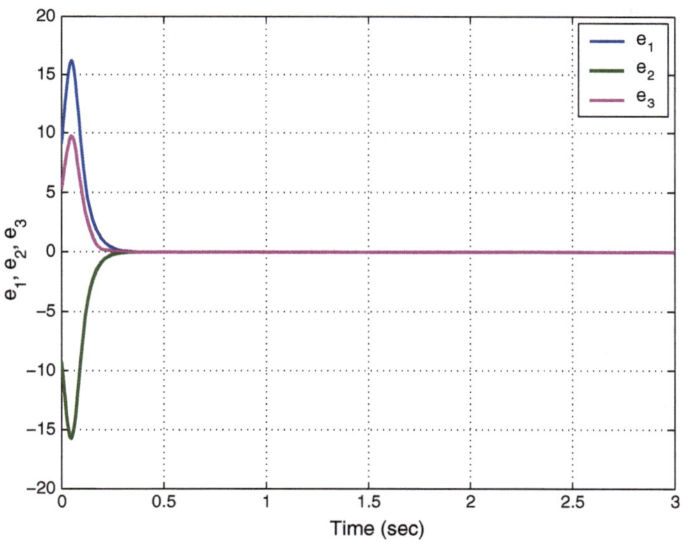

Fig. 10 Time-history of the synchronization errors e_1, e_2, e_3

6 Conclusions

In this work, we announced a novel 3-D dissipative circulant chaotic system with Labyrinth chaos. The novel chaotic system is a nine-term polynomial system with six sinusoidal nonlinearities. The phase portraits of the novel circulant chaotic system were illustrated and the qualitative properties of the novel circulant chaotic system were discussed. The novel circulant chaotic system has infinitely many equilibrium points and it exhibits labyrinth chaos. We showed that all the equilibrium points of the novel circulant chaotic system are saddle-foci and hence they are unstable. The Lyapunov exponents of the novel circulant chaotic system were obtained as $L_1 = 2.1714$, $L_2 = 0$ and $L_3 = -2.2373$, while the Kaplan–Yorke dimension of the novel circulant chaotic system was derived as $D_{KY} = 2.9705$. Since the Kaplan–Yorke dimension of the the novel circulant chaotic system has a very high value and close to three, the novel circulant chaotic system with labyrinth chaos exhibits highly complex behaviour. Hence, it is suitable for engineering applications such as cryptosystems, secure communications, etc. In this work, we also derived new results for the global chaos control of the novel circulant chaotic system with unknown parameters using adaptive control method. Finally, we also derived new results for the global chaos synchronization of the identical novel circulant chaotic systems with unknown parameters using adaptive control method. The main control results were established using Lyapunov stability theory. MATLAB simulations were depicted to illustrate all the main results of this work.

References

1. Abdurrahman A, Jiang H, Teng Z (2015) Finite-time synchronization for memristor-based neural networks with time-varying delays. Neural Netw 69:20–28
2. Arneodo A, Coullet P, Tresser C (1981) Possible new strange attractors with spiral structure. Commun Math Phys 79(4):573–576
3. Azar AT, Vaidyanathan S (2015) Chaos Modeling and Control Systems Design, vol 581. Springer, Germany
4. Behnia S, Afrang S, Akhshani A, Mabhouti K (2013) A novel method for controlling chaos in external cavity semiconductor laser. Optik 124(8):757–764
5. Cai G, Tan Z (2007) Chaos synchronization of a new chaotic system via nonlinear control. J Uncertain Syst 1(3):235–240
6. Chen G, Ueta T (1999) Yet another chaotic attractor. Int J Bifurc Chaos 9(7):1465–1466
7. Islam MM, Murase K (2005) Chaotic dynamics of a behaviour-based miniature mobile robot: effects of environment and control structure. Neural Netw 18(2):123–144
8. Karthikeyan R, Sundarapandian V (2014) Hybrid chaos synchronization of four-scroll systems via active control. J Electr Eng 65(2):97–103
9. Khalil HK (2001) Nonlinear Systems, 3rd edn. Prentice Hall, New Jersey
10. Li D (2008) A three-scroll chaotic attractor. Phys Lett A 372(4):387–393
11. Lorenz EN (1963) Deterministic periodic flow. J Atmos Sci 20(2):130–141
12. Lü J, Chen G (2002) A new chaotic attractor coined. Int J Bifurc Chaos 12(3):659–661
13. Matouk AE (2011) Chaos, feedback control and synchronization of a fractional-order modified autonomous Van der Pol-Duffing circuit. Commun Nonlinear Sci Numer Simul 16(2):975–986

14. Njah AN, Sunday OD (2009) Generalization on the chaos control of 4-D chaotic systems using recursive backstepping nonlinear controller. Chaos, Solitons Fractals 41(5):2371–2376
15. Pehlivan I, Moroz IM, Vaidyanathan S (2014) Analysis, synchronization and circuit design of a novel butterfly attractor. J Sound Vib 333(20):5077–5096
16. Pham VT, Volos CK, Vaidyanathan S, Le TP, Vu VY (2015) A memristor-based hyperchaotic system with hidden attractors: dynamics, synchronization and circuital emulating. J Eng Sci Technol Rev 8(2):205–214
17. Rasappan S, Vaidyanathan S (2013) Hybrid synchronization of *n*-scroll Chua circuits using adaptive backstepping control design with recursive feedback. Malays J Math Sci 73(1):73–95
18. Rasappan S, Vaidyanathan S (2014) Global chaos synchronization of WINDMI and Coullet chaotic systems using adaptive backstepping control design. Kyungpook Math J 54(1):293–320
19. Rhouma R, Belghith S (2011) Cryptanalysis of a chaos-based cryptosystem. Commun Nonlinear Sci Numer Simul 16(2):876–884
20. Rössler OE (1976) An equation for continuous chaos. Phys Lett A 57(5):397–398
21. Sampath S, Vaidyanathan S, Volos CK, Pham VT (2015) An eight-term novel four-scroll chaotic system with cubic nonlinearity and its circuit simulation. J Eng Sci Technol Rev 8(2):1–6
22. Sarasu P, Sundarapandian V (2011a) Active controller design for generalized projective synchronization of four-scroll chaotic systems. Int J Syst Signal Control Eng Appl 4(2):26–33
23. Sarasu P, Sundarapandian V (2011b) The generalized projective synchronization of hyperchaotic Lorenz and hyperchaotic Qi systems via active control. Int J Soft Comput 6(5):216–223
24. Sarasu P, Sundarapandian V (2012) Generalized projective synchronization of two-scroll systems via adaptive control. Int J Soft Comput 7(4):146–156
25. Sprott JC (1994) Some simple chaotic flows. Phys Rev E 50(2):647–650
26. Sprott JC (2010) Elegant Chaos. World Scientific, Singapore
27. Sundarapandian V (2010) Output regulation of the Lorenz attractor. Far East J Math Sci 42(2):289–299
28. Sundarapandian V (2013a) Adaptive control and synchronization design for the Lu-Xiao chaotic system. Lect Notes Electr Eng 131:319–327
29. Sundarapandian V (2013b) Analysis and anti-synchronization of a novel chaotic system via active and adaptive controllers. J Eng Sci Technol Rev 6(4):45–52
30. Sundarapandian V, Karthikeyan R (2011a) Anti-synchronization of hyperchaotic Lorenz and hyperchaotic Chen systems by adaptive control. Int J Syst Signal Control Eng Appl 4(2):18–25
31. Sundarapandian V, Karthikeyan R (2011b) Anti-synchronization of Lü and Pan chaotic systems by adaptive nonlinear control. Eur J Sci Res 64(1):94–106
32. Sundarapandian V, Karthikeyan R (2012) Adaptive anti-synchronization of uncertain Tigan and Li systems. J Eng Appl Sci 7(1):45–52
33. Sundarapandian V, Pehlivan I (2012) Analysis, control, synchronization, and circuit design of a novel chaotic system. Math Comput Model 55(7–8):1904–1915
34. Sundarapandian V, Sivaperumal S (2011) Sliding controller design of hybrid synchronization of four-wing chaotic systems. Int J Soft Comput 6(5):224–231
35. Suresh R, Sundarapandian V (2013) Global chaos synchronization of a family of *n*-scroll hyperchaotic Chua circuits using backstepping control with recursive feedback. Far East J Math Sci 7(2):219–246
36. Thomas R (1999) Deterministic chaos seen in terms of feedback circuits: analysis, synthesis, 'labyrnth chaos'. Int J Bifurc Chaos 9:1889–1905
37. Tigan G, Opris D (2008) Analysis of a 3D chaotic system. Chaos, Solitons Fractals 36: 1315–1319
38. Tuwankotta JM (2006) Chaos in a coupled oscillators system with widely spaced frequencies and energy-preserving non-linearity. Int J Non-Linear Mech 41(2):180–191
39. Usama M, Khan MK, Alghathbar K, Lee C (2010) Chaos-based secure satellite imagery cryptosystem. Comput Math Appl 60(2):326–337

40. Vaidyanathan S (2011a) Output regulation of Arneodo-Coullet chaotic system. Commun Comput Inf Sci 133:98–107
41. Vaidyanathan S (2011b) Output regulation of the unified chaotic system. Commun Comput Inf Sci 198:1–9
42. Vaidyanathan S (2012a) Adaptive controller and syncrhonizer design for the Qi-Chen chaotic system. Lect Notes Inst Comput Sci Social-Inform Telecommun Eng 84:73–82
43. Vaidyanathan S (2012b) Anti-synchronization of Sprott-L and Sprott-M chaotic systems via adaptive control. Int J Control Theory Appl 5(1):41–59
44. Vaidyanathan S (2012c) Global chaos control of hyperchaotic Liu system via sliding control method. Int J Control Theory Appl 5(2):117–123
45. Vaidyanathan S (2012d) Sliding mode control based global chaos control of Liu-Liu-Liu-Su chaotic system. Int J Control Theory and Appl 5(1):15–20
46. Vaidyanathan S (2013a) A new six-term 3-D chaotic system with an exponential nonlinearity. Far East J Math Sci 79(1):135–143
47. Vaidyanathan S (2013b) A ten-term novel 4-D hyperchaotic system with three quadratic nonlinearities and its control. Int J Control Theory Appl 6(2):97–109
48. Vaidyanathan S (2013c) Analysis and adaptive synchronization of two novel chaotic systems with hyperbolic sinusoidal and cosinusoidal nonlinearity and unknown parameters. J Eng Sci Technol Rev 6(4):53–65
49. Vaidyanathan S (2013d) Analysis, control and synchronization of hyperchaotic Zhou system via adaptive control. Adv Intell Syst Comput 177:1–10
50. Vaidyanathan S (2014a) A new eight-term 3-D polynomial chaotic system with three quadratic nonlinearities. Far East J Math Sci 84(2):219–226
51. Vaidyanathan S (2014b) Analysis and adaptive synchronization of eight-term 3-D polynomial chaotic systems with three quadratic nonlinearities. Eur Phys J: Spec Top 223(8):1519–1529
52. Vaidyanathan S (2014c) Analysis, control and synchronisation of a six-term novel chaotic system with three quadratic nonlinearities. Int J Model Identif Control 22(1):41–53
53. Vaidyanathan S (2014d) Generalised projective synchronisation of novel 3-D chaotic systems with an exponential non-linearity via active and adaptive control. Int J Model Identif Control 22(3):207–217
54. Vaidyanathan S (2014e) Global chaos synchronisation of identical Li-Wu chaotic systems via sliding mode control. Int J Model Identif Control 22(2):170–177
55. Vaidyanathan S (2014f) Qualitative analysis and control of an eleven-term novel 4-D hyperchaotic system with two quadratic nonlinearities. Int J Control Theory Appl 7:35–47
56. Vaidyanathan S (2015a) 3-cells Cellular Neural Network (CNN) attractor and its adaptive biological control. Int J PharmTech Res 8(4):632–640
57. Vaidyanathan S (2015b) A 3-D novel highly chaotic system with four quadratic nonlinearities, its adaptive control and anti-synchronization with unknown parameters. J Eng Sci Technol Rev 8(2):106–115
58. Vaidyanathan S (2015c) A novel chemical chaotic reactor system and its adaptive control. Int J ChemTech Res 8(7):146–158
59. Vaidyanathan S (2015d) Adaptive backstepping control of enzymes-substrates system with ferroelectric behaviour in brain waves. Int J PharmTech Res 8(2):256–261
60. Vaidyanathan S (2015e) Adaptive biological control of generalized Lotka-Volterra three-species biological system. Int J PharmTech Res 8(4):622–631
61. Vaidyanathan S (2015f) Adaptive chaotic synchronization of enzymes-substrates system with ferroelectric behaviour in brain waves. Int J PharmTech Res 8(5):964–973
62. Vaidyanathan S (2015g) Adaptive control of a chemical chaotic reactor. Int J PharmTech Res 8(3):377–382
63. Vaidyanathan S (2015h) Adaptive control of the FitzHugh-Nagumo chaotic neuron model. Int J PharmTech Res 8(6):117–127
64. Vaidyanathan S (2015i) Adaptive synchronization of chemical chaotic reactors. Int J ChemTech Res 8(2):612–621

65. Vaidyanathan S (2015j) Adaptive synchronization of generalized Lotka-Volterra three-species biological systems. Int J PharmTech Res 8(5):928–937
66. Vaidyanathan S (2015k) Adaptive synchronization of novel 3-D chemical chaotic reactor systems. Int J ChemTech Res 8(7):159–171
67. Vaidyanathan S (2015l) Adaptive synchronization of the identical FitzHugh-Nagumo chaotic neuron models. Int J PharmTech Res 8(6):167–177
68. Vaidyanathan S (2015m) Analysis, control and synchronization of a 3-D novel jerk chaotic system with two quadratic nonlinearities. Kyungpook Math J 55:563–586
69. Vaidyanathan S (2015n) Analysis, properties and control of an eight-term 3-D chaotic system with an exponential nonlinearity. Int J Model Identif and Control 23(2):164–172
70. Vaidyanathan S (2015o) Anti-synchronization of brusselator chemical reaction systems via adaptive control. Int J ChemTech Res 8(6):759–768
71. Vaidyanathan S (2015p) Chaos in neurons and adaptive control of Birkhoff-Shaw strange chaotic attractor. Int J PharmTech Res 8(5):956–963
72. Vaidyanathan S (2015q) Chaos in neurons and synchronization of Birkhoff-Shaw strange chaotic attractors via adaptive control. Int J PharmTech Res 8(6):1–11
73. Vaidyanathan S (2015r) Coleman-Gomatam logarithmic competitive biology models and their ecological monitoring. Int J PharmTech Res 8(6):94–105
74. Vaidyanathan S (2015s) Dynamics and control of brusselator chemical reaction. Int J ChemTech Res 8(6):740–749
75. Vaidyanathan S (2015t) Dynamics and control of tokamak system with symmetric and magnetically confined plasma. Int J ChemTech Res 8(6):795–803
76. Vaidyanathan S (2015u) Global chaos synchronization of chemical chaotic reactors via novel sliding mode control method. Int J ChemTech Res 8(7):209–221
77. Vaidyanathan S (2015v) Global chaos synchronization of the forced Van der Pol chaotic oscillators via adaptive control method. Int J PharmTech Res 8(6):156–166
78. Vaidyanathan S (2015w) Global chaos synchronization of the Lotka-Volterra biological systems with four competitive species via active control. Int J PharmTech Res 8(6):206–217
79. Vaidyanathan S (2015x) Lotka-Volterra population biology models with negative feedback and their ecological monitoring. Int J PharmTech Res 8(5):974–981
80. Vaidyanathan S (2015y) Lotka-Volterra two species competitive biology models and their ecological monitoring. Int J PharmTech Res 8(6):32–44
81. Vaidyanathan S (2015z) Output regulation of the forced Van der Pol chaotic oscillator via adaptive control method. Int J PharmTech Res 8(6):106–116
82. Vaidyanathan S, Azar AT (2015a) Analysis and control of a 4-D novel hyperchaotic system. Stud Comput Intell 581:3–17
83. Vaidyanathan S, Azar AT (2015b) Analysis, control and synchronization of a nine-term 3-D novel chaotic system. In: Azar AT, Vaidyanathan S (eds) Chaos Modelling and Control Systems Design, Studies in Computational Intelligence, vol 581. Springer, Germany, pp 19–38
84. Vaidyanathan S, Madhavan K (2013) Analysis, adaptive control and synchronization of a seven-term novel 3-D chaotic system. Int J Control Theory Appl 6(2):121–137
85. Vaidyanathan S, Pakiriswamy S (2013) Generalized projective synchronization of six-term Sundarapandian chaotic systems by adaptive control. Int J Control Theory Appl 6(2):153–163
86. Vaidyanathan S, Pakiriswamy S (2015) A 3-D novel conservative chaotic system and its generalized projective synchronization via adaptive control. J Eng Sci Technol Rev 8(2):52–60
87. Vaidyanathan S, Rajagopal K (2011) Hybrid synchronization of hyperchaotic Wang-Chen and hyperchaotic Lorenz systems by active non-linear control. Int J Syst Signal Control Eng Appl 4(3):55–61
88. Vaidyanathan S, Rajagopal K (2012) Global chaos synchronization of hyperchaotic Pang and hyperchaotic Wang systems via adaptive control. Int J Soft Comput 7(1):28–37
89. Vaidyanathan S, Rasappan S (2011) Global chaos synchronization of hyperchaotic Bao and Xu systems by active nonlinear control. Commun Comput Inform Sci 198:10–17

90. Vaidyanathan S, Rasappan S (2014) Global chaos synchronization of n-scroll Chua circuit and Lur'e system using backstepping control design with recursive feedback. Arab J Sci Eng 39(4):3351–3364
91. Vaidyanathan S, Sampath S (2012) Anti-synchronization of four-wing chaotic systems via sliding mode control. Int J Autom Comput 9(3):274–279
92. Vaidyanathan S, Volos C (2015) Analysis and adaptive control of a novel 3-D conservative no-equilibrium chaotic system. Archiv Control Sci 25(3):333–353
93. Vaidyanathan S, Volos C, Pham VT (2014a) Hyperchaos, adaptive control and synchronization of a novel 5-D hyperchaotic system with three positive Lyapunov exponents and its SPICE implementation. Archiv Control Sci 24(4):409–446
94. Vaidyanathan S, Volos C, Pham VT, Madhavan K, Idowu BA (2014b) Adaptive backstepping control, synchronization and circuit simulation of a 3-D novel jerk chaotic system with two hyperbolic sinusoidal nonlinearities. Arch Control Sci 24(3):375–403
95. Vaidyanathan S, Azar AT, Rajagopal K, Alexander P (2015a) Design and SPICE implementation of a 12-term novel hyperchaotic system and its synchronisation via active control. Int J Model Identif Control 23(3):267–277
96. Vaidyanathan S, Idowu BA, Azar AT (2015b) Backstepping controller design for the global chaos synchronization of Sprott's jerk systems. Stud Comput Intell 581:39–58
97. Vaidyanathan S, Rajagopal K, Volos CK, Kyprianidis IM, Stouboulos IN (2015c) Analysis, adaptive control and synchronization of a seven-term novel 3-D chaotic system with three quadratic nonlinearities and its digital implementation in LabVIEW. J Eng Sci Technol Rev 8(2):130–141
98. Vaidyanathan S, Sampath S, Azar AT (2015d) Global chaos synchronisation of identical chaotic systems via novel sliding mode control method and its application to Zhu system. Int J Model Identif Control 23(1):92–100
99. Vaidyanathan S, Volos C, Pham VT, Madhavan K (2015e) Analysis, adaptive control and synchronization of a novel 4-D hyperchaotic hyperjerk system and its SPICE implementation. Nonlinear Dyn 25(1):135–158
100. Vaidyanathan S, Volos CK, Kyprianidis IM, Stouboulos IN, Pham VT (2015f) Analysis, adaptive control and anti-synchronization of a six-term novel jerk chaotic system with two exponential nonlinearities and its circuit simulation. J Eng Sci Technol Rev 8(2):24–36
101. Vaidyanathan S, Volos CK, Pham VT (2015g) Analysis, adaptive control and adaptive synchronization of a nine-term novel 3-D chaotic system with four quadratic nonlinearities and its circuit simulation. J Eng Sci Technol Rev 8(2):181–191
102. Vaidyanathan S, Volos CK, Pham VT (2015h) Analysis, control, synchronization and SPICE implementation of a novel 4-D hyperchaotic Rikitake dynamo system without equilibrium. J Eng Sci Technol Rev 8(2):232–244
103. Vaidyanathan S, Volos CK, Pham VT (2015i) Global chaos control of a novel nine-term chaotic system via sliding mode control. In: Azar AT, Zhu Q (eds) Advances and Applications in Sliding Mode Control Systems, Studies in Computational Intelligence, vol 576. Springer, Germany, pp 571–590
104. Vincent UE, Njah AN, Laoye JA (2007) Controlling chaos and deterministic directed transport in inertia ratchets using backstepping control. Phys D 231(2):130–136
105. Volos CK, Kyprianidis IM, Stouboulos IN, Anagnostopoulos AN (2009) Experimental study of the dynamic behavior of a double scroll circuit. J Appl Funct Anal 4:703–711
106. Volos CK, Kyprianidis IM, Stouboulos IN (2013) Experimental investigation on coverage performance of a chaotic autonomous mobile robot. Robot Auton Syst 61(12):1314–1322
107. Volos CK, Kyprianidis IM, Stouboulos IN, Tlelo-Cuautle E, Vaidyanathan S (2015) Memristor: a new concept in synchronization of coupled neuromorphic circuits. J Eng Sci Technol Rev 8(2):157–173
108. Yang J, Zhu F (2013) Synchronization for chaotic systems and chaos-based secure communications via both reduced-order and step-by-step sliding mode observers. Commun Nonlinear Sci Numer Simul 18(4):926–937

109. Yang J, Chen Y, Zhu F (2014) Singular reduced-order observer-based synchronization for uncertain chaotic systems subject to channel disturbance and chaos-based secure communication. Appl Math Comput 229:227–238
110. Zhou W, Xu Y, Lu H, Pan L (2008) On dynamics analysis of a new chaotic attractor. Phys Lett A 372(36):5773–5777
111. Zhu C, Liu Y, Guo Y (2010) Theoretic and numerical study of a new chaotic system. Intell Inform Manag 2:104–109

A 3-D Novel Jerk Chaotic System and Its Application in Secure Communication System and Mobile Robot Navigation

Aceng Sambas, Sundarapandian Vaidyanathan, Mustafa Mamat, W.S. Mada Sanjaya and Darmawan Setia Rahayu

Abstract In this work, we study the complex dynamics of a six-term 3-D novel jerk chaotic system with two hyperbolic sinusoidal nonlinearities, which was proposed by Vaidyanathan et al. Arch Control Sci 24(3):375–403, 2014, [98]. The initial study in this chapter is to analyze the eigenvalue structures, various attractors, Lyapunov exponent analysis, Kaplan Yorke dimension, FFT analysis and Poincaré map analysis. We have studied the dynamic behavior of the system in the case of the bidirectional coupling via a linear resistor. Both experimental and simulation results have shown that chaotic synchronization is possible. Furthermore, the effectiveness of the bidirectional coupling method scheme between two identical 3-D novel jerk chaotic systems in a secure communication system is presented in detail. Also, the driving strategy of a mobile robot is studied, in order to generate the most unpredictable trajectory. Kinematics model of the robot's movement has been created and combined with the networking system of a 3-D novel jerk circuit so that the movement of the robot is very difficult to predict. Finally, numerical simulations by using MATLAB

A. Sambas (✉)
Department of Mechanical Engineering, Universitas Muhammadiyah Tasikmalaya, Tasikmalaya, Indonesia
e-mail: acenx.bts@gmail.com

M. Mamat
Faculty of Informatics and Computing, Universiti Sultan Zainal Abidin, Kuala Terengganu, Malaysia
e-mail: must@unisza.edu.my

S. Vaidyanathan
Research and Development Centre, Vel Tech University, Chennai, Tamil Nadu, India
e-mail: sundarvtu@gmail.com

W.S.M. Sanjaya
Department of Physics, Universitas Islam Negeri Sunan Gunung Djati Bandung, Bandung, Indonesia
e-mail: madasws@gmail.com

D.S. Rahayu
Bolabot Techno Robotic School, CV Sanjaya Star Group, Bandung, Indonesia
e-mail: darmawan.sr@gmail.com

S. Vaidyanathan and C. Volos (eds.), *Advances and Applications in Chaotic Systems*, Studies in Computational Intelligence 636,
DOI 10.1007/978-3-319-30279-9_12

2010, experimental results, as well as the implementation of circuit simulations by using Proteus has been performed in this study.

Keywords Chaos · Jerk systems · Bidirectional coupling · Chaos synchronization · Secure communication · Mobile robot

1 Introduction

Chaotic systems are defined as nonlinear dynamical systems which are sensitive to initial conditions, topologically mixing and with dense periodic orbits [5]. Sensitivity to initial conditions of chaotic systems is popularly known as the *butterfly effect*. Small changes in an initial state will make a very large difference in the behavior of the system at future states.

Some classical paradigms of 3-D chaotic systems in the literature are Lorenz system [16], Moore–Spiegel system [20], Rössler system [33], ACT system [4], Sprott systems [52], Chen system [7], Lü system [10], Cai system [6], Tigan system [64], Malasoma system [18], etc.

Many new chaotic systems have been discovered in the recent years such as Zhou system [118], Zhu system [119], Li system [12], Wei–Yang system [114], Han system [11], Sundarapandian systems [57, 61], Vaidyanathan systems [68, 70, 72–76, 80, 81, 84, 85, 98, 100, 102, 103, 105], Pehlivan system [28], Sampath–Vaidyanathan system [42], Pandey system [25], Pham system [31], etc.

The synchronization of chaotic systems is a phenomenon that occurs when two or more chaotic systems are coupled or when a chaotic system drives another chaotic system [110, 113]. Because of the butterfly effect which causes exponential divergence of the trajectories of two identical chaotic systems started with nearly the same initial conditions, the synchronization of chaotic systems is a challenging research problem in the chaos literature [5].

Major works on synchronization of chaotic systems deal with the complete synchronization (CS) of a pair of chaotic systems called the *master* and *slave* systems. The design goal of the complete synchronization is to apply the output of the master system to control the slave system so that the output of the slave system tracks the output of the master system asymptotically with time.

Chaos theory has important applications in several branches of science and engineering like physics [51, 104], economics [106], biology [43, 77], ecology [44], psychology [54], chemical reactions [21, 78, 79], robotics [19, 24, 108], random bits generator [107], encryption [1, 2, 14, 109], communication devices [3, 8], secure communication system [39–41, 117], circuits [13, 112], memristors [30, 32, 111], neural networks [15], fuzzy systems [23], etc.

Synchronization of chaotic systems deals with the problem of synchronizing the respective states of two chaotic systems called as *master* and *slave* systems asymptotically with time. Because of the butterfly effect, which causes the exponential divergence of the trajectories of two identical chaotic systems started with nearly the same initial conditions, synchronizing two chaotic systems is seemingly a very challenging problem in the chaos literature. Pecora and Carroll pioneered research on the chaos synchronization problem with a seminal work [27].

Over the last three decades, various types of chaos synchronization problems have been proposed in the chaos literature, which include complete synchronization [26, 34, 36, 56, 63, 66, 69, 71, 88, 89, 91–93, 95, 97, 99, 101], anti-synchronization [58, 59, 67, 82, 86, 87, 96], hybrid synchronization [9, 35, 37, 60, 62, 65, 83, 90, 94], generalized synchronization [38], impulsive synchronization [110], unidirectional synchronization [40], bidirectional synchronization [41], projective synchronization [17], generalized projective synchronization [45–49], etc.

Synchronization of chaotic systems is a key issue in symmetric chaos based secure communication schemes. Many researchers demonstrated, using simulation, that chaos can be synchronized and applied to secure communication schemes such as secure fiber optical communication scheme using chaos [3], secure communication based on chaotic stream cipher [117], secure communication with chaotic lasers [50], chaotic communication on a satellite formation flying [8] and satellite communications using transiently chaotic neural networks [15].

In mobile robot navigation, recent efforts to apply the theory of dynamical systems to robotic include the study of chaos in patrol mobile robot [19], floor-cleaning operation robot [24], motion control of robots using a chaotic truly random bits generator [107], chaotic mobile robot using fuzzy logic [23] and chaotic robot prediction by neuro-fuzzy control [22].

In this work, we investigate the novel Vaidyanathan jerk chaotic system with two hyperbolic sinusoidal nonlinearities [98]. Basic dynamical properties of the novel Vaidyanathan novel jerk chaotic system are analyzed by means of equilibrium points, eigenvalue structures, Lyapunov exponents. Kaplan Yorke dimension, Poincaré map and FFT analysis. To confirm the validity of the theoretical model, we also discuss the implementation of the novel Vaidyanathan jerk chaotic system by circuit design and simulation in MultiSIM. Furthermore, simulation results are used to visualize and illustrate the effectiveness of Vaidyanathan jerk chaotic system in synchronization and secure communication. In mobile robot navigation, kinematic model, numerical simulation and experimental results have been performed in this study.

In Sect. 2, a 3-D novel Vaidyanathan jerk chaotic system [98] is introduced and its qualitative properties are discussed in detail. In Sect. 3, the circuital realization of the Vaidyanathan jerk chaotic system is described. In Sect. 4, the bidirectional method is applied to synchronize identical novel Vaidyanathan jerk chaotic systems. In Sect. 5, the chaotic masking communication method of the 3-D novel Vaidyanathan jerk chaotic system is realized using MATLAB and MultiSIM programs. In Sect. 6, chaotic navigation mobile robot using 3-D novel Vaidyanathan jerk chaotic system is presented in detail. Section 7 summarizes the main results derived in this work.

2 3-D Novel Vaidyanathan Jerk Chaotic System

Moore and Spiegel found a model for the irregular variability in the luminosity of stars [20]. Moore–Spiegel chaotic system is given by the third order differential equation

$$\dddot{x} = -\ddot{x} + 9\dot{x} - x^2\dot{x} - 5x \tag{1}$$

In mechanics, if x represents the displacement of a body, then $\dot{x}$ and $\ddot{x}$ represent the *velocity* and *acceleration* of the body, respectively. The third order derivative $\dddot{x}$ is called the *jerk* of the body. Thus, the Moore–Spiegel system (1) represents a jerk dynamics.

By defining state variables $x_1 = x$, $x_2 = \dot{x}$ and $x_3 = \ddot{x}$, we can also represent the Moore–Spiegel jerk dynamics (1) in system form as

$$\begin{aligned} \dot{x}_1 &= x_2 \\ \dot{x}_2 &= x_3 \\ \dot{x}_3 &= -5x_1 + (9 - x_1^2)x_2 - x_3 \end{aligned} \tag{2}$$

In system form, the Moore–Spiegel chaotic system (2) is a six-term polynomial chaotic system with a cubic nonlinearity.

The simplest cubic case of a jerk chaotic dynamics was proposed by Malasoma [18]. Malasoma cubic jerk dynamics is given by

$$\dddot{x} = -a\ddot{x} + x\dot{x}^2 - x, \tag{3}$$

which is chaotic for $a = 2.03$.

In system form, Malasoma chaotic dynamics (3) can be represented as

$$\begin{aligned} \dot{x}_1 &= x_2 \\ \dot{x}_2 &= x_3 \\ \dot{x}_3 &= -x_1 + x_1 x_2^2 - a x_3 \end{aligned} \tag{4}$$

In 2000, Sprott found a new class of chaotic circuits using piecewise linear functions [53]. In 2009, Sun and Sprott found a simple jerk system with piecewise exponential nonlinearity [55]. There is also some interest in the literature in finding hyperjerk hyperchaotic systems and a recent work [101] discusses the properties of the Vaidyanathan hyperjerk system. In this work, we investigate the properties of the 3-D novel Vaidyanathan jerk chaotic system [98].

The 3-D novel Vaidyanathan jerk chaotic system is described by

$$\begin{cases} \dot{x} = y \\ \dot{y} = z \\ \dot{z} = x - a\left[\sinh(x) + \sinh(y)\right] - bz \end{cases} \tag{5}$$

where a and b are positive parameters.

In [101], it was shown that the Vaidyanathan jerk system (5) is *chaotic* when $a = 0.4$ and $b = 0.8$.

In this work, we take the parameters of the Vaidyanathan jerk system (5) as $a = 0.4$ and $b = 0.67$. For numerical simulations, we take the initial condition as $(x_0, y_0, z_0) = (0.8, 1.2, 0.5)$.

2.1 Equilibrium Points

The equilibrium points of the 3-D novel Vaidyanathan jerk system (5) for the chaotic case $a = 0.4$ and $b = 0.67$ are obtained by solving the system of equations

$$\begin{cases} y = 0 \\ z = 0 \\ x - a\left[\sinh(x) + \sinh(y)\right] - bz = 0 \end{cases} \tag{6}$$

Solving the system (6), we obtain

$$x - 0.4\sinh(x) = 0, \quad y = 0, \quad z = 0 \tag{7}$$

Thus, we find that the Vaidyanathan jerk system (5) has three equilibrium points given by

$$E_0 = \begin{bmatrix} 0 \\ 0 \\ 0 \end{bmatrix}, \quad E_+ = \begin{bmatrix} 2.5527 \\ 0 \\ 0 \end{bmatrix}, \quad E_- = \begin{bmatrix} -2.5527 \\ 0 \\ 0 \end{bmatrix} \tag{8}$$

To test the stability type of the equilibrium points E_0, E_+, E_-, we calculate the Jacobian matrix of the novel Vaidyanathan jerk chaotic system (5) at any point x (with $a = 0.4$ and $b = 0.67$):

$$J(x) = \begin{bmatrix} 0 & 1 & 0 \\ 0 & 0 & 1 \\ 1 - 0.4\cosh(x) & -0.4\cosh(y) & -0.67 \end{bmatrix} \tag{9}$$

We find that

$$J_0 = J(E_0) = \begin{bmatrix} 0 & 1 & 0 \\ 0 & 0 & 1 \\ 0.6 & -0.4 & -0.67 \end{bmatrix} \tag{10}$$

which has the eigenvalues

$$\lambda_1 = 0.5553, \quad \lambda_{2,3} = -0.6127 \pm 0.8397i \tag{11}$$

This shows that the equilibrium point E_0 is a saddle-focus, which is unstable. Next, we find that

$$J_+ = J(E_+) = \begin{bmatrix} 0 & 1 & 0 \\ 0 & 0 & 1 \\ -1.5839 & -0.4 & -0.67 \end{bmatrix} \tag{12}$$

which has the eigenvalues

$$\lambda_1 = -1.2998, \quad \lambda_{2,3} = 0.3149 \pm 1.0580i \tag{13}$$

This shows that the equilibrium point E_+ is a saddle-focus, which is unstable. Since $J(E_-) = J(E_+)$, the matrices $J(E_-)$ and $J(E_+)$ have the same eigenvalues. This shows that the equilibrium point E_- is also a saddle-focus, which is unstable.

2.2 Numerical Simulation Using MATLAB

Numerical simulations of the Vaidyanathan jerk chaotic system (5) are carried out by using MATLAB 2010. The fourth-order Runge–Kutta method is used to solve the system of differential equations (5).

We consider the parameter values as $a = 0.4$ and $b = 0.67$. We take the initial condition as $(x_0, y_0, z_0) = (0.8, 1.2, 0.5)$.

Figure 1a–c show the 2-D projections of the strange attractor of the Vaidyanathan jerk system (5) on to the $x{-}y$ plane, the $y{-}z$ plane and the $x{-}z$ plane, respectively. The time series of the Vaidyanathan jerk circuit's variables are shown in Fig. 1d.

2.3 Lyapunov Exponents and Lyapunov Dimension

Lyapunov exponent of a dynamical system is a measure of exponential divergence of orbits, which characterizes the rate of separation of very close trajectories [5]. In a three dimensional system, like the Vaidyanathan jerk system (5), there are three Lyapunov exponents (LE_1, LE_2, LE_3). In more details, for a 3-D continuous dissipative system, the values of the Lyapunov exponents are useful for distinguishing among the various types of phase orbits. So, the possible spectra of attractors, of this class of dynamical systems, can be classified into four groups, based on the Lyapunov exponents [115].

- $(\lambda_1, \lambda_2, \lambda_3) \longrightarrow (-, -, -)$: a fixed point
- $(\lambda_1, \lambda_2, \lambda_3) \longrightarrow (0, -, -)$: a limit point.
- $(\lambda_1, \lambda_2, \lambda_3) \longrightarrow (0, 0, -)$: a 2-torus.
- $(\lambda_1, \lambda_2, \lambda_3) \longrightarrow (+, 0, -)$: a strange attractor.

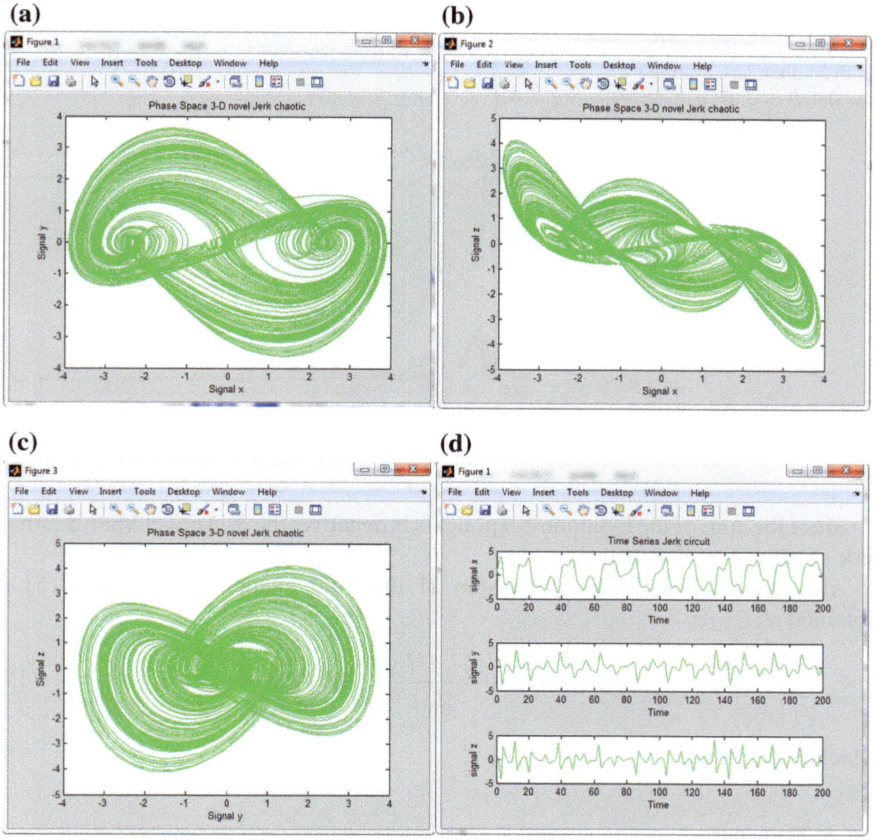

Fig. 1 Numerical simulation results using MATLAB 2010, for $a = 0.4$, $b = 0.67$, in **a** $x-y$ plane, **b** $x-z$ plane, **c** $y-z$ plane, **d** time series of x, y, z signals

Lyapunov dimension of a chaotic system is also called Kaplan–Yorke dimension, which gives an estimate of the rate of entropy production and of the fractal dimension of the chaotic dynamical system [5].

We take the parameter values as $a = 0.4$ and $b = 0.67$. We take the initial condition as $(x_0, y_0, z_0) = (0.8, 1.2.0.5)$. Then the values of the Lyapunov exponents of the Vaidyanathan jerk system (5) are numerically obtained as

$$L_1 = 0.0905, \quad L_2 = 0, \quad L_3 = -0.7576 \tag{14}$$

The dynamics of the Lyapunov exponents of the Vaidyanathan jerk system (5) is shown in Fig. 2.

Thus, the maximal Lyapunov exponent (MLE) of the 3-D novel Vaidyanathan jerk system (5) is $L_1 = 0.0905 > 0$, which means that the Vaidyanathan jerk system (5) has a chaotic behavior for the parameter values $a = 0.4$, $b = 0.67$.

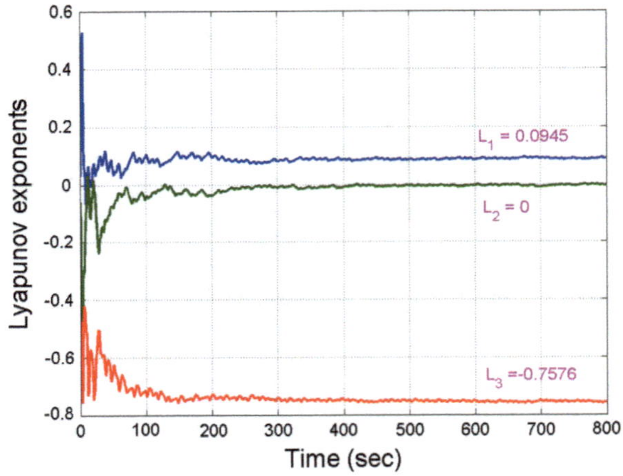

Fig. 2 The dynamics of Lyapunov exponents of the Vaidyanathan jerk system for $a = 0.4, b = 0.67$

Since the sum of the Lyapunov exponents is negative, the 3-D novel Vaidyanathan jerk system (5) is dissipative.

Also, the Kaplan–Yorke dimension of the Vaidyanathan jerk system (5) is obtained as

$$D_{KY} = 2 + \frac{L_1 + L_2}{|L_3|} = 2.1195, \tag{15}$$

which is fractional.

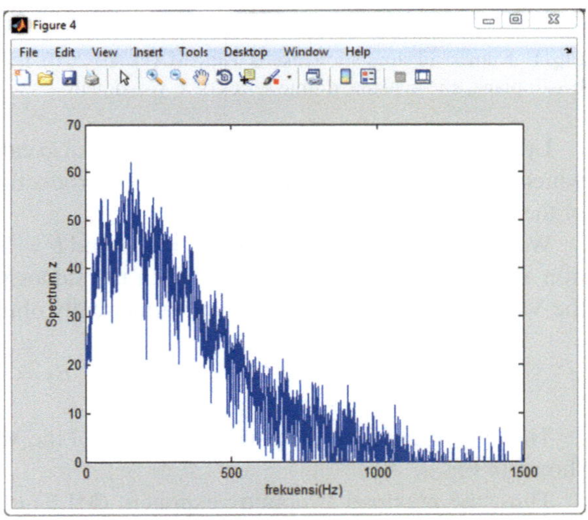

Fig. 3 The frequency spectrum generated numerically from chaotic system, Spectrum z versus frequency, for $a = 0.4, b = 0.67$

(a)

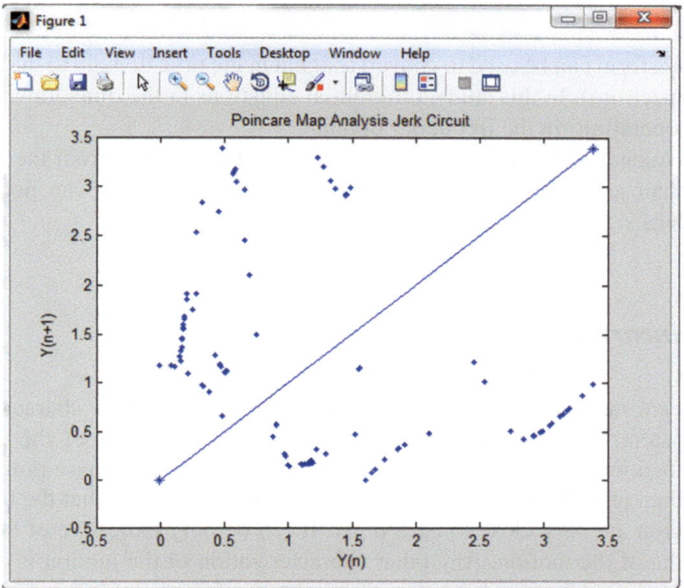

(b)

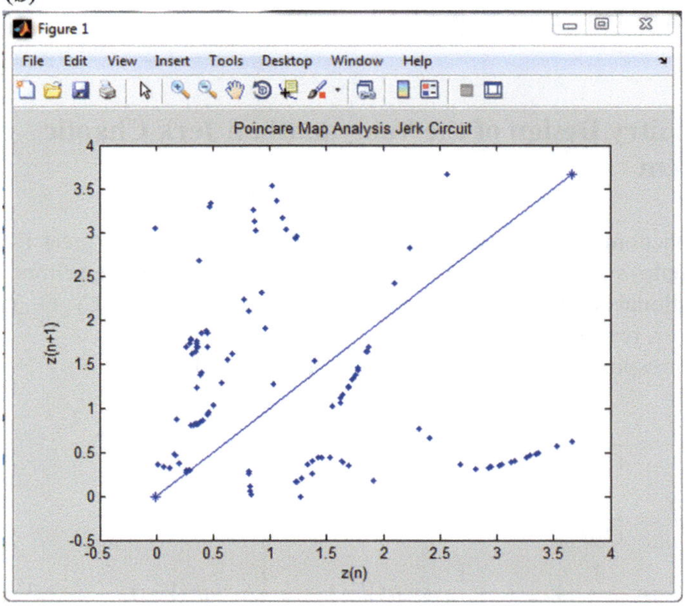

Fig. 4 A gallery of Poincaré maps for the Vaidyanathan jerk chaotic system for $a = 0.4$, $b = 0.67$. The plots give the maxima of **a** $y(n+1)$ versus those of $y(n)$; **b** $z(n+1)$ versus those of $z(n)$

2.4 FFT Analysis

Fourier transform can be used to convert a signal in the time domain to the frequency domain (spectrum). In this study, differential equations in the time domain become algebraic operations in the frequency domain.

The frequency spectra of signal z is generated numerically from the 3-D novel Vaidyanathan jerk chaotic system (5), is shown in Fig. 3. One can note that the bandwidth of signal z is in about 0–1.5 kHz.

2.5 Poincaré Map Analysis

The Poincaré map is a useful tool for analysing the dynamical characteristics of chaotic system. It is used to discuss the dynamical behaviour of the 3-D novel Vaidyanathan jerk chaotic system (5). In the chaotic case, the phase portrait of the Vaidyanathan jerk chaotic system (5) is very dense in the sense that the trajectories of the motion are very close to each other. It can be only indicative of the minima and maxima of the motion. Any other characterization of the motion is difficult to be interpreted. So, one way to capture the qualitative features of the strange attractor is to obtain the Poincaré map [11]. Figure 4a, b show the Poincaré section map by using MATLAB for $a = 0.4$, $b = 0.67$.

3 Circuitry Design of the Vaidyanathan Jerk Chaotic System

Chaotic phenomena in electric circuits have been studied with great interest. The circuit employs simple electronic elements, such as resistors, capacitors, multiplier and operational amplifiers. In this circuit (Fig. 5), the voltages of C_1, C_2, C_3 are used as x, y and z, respectively.

The corresponding circuit equation can be described as

$$\begin{aligned}
\frac{dV_{C1}}{dt} &= \frac{1}{C_1R_1}V_{C2} \\
\frac{dV_{C2}}{dt} &= \frac{1}{C_2R_2}V_{C3} \\
\frac{dV_{C3}}{dt} &= \frac{1}{C_3R_3}V_{C1} - \frac{1}{C_3R_4}sinh(V_{C1}) - \frac{1}{C_3R_5}sinh(V_{C2}) - \frac{1}{C_3R_6}V_{C3}
\end{aligned} \tag{16}$$

We choose $R_1 = R_2 = R_3 = R_7 = R_8 = R_9 = R_{10} = R_{11} = R_{12} = R_{17} = 10\,\text{K}\Omega$, $R_4 = R_5 = 250\,\Omega$, $R_6 = 15\,\text{K}\Omega$, $R_{16} = R_{20} = 1\,\text{M}\Omega$, $R_{15} = R_{19} = 60\,\text{K}\Omega$.

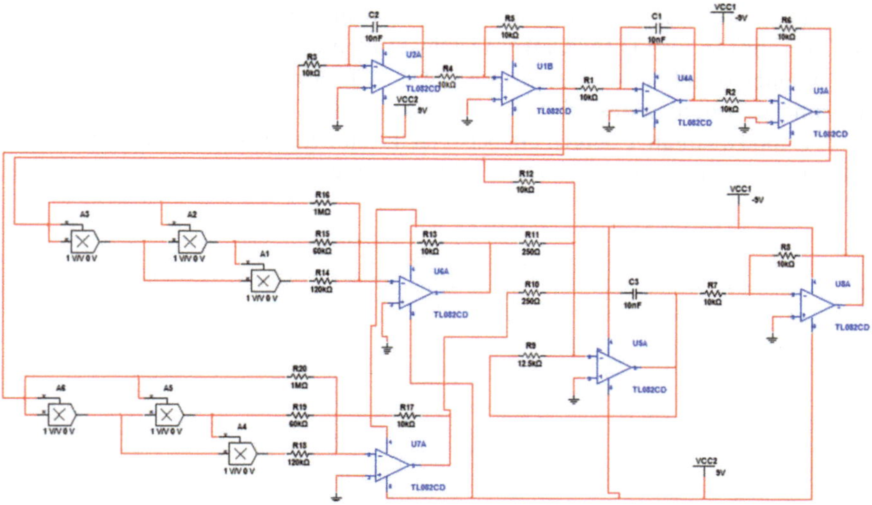

Fig. 5 Schematic of the Vaidyanathan jerk chaotic system by using MultiSIM 10.0

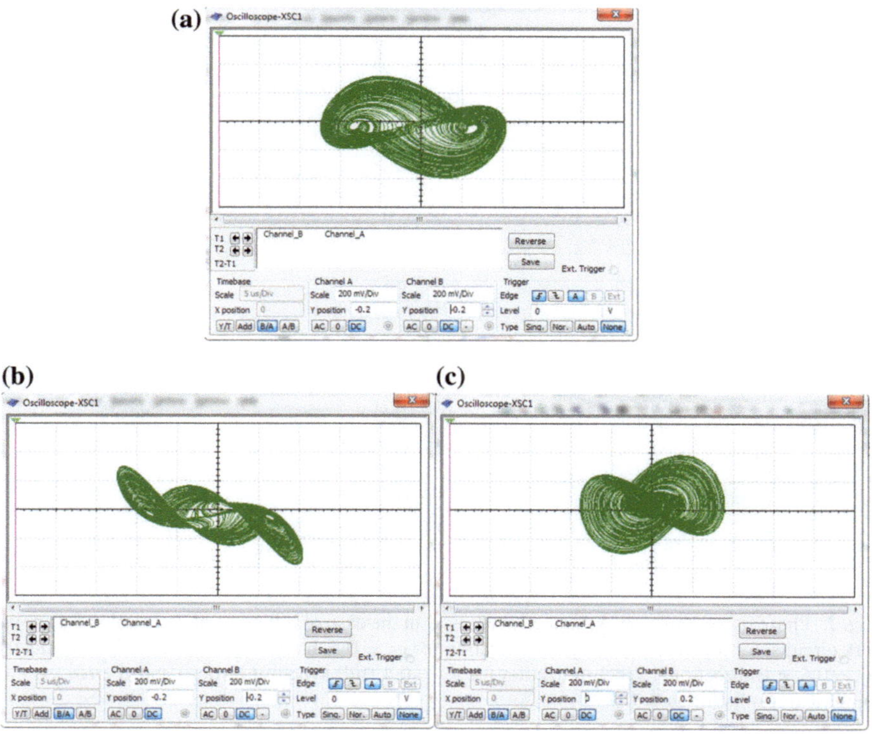

Fig. 6 Various projections of the Vaidyanathan chaotic attractor using MultiSIM 10.0, for $a = 0.4$, $b = 0.67$, **a** $x{-}y$ plane, **b** $x{-}z$ plane, **c** $y{-}z$ plane

$R_{14} = R_{18} = 120\,\text{K}\Omega$, $C_1 = C_2 = C_3 = C_4 = 10\,\text{nF}$. The circuit has three integrators (by using Op-amp TL082CD) in a feedback loop and six multipliers (IC AD633). The supplies of all active devices are $\pm 9\,\text{V}$. With MultiSIM 10.0, we obtain the experimental observations of system (16) as shown in Fig. 6. As compared with Fig. 1, a good qualitative agreement between the numerical simulations and the MultiSIM 10.0 results of the 3-D novel Vaidyanathan jerk chaotic system is confirmed.

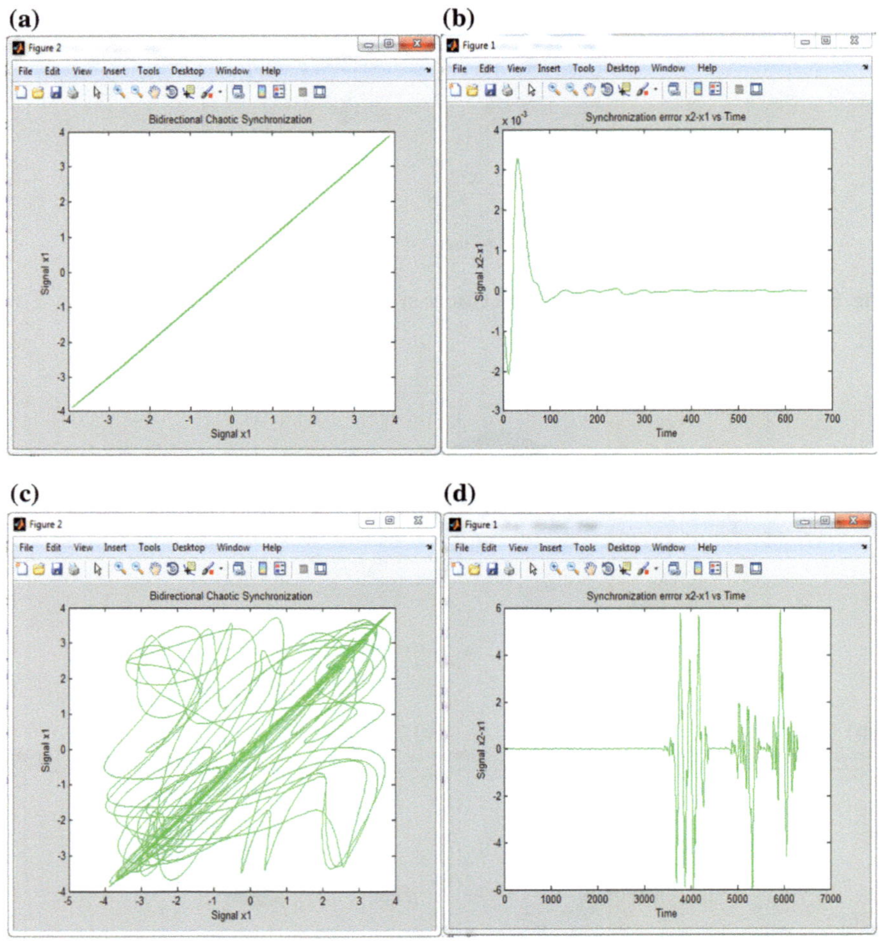

Fig. 7 Phase portrait of x_2 *vs* x_1 and error $x_2 - x_1$ in the case of bidirectionally coupled 3-D novel Jerk circuits, for **a** Bidirectional synchronization for $g_c = 0.5$, **b** error $x_2 - x_1$ numerical results for $g_c = 0.5$, **c** Bidirectional synchronization for $g_c = 0.4$ (full synchronization) and **d** error $x_2 - x_1$ numerical results for $g_c = 0.4$

4 Bidirectional Chaotic Synchronization

4.1 Mathematical Model of Bidirectional Synchronization

Synchronization between coupled chaotic systems has received considerable attention and led to communication applications. With coupling method and synchronizing identical chaotic systems, a message signal sent by a transmitter system can be reproduced at a receiver under the influence of a single chaotic signal through synchronization. This work presents the study of numerical simulation of

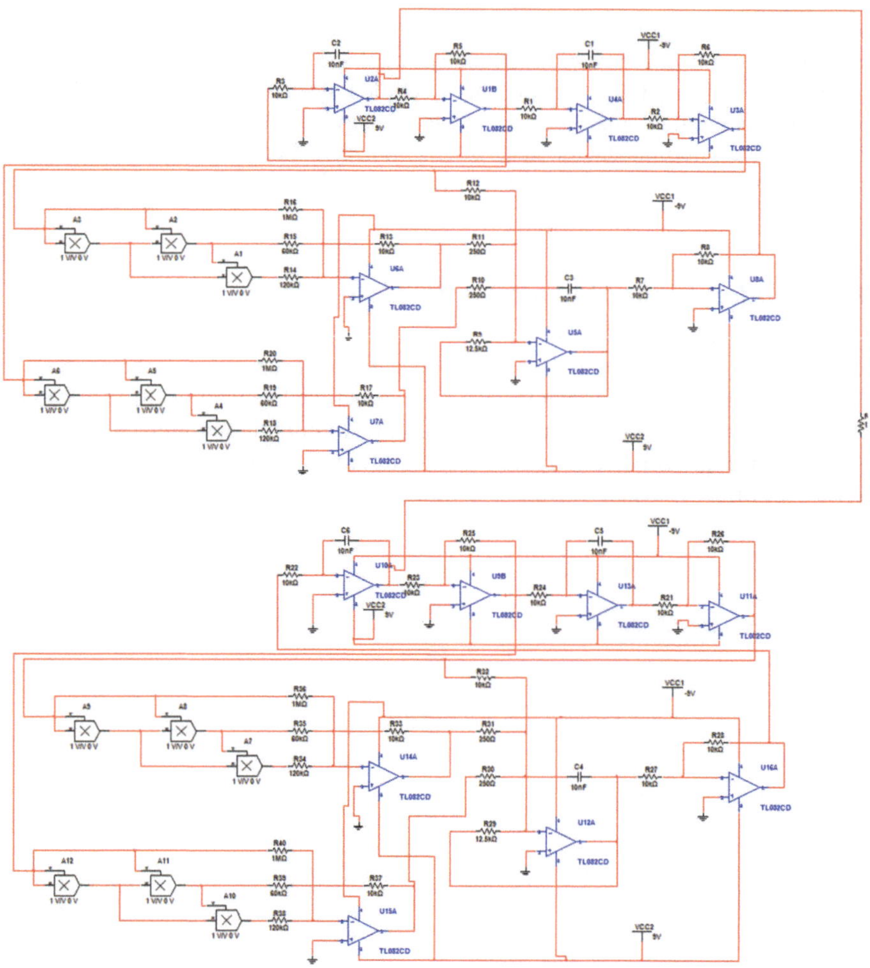

Fig. 8 Bidirectional chaotic synchronization 3-D novel Jerk circuit by using MultiSIM 10.0

chaos synchronization of coupled chaotic 3-D novel Jerk chaotic circuits. The following (bidirectional coupling) configuration is described below:

$$\begin{aligned}
\dot{x_1} &= y_1 \\
\dot{y_1} &= z_1 + g_c(y_2 - y_1) \\
\dot{z_1} &= x_1 - a[\sinh(x_1) + \sinh(y_1)] - bz_1 \\
\dot{x_2} &= y_2 \\
\dot{y_2} &= z_2 + g_c(y_1 - y_2) \\
\dot{z_2} &= x_2 - a[\sinh(x_2) + \sinh(y_2)] - bz_2
\end{aligned} \tag{17}$$

The coupling coefficient g_c is present in the equations of both systems, since the coupling between them is mutual. Numerical simulations of system (17), by using the fourth-order Runge–Kutta method, are used to describe the dynamics of chaotic synchronization of bidirectionally coupled 3-D novel Jerk chaotic systems. In bidirectional (mutual) coupling, both coupled systems are connected in such a way that they mutually influence each other's behavior. Synchronization numerically appears for a coupling factor $g_c \geq 0.5$ as shown in Fig. 7a, with error $e_x = x_1 - x_2 \longrightarrow 0$, which implies the complete synchronization.

4.2 Analog Circuit Simulation in MultiSIM

Simulation results show that the two systems are synchronized well. Figure 8 shows the circuit schematic for implementing the bidirectional synchronization of coupled Jerk systems. Chaotic synchronization appears for a coupling strength $R_{21} \leq 100\,\text{m}\Omega$, as shown in Fig. 9a. For different initial conditions or resistance coupling strength $R_{21} > 100\,\text{m}\Omega$, the synchronization cannot occur as shown in Fig. 9b.

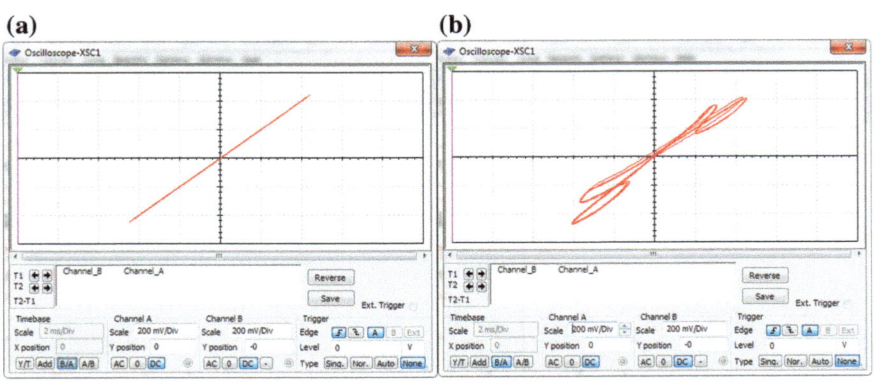

Fig. 9 Synchronization phase portrait of x_2 versus x_1, for **a** $R_{21} = 100\,\text{m}\Omega$ and **b** $R_{21} = 1\,\Omega$ MultiSIM 10.0, for $a = 0.4$, $b = 0.67$

5 Application in Secure Communication System

5.1 Mathematical Model of Secure Communication Systems

To study the effectiveness of signal masking approach in the 3-D novel Jerk circuit, we first set the information-bearing signal $m_s(t)$ in the form of square wave:

$$m_s(t) = \frac{4}{\pi}\sum_{n=1}^{\infty}\frac{1}{2n-1}sin[(2n-1)2\pi ft] \tag{18}$$

The sum of the signal $m_s(t)$ and the chaotic signal $m_{3-DnovelJerkcircuit}(t)$, produced by the 3-D novel Jerk circuit, is the new encryption signal $m_{encryption}(t)$, which is given by Eq. (19).

$$m_{encryption}(t) = m_s(t) + m_{3-DnovelJerkcircuit}(t) \tag{19}$$

The signal $m_{3-DnovelJerkcircuit}(t)$ is one of the parameters of Eq. 5. After finishing the encryption process the original signal can be recovered with the following procedure.

$$m_{decryption}(t) = m_{encryption}(t) - m_{3-DnovelJerkcircuit}(t) \tag{20}$$

So, $m_{decryption}(t)$is the original signal and must be the same with $m_s(t)$. Due to the fact that the input signal can be recovered from the output signal, it turns out that it is possible to implement a secure communication system using the proposed chaotic system.

5.2 Numerical Simulation of Secure Communication Systems

In chaos-based secure communication scheme, chaos synchronization is the critical issue, because the two identical chaos generators in the transmitter and the receiver end need to be synchronized. Information signal is added to the chaotic signal at transmitter and at receiver the masking signal is regenerated and subtracted from the receiver signal [116]. Figure 10a–c show the MATLAB 2010 numerical simulation results for the proposed chaotic masking communication scheme.

5.3 Analog Circuit Simulation of Secure Communication System

Information is masked by chaotic signals at the transmitter, and then sent to the receiver by the public channel. Finally the encrypted signals are decrypted at the receiver. In this system, the key issue is that the two identical chaos generators

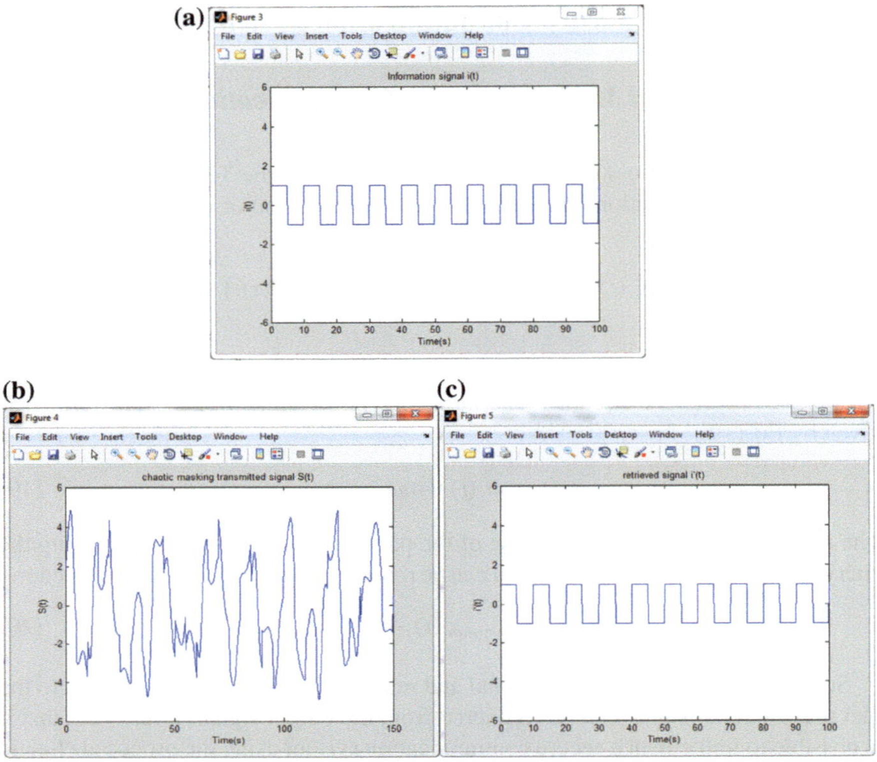

Fig. 10 MATLAB 2010 simulation of 3-D novel Jerk circuit masking communication system when amplitude is 1 V and frequency 1 kHz: **a** Information signal. **b** Chaotic masking transmitted signal. **c** Retrieved signal

Fig. 11 The principle diagram of chaos-based secure communication

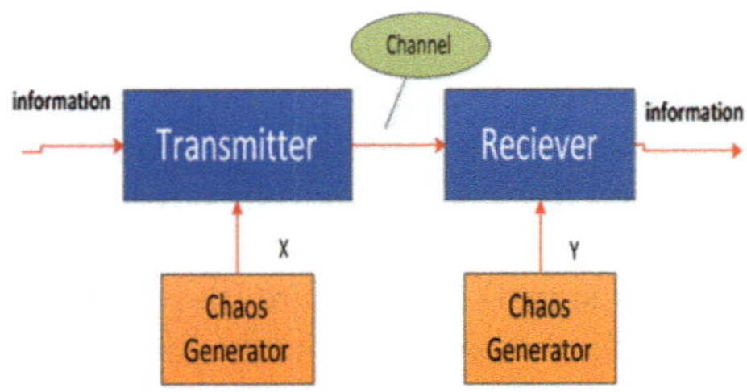

in the transmitter end and the receiver end need to be synchronized. Thus, chaos synchronization is the key technique throughout this whole process [116]. The principle diagram of symmetric chaos-based secure communication schemes is shown in Fig. 11.

(a)
(b)
(c)

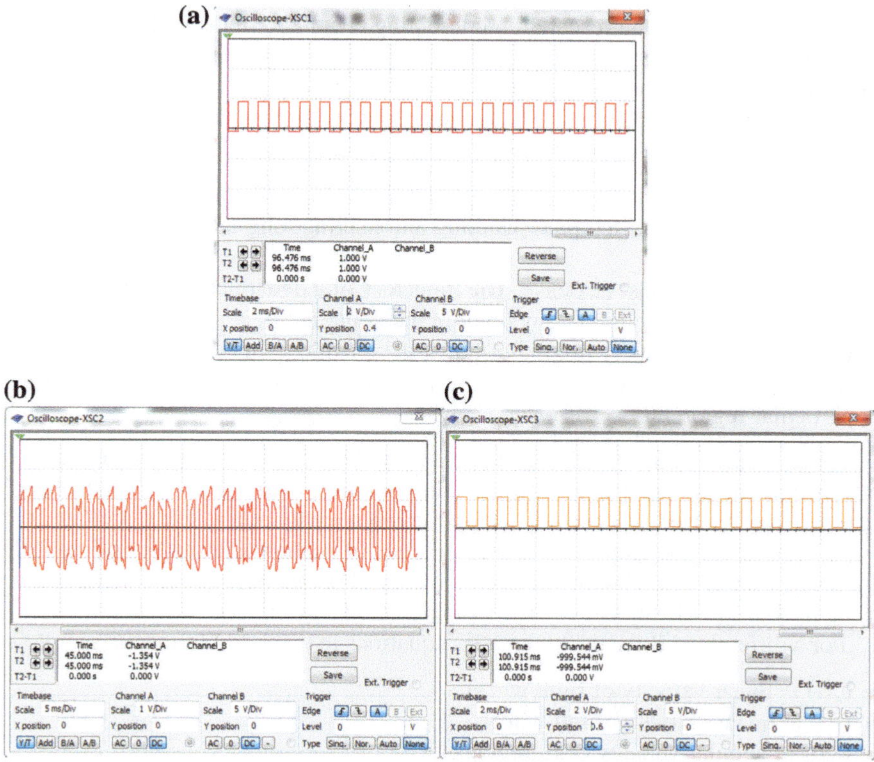

Fig. 12 MultiSIM 10.0 outputs of Jerk circuit masking communication systems, when amplitude is 1 V and frequency 1 kHz: **a** Information signal. **b** Chaotic masking transmitted signal. **c** Retrieved signal

The information signal is a sinusoidal pulse signal of frequency $f = 1\,\text{kHz}$ and amplitude $A = 1\,\text{V}$ (Fig. 12a), provided by an external source. In Fig. 12c we can see that the recovered signal is exactly the same with the information signal, while the transmitted signal (Fig. 12b), where the chaotic masking transmitted signal. Simulation results with MultiSIM 10.0 have shown that the performance of chaotic 3-D novel Jerk circuits in chaotic masking and message recovery is very satisfactory.

6 Application in Mobile Robot Navigation

6.1 Mathematical Models of Mobile Robot Navigation

In a chaos-based navigation system, chaos is used to control the movement of the system. Most of the chaos is used to control the movement of the system Arduino.

The main purpose of the research in combining chaotic systems with robots is that a chaotic system is very sensitive to initial conditions and hence it is very hard to predict the trajectory of the system.

There are three conditions in a dynamic system to produce chaos-based robot trajectory [108].

- It must be very sensitive on initial conditions. This feature contributes to the desired robot unexpected path planning and making long-term chaotic trajectory prediction
- Its chaotic orbits must be dense. The trajectory of a dynamical system is dense, if it comes arbitrarily close to any point in the domain.
- It must be topologically mixing. That is, the chaotic system will move over time so that each designated area of the trajectory will eventually cover part of any particular region.

An advantage of using chaos is that the behavior of the robot can be predicted in advance by the system designer. So, an autonomous mobile robot with such characteristics may be used successfully as mobile robot navigation.

Kinematics is the study of the mathematics of motion without considering the forces that develop a relationship between control parameters and the behavior of a system in space. The model of the robot is as shown in Fig. 13.

For a differential drive the kinematics equations in the world frame are as follows

- v_r(t) = linear velocity of right wheel
- v_l(t) = linear velocity of left wheel
- w_r(t) = angular velocity of right wheel
- w_l(t) = angular velocity of left wheel
- r = nominal radius of each wheel
- L = distance between the two wheels
- R = Instantaneous curvature radius of the robot trajectory, relative to the mid-point axis
- ICC = Instantaneous Center of Curvature
- $R - \frac{L}{2}$ = Curvature radius of trajectory described by left wheel
- $R + \frac{L}{2}$ = Curvature radius of trajectory described by right wheel

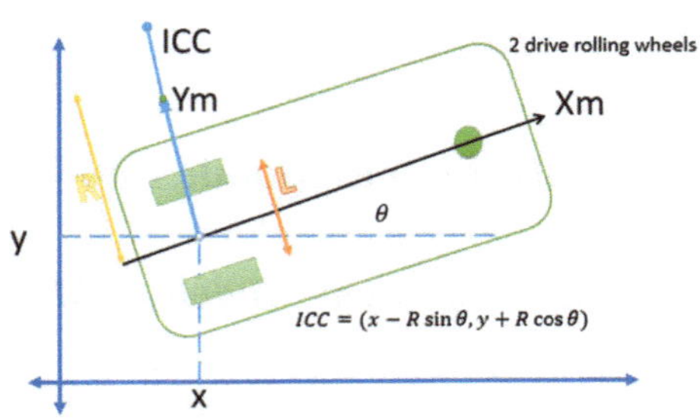

Fig. 13 Kinematic model of mobile robot

With respect to ICC the angular velocity of the robot is given as follows:

$$
\begin{aligned}
w(t) &= \frac{v_r(t)}{R+\frac{L}{2}} \\
w(t) &= \frac{v_l(t)}{R+\frac{L}{2}} \\
w(t) &= \frac{v_r(t)-v_l(t)}{L}
\end{aligned}
\tag{21}
$$

The instantaneous curvature radius of the robot trajectory relative to the mid-point axis is given as:

$$
R=\frac{L(v_l(t)+v_r(t))}{2(v_l(t)+v_r(t))} \tag{22}
$$

Therefore the linear velocity of the robot is given as:

$$
v(t)=w(t)R=\frac{1}{2}(v_r(t)+v_l(t)) \tag{23}
$$

The kinematics equations in the world frame can be represented as follows [29]:

$$
\begin{aligned}
\dot{X} &= v(t)\cos\theta(t) \\
\dot{Y} &= v(t)\sin\theta(t) \\
\dot{\theta} &= w(t)
\end{aligned}
\tag{24}
$$

These are the equations that are used to build a model of the robot. These equations were used to simulate the robot in MATLAB. The 3-D novel Vaidyanathan jerk system controller was tested and fine-tuned on this model as well as compared with other controllers for optimum results.

6.2 *Numerical Simulation of Mobile Robot Navigation*

In this work, the three proposed dynamical systems were solved numerically by using the fourth order Runge–Kutta algorithm. Searching for sets of optimal parameters for the 3D novel jerk system for generating the best possible patterns is very time-consuming task. Therefore, for convenience we retain their original parameters of these systems as used in the literature. Here is the equation of linear velocity and angular velocity, where in the linear velocity of right wheel and linear velocity of left wheel are replaced by a chaotic signal x and y:

$$
\begin{aligned}
v(t) &= \frac{1}{2}(\dot{x}(t)+\dot{y}(t)) \\
w(t) &= \frac{\dot{x}(t)-\dot{y}(t)}{L}
\end{aligned}
\tag{25}
$$

By combining Eqs. (5) and (24), the following dynamics is obtained.

$$\begin{aligned} \dot{x} &= y \\ \dot{y} &= z \\ \dot{z} &= x - a[\sinh(x) + \sinh(y)] - bz \\ \dot{X} &= v(t)\cos\theta(t) \\ \dot{Y} &= v(t)\sin\theta(t)z \\ \dot{\theta} &= w(t) \end{aligned} \tag{26}$$

The system (26) describes the mobile robot navigation based on the novel Vaidyanathan jerk system. The behavior of the 3D novel Vaidyanathan jerk system is chaotic. The chaotic mobile robot trajectory shown in Fig. 14 may be obtained by solving the system dynamics (26) by taking the parameter values and initial conditions as

$$a = 0.4, b = 0.67, L = 0.05, (x_0, y_0, z_0) = (0.8, 1, 2, 0.5), (X_0, Y_0, \theta_0) = (0, 0, 0)$$

The duration for run-time for simulation was taken as $T = 10{,}000$ s.

Figure 14 shows the motion of the mobile robot with the proposed controller. It is observed that the motion of the robot is unpredictable and with sensitive dependence on the initial condition. The trajectories generated by (26) scanned the whole workspace regardless of the shape of workspace.

In Fig. 14, the mobile robot motion path with equation x and y for PWM shows significantly higher value of coverage rate in regard to the other systems, where the 72 % of the terrain shows to be covered by the robot. The results show that the mobile robot navigation systems with combining the equation $x - y$ have a better level of coverage rate. The robot's workplace is supposed to be a square terrain with dimensions $M = 5\,\text{m} \times 5\,\text{m} = 25\,\text{m}^2$ in normalized unit cells. Furthermore, a

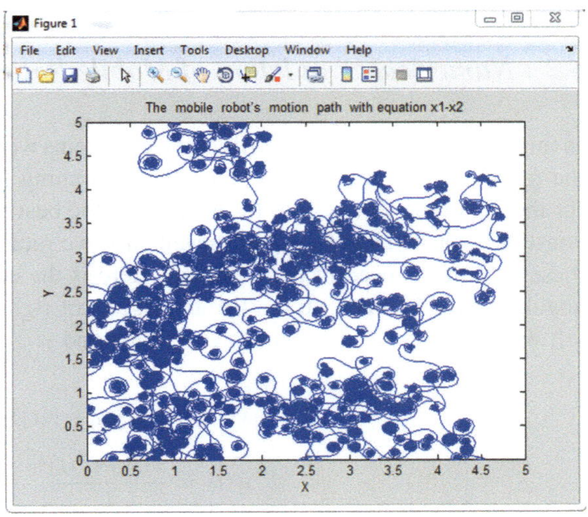

Fig. 14 Trajectory mobile robot navigation of 3D novel Jerk system using MATLAB

second interesting evaluation criterion is the coverage time of the system, which is the total time for the system to cover the entire terrain.

6.3 Experimental Results of Mobile Robot Navigation

Embedded controller realizations are generally subject to stronger constraints than their PC-based counterparts regarding speed, memory, and the environment in which the embedded system operates. In this work, the hardware used in building the robot along with the software implementation are given. The basic hardware components used are, motor driver type L293D and Arduino Uno ATMEGA 328P.

In this work the experimental results concerning the coverage performance of an autonomous mobile robot, by using equation 3-D novel Jerk system the chaotic motion controller, are presented. Chaos signal from the computer via serial communication is sent to the Arduino. Furthermore, the Arduino sends a command in the form of PWM to drive the linear velocity of right wheel and linear velocity of left wheel. Figure 15 shows the circuit schematic of principle mobile robot navigation. The autonomous mobile robot of this work is a four wheels platform, in which only the two wheels are independently controlled on velocity and rotation sense by using two gear motors Fig. 16.

The 3-D novel Vaidyanathan jerk system shows significantly higher value of coverage rate in regard to the other systems. The 3-D novel Vaidyanathan jerk system has the better performance which is obvious in Fig. 17, where the terrain shows to be covered by the robot. This happens because the 3-D novel Vaidyanathan jerk system produces a mobile robot's orbit which is constituted by spiral curves that abstain longer distances concerning the other two systems.

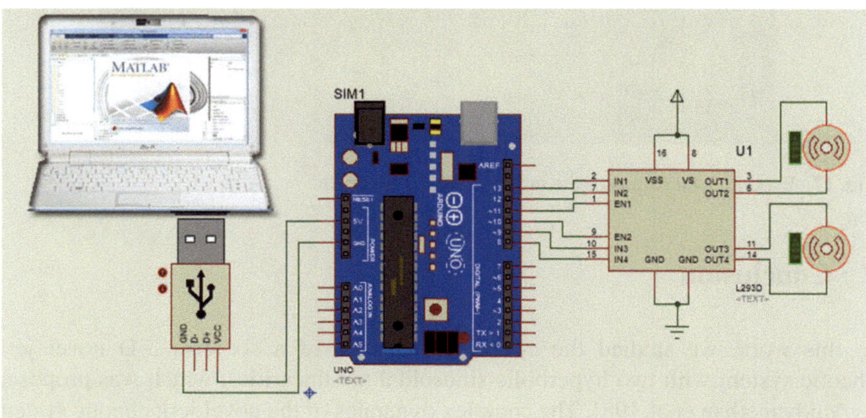

Fig. 15 Schematic of principle mobile robot navigation using Proteus

Fig. 16 The chaotic autonomous mobile robot

Fig. 17 Experimental results of the mobile robot motion path

7 Conclusion

In this work, we studied the complex dynamics of a six-term 3-D novel jerk chaotic system with two hyperbolic sinusoidal nonlinearities, which was proposed by Vaidyanathan et al. [98]. The complex dynamics of the novel jerk chaotic system has also been explored in detail including eigenvalue structures, various attractors, Lyapunov exponent analysis, Kaplan–Yorke dimension, FFT analysis and Poincaré map analysis. Moreover, it is implemented via a designed circuit and tested experimentally with MultiSIM. The MultiSIM results of the 3D novel Vaidyanathan jerk system were well agreed with the simulation results.

The chaotic synchronization of two identical 3-D novel Vaidyanathan jerk system has been investigated by implementing bidirectional method technique. The proposed method of synchronization between chaotic circuits can be applied successfully to a secure communication scheme. Chaos synchronization and chaos masking were realized by using MATLAB 2010 and MultiSIM 10.0 programs. The comparison between MATLAB 2010 and MultiSIM 10.0 simulation results demonstrate the effectiveness of the proposed secure communication scheme. Finally, the effectiveness of the bidirectional scheme between two identical 3-D novel Vaidyanathan jerk circuits in a secure communication system is presented in details. Integration of theoretical physics, the numerical simulation by using MATLAB 2010, as well as the implementation of circuit simulations by using MultiSIM 10.0 have been performed in this study.

In mobile robot navigation, the driving strategy of a mobile robot is presented, in order to generate the most unpredictable trajectory, as well as the trajectory with the higher coverage rate of a specific terrain. 3D novel Jerk system shows significantly higher value of coverage rate in regard to the other systems, which is the criterion of success of such robot's mission, among the proposed dynamical systems. This is due to the nature of the jerk chaotic attractors produced by the 3-D novel Vaidyanathan jerk chaotic system. This study has some flaws, especially in the delivery of data from the computer to the Arduino still using USB. It is possible to use the wireless mobile robot navigation system in future research.

References

1. Abdulkareem A, Abduljaleel IQ (2013) Speech encryption using chaotic map and blowfish algorithms. J Basrah Res—Sci 39(2):68–76
2. Andreatos AS, Leros AP (2013) Secure image encryption based on a Chua chaotic noise generator. J Eng Sci Technol Rev 6(4):90–103
3. Argyris A, Syvridis D, Larger L, Annovazzi-Lodi V, Colet P, Fischer I, Garcia-Ojalvo J, Mirasso CR, Pesquera L, Shore KA (2005) Chaos-based communications at high bit rates using commercial fibre-optic links. Nature 438:343–346
4. Arneodo A, Coullet P, Tresser C (1981) Possible new strange attractors with spiral structure. Commun Math Phys 79(4):573–576
5. Azar AT, Vaidyanathan S (2015) Chaos modeling and control systems design, vol 581. Springer, Germany
6. Cai G, Tan Z (2007) Chaos synchronization of a new chaotic system via nonlinear control. J Uncertain Syst 1(3):235–240
7. Chen G, Ueta T (1999) Yet another chaotic attractor. Int J Bifurc Chaos 9(7):1465–1466
8. Grzybowski JMV, Rafikov M, Macau EEN (2010) Chaotic communication on a satellite formation flying—the synchronization issue in a scenario with transmission delays. Acta Astronaut 66(7–8):1160–1168
9. Karthikeyan R, Sundarapandian V (2014) Hybrid synchronization of four-scroll systems via active control. J Electr Eng 65(2):97–103

10. Lü J, Chen G (2002) A new chaotic attractor coined. Int J Bifurc Chaos 12(3):659–661
11. Lü J, Han F, Yu X, Chen G (2004) Generating 3-D multi-scroll chaotic attractors: a hysteresis series switching method. Automatica 40(10):1677–1687
12. Li D (2008) A three-scroll chaotic attractor. Phys Lett A 372(4):387–393
13. Li XF, Chlouverakis KE, Xu DL (2009) Nonlinear dynamics and circuit realization of a new chaotic flow: a variant of Lorenz, Chen and Lü. Nonlinear Anal-Real World Appl 10(4):2357–2368
14. Lian S, Sun J, Liu G, Wang Z (2008) Efficient video encryption scheme based on advanced video coding. Multimed Tools Appl 38(1):75–89
15. Liu W, Wang L (2009) Minimizing interference in satellite communications using transiently chaotic neural networks. Comput Math Appl 57(6):1024–1029
16. Lorenz EN (1963) Deterministic periodic flow. J Atmos Sci 20(2):130–141
17. Mainieri R, Rehacek J (1999) Projective synchronization in three-dimensional chaotic systems. Phys Rev Lett 82:3042–3045
18. Malasoma JM (2000) What is the simplest dissipative chaotic jerk equation which is parity invariant? Phys Lett A 264:383–389
19. Martins-Filho LS, Macau EEN (2007) Patrol mobile robots and chaotic trajectories. Math Probl Eng 2007:1–10
20. Moore DW, Spiegel EA (1966) A thermally excited nonlinear oscillator. Astrophys J 143: 871–887
21. Nakajima K, Sawada Y (1980) Experimental studies on the weak coupling of oscillatory chemical reaction systems. Astrophys J 72(4):2231–2234
22. Ng KC, Trivedi MM (1998) A neuro-fuzzy controller for mobile robot navigation and multi-robot convoying. IEEE Trans Syst Man Cybern—Part B: Cybern 28(6):829–840
23. Ni J, Yang SX (2012) A fuzzy-logic based chaos GA for cooperative foraging of multi-robots in unknown environments. Int J Robot Autom 27(1):15–30
24. Palacin J, Salde JA, Valganon I, Clua X (2004) Building a mobile robot for a floor-cleaning operation in domestic environments. IEEE Trans Instrum Meas 53(5):1418–1424
25. Pandey A, Baghel RK, Singh RP (2013) An autonomous chaotic circuit for wideband secure communication. Int J Eng Bus Enterp Appl 4(1):44–47
26. Park JH (2006) Chaos synchronization between two different chaotic dynamical systems. Chaos Solitons Fractals 27:549–554
27. Pecora LM, Carroll TL (1990) Synchronization in chaotic systems. Phys Rev Lett 64:821–825
28. Pehlivan I, Moroz IM, Vaidyanathan S (2014) Analysis, synchronization and circuit design of a novel butterfly attractor. J Sound Vib 333(20):5077–5096
29. Peri VM (2005) Fuzzy Logic Controller for an Autonomous Mobile Robot. In: PhD Thesis, Jawaharlal Nehru Technological University
30. Pham V-T, Volos C, Jafari S, Wang X, Vaidyanathan S (2014) Hidden hyperchaotic attractor in a novel simple memristive neural network. Optoelectr Adv Mater Rapid Commun 8(11–12):1157–1163
31. Pham V-T, Vaidyanathan S, Volos CK, Jafari S (2015) Hidden attractors in a chaotic system with an exponential nonlinear term. Eur Phys J: Spec Top 224(8):1507–1517
32. Pham V-T, Volos CK, Vaidyanathan S, Le TP, Vu VY (2015) A memristor-based hyperchaotic system with hidden attractors: dynamics, synchronization and circuital emulating. J Eng Sci Technol Rev 8(2):205–214
33. Rössler OE (1976) An equation for continuous chaos. Phys Lett A 57(5):397–398
34. Rasappan S, Vaidyanathan S (2012) Global chaos synchronization of WINDMI and Coullet chaotic systems by backstepping control. Far East J Math Sci 67(2):265–287
35. Rasappan S, Vaidyanathan S (2012) Hybrid synchronization of n-scroll Chua and Lur'e chaotic systems via backstepping control with novel feedback. Arch Control Sci 22(3):343–365
36. Rasappan S, Vaidyanathan S (2012) Synchronization of hyperchaotic Liu system via backstepping control with recursive feedback. Commun Comput Inf Sci 305:212–221
37. Rasappan S, Vaidyanathan S (2013) Hybrid synchronization of n-scroll Chua circuits using adaptive backstepping control design with recursive feedback. Malays J Math Sci 7(2):219–246

38. Rulkov NF, Sushchik MM, Tsimring LS, Abarbanel HDI (1995) Generalized synchronization of chaos in directionally coupled chaotic systems. Phys Rev E 51(2):980–994
39. Sambas A, Sanjaya WSM, Halimatussadiyah (2012) Unidirectional chaotic synchronization of Rossler circuit and its application for secure sommunication. WSEAS Trans Syst 11(9):506–515
40. Sambas A, Sanjaya WSM, Mamat M (2013) Design and numerical simulation of unidirectional chaotic synchronization and its application in secure communication system. J Eng Sci Technol Rev 6(4):66–73
41. Sambas A, Sanjaya WSM, Mamat M, Halimatussadiyah (2013) Design and analysis bidirectional chaotic synchronization of Rossler circuit and its application for secure communication. Appl Math Sci 7(1):11–21
42. Sampath S, Vaidyanathan S, Volos CK, Pham V-T (2015) An eight-term novel four-scroll chaotic system with cubic nonlinearity and its circuit simulation. J Eng Sci Technol Rev 8(2):1–6
43. Sanjaya WSM, Mamat M, Salleh Z, Mohd I (2011) Bidirectional chaotic synchronization of Hindmarsh-Rose neuron model. Appl Math Sci 5:2685–2695
44. Sanjaya WSM, Mohd I, Mamat M, Salleh Z (2012) Mathematical model of three species food chain interaction with mixed functional response. Int J Mod Phys 9:334–340
45. Sarasu P, Sundarapandian V (2011) Active controller design for generalized projective synchronization of four-scroll chaotic systems. Int J Syst Signal Control Eng Appl 4(2):26–33
46. Sarasu P, Sundarapandian V (2011) The generalized projective synchronization of hyperchaotic Lorenz and hyperchaotic Qi systems via active control. Int J Soft Comput 6(5):216–223
47. Sarasu P, Sundarapandian V (2012) Adaptive controller design for the generalized projective synchronization of 4-scroll systems. Int J Syst Signal Control Eng Appl 5(2):21–30
48. Sarasu P, Sundarapandian V (2012) Generalized projective synchronization of three-scroll chaotic systems via adaptive control. Eur J Sci Res 72(4):504–522
49. Sarasu P, Sundarapandian V (2012) Generalized projective synchronization of two-scroll systems via adaptive control. Int J Soft Comput 7(4):146–156
50. Sciamanna M, Shore KA (2015) Chaos synchronization and chaos communication using laser diodes. Nat Photon 9:151–162
51. Shinbrot T, Grebogi C, Wisdom J, Yorke JA (1992) Chaos in a double pendulum. Am J Phys 60:491–499
52. Sprott JC (1994) Some simple chaotic flows. Phys Rev E 50(2):647–650
53. Sprott JC (2000) Simple chaotic systems and circuits. Am J Phys 68:758–763
54. Sprott JC (2004) Dynamical models of love. Nonlinear Dyn Psychol Life Sci 8:303–314
55. Sun KH, Sprott JC (2009) A simple jerk system with piecewise exponential nonlinearity. Int J Nonlinear Sci Numer Simul 10(11–12):1443–1450
56. Sundarapandian V (2013) Adaptive control and synchronization design for the Lu-Xiao chaotic system. Lect Notes Electr Eng 131:319–327
57. Sundarapandian V (2013) Analysis and anti-synchronization of a novel chaotic system via active and adaptive controllers. J Eng Sci Technol Rev 6(4):45–52
58. Sundarapandian V, Karthikeyan R (2011) Anti-synchronization of hyperchaotic Lorenz and hyperchaotic Chen systems by adaptive control. Int J Syst Signal Control Eng Appl 4(2):18–25
59. Sundarapandian V, Karthikeyan R (2012) Adaptive anti-synchronization of uncertain Tigan and Li systems. J Eng Appl Sci 7(1):45–52
60. Sundarapandian V, Karthikeyan R (2012) Hybrid synchronization of hyperchaotic Lorenz and hyperchaotic Chen systems via active control. J Eng Appl Sci 7(3):254–264
61. Sundarapandian V, Pehlivan I (2012) Analysis, control, synchronization, and circuit design of a novel chaotic system. Math Comput Modell 55(7–8):1904–1915
62. Sundarapandian V, Sivaperumal S (2011) Sliding controller design of hybrid synchronization of four-wing chaotic systems. Int J Soft Comput 6(5):224–231
63. Suresh R, Sundarapandian V (2013) Global chaos synchronization of a family of n-scroll hyperchaotic Chua circuits using backstepping control with recursive feedback. Far East J Math Sci 73(1):73–95

64. Tigan G, Opris D (2008) Analysis of a 3D chaotic system. Chaos Solitons Fractals 36:1315–1319
65. Vaidyanathan S (2011) Hybrid chaos synchronization of Liu and Lü systems by active nonlinear control. Commun Comput Inf Sci 204:1–10
66. Vaidyanathan S (2012) Analysis and synchronization of the hyperchaotic Yujun systems via sliding mode control. Adv Intell Syst Comput 176:329–337
67. Vaidyanathan S (2012) Anti-synchronization of Sprott-L and Sprott-M chaotic systems via adaptive control. Int J Control Theory Appl 5(1):41–59
68. Vaidyanathan S (2013) A new six-term 3-D chaotic system with an exponential nonlinearity. Far East J Math Sci 79(1):135–143
69. Vaidyanathan S (2013) A ten-term novel 4-D hyperchaotic system with three quadratic nonlinearities and its control. Int J Control Theory Appl 6(2):97–109
70. Vaidyanathan S (2013) Analysis and adaptive synchronization of two novel chaotic systems with hyperbolic sinusoidal and cosinusoidal nonlinearity and unknown parameters. J Eng Sci Technol Rev 6(4):53–65
71. Vaidyanathan S (2013) Analysis, control and synchronization of hyperchaotic Zhou system via adaptive control. Adv Intell Syst Comput 177:1–10
72. Vaidyanathan S (2014) A new eight-term 3-D polynomial chaotic system with three quadratic nonlinearities. Far East J Math Sci 84(2):219–226
73. Vaidyanathan S (2014) Analysis and adaptive synchronization of eight-term 3-D polynomial chaotic systems with three quadratic nonlinearities. Eur Phys J: Spec Top 223(8):1519–1529
74. Vaidyanathan S (2014) Analysis, control and synchronisation of a six-term novel chaotic system with three quadratic nonlinearities. Int J Modell Identif Control 22(1):41–53
75. Vaidyanathan S (2014) Generalized projective synchronisation of novel 3-D chaotic systems with an exponential non-linearity via active and adaptive control. Int J Modell Identif Control 22(3):207–217
76. Vaidyanathan S (2015) A 3-D novel highly chaotic system with four quadratic nonlinearities, its adaptive control and anti-synchronization with unknown parameters. J Eng Sci Technol Rev 8(2):106–115
77. Vaidyanathan S (2015) Adaptive backstepping control of enzymes-substrates system with ferroelectric behaviour in brain waves. Int J PharmTech Res 8(2):256–261
78. Vaidyanathan S (2015) Adaptive control of a chemical chaotic reactor. Int J PharmTech Res 8(3):377–382
79. Vaidyanathan S (2015) Adaptive synchronization of chemical chaotic reactors. Int J ChemTech Res 8(2):612–621
80. Vaidyanathan S (2015) Analysis, properties and control of an eight-term 3-D chaotic system with an exponential nonlinearity. Int J Modell Identif Control 23(2):164–172
81. Vaidyanathan S, Azar AT (2015) Analysis, control and synchronization of a nine-term 3-D novel chaotic system. In: Azar AT, Vaidyanathan S (eds) Chaos Modelling and Control Systems Design, Studies in Computational Intelligence, vol 581. Springer, Germany, pp 19–38
82. Vaidyanathan S, Azar AT (2015) Anti-synchronization of identical chaotic systems using sliding mode control and an application to Vaidyanathan-Madhavan chaotic systems. Stud Comput Intell 576:527–547
83. Vaidyanathan S, Azar AT (2015) Hybrid synchronization of identical chaotic systems using sliding mode control and an application to Vaidyanathan chaotic systems. Stud Comput Intell 576:549–569
84. Vaidyanathan S, Madhavan K (2013) Analysis, adaptive control and synchronization of a seven-term novel 3-D chaotic system. Int J Control Theory Appl 6(2):121–137
85. Vaidyanathan S, Pakiriswamy S (2015) A 3-D novel conservative chaotic System and its generalized projective synchronization via adaptive control. J Eng Sci Technol Rev 8(2):52–60
86. Vaidyanathan S, Rajagopal K (2011) Anti-synchronization of Lü and Pan chaotic systems by adaptive nonlinear control. Eur J Sci Res 64(1):94–106

87. Vaidyanathan S, Rajagopal K (2011) Anti-synchronization of Li and T chaotic systems by active nonlinear control. Commun Comput Inf Sci 198:175–184
88. Vaidyanathan S, Rajagopal K (2011) Global chaos synchronization of hyperchaotic Pang and Wang systems by active nonlinear control. Commun Comput Inf Sci 204:84–93
89. Vaidyanathan S, Rajagopal K (2011) Global chaos synchronization of Lü and Pan systems by adaptive nonlinear control. Commun Comput Inf Sci 205:193–202
90. Vaidyanathan S, Rajagopal K (2011) Hybrid synchronization of hyperchaotic Wang-Chen and hyperchaotic Lorenz systems by active non-linear control. Int J Syst Signal Control Eng Appl 4(3):55–61
91. Vaidyanathan S, Rajagopal K (2012) Global chaos synchronization of hyperchaotic Pang and hyperchaotic Wang systems via adaptive control. Int J Soft Comput 7(1):28–37
92. Vaidyanathan S, Rasappan S (2010) New results on the global chaos synchronization for Liu-Chen-Liu and Lü chaotic systems. Commun Comput Inf Sci 102:20–27
93. Vaidyanathan S, Rasappan S (2011) Global chaos synchronization of hyperchaotic Bao and Xu systems by active nonlinear control. Commun Comput Inf Sci 198:10–17
94. Vaidyanathan S, Rasappan S (2011) Hybrid synchronization of hyperchaotic Qi and Lü systems by nonlinear control. Commun Comput Inf Sci 131:585–593
95. Vaidyanathan S, Sampath S (2011) Global chaos synchronization of hyperchaotic Lorenz systems by sliding mode control. Commun Comput Inf Sci 205:156–164
96. Vaidyanathan S, Sampath S (2012) Anti-synchronization of four-wing chaotic systems via sliding mode control. Int J Autom Comput 9(3):274–279
97. Vaidyanathan S, Volos C, Pham V-T (2014) Hyperchaos, adaptive control and synchronization of a novel 5-D hyperchaotic system with three positive Lyapunov exponents and its SPICE implementation. Arch Control Sci 24(4):409–446
98. Vaidyanathan S, Volos C, Pham V-T, Madhavan K, Idowu BA (2014) Adaptive backstepping control, synchronization and circuit simulation of a 3-D novel jerk chaotic system with two hyperbolic sinusoidal nonlinearities. Arch Control Sci 24(3):375–403
99. Vaidyanathan S, Pham V-T, Volos CK (2015) A 5-D hyperchaotic Rikitake dynamo system with hidden attractors. Eur Phys J: Spec Top 224(8):1575–1592
100. Vaidyanathan S, Rajagopal K, Volos CK, Kyprianidis IM, Stouboulos IN (2015) Analysis, adaptive control and synchronization of a seven-term novel 3-D chaotic system with three quadratic nonlinearities and its digital implementation in LabVIEW. J Eng Sci Technol Rev 8(2):130–141
101. Vaidyanathan S, Volos C, Pham V-T, Madhavan K (2015) Analysis, adaptive control and synchronization of a novel 4-D hyperchaotic hyperjerk system and its SPICE implementation. Arch Control Sci 25(1):135–158
102. Vaidyanathan S, Volos CK, Kyprianidis IM, Stouboulos IN, Pham V-T (2015) Analysis, adaptive control and anti-synchronization of a six-term novel jerk chaotic system with two exponential nonlinearities and its circuit simulation. J Eng Sci Technol Rev 8(2):24–36
103. Vaidyanathan S, Volos CK, Pham V-T (2015) Analysis, adaptive control and adaptive synchronization of a nine-term novel 3-D chaotic system with four quadratic nonlinearities and its circuit simulation. J Eng Sci Technol Rev 8(2):181–191
104. Vaidyanathan S, Volos CK, Pham V-T (2015) Analysis, control, synchronization and SPICE implementation of a novel 4-D hyperchaotic Rikitake dynamo system without equilibrium. J Eng Sci Technol Rev 8(2):232–244
105. Vaidyanathan S, Volos CK, Pham V-T (2015) Global chaos control of a novel nine-term chaotic system via sliding mode control. In: Azar AT, Zhu Q (eds) Advances and Applications in Sliding Mode Control Systems, Studies in Computational Intelligence, vol 576. Springer, Germany, pp 571–590
106. Volos CK, Kyprianidis IM, Stavrinides SG, Stouboulos IN, Magafas I, Anagnostopoulos AN (2011) Nonlinear financial dynamics from an engineer's point of view. J Eng Sci Technol Rev 4(3):281–285
107. Volos CK, Kyprianidis IM, Stouboulos IN (2012) Motion control of robots using a chaotic truly random bits generator. J Eng Sci Technol Rev 5(2):6–11

108. Volos CK, Kyprianidis IM, Stouboulos IN (2013) Experimental investigation on coverage performance of a chaotic autonomous mobile robot. Robot Auton Syst 61:1314–1322
109. Volos CK, Kyprianidis IM, Stouboulos IN (2013) Text encryption scheme realized with a chaotic pseudo-random bit generator. J Eng Sci Technol Rev 6:9–14
110. Volos CK, Stavrinides SG, Kyprianidis IM, Stouboulos IN, Ozer M, Anagnostopoulos AN (2013) Impulsive synchronization between double-scroll circuits. In: Stavrinides SG, Banerjee S, Caglar SH, Ozer M (eds) Chaos and Complex Systems. Springer, New York, pp 469–474
111. Volos CK, Kyprianidis IM, Stouboulos IN, Tlelo-Cuautle E, Vaidyanathan S (2015) Memristor: A new concept in synchronization of coupled neuromorphic circuits. J Eng Sci Technol Rev 8(2):157–173
112. Volos CK, Pham V-T, Vaidyanathan S, Kyprianidis IM, Stouboulos IN (2015) Synchronization phenomena in coupled Colpitts circuits. J Eng Sci Technol Rev 8(2):142–151
113. Voss HU (2000) Anticipating chaotic synchronization. Phys Rev E 61:5115–5119
114. Wei Z, Yang Q (2010) Anti-control of Hopf bifurcation in the new chaotic system with two stable node-foci. Appl Math Comput 217(1):422–429
115. Wolf A, Swift JB, Swinney HL, Vastano JA (1985) Determining Lyapunov exponents from a time series. Phys D 16:285–317
116. Zhang H (2010) Chaos Synchronization and Its Application to Secure Communication. In: PhD thesis, University of Waterloo, Canada
117. Zhang W, Tang S, Zhang L, Ma Z, Song J (2015) Chaotic stream cipher-based secure d ata communications over intelligent transportation network. Int J Antennas Propag 2015:1–13
118. Zhou W, Xu Y, Lu H, Pan L (2008) On dynamics analysis of a new chaotic attractor. Phys Lett A 372(36):5773–5777
119. Zhu C, Liu Y, Guo Y (2010) Theoretic and numerical study of a new chaotic system. Intell Inf Manag 2:104–109

On the Verification for Realizing Multi-scroll Chaotic Attractors with High Maximum Lyapunov Exponent and Entropy

E. Tlelo-Cuautle, M. Sánchez-Sánchez, V.H. Carbajal-Gómez, A.D. Pano-Azucena, L.G. de la Fraga and G. Rodriguez-Gómez

Abstract Nowadays, many works have been presented regarding the modeling, simulation and circuit realization of different kinds of continuous-time multi-scroll chaotic attractors. However, very few works describe the experimental realization of attractors having high maximum Lyapunov exponent (MLE) and high entropy, which are desirable characteristics to guarantee better chaotic unpredictability. For instance, two chaotic oscillators having the same MLE values can behave in a very different way, e.g. showing different entropy values. That way, we describe the experimental realization of an optimized multi-scroll chaotic oscillator with both high MLE and entropy. First, the MLE is optimized by applying an evolutionary algorithm, which provides a set of feasible solutions. Second, the associated entropy is evaluated for each feasible solution. In this chapter, experimental results are shown for the electronic implementation of a chaotic oscillator generating 2-, 5- and 10-scrolls. Finally, the experimental results show that by increasing the number of scrolls both the MLE and its associated entropy increase in a similar proportion, thus guaranteeing better unpredictability.

E. Tlelo-Cuautle (✉) · V.H. Carbajal-Gómez · A.D. Pano-Azucena · G. Rodriguez-Gómez
INAOE, Luis Enrique Erro No. 1, Tonantzintla, Puebla, Mexico
e-mail: etlelo@inaoep.mx

V.H. Carbajal-Gómez
e-mail: victhug26@gmail.com

A.D. Pano-Azucena
e-mail: ana.dalia.p.a@gmail.com

G. Rodriguez-Gómez
e-mail: grodrig@inaoep.mx

M. Sánchez-Sánchez
Institute of Agricultural Engineering, Universidad del Papaloapan, Loma Bonita, Oaxaca, Mexico
e-mail: masanchez@unpa.edu.mx

L.G. de la Fraga
CINVESTAV, Mexico City, Mexico
e-mail: fraga@cs.cinvestav.mx

S. Vaidyanathan and C. Volos (eds.), *Advances and Applications in Chaotic Systems*, Studies in Computational Intelligence 636,
DOI 10.1007/978-3-319-30279-9_13

Keywords Multi-scroll chaotic oscillator · Maximum Lyapunov exponent · Entropy · Evolutionary algorithm · Operational amplifier

1 Introduction

In Electronics, a great variety of chaotic oscillators has been implemented with different kinds of electronic devices [7, 15, 19], using integrated circuits technology [20], and more recently by using field programmable gate arrays [17]. In addition, recent works show a relationship between the number of scrolls and the value of the maximum Lyapunov exponent (MLE) [2, 18], and also a relationship between the number of scrolls and the associated entropy [23]. Both characteristics associated to MLE and entropy are quite desirable to improve for the development of enhanced applications in nonlinear systems, like for example: the implementation of random number generators [1, 4, 10, 23] that are quite useful in robotics [16, 21].

In this chapter we show the experimental realization of multi-scroll chaotic attractors that are optimized to provide a high value of the MLE, and for each attractor it is also guaranteed to have a good distribution of the trajectories that is visualized in the phase-space portraits. As concluded in [23], the next sections show that the MLE increases by increasing the number of scrolls, indicating a better unpredictability of the dynamical system due to the increment of its associated entropy. In addition, we also show the experimental realization of multi-scroll chaotic attractors having a uniform distribution of its trajectories in the phase-space portrait, because when the phases are not well distributed among all the scrolls, some scrolls cannot be formed, thus leading to a pretty difficult problem for the electronic implementation.

The optimization algorithm proposed in [2], is applied herein. It is based on the evolutionary algorithm known as non-dominated sorting genetic algorithm (NSGA-II) [3], and it optimizes two characteristics, namely: (a) Maximizing the positive Lyapunov exponent, and (b) minimizing the dispersions of the phase transitions among all scrolls in an attractor. The results of the optimization algorithm show that both characteristics are in conflict, so that a feasible set of solutions is provided to select the best one according to the problem at hand.

From the set of feasible solutions provided by applying [2], we select some multi-scroll chaotic attractors having high MLE, and then we realize experiments by implementing the chaotic oscillators with commercially available operational amplifiers. The purpose of the experiments are oriented to verify the relationship on the number of scrolls, their MLE value and the associated entropy.

The case of study is the multi-scroll chaotic oscillator based on saturated nonlinear function series, already described in [7]. It is a third order continuous-time dynamical system, and then it has three Lyapunov exponents, one being positive, which is known as MLE and its value indicates the degree of chaotic behavior. It is optimized by applying the evolutionary algorithm given in [2], and its associated entropy is evaluated from both numerical simulations and experimental data for generating 2-, 5- and 10-scrolls. The goal of performing the experiments is to verify that the higher the number of scrolls the higher the MLE and its associated entropy [9]. Further, optimized chaotic oscillators should improve applications like random number generators [1, 4, 10, 16, 21, 23], synchronization [6], and secure communication systems [5], for instance.

2 Multi-scroll Chaotic Oscillators

The study of chaotic dynamical systems traces its origin to the findings of E.N. Lorenz in the early 1960s. His interest in weather forecasting led him to the discovery of a nonlinear dynamical system that displayed high sensitivity to the initial conditions, an essential property of chaotic systems. Ever since then, decades of research have expanded the understanding of this ubiquitous phenomena and produced several applications in different areas of engineering.

Dynamical systems describe motion in nature that can be modeled by equations of the following forms:

$$\dot{x}(t) = f(a, x(t)) \quad t > 0 \tag{1}$$

and

$$x(n+1) = f(a, x(n)) \quad n \in \mathbb{N} \tag{2}$$

In these equations, the state variables of the dynamical system are represented by $x(t)$ and $x(n)$. The possible values of the state variables imply that (1) is a differential equation while (2) is a difference equation or map. In both cases, x represents a q-dimensional state vector, i.e. $x \in R^q$. The control parameter a (also called bifurcation parameter) has m components such that $a \in R^m$. The control parameter affects the evolution of the state variables, and the relationship between the parameters and the state variables is defined by a function f. The range of f is in the same space of the state vector x. Chaos, defined colloquially as irregular an unpredictable behavior, can stem from dynamical systems as long as f has the suitable properties. Likewise, a must be set to the appropriate values in order to precipitate a transition to chaos. The chaotic behavior product of dynamical systems is known as deterministic chaos.

In electronics, chaotic oscillators have been implemented to generate double or multi-scroll chaotic attractors. In the last case, multi-scrolls have been generated by using different kinds of electronic devices [7, 15, 19], as well as by designing integrated circuits [20] or by using configurable digital architectures [17]. However, analog realizations suffer the limitations of the electronic devices [11], or suffer the variations problems from integrated circuit fabrication technologies as shown in [20], where a variation in process, voltage or temperature (PVT) may degrade or even eliminate the normal behavior of a chaotic oscillator, rendering it useless. That way, still more research is done regarding the design of chaotic oscillators using analog devices. In this chapter, the chaotic oscillator is realized by using commercially available operational amplifiers.

The case of study is the multi-scroll chaotic oscillator based on saturated nonlinear function series [7], which can be described by the system of differential equations given by (3), where a, b, c, and d_1 are positive constants that can have values in the interval [0, 1]. In (3), the dynamical system is controlled by a saturated nonlinear function series f that is approximated by piecewise-linear functions.

$$\begin{aligned} \dot{x} &= y \\ \dot{y} &= z \\ \dot{z} &= -ax - by - cz + d_1 f(x; m) \end{aligned} \tag{3}$$

In the following, we describe in detail how the saturated function f in (3) is obtained. Let f_0 be the saturated function:

$$f_0(x; m) = \begin{cases} 1, & \text{if } x > m \\ \frac{x}{m}, & \text{if } |x| \leq m \\ -1, & \text{if } x < -m, \end{cases} \tag{4}$$

where $1/m$ is the slope of the middle segment and $m > 0$; the upper radial $\{f_0(x; m) = 1 \mid x > m\}$, and the lower radial $\{f_0(x; m) = -1 \mid x < -m\}$ are called *saturated plateaus*, and the segment $\{f_0(x; m) = x/m \mid |x| \leq m\}$ between the two saturated plateaus is called *saturated slope*.

Lets us consider now the saturated functions f_h and f_{-h} defined as:

$$f_h(x; m, h) = \begin{cases} 2, & \text{if } x > h + m \\ \frac{x-k}{m} + 1, & \text{if } |x - h| \leq m \\ 0, & \text{if } x < h - m, \end{cases} \tag{5}$$

and

$$f_{-h}(x; m, -h) = \begin{cases} 0, & \text{if } x > h + m \\ \frac{x-k}{m} - 1, & \text{if } |x - h| \leq m \\ -2, & \text{if } x < h - m, \end{cases} \tag{6}$$

where h is called the *saturated delay time* and $h > m$. Therefore, a saturated function series for a chaotic oscillator with s scrolls is defined as the function:

$$f(x; m) = \sum_{i=0}^{s-2} f_{2i-s+2}(x; m, 2i - s + 2) \tag{7}$$

where $s > 2$.

For example, using $f = f_0$ in (3), a 2-scrolls chaotic oscillator can be generated. Therefore, the saturated function series to generate 3-scrolls is $f(x; m) = f_{-1}(x; m, -1) + f_1(x; m, 1)$. To generate a 4-scrolls attractor it will be $f(x; m) = f_{-2}(x; m, -2) + f_0(x; m) + f_2(x; m, 2)$, and so on.

In [2], the optimization of the MLE requires as input data, the number of scrolls to be generated. Then, a bi-objective optimization problem is encoded: (i) to maximize MLE, and (ii) to minimize the variability in the oscillator's phase-space transitions or the trajectories. From (3), the optimization problem is devoted to find the values of the four coefficient variables a, b, c and d_1 that solve both objectives (i) and (ii). Those four coefficients can take values within the range [0.0, 1.0], and one decides how many decimal numbers to use.

3 Computing Lyapunov Exponents and Entropy

Lyapunov exponents are asymptotic measures that characterize the average rate of growth (or shrinking) of small perturbations to the solutions of a dynamical system. Lyapunov exponents provide quantitative measures of response sensitivity of a dynamical system to small changes in initial conditions. The number of Lyapunov exponents is equal to the number of states variables in the dynamical system, and at least three state variables are required to generate chaotic behavior. In this chapter, the case of study is a multi scroll chaotic oscillator having three state variables, described by (3). The experimental results presented in the next sections will verify what is already known that by increasing the number of scrolls both the MLE and its associated entropy increase in a similar proportion [23].

3.1 Lyapunov Exponents

Lets us consider an n-dimensional dynamical system:

$$\dot{x} = f(x), \quad t > 0, \quad x(0) = x_0 \in \mathbb{R}^n \tag{8}$$

where x and f are n-dimensional vector fields. To determine the n-Lyapunov exponents of the system one have to find the long term evolution of small perturbations to a trajectory, which are determined by the variational equation of (8),

$$\dot{y} = \frac{\partial f}{\partial x}\big(x(t)\big)y = J\big(x(t)\big)y \tag{9}$$

where J is the $n \times n$ Jacobian matrix of f. A solution to (9) with a given initial perturbation $y(0)$ can be written as

$$y(t) = Y(t)y(0) \tag{10}$$

with $Y(t)$ as the fundamental solution satisfying

$$\dot{Y} = J\big(x(t)\big)Y, \quad Y(0) = I_n \tag{11}$$

Here I_n denotes the $n \times n$ identity matrix. If we consider the evolution of an infinitesimal n-parallelepiped $[p_1(t), \ldots, p_n(t)]$ with the axis $p_i(t) = Y(t)p_i(0)$ for $i = 1, \ldots, n$, where $p_i(0)$ denotes an orthogonal basis of $\mathbb{R}^n$. The ith Lyapunov exponent, which measures the long-time sensitivity of the flow $x(t)$ with respect to the initial data $x(0)$ at the direction $p_i(t)$, is defined by the expansion rate of the length of the ith axis $p_i(t)$ and is given by

$$\lambda_i = \lim_{t\to\infty} \frac{1}{t} \ln \left\| p_i(t) \right\| \tag{12}$$

In summary, the Lyapunov exponents can be computed as follows [2, 13, 18, 22]:

1. Initial conditions of the system and the variational system are set to $\mathbf{X_0}$ and $\mathbf{I_{n\times n}}$, respectively.
2. The systems are integrated by several steps until an orthonormalization period TO is reached. The integration of the variational system $\mathbf{Y} = [y_1, y_2, y_3]$ depends on the specific Jacobian that the original system $\mathbf{X}$ is using in the current step.
3. The variational system is orthonormalized by using the standard Gram-Schmidt method [12], and the logarithm of the norm of each Lyapunov vector contained in $\mathbf{Y}$ is obtained and accumulated in time.
4. The next integration is carried out by using the new orthonormalized vectors as initial conditions. This process is repeated until the full integration period T is reached.
5. The Lyapunov exponents are obtained by

$$\lambda_\mathrm{i} \approx \frac{\mathbf{1}}{\mathbf{T}} \sum_{\mathrm{j=TO}}^{\mathbf{T}} \ln \left\| \mathbf{y_i} \right\|$$

The time-step selection was set as in [18], by using the minimum absolute value of all the eigenvalues of the system $\lambda_{\min}$, and ψ was chosen well above the sample theorem as 50.

$$t_{\text{step}} = \frac{1}{\lambda_{\min}\psi}$$

The orthogonalization period TO was chosen about 50 t_{step}. This procedure is used herein as in [2] to optimize the MLE.

3.2 Evaluation of Entropy

For chaotic oscillators, the entropy is an alternative choice to Lyapunov exponents because it reveals aspects of the underlying dynamical system (i.e., it quantifies the stretching and the folding aspects at the same time). In this manner, in this chapter the entropy is evaluated because its rate of growth is an interesting parameter to quantify disorder in chaotic oscillators. In the same direction, as chaotic attractors can be recognized by visual inspection in their phase-space portraits, we perform a numerical quantification of chaos by optimizing the MLE of the chaotic oscillator described by (3). The entropy has also some relationships of interest as for the sum of Lyapunov exponents [13, 14], which measure the instability of nearby trajectories.

The entropy is computed herein by applying the algorithm presented by Moddemeijer, which is online available at http://www.cs.rug.nl/~rudy/matlab/. That way, in Sect. 5, we list 10 values of the MLE and their associated entropy that is evaluated from both numerical simulation and experimental data for generating 2-, 5- and 10-scrolls attractors.

4 Circuit Realization with Commercially Available Operational Amplifiers

The multi-scroll chaotic oscillator based on saturated nonlinear function series f is described by (3). For the circuit realization, one should approximate function f by piecewise-linear (PWL) segments as follows:

$$f(x;k,h,p,q) = \begin{cases} (2q+1)k, \\ k(x-ih)+2ik, \\ (2i+1)k, \\ -(2p+1)k. \end{cases} \tag{13}$$

with

$$\begin{array}{l} x > qh+1 \\ |x-ih| \leq 1, \quad -p \leq i \leq q \\ ih+1 < x < (i+1)h-1, \quad -p \leq i \leq q-1 \\ x < -ph-1. \end{array}$$

For instance, Fig. 1 shows two kinds of saturated functions. The one with 5 linear segments is used to generate odd number of scrolls, and the one with 7 segments is used to generate even number of scrolls. The difference is that the one on the right has an slope crossing the origin of the plane. Thus, by increasing the number of segments from the three near the origin in Fig. 1b, one generates as many even number of scrolls as the number of saturated levels, which are the linear segments with slope = 0. In a similar way, starting from five segments, as the PWL description

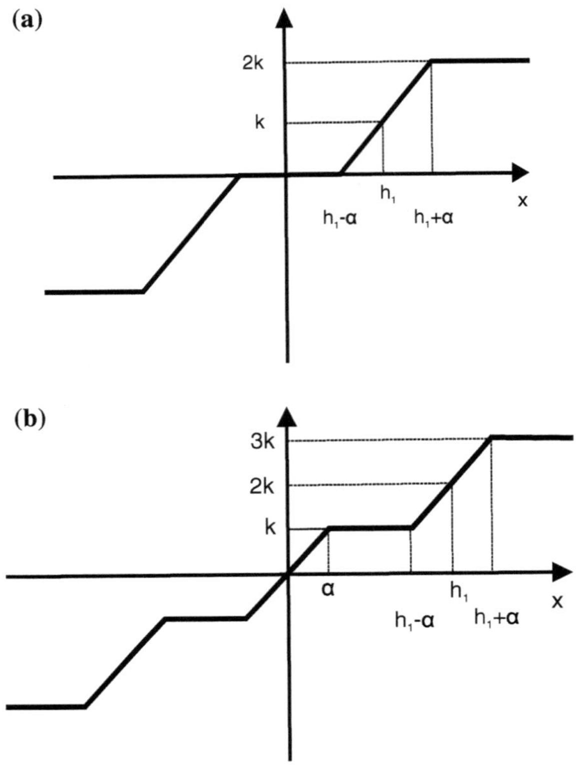

Fig. 1 PWL descriptions of a saturated nonlinear function series to generate **a** 3-scrolls, and **b** 4-scrolls

in Fig. 1a, one can generate as many odd number of scrolls as the number of saturated levels.

In simulating multi-scroll chaotic oscillators, one should scale the values to be realized with electronic devices. For example, by simulating 6-scrolls one needs a PWL description like the one in Fig. 1b but with 11 segments (6 saturated levels plus 5 slopes). Therefore, by setting $a, b, c, d_1 = 0.7$, $k = 10$, $h = 20$, $p = q = 2$, the simulation result is shown in Fig. 2. As one sees, the ranges for the vertical and horizontal axes are around ± 12 and ± 60, respectively. It is pretty clear that the horizontal range cannot be realized using commercially available operational amplifiers because they can be biased only up to ± 18 V.

To cope with this problem one can scale the PWL description by modifying (13) by α. Now, the saturated nonlinear function series is redefined by (14), where α allows that $k < 1$, because the chaos condition now applies on $s = \frac{k}{\alpha}$, the new slope. In this manner, k and α can be selected to allow $k < 1$, so that the ranges in (13) can be scaled. As a result, now the generation of a 6-scrolls attractor with $a = b = c = d_1 = 0.7$, $k = 1$, $\alpha = 0.1$, $s = 10$, $h = 2$, and $p = q = 2$, is shown in Fig. 3. As one sees, the ranges of the attractor are within the ranges that can be processed by commercially available operational amplifiers. Besides, it is possible to compute small ranges for

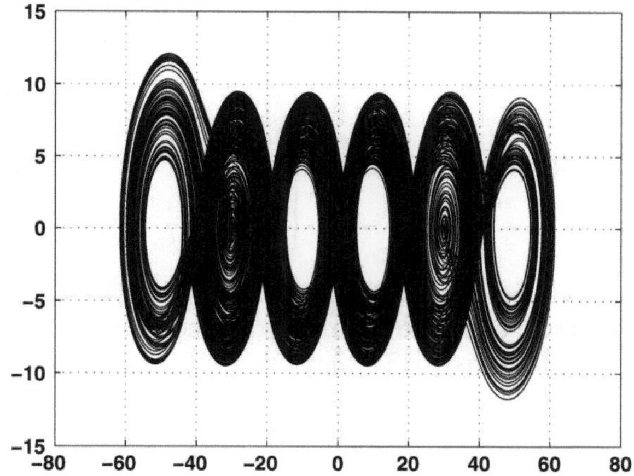

Fig. 2 Generating a 6-scrolls attractor without scaling

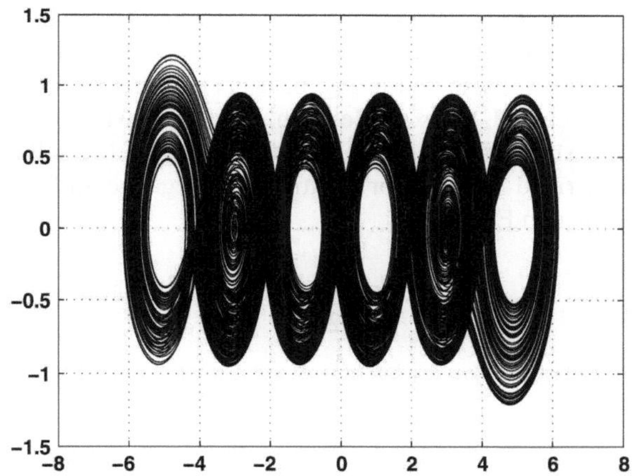

Fig. 3 Generating 6-scrolls with ranges that can be processed by commercially available operational amplifiers

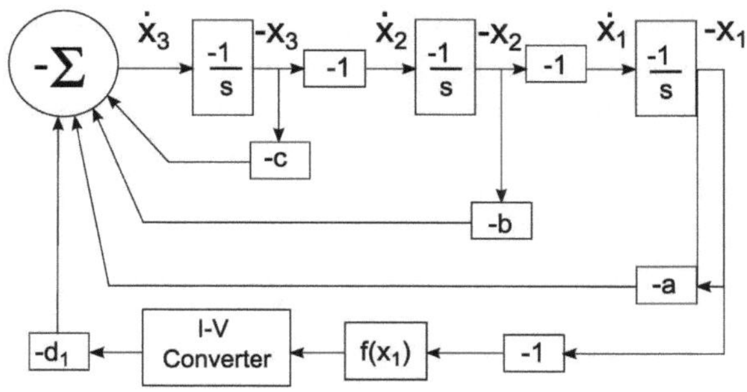

Fig. 4 Block diagram description of (3)

realizing attractors with integrated circuit technology [20], it just depends on setting the values of k and α.

$$f(x_1; k, h, p, q) = \begin{cases} (2q+1)k & x_1 > qh + \alpha \\ \frac{k}{\alpha}(x_1 - ih) + 2ik & \mid x_1 - ih \mid \leq \alpha \\ & -p \leq i \leq q \\ (2i+1)k & ih + \alpha < x_1 < (i+1)h - \alpha \\ & -p \leq i \leq q-1 \\ -(2p+1)k & x_1 < -ph - \alpha \end{cases} \tag{14}$$

By applying flow diagrams from linear systems, the dynamical system in (3) can be described by the block diagram shown in Fig. 4. An analogy to electronics, the diagram consists of 3 integrators, 1 adder, one current-to-voltage (I/V) converter, one block for the saturated nonlinear function series ($f(x_1)$) and amplifiers. In this manner, each block can be realized with commercially available operational amplifiers. One of the realizations is shown in Fig. 5, where the block for ($f(x_1)$) is labeled SNLF.

For realizing the nonlinear saturated function series, one can take advantage of the saturation properties of the operational amplifiers. In this manner, two saturated levels can be implemented in voltage mode by using the finite-gain model of the operational amplifier, as shown in Fig. 6. It is clear that by simulation, several limitations can be included, for example: gain, bandwidth, slew rate and saturation [11]. Therefore, if a shift-voltage ($\pm E$) is added, one gets the shifted-voltage saturated functions described by (15) for positive and negative shifts, respectively, these effects are shown in Fig. 7.

$$V_o = \frac{A_v}{2}\left(|V_i + \frac{V_{sat}}{A_v} - E| - |V_i - \frac{V_{sat}}{A_v} - E|\right) \tag{15}$$
$$V_o = \frac{A_v}{2}\left(|V_i + \frac{V_{sat}}{A_v} + E| - |V_i - \frac{V_{sat}}{A_v} + E|\right)$$

The saturated nonlinear function series can now be implemented as shown in Fig. 8, where the number of operational amplifiers equals the number of scrolls to be generated, minus one. The reason is that one operational amplifier can generate 2-scrolls, then one needs three amplifiers to generate 3-scrolls, and so on. In the same manner, to generate a saturated nonlinear function with different voltage-shifts, then E takes different values in (15). On the other hand, the values of the plateaus k, in voltage and current, the breakpoints α, the slope s and the saturated delays h are evaluated by (16) [15, 19].

$$k = R_{ix}I_{sat}, \quad I_{sat} = \frac{V_{sat}}{R_c}, \quad \alpha = \frac{R_{iz}|V_{sat}|}{R_{fz}}, \quad h = \frac{E_i}{(1 + \frac{R_{iz}}{R_{fz}})} \tag{16}$$

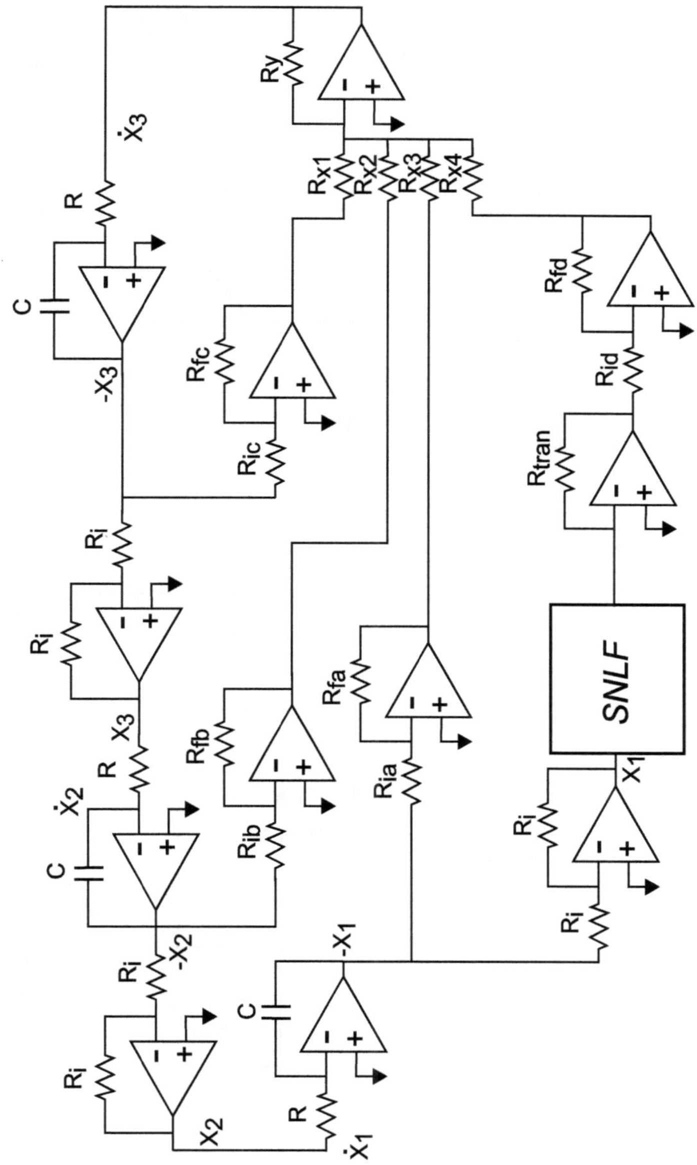

Fig. 5 Circuit realization of (3) by using commercial operational amplifiers

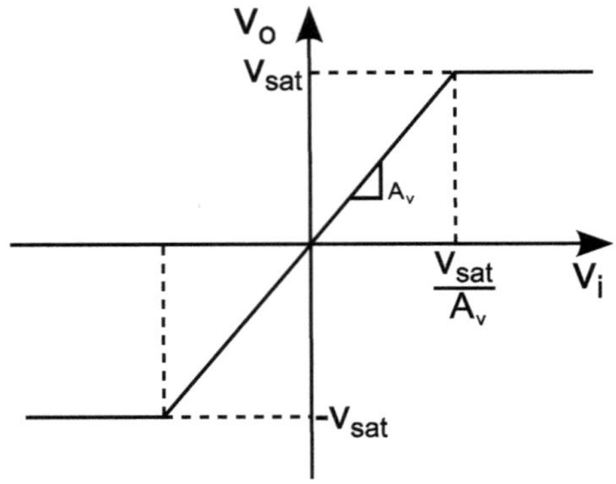

Fig. 6 Finite gain model of the operational amplifier

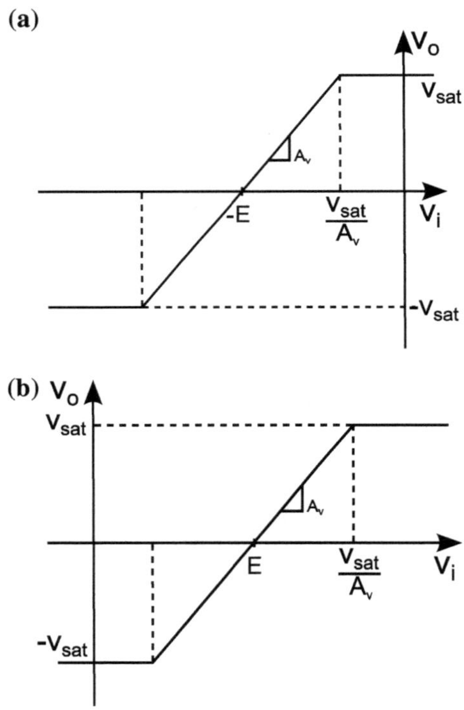

Fig. 7 Shift of the voltage when E takes **a** Negative and **b** Positive values

5 Experimental Verification Results

To have control on varying the coefficients a, b, c and d_1 in (3), the multi-scroll chaotic oscillator was implemented as shown in Fig. 5, where the block sketching the saturated nonlinear function (SNLF) is shown in Fig. 8. The values of the circuit elements are: $C = 1\,\text{nF}$, $R = 1\,\text{M}\Omega$, $R_{ia} = R_{ib} = R_{ic} = R_{id} = 10\,\text{k}\Omega$, $R_i = R_f$, with $V_{sat} = \pm 16$ to $\pm 18\,\text{V}$. Besides, to set the corresponding values of the coefficients a,

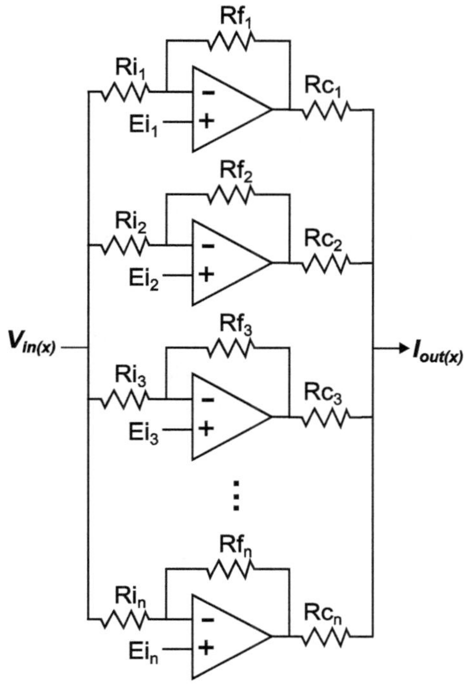

Fig. 8 Realization of the saturated nonlinear function series using operational amplifiers

Table 1 Optimized MLE and its associated entropy for generating 2-scrolls

Case	a	b	c	d_1	MLE	Simulated entropy
1	1.0000	1.0000	0.4997	1.0000	0.3761	1.4742
2	1.0000	0.7884	0.6435	0.6665	0.3713	1.0709
3	0.8661	1.0000	0.3934	0.9903	0.3607	1.15806
4	0.7746	0.6588	0.5846	0.4931	0.3460	1.1133
5	1.0000	0.7000	0.6780	0.1069	0.3437	0.7281
6	1.0000	0.7000	0.7000	0.2542	0.3425	1.16843
7	0.7743	0.6716	0.5892	1.8469	0.3391	1.5712
8	0.9248	0.7491	0.6686	0.6814	0.3385	1.1628
9	0.7178	0.6593	0.5546	0.2247	0.3376	0.2925
10	0.7060	0.6451	0.5523	0.2181	0.3320	0.2765
11	0.7060	0.7000	0.7000	0.7000	0.2658	1.3312

b, c and d_1, associated to the optimized values for MLE listed in Tables 1, 2 and 3, linear precision potentiometers were used to tune the four decimals. In Fig. 5, the resistances associated to the four coefficients are labeled as: R_{fa}, R_{fb}, R_{fc}, R_{fd}.

Table 2 Optimized MLE and its associated entropy for generating 5-scrolls

Case	*a*	*b*	*c*	*d*	MLE	Entropy simulated	Entropy experiment
1	1.0000	0.7250	0.2250	1.0000	0.6919	2.2481	2.0131
2	0.9880	0.7140	0.2050	1.0000	0.6914	2.2962	2.1472
3	0.9890	0.7300	0.2070	1.0000	0.6908	2.2708	2.0779
4	0.9910	0.6810	0.2300	0.9810	0.6814	2.2906	2.1175
5	0.9880	0.7480	0.1890	1.0000	0.6663	1.3800	1.9619
6	0.9840	0.6810	0.2270	0.9830	0.6651	2.3365	2.0757
7	0.9890	0.6810	0.2040	0.9790	0.6645	2.1736	2.3032
8	1.0000	0.7840	0.2000	1.0000	0.6533	2.2628	2.3024
9	0.9800	0.7960	0.1570	1.0000	0.6523	1.3214	2.1260
10	1.0000	0.7330	0.2050	1.0000	0.6471	2.2560	2.0287
11	0.7000	0.7000	0.7000	0.7000	0.2840	2.2352	1.9403

Table 3 Optimized MLE and its associated entropy for generating 10-scrolls

Case	*a*	*b*	*c*	*d*	MLE	Entropy simulated	Entropy experiment
1	1.0000	0.5160	0.1190	1.0000	0.8853	2.8882	2.6302
2	1.0000	0.5054	0.1140	1.0000	0.8826	2.9032	2.6152
3	1.0000	0.5130	0.1180	1.0000	0.8792	2.8863	2.6193
4	1.0000	0.5410	0.1060	1.0000	0.8712	2.8874	2.5166
5	1.0000	0.5930	0.0840	1.0000	0.8545	2.8664	2.4594
6	1.0000	0.5160	0.1580	1.0000	0.8438	2.9273	2.6874
7	1.0000	0.6430	0.0975	1.0000	0.8314	2.8957	2.4891
8	1.0000	0.7000	0.1160	1.0000	0.7825	2.8788	2.6890
9	1.0000	0.7995	0.2127	0.9831	0.7249	2.6036	1.8740
10	1.0000	0.7200	0.4195	1.0000	0.6177	2.8748	2.6213
11	0.7000	0.7000	0.7000	0.7000	0.3026	2.8956	2.6157

The measurements were performed using a 200 MHz oscilloscope with a sampling frequency of 1 G/s. This equipment introduces errors in saving the samples, so that it is reflected in the differences when computing the entropy from simulated and experimental results, as shown in Tables 2 and 3.

The experimental results for the realization of the saturated nonlinear function series with 3 and 19 segments, to generate 2- and 10-scrolls, respectively, are shown in Fig. 9. Other saturated nonlinear function series can be generated as already shown in [8, 11].

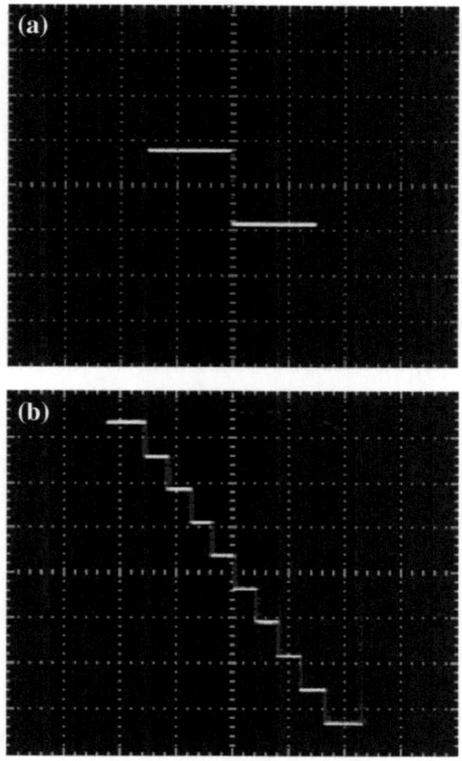

Fig. 9 Experimental results for the saturated nonlinear function series with **a** 3, and **b** 19 segments

Figure 10 shows the simulation results for six cases from Table 1. As one sees, case 11 has the lowest MLE, where all coefficients are set to traditional values of 0.7, as used in [8]. Furthermore, when applying [2], the other 10 cases provide higher MLEs, because the coefficients a, b, c and d_1 were varied (optimized). The simulated entropy also shown in Table 1, shows slight variations when the MLE increases, but it can be appreciated that in general it increases as MLE does it.

Figure 11 shows experimental results for generating two-scrolls for six cases in Table 1. As one sees, the more complex behavior appears for the highest MLE. This is better appreciated when incrementing the number of scrolls, as shown in the following cases for generating 5- and 10-scrolls.

Figure 12 shows six cases from Table 2. In these cases it is better appreciated that the higher the value of the MLE, the better the complex behavior of the 5-scrolls attractor, i.e. the scrolls are less defined in the phase-space portraits, as already shown in [2]. This is indeed confirmed in Fig. 13 for generating a 5-scrolls attractor, for which we list the entropy computed from experimental data. It can be appreciated that the entropy increases as MLE does it.

Figure 14 shows six simulation results for generating 10-scrolls from Table 3. As one sees, case 11 is generated when the four coefficients have the same value of 0.7,

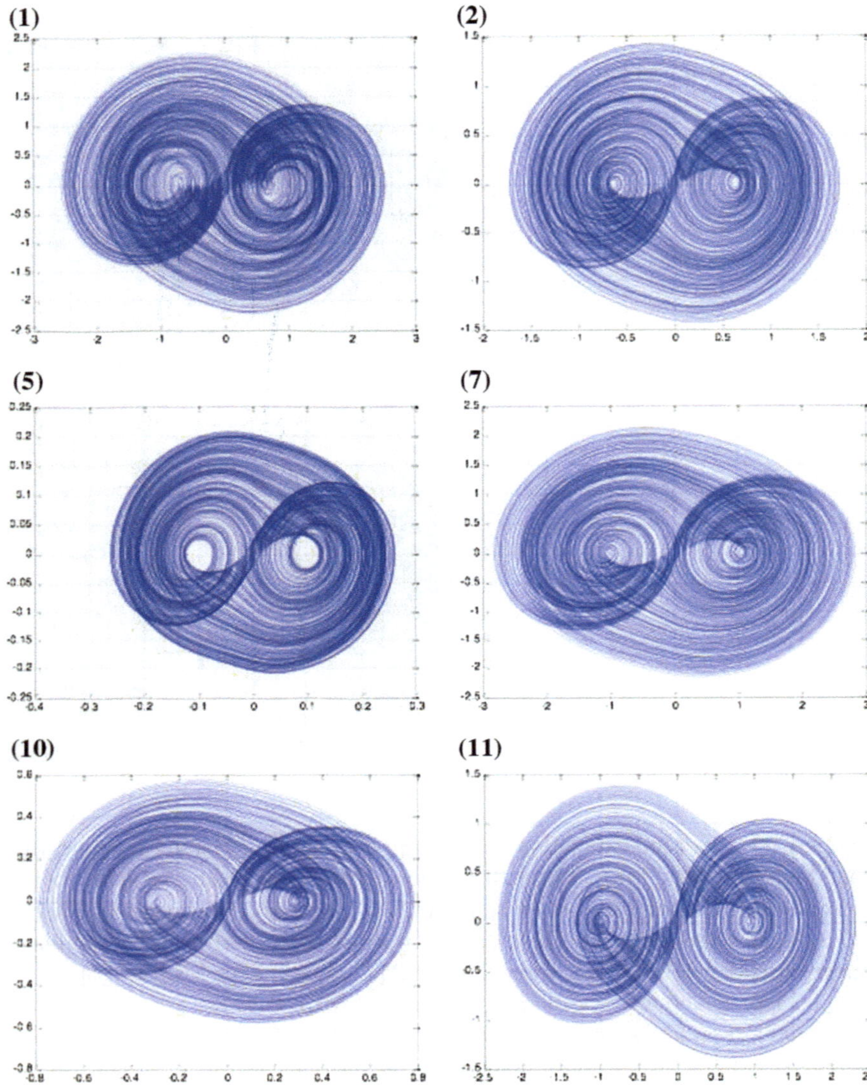

Fig. 10 Simulation results generating 2-scrolls for six cases in Table 1

for which the 10-scrolls are pretty good appreciated. However, the scrolls become more complex as the MLE increases, so that case 2 in Fig. 14 shows a more complex attractor.

The simulated entropy in Table 3 shows a little bit difference for the 11 cases, where cases 2 and 6 have the higher simulated entropy value.

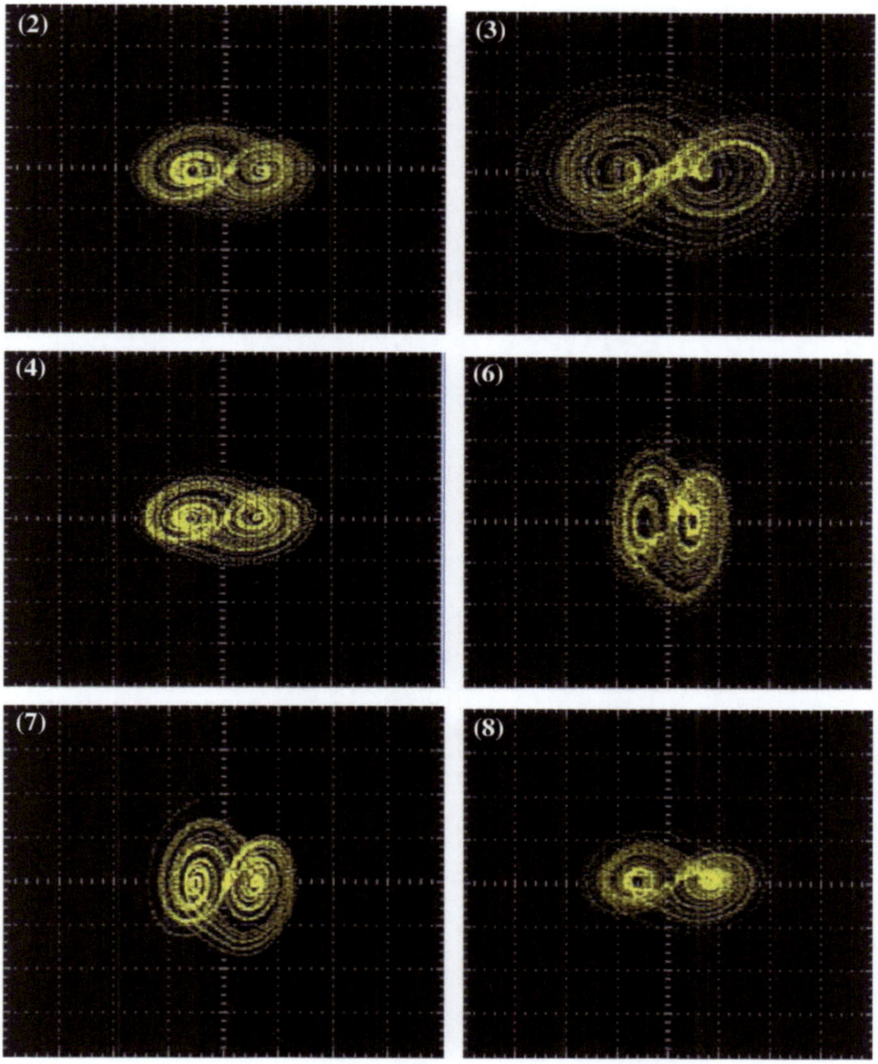

Fig. 11 Experimental results generating 2-scrolls for six cases in Table 1

Figure 15 shows six experimental cases from Table 3. The 10-scrolls attractors for those cases were generated using the saturated nonlinear function series shown in Fig. 9b, with 19 segments. As supposed, case 1 has the more complex chaotic behavior because it has the highest MLE. The other 5 cases shown in Fig. 15 are also complex because MLE is higher than when using traditional coefficient values of 0.7 [8]. In that case, the scrolls are more defined in the phase space diagram, as for the simulated case 11 in Fig. 14. As it was done for the 5-scrolls attractor, we list the

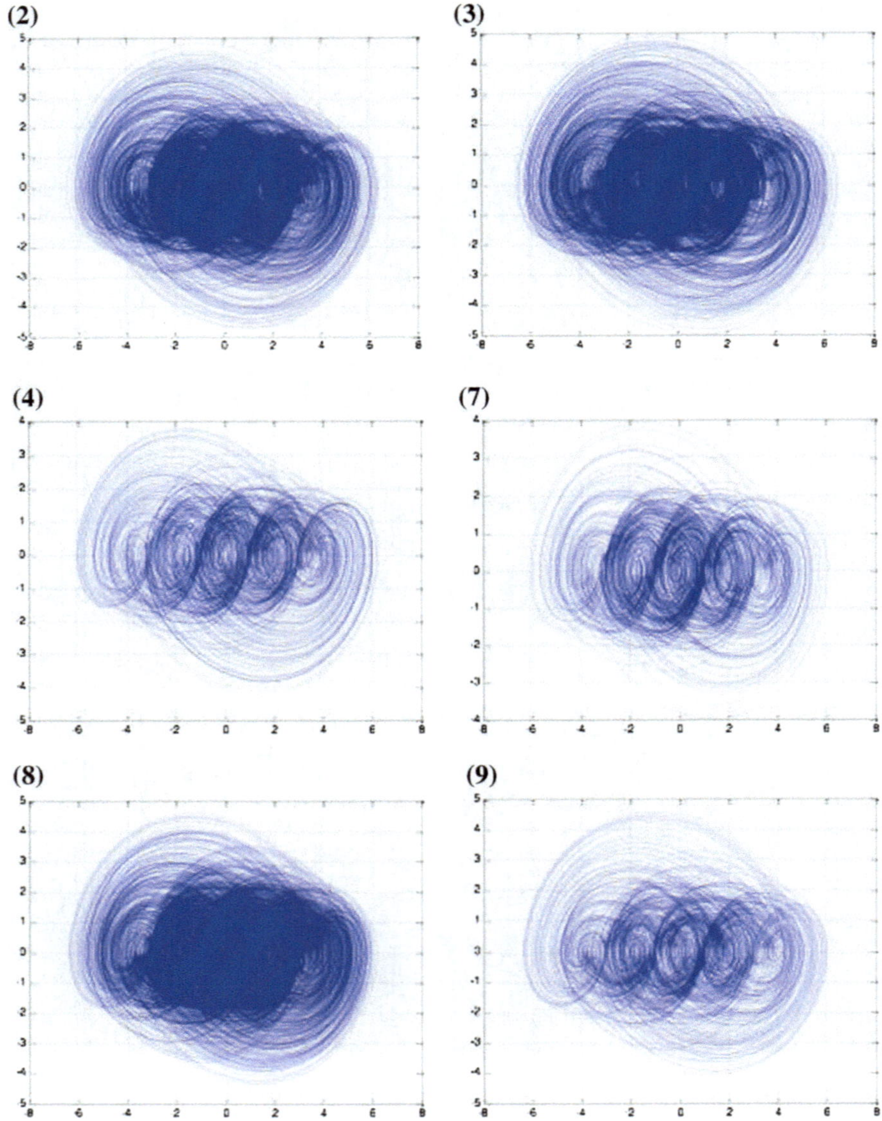

Fig. 12 Simulation results generating 5-scrolls for six cases in Table 2

entropy computed from experimental data in Table 3. Again, it can be appreciated that the entropy varies as MLE does it.

Figure 16 shows the state variable x for case 11, where one can count quite clearly the 10 levels that are associated to the 10 saturated levels of the saturated nonlinear function series shown in Fig. 9b. A similar behavior is for the state variable x for the

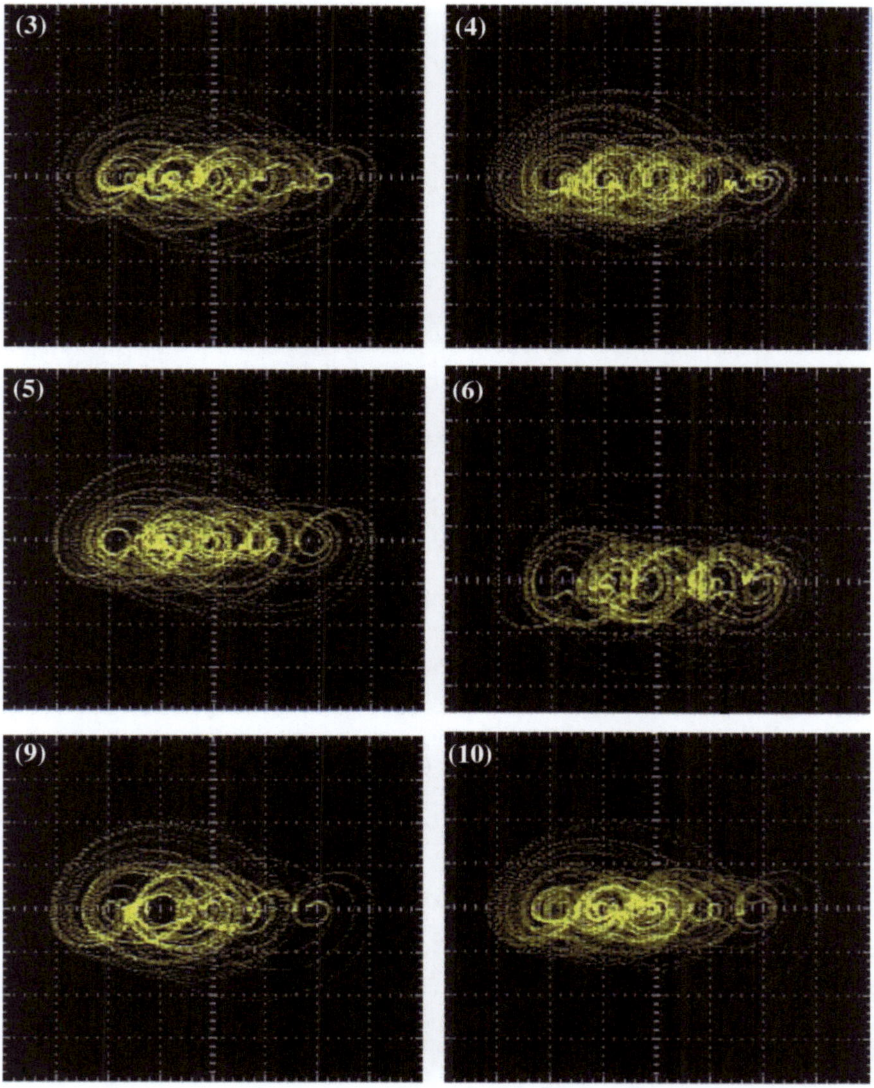

Fig. 13 Experimental results generating 5-scrolls for six cases in Table 2

other cases, but the phase space portraits are more complex as MLE being increased, as shown in Figs. 14 and 15.

From the simulated and experimental data, it can be concluded that the more scrolls are generated the higher the values for the entropy and MLE.

It is worth mentioning that because we used a 200 MHz oscilloscope with a sampling frequency of 1 G/s, the saved experimental data is contaminated with undesired

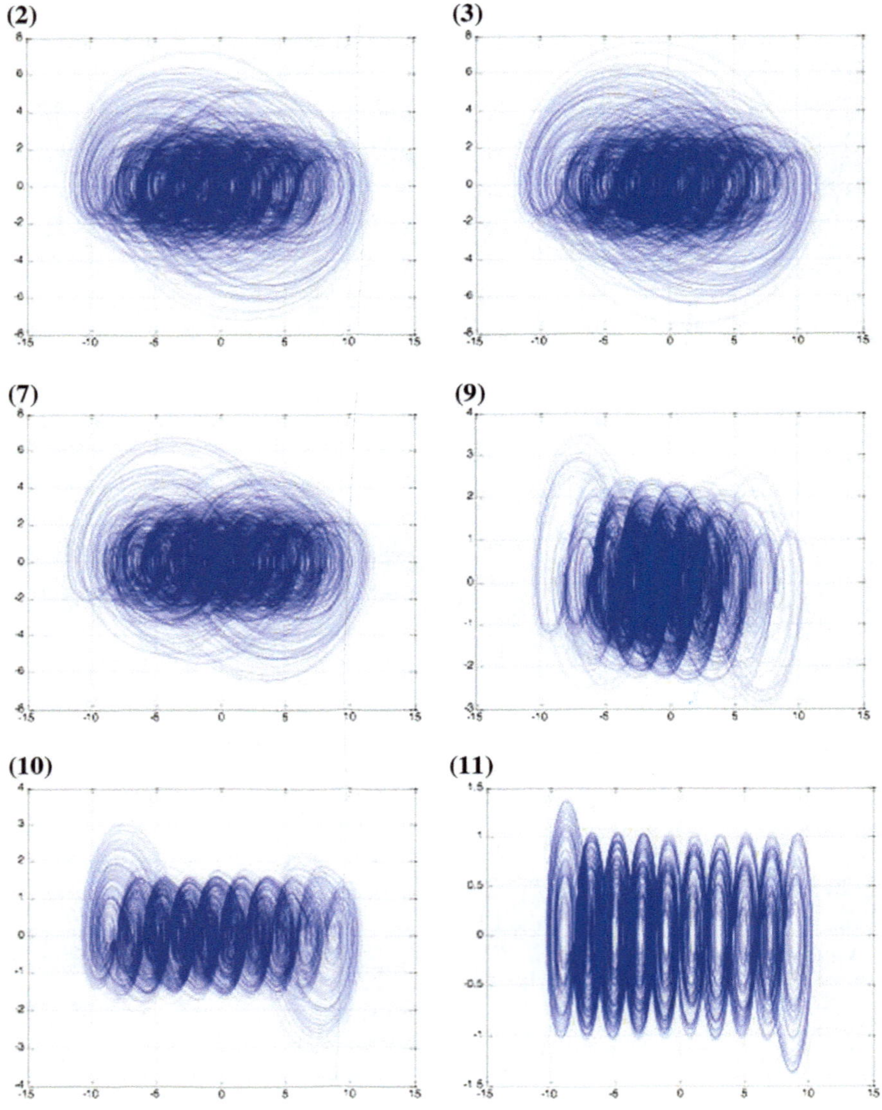

Fig. 14 Simulation results generating 10-scrolls for six cases in Table 3

frequencies, as shown in Fig. 17. It means that one should filter the experimental signal to avoid aliasing and then recover the chaotic signal. Figure 18 shows the comparison of the signals in the phase space portraits, when they are plotted directly from the experimental data, and after the signal is filtered.

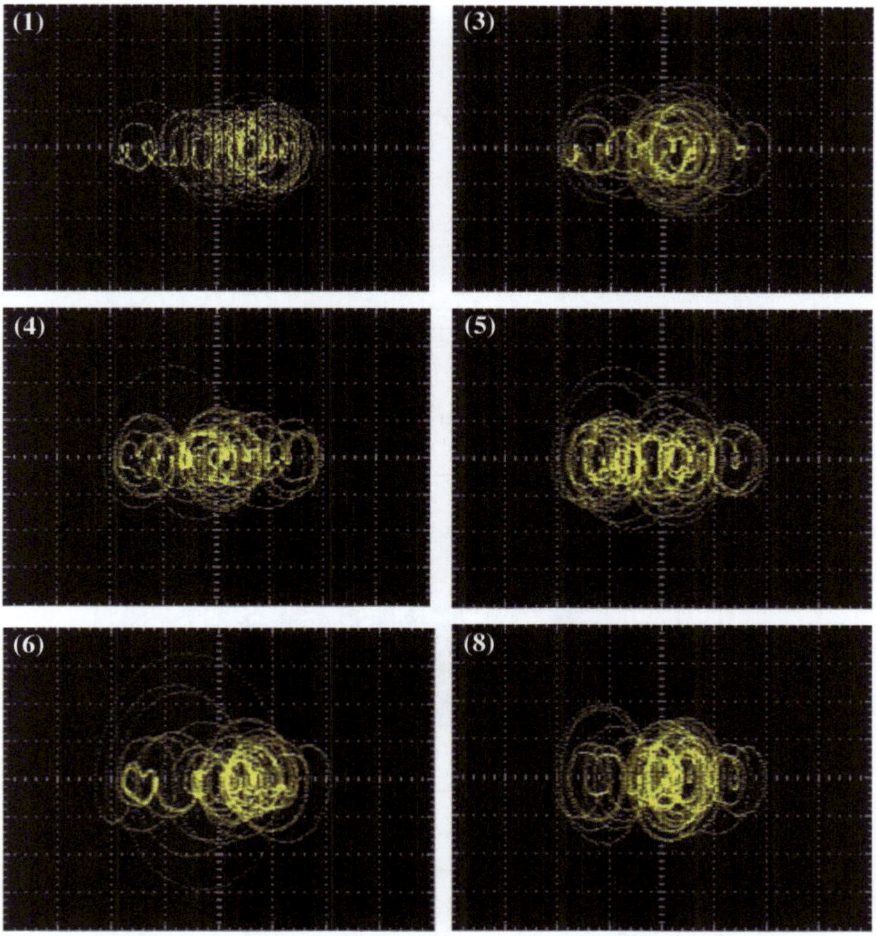

Fig. 15 Experimental results generating 10-scrolls for six cases in Table 3

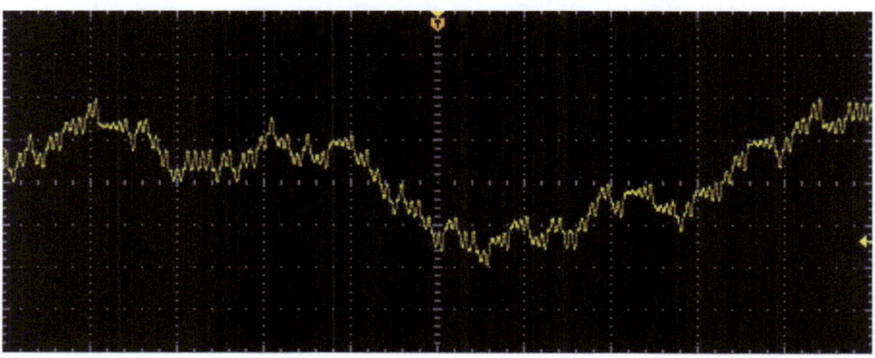

Fig. 16 Counting 10 levels when generating a 10-scrolls attractor

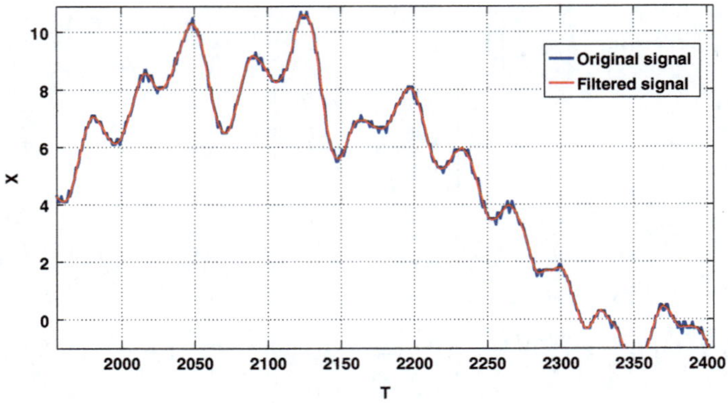

Fig. 17 Signal from experimental data and after it is filtered in MATLAB^{TM}

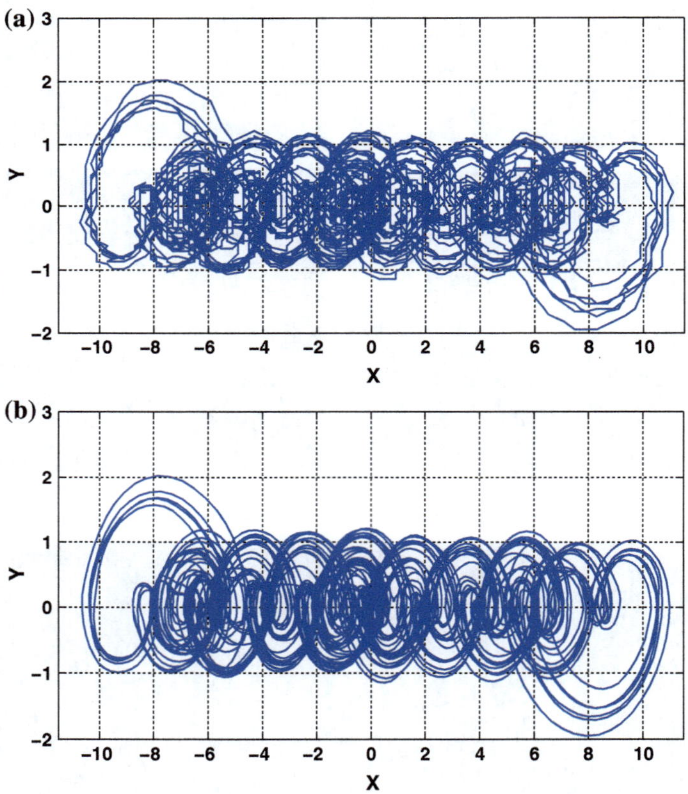

Fig. 18 Phase space portraits for the 10-scrolls attractor from: **a** experimental data and, **b** after it is filtered

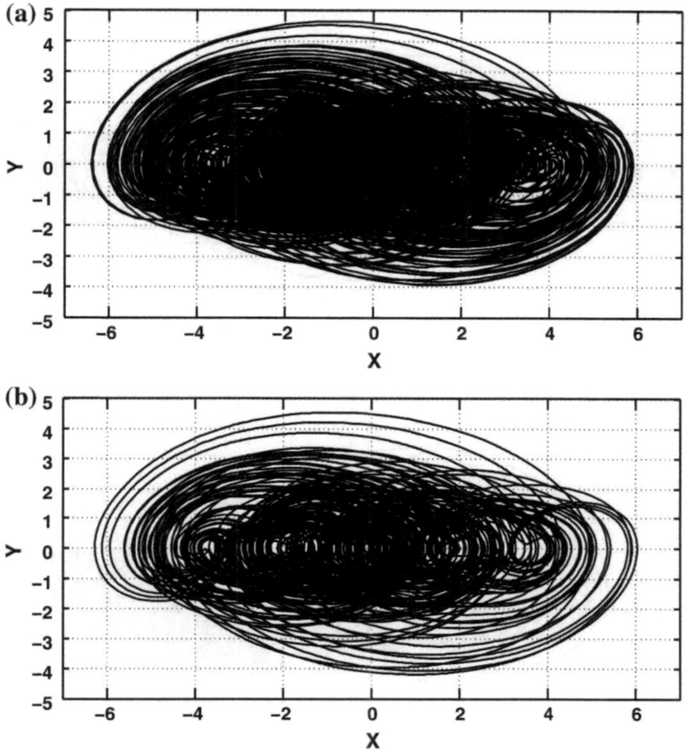

Fig. 19 Phase space portraits from experimental data showing 5-scrolls with the highest **a** MLE and, **b** entropy

The signals were filtered in MATLAB using y = sgolayfilt(x, k, f) (a Savitzky–Golay Finite Impulse Response smoothing filter). If x is a matrix, sgolayfilt operates on each column. The polynomial order k must be less than the frame size f, which must be odd. In our experiments, we used $k = 9$ and $f = 31$, to approximate the experimental data to the observed signals in the oscilloscope.

Finally, from the filtered data and from Tables 2 and 3, we selected the cases with the highest values for the MLE and their associated entropy computed from experimental data, so that they are shown in Figs. 19 and 20 for generating 5- and 10-scrolls, respectively.

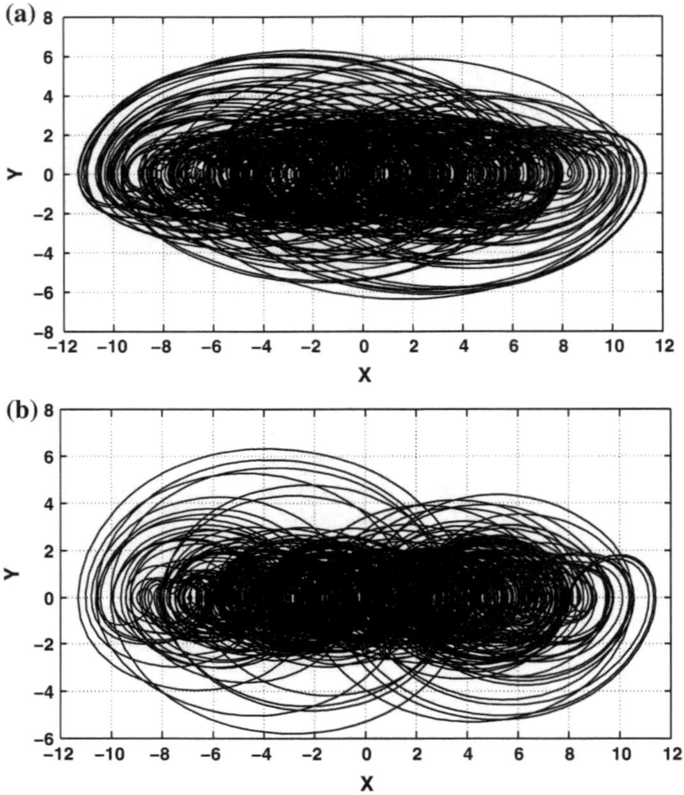

Fig. 20 Phase space portraits from experimental data showing 10-scrolls with the highest **a** MLE and, **b** entropy

6 Conclusion

This article showed the experimental verification on optimizing the MLE in a multi-scroll chaotic oscillator based on saturated function series, and its associated entropy. The optimization of MLE was performed by applying an evolutionary algorithm for generating 2-, 5- and 10-scrolls.

The laboratory experiments confirmed that the chaotic behavior becomes more complex as MLE is maximized. Furthermore, to better confirm the chaotic complexity associated to the value of MLE, we listed the associated entropy from simulated and experimental data for generating 2-, 5- and 10-scrolls attractors.

It was also discussed that to eliminate the undesired frequencies introduced by the poor sampling of the oscilloscope, the experimental data (signal) should be filtered.

As a final conclusion, the experiments showed that multi-scroll chaotic oscillators have a more complex chaotic behavior when the number of scrolls increases. For instance, Tables 1, 2 and 3 clearly show that by increasing the number of scrolls, when the chaotic oscillator is optimized, both MLE and the entropy increases.

Acknowledgments This work has been partially supported by CONACyT-Mexico under grants 168357 and 237991.

References

1. Cicek I, Pusane AE, Dundar G (2014) A novel design method for discrete time chaos based true random number generators. Integr VLSI J 47(1):38–47
2. de la Fraga LG, Tlelo-Cuautle E (2014) Optimizing the maximum lyapunov exponent and phase space portraits in multi-scroll chaotic oscillators. Nonlinear Dyn 76(2):1503–1515
3. Deb K, Pratap A, Agarwal S, Meyarivan T (2002) A fast and elitist multiobjective genetic algorithm: NSGA-II. IEEE Trans Evol Comput 6(2):182–197
4. Ergün S, Özogez S (2010) Truly random number generators based on non-autonomous continuous-time chaos. Int J Circuit Theory Appl 38(1):1–24
5. Gamez-Guzman L, Cruz-Hernandez C, Lopez-Gutierrez R, Garcia-Guerrero E (2009) Synchronization of chua's circuits with multi-scroll attractors: Application to communication. Commun Nonlinear Sci Numer Simul 14(6):2765–2775. doi:10.1016/j.cnsns.2008.10.009, http://www.sciencedirect.com/science/article/pii/S1007570408003298
6. Karthikeyan R, Sundarapandian V (2014) Hybrid chaos synchronization of four-scroll systems via active control. J Electr Eng 65(2):97–103
7. Lü J, Chen G (2006) Generating multiscroll chaotic attractors: theories, methods and applications. Int J Bifurc Chaos 16(4):775–858
8. Lu J, Chen G, Yu X, Leung H (2004) Design and analysis of multiscroll chaotic attractors from saturated function series. IEEE Trans Circuits Syst 51:2476–2490
9. Moddemeijer R (1989) On estimation of entropy and mutual information of continuous distributions. Signal Process 16(3):233–248
10. Nejati H, Beirami A, Ali WH (2012) Discrete-time chaotic-map truly random number generators: design, implementation, and variability analysis of the zigzag map. Analog Integr Circuits Signal Process 73(1):363–374. doi:10.1007/s10470-012-9893-9
11. Ortega-Torres E, Sanchez-Lopez C, Mendoza-Lopez J (2013) Frequency behavior of saturated nonlinear function series based on opamps. Revista Mexicana De Fiscia 59(6):504–510
12. Parker T, Chua L (1989) Practical numerical algorithms for chaotic systems. Springer, New York
13. Pesis YB (1977) Characteristic Lyapunov exponents and smooth ergodic theory. Russ Math Surv 32(4):55–112
14. Ruelle D (1979) Bifurcation theory and its application in scientific disciplines. New York Academy of Science, New York
15. Sánchez-López C, Trejo-Guerra R, Munoz-Pacheco JM, Tlelo-Cuautle E (2010) N-scroll chaotic attractors from saturated function series employing CCII+s. Nonlinear Dyn 61(1–2):331–341
16. Tlelo-Cuautle E, Ramos-López HC, Sánchez-Sánchez M, Pano-Azucena AD, Sánchez-Gaspariano LA, Nunez-Perez JC, Camas-Anzueto JL (2014) Application of a chaotic oscillator in an autonomous mobile robot. J Electr Eng 65(3):157–162
17. Tlelo-Cuautle E, Rangel-Magdaleno J, Pano-Azucena A, Obeso-Rodelo P, Nunez-Perez J (2015) FPGA realization of multi-scroll chaotic oscillators. Commun Nonlinear Sci Numer Simul 27(1–3):66–80. doi:10.1016/j.cnsns.2015.03.003, http://www.sciencedirect.com/science/article/pii/S1007570415000878

18. Trejo-Guerra R, Tlelo-Cuautle E, Munoz-Pacheco JM, Sánchez-López C, Cruz-Hernández C (2010) On the relation between the number of scrolls and the lyapunov exponents in PWL-functions-based η-scroll chaotic oscillators. Int J Nonlinear Sci Numer Simul 11(11):903–910. doi:10.1515/IJNSNS.2010.11.11.903
19. Trejo-Guerra R, Tlelo-Cuautle E, Sánchez-López C, Muñoz-Pacheco J, Cruz-Hernández C (2010) Realization of multiscroll chaotic attractors by using current-feedback operational amplifiers. Revista Mexicana de Fisica 54(4):268–274
20. Trejo-Guerra R, Tlelo-Cuautle E, Jiménez-Fuentes JM, Sánchez-López C, Muñoz-Pacheco JM, Espinosa-Flores-Verdad G, Rocha-Pérez JM (2012) Integrated circuit generating 3-and 5-scroll attractors. Commun Nonlinear Sci Numer Simul 17(11):4328–4335
21. Volos CK, Kyprianidis IM, Stouboulos I (2013) Experimental investigation on coverage performance of a chaotic autonomous mobile robot. Robot Auton Syst 61(12):1314–1322
22. Wolf A, Swift JB, Swinney HL, Vastano JA (1985) Determining lyapunov exponents from a time series. Phys D: Nonlinear Phenom 16(3):285–317. doi:10.1016/0167-2789(85)90011-9, http://www.sciencedirect.com/science/article/pii/0167278985900119
23. Yalcin ME (2007) Increasing the entropy of a random number generator using n-scroll chaotic attractors. Int J Bifurc Chaos 17(12):4471–4479. doi:10.1142/S0218127407020130

Chaotic Synchronization of CNNs in Small-World Topology Applied to Data Encryption

A.G. Soriano-Sánchez, C. Posadas-Castillo, M.A. Platas-Garza and C. Elizondo-González

Abstract This work addresses the synchronization of chaos and its main application. Three different models of CNN are considered as generators of chaotic behavior. The Watts–Strogatz and Newman–Watts algorithms are used to arrange the chaotic CNN models in small-world topology. The resulting complex network is carried synchrony with the Complex Systems Theory. Prior to encryption, recently established selection criteria, which consider energy and frequency characteristics, are used to choose the chaotic signal that best hides the message. After selecting the signal, encryption, transmission and retrieval of a confidential message are performed. When considering the message requirements by using the selection criteria, the security level of encryption is improved in the time and frequency domain.

Keywords Chaotic CNNs · Chaos synchronization · Small-world networks · Secret communications

1 Introduction

In this work, chaos synchronization and chaotic encryption are discussed. Both fields have been extensively studied due to their importance for engineering. As a result, a big part of their body of knowledge has been established in the last two decades.

A.G. Soriano-Sánchez · C. Posadas-Castillo (✉) · M.A. Platas-Garza
C. Elizondo-González
Facultad de Ingeniería Mecánica y Eléctrica, Universidad Autónoma de Nuevo León, San Nicolas de los Garza, Mexico
e-mail: cornelio.posadascs@uanl.edu.mx

A.G. Soriano-Sánchez
e-mail: allansori@gmail.com

M.A. Platas-Garza
e-mail: miguel.platasg@uanl.mx

C. Elizondo-González
e-mail: celizond@yahoo.com

S. Vaidyanathan and C. Volos (eds.), *Advances and Applications in Chaotic Systems*, Studies in Computational Intelligence 636,
DOI 10.1007/978-3-319-30279-9_14

Chaos as we know it today has its beginning in the early 1960s. With the appearance of computers, it was possible to visualize the behavior of some systems from the solution of their differential equations. That was when Edward N. Lorenz published his historical article on deterministic nonperiodic flow [27]. By using numerical methods, Lorenz obtained the trajectories of equations describing the forced dissipative hydrodynamic flow to be identified in phase space. As a result, a lot of information on chaos subsequently emerged. Chaos turned out to be such an interesting phenomenon that many systems have been created solely to generate it and deepen its study.

Among the best known chaos generators can be found conventional chaotic oscillators like Chua, Lorenz and Rössler models [10, 27, 31, 41], multi-scroll attractor chaotic oscillators [29, 30, 55, 64], fractional-order chaotic oscillators [19, 38, 54] and Cellular Neural Networks [4, 39, 67, 68], for instance.

The so-called Cellular Neural Networks (CNNs) have been extensively studied since their beginning three decades ago. The acronym CNN for Cellular Neural Network was first introduced by L.O. Chua and L. Yang in 1988 [11]. The best features of these systems are: on one hand, their ability of real-time signal processing; on the other hand, their local interconnection makes them tailor-made for monolithic implementation [11, 65]. We are interested in the application that interpreted CNN as Cellular Nonlinear Networks. In this way, CNNs are considered for generating chaotic signals. We are interested in three different CNNs that exhibit chaotic behavior: the standard CNN model (Chua–Yang) [11], the chaotic 3D CNN [18] and the Lu–He CNN model [28]. In addition to the complexity generated by locally coupled nonlinear dynamical systems to generate a specific behavior, we will arrange the CNNs in small-world topology to generate a network to be synchronized.

Chaos synchronization has its beginning in the early 1990s when L.M. Pecora and T.L. Carroll synchronized, for the first time, two identical chaotic oscillators with different initial conditions [35]. The basic concepts and applications of chaos synchronization were established some years later.

In 1994, C.W. Wu and L.O. Chua defined concepts such as asymptotic and partial synchronization. They established the relation between asymptotic synchronization and asymptotic stability [62]. The role of unstable periodic orbits in synchronous chaotic behavior was investigated by J.F. Heagy et al. in 1995 [17]. They proved how desynchronized bursting behavior is initiated, and suggested taking this phenomenon into account to yield high quality chaotic synchronization.

In 1996, N.F. Rulkov discussed the cooperative behavior related to the regimes of synchronized chaos and outlined some examples that illustrate different types of identical chaotic oscillations [42]. One year later, L.M. Pecora et al. reviewed the basics of chaotic synchronization and examined coupling configurations as well as secure communication schemes [36]. G. Kulumbán et al. provided a unified approach for the analysis and comparison of conventional and chaotic communications systems, clarifying the role of synchronization for chaotic communications and describing chaotic synchronization schemes [21].

Since then, chaotic encryption has attracted the interest of researchers. As a result, chaotic communications have been implemented extensively, this because non-periodicity and apparent randomness of chaotic signals seem to be their main benefits [22].

Three major encryption schemes have been proposed for private communications [13]: chaotic masking scheme [3, 37, 56], chaos shift keying scheme [20, 63, 66] and chaotic modulation scheme [5, 8, 58]. Recently, new encryption techniques and communication schemes have been formulated. Some of them consider generating pseudo-random sequences by iterating chaotic maps to encrypt image blocks [25]. The aim was to generate pseudo-random sequences with high initial-value sensitivity and good randomness. Some others focus on the robustness and security [7]. In order to increase the security of the algorithm, the size of the key space and the complexity of the coupling parameters are increased.

On the other hand, several methods have been proposed to achieve chaotic synchronization, which is a key step in the private communication process. The most important results that have been derived from this research are: firstly, the discovery that the behavior of many biological and non-biological systems can be modeled by the dynamics of complex networks: modeling of the human brain [49, 52], the spread of epidemics in a population [44] and modeling of economic systems [23]. Secondly, the effect of topology on the realization of system processes: synchronization of networked nonlinear systems [2, 50], synchronization of pyloric central pattern generator of the lobster [12], the presence of Alzheimer in a human brain [53] and stable growth of a neurons population [14], for instance.

The above quoted papers are some results of the investigation of complex networks arranged in a particular way, *small-world topology*. The small-world networks have their beginning in the 1960s when Stanley Milgram performed an experiment that led to the well-known concept of six degrees of separation [15]. According to [61] and [32], Milgram's experiment consisted of randomly distributed letters to people in Nebraska to be sent to Boston by people who might know the consignee. Milgram found that it had only taken an average of six steps for a letter to get from Nebraska to Boston. He concluded that six was the average number of acquaintances separating people in the entire world.

Such networks became popular after D.J. Watts and S.H. Strogatz published the algorithm to introduce the small-world property into a regular network. They showed that the resulting network fulfilled two main characteristics: high clustering coefficient and short average path length [61]. After this pioneering work, many researcher turned their attention to those types of networks during the next years.

This work carries out a process of secret communication by using chaos. The chaotic signal, which encrypts the message, is chosen based on its energy and frequency characteristics to ensure good encryption. Three models of CNNs are used as chaos generators arranged in small-world topology. The resulting network is synchronized by the Complex System Theory.

The remainder of the chapter is organized as follows: a brief review on complex dynamical networks and their synchronization is given in Sect. 2. Section 3 provides the explanation of the small-world algorithms and the description of some basic network characteristics. Section 4 provides the definition and necessary concepts of CNNs as well as the model descriptions and the corresponding chaotic attractors.

Section 5 provides the necessary conditions to achieve synchronization and examples of small-world network synchronization. Section 6 provides two selection criteria to evaluate the masking signal as well as the chaotic encryption, transmission and retrieval of a confidential message. Section 7 provides the results interpretation and some remarks on the chaotic encryption and small-world algorithms. Some conclusions and the possible direction of the research are given in Sect. 8.

2 Complex Networks

In the present section we will address the topic of complex networks and their synchronization. We will provide the definition of a complex network and the coupling matrix technique, which is used to achieve synchrony.

Among the possible definitions of a complex network, we will use the one suggested by Wang [59].

Definition 1 A complex network is defined as an interconnected set of oscillators (two or more), where each oscillator is a fundamental unit, with its dynamic depending of the nature of the network.

Each oscillator is defined as follows

$$\dot{\mathbf{x}}_i = f(\mathbf{x}_i) + \mathbf{u}_i, \qquad \mathbf{x}_i(0), \qquad i = 1,\ 2, \ldots,\ N, \tag{1}$$

where N is the network's size, $\mathbf{x}_i = [x_{i1},\ x_{i2},\ \ldots,\ x_{in}] \in \mathbb{R}^{\mathrm{n}}$ represents the state variables of the oscillator i. $\mathbf{x}_i(0) \in \mathbb{R}^{\mathrm{n}}$ are the initial conditions for oscillator i. $\mathbf{u}_i \in \mathbb{R}^{\mathrm{n}}$ establishes the synchronization between two or more oscillators and is defined as follows [60]

$$\mathbf{u}_i = c \sum_{j=1}^{N} a_{ij} \Gamma \mathbf{x}_j, \qquad i = 1,\ 2, \ldots,\ N. \tag{2}$$

The constant $c > 0$ represents the coupling strength. $\Gamma \in \mathbb{R}^{\mathrm{n}\times\mathrm{n}}$ is a constant matrix to determine the coupled state variable of each oscillator. Assume that $\Gamma = \mathrm{diag}(r_1,\ r_2, \ldots,\ r_n)$ is a diagonal matrix. If two oscillators are linked through their kth state variables, then, the diagonal element $r_k = 1$ for a particular k and $r_j = 0$ for $j \neq k$.

Synchronization is achieve through (2), where a_{ij} are the elements of $A \in \mathbb{R}^{\mathrm{N}\times\mathrm{N}}$ which is the coupling matrix. The matrix A shows the connections between oscillators; if the oscillator ith is connected to the oscillator jth, then $a_{ij} = 1$, otherwise $a_{ij} = 0$ for $i \neq j$. The diagonal elements of A matrix are defined as

$$a_{ii} = -\sum_{j=1, j\neq i}^{N} a_{ij} = -\sum_{j=1, j\neq i}^{N} a_{ji} \qquad i = 1, 2, \ldots, N. \tag{3}$$

The dynamical complex network (1) and (2) is said to achieve synchronization if

$$\mathbf{x}_1(t) = \mathbf{x}_2(t) = \cdots = \mathbf{x}_\mathrm{N}(t) \text{ as } t \to \infty. \tag{4}$$

In this chapter we will synchronize N-coupled CNNs arranged in small-world topology.

3 Small-World Networks

Even though the small-world property was derived from a research social in nature, numerous applications have been found in different fields ranging from engineering to biology [1, 16, 26, 40], therefore it has become a topic widely studied.

The small-world property consists in the existence of long-range links connecting pairs of nodes distant from each other. The concept of the six degrees of separation is implied due to it is needed a small number of steps (acquaintances) to reach any node in this type of networks.

Two of the most important features of complex networks that will be affected by the small-world property are the following: on one hand *the clustering coefficient* C, which is defined as the average fraction of pairs of neighbors of an oscillator that are also neighbors of each other, the clustering coefficient C_i of the oscillator i is defined as the ratio between the actual number E_i of edges that exist between k_i oscillators and the total number $k_i(k_i - 1)/2$, so

$$C_i = \frac{2E_i}{k_i(k_i - 1)}. \tag{5}$$

The clustering coefficient C of the whole network is the averaged of C_i over all i. On the other hand *the average path length* L, which is defined as the distance between two oscillators averaged over all pairs of oscillators [6, 59]

$$L = \frac{1}{N(N-1)} \sum_{i\neq j} d_{ij}, \qquad 1 \leq i, j \leq N \tag{6}$$

where d_{ij} is the distance between node i and node j. Due to the existence of long-range links, the small-world network has high clustering coefficient $C(N, p)$ and short average path length $L(N, p)$.

In the following, the algorithms to introduce the small-world property will be described.

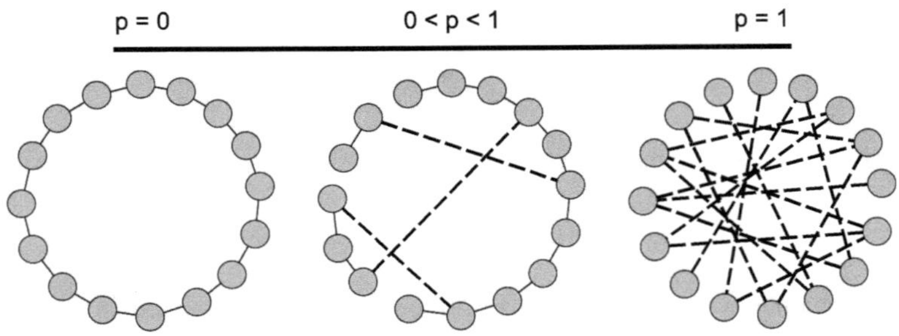

Fig. 1 Evolution of the Watts–Strogatz small-world algorithm. The *solid lines* are the links in their original positions. The *dash–dot line* is the rewired link to a randomly chosen position

3.1 Watts–Strogatz Small-World Algorithm

In 1998 D.J. Watts and S.H. Strogatz proposed an algorithm to introduce the small-world property into a regular network. Such network is arranged in the nearest neighbor topology, which is a ring lattice with periodic boundary conditions [34].

The Watts–Strogatz model is created by rewiring a certain amount of the existing links to new randomly chosen positions. Restrictions are:

1. The size of the networks remains unchanged.
2. The amount of links remains unchanged.
3. No CNN is allowed to have multiple links with other CNN.
4. No CNN is allowed to have links with itself.
5. The relation $N \gg k$ must hold,

where N is the size of the network, k is the periodic boundary condition, i.e., CNN i is connected with $i \pm 1, i \pm 2, \ldots, i \pm k$ neighboring CNNs; p is the probability to rewire a link, thus Nkp are the long-range links that can be created. Figure 1 shows the evolution of the Watts–Strogatz small-world algorithm.

When $p = 0$ the topology remains unchanged and the network is considered regular. When $0 < p < 1$ one obtains a small-world network. At the point where $p = 1$ all the links have been rewired and the network topology has become random. An important issue to be taken under consideration is the fact that this algorithm can lead us to generate isolated cluster since it breaks the existing links.

3.2 Newman–Watts Small-World Algorithm

After Watts and Strogatz published their pioneering algorithm to generate small-world networks, a revised version emerged one year later. In 1999, M.E.J. Newman and D.J. Watts proposed a modified version of the original small-world model [33, 34].

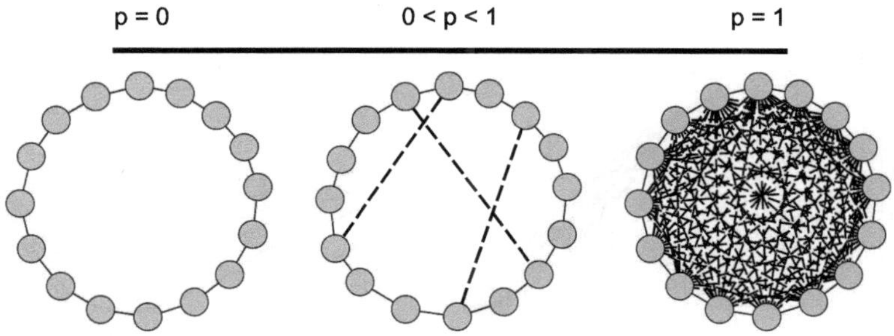

Fig. 2 Evolution of the Newman–Watts small-world algorithm. The *solid lines* are the original links. The *dash–dot lines* are the links randomly added

Newman–Watts algorithm also starts from the nearest neighbor topology, but unlike the previous case, the algorithm introduces the small-world property by adding links to pairs of CNN randomly chosen. Restrictions remain almost the same except for the second one which has been removed.

To determine the amount of links to be added we consider the following: a CNN i is already connected with its $2k$ neighboring CNNs. The fourth restriction does not allow the CNN i to have links with itself, thus, it can connect $N - (2k + 1)$ CNNs more; therefore, for the whole network we have $N(N - (2k + 1))$ links. However, since the considered network is undirected and the third restriction does not allow multiple links between pairs of CNNs, the connection from CNN i to CNN j is the same as the connection from CNN j to CNN i; therefore we have $N(N - (2k + 1))/2$ possible links. As the Newman–Watts algorithm is applied, $N(N - (2k + 1))p/2$ links are introduced.

Figure 2 shows the evolution of the Newman–Watts small-world algorithm. When $p = 0$, as in the previous case, the topology remains unchanged and the network is considered regular. When $0 < p < 1$ one obtains a small-world network by adding links to randomly pairs of CNNs. At the point where $p = 1$ all the possible links have been added and the network has become globally coupled.

Figure 3 depicts the evolution of the clustering coefficient for a network with different values of k when applying the small-world algorithms. As above mentioned, both algorithms have led us to a different type of network. When $p = 1$, Watts–Strogatz algorithm generates a random network whose clustering coefficient is small (Fig. 3a), i.e. the network has a low connectivity. On the other hand, Newman–Watts algorithm generates a globally coupled network whose clustering coefficient is the highest possible (Fig. 3b), i.e., every pair of CNN has a link.

In this chapter we will use both small-world algorithms in order to determine the advantages that these could bring to achieve synchronization.

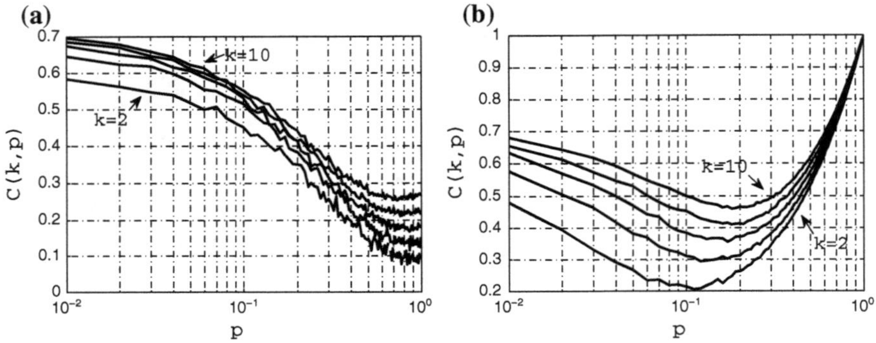

Fig. 3 Evolution of the clustering coefficient of a network where $N = 100$ and $k = 2, 4, 6, 8, 10$ when applying small-world algorithms: **a** Clustering coefficient when links are rewired (*Watts–Strogatz algorithm*). **b** Clustering coefficient when links are added (*Newman–Watts algorithm*)

4 CNNs as Chaos Generators

The present section focuses on the definition and description of a CNN. Authors will first provide the definition of a CNN and its general characteristics. Then, the mathematical models to be used will be described and examples of their chaotic attractors will be shown.

4.1 Cellular Neural Network

Among the existing definitions of CNNs, authors resort to the following one given in [11]:

Definition 2 A Cellular Neural Network is an spatial arrangement consisting of locally-coupled cells. Each cell is a dynamical system which has an input, an output and a state evolving according to some prescribed dynamical laws. The cell in row i and column j is denoted as $C(i, j)$ and it is said to be isolated if it is not coupled to any other cell.

The variables for an isolated cell are [11]: input $u(t) \in \mathbb{R}^u$, threshold $z(t) \in \mathbb{R}^z$ which is usually assumed to be scalar, state $x(t) \in \mathbb{R}^x$ and output $y(t) \in \mathbb{R}^y$. An example of an isolated cell is given in Fig. 4.

Each cell is coupled only among the neighboring cells lying within some prescribed sphere of influence with radius r, i.e., the r-neighborhood of the cell, which is defined as follows [11]

$$S_r(i, j) = \{C(k, l) | \max\{|k - i|, |l - j|\} \leq r, \quad 1 \leq k \leq M, 1 \leq l \leq N\}. \tag{7}$$

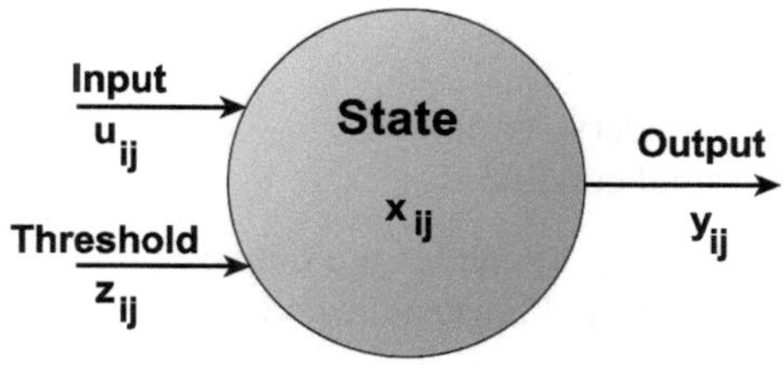

Fig. 4 A two-dimensional isolated cell: input $u_{ij}(t) \in \mathbb{R}^u$, threshold $z_{ij}(t) \in \mathbb{R}^z$, state $x_{ij}(t) \in \mathbb{R}^x$ and output $y_{ij}(t) \in \mathbb{R}^y$

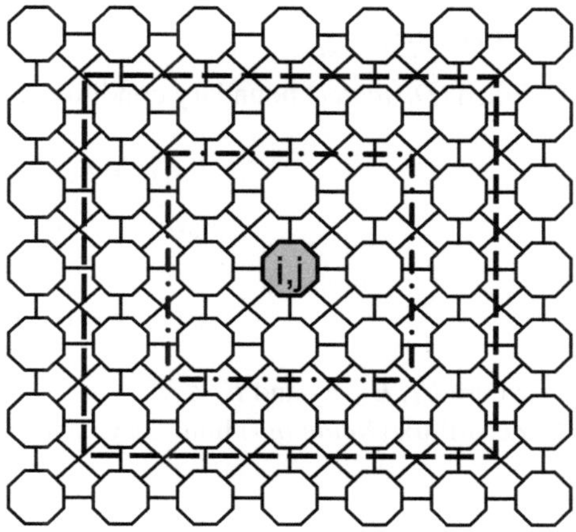

Fig. 5 $M \times N$ arrays of the cell $C(i, j)$: 3×3 array obtained with $r = 1$ (*dash–dot line*) and 5×5 array by using $r = 2$ (*dash line*)

Depending on the magnitude of the r-neighborhood, one can build $M \times N$ rectangular arrays of cells. Figure 5 shows two possible arrangements formed by the sphere of influence of cell $C(i, j)$: a 3×3 array obtained by using $r = 1$ and a 5×5 array obtained by using $r = 2$. The resulting network, which can be 1-D as shown in Fig. 6a or 2-D as shown in Fig. 6b, is governed mathematically by four specifications [9]:

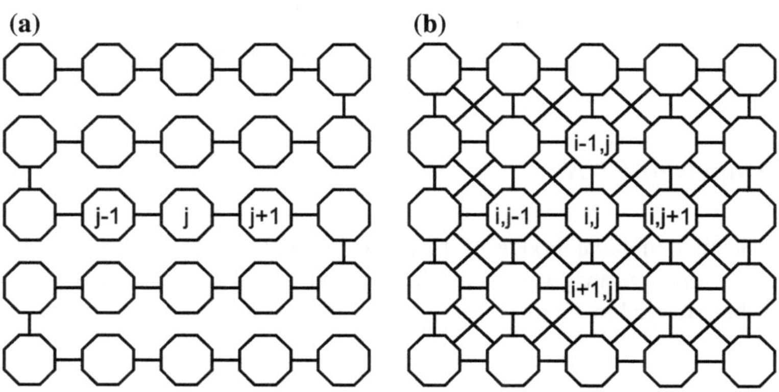

Fig. 6 Cells (*represented by octagons*) forming CNNs of different dimensions: **a** 1-D CNN. **b** 2-D CNN

- Cell dynamics,
- Synaptic law,
- Boundary conditions,
- Initial conditions.

The cell dynamics are defined by the state equations. The synaptic law defines the coupling between the considered cell C_i and the cells included within the sphere of influence S_i. The boundary conditions define the neighborhood of those cells that are located on an edge of the array. The most common boundary conditions for 2-D CNNs are [65]:

1. Fixed (*Dirichlet*) boundary conditions:

$$x_{i,0} = E_1, \quad i \in \{1, 2, \ldots, M\}, \tag{8a}$$
$$x_{i,N+1} = E_2, \quad i \in \{1, 2, \ldots, M\}, \tag{8b}$$
$$x_{0,i} = E_3, \quad j \in \{1, 2, \ldots, N\}, \tag{8c}$$
$$x_{M+1,j} = E_4, \quad j \in \{1, 2, \ldots, N\}, \tag{8d}$$

   where E_1, E_2, E_3 and $E_4 \in \mathbb{R}$.
2. Zero-Flux (*Neumann*) boundary conditions:

$$x_{i,0} = x_{i,1}, \quad i \in \{1, 2, \ldots, M\}, \tag{9a}$$
$$x_{i,N+1} = x_{i,M}, \quad i \in \{1, 2, \ldots, M\}, \tag{9b}$$
$$x_{0,i} = x_{1,j}, \quad j \in \{1, 2, \ldots, N\}, \tag{9c}$$
$$x_{M+1,j} = x_{N,j}, \quad j \in \{1, 2, \ldots, N\}. \tag{9d}$$

3. Periodic (*Toroidal*) boundary conditions:

$$x_{i,0} = x_{i,M}, \quad i \in \{1, 2, \ldots, M\}, \tag{10a}$$
$$x_{i,N+1} = x_{i,1}, \quad i \in \{1, 2, \ldots, M\}, \tag{10b}$$
$$x_{0,i} = x_{N,j}, \quad j \in \{1, 2, \ldots, N\}, \tag{10c}$$
$$x_{M+1,j} = x_{1,j}, \quad j \in \{1, 2, \ldots, N\}. \tag{10d}$$

4.2 Models of CNNs

Below, the mathematical models of the CNNs are provided. Examples of their states and chaotic attractors are given as well. These models will be used as chaos generators and will be arranged in small-world topology to be synchronized.

4.2.1 The Standard CNN

We will describe first the Chua–Yang model which is widely known as the standard model because it has been extensively studied. The standard CNN is described as follows [11]

$$\dot{x}_{ij} = -x_{ij} + z_{ij} + \sum_{kl \in S_r(i,j)} a_{kl} y_{kl} + \sum_{kl \in S_r(i,j)} b_{kl} v_{kl}, \quad i = 1, \ldots, M\,; j = 1, \ldots, N, \quad (11)$$

$$y_{ij} = f(x_{ij}), \quad (12)$$

where z_{ij} is a scalar for simplicity, $S_r(i,j)$ is the sphere of influence with radius r, i.e., the r-neighborhood of the cell. $\sum_{kl \in S_r(i,j)} a_{kl} y_{kl}$ and $\sum_{kl \in S_r(i,j)} b_{kl} v_{kl}$ are the local couplings, and

$$f(x_{ij}) = \frac{1}{2}\left(|x_{ij}+1| - |x_{ij}-1|\right). \quad (13)$$

For the particular case where $M = 1$ and $N = 2$, (11) and (12) assume the simpler form of a 1×2 array [71]

$$\begin{cases} \dot{x}_1 = -x_1 + a_{00} f(x_1) + a_{01} f(x_2) + b_{00} v_1(t), \\ \dot{x}_2 = -x_2 + a_{0,-1} f(x_1) + a_{00} f(x_2) + b_{00} v_2(t), \\ y_1 = f(x_1), \\ y_2 = f(x_2), \end{cases} \quad (14)$$

where by using $a_{00} = 2$, $a_{0,-1} = 1.2$, $a_{01} = -1.2$, $b_{00} = 1$, $v_1(t) = 4.04 \sin(\pi t/2)$ and $v_2(t) = 0$, we simplify (14) to

$$\begin{cases} \dot{x}_1 = -x_1 + 2f(x_1) - 1.2 f(x_2) + 4.04 \sin\left(\frac{\pi}{2} t\right), \\ \dot{x}_2 = -x_2 + 1.2 f(x_1) + 2 f(x_2), \end{cases} \quad (15)$$

with the nonlinear function

$$f(x_{1,2}) = \frac{1}{2}\left(|x_{1,2}+1| - |x_{1,2}-1|\right). \quad (16)$$

It was first shown in [71] that (15) and (16) exhibited chaotic behavior for the parameters given. Figure 7a, b shows an example of the chaotic attractor and the state variables respectively generated from (15) and (16).

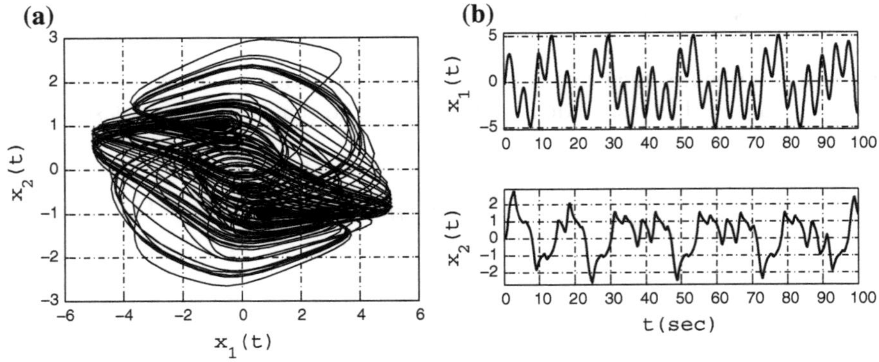

Fig. 7 **a** View (x_1, x_2)-plane for the chaotic attractor of the standard CNN (15) and (16). **b** $x_1(t)$ and $x_2(t)$ state variables obtained with $\mathbf{x}(0) = [-0.2,\ 0.1]^T$

4.2.2 Chaotic 3D CNN

The second model taken under consideration is the so-called chaotic 3D CNN. This model is a simpler form of the biological one proposed by J. J. Hopfield which was described as follows [18]

$$C_i\left(\frac{\mathrm{d}u_i}{\mathrm{dt}}\right) = -\frac{u_i}{R_i} + \sum_j T_{ij}V_j + I_i, \tag{17}$$

$$V_i = g_i(u_i), \tag{18}$$

to model the graded response and the effect of the action potentials. T_{ij} was biologically viewed as a description of the synaptic interconnection strength from neuron j to neuron i. V_i is the output of the neuron j and it is bounded above and below $V_i^0 \leq V_i \leq V_i^1$. The simplified model of the Hopfield system is described as follows [69]

$$\dot{\mathbf{x}} = -\mathbf{x} + T\tanh(\mathbf{x}), \tag{19}$$

where the state vector $\mathbf{x} \in \mathbb{R}^3$, $\tanh(\mathbf{x}) = [\tanh(x_1)\ \tanh(x_2)\ \tanh(x_3)]^T$ and

$$T = \begin{bmatrix} T_{11} & T_{12} & T_{13} \\ T_{21} & T_{22} & T_{23} \\ T_{31} & T_{32} & T_{33} \end{bmatrix} = \begin{bmatrix} 1.49 & 2 & 1 \\ -2 & 1.7 & 0 \\ 4 & -4 & 2 \end{bmatrix}. \tag{20}$$

The chaotic 3D CNN described by (19) exhibits chaotic behavior for parameters given in (20) [69]. References [24, 70] give alternative parameters for (19) to exhibit chaotic or hyperchaotic behavior. An example of the chaotic attractor and state variables are shown in Fig. 8a, b respectively. They were obtained by using the initial conditions $\mathbf{x}(0) = [-0.2,\ 0.1,\ 0.1]^T$.

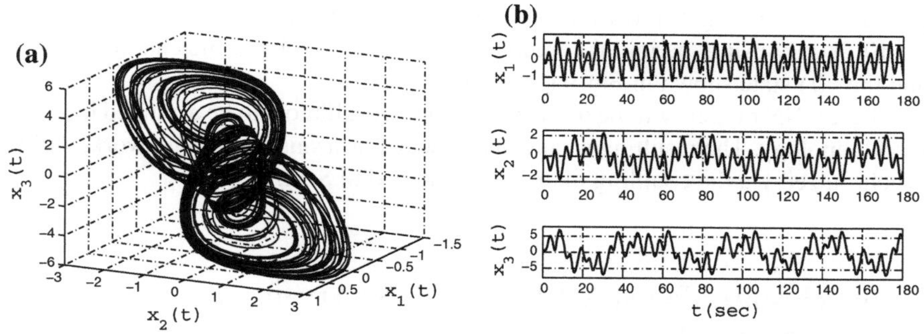

Fig. 8 **a** View (x, y, z)-plane for the chaotic attractor of the 3D CNN in (19). **b** $x_1(t)$, $x_2(t)$ and $x_3(t)$ states variables obtained with $\mathbf{x}(0) = [0.1,\ -0.2,\ -0.2]^T$

4.2.3 Lu–He CNN

The third model under consideration is the one-dimensional Lu–He system with time-delay. The importance of the model is due to it is well-know that chaotic behavior cannot be found in continuous systems with less than two dimensions. It does not seem to be true for systems with delay. That is what H. Lu et al. claimed and proved when first proposed a simple equation with only one variable as chaos generator [28]. This model has been widely studied and different modalities have emerged [39, 45–48, 57]. The time-delay Lu–He system is described as follows [28]

$$\dot{x}(t) = 0.001x(t) - 3.8\left(|x_\tau + 1| - |x_\tau - 1|\right) + 2.85\left(\left|x_\tau + \frac{4}{3}\right| - \left|x_\tau - \frac{4}{3}\right|\right), \tag{21}$$

where $x_\tau = x(t - \tau)$. Equation (21) has an infinite-dimensional solution space, with initial condition as any continuous function defined on the closed interval $[-\tau,\ 0]$

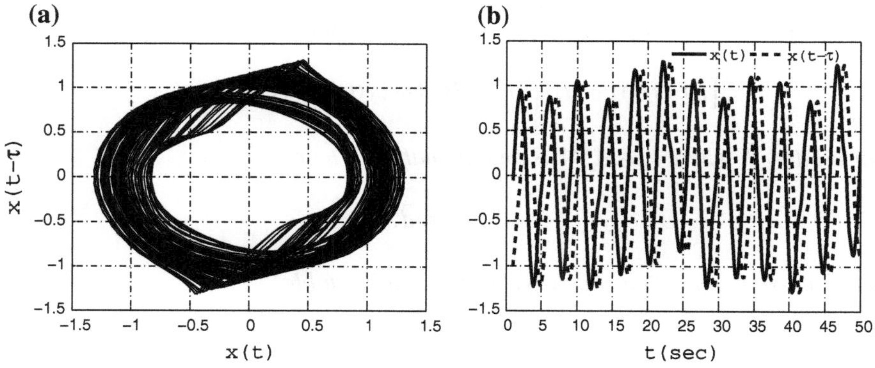

Fig. 9 **a** View $(x(t),\ x(t-\tau))$-plane for the chaotic attractor of the Lu–He system in (21). **b** $x(t)$ (*solid line*) and $x(t-\tau)$ (*dash line*) states variables obtained with $x(0) = -1$

[28]. By considering $\tau = 1$, we choose the arbitrary function $\phi_0 = 0.5 \in [-\tau,\ 0]$ for $t \in [0,\ \tau]$. By using the initial condition $x(0) = -1$ we obtain the chaotic attractor and the $x(t)$ state shown in Fig. 9a, b respectively.

These CNN models will be used as the chaotic oscillators to be arranged in small-world topology. The resulting arrangement, which is a small-world network, will be synchronized via the coupling matrix as explained in Sect. 2.

5 Synchronization of *N* Chaotic CNNs Arranged in Small-World Topology

Section 5 will be devoted to achieve synchronization between chaotic CNNs arranged in a specific topology. The CNN models described in Sect. 4 will be used to compose the small-world networks. The small-world property will be introduced by the algorithms described in Sect. 3. In the following, conditions to achieve synchronization are provided.

5.1 Conditions for Synchronization by Using the Coupling Matrix

Suppose there are no isolated clusters in the complex network, then the coupling matrix A is a symmetric irreducible matrix, so one eigenvalue of A is zero and all the other eigenvalues are strictly negative, this means, $\lambda_{2,\ldots,\mathrm{N}}(A) < 0$.

Theorem 1 [59] *Consider the dynamical network (1). Let*

$$0 = \lambda_1 > \lambda_2 \geq \lambda_3 \cdots \geq \lambda_N, \tag{22}$$

be the eigenvalues of its coupling matrix A. Suppose that there exist an $n \times n$ *diagonal matrix* $H > 0$ *and two constants* $\bar{d} < 0$ *and* $\tau > 0$, *such that*

$$\left[Df\left(s\left(t\right)\right) + d\Gamma\right]^T H + H\left[Df\left(s\left(t\right)\right) + d\Gamma\right] \leq -\tau I_n, \tag{23}$$

for all $d \leq \bar{d}$, *where* $I_n \in \mathbb{R}^{n \times n}$ *is an unit matrix. If moreover,*

$$c\lambda_2 \leq \bar{d}, \tag{24}$$

then the synchronization state (4) is exponentially stable.

The coupling strength is computed as follows

$$c \geq \left| \frac{\bar{d}}{\lambda_2} \right|, \tag{25}$$

which will affect the stability of the synchronization state (4) through the control law (2).

5.2 Synchronization Results

In the following, complex networks of identical chaotic CNNs will be synchronized. The CNNs considered can be either the standard CNN (Chua–Yang model) described by (15) and (16), the chaotic 3D CNN described by (19) and (20) or the Lu–He CNN model described by (21).

Considering a synchronization scheme of N-coupled chaotic CNN, the coupling matrix of the original topology (*nearest neighbor*) is obtained as explained in Sect. 2. All its eigenvalues are $0 = \lambda_1 > \lambda_2 \geq \lambda_3 \cdots \geq \lambda_N$. As the small-world property is introduced, the largest nonzero eigenvalue $\lambda_2(p)$ will vary, thus, the coupling strength given by (25) becomes

$$c(p) \geq \left| \frac{\bar{d}}{\lambda_2(p)} \right|. \tag{26}$$

The coupling strength $c(p)$ is computed for each $\lambda_2(p)$, which varies depending on the small-world algorithm. The obtained ratio is the lowest boundary necessary for each type of CNN to reach synchrony and it decreases as the probability increases.

The following characteristics apply to Case 1, Case 2 and Case 3: initial conditions were randomly generated for each chaotic CNN within the range $[-16, 10]$ without repeating them. The Gamma matrix is defined such that synchronization is achieved by the first state variable, i.e., for the standard CNN (Case 1) $\Gamma = \text{diag}(1, 0)$, for the chaotic 3D CNN (Case 2) $\Gamma = \text{diag}(1, 0, 0)$ and finally, for the Lu–He CNN (Case 3) $\Gamma = 1$. The size of the complex network is $N = 300$ for all cases. The periodic boundary conditions are $k = 3$ (Case 1 and Case 3) and $k = 10$ (Case 2). The small-world property is introduced by Newman–Watts algorithm (Case 1 and Case 3) and Watts–Strogatz algorithm (Case 2). The long-range connections will be either added or created as p grows.

The coupling strength is computed according to (26) and the resulting values are depicted in Fig. 10, where $\bar{d} = -10$ and $\bar{d} = -1$ were used for the standard CNN and the Lu–He CNN models respectively when using Newman–Watts algorithm. According to (2), the control laws u_{i1} for $i = 1, \ldots, N$ are given by the A matrix nonzero elements for all cases.

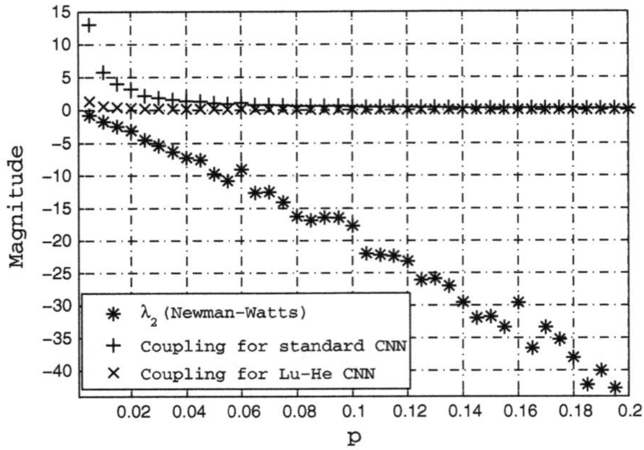

Fig. 10 Evolution of the coupling strength lower boundary for two types of CNN as function of the probability when using the Newman–Watts small-world algorithm. The size of the network $N = 300$ with a periodic boundary condition $k = 3$

5.2.1 Case 1: Synchronization of *N* Standard CNNs

By using $\bar{d} = -10$, the chaotic state variables $x_1(t)$ and $x_2(t)$ of a single standard CNN described by (15)–(16) are stabilized. The coupling strength c is computed from (26), thus, $c(p) \geq |-10/\lambda_2(p)|$ which lower boundary is given in Fig. 10 with "+"-mark.

The set of equations that describes the complex network is given as follows

$$\begin{cases} \dot{x}_{i1} = -x_{i1} + 2f(x_{i1}) - 1.2f(x_{i2}) + 4.04 \sin\left(\frac{\pi}{2}t\right) + c\sum_{j=1}^{N} a_{ij}x_{j1}, \\ \dot{x}_{i2} = -x_{i2} + 1.2f(x_{i1}) + 2f(x_{i2}), \end{cases} \tag{27}$$

with the nonlinear function

$$f(x_{i1,2}) = \frac{1}{2}\left(|x_{i1,2} + 1| - |x_{i1,2} - 1|\right), \tag{28}$$

where $1 \leq i \leq 300$. By using $p = 0.2$, the coupling strength was set in $c = 1$ and the following synchronization results were obtained. Figure 11a shows the time evolution of some chaotic state variables x_{j1} and x_{j2} for $j = 26, 53, 249, 130$. Figure 11b shows the phase portraits between some arbitrary chosen states and the chaotic attractor. Synchronization is confirmed by the phase portraits, therefore, condition $\mathbf{x}_1(t) = \mathbf{x}_2(t) = \cdots = \mathbf{x}_N(t)$ as $t \to \infty$ holds.

5.2.2 Case 2: Synchronization of *N* Chaotic 3D CNN

By using $\bar{d} = -1$ the chaotic state variables $x_1(t)$, $x_2(t)$ and $x_3(t)$ of a single chaotic 3D CNN described by (19) and (20) are stabilized. The coupling strength is obtained by using (26) as follows $c(p) \geq |-1/\lambda_2(p)|$. Note that we only have to determine

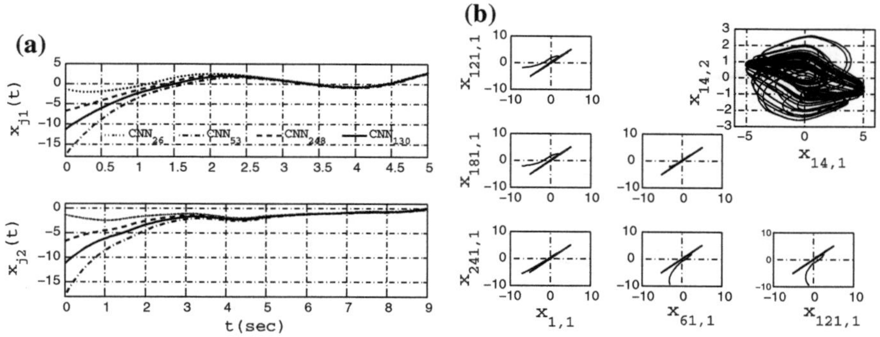

Fig. 11 **a** Time evolution of state variables x_{j1} and x_{j2} for $j = 26,\ 53,\ 249,\ 130$. **b** Phase portraits to corroborate synchronization between x_{i1} versus x_{j1} for $i = 1,\ 61,\ 121; j = 121,\ 181,\ 241$ and view $(x_{14,1}, x_{14,2})$-plane of the chaotic attractor

$\lambda_2(p)$ which will vary as the network varies through the probability p. For this case, the small-world property will be introduced by the Watts–Strogatz algorithm.

The equations that describe the complex network are given as follows

$$\dot{\mathbf{x}}_i = -\mathbf{x}_i + T\tanh(\mathbf{x}_i) + c\sum_{j=1}^{N} a_{ij}\Gamma\mathbf{x}_j, \qquad 1 \le i \le 300, \tag{29}$$

where

$$T = \begin{bmatrix} T_{11} & T_{12} & T_{13} \\ T_{21} & T_{22} & T_{23} \\ T_{31} & T_{32} & T_{33} \end{bmatrix} = \begin{bmatrix} 1.49 & 2 & 1 \\ -2 & 1.7 & 0 \\ 4 & -4 & 2 \end{bmatrix}. \tag{30}$$

For this case by using $p = 0.4$, the corresponding coupling strength is $c = 0.5$. Synchronization results are the following. Figure 12a shows the time evolution of some state variables x_{j1}, x_{j2} and x_{j3} for $j = 55,\ 242,\ 72,\ 174$. Figure 12b provides the phase portraits between some arbitrary chosen states and the corresponding chaotic attractor for the network (29) and (30). Synchronization is verified by the phase portraits, thus, the condition $\mathbf{x}_1(t) = \mathbf{x}_2(t) = \cdots = \mathbf{x}_N(t)$ as $t \to \infty$ holds as well.

5.2.3 Case 3: Synchronization of *N* Lu–He CNN

Similarly as in the previous cases, we proceeded to meet the conditions of Theorem 1. By using $\bar{d} = -1$ the chaotic state variables $x(t)$ and $x(t-\tau)$ of a single Lu–He CNN described by (21) are stabilized. The coupling strength is obtained by using (26) as follows $c(p) \ge |-1/\lambda_2(p)|$ which lower boundary is given in Fig. 10 with "×"-mark. The small-world property will be introduced by the Newman–Watts algorithm as in Case 1.

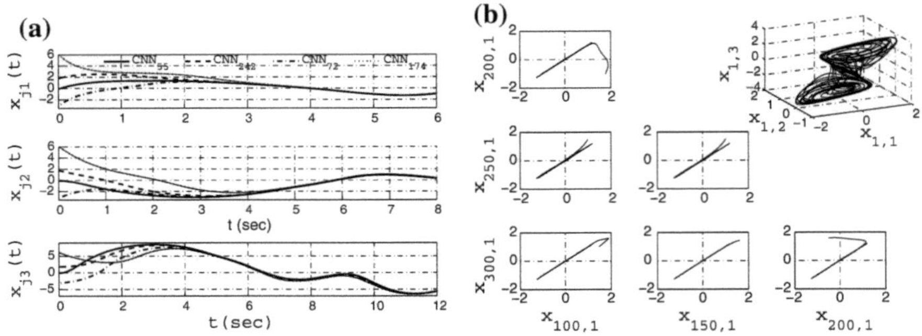

Fig. 12 **a** Time evolution of state variables x_{j1}, x_{j2} and x_{j3} for $j = 55, 242, 72, 174$. **b** Phase portraits to corroborate synchronization between x_{i1} versus x_{j1} for $i = 100, 120, 200$; $j = 200, 250, 300$ and view $(x_{1,1}, x_{1,2}, x_{1,3})$-plane of the chaotic attractor

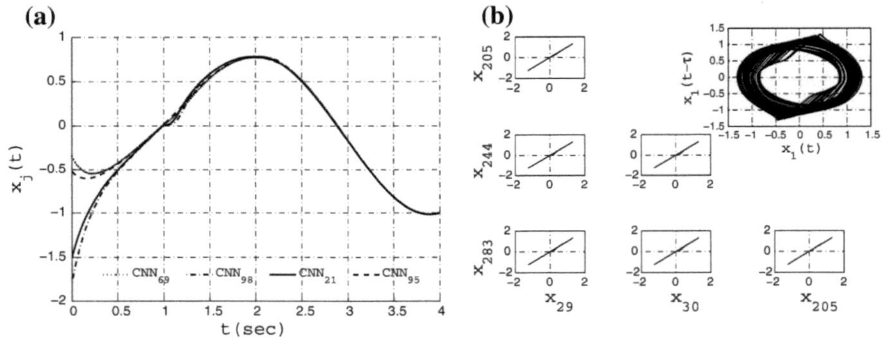

Fig. 13 **a** Time evolution of state variables x_{j1} for $j = 69, 98, 21, 95$. **b** Phase portraits to corroborate synchronization between x_{i1} versus x_{j1} for $i = 29, 30, 205$; $j = 205, 244, 283$ and view $(x_1(t), x_1(t-\tau))$-plane of the chaotic attractor

Equations describing the complex network are given as follows

$$\dot{x}_i(t) = 0.001x_i(t) - 3.8\left(\left|x_{i,\tau} + 1\right| - \left|x_{i,\tau} - 1\right|\right) \\ +2.85\left(\left|x_{i,\tau} + \frac{4}{3}\right| - \left|x_{i,\tau} - \frac{4}{3}\right|\right) + c\sum_{j=1}^{N} a_{ij}x_j, \tag{31}$$

where $x_{i,\tau} = x_i(t-\tau)$ and $1 \leq i \leq 300$. By using $p = 0.3$, the coupling strength was set in $c = 1$. The following synchronization results were obtained. Figure 13a shows the time evolution of some state variables arbitrary chosen x_{j1} for $j = 69, 98, 21, 95$. Figure 13b depicts the phase portraits between some arbitrary chosen states and the corresponding chaotic attractor. Synchronization is confirmed by the phase portraits, therefore, condition $\mathbf{x}_1(t) = \mathbf{x}_2(t) = \cdots = \mathbf{x}_N(t)$ as $t \to \infty$ holds.

In the remainder of this work, one of the most important application of chaotic signals will be described. A process of secret communication will be performed by using the presented chaotic signals.

6 Chaotic Encryption

One of the most important application of chaos in engineering is secret communications. Chaotic encryption has become one of the most feasible alternatives to hide a confidential message; as a result, it has been deeply studied for the last decade or two which resulted in the appearance of several ways to encrypt [3, 8, 37, 58, 63, 66]. However, chaotic encryption seems to be vulnerable when omitting the energy and spectral characteristics of signals involved. For this reason, the chaotic signals considered will be evaluated to determine a suitable one to be used when encrypting the confidential message.

6.1 Chaotic Signals Evaluation

In the following, the criteria for assessing the quality of the candidate signals to encrypt will be briefly described. The resulting indices, which were applied to the signals showed in the synchronization process, will be given as well. These data will be used to select the best chaotic signal to encrypt a confidential message.

The first selection criterion considers the energy level of the chaotic signal, which suggests the bigger the energy level, the better the signal to encrypt. The criterion is described as follows [51]

$$\sum_{n=0}^{N-1} |x_c(n)|^2 \gg \sum_{n=0}^{N-1} |x_m(n)|^2, \tag{32}$$

where $x_c(n)$ is the samples set of the chaotic signal and $x_m(n)$ is the samples set of the message (digital audio). The criterion, J_1 is defined as the ratio between the right and the left parts of (32), i.e., J_1 yields the times the chaotic signal energy exceeds the message energy, therefore, $J_1 \gg 1$ results in a good encryption.

The second criterion considers the chaotic signal energy in a frequency range which result to be the range where the most energy of the message is located. This is done to ensure good encryption from the frequency domain. The criterion is described as follows [51]

$$\sum_{k=0}^{N-1} \alpha(k)|X_c(k)|^2 \gg \sum_{k=0}^{N-1} \alpha(k)|X_m(k)|^2, \tag{33}$$

where $X_c(k)$ are the chaotic signal spectrum samples, $X_m(k)$ are the message spectrum samples and $\alpha(k)$ is a frequency weighting function that selects the range where the message is located. J_2 is defined as the ratio between right and left parts of (33), which gives the relation between chaotic signal energy and message energy in a specific frequency band if $\alpha(k) = 1$ in $K \in [k_1, k_2]$. Criterion J_2 will yield how many times the weighted chaotic signal energy exceeds the weighted energy of the message, thus, $J_2 \gg 1$ results in a good encryption.

As an additional tool, the index of correlation between the message and its encryption will allow us to support the choice of the chaotic signal. The correlation index is defined as follows [43]

$$r_{xy} = \frac{\text{cov}(X, Y)}{\sigma_X \sigma_Y}, \tag{34}$$

where σ is the standard deviation. In terms of mean and expectation, the covariance $\text{cov}(X, Y)$ can be expressed as $E\left[(X - \mu_X)(Y - \mu_Y)\right]$, where μ is the mean of X or Y and E is the expectation. The resulting index is within $[-1, 1]$. It is said that x and y are weakly (strongly) correlated if $r_{xy} \approx 0$ ($r_{xy} \approx \pm 1$). When $r_{xy} = -1$ there is a total negative correlation while $r_{xy} = 0$ means there is no correlation.

The indices corresponding to each type of CNN are given in Table 1. These data were obtained from the complex networks previously synchronized. The weighting function $\alpha(k)$ has a unity gain in the message's frequency range 0.01–5 kHz and null elsewhere. The total energy of the message is 3.6085×10^3.

Two things are worth mentioning from Table 1: in the first place, the fact that the chaotic states generated by the CNN standard model seem to be better candidates since all their values resulted to be $J_1 > 1$ and $J_2 > 1$. In the second place, we highlight the resulting values of J_1 and J_2 corresponding to the $x_{1,2}$ state of the standard CNN model. It may be noticed that, according to J_2, almost all the chaotic

Table 1 Data obtained from the state vector of the CNN_1 of each complex network in Sect. 5: resulting indices J_1, J_2 after applying selection criteria (32) and (33) and the correlation between the message $m(t)$ and its encryption $s(t)$ in time and frequency domain $r_{m,s}$ and $R_{m,s}$ respectively

CNN model	State	E^a	J_1	J_2	$r_{m,s}$	$R_{m,s}$
CNN standard	$x_{1,1}$	3.6447^b	10.1004	1.4491	0.0046	0.0118
	$x_{1,2}$	0.7651^b	2.1204	2.2055	0.0067	−0.0021
Chaotic 3D CNN	$x_{1,1}$	0.2159^b	0.5984	1.7435	0.0023	0.0049
	$x_{1,2}$	0.5816^b	1.6118	0.4302	0.0012	−0.0138
	$x_{1,3}$	6.6921^b	18.5455	0.6181	0.0041	0.025
Lu–He CNN	x_1	0.022	6.1022^c	0.0025	0.1716^d	0.1148^d

[a] Chaotic signal energy
[b] $\times 10^4$
[c] $\times 10^{-6}$
[d] $\times 10^{-3}$

signal energy is located in the message's frequency range. The reader has to notice that even when the $x_{1,3}$ state corresponding to the chaotic 3D CNN model has a big J_1, the corresponding $J_2 < 1$ which means that almost all the chaotic signal energy is outside the range of interest.

Columns $r_{m,s}$ and $R_{m,s}$ give us the correlation between the message and its encryption, $r_{m,s}$ is the correlation index in time domain while $R_{m,s}$ was obtained from their frequency spectrum when using all available signals. The lower the index, the weaker the correlation between the message and its encryption. With these data we can confirm the election of the chaotic signal to encrypt the message.

6.2 Encryption Results

The end result of this research is to apply the apparent randomness of chaotic systems to hide information, and use their ability to be synchronized to retrieve it. For this purpose, the chaotic systems are CNNs arranged in small-world topology. The chaotic signal to mask the message is selected according to its encryption capability determined by (32) and (33). It is highly recommended to choose the chaotic signal based on both criteria J_1 and J_2 to prevent a bad encryption.

In the following, the encryption, transmission and retrieval of a confidential message will be carried out. Figure 14 shows the two-channel communication diagram with multi-user modality that will be used. The additive chaotic encryption will be utilized to hide the confidential message. It consist on the application of autonomous chaotic oscillators whose output signal is added to the information signal. This sum is sent over a communication channel. Another chaotic signal of the encoder is also transmitted and used by the decoder to synchronize an equivalent chaotic oscillator

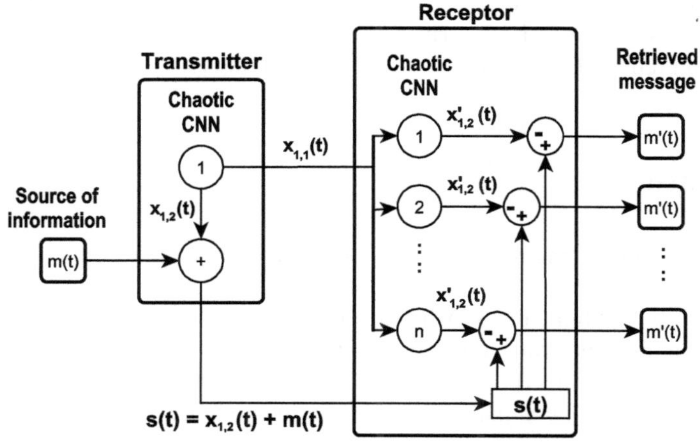

Fig. 14 Two-channel communication diagram with multi-user modality

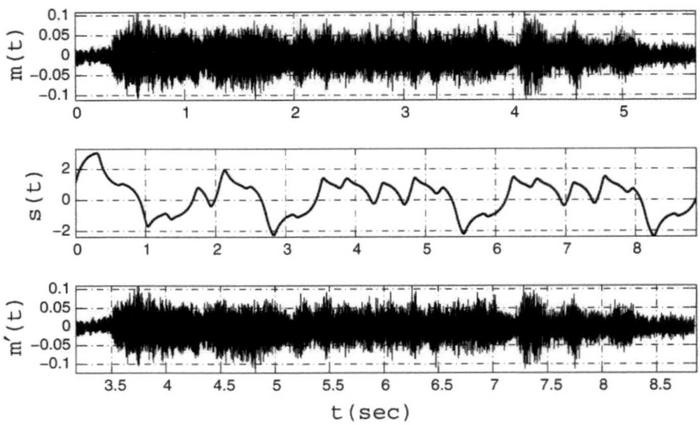

Fig. 15 Resulting signals of the communication process. At the *top*, the message $m(t)$ to be encrypted. In the *middle* the transmitted signal $s(t)$ which is the sum of the $x_{1,2}$ chaotic state and the confidential message $m(t)$. At the *bottom*, the retrieved message $m'(t)$

with the encoder system. The reconstructed chaotic signal is then subtracted from the sum transmitted which finally restores the information [13].

We will consider as a confidential message to be transmitted a short part of the classical song *Le nozze di Figaro (The Marriage of Figaro)* by W. A. Mozart. Figure 15 shows the resulting signals of the communication process. At the top, the message to be encrypted $m(t)$; in the middle, the encrypted message $s(t) = x_{1,2}(t) + m(t)$ which is transmitted by the second channel; finally at the bottom, the retrieved message $m'(t) = s(t) - x'_{1,2}(t)$ for every user.

The encryption was made by using the $x_{1,2}$ state of the standard CNN model, since as explained, almost all its energy is located in the message's frequency band. Without hesitation, selecting the chaotic signal based on its energy and bandwidth characteristics improves the security level of the encrypted message, since it is prevented from retrieval by conventional filtering techniques.

7 Discussion

In this section the strengths and weaknesses of the three main results of this work are discussed.

Firstly, we highlight the synchronization of complex networks composed of chaotic CNN models arranged in small world topology.

For purposes of encryption, the use of such systems allows us to generate a wide variety of chaotic signals, since due to its structure, the number of chaotic signals generated by a CNN varies according to its dimension, which makes the selection process more effective, since the greater the number of signals evaluated, more extensive is the field of requirements that can be met.

Secondly, we highlight the use of criteria, which consider energy and frequency characteristics, to select the chaotic signal to hide the message. This brings an improvement in the security level of the encrypted message, since knowing its bandwidth allows us to choose a suitable chaotic signal to cover those frequencies. As a result, the encrypted message is prevented from being retrieved by using filtering techniques. However, the proposed criteria do not provide local information of the chaotic signal, i.e., the criteria do not allow us to know if its magnitude is insufficient to mask the message in a given time interval.

Correlation indices (Table 1 columns $r_{m,s}$ and $R_{m,s}$) supported the choice of the chaotic signal since they are very close to zero. This ensures that the encrypted message is weakly correlated with the original one, therefore, the chosen chaotic signal properly encrypts the confidential information.

As the third and final result, we have the encryption, transmission and retrieval of a confidential message. This was possible by making use of the ability of chaotic systems to be synchronized. The chaotic system, shown in the communication diagram (Fig. 14), is a complex network with small-world topology. This turns out to be novel, since usually the system that generates the chaotic signal is a single oscillator, so emerging dynamics, present in complex systems, can be exploited. Because it was possible to achieve identical synchronization, the confidential message was fully retrieved.

8 Conclusion

In this work the synchronization of complex networks with small-world topology composed of chaotic CNNs models was performed. Synchronization was accomplished by using the technique of coupling matrix. The simulations showed the importance of the probability when computing the coupling strength. We concluded that the bigger the probability, the smaller the coupling strength.

Another important aspect that involves the probability is the clustering coefficient, which varies considerably as a function of probability p (Fig. 3a, b). For purposes of synchronization, both algorithms could be merged into a combination to keep the network clustering coefficient as big as possible, i.e., by choosing Watts–Strogatz algorithm for small values of p and Newman–Watts algorithm for large values of p and thus promote the status of synchrony.

The encryption results allow us to determine that selecting the chaotic signal according to indices delivered by the criteria (Table 1), the message will be encrypted properly and there will be no need to attenuate the message intensity. Once again, we emphasize the importance of considering the message requirements, especially its frequency location, when selecting the chaotic signal to hide it.

Private communications using chaos have still issues to be addressed to improve the security of the encrypted message. Adding a third selection criterion that considers the modulation of chaotic signal energy to the frequency band of interest, or perform multiple encryption using different frequency bands surely will impact on the effectiveness of the encryption.

Acknowledgments This work was supported by CONACYT México under Research Grant No. 166654, PAICYT under Research Grant IT956-11 and Facultad de Ingeniería Mecánica y Eléctrica (FIME-UANL).

References

1. Anishchenko A, Treves A (2006) Autoassociative memory retrieval and spontaneous activity bumps in small-world networks of integrate-and-fire neurons. J Physiol 100:225–236
2. Arellano-Delgado A, Cruz-Hernández C, López-Gutiérrez RM, Posadas-Castillo C (2015) Outer synchronization of simple firefly discrete models in coupled networks. Math Probl Eng 2015:1–14
3. Arellano-Delgado A, López-Gutiérrez R, Cruz-Hernández C, Posadas-Castillo C, Cardoza-Avendano L, Serrano-Guerrero H (2013) Experimental network synchronization via plastic optical fiber. Opt Fiber Technol 19(2):93–108
4. Arena P, Fortuna L, Porto D (2000) Chaotic behavior in noninteger-order cellular neural networks. Phys Rev E 61(1):776–781
5. Banerjee S (2009) Synchronization of time-delayed systems with chaotic modulation and cryptography. Chaos Soliton Fractals 42(3):745–750
6. Barrat A, Weigt M (2000) On the properties of small-world network models. Eur Phys J B 13:547–560
7. Behnia S, Akhshani A, Akhavan A, Mahmodi H (2009) Applications of tripled chaotic maps in cryptography. Chaos Solitons Fractals 40:505–519
8. Bhat L, Sudha KL (2012) Performance analysis of chaotic DS-CDMA with CSK modulation. Int J Adv Eng Res Stud 1(3):133–136
9. Chua LO, Hasler M, Moschytz GS, Neirynck J (1995) Autonomous cellular neural networks: a unified paradigm for pattern formation and active wave propagation. IEEE Trans Circuits Sys I Fundam Theory Appl 42(10):559–577
10. Chua LO, Komuro M, Matsumoto T (1986) The double scroll family. IEEE Trans Circuits Syst 11:1072–1118
11. Chua LO, Yang L (1998) Cellular neural networks: theory and applications. IEEE Trans Circuits Syst 35:1257–1290
12. Cornejo-Pérez O, Solís-Perales GC, Arenas-Prado JA (2012) Synchronization dynamics in a small pacemaker neuronal ensemble via a robust adaptive controller. Chaos Solitons Fractal 45:861–868
13. Dachselt F, Schwarz W (2001) Chaos and cryptography. IEEE Trans Circuits Syst I Fundam Theory Appl 48:1498–1509
14. Gray RT, Fung CKC, Robinson PA (2009) Stability of small-world networks of neural populations. Neurocomputing 72:1565–1574
15. Guare J (ed) (1990) Six degrees of separation: a play. Vintage, New York
16. Guzmán-Vargas L, Hernández-Pérez R (2006) Small-world topology and memory effects on decision time in opinion dynamics. Physica A 372:326–332
17. Heagy JF, Carroll TL, Pecora LM (1995) Desynchronization by periodic orbits. Phys Rev E 52:R1253–R1256
18. Hopfield JJ (1984) Neurons with graded response have collective computational properties like those of two-state neurons. Proc Natl Acad Sci USA 81:3088–3092
19. Huang X, Zhao Z, Wang Z, Li Y (2012) Chaos and hyperchaos in fractional-order cellular neural networks. Neurocomputing 94:13–21

20. Kaddoum G, Gagnon F (2012) Design of a high-data-rate differential chaos-shift keying system. IEEE Trans Circuits Syst II Express Briefs 59(7):448–452
21. Kolumbán G, Kennedy MP, Chua LO (1997) The role of synchronization in digital communication. IEEE Trans Circuits Syst I Fundam Theory Appl 44:927–935
22. Larson LE, Liu JM, Tsimring LS (eds) (2006) Digital communications using chaos and nonlinear dynamics. Springer, New York
23. Latora V, Marchiori M (2003) Economic small-world behavior in weighted networks. Eur Phys J B 32:249–263
24. Li Q, Yang XS (eds) (2007) Horseshoe dynamics in a small hyperchaotic neural network. Springer, Berlin
25. Lian S (2009) Efficient image or video encryption based on spatiotemporal chaos system. Chaos Solitons Fractals 40:2509–2519
26. Liu Y, Liang M, Zhou Y, He Y, Hao Y, Song M, Yu C, Liu H, Liu Z, Jiang T (2008) Disrupted small-world networks in schizophrenia. Brain 131:945–961
27. Lorenz EN (1963) Deterministic nonperiodic flow. J Atmos Sci 20:130–141
28. Lu H, He Y, He X (1998) A chaos-generator: analyses of complex dynamics of a cell equation in delayed cellular neural networks. IEEE Trans Circuits Syst I Fundam Theory Appl 45(2):178–181
29. Lu J, Chen G (2006) Generating multiscroll chaotic attractors: theories, methods and applications. Int J Bifurc Chaos 16(4):775–858
30. Lu J, Yu S, Leung H, Chen G (2006) Experimental verification of multidirectional multiscroll chaotic attractors. IEEE Trans Circuits Syst I Regul Pap 53(1):149–165
31. Matsumoto T (1984) A chaotic attractor from chua's circuit. IEEE Trans Circuits Syst 12:1055–1058
32. Milgram S (1967) The small-world problem. Psychol Today 1(1):61–67
33. Newman MEJ, Watts DJ (1999) Renormalization group analysis of the small-world network model. Phys Lett A 263:341–346
34. Newman MEJ, Watts DJ (1999) Scaling and percolation in the small-world network model. Phys Rev E 60:7332–7342
35. Pecora LM, Carroll TL (1990) Synchronization in chaotic systems. Phys Rev Lett 64:821–824
36. Pecora LM, Carroll TL, Johnson GA, Mar DJ, Heagy JF (1997) Fundamentals of synchronization in chaotic systems, concepts, and applications. Chaos 7:520–543
37. Pehlivan I, Uyaroğlu Y, Yoğun M (2010) Chaotic oscillator design and realizations of the Rucklidge attractor and its synchronization and masking simulations. Sci Res Essays 5(16):2210–2219
38. Petráš I (2008) A note on the fractional-order Chua's system. Chaos Soliton Fractals 38:140–147
39. Posadas-Castillo C, Cruz-Hernández C, López-Gutiérrez R (2008) Synchronization of chaotic neural networks with delay in irregular networks. Appl Math Comput 205:487–496
40. Ramani A, Grammaticos B, Satsuma J (2009) Modelling the dynamics of nonendemic epidemics. Chaos Soliton Fractals 40:491–496
41. Rossler OE (1976) An equation for continuous chaos. Phys Lett A 57(5):397–398
42. Rulkov NF (1996) Images of synchronized chaos: experiments with circuits. Chaos 6:262–279
43. Salkind N (ed) (2007) Pearson product-moment correlation coefficient. Sage Publications, Thousand Oaks
44. Santos FC, Rodrigues JF, Pacheco JM (2005) Epidemic spreading and cooperation dynamics on homogeneous small-world networks. Phys Rev E 72:056–128
45. Senthilkumar DV, Kurths J (2010) Characteristics and synchronization of time-delay systems driven by a common noise. Eur Phys J Special Topics 187:87–93
46. Senthilkumar DV, Lakshmanan M (2007) Delay time modulation induced oscillating synchronization and intermittent anticipatory/lag and complete synchronizations in time-delay nonlinear dynamical systems. 17(013112)
47. Senthilkumar DV, Lakshmanan M (2007) Intermittency transition to generalized synchronization in coupled time-delay systems. Phys Rev E 76(066210)

48. Senthilkumar DV, Lakshmanan M, Kurths J (2008) Transition from phase to generalized synchronization in time-delay systems. Chaos 18(023118)
49. Shibata S, Komaki Y, Seki F, Inouye MO, Nagai T, Okano H (2015) Connectomics: comprehensive approaches for whole-brain mapping. Microscopy (Tokyo) 64(1):57–67
50. Soriano-Sánchez AG, Posadas-Castillo C, Platas-Garza MA (2015) Synchronization of generalized chua's chaotic oscillators in small-world topology. J Eng Sci Technol Rev 8(2):185–191
51. Soriano-Sánchez AG, Posadas-Castillo C, Platas-Garza MA, Diaz-Romero DA (2015) Performance improvement of chaotic encryption via energy and frequency location criteria. Math Comput Simul 112:14–27
52. Sporns O (2011) The human connectome: a complex network. Ann N Y Acad Sci 1224: 109–125
53. Stam CJ, Jones BF, Nolte G, Breakspear M, Scheltens P (2007) Small-world networks and functional connectivity in Alzheimer's disease. Cereb Cortex 17:92–99
54. Tavazoei MS, Haeri M (2008) Chaotic attractors in incommensurate fractional order systems. Physica D 237:2628–2637
55. Tlelo-Cuautle E, Rangel-Magdaleno JJ, Pano-Azucena AD, Obeso-Rodelo PJ, Nunez-Perez JC (2015) FPGA realization of multi-scroll chaotic oscillators. Commun Nonlinear Sci Numer Simul 27(1):66–80
56. Uyaroğlu Y, Pehlivan I (2010) Nonlinear Sprott94 Case A chaotic equation: synchronization and masking communication applications. Comput Electr Eng 36(6):1093–1100
57. Vasegh N, Sedigh AK (2009) Chaos control in delayed chaotic systems via sliding mode based delayed feedback. Chaos Soliton Fractals 40:159–165
58. Wang MJ, Wang XY, Pei BN (2012) A new digital communication scheme based on chaotic modulation. Nonlinear Dyn 67(2):1097–1104
59. Wang XF (2002) Complex networks: topology, dynamics and synchronization. Int J Bifurc Chaos 12(5):885–916
60. Wang XF, Chen G (2002) Synchronization in small-world dynamical networks. Int J Bifurc Chaos 12(1):187–192
61. Watts DJ, Strogatz SH (1998) Collective dynamics of small-world networks. Nature 393:440–442
62. Wu CW, Chua LO (1994) A unified framework for synchronization and control of dynamical systems. Int J Bifurc Chaos 4:979–998
63. Xu WK, Wang L, Kolumbán G (2011) A novel differential chaos shift keying modulation scheme. Int J Bifurc Chaos 21(3):799–814
64. Yalçin ME (2007) Multi-scroll and hypercube attractors from a general jerk circuit using Josephson junctions. Chaos Soliton Fractals 34:1659–1666
65. Yalçin ME, Suykens JAK, Vandewalle JPL (eds) (2004) Cellular neural networks, multi-scroll chaos and synchronization, serie A, vol 50. World Scientific, Singapore
66. Yang H, Jiang GP (2012) High-efficiency differential-chaos-shift-keying scheme for chaos-based noncoherent communication. IEEE Trans Circuits Syst II Express Briefs 59(5):312–316
67. Yang W, Sun J (2010) Function projective synchronization of two-cell quantum-CNN chaotic oscillators by nonlinear adaptive controller. Phys Lett A 374(4):557–561
68. Yang XS, Huang Y (2006) A new class of chaotic simple three-neuron cellular neural networks. Int J Bifurc Chaos 16:1019–1021
69. Yang XS, Li Q (2006) Horseshoe chaos in cellular neural networks. Int J Bifurc Chaos 16:157–161
70. Yuan Q, Li Q, Yang XS (2009) Horseshoe chaos in a class of simple Hopfield neural networks. Chaos Soliton Fractals 39:1522–1529
71. Zou F, Nossek JA (1991) A chaotic attractor with cellular neural networks. IEEE Trans Circuits Syst 38(7):811–812

Fuzzy Adaptive Synchronization of Incommensurate Fractional-Order Chaotic Systems

A. Bouzeriba, A. Boulkroune, T. Bouden and S. Vaidyanathan

Abstract This chapter presents a fuzzy adaptive control scheme for achieving a generalized projective synchronization (GPS) of two incommensurate fractional-order chaotic systems. The master system and the slave system are assumed to be with non-identical structure, external dynamical disturbances, uncertain models and distinct fractional-orders. The adaptive fuzzy systems are employed for approximating some unknown nonlinear functions. Lyapunov method is adopted for deriving the adaptation laws and proving the stability of the closed-loop system. Under mild assumptions, the proposed control scheme can guarantee all the signals in the closed-loop system remain bounded and the synchronization errors converge asymptotically towards a small of neighbourhood of the origin. Finally, numerical experiment results are presented to show the effectiveness of the proposed synchronization scheme.

Keywords Fuzzy adaptive control · Incommensurate fractional-order systems · Chaotic systems · Chaos synchronization

1 Introduction

Fractional calculus is an area of mathematics that deals with differentiation and integration of arbitrary orders. Recently, the fractional calculus has been studied with increasing interest from chemists, physicians and engineers. In fact, it was

A. Bouzeriba · A. Boulkroune (✉)
LAJ, University of Jijel, BP. 98, Ouled-Aissa, 18000 Jijel, Algeria
e-mail: boulkroune2002@yahoo.fr

A. Bouzeriba
e-mail: bouzeriba.amel@yahoo.fr

T. Bouden
NDT, University of Jijel, BP. 98, Ouled-Aissa, 18000 Jijel, Algeria
e-mail: bouden_toufik@yahoo.com

S. Vaidyanathan
R & D Centre, Vel Tech University, Avadi, Chennai, India
e-mail: sundarvtu@gmail.com

S. Vaidyanathan and C. Volos (eds.), *Advances and Applications in Chaotic Systems*, Studies in Computational Intelligence 636,
DOI 10.1007/978-3-319-30279-9_15

found that various systems in interdisciplinary fields can be precisely modelled by fractional-order differential equations such as [5, 39]: dielectric polarization, electrode-electrolyte polarization, viscoelastic systems, electromagnetic waves, heat diffusion systems, finance systems, batteries, neurons, and so on. That is to say, fractional derivatives give a superb instrument for a precise description of memory and heredity features of several material and processes.

Chaotic systems are nonlinear and deterministic rather than probabilistic. They are characterized by the self similarity of the strange attractor and unusual sensitivity to initial conditions quantified by fractal dimension and the existence of a positive Lyapunov exponent, respectively, [2]. Recently, many works have shown that some fractional-order systems can behave chaotically, namely fractional-order Rössler system [23], fractional-order Lorenz system [20], fractional-order Arneodo system [32], fractional-order Lü system [16].

Synchronization problem consists in designing a slave system whose behavior mimics another system (i.e. master system). The latter drives the slave system via the transmitted signals. In the literature, various types of the chaos synchronization have been revealed, such as complete synchronization (CS) [11, 14, 45], phase synchronization (PHS) [33, 37], lag synchronization (LS) [12, 24], generalized synchronization (GS) [51], generalized projective synchronization (GPS) [25, 27, 40–44, 47, 55]. However, all these synchronization methods focus on integer-order chaotic systems, which is a very special case of the non-integer-order (i.e. fractional-order) chaotic systems. In addition, it has been assumed in [11, 12, 14, 24, 25, 27, 33, 37, 45, 51, 55] that models of the chaotic systems are almost known. Therefore, it is very interesting to extend these fundamental results to uncertain fractional-order chaotic systems and to incorporate an online function approximator (such as adaptive fuzzy system) to deal with model uncertainties.

Based on the universal approximation feature of the fuzzy systems [50], fuzzy adaptive control schemes [3, 4, 6, 15, 21, 35, 36, 48, 49] have been developed for a class of uncertain chaotic systems with integer-order. In these schemes, the fuzzy logic systems are employed to online estimate the uncertain nonlinear functions. The stability of the closed-loop system has been analyzed in Lyapunov sense. In order to deal with the bounded external disturbances and fuzzy approximation errors, a robust control term has been added to the dominate fuzzy adaptive control term. This robust control term can be designed by a sliding mode control (SMC) approach [6, 15, 35, 36] and an $H\infty$ control approach [21, 48, 49]. However, these synchronization schemes [6, 15, 21, 35, 36, 48, 49] are limited to uncertain chaotic integer-order systems.

Synchronization of fractional-order chaotic systems is yet considered as a challenging research topic [22, 28–31, 52–54]. In [54], chaos synchronization of variable-order fractional financial system based on active control method has been studied. In [52], a local stability criterion for synchronization of incommensurate fractional-order chaotic systems has been derived based on the stability theory of linear incommensurate fractional-order differential equations. A modified projective synchronization via active SMC of two different fractional-order systems has been proposed in [53]. The author of [22] has designed an impulsive synchronization system for fractional order chaotic systems. Chaos synchronization between two different uncertain fractional order chaotic systems has been studied based on adaptive fuzzy sliding mode control in [31]. A generalized projective synchronization of fractional order chaotic system has been developed in [28] using an adaptive fuzzy sliding mode control strategy. In [30], an adaptive fuzzy control based synchronization of uncertain fractional-order chaotic systems with time delay has been proposed. In [29]. An adaptive fuzzy logic controller has been designed for achieving an H_∞ synchronizing for a class of uncertain fractional-order chaotic systems. However, the fundamental results of [29–31] are already questionable, because the stability analysis has not been derived rigorously in mathematics, as stated in [1, 46].

In the current chapter, a direct adaptive fuzzy control is designed to achieve a GPS of two different incommensurate fractional-order chaotic systems with uncertain dynamics and external disturbances. Lyapunov method is adopted to carry out the design of the adaptation laws and the stability analysis of the corresponding closed-loop system. To show the effectiveness of the proposed synchronization scheme, an illustrative example will be presented. Compared to the previous works [22, 28–31, 52–54], the main contributions of this chapter are:

(1) The master and slave systems are assumed to be with non-identical structure, external dynamical disturbances, uncertain models and distinct fractional-orders. To the best of our knowledge, the design of a direct adaptive fuzzy control for such a class of fractional-order chaotic systems has not been formerly considered in the control literature.
(2) The conditions imposed in the previous literature [22, 52–54] on full or partial knowledge of the models of the master and slave systems are neglected here. In fact, the adaptive fuzzy systems incorporated in the proposed controller permits to online approximate the uncertain functions.
(3) Unlike the closely related works [29–31], the stability analysis of the underlying closed-loop system is rigorously established in this paper, through the use of some properties of the Caputo fractional-order derivative [13, 19, 26, 34, 38, 56].
(4) Compared to [28], the proposed control law is very simple, continuous and free of singularity problem. Moreover, one does not use the fractional-order derivatives of the master state vector as input for the designed fuzzy systems.

2 Basic Definitions and Preliminaries for Fractional-Order Systems

The most frequently used definitions for fractional derivatives are: Riemann–Liouville, Grünwald–Letnikov, and Caputo definitions [34]. As the Caputo fractional operator is more consistent than another ones [13, 19, 26, 34, 38, 56], then this operator will be employed in the rest of this paper. Also, a modification of

Adams–Bashforth–Moulton algorithm proposed in [17, 18] will be used for computer numerical simulation of the Caputo fractional-order differential equations.

The Caputo fractional derivative of a function $x(t)$ with respect to time is defined as follows [34]:

$$D_t^{\alpha} x(t) = \frac{1}{\Gamma(m-\alpha)} \int_0^t (t-\tau)^{-\alpha+m-1} x^{(m)}(\tau) d\tau, \tag{1}$$

where $m = [\alpha] + 1$, $[\alpha]$ is the integer part of α, D_t^{α} is called the α-*order* Caputo differential operator, and $\Gamma(.)$ is the well-known Euler's gamma function:

$$\Gamma(P) = \int_0^{\infty} t^{p-1} e^{-t} dt; \quad \text{with } \Gamma(P+1) = P\Gamma(P) \tag{2}$$

This function can be seen as an extension of the factorial to real number arguments.

The following properties of the Caputo fractional-order derivative will be employed in the sequent sections [13, 26, 34]:

Property 1 *Let* $0 < q < 1$, *then*

$$Dx(t) = D_t^{1-q} D_t^q x(t), \quad where \;\; D = \frac{d}{dt}. \tag{3}$$

Property 2 *The Caputo fractional derivative operator is a linear operator:*

$$D_t^q(\nu x(t) + \mu y(t)) = \nu D_t^q x(t) + \mu D_t^q y(t), \tag{4}$$

where ν *and* μ *are real constants.*

Especially, $D_t^q x(t) = D_t^q(x(t) + 0) = D_t^q x(t) + D_t^q 0$, *then, we have* $D_t^q 0 = 0$.

Property 3 *Consider a Caputo fractional nonlinear system [19, 38, 56]:*

$$D_t^q x(t) = f(x(t)), \quad with \;\; 0 < q < 1 \tag{5}$$

If one assumes that $f(x(t))$*satisfies the Lipchitz condition with respect to* x, *i.e.,*

$$\| f(x(t)) - f(x_1(t)) \| \le \ell \, \|x(t) - x_1(t)\| \,, \tag{6}$$

where ℓ *is a positive constant. Without loss of generality, one also assumes that* $f(x)$*satisfies* $f(x) = 0$ *at* $x = 0$.

It follows that:

$$\| f(x(t)) \| \le \ell \, \|x(t)\| \,, \tag{7}$$

3 Problem Statement and Fuzzy Logic Systems

3.1 Problem Statement

Our main motivation consists in designing a fuzzy adaptive control system achieving a GPS between two different fractional-order chaotic systems. Figure 1 presents this proposed synchronization scheme.

The uncertain fractional-order chaotic master system can be given in the following form:

$$D_t^{\alpha_i} x_i = f_{mi}(x), \quad \text{for } i = 1, \ldots, n \tag{8}$$

where $D_t^{\alpha_i} = \frac{d^{\alpha_i}}{dt^{\alpha_i}}$, $0 < \alpha_i < 1$is the fractional-order, $x = [x_1, \ldots, x_n]^T \in R^n$ is the state vector that is measurable, and $f_{mi}(x)$ is an unknown continuous nonlinear function.

The slave system is described by

$$D_t^{\beta_i} y_i = f_{si}(y) + u_i + d_i(t, y), \quad \text{for } i = 1, \ldots, n \tag{9}$$

where $0 < \beta_i < 1$ is the fractional-order of the states of the slave system, $f_{si}(y)$ is an unknown continuous nonlinear function, $y = [y_1, \ldots, y_n]^T \in R^n$ is its state vector which is also assumed to be measurable. $u$is the control input to be determined, and $d_i(t, y)$denotes the lumped disturbances that can include unmodeled dynamics, parametric variations, and external and noise disturbances.

Remark 1 Compared to [22, 28–31, 52–54], the systems (8) and (9) considered in this paper represent a relatively large class of uncertain fractional-order chaotic systems, for the following reasons:

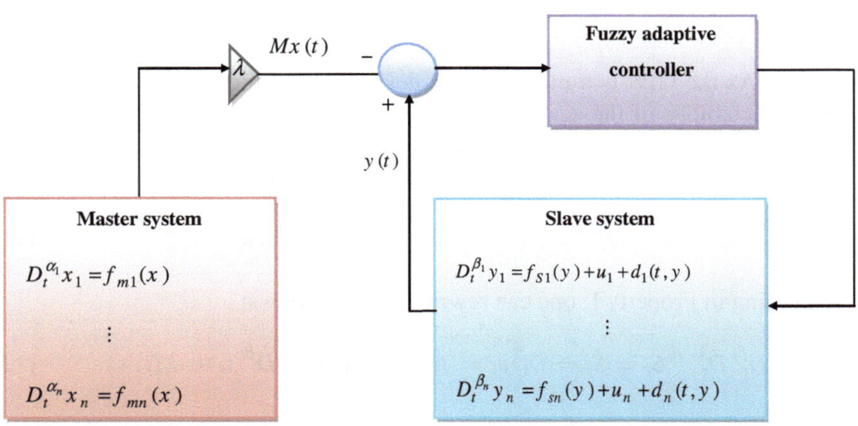

Fig. 1 The proposed chaos synchronization scheme

(a) Many chaotic systems can be described in this considered form, such as fractional-order Chen system, fractional-order Lorenz system, fractional-order Lü system, fractional-order unified chaotic system, and so on.
(b) The master system and slave system are assumed to be completely different, with incommensurate (or inhomogeneous) fractional-orders (i.e. $\alpha_1 \neq \alpha_2 \neq \cdots \neq \alpha_n$, and $\beta_1 \neq \beta_2 \neq \cdots \neq \beta_n$)and with different orders (i.e. $\alpha_i \neq \beta_i$, and $\sum_1^n \alpha_i \neq \sum_1^n \beta_i$).
(c) Both systems are assumed to be with uncertain dynamics (i.e. $f_{mi}(x)$ and $f_{si}(y)$ are unknown). The slave system is affected by external dynamic disturbances.

Remark 2 If ($\alpha_1 = \alpha_2 = \cdots = \alpha_n$) and ($\beta_1 = \beta_2 = \cdots = \beta_n$), the GPS of different incommensurate fractional-order chaotic systems will be reduced to a GPS of different commensurate fractional-order chaotic systems. Especially, if $\alpha_1 = \alpha_2 = \cdots = \alpha_n = 1$ and $\beta_1 = \beta_2 = \cdots = \beta_n = 1$, the GPS of different commensurate fractional-order chaotic systems will become a GPS of integer-order chaotic systems.

Our objective is to design an appropriate fuzzy adaptive control law u_i (for all $i = 1, \ldots, n$)such that a GPS between the master system (8) and the slave system (9) is practically realized, while ensuring the boundedness of all closed-loop system signals.

The synchronization error variables between the two systems are defined as follows:

$$e_i = y_i - \lambda_i x_i, \quad \text{for } i = 1, \ldots, n \tag{10}$$

where $\lambda_i$is the scaling factor that defines a proportional relation between the synchronized systems.

Remark 3 If $\lambda_1 = \lambda_2 \cdots = \lambda_n = \lambda$, the GPS problem becomes a projective synchronization (PS) problem. In particular, if $\lambda = 1$ and -1, this problem is further reduced to CS and anti-phase synchronization (AS), respectively. If the scaling factors $\lambda_1 = \lambda_2 \cdots = \lambda_n = 0$, the synchronization problem is turned into a chaos control problem.

Remark 4 As the GPS is a more general definition that the PS, it is obvious that the unpredictability of the scaling matrix $M = diag(\lambda_1, \lambda_2, \ldots, \lambda_n)$, in GPS can additionally enhance the security in communication applications.

Now, as in [38], one defines a new error variable S_i such that:

$$D_t^{1-\beta_i} S_i = e_i, \quad \text{for } i = 1, 2, \ldots, n \tag{11}$$

According to Property 1, one can rewrite (11) as follows:

$$D_t^{\beta_i} D_t^{1-\beta_i} S_i = \dot{S}_i = D_t^{\beta_i} e_i = f_{si}(y) + u_i - \lambda_i D_t^{\beta_i} x_i + d_i(t, y). \tag{12}$$

or

$$\dot{S}_i = \phi_i(x, y, d_i) + u_i, \tag{13}$$

with

$$\phi_i(x, y, d_i) = \left[f_{si}(y) - \lambda_i D_t^{\beta_i} x_i \right] + d_i(t, y) \tag{14}$$

Remark 5 By dynamics (13), we know that the GPS between the master system (8) and the slave system (9) is transformed into the following problem: designing a control law u_i such that the dynamics (13) is practically stabilized. In short, this GPS problem becomes stabilization one.

Remark 6 Since the nonlinear function $\phi_i(x, y, d_i)$ is uncertain, the design of a control law to stabilize the dynamics (13) is not easy. To overcome such a problem, one will use later an adaptive fuzzy system to estimate the functional upper-bound of $\phi_i(x, y, d_i)$.

Remark 7 From (11) and according to Properties 1–3 of the Caputo fractional derivative operator [13, 19, 26, 34, 38, 56], one can easily show the existence of a positive real number l_i, such that $|e_i| \leq l_i |S_i|$. Hence, $S_i = 0$ implies that $e_i = 0$ and the boundedness of S_i implies that of e_i.

3.2 Fuzzy Approximator

The basic components of a fuzzy logic system include a fuzzifier, a fuzzy knowledge-base, an inference engine and a defuzzifier [7–10, 50]. The fuzzy inference engine uses a set of fuzzy If–Then rules to perform a mapping from an input $x^T = [x_1, \ldots, x_n] \in R^n$ to an output $\hat{f} \in R$. The ith fuzzy rule is written as:

$$R^{(i)}: if\ x_1\ is\ A_1^i\ and \ldots and\ x_n\ is\ A_n^i\ then\ \hat{f}\ is\ f^i, \tag{15}$$

where $A_1^i, A_2^i, \ldots, and\ A_n^i$ are fuzzy sets and $f^i$is a fuzzy singleton for the output in the ith rule.

If one uses a singleton fuzzifier, product inference, and center-average defuzzifier, the output of this fuzzy logic system can be simply formulated as

$$\hat{f}(x) = \frac{\sum_{i=1}^{m} f^i \left(\prod_{j=1}^{n} \mu_{A_j^i}(x_j) \right)}{\sum_{i=1}^{m} \left(\prod_{j=1}^{n} \mu_{A_j^i}(x_j) \right)} = \theta^T \psi(\underline{x}), \tag{16}$$

where $\mu_{A_j^i}(x_j)$ is the membership function of the fuzzy set A_j^i, m is the number of fuzzy rules, $\theta^T = \left[f^1, \ldots, f^m \right]$ is the adaptive parameter (consequent parameters) vector, and $\psi^T = \left[\psi^1, \psi^2, \ldots, \psi^m \right]$ where

$$\psi^i(x) = \frac{\prod_{j=1}^{n} \mu_{A_j^i}(x_j)}{\sum_{i=1}^{m} \left(\prod_{j=1}^{n} \mu_{A_j^i}(x_j) \right)} \tag{17}$$

is the fuzzy basis function (FBF).

The fuzzy logic system (16) is able to uniformly approximate any real continuous function $f(\underline{x})$ defined on a compact set to any arbitrary accuracy [7–10, 50]. Of particular importance, one assumes that the FBFs, i.e. $\psi(\underline{x})$, are properly specified in advance by designer. However, the vector θ is determined by some adequate adaptive laws that will be designed later.

4 Design of Fuzzy Adaptive Controller

In the sequel, the following mild assumptions are required:

Assumption 1 The unknown external disturbance satisfies:

$$|d_i(t, y)| \leq \bar{d}_i(y), \tag{18}$$

where $\bar{d}_i(y)$ is an unknown continuous positive function.

Assumption 2 There exists an unknown continuous positive function $\bar{\phi}_i(y)$ such that:

$$|\phi_i(x, y, d_i)| \leq \bar{\phi}_i(y), \quad \text{for } i = 1, \ldots, n \tag{19}$$

The unknown function $\bar{\phi}_i(y)$ can be approximated, on a compact set Ω_y, by the linearly parameterized fuzzy systems (16) as follows:

$$\hat{\bar{\phi}}_i(y, \theta_i) = \theta_i^T \psi_i(y), \quad \text{with } i = 1, \ldots, n \tag{20}$$

where $\psi_i(y)$ is the FBF vector, which is determined a priori by the designer, and θ_i is the vector of the adjustable parameters of this fuzzy system.

Without loss of generality, we assume that there exists an optimal fuzzy approximator with m^* fuzzy rules that can identify the nonlinear function $\bar{\phi}_i(y)$ with a minimal approximation error, i.e.

$$\bar{\phi}_i(y) = \hat{\bar{\phi}}_i(y, \theta_i^*) + \delta_i(y) = \theta_i^{*T} \psi_i(y) + \delta_i(y) \tag{21}$$

where $\delta_i(y)$ is the minimal approximation error and usually assumed to be bounded for all $y \in \Omega_y$, *i.e.* $|\delta_i(y)| \leq \bar{\delta}_i$, with $\bar{\delta}_i$ is an unknown constant [7–10, 50], and

$$\theta_i^* = \arg\min_{\theta_i} \left[\sup_{y \in \Omega_y} \left| \bar{\phi}_i(y) - \hat{\bar{\phi}}_i(y, \theta_i) \right| \right] \tag{22}$$

Notice that θ_i^* is the optimal value of θ_i [7–10, 50] and mainly introduced for analysis purposes. Its value is not needed when implementing the controller.

From the previous analysis, one can get the following expressions:

$$\begin{aligned}\hat{\bar{\phi}}_i(y,\theta_i)-\bar{\phi}_i(y) &= \hat{\bar{\phi}}_i(y,\theta_i)-\hat{\bar{\phi}}_i(y,\theta_i^*)+\hat{\bar{\phi}}_i(y,\theta_i^*)-\bar{\phi}_i(y),\\ &= \theta_i^T\psi_i(y)-\theta_i^{*T}\psi_i(y)-\delta_i(y),\\ &= \tilde{\theta}_i^T\psi_i(y)-\delta_i(y).\end{aligned} \tag{23}$$

with $\tilde{\theta}_i=\theta_i-\theta_i^*$, for $i=1,\ldots,n$.

For the master-slave system (8) and (9), the control law can be designed as:

$$u_i=\frac{-\rho_i^2(t)S_i}{(\rho_i\,|S_i|+\varepsilon_i)} \quad \text{for } i=1,\ldots,n \tag{24}$$

with $\rho_i(t)=\theta_i^T\psi_i(y)+k_{0i}+k_{1i}\,|S_i|$, where k_{0i} is an adaptive parameter which will be designed later, and $k_{1i}>0$ is a positive design parameter. $\varepsilon_i>0$ is a positive and small design parameter.

From (13), we have

$$S_i\dot{S}_i=S_i\phi_i(x,y,d_i)+S_iu_i\le|S_i|\,\bar{\phi}_i(y)+S_iu_i \tag{25}$$

Using (23) and substituting the control law (24) into (25) yields

$$\begin{aligned}S_i\dot{S}_i &\le |S_i|\,\bar{\phi}_i(y)-\rho_i^2(t)S_i^2/(\rho_i\,|S_i|+\varepsilon_i)+\rho_i\,|S_i|-\rho_i\,|S_i|\\ &\le -|S_i|\,\tilde{\theta}_i^T\psi_i(y)+|S_i|\,|\delta_i(y)|-k_{0i}\,|S_i|-k_{1i}S_i^2+\varepsilon_i.\end{aligned} \tag{26}$$

Adaptation laws are designed as follows:

$$\dot{\theta}_i=\gamma_{\theta i}(|S_i|\,\psi_i(y)-\sigma_{\theta i}\theta_i), \quad \text{with } \theta_{ij}(0)>0 \tag{27}$$

$$\dot{k}_{0i}=\gamma_{ki}(|S_i|-\sigma_{ki}k_{0i}), \quad \text{with } k_{0i}(0)>0 \tag{28}$$

where $\gamma_{\theta i}$, $\sigma_{\theta i}$, σ_{ki} and γ_{ki} are strictly positive design parameters.

Now, we are in a position to present our main result.

Theorem 1 *For the master-slave system (8) and (9), if Assumptions 1 and 2 are valid, the control law (24) together with its adaptation laws (27) and (28) can guarantee the following properties:*

- *All the signals in the closed-loop system are bounded.*
- *Signals S_i and e_i asymptotically converge to a residual set that can be made small by properly adjusting the design parameters.*

Proof Consider the following Lyapunov function candidate:

$$V_i=\frac{1}{2}S_i^2+\frac{1}{2\gamma_{\theta i}}\left\|\tilde{\theta}_i\right\|^2+\frac{1}{2\gamma_{ki}}\tilde{k}_{0i}^2, \tag{29}$$

with $\tilde{k}_{0i} = k_{0i} - k_{0i}^*$, where $k_{0i}^* = \bar{\delta}_i$.

Differentiating V_i with respect to time yields

$$\dot{V}_i = S_i \dot{S}_i + \frac{1}{\gamma_{\theta i}} \tilde{\theta}_i^T \dot{\theta}_i + \frac{1}{\gamma_{ki}} \dot{k}_{0i} \tilde{k}_{0i} \tag{30}$$

Using (26)–(28), $\dot{V}_i$ becomes

$$\dot{V}_i \leq -k_{1i} S_i^2 + \varepsilon_i - \sigma_{\theta i} \tilde{\theta}_i^T \theta_i - \sigma_{ki} k_{0i} \tilde{k}_{0i}. \tag{31}$$

It is clear that

$$-\sigma_{\theta i} \tilde{\theta}_i^T \theta_i \leq -\frac{\sigma_{\theta i}}{2} \left\| \tilde{\theta}_i \right\|^2 + \frac{\sigma_{\theta i}}{2} \left\| \theta_i^* \right\|^2 \tag{32}$$

$$-\sigma_{ki} k_{0i} \tilde{k}_{0i} \leq -\frac{\sigma_{ki}}{2} \tilde{k}_{0i}^2 + \frac{\sigma_{ki}}{2} k_{0i}^{*2} \tag{33}$$

Substituting (32) and (33) into (31), we get

$$\begin{aligned} \dot{V}_i &\leq -k_{1i} S_i^2 - \frac{\sigma_{\theta_i}}{2} \left\| \tilde{\theta}_i \right\|^2 - \frac{\sigma_{ki}}{2} \tilde{k}_{0i}^2 + \frac{\sigma_{ki}}{2} k_{0i}^{*2} + \frac{\sigma_{\theta i}}{2} \left\| \theta_i^* \right\|^2 + \varepsilon_i \\ &\leq -\eta_i V_i + \mu_i \end{aligned} \tag{34}$$

with $\mu_i = \frac{\sigma_{ki}}{2} k_{0i}^{*2} + \frac{\sigma_{\theta i}}{2} \left\| \theta_i^* \right\|^2 + \varepsilon_i$ and $\eta_i = \min\left\{2k_{1i}, \sigma_{\theta_i} \gamma_{\theta_i}, \sigma_{ki} \gamma_{ki}\right\}$.

Therefore, we get

$$0 \leq V_i(t) \leq \frac{\mu_i}{\eta_i} + \left(V_i(0) - \frac{\mu_i}{\eta_i} \right) e^{-\eta_i t} \tag{35}$$

which means that all the signals of the closed-loop system are bounded. Especially, we have

$$|S_i| \leq \sqrt{\frac{2\mu_i}{\eta_i}}, \quad \text{when } t \to \infty \tag{36}$$

Therefore, S_i converges to an adjustable residual set (defined by (36)). Since μ_i can be selected arbitrary and η_i only depends on the design parameters (that can be selected sufficiently large), the ultimate bounds of the error S_i can be made arbitrary small.

Based on Remark 7 and Properties 1–3, the synchronization errors are stable and also converge to an adjustable residual set. □

Remark 8 It is worth noting that the synchronization of the fractional-order chaotic systems with known or partially known dynamics has been comprehensively studied in many works [22, 29–31, 52–54]. But, to the best of our knowledge, there are no theoretical or applied works in the literature on the synchronization of incommensurate fractional-order chaotic systems with completely unknown and different models and dynamic disturbances, except the works of [29–31] which are already questionable (as stated in [1, 46]). Hence, the results of [22, 29–31, 52–54] cannot be directly applied to the considered master-slave system.

5 Simulation Results

The proposed controller will be applied to synchronize two different chaotic systems with distinct fractional-orders. We will consider three simulation cases according to the value of λ_i.

We consider the fractional-order Chua's oscillator system [57] as the master system, which is described by

$$\begin{cases} D_t^{\alpha_1} x_1 = a(x_2 - x_1 - f(x_1)), \\ D_t^{\alpha_2} x_2 = x_1 - x_2 + x_3, \\ D_t^{\alpha_3} x_3 = -bx_2 - cx_3, \end{cases} \tag{37}$$

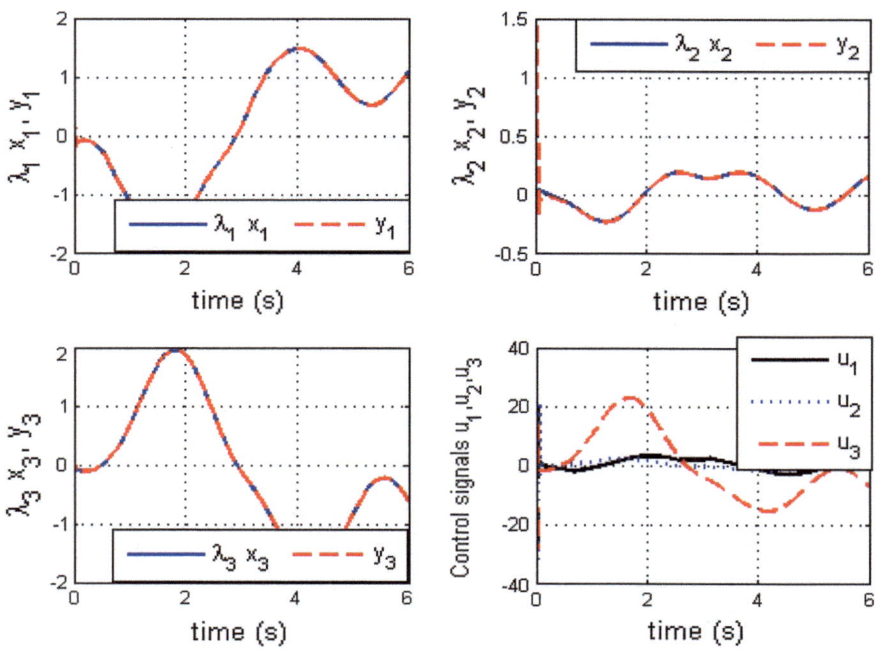

Fig. 2 Simulation results of an anti-phase projective synchronization

where $f(x_1) = m_1 x_1 + \frac{1}{2}(m_0 - m_1)(|x_1 + 1| - |x_1 - 1|)$, $m_0 = -1.1726$, $m_1 = -0.7872$, $a = 10, 725$, $b = 10.593$, $c = 0.268$. The system (37) is a chaotic attractor for $\alpha_1 = 0.93$, $\alpha_2 = 0.99$ and $\alpha_3 = 0.92$, [57].

The controlled fractional-order Rössler system in [23] is considered as the slave system, which is described by:

$$\begin{cases} D_t^{\beta_1} y_1 = -(y_2 + y_3) + u_1 + d_1(t), \\ D_t^{\beta_2} y_2 = y_1 + a y_2 + u_2 + d_2(t), \\ D_t^{\beta_3} y_3 = b + y_3(y_1 - c) + u_3 + d_3(t), \end{cases} \tag{38}$$

where $a = 0.5$, $b = 0.2$ and $c = 10$. For $u_i = 0$ and $d_i(.) = 0$, the system (38) has a chaotic behavior for $\beta_1 = 0.9$, $\beta_2 = 0.85$ and $\beta_3 = 0.95$.

The external dynamic disturbances are selected as follows: $d_1(t) = d_2(t) = d_3(t) = 0.2\sin(3t) + 0.2\cos(3t)$.

The initial conditions are: $x(0) = [x_1(0), x_2(0), x_3(0)]^T = [0.2, -0.1, 0.1]^T$ and $y(0) = [y_1(0), y_2(0), y_3(0)]^T = [0.5, 1.5, 0.1]^T$.

The adaptive fuzzy systems, $\theta_i^T \psi_i(y)$, with $i = 1, 2, 3$, have the vector $y = [y_1, y_2, y_3]^T$ as input. For each input variable of these fuzzy systems, as in [8], we define three (one triangular and two trapezoidal) membership functions uniformly distributed on the intervals $[-2, 2]$. The design parameters are chosen as follows:

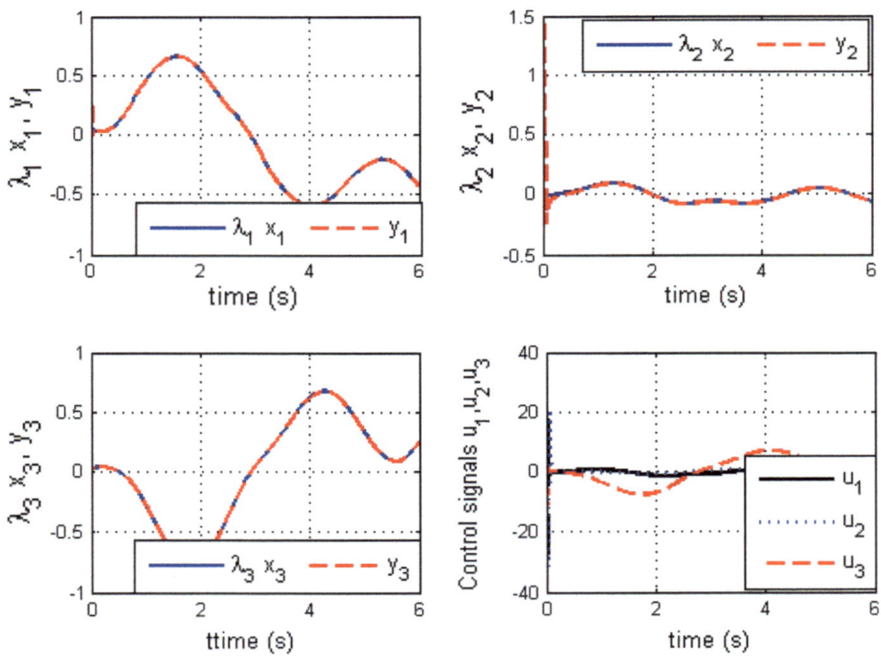

Fig. 3 Simulation results of a complete projective synchronization

$k_{11} = k_{12} = k_{13} = 10$, $\gamma_{\theta 1} = \gamma_{\theta 2} = \gamma_{\theta 3} = 50$, $\sigma_{\theta 1} = \sigma_{\theta 2} = \sigma_{\theta 3} = 0.001$, $\sigma_{k1} = \sigma_{k2} = \sigma_{k3} = 0.05$.

The initial conditions for the adaptive parameters are selected as: $\theta_{1j}(0) = \theta_{2j}(0) = \theta_{3j}(0) = 0.001$and $k_{0i}(0) = 0.001$.

Three simulation cases will be considered.

Case 1: Anti-phase projective synchronization ($\lambda_i = -0.5$).
The obtained results are given in Fig. 2. It is clear that the trajectories of slave system (y_1, y_2, y_3) practically and quickly converge to that of the master system $(\lambda_1 x_1, \lambda_2 x_2, \lambda_3 x_3)$. Hence, the projective anti-phase synchronization between the master and slave systems is effectively realized.

Case 2: Complete projective synchronization ($\lambda_i = 0.2$).
The obtained simulation results for this complete projective synchronization are given in Fig. 3. From this figure, it is clear that the trajectories of slave system (y_1, y_2, y_3) effectively track to that of the master system $(\lambda_1 x_1, \lambda_2 x_2, \lambda_3 x_3)$. The corresponding control signals are also bounded, continuous and admissible.

Case 3: Generalized projective synchronization ($\lambda_1 = 1, \lambda_2 = -1, \lambda_3 = 0.5$).
Figure 4 depicts the generalized projective synchronization between the master system (37) and the slave system (38).

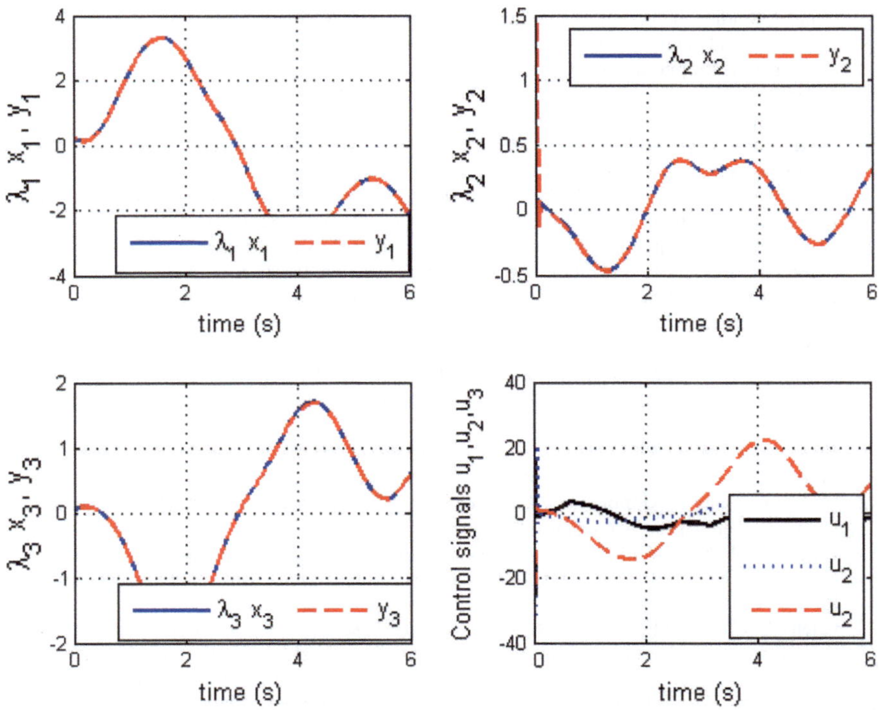

Fig. 4 Simulation results of a generalized projective synchronization

6 Conclusion

This chapter has been conducted to investigate the problem of GPS of two incommensurate fractional-order chaotic systems with uncertain dynamics and external disturbances. The master and slave system are assumed to be with different fractional-orders and different structure. This GPS has been practically achieved by using a smooth fuzzy adaptive control law. The adaptive fuzzy systems, incorporated in the controller, are employed for approximating unknown nonlinear functions. Furthermore, a Lyapunov based analysis has been carried out to conclude about the practical stability as well as the asymptotic convergence of the underlying synchronization errors towards a small of neighborhood of the origin. Finally, numerical simulations have been given to verify the effectiveness of our synchronization scheme.

References

1. Aghababa MP (2012) Comments on H∞ synchronization of uncertain fractional order chaotic systems: adaptive fuzzy approach. ISA Trans 51(1):11–12
2. Azar AT, Vaidyanathan S (2015) Chaos modeling and control systems design, Studies in computational intelligence, vol 581. Springer, Germany
3. Azar AT, Vaidyanathan S (2015) Computational intelligence applications in modeling and control, Studies in computational intelligence, vol 575. Springer, Germany
4. Azar AT, Vaidyanathan S (2015) Handbook of research on advanced intelligent control engineering and automation. IGI-Global, USA
5. Baleanu D, Güvenç ZB, Tenreiro Machado JA (2010) New trends in nanotechnology and fractional calculus applications. Springer, New York
6. Boulkroune A, Msaad M (2012) Fuzzy adaptive observer-based projective synchronization for nonlinear systems with input nonlinearity. J Vib Control 18:437–450
7. Boulkroune A, M'saad M (2012) On the design of observer-based fuzzy adaptive controller for nonlinear systems with unknown control gain sign. Fuzzy Sets Syst 201:71–85
8. Boulkroune A, Tadjine M, M'saad M, Farza M (2008) How to design a fuzzy adaptive control based on observers for uncertain affine nonlinear systems. Fuzzy Sets Syst 159:926–948
9. Boulkroune A, M'Saad M, Farza M (2011) Adaptive fuzzy controller for multivariable nonlinear state time-varying delay systems subject to input nonlinearities. Fuzzy Sets Syst 164(1):45–65
10. Boulkroune A, Bouzeriba A, Hamel S, Bouden T (2014) A projective synchronization scheme based on fuzzy adaptive control for unknown multivariable chaotic systems. Nonlinear Dyn 78:433–447
11. Bowonga S, Kakmenib M, Koinac R (2006) Chaos synchronization and duration time of a class of uncertain systems. Math Comput Simul 71:212–228
12. Cailian C, Gang F, Xinping G (2005) An adaptive lag-synchronization method for time-delay chaotic systems. Proc Am Control Conf 6:4277–4282
13. Carpinteri A, Mainardi F (1997) Fractals and fractional calculus. Springer, New York
14. Carroll TL, Heagy JF, Pecora LM (1996) Transforming signals with chaotic synchronization. Phys Rev E 54:4676–4680
15. Chen CS, Chen HH (2009) Robust adaptive neural-fuzzy-network control for the synchronization of uncertain chaotic systems. Nonlinear Anal Real World Appl 10(3):1466–1479
16. Deng W, Li C (2005) Chaos synchronization of the fractional Lü system. Phys A 353:61–72
17. Diethelm K, Ford NJ (2002) Analysis of fractional differential equations. J Math Anal Appl 265:229–248

18. Diethelm K, Ford NJ, Freed AD (2002) A predictor-corrector approach for the numerical solution of fractional differential equations. Nonlinear Dyn 29:3–22
19. Dong-Feng W, Jin-Ying Z, Xiao-Yan W (2011) Synchronization of uncertain fractional-order chaotic systems with disturbance based on a fractional terminal sliding mode controller. Chin Phys B 20(11):110506
20. Grigorenko I, Grigorenko E (2003) Chaotic dynamics of the fractional Lorenz system. Phys Rev Lett 91:034101
21. Hwang EJ, Hyun CH, Kim E, Park M (2009) Fuzzy model based adaptive synchronization of uncertain chaotic systems: robust tracking control approach. Phys Lett A 373(22):1935–1939
22. Jin-Gui L (2013) A novel study on the impulsive synchronization of fractional-order chaotic systems. Chin Phys B 22:060510
23. Li C, Chen G (2004) Chaos and hyperchaos in the fractional-order Rössler equations. Phys A 341:55–61
24. Li C, Liao X, Wong KW (2005) Lag synchronization of hyperchaos with application to secure communications. Chaos Solitons Fractals 23(1):183–193
25. Li GH (2006) Projective synchronization of chaotic system using backstepping control. Chaos Solitons Fractals 29(2):490–494
26. Li Y, Chen YQ, Podlubny I (2010) Stability of fractional-order nonlinear dynamic systems: Lyapunov direct method and generalized Mittag-Leffler stability. Comput Math Appl 59(5):1810–1821
27. Li Z, Xu D (2004) A secure communication scheme using projective chaos synchronization. Chaos Solitons Fractals 22(2):477–481
28. Li-Ming W, Yong-Guang T, Yong-Quan C, Feng W (2014) Generalized projective synchronization of the fractional-order chaotic system using adaptive fuzzy sliding mode control. Chin Phys B 23:100501
29. Lin TC, Kuo CH (2011) synchronization of uncertain fractional order chaotic systems: adaptive fuzzy approach. ISA Trans 50:548–556
30. Lin TC, Lee TY (2011) Chaos synchronization of uncertain fractional-order chaotic systems with time delay based on adaptive fuzzy sliding mode control. IEEE Trans Fuzzy Syst 19: 623–635
31. Lin TC, Lee TY, Balas VE (2011) Adaptive fuzzy sliding mode control for synchronization of uncertain fractional order chaotic systems. Chaos Solitons Fractals 44:791–801
32. Lu J (2005) Chaotic dynamics and synchronization of fractional order Arneodo's systems. Chaos Solitons Fractals 26(4):1125–1133
33. Pikovsky AS, Rosenblum MG, Osipov GV, Kurths J (1997) Phase synchronization of chaotic oscillators by external driving. Phys D Nonlinear Phenom 104:219–238
34. Podlubny I (1999) Fractional differential equations. Academic Press, New York
35. Poursamad A, Davaie-Markazi AH (2009) Robust adaptive fuzzy control of unknown chaotic systems. Appl Soft Comput 9(3):970–976
36. Roopaei M, Jahromi MZ (2008) Synchronization of two different chaotic systems using novel adaptive fuzzy sliding mode control. Chaos 18:033133
37. Rosenblum MG, Pikovsky AS, Kurths J (1996) Phase synchronization of chaotic oscillators. Phys Rev Lett 76:1804–1807
38. Ruo-Xun Z, Shi-Ping Y (2011) Adaptive stabilization of an incommensurate fractional order chaotic system via a single state controller. Chin Phys B 20(11):110506
39. Sabatier J, Agrawal OP, Machado JAT (2007) Advances in fractional calculus. Springer, London
40. Sarasu P, Sundarapandian V (2011) Active controller design for generalized projective synchronization of four-scroll chaotic systems. Int J Syst Signal Control Eng Appl 4(2):26–33
41. Sarasu P, Sundarapandian V (2011) The generalized projective synchronization of hyperchaotic Lorenz and hyperchaotic Qi systems via active control. Int J Soft Comput 6(5):216–223
42. Sarasu P, Sundarapandian V (2012) Adaptive controller design for the generalized projective synchronization of 4-scroll systems. Int J Syst Signal Control Eng Appl 5(2):21–30
43. Sarasu P, Sundarapandian V (2012) Generalized projective synchronization of three-scroll chaotic systems via adaptive control. Eur J Sci Res 72(4):504–522

44. Sarasu P, Sundarapandian V (2012) Generalized projective synchronization of two-scroll systems via adaptive control. Int J Soft Comput 7(4):146–156
45. Sun J, Zhang Y (2004) Impulsive control and synchronization of Chua's oscillators. Math Comput Simul 66:499–508
46. Tavazoei MS (2012) Comments on chaos synchronization of uncertain fractional-order chaotic systems with time delay based on adaptive fuzzy sliding mode control. IEEE Trans Fuzzy Syst 20(5):993–995
47. Vaidyanathan S, Pakiriswamy S (2013) Generalized projective synchronization of six-term Sundarapandian chaotic systems by adaptive control. Int J Control Theory Appl 6(2):153–163
48. Wang J, Zhang Z, Li H (2008) Synchronization of FitzHugh-Nagumo systems in EES via H8 variable universe adaptive fuzzy control. Chaos Solitons Fractals 36:1332–1339
49. Wang J, Chen L, Deng B (2009) Synchronization of Ghostburster neuron in external electrical stimulation via H8 variable universe fuzzy adaptive control. Chaos Solitons Fractals 39(5):2076–2085
50. Wang LX (1994) Adaptive fuzzy systems and control: design and stability analysis. Prentice-Hall, Englewood Cliffs, New Jersey
51. Wang YW, Guan ZH (2006) Generalized synchronization of continuous chaotic systems. Chaos Solitons Fractals 27:97–101
52. Wang JW, Zhang YB (2009) Synchronization in coupled nonidentical incommensurate fractional-order systems. Phys Lett A 374:202–207
53. Wang X, Zhang X, Ma C (2012) Modified projective synchronization of fractional order chaotic systems via active sliding mode control. Nonlinear Dyn 69:511–517
54. Xu Y, He Z (2013) Synchronization of variable-order fractional financial system via active control method. Cent Eur J Phys 11:824–835
55. Yan J, Li C (2005) Generalized projective synchronization of a unified chaotic system. Chaos Solitons Fractals 26(4):1119–1124
56. Zhang R, Yang S (2012) Robust chaos synchronization of fractional-order chaotic systems with unknown parameters and uncertain perturbations. Nonlinear Dyn 69(3):983–992
57. Zhu H, Zhou S, Zhang J (2009) Chaos and synchronization of the fractional order Chua's system. Chaos Solitons Fractals 39:1595–1603

Implementation of a Laboratory-Based Educational Tool for Teaching Nonlinear Circuits and Chaos

A.E. Giakoumis, Ch.K. Volos, I.N. Stouboulos, I.M. Kyprianidis, H.E. Nistazakis and G.S. Tombras

Abstract The last three decades the subject of nonlinear circuits has become an interesting topic not only due to its applications in various fields but also for educational aims. In this direction, Chua's circuit is considered a cornerstone because it is a unique platform both for the understanding of nonlinear phenomena and the study of experimental chaos as well. So, in this chapter, a new laboratory setup of Chua's oscillator circuit is presented. The proposed realization is suitable for studying, in the laboratory, the design of a nonlinear circuit step by step. It is also a very useful tool for illustrating in the oscilloscope well-known phenomena related with chaos theory, such as period doubling route to chaos, crisis phenomena, intermittency, and attractors' coexistence. The proposed platform could be a useful laboratory-based educational tool for teaching nonlinear circuits in courses related with nonlinear dynamics and chaos for undergraduate, postgraduate and Ph.D. students.

Keywords Nonlinear circuit · Chaos · Chua's oscillator · Chua's diode · Period doubling route to chaos · Crisis phenomena · Intermittency · Attractor's coexistence

---

A.E. Giakoumis · Ch.K. Volos (✉) · I.N. Stouboulos · I.M. Kyprianidis
Physics Department, Aristotle University of Thessaloniki, GR-54124 Thessaloniki, Greece
e-mail: volos@physics.auth.com

A.E. Giakoumis
e-mail: ang1960@el.teithe.gr

I.N. Stouboulos
e-mail: stouboulos@physics.auth.com

I.M. Kyprianidis
e-mail: imkypr@auth.gr

H.E. Nistazakis · G.S. Tombras
Faculty of Physics, Department of Electronics, Computers, Telecommunications
and Control, National and Kapodistrian University of Athens, GR-15784 Athens, Greece
e-mail: enistaz@phys.uoa.gr

G.S. Tombras
e-mail: gtombras@phys.uoa.gr

S. Vaidyanathan and C. Volos (eds.), *Advances and Applications in Chaotic Systems*, Studies in Computational Intelligence 636,
DOI 10.1007/978-3-319-30279-9_16

1 Introduction

For decades, the engineering undergraduate education in the area of circuit and design has been mainly focused in linear models. The reason is that linear system theory has been thoroughly developed and mathematical tools are available to analyze such systems. This philosophy has led many scientists and experimentalists to disregard many observed phenomena because linear system theory cannot explain them. Furthermore, in the last decades, there is a strong interest in exploring systems that display unusual complicated waveforms, commonly known as strange attractors. These attractors have been increasingly observed in several nonlinear deterministic systems.

Therefore, it is important for today's students to be exposed to these complex chaos phenomena. As a consequence, electronic oscillators generating chaotic waveforms are the most convenient tools for practical training of students taking courses on nonlinear dynamics and chaos [1, 4, 34–36, 40]. From a didactical point of view, the oscillator should not be higher than a third-order system and preferably autonomous. In more details, the system should have at least three state variables in order to be chaotic, according to Poincaré–Bendixson theorem, while the choice of an autonomous system is done because this will eliminate the need for an external driving input such as a sinusoidal source. Also, it should contain a single, simply defined and common nonlinear unit. Smooth, monotonous and unambiguous nonlinear functions are preferred to piecewise linear, non-monotonous and ambiguous ones. From a technical point of view the circuit should contain as few elements as possible. All the devices should be commercially available and cheap. The circuit should be easy to build and tune up. Also, the oscillator should operate at kilohertz frequencies to simplify the measuring procedures.

For this reason a number of chaotic oscillators have been described in literature [7]. However, the first nonlinear circuit, the Chua's oscillator, which is designed for exploring chaos in the laboratory, is the most suitable from all, because it shows a very rich dynamical behavior [17, 23]. Until now, a great variety of nonlinear phenomena, such as routes to chaos, stochastic resonance, signal amplification via chaos, 1/f noise generation, antimonotonicity, period adding, crisis phenomena etc., have been discovered in Chua's oscillator circuit.

In this chapter, a new laboratory setup of Chua's oscillator circuit is presented. The proposed realization is suitable for studying, in the laboratory, the design of a nonlinear circuit step by step. It is also very useful for illustrating in the oscilloscope the aforementioned phenomena that a nonlinear circuit can produce. So, this work presents a laboratory-based educational tool for teaching nonlinear circuits in courses related with nonlinear dynamics and chaos for undergraduate, postgraduate and Ph.D. students.

This chapter is organized as follows. In Sect. 2 the history of Chua's oscillator circuit as well as its detailed description are reported in details. The design procedure of the proposed laboratory setup of this circuit is presented step by step in Sect. 3. In Sect. 4 the experiments, which could be done with this laboratory-based educational tool are described and the obtained results are illustrated. Finally, Sect. 5 includes the conclusion remarks of this chapter and some thoughts for future work.

2 The Circuit of Chua's Oscillator

2.1 History

Chaos has preoccupied scientists since the late of 19th century. In more details, the study began in 1889 by Henry Poincaré [38], who was the first that observed this kind of dynamical behavior. However, chaos theory begins to take form in the second half of the 20th century after observations of the evolution of different physical systems [29, 30], while many scientists confirmed that chaos finds application in all disciplines, including biology, chemistry, mechanics, economics, etc. [15, 18, 27, 33, 34]. Chaos results from the exceptional sensitivity of a system to initial conditions, an effect which is popularly referred to as the "*Butterfly Effect*" [21].

From 1983 onwards there is a surge in the study of chaos in electronic circuits. It started with the design of the first autonomous chaotic electronic circuit by Chua in 1983 [8, 10, 11, 17]. In order to reach, however, the implementation of chaotic circuits, the design of nonlinear and negative resistors, including dynatron vacuum tubes [22], kallirotron, transitron, etc. preceded [6, 41]. The discovery of the transistor in 1947 led to the investigation of solid state negative resistors [3]. In 1958 Esaki discovered the tunnel diode and eight years later the Gunn diode [14, 42]. However, the discovery of operational amplifiers gave scientists the possibility to easily and quickly build nonlinear elements with piecewise-linear functions of current versus voltage.

The Chua circuit is the first nonlinear circuit designed to confirm experimentally the appearance of chaos. This circuit is considered a paradigm because it is a fact that Chua's circuit is the first chaotic system in which chaos was systematically derived, physically confirmed and rigorously analyzed. In more details, Chua's circuit constituted for three decades the simplest chaotic circuit and it consists of only five components: two capacitors, an inductor, a resistance and a nonlinear element, known as a Chua diode (Fig. 1). Although the capacitors, the inductor and the resistor are conventional parts, the nonlinear element can be designed and implemented in many ways, depending on the use that it is intended for.

Historically, the birth of the Chua circuit emerged in 1983 during the visit of Leon Chua at the lab of Takashi Matsumoto at the University of Waseda in Tokyo [10, 11]. At that time there was a strong desire to implement circuits, which would provide experimental confirmation of chaos, a phenomenon that until then had been

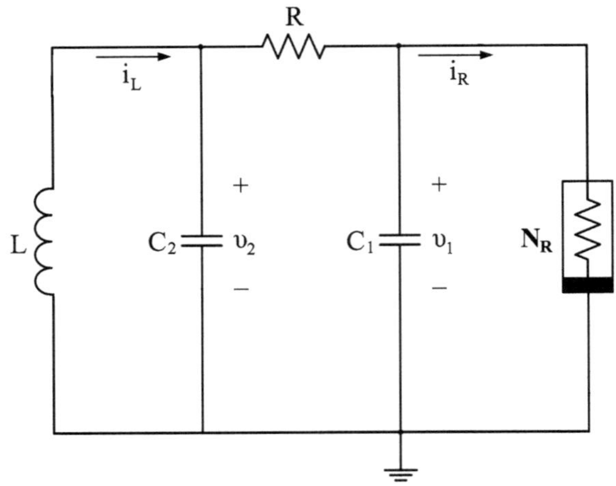

Fig. 1 The Chua circuit

observed and studied in simulation level only. This desire was for Chua the motivation to explore the possibility of drawing an autonomous chaotic circuit. However, the starting point of this effort was different from that which has been followed in previous years by various researchers, and rather than starting from known chaotic dynamical systems, such as the Lorenz or the Rössler system [29, 39], he attempted to design an autonomous nonlinear circuit that exhibits chaotic behavior.

For this reason, Chua decided that the circuit should have three equilibria, as few passive components as possible and a single nonlinear resistor with a split linear voltage-current characteristic. So, the characteristic of the nonlinear resistor had to consist of three sections, each with a negative slope. Also, because all the other components of the circuit were passive, the nonlinear resistor N_R had to be an active element, in order to ensure the instability of the equilibrium points.

This observation combined with the fact that the characteristic of the nonlinear resistor N_R should be a function controlled by voltage, as it was easier to be designed, led Chua to the design of the voltage-current characteristic of Fig. 2. However, as all physical resistors are ultimately passive elements, which means that for large values of voltage and current the power consumed, $P = i\upsilon$, is positive, Chua settled on the characteristic of Fig. 3, where the two additional outer sections do not affect the circuit equilibria but ensure the above requirements.

After the design of the circuit, Chua cooperated with Matsumoto and Komuro [9] and chose the values of each element via simulations while taking into consideration the fact that the line load should intersect with the characteristic of the nonlinear resistor N_R at three points. After this procedure they eventually discovered that the circuit can produce double-scroll chaotic attractors, thus achieving the mission for which it was designed.

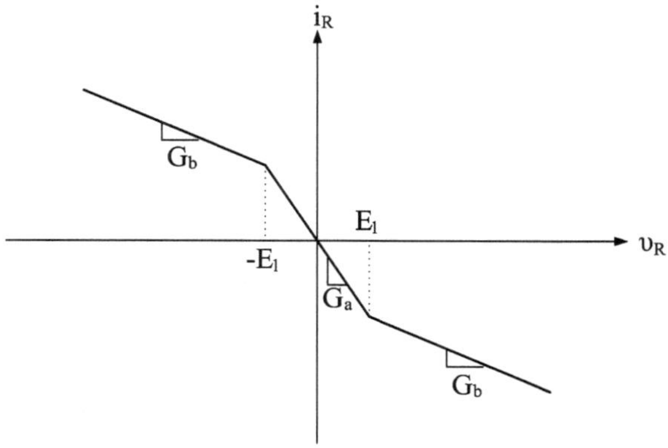

Fig. 2 The current-voltage characteristic of the non-linear resistor N_R of the circuit, as originally designed by Chua

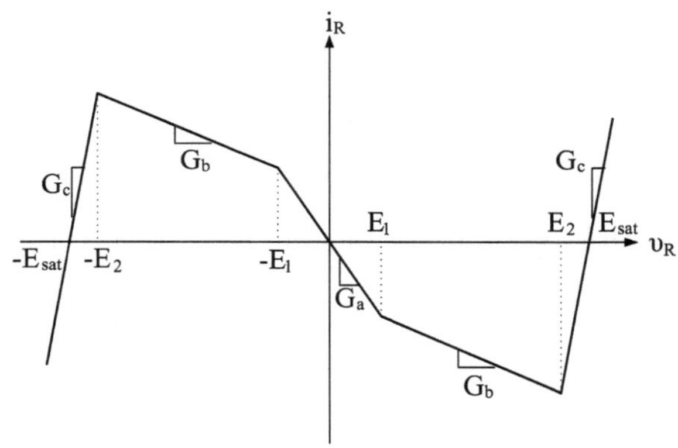

Fig. 3 The current-voltage characteristic of the non-linear resistor N_R of the circuit, as the final result

2.2 *The Description of Chua's Oscillator*

The Chua circuit of Fig. 1 is described mathematically by the following set of differential equations:

$$\begin{cases} C_1 \frac{d\upsilon_1}{dt} = G\,(\upsilon_2 - \upsilon_1) - f(\upsilon_1) \\ C_2 \frac{d\upsilon_2}{dt} = G\,(\upsilon_1 - \upsilon_2) + i_L \\ L \frac{di_L}{dt} = -\upsilon_2 \end{cases} \tag{1}$$

where $G = 1/R$ and $f(.) = i_R$ is the voltage-current function of the nonlinear element.

However, given the small parasitic ohmic resistance R_0 of the inductor, the so-called Chua's oscillator circuit arises (Fig. 4).

The Chua's oscillator circuit is described respectively by the following system of differential equations:

Fig. 4 The Chua's oscillator circuit

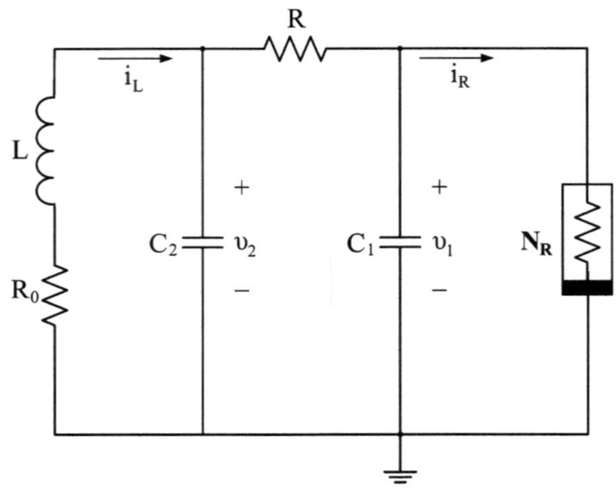

$$\begin{cases} C_1\frac{d\upsilon_1}{dt} = G\,(\upsilon_2 - \upsilon_1) - f(\upsilon_1) \\ C_2\frac{d\upsilon_2}{dt} = G\,(\upsilon_1 - \upsilon_2) + i_L \\ L\frac{di_L}{dt} = -\upsilon_2 - R_0 i_L \end{cases} \tag{2}$$

or, alternatively, by the following normalized system:

$$\begin{cases} \frac{dx}{d\tau} = \alpha\,[y - x - f(x)] \\ \frac{dy}{d\tau} = x - y + z \\ \frac{dz}{d\tau} = -\beta y - \gamma z \end{cases} \tag{3}$$

where constitutive variables $x = \upsilon_1/E_1$ and $y = \upsilon_2/E_1$ represent the voltages across the capacitors C_1 and C_2, while the variable $z = (i_L\,R)/E_1$ represents the current flowing through the inductor L. The dimensionless time τ is defined as $\tau = t/RC_2$ and the normalized parameters α, β and γ as: $\alpha = C_2/C_1$, $\beta = R^2C_2/L$ and $\gamma = RR_0C_2/L$, respectively.

In addition, the dimensionless form of the nonlinear function $f(x)$ of the N_R Chua diode is given by the following equation:

$$\begin{aligned} f(x) = m_c x + 0.5\,(m_a - m_b)\,(|x + 1| - |x - 1|) \\ + 0.5\,(m_b - m_c)\,(|x + E_2/E_1| - |x - E_2/E_1|) \end{aligned} \tag{4}$$

where $m_a = RG_a$, $m_b = RG_b$ and $m_c = RG_c$.

As it is mentioned, the circuit of Chua's oscillator can produce double-scroll attractors (Fig. 5), which are an indication of the generated chaotic behavior of the circuit.

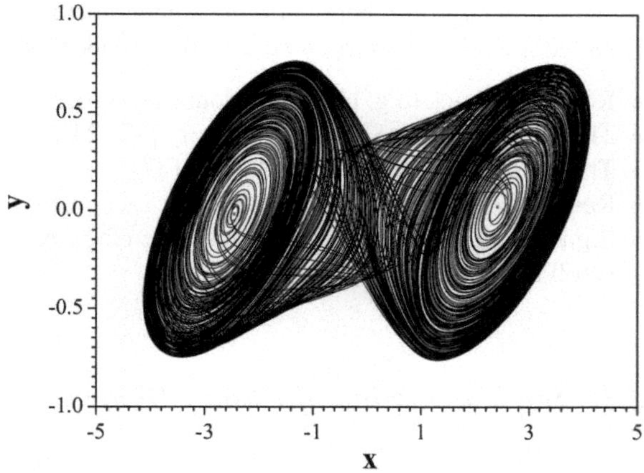

Fig. 5 The double-scroll chaotic attractor generated by the Chua's oscillator circuit

More specifically, the Chua oscillator circuit has three equilibria. The first of these is at the origin, $(x, y, z) = (0, 0, 0)$, while the other two are at the points:

$$P^{+}(x, y, z) = \left(-\frac{1}{G+G_b}(G_a - G_b)E_1, 0, \frac{G}{G+G_b}(G_a - G_b)E_1\right)$$

and

$$P^{-}(x, y, z) = \left(\frac{1}{G+G_b}(G_a - G_b)E_1, 0, -\frac{G}{G+G_b}(G_a - G_b)E_1\right),$$

where $G = \frac{1}{R+R_0}$.

The equilibrium points P^+ and P^- are located at the centers of the holes of the created scrolls (Fig. 5). A typical orbit that produces double-scroll chaotic attractors rotates around one of the two equilibrium points in a random way and when the orbit moves away from it, it is attracted by the other equilibrium point and repeats the same procedure producing the double-scroll chaotic attractor of Fig. 5.

The Chua oscillator circuit is the most typical representative of a larger class of nonlinear circuits, known as Chua's Circuit Family. A circuit, described by the constitutive equation $\dot{\boldsymbol{x}} = \boldsymbol{f}(\boldsymbol{x})$ belongs to this class if and only if [17]:

1. The function $\boldsymbol{f}(\cdot)$ is continuous,
2. The function $\boldsymbol{f}(\cdot)$ is characterized by odd symmetry, i.e. $\boldsymbol{f}(-\boldsymbol{x}) = -\boldsymbol{f}(\boldsymbol{x})$ and
3. The state space can be divided by two parallel planes U_1 and U_{-1} into three sections, D_1, D_0 and D_{-1}.

According to Chua:

"*The Chua oscillator circuit is structurally the simplest and dynamically the most complex member of Chua's circuit family.*" [8]

To date, a large number of nonlinear phenomena associated with chaos theory, has been recorded in literature, produced by members of Chua's circuit family, such as:

- Routes to chaos (e.g. by period doubling) [16].
- The phenomenon of antimonotonicity [13, 24].
- The phenomenon of period addition [37].
- Resonance phenomena [2].
- Signal amplification through chaos, when the circuit operates in area of the simple scroll attractor [20].

2.3 Implementation of Chua's Diode

The Chua's oscillator circuit can be circuitally implemented in many ways. As all linear elements (resistors, capacitors, inductor), that comprise the circuit are ready to find in commerce, the interest focuses on how to implement the nonlinear element N_R. Today various realizations based on the use of operational amplifiers [23, 43] diode [31], transistor [32] and transconductance amplifiers [12] have been proposed.

In this work, the typical voltage-current nonlinear element (Chua diode) of the circuit of Chua's oscillator derived from the parallel connection of two negative resistance converters N_{R1} and N_{R2}. The first N_{R1} consists of three sections with slopes G_{a1} and G_{b1}, while the breakpoints are $\pm E_2$, as shown in Fig. 6b. Correspondingly, the second negative resistance N_{R2} converter consists of three sections with slopes G_{a2} and G_{b2}, and the breakpoints are $\pm E_1$, as illustrated in Fig. 6a. The final characteristic of the diode Chua, resulting from the parallel connection of the two negative resistance converters N_{R1} and N_{R2}, consists of five sections with slopes G_a, G_b and G_c, while the two breakpoints are $\pm E_1$and $\pm E_2$ (Fig. 6c).

Figure 7 shows the complete Chua's oscillator circuit, where the nonlinear element N_R has been implemented using two parallel-connected negative resistance converters N_{R1} and N_{R2}, implemented with op-amps. Assuming that $R_1 = R_2$, the slopes of the characteristic of N_{R1} are $G_{b1} = 1/R_1$ and $G_{\mathrm{a}1} = -1/R_3$, while the breakpoints are $E_2 = \pm\frac{R_3}{R_2+R_3}E_{sat}$. Respectively, assuming $R_4 = R_5$ the slopes of characteristic of N_{R2} is $G_{\mathrm{b}2} = 1/R_4$ and $G_{\mathrm{a}2} = -1/R_6$, while the breakpoints are $E_1 = \pm\frac{R_6}{R_5+R_6}E_{sat}$. Therefore, as shown graphically in Fig. 12, we get: $G_{\mathrm{a}} = G_{\mathrm{a}1} + G_{\mathrm{a}2}$ and $G_{\mathrm{b}} = G_{\mathrm{a}1} + G_{\mathrm{b}2}$.

3 The Proposed Educational Tool Based on Chua's Oscillator Circuit

In this section the proposed laboratory setup of Chua's oscillator circuit for studying, in the laboratory, is presented. This laboratory setup consists of two independent circuits (A) and (B) (Fig. 8), which when they are connected, as it will be described

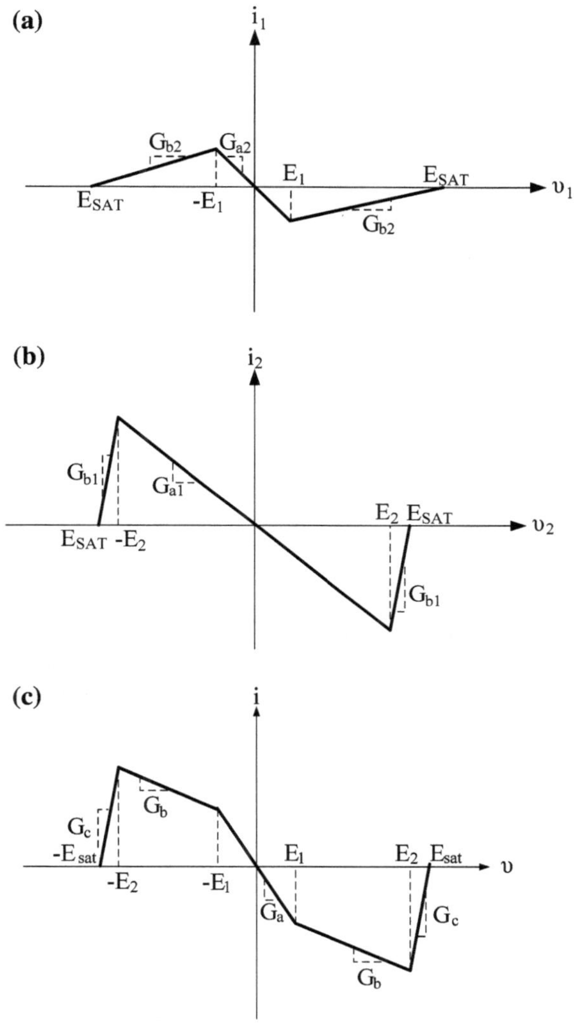

Fig. 6 Finding the voltage-current characteristic of the Chua diode graphically. **a** The υ-i characteristic of N_{R2}, **b** the υ-i characteristic of N_{R1} and **c** the υ-i characteristic of the parallel combination of N_{R1} and N_{R2}

in details below, provide the user with the Chua's oscillator circuit. These two circuits (A) and (B) are inside plastic transparent boxes, in order to be more robust for teaching purposes in undergraduate or postgraduate laboratories.

The circuit A carries out the portion of the Chua's oscillator circuit, which includes the elements L, R_0, and C_1, while at the top of the device there are connectors for the connection of the variable resistor R and the variable capacitor C_2, through which a greater degree of freedom regarding the regulation of the circuit configuration and hence the dynamic behavior can be achieved.

The circuit B implements the nonlinear resistor (Chua diode) of the Chua's oscillator circuit. For its implementation two potentiometers R_{3P} and R_{6P} are used, as

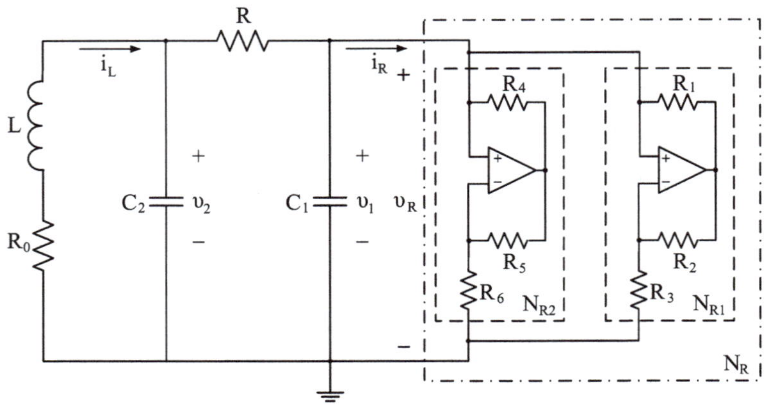

Fig. 7 Chua's oscillator circuit as implemented using operational amplifiers

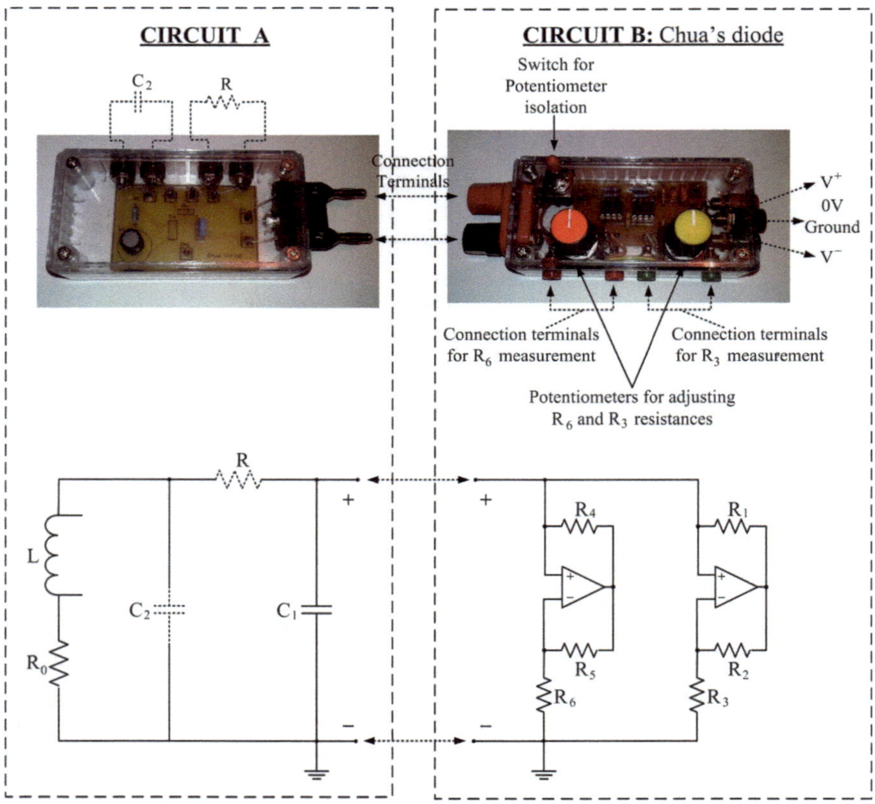

Fig. 8 The circuits (**A**) and (**B**) as parts of the Chua's oscillator circuit

Table 1 Values of the components of circuit A

Components	Values
L	18 mH
R_0	10 Ω
R	Variable resistor
C_1	8.2 nF
C_2	Variable capacitor

Table 2 Values of the components of circuit B

Components	Values
R_1	1.5 kΩ
R_2	1.5 kΩ
R_3	2 kΩ ($+R_{3P} = 1\,\text{k}\Omega \rightarrow$ potentiometer)
R_4	22 kΩ
R_5	22 kΩ
R_6	3 kΩ ($+R_{6P} = 1\,\text{k}\Omega \rightarrow$ potentiometer)
Operational amplifier	OP07

shown in Fig. 8, for adjusting the resistance values of $R_3 = 2\,\text{k}\Omega$ ($+R_{3P} = 1\,\text{k}\Omega \rightarrow$ potentiometer) and $R_6 = 3\,\text{k}\Omega$ ($+R_{6P} = 1\,\text{k}\Omega \rightarrow$ potentiometer). At the top there is a switch, which when is in position off (right) isolates the two potentiometers from the circuit, in order to be adjusted to the chosen values. Therefore, through the setting of these two potentiometers a greater degree of freedom in terms of configuration of the i-υ characteristic of the Chua's diode and therefore the dynamic behavior of the Chua's oscillator circuit, is achieved. Moreover, in the right part of the device there are connector terminals for the positive ($V+$), the negative (V^-) supply and the ground. Finally, the two LEDs, when lit, confirm that the operating voltage of the device is in the permissible operating limits of the circuit.

In Tables 1 and 2 the values of the elements of the two circuits (A) and (B) are displayed.

4 Experimental Procedure

In this section the experimental results of the use of Chua's oscillator circuit are presented. In more details, the experimental characteristic of the nonlinear resistor with two different methods as well as the experimental observation of circuit's dynamic behavior are studied by using the proposed laboratory setup of the Chua's oscillator circuit.

Table 3 Values of potentiometers of the non-linear resistor circuit

Potentiometer	Value (Ω)
R_{3P}	200
R_{6P}	300

4.1 Experimental Characteristic of the Nonlinear Resistor of Chua's Oscillator Circuit

In order to find the experimental characteristic of the nonlinear resistor of the Chua oscillator circuit, circuit (B) is used. For this purpose we followed two different methods. The first method allows the supervisory observation of the desired characteristic by using the oscilloscope, while the second method is the analytical finding of the experimental characteristic through measurement of voltages and currents (υ_R, i_R) of the nonlinear resistor with the help of the multimeter. For the aim of this work, the potentiometers are adjusted to the values as shown in Table 3.

4.1.1 Supervisory Observation of Circuit's Characteristic

For the supervisory observation of the experimental characteristic of the nonlinear resistor of Chua oscillator circuit, we used the circuit of Fig. 9, in which the resistor values are determined in the previous measurement.

The voltage υ_S is a triangular low-frequency voltage signal (200 Hz) with amplitude $V_S = 10$ V, which is applied via the resistor R_S to the nonlinear resistor N_R, which is part of the circuit B. Also, the supplied voltages of op-amps are ± 10 V.

The resistor R_S is used for measuring the current i_R flowing through the nonlinear resistor when a voltage υ_R is applied at its ends. A suitable for this purpose resistance value of R_S is 100 Ω. Then the voltage across the resistor R_S is:

$$\upsilon_{Rs} = -i_R R_s = -100\, i_R \tag{5}$$

Therefore, with the setup of Fig. 9 the voltage υ_{R_S} can be displayed on an oscilloscope, and indirectly the current i_R versus the voltage υ_R (Fig. 10). For this purpose, the channel Y of the oscilloscope is connected to the ends of the resistor R_S and the signal is inverted, while the channel X is connected to the ends of the nonlinear resistor, by setting the oscilloscope at XY mode.

4.1.2 Analytical Measurement of Circuit's Characteristic

At this section the analytical finding of the experimental characteristic of the nonlinear resistor through the measurement of voltages and currents (υ_R, i_R) using the multimeter, is presented. For this purpose, the circuit of Fig. 11 is implemented,

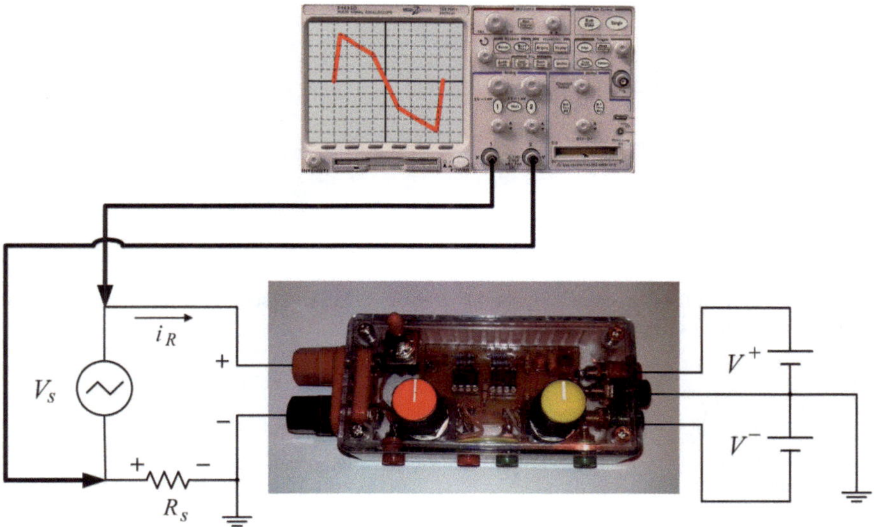

Fig. 9 The circuit for the supervisory observation of the experimental characteristic of the nonlinear resistor of the Chua's oscillator circuit

Fig. 10 The experimental i-υ characteristic of the nonlinear resistor N_R (*horizontal axes* 2 V/div, *vertical axes* 2 V/div)

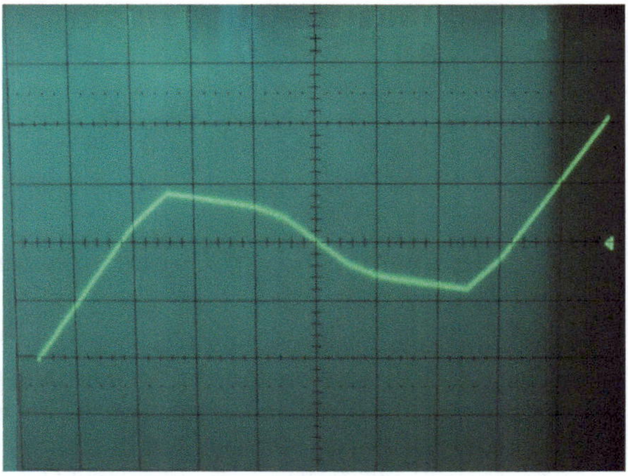

where the nonlinear resistor N_R is supplied by a DC voltage source $\upsilon_R = \upsilon_{DC}$, producing a current i_R. In Fig. 12 the i-υ characteristic of the nonlinear resistor N_R, which is obtained with this method, is shown. With this procedure, the exact values of the experimental breakpoints, saturation voltages and slopes of the five-segments of the i-υ characteristic are measured as they are presented in Table 4.

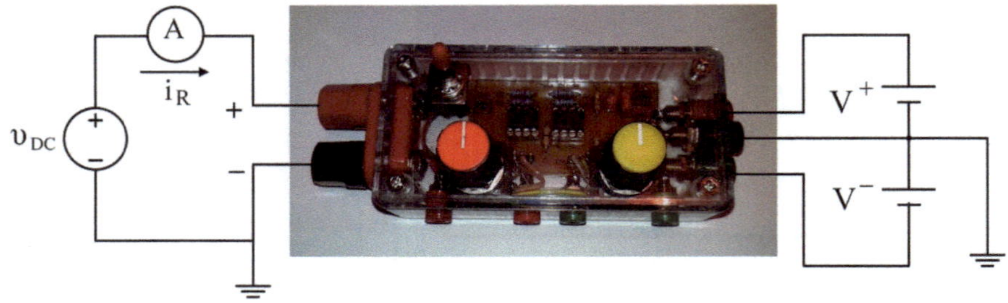

Fig. 11 The circuit for finding in analytical way the i-υ characteristic of the nonlinear resistor of the Chua's oscillator circuit

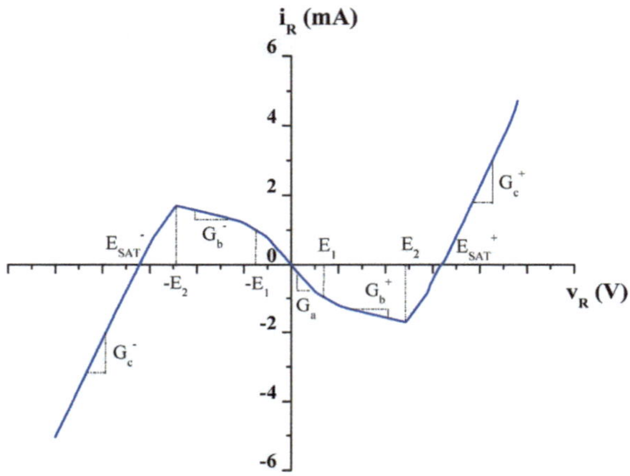

Fig. 12 The i-υ characteristic of the nonlinear resistor N_R, through the measurement of voltages and currents using the multimeter

4.2 *Experimental Study of Circuit's Dynamics*

In order to study the dynamics of the Chua's oscillator circuit, the two circuits A and B are connected, while a resistor box, for adjusting the resistor R and a capacitor box, for adjusting the capacitor C_2, are used, as shown in Fig. 13. Also the channels X and Y of the oscilloscope are connected to the ends of the capacitors C_2 and C_1, respectively, so that the voltages υ_2 and υ_1 could be displayed on the oscilloscope.

Next, the two possible way of studying the circuit's dynamics is presented in details. Keeping each time the value of the element R (or C_2) constant and varying the value of the other C_2 (or R), the phenomena related to the dynamic behavior of the circuit (e.g., path to the chaos through period doubling, attractors crisis, antimonotonicity, coexistence of attractors) are observed at the oscilloscope.

Table 4 Experimental breakpoints, saturation voltages and slopes of the five-segments of the i-υ characteristic

Parameters	Values
$-E_1$	−1.2116 V
E_1	1.2011 V
$-E_2$	−4.84 V
E_2	4.84 V
$-E_{sat}$	−6.43 V
E_{sat}	6.43 V
G_a	−0.7849 mS
G_b^+	−0.1650 mS
G_b^-	−0.1659 mS
G_c^-	1.3694 mS
G_c^+	1.3729 mS

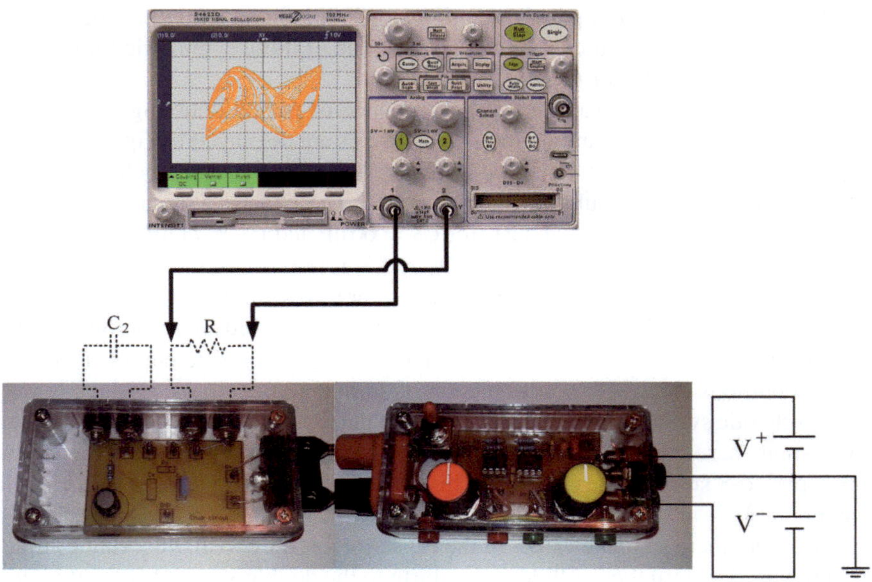

Fig. 13 Setup for the experimental study of the dynamic behavior of the Chua's oscillator circuit

4.2.1 Chua's Oscillator Dynamics by Varying the Capacitor C_2

In the first case, the resistor R is kept constant ($R = 1.5\,\mathrm{k}\Omega$), while the capacitor C_2 play the role of the bifurcation parameter. Starting from low values of C_2 ($C_2 = 20\,\mathrm{nF}$) the system is in the stable equilibrium point P^+ or P^- (Fig. 14). However, by increasing the value of C_2 the first bifurcation that can be observed is the loss of

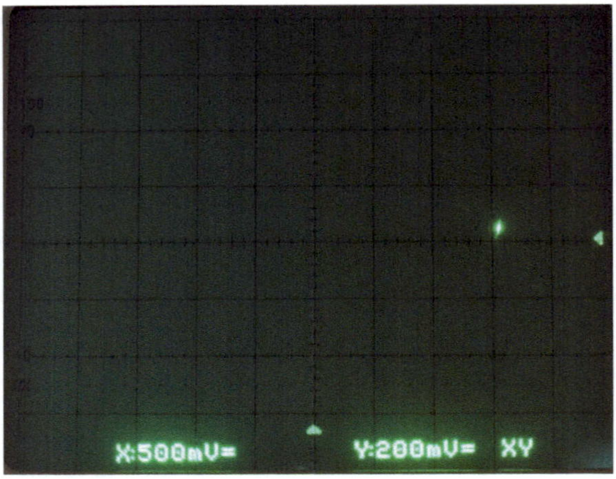

Fig. 14 Experimental phase portrait of υ_2 versus υ_1, for $R = 1.5\,\text{k}\Omega$ and $C_2 = 20\,\text{nF}$. (Stable equilibrium point P^+) *Horizontal axes* 0.5 V/div, *vertical axes* 0.2 V/div

stability of the equilibrium point through a Hopf bifurcation and, thus, the birth of two symmetric stable limit cycles of period-1. One of these symmetric limit cycles can be shown in Fig. 15 for $C_2 = 40\,\text{nF}$.

Increasing further the value of C_2, a sequence of period-doubling bifurcations can be observed. Period-2 and period-4 limit cycles are displayed in Figs. 16 and 17, respectively. This sequence of period-doubling bifurcations leads to chaos through the well-known period-doubling route to chaos. The chaotic attractor that can be observed is shown in Figs. 18 and 21, for $C_2 = 66\,\text{nF}$ and $C_2 = 71\,\text{nF}$. This attractor is confined to the two regions D_1 and D_0 and is referred to either as single-scroll chaotic attractor or Rössler screw-type attractor for its resemblance to the structure of the Rössler attractor. Since the circuit is symmetric, a mirrored single-scroll attractor lies in the regions D_{-1} and D_0 (Fig. 19), which is produced for different set of circuit's initial conditions. This is the phenomenon of attractors' coexistence, which is very usual in chaotic systems. The region of values of the capacitor C_2, in which the circuit is in a single-scroll chaotic state, interrupted by a small window, in which the circuit is in a period-3 steady state (Fig. 20). This phenomenon is very common in chaotic systems [28].

For further increasing the values of the capacitor C_2, the two distinct single attractors grow in size until they collide giving birth to the double-scroll chaotic attractor, which spans all the three regions D_1, D_0 and D_{-1}, as shown in Figs. 22 and 23 for $C_2 = 100\,\text{nF}$ and $C_2 = 200\,\text{nF}$, respectively.

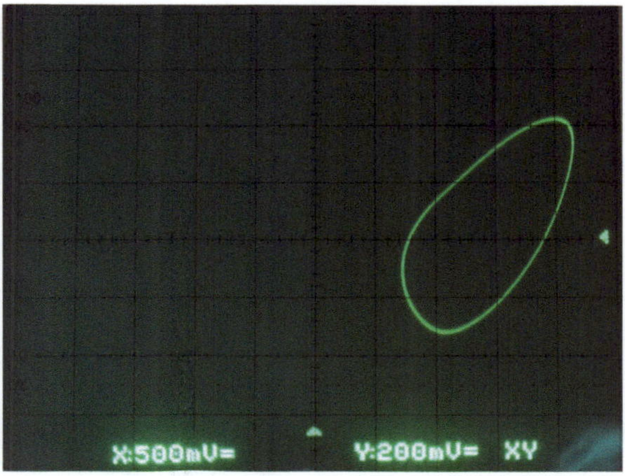

Fig. 15 Experimental phase portrait of υ_2 versus υ_1, for $R = 1.5\,\text{k}\Omega$ and $C_2 = 40\,\text{nF}$. (Period-1 steady state) *Horizontal axes* 0.5 V/div, *vertical axes* 0.2 V/div

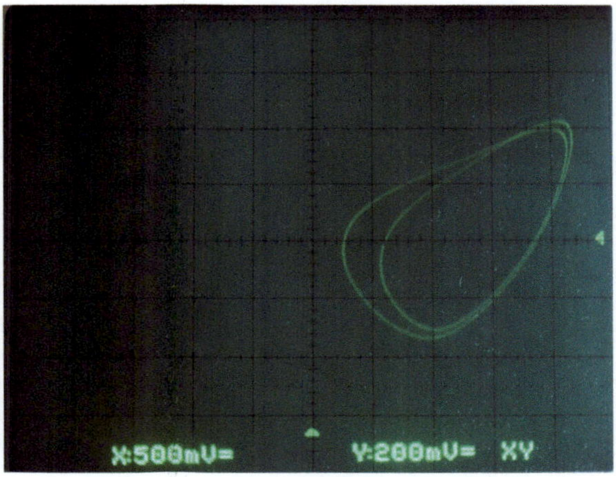

Fig. 16 Experimental phase portrait of υ_2 versus υ_1, for $R = 1.5\,\text{k}\Omega$ and $C_2 = 60\,\text{nF}$. (Period-2 steady state) *Horizontal axes* 0.5 V/div, *vertical axes* 0.2 V/div

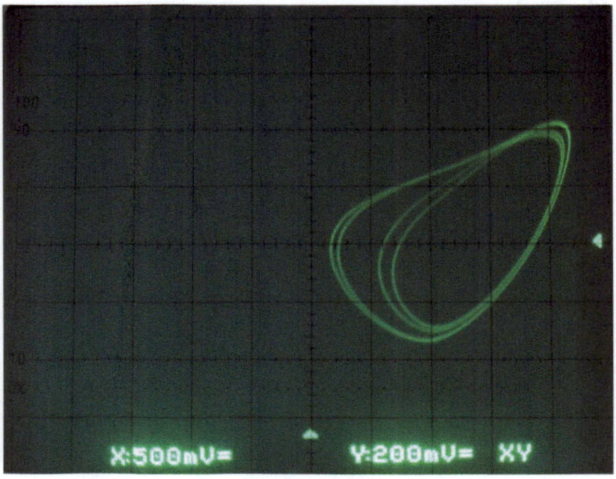

Fig. 17 Experimental phase portrait of υ_2 versus υ_1, for $R = 1.5\,\text{k}\Omega$ and $C_2 = 64\,\text{nF}$. (Period-4 steady state) *Horizontal axes* 0.5 V/div, *vertical axes* 0.2 V/div

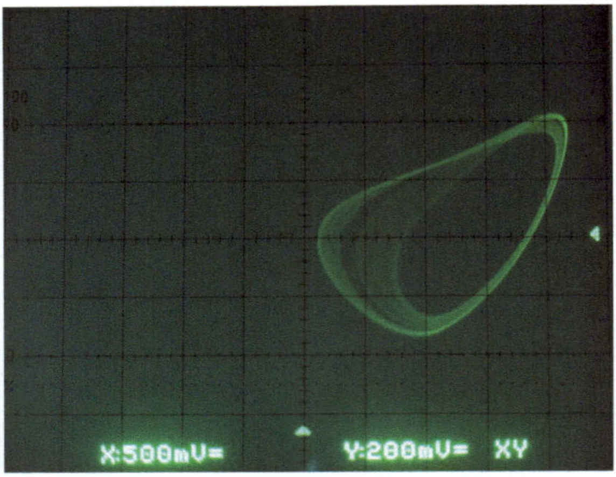

Fig. 18 Experimental phase portrait of v_2 versus v_1, for $R = 1.5\,\mathrm{k}\Omega$ and $C_2 = 66\,\mathrm{nF}$. (Single-scroll chaotic attractor) *Horizontal axes* 0.5 V/div, *vertical axes* 0.2 V/div

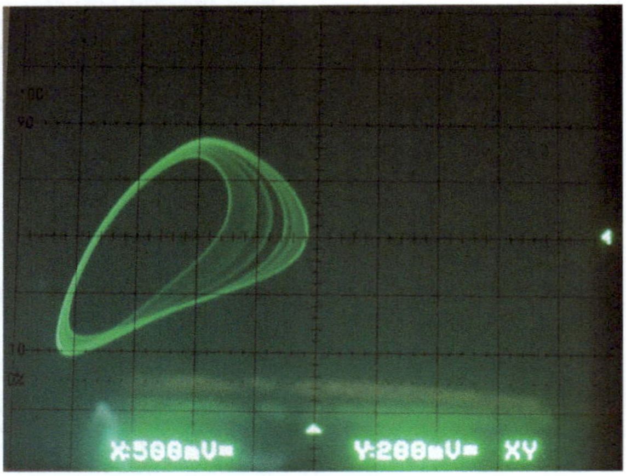

Fig. 19 Experimental phase portrait of v_2 versus v_1, for $R = 1.5\,\mathrm{k}\Omega$ and $C_2 = 66\,\mathrm{nF}$. (Single-scroll chaotic attractor) *Horizontal axes* 0.5 V/div, *vertical axes* 0.2 V/div

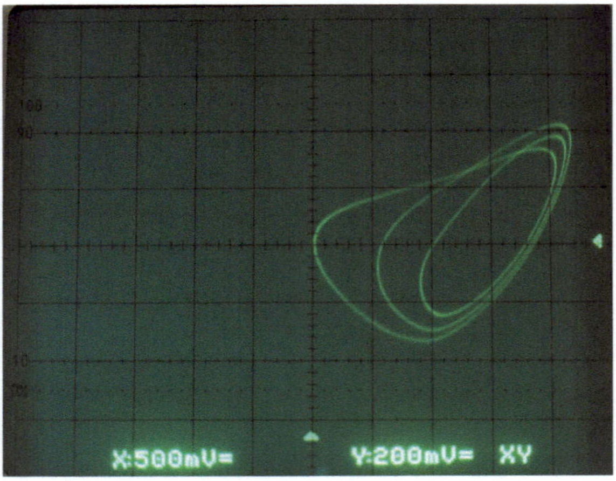

Fig. 20 Experimental phase portrait of v_2 versus v_1, for $R = 1.5\,\mathrm{k}\Omega$ and $C_2 = 70\,\mathrm{nF}$. (Period-3 steady state) *Horizontal axes* 0.5 V/div, *vertical axes* 0.2 V/div

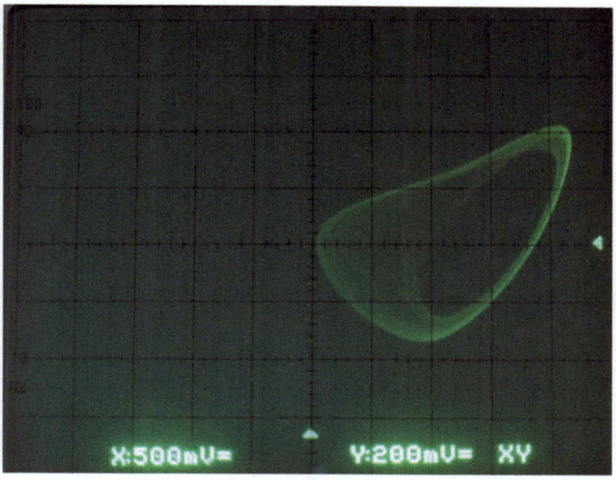

Fig. 21 Experimental phase portrait of υ_2 versus υ_1, for $R = 1.5\,\mathrm{k}\Omega$ and $C_2 = 71\,\mathrm{nF}$. (Single-scroll chaotic attractor) *Horizontal axes* 0.5 V/div, *vertical axes* 0.2 V/div

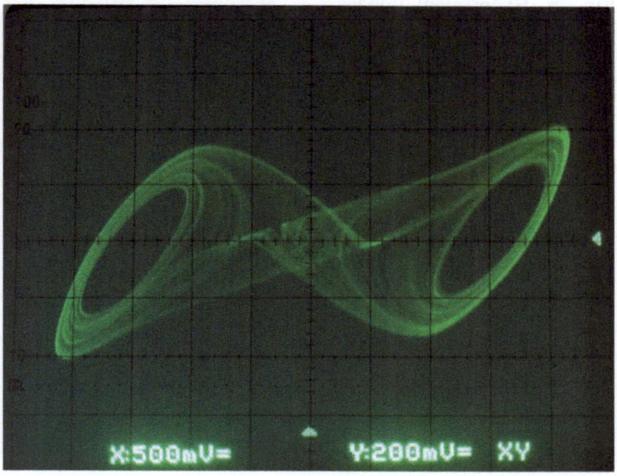

Fig. 22 Experimental phase portrait of υ_2 versus υ_1, for $R = 1.5\,\mathrm{k}\Omega$ and $C_2 = 100\,\mathrm{nF}$. (Double-scroll chaotic attractor) *Horizontal axes* 0.5 V/div, *vertical axes* 0.2 V/div

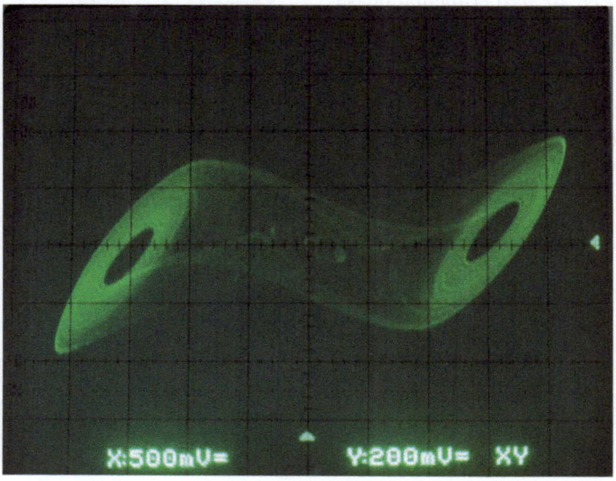

Fig. 23 Experimental phase portrait of υ_2 versus υ_1, for $R = 1.5\,\mathrm{k}\Omega$ and $C_2 = 200\,\mathrm{nF}$. (Double-scroll chaotic attractor) *Horizontal axes* 0.5 V/div, *vertical axes* 0.2 V/div

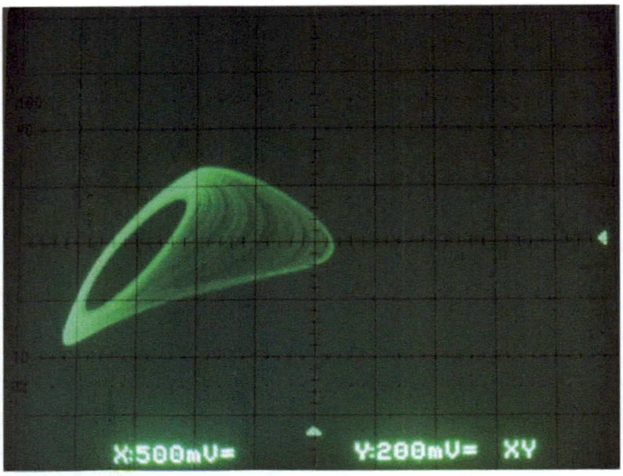

Fig. 24 Experimental phase portrait of υ_2 versus υ_1, for $R = 1.5\,\mathrm{k}\Omega$ and $C_2 = 210\,\mathrm{nF}$. (Single-scroll chaotic attractor) *Horizontal axes* 0.5 V/div, *vertical axes* 0.2 V/div

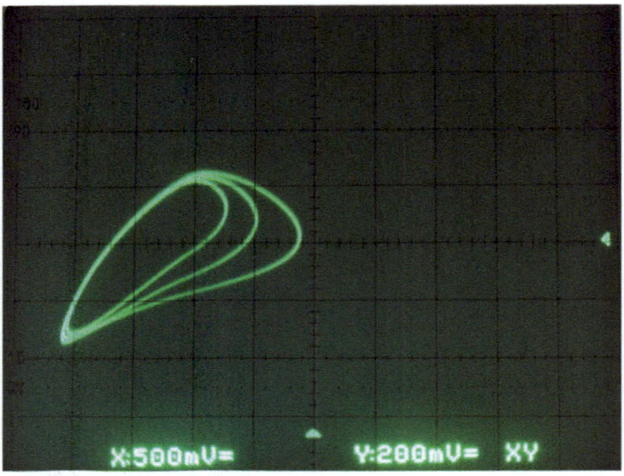

Fig. 25 Experimental phase portrait of υ_2 versus υ_1, for $R = 1.5\,\mathrm{k}\Omega$ and $C_2 = 230\,\mathrm{nF}$. (Period-3 steady state) *Horizontal axes* 0.5 V/div, *vertical axes* 0.2 V/div

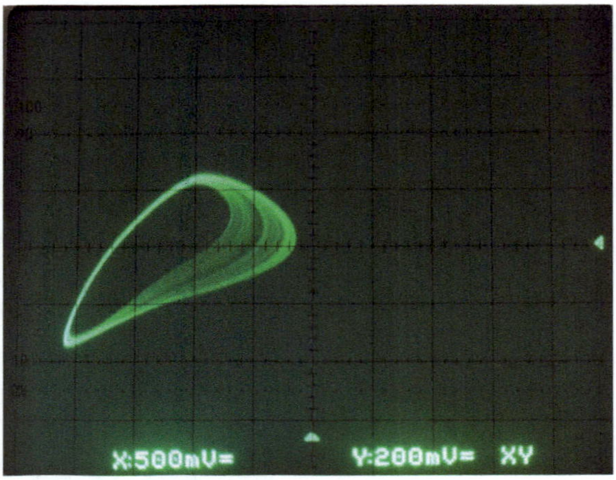

Fig. 26 Experimental phase portrait of υ_2 versus υ_1, for $R = 1.5\,\mathrm{k}\Omega$ and $C_2 = 232\,\mathrm{nF}$. (Single-scroll chaotic attractor) *Horizontal axes* 0.5 V/div, *vertical axes* 0.2 V/div

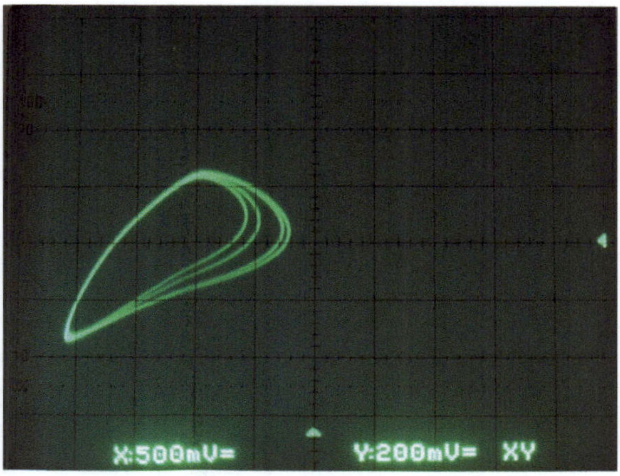

Fig. 27 Experimental phase portrait of υ_2 versus υ_1, for $R = 1.5\,\mathrm{k\Omega}$ and $C_2 = 238\,\mathrm{nF}$. (Period-4 steady state) *Horizontal axes* 0.5 V/div, *vertical axes* 0.2 V/div

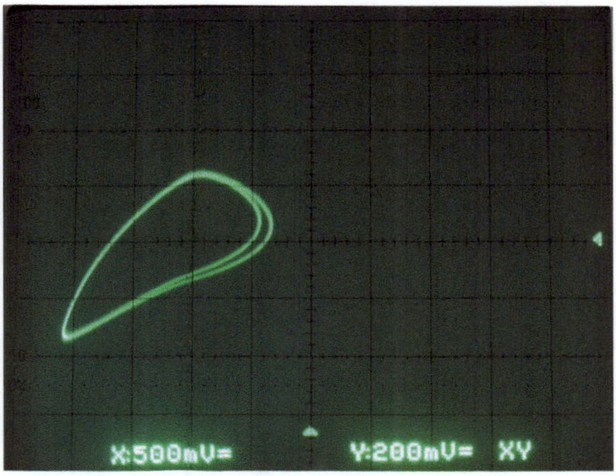

Fig. 28 Experimental phase portrait of υ_2 versus υ_1, for $R = 1.5\,\mathrm{k\Omega}$ and $C_2 = 250\,\mathrm{nF}$. (Period-2 steady state) *Horizontal axes* 0.5 V/div, *vertical axes* 0.2 V/div

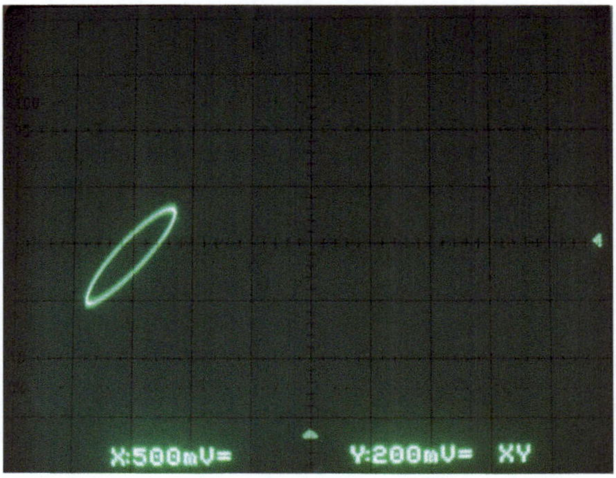

Fig. 29 Experimental phase portrait of υ_2 versus υ_1, for $R = 1.5\,\mathrm{k\Omega}$ and $C_2 = 260\,\mathrm{nF}$. (Period-1 steady state) *Horizontal axes* 0.5 V/div, *vertical axes* 0.2 V/div

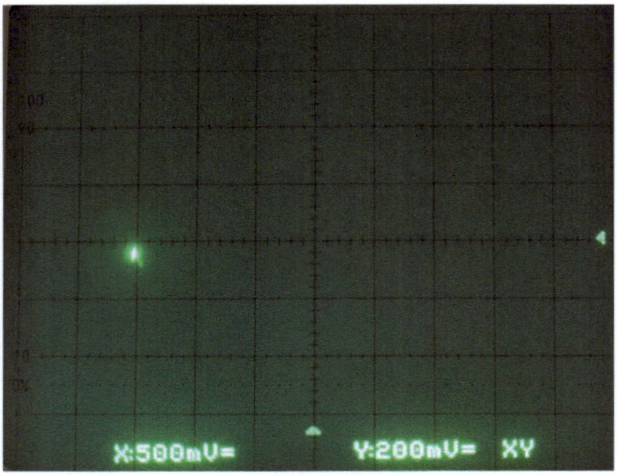

Fig. 30 Experimental phase portrait of υ_2 versus υ_1, for $R = 1.5\,\text{k}\Omega$ and $C_2 = 270\,\text{nF}$. (Stable equilibrium point P^+) *Horizontal axes* 0.5 V/div, *vertical axes* 0.2 V/div

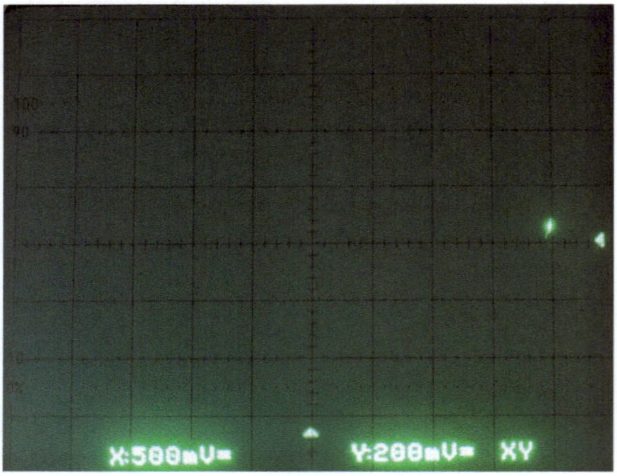

Fig. 31 Experimental phase portrait of υ_2 versus υ_1, for $C_2 = 80\,\text{nF}$ and $R = 1.65\,\text{k}\Omega$. (Stable equilibrium point P^+) *Horizontal axes* 0.5 V/div, *vertical axes* 0.2 V/div

For values of C_2 greater than 200 nF a reverse route from a double-scroll chaotic attractor to a single-scroll chaotic attractor (Figs. 24 and 26), which is interrupted again by a period-3 steady state (Fig. 25) and finally to a period-1 steady state (Fig. 29) through a reverse sequence of period-doubling (Figs. 27 and 28) is observed. In Fig. 30, for $C_2 = 270\,\text{nF}$ the circuit results to a stable equilibrium point.

All this route of system's dynamic behavior, by following the forward and reverse period doubling sequences, e.g. period-1 $\rightarrow$ period-2 $\rightarrow$ period-4 $\rightarrow \cdots \rightarrow$ single-scroll attractor $\rightarrow$ double-scroll attractor $\rightarrow$ single-scroll attractor $\rightarrow \cdots \rightarrow$ period-4 $\rightarrow$ period-2 $\rightarrow$ period-1, is the well-known phenomenon of antimonotonicity [5, 25, 26].

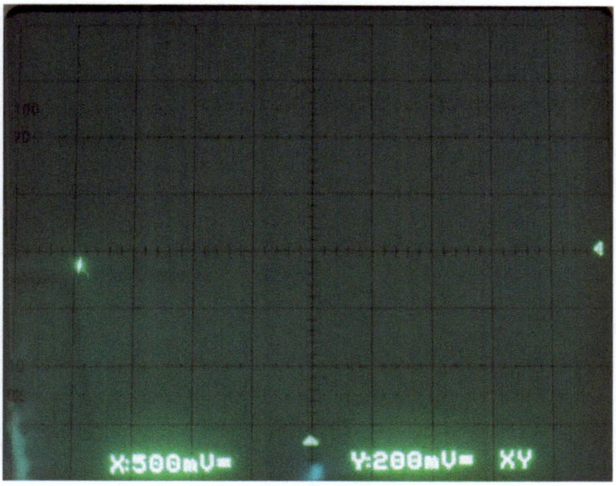

Fig. 32 Experimental phase portrait of υ_2 versus υ_1, for $C_2 = 80\,\text{nF}$ and $R = 1.65\,\text{k}\Omega$. (Stable equilibrium point P^-) *Horizontal axes* 0.5 V/div, *vertical axes* 0.2 V/div

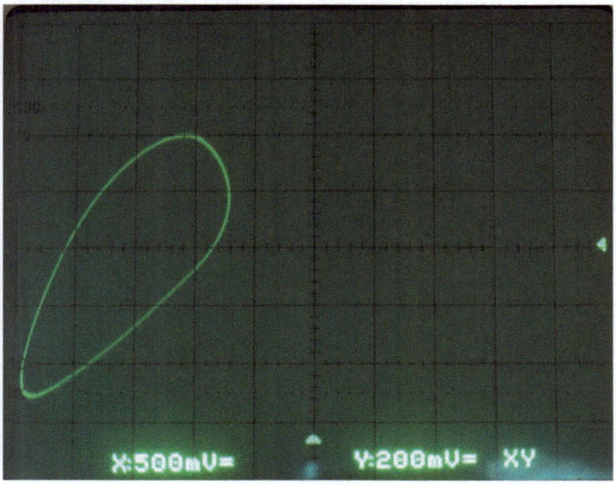

Fig. 33 Experimental phase portrait of υ_2 versus υ_1, for $C_2 = 80\,\text{nF}$ and $R = 1.6\,\text{k}\Omega$. (Period-1 steady state) *Horizontal axes* 0.5 V/div, *vertical axes* 0.2 V/div

4.2.2 Chua's Oscillator Dynamics by Varying the Resistor R

In the second case, the capacitor C_2 is kept constant ($C_2 = 80\,\text{nF}$), while the resistor plays the role of the bifurcation parameter. For large values of R ($R = 1.65\,\text{k}\Omega$) whether the system is in the stable equilibrium point P^+ or P^-, depends on circuit's initial conditions (Figs. 31 and 32). However, by decreasing the value of R the first bifurcation that can be observed is the loss of stability of the equilibrium point through a Hopf bifurcation, as in the previous case, and thus, the birth of a symmetric stable limit cycles of period-1 is produced (Fig. 33).

Decreasing further the value of R, a sequence of period-doubling bifurcations can be observed. Period-2 and period-4 limit cycles are displayed in Figs. 34 and 35, respectively. This sequence of period-doubling bifurcations leads to a single-scroll

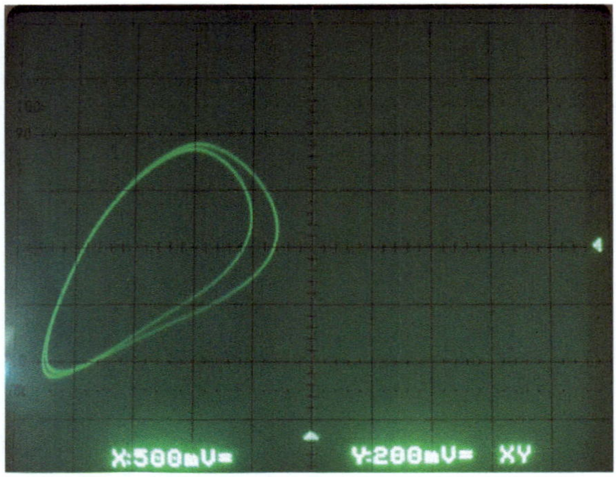

Fig. 34 Experimental phase portrait of υ_2 versus υ_1, for $C_2 = 80\,\text{nF}$ and $R = 1.54\,\text{k}\Omega$. (Period-2 steady state) *Horizontal axes* 0.5 V/div, *vertical axes* 0.2 V/div

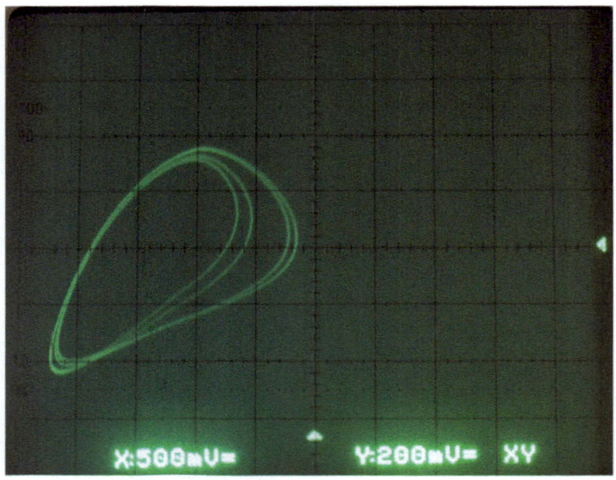

Fig. 35 Experimental phase portrait of υ_2 versus υ_1, for $C_2 = 80\,\text{nF}$ and $R = 1.526\,\text{k}\Omega$. (Period-4 steady state) *Horizontal axes* 0.5 V/div, *vertical axes* 0.2 V/div

chaotic attractor through the well-known period-doubling route to chaos (Fig. 36). For further decreasing of the value of R, a double-scroll chaotic attractor is observed, as shown in Figs. 37, 39 and 41 for $R = 1.49\,\text{k}\Omega$, $R = 1.42\,\text{k}\Omega$ and $R = 1.35\,\text{k}\Omega$, respectively. This region of values of R, in which the circuit is in a double-scroll chaotic attractor's behavior, interrupted by windows, in which the circuit is in symmetric periodic behavior (Fig. 38) or in asymmetric periodic behavior (Fig. 40).

Finally, the circuit, for further decreasing the value of R, results to a stable external limit cycle of period-1 (Fig. 42). This behavior coexists with the double-scroll chaotic attractor and is a consequence of the form of i-υ characteristic of the nonlinear resistor N_R. In fact, when the three segment nonlinearity of Fig. 2 is considered, for large initial conditions the behavior of the circuit may be unstable, which is clearly not the case of a real circuit. However, when the five segment nonlinearity of Fig. 3 is taken into account, it can be demonstrated that the double-scroll attractor coexists

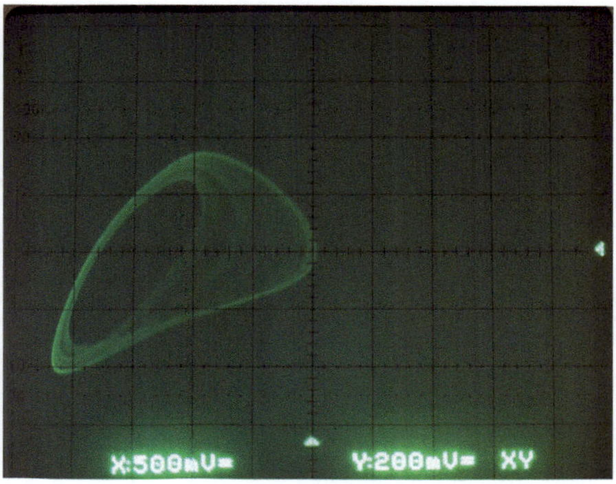

Fig. 36 Experimental phase portrait of υ_2 versus υ_1, for $C_2 = 80\,\text{nF}$ and $R = 1.51\,\text{k}\Omega$. (Single-scroll chaotic attractor) *Horizontal axes* 0.5 V/div, *vertical axes* 0.2 V/div

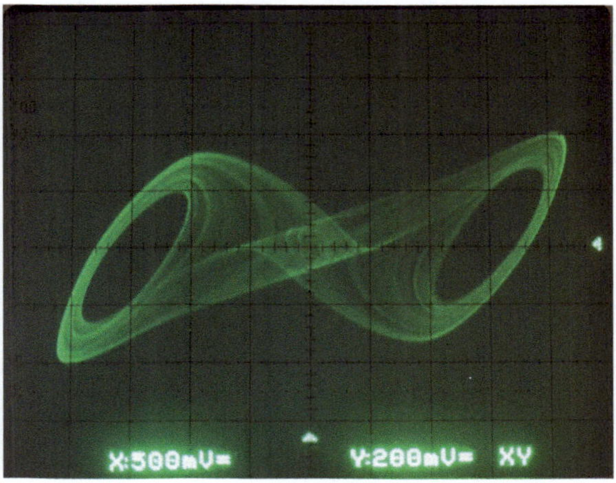

Fig. 37 Experimental phase portrait of υ_2 versus υ_1, for $C_2 = 80\,\text{nF}$ and $R = 1.49\,\text{k}\Omega$. (Double-scroll chaotic attractor) *Horizontal axes* 0.5 V/div, *vertical axes* 0.2 V/div

with a stable external limit cycle and that an unstable saddle-type periodic orbit separates the basins of attraction of the two attractors. This phenomenon is known as a boundary crisis [19].

5 Conclusion

In this chapter, a new laboratory setup of Chua's oscillator circuit was presented. The proposed realization is suitable for teaching, in the laboratory, courses related with nonlinear circuits and chaos for undergraduate, postgraduate and Ph.D. students. It constituted of two independent circuits, which the first one is the nonlinear resistor while the second is the rest of the Chua's oscillator circuit. In this way, experiments

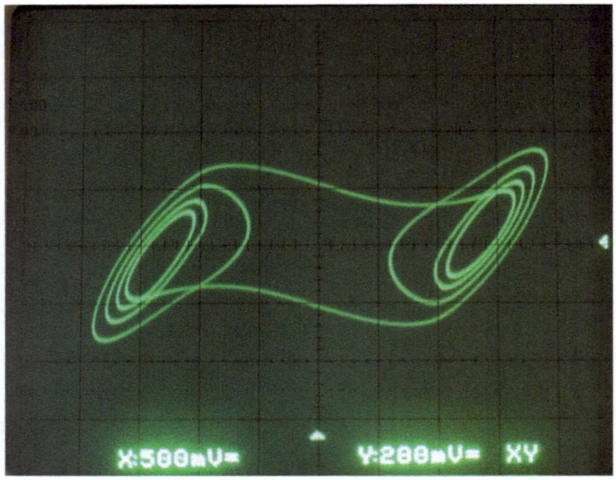

Fig. 38 Experimental phase portrait of υ_2 versus υ_1, for $C_2 = 80\,\text{nF}$ and $R = 1.43\,\text{k}\Omega$. (Periodic steady state) *Horizontal axes* 0.5 V/div, *vertical axes* 0.2 V/div

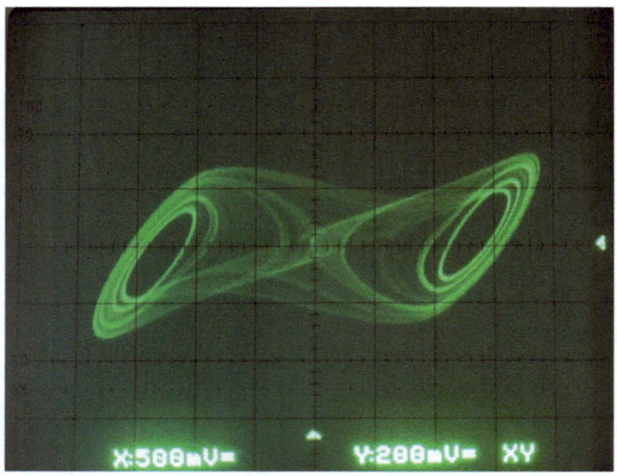

Fig. 39 Experimental phase portrait of υ_2 versus υ_1, for $C_2 = 80\,\text{nF}$ and $R = 1.42\,\text{k}\Omega$. (Double-scroll chaotic attractor) *Horizontal axes* 0.5 V/div, *vertical axes* 0.2 V/div

related with the finding of the i-υ characteristic curve, for different values of the resistors R_3 and R_6, could be done in the laboratory. Also, the dynamic behavior of Chua's oscillator circuit could be studied by changing either the value of the resistor R or the value of the capacitor C_2.

So, the proposed laboratory setup of Chua's oscillator circuit offers to the student a great variety of exercises through which well-known phenomena related with chaos theory, such as period-doubling route to chaos, crisis phenomena, intermittency, and attractors' coexistence, can be experimentally observed. As a future thought, the realization of other interesting nonlinear circuits and memristor emulators, in such way, could be done.

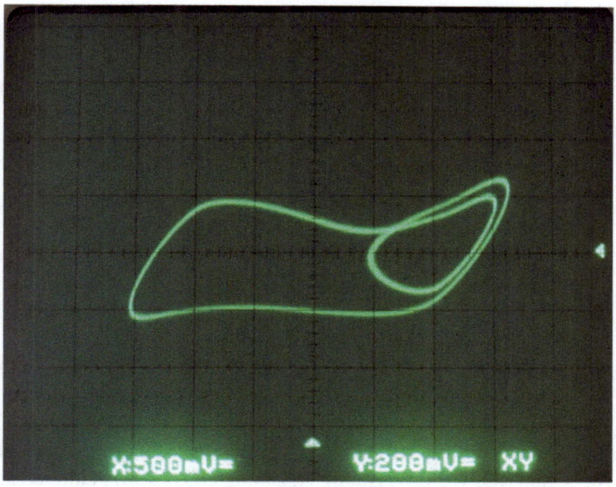

Fig. 40 Experimental phase portrait of υ_2 versus υ_1, for $C_2 = 80\,\text{nF}$ and $R = 1.36\,\text{k}\Omega$. (Periodic asymmetric attractor) *Horizontal axes* 0.5 V/div, *vertical axes* 0.2 V/div

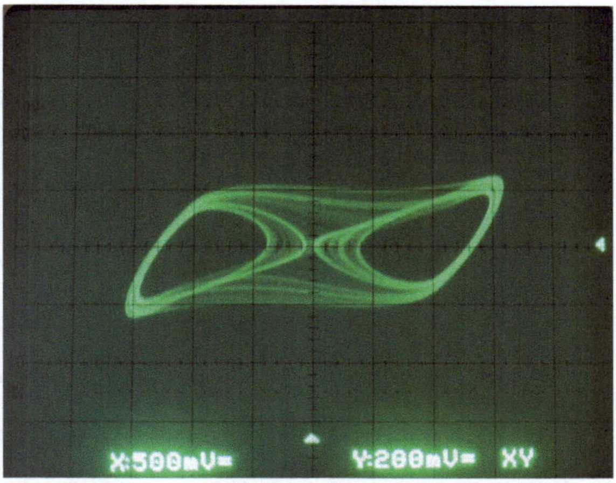

Fig. 41 Experimental phase portrait of υ_2 versus υ_1, for $C_2 = 80\,\text{nF}$ and $R = 1.35\,\text{k}\Omega$. (Double-scroll chaotic attractor) *Horizontal axes* 0.5 V/div, *vertical axes* 0.2 V/div

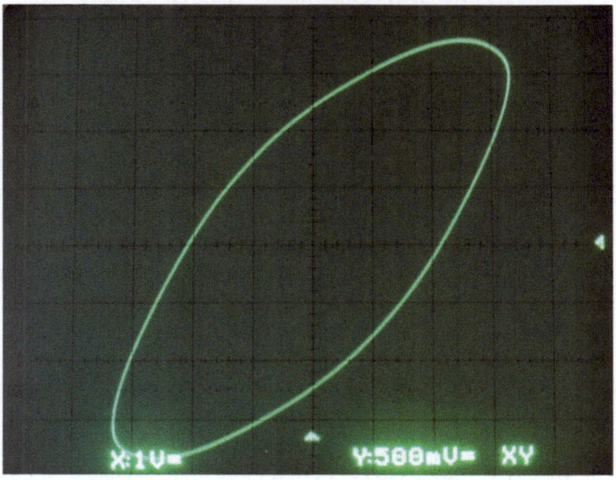

Fig. 42 Experimental phase portrait of υ_2 versus υ_1, for $C_2 = 80\,\text{nF}$ and $R = 1.32\,\text{k}\Omega$. (Period-1 steady state) *Horizontal axes* 1 V/div, *vertical axes* 0.5 V/div

References

1. Alligood KT, Sauer TD, Yorke JA (2000) Chaos: an introduction to dynamical systems. Springer, New York
2. Anishchenko V, Safonova M, Chua LO (1992) Stochastic resonance in Chua's circuit. Int J Bifurc Chaos 2:397–401
3. Arns RG (1998) The other transistor: early history of the metal-oxide semiconductor field-effect transistor. Eng Sci Educ J 7(5):233–240
4. Baker GL, Gollub JP (1990) Chaotic dynamics: an introduction. Cambridge University Press, Cambridge
5. Bier M, Bountis TC (1984) Remerging Feigenbaum trees in dynamical systems. Phys Lett A 104:239–244
6. Brunetti C (1939) The transitron oscillator. Proc IRE 27(2):88–94
7. Chen G, Ueta T (2002) Chaos in circuits and systems. World Scientific, Singapore
8. Chua LO (1994) Chua's circuit 10 year later. Int J Bifurc Chaos 22:279–305
9. Chua LO, Yu J, Yu Y (1983) Negative resistance devices. Int J Circuit Theory Appl 11:161–186
10. Chua LO, Wu CW, Huang A, Zhong GQ (1993) A universal circuit for studying and generating chaos—part I: routes to chaos. IEEE Trans Circuits Syst I 40(10):732–744
11. Chua LO, Wu CW, Huang A, Zhong GQ (1993) A universal circuit for studying and generating chaos—part II: strange attractors. IEEE Trans Circuits Syst I 40(10):745–761
12. Cruz JM, Chua LO (1992) A CMOS IC nonlinear resistor for Chua's circuit. ERL Memorandum, Electronics Research Laboratory, University of California, Berkeley
13. Dawson P, Grebogi C, Yorke J, Kan I (1992) Antimonotonicity-inevitable reversal of period doubling cascades. Phys Lett A 162:249–252
14. Esaki L (1958) New phenomenon in narrow germanium p-n junctions. Phys Rev 109(2):603
15. Field RJ, Györgyi L (1993) Chaos in chemistry and biochemistry. World Scientific Publishing, Singapore
16. Feigenbaum MJ (1979) The universal metric properties of nonlinear transformations. J Stat Phys 21:669–706
17. Fortuna L, Frasca M, Xibilia MG (2009) Chua's circuit implementations: yesterday, today and tomorrow. World Scientific, Singapore
18. Grebogi C, Yorke J (1997) The impact of chaos on science and society. United Nations University Press, Tokyo
19. Grebogi C, Ott E, Yorke JA (1983) Crises: sudden changes in chaotic attractors and chaotic transients. Phys D 7:181–200
20. Halle K, Chua LO, Anishchenko V, Safonova M (1992) Signal amplification via chaos: experimental evidence. Int J Bifurc Chaos 2:1011–1020
21. Hasselblatt B, Katok A (2003) A first course in dynamics: with a panorama of recent developments. Cambridge University Press, Cambridge
22. Hull AW (1918) The dynatron: a vacuum tube possessing negative electric resistance. Proc Inst Radio Eng 6(1):5–35
23. Kennedy MP (1992) Robust op amp realization of Chua's circuit. Frequenz 46(3–4):66–80
24. Kocarev L, Halle K, Eckert K, Chua LO (1993) Experimental observations of antimonotonicity in Chua's circuit. Int J Bifurc Chaos 3:1051–1055
25. Kyprianidis IM, Fotiadou ME (2006) Complex dynamics in Chua's canonical circuit with a cubic nonlinearity. WSEAS Trans Circuits Syst 5:1036–1043
26. Kyprianidis IM, Haralabidis P, Stouboulos IN, Bountis T (2000) Antimonotonicity and chaotic dynamics in a fourth order autonomous nonlinear electric circuit. Int J Bifurc Chaos 10:1903–1915
27. Kyrtsou C, Vorlow C (2005) Complex dynamics in macroeconomics: a novel approach. In: Diebolt C, Kyrtsou C (eds) New trends in macroeconomics. Springer, Berlin, pp 223–245. ISBN-13: 978-3-540-21448-9
28. Li TY, Yorke JA (1975) Period three implies chaos. Am Math Mon 82(10):985–992

29. Lorenz EN (1963) Deterministic non-periodic flow. J Atmos Sci 20:130–141
30. Mandelbrot B (1977) The fractal geometry of nature. W.H. Freeman Company, New York
31. Matsumoto T (1984) A chaotic attractor from Chua's circuit. IEEE Trans Circuits Syst CAS–31(12):1055–1058
32. Matsumoto T, Chua LO, Tokumasu K (1986) Double scroll via a two-transistor circuit. IEEE Trans Circuits Syst 33(8):828–835
33. May RM (1976) Theoretical ecology: principles and applications. W.B. Saunders Company, Philadelphia
34. Moon FC (1987) Chaotic vibrations: an introduction for applied scientists and engineers. Wiley, New York
35. Nicolis G (1995) Introduction to nonlinear science. Cambridge University Press, Cambridge
36. Ott E (1993) Chaos in dynamical systems. Cambridge University Press, Cambridge
37. Pivka L, Spany V (1993) Boundary surfaces and basin bifurcations in Chua's circuit. J Circuits Syst Comput 3:441–470
38. Poincaré JH (1890) Sur le probleme des trois corps et les equations de la dynamique. Divergence des series de M. Lindstedt. Acta Math 13:1–270
39. Rössler OE (1976) An equation for continuous chaos. Phys Lett 57A(5):397–398
40. Strogatz SH (1994) Nonlinear dynamics and chaos. Addison-Wesley, New York
41. Turner LB (1920) The Kallirotron. An aperiodic negative-resitance triode combination. Radio Rev 1:317–329
42. Voelcker J (1989) The Gunn effect. IEEE Spectr 26(7). doi:10.1109/6.29344
43. Zhong GQ, Ayron F (1985) Experimental confirmation of chaos from Chua's circuit. Int J Circuit Theory Appl 13(11):93–98

Control of Shimizu–Morioka Chaotic System with Passive Control, Sliding Mode Control and Backstepping Design Methods: A Comparative Analysis

Uğur Erkin Kocamaz, Yilmaz Uyaroğlu and Sundarapandian Vaidyanathan

Abstract This chapter investigates the control of continuous time Shimizu–Morioka chaotic system with unknown system parameters by means of three different control approaches, namely passive control, sliding mode control and backstepping design. Based on the properties of sliding mode control theory, the appropriate surfaces are designed. Lyapunov functions are used to realize that the passive controller and backstepping controllers ensure the global asymptotic stability of the system. Owing to the controllers, the Shimizu–Morioka chaotic system stabilizes towards its equilibrium points in the state space. Numerical simulations are performed to show and compare the efficiency of the proposed control methods.

Keywords Shimizu–Morioka chaotic system · Passive control · Sliding mode control · Backstepping design · Chaos control

1 Introduction

Since Lorenz found the first chaotic attractor [21], plenty of new chaotic attractors have been discovered. Henon map [10], Rössler attractor [33], Chua's double scroll [24], Chen [5] and Lü [22] chaotic systems are the well-known and most studied ones. Shimizu and Morioka [35] introduced a continuous time chaotic system and Shimizu–Morioka chaotic system has become one of the important chaotic systems.

U.E. Kocamaz (✉)
Bahcelievler, Guvercin, Serdivan, Sakarya, Turkey
e-mail: ugurkocamaz@gmail.com

Y. Uyaroğlu
Faculty of Engineering, Department of Electrical & Electronics Engineering,
Sakarya University, Serdivan, 54187 Sakarya, Turkey
e-mail: uyaroglu@sakarya.edu.tr

S. Vaidyanathan
Research and Development Centre, Vel Tech Dr. RR and Dr. SR Technical University,
Avadi, Chennai 600 062, Tamil Nadu, India
e-mail: sundarvtu@gmail.com

S. Vaidyanathan and C. Volos (eds.), *Advances and Applications in Chaotic Systems*, Studies in Computational Intelligence 636,
DOI 10.1007/978-3-319-30279-9_17

This system shows similar bifurcation as in the Lorenz chaotic attractor, but it is not equivalent to the Lorenz chaotic system in the topological structure. All of the chaotic systems have different characteristics and the investigations on the dynamic behaviours of them such as Lyapunov exponents, Hopf bifurcation analysis, topological structure, stability, control and synchronization are realized separately in many papers. Recently, several novel chaotic attractors have also been presented [18, 32], and many more will be revealed due to their potential applications especially in data encryption and secure communication [1, 30].

A nonlinear system under chaotic behaviour might have undesired trajectories, so the control of chaotic systems has become one of the major research areas for the nonlinear systems. When the control of a chaotic system is achieved, the system stabilizes towards its equilibrium points. At first, Hubler [11] introduced an adaptive control for chaotic systems. Then, Ott, Grebogi and Yorke [29] proposed a method called OGY for controlling chaotic systems. After these pioneering studies, various effective methods have been applied for the control of chaotic systems such as linear feedback [17], nonlinear feedback [51], adaptive [16], sliding mode [2, 13, 26, 47, 48], passive [4, 8, 23, 31, 49], backstepping design [26, 27, 46, 50] and impulsive [44] controls. Among them, the passive control method has been gaining importance in the control of chaotic systems on account of using only one state controller which provides considerable significance in reducing the cost and complexity. The passivity theory keeps the system internally stable with implementing a controller which renders the closed loop system passive upon the properties of the system. Recently, the passive control method has been implemented for the control of Lorenz [49], Chen [31], unified [4], Rabinovich [8], n-dimensional [23] chaotic systems. The sliding mode control is one of the well-known control methods, and its dynamic performance is determined by the prescribed manifold or sliding surface where a switching structure gets the control. It maintains discontinuous control by enforcing the system states to stay on the sliding surface [36]. In recent years, the sliding mode control has been successfully applied in the control of Lorenz [48], Chua [13], Rössler [2], Duffing–Holmes [47], Arneodo [26] chaotic systems. Backstepping design is a systematic control method that combines the choice of a Lyapunov function with the design of a feedback controller. First, a small subsystem is only considered, for which a virtual control law is constructed. Then, the design is extended in several steps until the control laws for the full system have been constructed [9]. It has been successfully used to control chaotic Lorenz [46], Chen [46], Lü [46], Arneodo [26], hyperchaotic Rössler [50], hyperchaotic Liu [27] systems. The methodology of passive control, sliding mode control and backstepping design methods can be reached in number of papers [8, 27, 40, 46, 47, 49]. In some papers, the control of chaotic systems is investigated by using more than one method with giving comparative analyses [6, 26, 28, 43, 45].

Since the continuous time Shimizu–Morioka chaotic system was introduced [30], its dynamical behaviours and properties have been extensively investigated in some papers [7, 15, 19, 20, 25, 34, 37]. The synchronization of Shimizu–Morioka chaotic system is implemented with active control [14, 38, 40], feedback control [17], adaptive control [41], sliding mode control [40] and linear-nonlinear decomposition

[3] methods. Feedback controllers [17], adaptive controllers [41], delayed feedback controllers [7] and linearized feedback controllers [12] are employed for the control of continuous time Shimizu–Morioka chaotic system. According to the literature review, the control of Shimizu–Morioka chaotic system was not investigated with the passive control, sliding mode control and backstepping design approaches.

This chapter concerns on the further investigations on the control of Shimizu–Morioka chaotic system. First, a brief description of continuous time nonlinear Shimizu–Morioka system is given in Sect. 2. Then, three different control methods, namely passive control, sliding mode control and backstepping design, are applied for achieving the control of this nonlinear system to its equilibrium points in Sect. 3. Afterwards, numerical simulations are demonstrated to verify and compare the effectiveness of the control results for the Shimizu–Morioka chaotic system in Sect. 4. Finally, Sect. 5 concludes the chapter.

2 Shimizu–Morioka Chaotic System

The continuous time Shimizu–Morioka chaotic system is defined by a set of three first-order differential equations as

$$\begin{cases} \dot{x} = y, \\ \dot{y} = (1 - z)x - ay, \\ \dot{z} = -bz + x^2, \end{cases} \tag{1}$$

where x, y, z are state variables and a, b are positive constant parameters [35]. It exhibits chaotic behaviour when the parameter values are selected as $a = 0.75$ and $b = 0.45$ [41]. According to these parameters and the initial conditions $x(0) = 0$, $y(0) = 0.25$ and $z(0) = 1$, the time series of Shimizu–Morioka system are illustrated in Fig. 1, the 2D phase plots are illustrated in Fig. 2, and the 3D phase plane is illustrated in Fig. 3.

The equilibria of Shimizu–Morioka chaotic system (1) can be found with getting $\dot{x} = 0$, $\dot{y} = 0$ and $\dot{z} = 0$ as follows:

$$\begin{cases} y = 0, \\ (1 - z)x - ay = 0, \\ -bz + x^2 = 0. \end{cases} \tag{2}$$

Hence, the Shimizu–Morioka chaotic system has three equilibrium points; $E_0(0, 0, 0)$, $E_1(\sqrt{b}, 0, 1)$ and $E_2(-\sqrt{b}, 0, 1)$. When the b parameter is equal to 0.45, the equilibrium points become $E_1(0.67082, 0, 1)$ and $E_2(-0.67082, 0, 1)$.

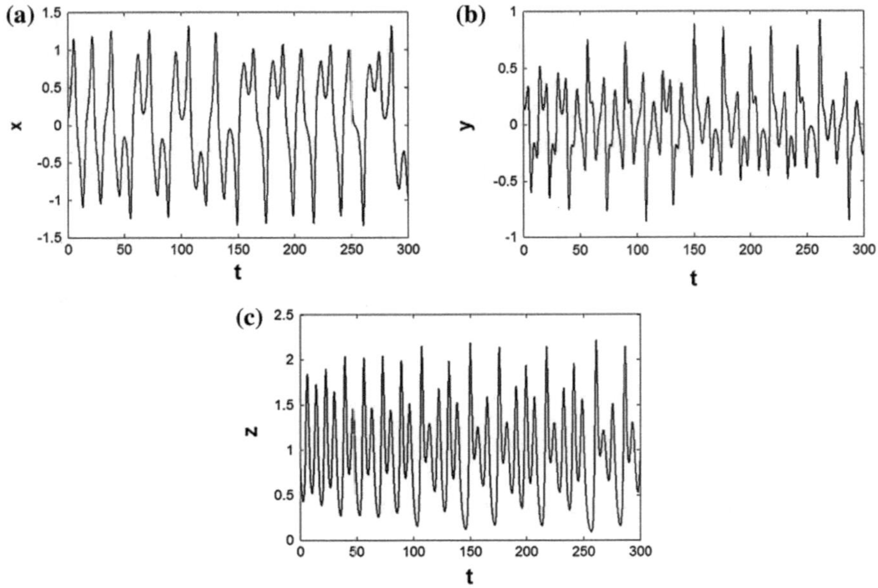

Fig. 1 Time series of Shimizu–Morioka chaotic system for **a** x signals, **b** y signals, **c** z signals

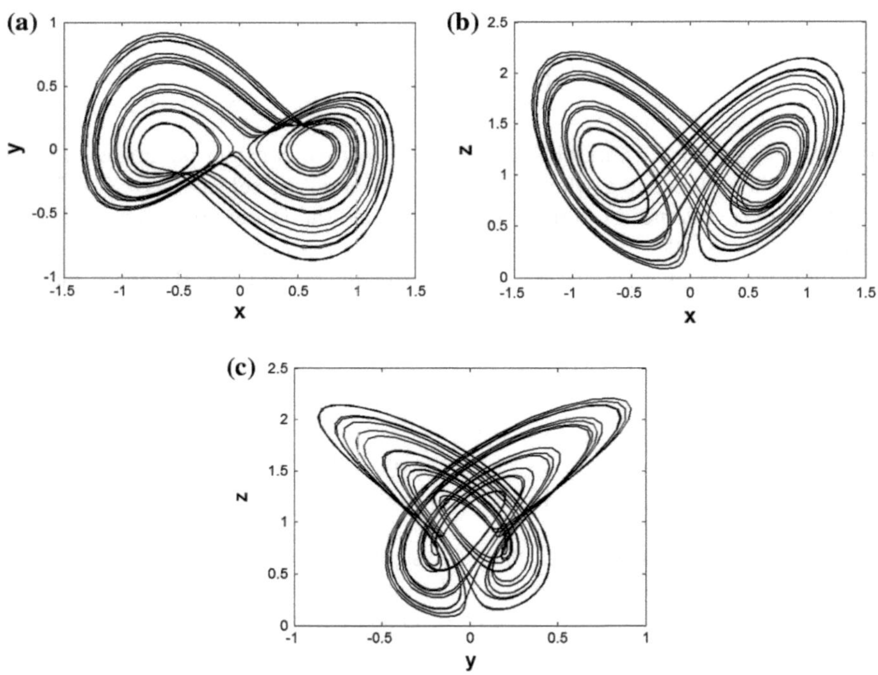

Fig. 2 Phase plots of Shimizu–Morioka chaotic system in **a** x–y phase plot, **b** x–z phase plot, **c** y–z phase plot

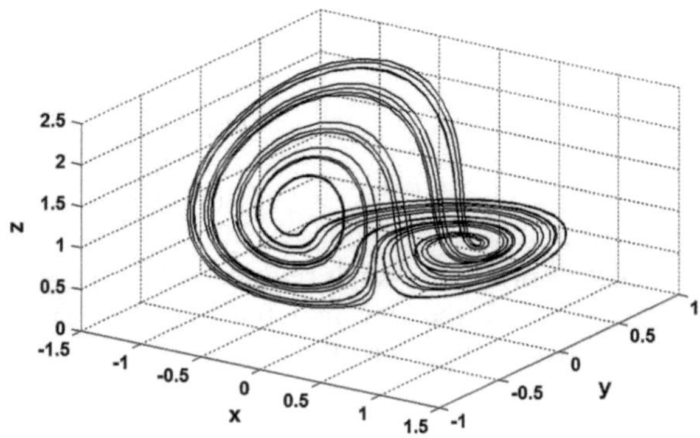

Fig. 3 3D phase plane of Shimizu–Morioka chaotic system

3 Control of Shimizu–Morioka Chaotic System

3.1 Passive Control

In this section, passive control method is applied to system (1) in order to control the Shimizu–Morioka chaotic system to equilibrium points. When the passive controller u is added to system (1), it changes into the following form:

$$\begin{cases} \dot{x}_1 = x_2 + u, \\ \dot{x}_2 = (1 - x_3)x_1 - ax_2, \\ \dot{x}_3 = -bx_3 + x_1{}^2. \end{cases} \tag{3}$$

The state variable x_1 is considered as the output of the system and assuming that $y = x_1$, $z_1 = x_2$, $z_2 = x_3$, $z = [z_1 z_2]^T$, then system (3) becomes

$$\begin{cases} \dot{z}_1 = (1 - z_2)y - az_1, \\ \dot{z}_2 = -bz_2 + y^2, \\ \dot{y} = z_1 + u. \end{cases} \tag{4}$$

The passivity theory has the following generalized form

$$\begin{cases} \dot{z} = f_0(z) + p(z, y)y, \\ \dot{y} = b(z, y) + a(z, y)u, \end{cases} \tag{5}$$

where system (4) can be expressed in the normal form of system (5) as follows:

$$f_0(z) = \begin{bmatrix} -az_1 \\ -bz_2 \end{bmatrix}, \tag{6}$$

$$p(z, y) = \begin{bmatrix} 1 - z_2 \\ 2y \end{bmatrix}, \tag{7}$$

$$b(z, y) = z_1, \tag{8}$$

$$a(z, y) = 1. \tag{9}$$

Let, the storage function is selected as $V(z, y) = W(z) + 0.5(y^2)$ where $W(z) = 0.5({z_1}^2 + {z_2}^2)$ is the Lyapunov function of $f_0(z)$ with $W(0) = 0$. Then, by taking the derivative of $W(z)$ with Eq. (6),

$$\begin{aligned} \dot{W}(z) &= \frac{\partial W(z)}{\partial z} f_0(z) = [z_1 \ z_2] \begin{bmatrix} -az_1 \\ -bz_2 \end{bmatrix} \\ &= -a{z_1}^2 - b{z_2}^2. \end{aligned} \tag{10}$$

Because $W(z) \geq 0$ and $\dot{W}(z) \leq 0$, it can be concluded that $W(z)$ is the Lyapunov function of $f_0(z)$ and the $f_0(z)$ is globally asymptotically stable [8]. So, the controlled Shimizu–Morioka chaotic system (3) is a minimum phase system based on the Lyapunov stability. According to the passivity theory, the controlled system can be equivalent to a passive system and globally asymptotically stabilized at its zero equilibrium point by the following state controller [49]:

$$\begin{aligned} u &= a(z, y)^{-1} \left[-b^T(z, y) - \frac{\partial W(z)}{\partial z} p(z, y) - \alpha y + v \right] \\ &= 1^{-1} \left[-z_1 - \left[z_1 \ z_2 \right] \begin{bmatrix} 1 - z_2 \\ 2y \end{bmatrix} - \alpha y + v \right] \\ &= -2z_1 + z_1 z_2 - 2z_2 y - \alpha y + v \end{aligned} \tag{11}$$

where $\alpha > 0$ is a real constant and v is an external input signal. Providing that the conversions $y = x_1$, $z_1 = x_2$ and $z_2 = x_3$ are taken back, the controller u can be denoted by the following form:

$$u = -2x_2 + x_2 x_3 - 2x_1 x_3 - \alpha x_1 + v. \tag{12}$$

The control of Shimizu–Morioka chaotic system (3) by using the passive control method is completed with Eq. (12). Hence, the control of Shimizu–Morioka chaotic system with uncertain parameters by means of passive control is achieved.

Substituting Eq. (12) into system (3) yields

$$\begin{cases} \dot{x}_1 = -x_2 + x_2 x_3 - 2x_1 x_3 - \alpha x_1 + v, \\ \dot{x}_2 = (1 - x_3) x_1 - a x_2, \\ \dot{x}_3 = -b x_3 + {x_1}^2. \end{cases} \tag{13}$$

Although the passive controlled Shimizu–Morioka chaotic system changes the equilibria of the system, the controller stabilizes the chaotic system at any equilibrium point with adjusting the v parameter. Let $\dot{x} = 0$ and then the controlled system (13) becomes

$$\begin{cases} 0 = -x_2 + x_2x_3 - 2x_1x_3 - \alpha x_1 + v, \\ 0 = (1 - x_3)x_1 - ax_2, \\ 0 = -bx_3 + {x_1}^2. \end{cases} \tag{14}$$

Thus,

$$\begin{cases} x_3 = x_1^2/b, \\ x_2 = (x_1 - {x_1}^3/b)/a, \\ v = -(x_1^2x_2 - 2{x_1}^3)/b + x_2 + \alpha x_1. \end{cases} \tag{15}$$

The conditions in Eq. (15) maintain the global asymptotical stability of Shimizu–Morioka chaotic system towards its $E_0(0, 0, 0)$, $E_1(\sqrt{b}, 0, 1)$ and $E_2(-\sqrt{b}, 0, 1)$ equilibrium points.

3.2 *Sliding Mode Control*

In this section, sliding mode control method is applied to system (1) in order to control the Shimizu–Morioka chaotic system to equilibrium points. When the sliding mode controllers u_1, u_2 and u_3 are added to system (1), it changes into the following form:

$$\begin{cases} \dot{x}_1 = x_2 + u_1, \\ \dot{x}_2 = (1 - x_3)x_1 - ax_2 + u_2, \\ \dot{x}_3 = -bx_3 + x_1^2 + u_3. \end{cases} \tag{16}$$

A fixed point of system (16) can be denoted as (x_d, y_d, z_d). After that, the trajectory error states are determined as $e_1 = x_1 - x_d$, $e_2 = x_2 - y_d$ and $e_3 = x_3 - z_d$. Then the state variables are obtained as $x_1 = e_1 + x_d$, $x_2 = e_2 + y_d$ and $x_3 = e_3 + z_d$, the error state dynamic equations of system (16) become

$$\begin{cases} \dot{e}_1 = e_2 + y_d + u_1, \\ \dot{e}_2 = (1 - (e_3 + z_d))(e_1 + x_d) - a(e_2 + y_d) + u_2, \\ \dot{e}_3 = -b(e_3 + z_d) + (e_1 + x_d)^2 + u_3. \end{cases} \tag{17}$$

The error dynamic equations of system (17) can be rewritten as follows:

$$\begin{cases} \dot{e}_1 = e_2 + y_d + u_1, \\ \dot{e}_2 = (1 - z_d)e_1 - ae_2 - x_de_3 - e_1e_3 + x_d - ay_d - x_dz_d + u_2, \\ \dot{e}_3 = {e_1}^2 + 2x_de_1 - be_3 + {x_d}^2 - bz_d + u_3. \end{cases} \tag{18}$$

The y_d, $x_d - ay_d - x_dz_d$ and $x_d^2 - bz_d$ equations in system (18) are stable values and do not affect the system dynamics when controlling to the zero equilibrium point. Therefore, the error state dynamic equations of system (18) can be simplified as

$$\begin{cases} \dot{e}_1 = e_2 + u_1, \\ \dot{e}_2 = (1 - z_d)e_1 - ae_2 - x_d e_3 - e_1 e_3 + u_2, \\ \dot{e}_3 = 2x_d e_1 - be_3 + e_1^2 + u_3. \end{cases} \tag{19}$$

The error dynamics (19) are regularized in the matrix notation as

$$\dot{e} = Ae + \eta(x, y) + u \tag{20}$$

where

$$A = \begin{bmatrix} 0 & 1 & 0 \\ 1 - z_d & -a & -x_d \\ 2x_d & 0 & -b \end{bmatrix}, \eta(x, y) = \begin{bmatrix} 0 \\ -e_1 e_3 \\ e_1^2 \end{bmatrix}, u = \begin{bmatrix} u_1 \\ u_2 \\ u_3 \end{bmatrix}. \tag{21}$$

Based on the sliding mode theory, the control signal u is defined as [40]:

$$u(t) = -\eta(x, y) + Bv(t) \tag{22}$$

where B is selected so that (A, B) is controllable. So, B is chosen as

$$B = [0\ 1\ 1]^{\mathrm{T}}. \tag{23}$$

According to the Hurwitz criterion, the vector C must be selected such that the system matrix of the controlled dynamics $[I - B(CB)^{-1}C]A$ has all eigenvalues with negative real parts. The sliding mode variable C is taken as [9.5 2.5 –1.5] for the E_0 and E_2 equilibrium points. Then, the sliding surface is constructed as

$$s = Ce = [\,9.5\ 2.5\ -1.5\,]e = 9.5e_1 + 2.5e_2 - 1.5e_3 \tag{24}$$

which makes the sliding mode state equation asymptotically stable. The selected C vector does not maintain the Hurwitz criterion for the E_1 equilibrium point, so it is selected differently as [–1.25 –2.5 3.5]. Then, the surface for E_1 equilibrium point becomes

$$s = Ce = [\,-1.25\ -2.5\ 3.5\,]e = -1.25e_1 - 2.5e_2 + 3.5e_3. \tag{25}$$

From the property of the sliding mode control methodology [40]:

$$v(t) = -(CB)^{-1}\left[C(kI + A)e + q\,\mathrm{sign}(s)\right] \tag{26}$$

where I is the identity matrix and k, q are positive real constants. A large value of k may cause chattering and an appropriate value of q reduces the chattering and also speeds up the reaching time to the sliding surface.

After that, the required sliding mode control signal is determined by Eq. (22) where $\eta(e)$ and B are obtained from Eqs. (21) and (23), respectively:

$$\begin{cases} u_1 = 0, \\ u_2 = (x - x_d)(z - z_d) + v(t), \\ u_3 = -(x - x_d)^2 + v(t). \end{cases} \tag{27}$$

The control of Shimizu–Morioka chaotic system (16) by using the sliding mode control method is completed with Eq. (27). Hence, the control of Shimizu–Morioka chaotic system with uncertain parameters by means of sliding mode control is achieved.

3.3 Backstepping Design

In this section, backstepping control method is applied to system (1) in order to control the Shimizu–Morioka chaotic system to equilibrium points. When the backstepping controllers u_1, u_2 and u_3 are added to system (1), it changes into the following form:

$$\begin{cases} \dot{x}_1 = x_2 + u_1, \\ \dot{x}_2 = (1 - x_3)x_1 - ax_2 + u_2, \\ \dot{x}_3 = -bx_3 + x_1^2 + u_3. \end{cases} \tag{28}$$

A fixed point of system (28) can be denoted as (x_d, y_d, z_d). After that, the trajectory error states are determined as $e_1 = x_1 - x_d$, $e_2 = x_2 - y_d$ and $e_3 = x_3 - z_d$. Then the state variables are obtained as $x_1 = e_1 + x_d$, $x_2 = e_2 + y_d$ and $x_3 = e_3 + z_d$, the error state dynamic equations of system (28) become

$$\begin{cases} \dot{e}_1 = e_2 + y_d + u_1, \\ \dot{e}_2 = (1 - (e_3 + z_d))(e_1 + x_d) - a(e_2 + y_d) + u_2, \\ \dot{e}_3 = -b(e_3 + z_d) + (e_1 + x_d)^2 + u_3. \end{cases} \tag{29}$$

System (29) can be considered as a control problem with control inputs u_1, u_2 and u_3, which are the functions of the error vectors e_1, e_2 and e_3.

There are three steps in the following backstepping design procedure. At the ith step, an intermediate control function α_i will be obtained by constructing an appropriate Lyapunov function V_i.

Step 1: Define $w_1 = e_1$. Then its derivative is

$$\dot{w}_1 = e_2 + y_d + u_1, \tag{30}$$

where $e_2 = \alpha_1(w_1)$ is regarded as a virtual controller.

For the design of $\alpha_1(w_1)$ to stabilize the w_1—subsystem (30), a Lyapunov function V_1 is chosen as

$$V_1 = \frac{1}{2}w_1^2. \tag{31}$$

The derivative of V_1 is

$$\dot{V}_1 = w_1\dot{w}_1 = w_1(\alpha_1(w_1) + y_d + u_1). \tag{32}$$

If $u_1 = -y_d$ and $\alpha_1(w_1) = -w_1$, then $\dot{V}_1$ is obtained as

$$\dot{V}_1 = -w_1^2, \tag{33}$$

which is negative definite. This implies that the w_1—subsystem (30) is globally asymptotically stable.

Since the virtual control function $\alpha_1(w_1)$ is estimative, the error variable can be defined as

$$w_2 = e_2 - \alpha_1(w_1) = e_2 + w_1. \tag{34}$$

Then, the following (w_1, w_2)—subsystem is obtained

$$\begin{cases} \dot{w}_1 = w_2 - w_1, \\ \dot{w}_2 = (1 - e_3 - z_d)(w_1 + x_d) + (1 - a)(w_2 - w_1) - ay_d + u_2, \end{cases} \tag{35}$$

where $e_3 = \alpha_2(w_1, w_2)$ is regarded as a virtual controller.

Step 2: In this step, the following Lyapunov function V_2 is constructed to stabilize the (w_1, w_2)—subsystem (35)

$$V_2 = \frac{1}{2}(w_1^2 + w_2^2). \tag{36}$$

Its derivative is

$$\begin{aligned} \dot{V}_2 = &- w_1^2 + w_2(w_1 + (1 - \alpha_2 - z_d)(w_1 + x_d) + (1 - a)(w_2 - w_1) \\ &- ay_d + u_2). \end{aligned} \tag{37}$$

If $\alpha_2(w_1, w_2) = 0$, then

$$u_2 = -2w_1 - x_d + z_d(w_1 + x_d) - (1 - a)(w_2 - w_1) + ay_d - k_1 w_2, \tag{38}$$

where $k_1 > 0$ is a gain constant.

Substituting Eq. (38) into Eq. (37) yields

$$\dot{V}_2 = -w_1^2 - k_1 w_2^2, \tag{39}$$

which is negative definite. This implies that the (w_1, w_2)—subsystem (35) is globally asymptotically stable.

Since the virtual control function $\alpha_2(w_1, w_2)$ is estimative, the error variable can be defined as

$$w_3 = e_3 - \alpha_2(w_1, w_2) = e_3. \tag{40}$$

Then, the following (w_1, w_2, w_3)—subsystem is obtained

$$\begin{cases} \dot{w}_1 = w_2 - w_1, \\ \dot{w}_2 = -w_1 - w_3(w_1 + x_d) - k_1 w_2, \\ \dot{w}_3 = -b(w_3 + z_d) + (w_1 + x_d)^2 + u_3. \end{cases} \tag{41}$$

Step 3: In order to stabilize the (w_1, w_2, w_3)—subsystem (41), the following Lyapunov function is constructed

$$V_3 = \frac{1}{2}(w_1^2 + w_2^2 + w_3^2). \tag{42}$$

The derivative of V_3 is

$$\dot{V}_3 = -w_1^2 - k_1 w_2^2 + w_3(-w_2(w_1 + x_d) - b(w_3 + z_d) + (w_1 + x_d)^2 + u_3). \tag{43}$$

Then, the controller u_3 can be taken as

$$u_3 = w_2(w_1 + x_d) + b(w_3 + z_d) - (w_1 + x_d)^2 - k_2 w_3, \tag{44}$$

where $k_2 > 0$ is a gain constant.

Substituting Eq. (44) into Eq. (43) yields

$$\dot{V}_3 = -w_1^2 - k_1 w_2^2 - k_2 w_3^2, \tag{45}$$

which is negative definite. This implies that the (w_1, w_2, w_3)—subsystem (41) is globally asymptotically stable.

Since $w_1 = e_1$, $w_2 = e_2 - \alpha_1(w_1)$ and $w_3 = e_3 - \alpha_2(w_1, w_2)$, this implies that $(e_1, e_2, e_3) \to (0, 0, 0)$ as $t \to \infty$. Hence, the control of Shimizu–Morioka chaotic system with uncertain parameters by means of backstepping design is achieved.

In summary, the backstepping control that achieves global asymptotic stability of the error dynamics (29) is

$$\begin{cases} u_1 = -y_d, \\ u_2 = -2w_1 - x_d + z_d(w_1 + x_d) - (1 - a)(w_2 - w_1) + a y_d - k_1 w_2, \\ u_3 = w_2(w_1 + x_d) + b(w_3 + z_d) - (w_1 + x_d)^2 - k_2 w_3, \end{cases} \tag{46}$$

where k_1, k_2 are positive gain constants and w_1, w_2, w_3 are defined by

$$\begin{cases} w_1 = e_1, \\ w_2 = e_1 + e_2, \\ w_3 = e_3. \end{cases} \tag{47}$$

4 Numerical Simulations

In this section, numerical simulations are performed to demonstrate the control of Shimizu–Morioka chaotic system in Eqs. (3), (16) and (28). The fourth-order Runge–Kutta method is used in all numerical simulations with variable time step. The parameter values of Shimizu–Morioka system are considered as $a = 0.75$ and $b = 0.45$ with the initial conditions $x(0) = 0$, $y(0) = 0.25$ and $z(0) = 1$ to ensure the chaotic behaviour. The passive control parameter is chosen as $\alpha = 10$, the sliding mode con-

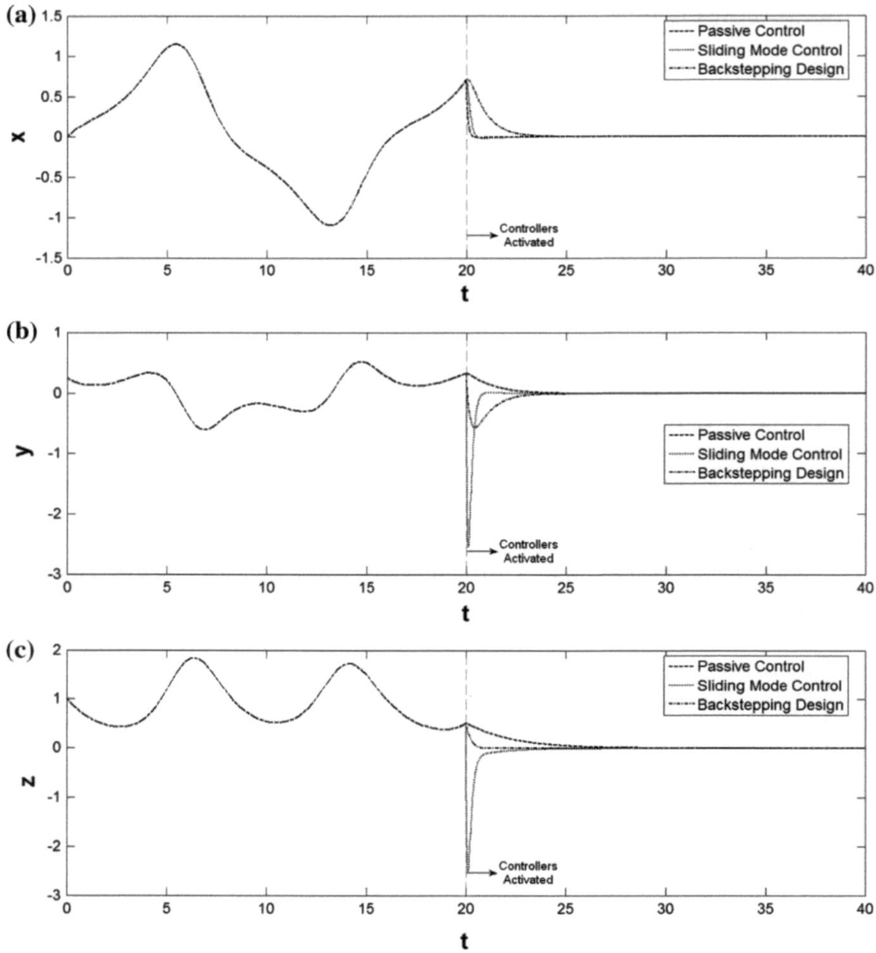

Fig. 4 The controlled Shimizu–Morioka chaotic system towards $E_0(0, 0, 0)$ equilibrium point when the controllers are activated at $t = 20$ for **a** x signals, **b** y signals, **c** z signals

trol parameters are selected as $k = 10, q = 0.1$ and the backstepping control gains are taken as $k_1 = 5, k_2 = 5$. The controllers are activated at $t = 20$ in the simulations.

A chaotic system stabilizes towards a desired equilibrium point with sliding mode and backstepping controllers by adjusting its x_d, y_d and z_d parameters. For instance, $x_d = \sqrt{b}$, $y_d = 0$ and $z_d = 1$ for the $E_1(\sqrt{b}, 0, 1)$ equilibrium point. A passive controller slightly changes the controlled system but it still preserves to stabilize towards the equilibrium points with adjusting the v parameter. According to the conditions in Eq. (15) with the parameter values $b = 0.45$ and $\alpha = 10$, the v passive control parameter is calculated as 0, 8.0498 and –8.0498 for the $E_0(0, 0, 0)$, $E_1(\sqrt{b}, 0, 1)$ and $E_2(-\sqrt{b}, 0, 1)$ equilibrium points, respectively.

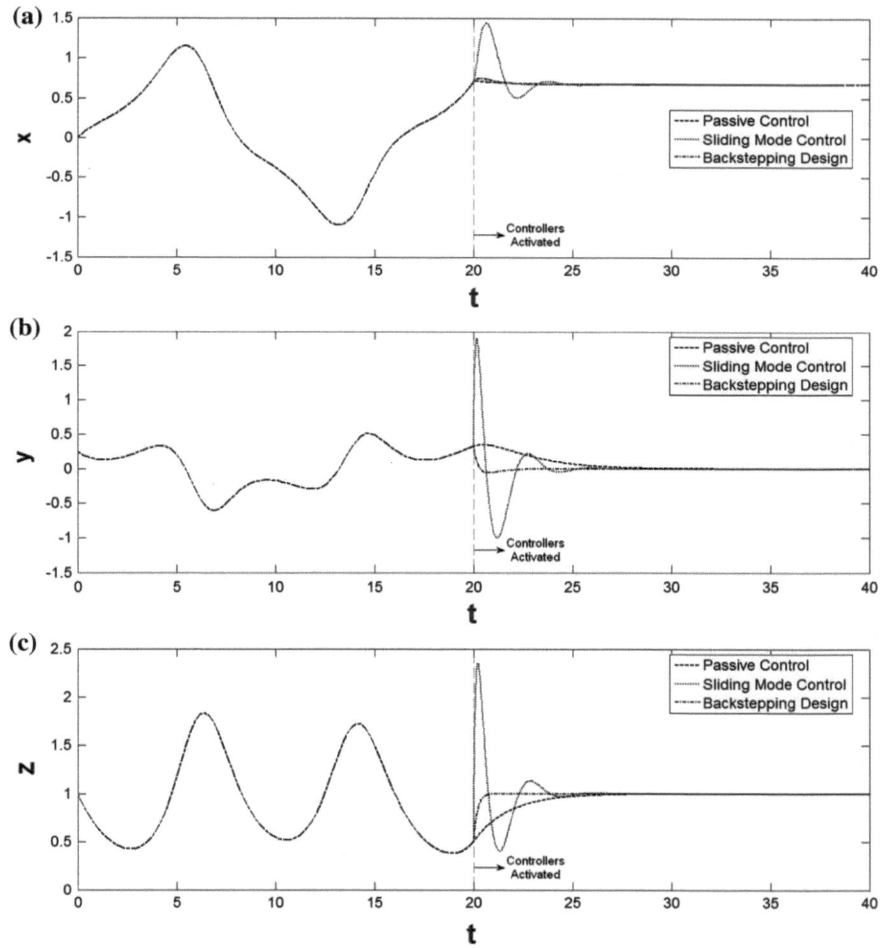

Fig. 5 The controlled Shimizu–Morioka chaotic system towards $E_1(\sqrt{b}, 0, 1)$ equilibrium point when the controllers are activated at $t = 20$ for **a** x signals, **b** y signals, **c** z signals

The simulation results of controlled Shimizu–Morioka chaotic system with the passive control, sliding mode control and backstepping design methods to E_0 (0, 0, 0),$E_1(\sqrt{b}, 0, 1)$ and $E_2(-\sqrt{b}, 0, 1)$ equilibrium points are shown in Figs. 4, 5 and 6, respectively.

As expected, the related figures show that the outputs of Shimizu–Morioka chaotic system converge to its equilibrium points in an appropriate time period, after the passive controller, the sliding mode controllers and backstepping controllers are activated. Thus, computer simulations have confirmed all the theoretical analyses of proposed passive control, sliding mode control and backstepping design meth-

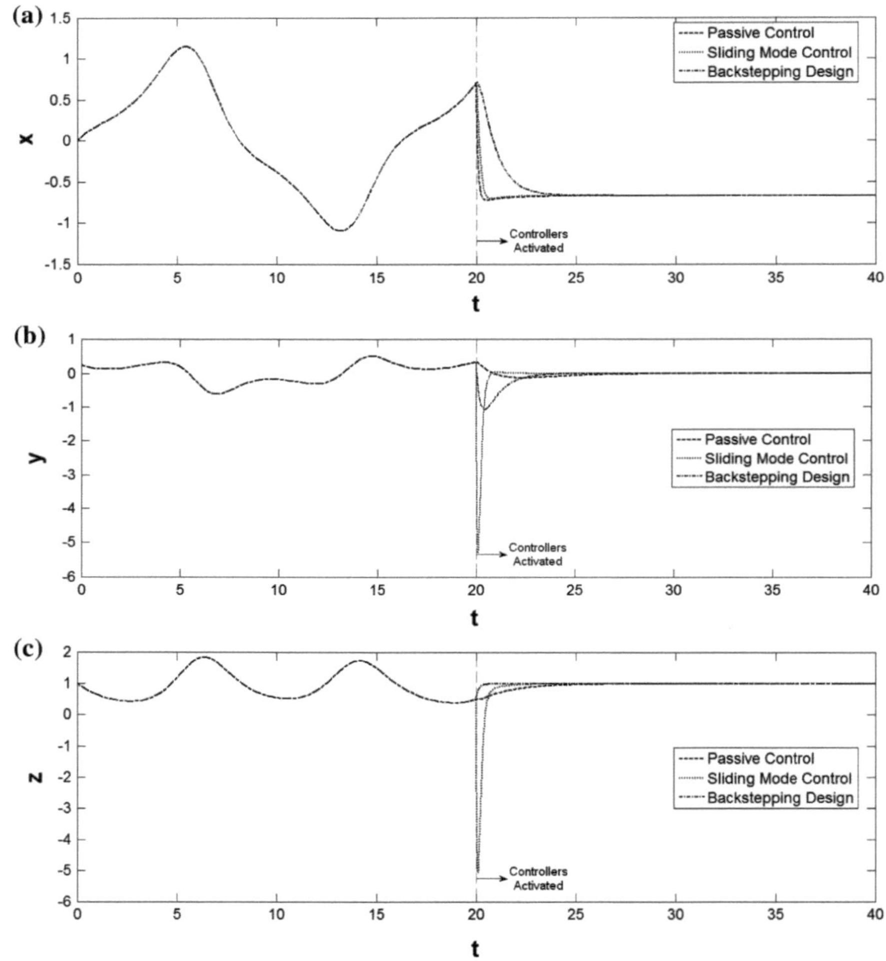

Fig. 6 The controlled Shimizu–Morioka chaotic system towards $E_2(-\sqrt{b}, 0, 1)$ equilibrium point when the controllers are activated at $t = 20$ for **a** x signals, **b** y signals, **c** z signals

ods. Figures 4, 5 and 6 include comparative results for the control of Shimizu–Morioka chaotic system. While the control is provided at $t \geq 24$ by using the sliding mode controllers, it is observed when $t \geq 25$ with the backstepping controllers and $t \geq 29$ with the passive controller for the $E_0(0, 0, 0)$ equilibrium point. Furthermore, the control is firstly observed owing to the sliding mode controllers for the $E_1(\sqrt{b}, 0, 1)$ and $E_2(-\sqrt{b}, 0, 1)$ equilibrium points. By comparison of the passive control, sliding mode control and backstepping design methods at the same specified values of Shimizu–Morioka chaotic system, it can be concluded that the sliding mode control method is more successfully on achieving the control of Shimizu–Morioka chaotic system.

5 Conclusion

In this chapter, a single passive controller, two sliding mode controllers and three backstepping controllers have been designed to realize the global asymptotical stability of continuous time Shimizu–Morioka chaotic system towards its equilibrium points with parameter uncertainties. Simulation results are demonstrated to verify all the theoretical analyses. In addition, the performances of the proposed control methods are compared with the numerical simulations. They have shown that the sliding mode controllers regulate the control of Shimizu–Morioka chaotic system in a better time period than the passive controller and backstepping controllers, which exposes the effectiveness of sliding mode control method. Backstepping design method also gives good control results. On the other hand, the advantage of passive control method is to achieve the control of Shimizu–Morioka chaotic system with only one state controller which provides easiness in implementation and low-cost production.

References

1. Banerjee S, Kurths J (2014) Chaos and cryptography: a new dimension in secure communications. Eur Phys J-Spec Topics 223(8):1441–1445
2. Chang JF, Hung ML, Yang YS, Liao TL, Yan JJ (2008) Controlling chaos of the family of Rössler systems using sliding mode control. Chaos, Solitons Fractals 37(2):609–622
3. Chantov D (2009) Chaotic synchronization methods based on stability analysis of linear systems. Trak Univ J Sci 10(2):165–171
4. Chen X, Liu C (2010) Passive control on a unified chaotic system. Nonlinear Anal: Real World Appl 11:683–687
5. Chen G, Ueta T (1999) Yet another chaotic attractor. Int J Bifurc Chaos 9(7):1465–1466
6. de Paula AS, Savi MA (2011) Comparative analysis of chaos control methods: a mechanical system case study. Int J Non-Linear Mech 46(8):1076–1089
7. El-Dessoky MM, Yassen MT, Aly ES (2014) Bifurcation analysis and chaos control in Shimizu-Morioka chaotic system with delayed feedback. Appl Math Comput 243:283–297
8. Emiroglu S, Uyaroglu Y (2010) Control of Rabinovich chaotic system based on passive control. Sci Res Essays 5(21):3298–3305

9. Guessas L, Benmahammed K (2011) Adaptive backstepping and PID optimized by genetic algorithm in control of chaotic. Int J In-novative Comput Inf Control 7(9):5299–5312
10. Henon M (1976) A two-dimensional mapping with a strange attractor. Commun Math Phys 50(1):69–77
11. Hubler A (1989) Adaptive control of chaotic systems. Helv Phys Acta 62:343–346
12. Islam N, Mazumdar HP, Das A (2009) On the stability and control of the Schimizu-Morioka system of dynamical equations. Differ Geom-Dyn Syst 11:135–143
13. Jang MJ, Chen CL, Chen CK (2002) Sliding mode control of chaos in the cubic Chua's circuit system. Int J Bifurc Chaos 12(6):1437–1449
14. Kareem SO, Ojo KS, Njah AN (2012) Function projective synchronization of identical and non-identical modified finance and Shimizu-Morioka systems. Pramana—J Phys 79(1):71–79
15. Leonov GA (2012) General existence conditions of homoclinic trajectories in dissipative systems. Lorenz, Shimizu-Morioka, Lu and Chen systems. Phys Lett A 376:3045–3050
16. Li CL, Tong YN (2013) Adaptive control and synchronization of a fractional-order chaotic system. Pramana—J Phys 80(4):583–592
17. Liao X, Xu F, Wang P, Yu P (2009) Chaos control and synchronization for a special generalized Lorenz canonical system–the SM system. Chaos, Solitons Fractals 39(5):2491–2508
18. Lin Y, Wang CH (2015) A novel grid multi-scroll chaotic oscillator. Electron World 121(1946):35–39
19. Liu L, Gao B (2011) Conditions for appearance and disappearance of limit cycles in the Shimizu-Morioka system. Int J Bifurc Chaos 21(9):2489–2503
20. Llibre J, Pessoa C (2015) The Hopf bifurcation in the Shimizu-Morioka system. Nonlinear Dyn 79(3):2197–2205
21. Lorenz EN (1963) Deterministic nonperiodic flow. J Atmos Sci 20:130–141
22. Lü J, Chen G, Zhang S (2002) The compound structure of a new chaotic attractor. Chaos, Solitons & Fractals 14(5):669–672
23. Mahmoud GM, Mahmoud EE, Arafa AA (2013) Passive control of n-dimensional chaotic complex nonlinear systems. J Vib Control 19(7):1061–1071
24. Matsumoto T (1984) A chaotic attractor from Chua's circuit. IEEE Trans Circuits Syst CAS 31(12):1055–1058
25. Messias M, Gouveia MRA, Pessoa C (2012) Dynamics at infinity and other global dynamical aspects of Shimizu-Morioka equations. Nonlinear Dyn 69(1–2):577–587
26. Motallebzadeh F, Dadras S, Motallebzadeh F, Ozgoli S (2009) Controlling chaos in Arneodo system. 17th Mediterranean Conference on Control & Automation, Thessaloniki, Greece, IEEE, vols 1–3, pp 314–319
27. Njah AN (2010) Tracking control and synchronization of the new hyperchaotic Liu system via backstepping techniques. Nonlinear Dyn 61(1–2):1–9
28. Ojo KS, Njah AN, Ogunjo ST (2013) Comparison of backstepping and modified active control in projective synchronization of chaos in an extended Bonhöffer-van der Pol oscillator. Pramana—J Phys 80(5):825–835
29. Ott E, Grebogi C, Yorke JA (1990) Controlling chaos. Phys Rev Lett 64(11):1196–1199
30. Peng ZP, Wang CH, Lin Y, Luo XW (2014) A novel four-dimensional multi-wing hyper-chaotic attractor and its application in image encryption. Acta Phys Sinica 63(24):240506
31. Qi D, Zhao G, Song Y (2004) Passive control of Chen chaotic system. Proceedings of the 5th World Congress on Intelligent Control and Automation, Hangzhou, China, pp 1284–1286
32. Qiao ZQ, Li XY (2014) Dynamical analysis and numerical simulation of a new Lorenz-type chaotic system. Math Comput Model Dyn Syst 20(3):264–283
33. Rössler OE (1976) An equation for continuous chaos. Phys Lett A 57(5):397–398
34. Shilnikov AL (1993) On bifurcations of the Lorenz attractor in the Shimizu-Morioka model. Physica D 62:338–346
35. Shimizu T, Morioka N (1980) On the bifurcation of a symmetric limit cycle to an asymmetric one in a simple model. Phys Lett A 76(3–4):201–204
36. Slotine JJ (1984) Sliding controller design for nonlinear systems. Int J Control 40(2):421–434

37. Tigan G, Turaev D (2011) Analytical search for homoclinic bifurcations in the Shimizu-Morioka model. Physica D 240:985–989
38. Vaidyanathan S (2011) Global chaos synchronization of Shimizu-Morioka and Liu-Chen chaotic systems by active nonlinear control. Int J Adv Sci Technol 2(4):11–20
39. Vaidyanathan S (2011) Global chaos synchronization of Arneodo and Shimizu-Morioka chaotic systems by active nonlinear control. Int J Adv Sci Technol 2(6):32–42
40. Vaidyanathan S (2011) Sliding mode controller design for synchronization of Shimizu-Morioka chaotic systems. Int J Inf Sci Tech 1(1):20–29
41. Vaidyanathan S (2012) Adaptive control and synchronization of Shimizu-Morioka chaotic system. Int J Found Comput Sci Technol 2(4):29–42
42. Wang X, Li X (2010) Feedback control of the Liu chaotic dynamical system. Int J Modern Phys B 24(3):397–404
43. Wang MJ, Wang XY (2009) Controlling Liu system with different methods. Modern Phys Lett B 23(14):1805–1818
44. Wu X, Lu J, Tse CK, Wang J, Liu J (2007) Impulsive control and synchronization of the Lorenz systems family. Chaos, Solitons Fractals 31(3):631–638
45. Yang C, Tao CH, Wang P (2010) Comparison of feedback control methods for a hyperchaotic Lorenz system. Phys Lett A 374(5):729–732
46. Yassen MT (2006) Chaos control of chaotic dynamical systems using backstepping design. Chaos, Solitons Fractals 27(2):537–548
47. Yau HT, Chen CK, Chen CL (2000) Sliding mode control of chaotic systems with uncertainties. Int J Bifurc Chaos 10(5):1139–1147
48. Yau HT, Yan JJ (2004) Design of sliding mode controller for Lorenz chaotic system with nonlinear input. Chaos, Solitons Fractals 19(4):891–898
49. Yu W (1999) Passive equivalence of chaos in Lorenz system. IEEE Trans Circuits Syst-I: Fundam Theory Appl 46(7):876–878
50. Zhang H, Ma XK, Li M, Zou JL (2005) Controlling and tracking hyperchaotic Rössler system via active backstepping design. Chaos, Solitons Fractals 26(2):353–361
51. Zhou WN, Pan L, Li Z, Halang WA (2009) Non-linear feedback control of a novel chaotic system. Int J Control, Autom Syst 7(6):939–944

Generalized Projective Synchronization of a Novel Chaotic System with a Quartic Nonlinearity via Adaptive Control

Sundarapandian Vaidyanathan and Sarasu Pakiriswamy

Abstract In this work, we announce a 3-D six-term novel chaotic system with a quartic nonlinearity. Next, the qualitative properties of the novel chaotic system are discussed in detail. We show that the novel chaotic system has three unstable equilibrium points. The Lyapunov exponents of the novel chaotic system are obtained as $L_1 = 0.1507$, $L_2 = 0$ and $L_3 = -0.9521$, while the Kaplan–Yorke dimension of the novel chaotic system is obtained as $D_{KY} = 2.1583$. The maximal Lyapunov exponent (MLE) of the novel chaotic system is obtained as $L_1 = 0.1507$. Using Lyapunov stability theory, this work also derives an adaptive controller for the generalized projective synchronization (GPS) of identical novel chaotic systems with unknown parameters. In the chaos literature, many types of synchronization such as complete synchronization (CS), anti-synchronization (AS), hybrid synchronization (HS), projective synchronization (PS) and generalized synchronization (GS) are considered for the synchronization of a pair of chaotic systems called *master* and *slave* systems. All these types of synchronization are special cases of the generalized projective synchronization (GPS) of chaotic systems. MATLAB plots have been depicted to illustrate the phase portraits of the novel chaotic system and also the GPS results for the novel chaotic systems using adaptive controllers.

Keywords Chaos · Chaotic systems · Synchronization · Active control · Adaptive control

---

S. Vaidyanathan (✉)
Research and Development Centre, Vel Tech University, Avadi,
Chennai 600062, Tamil Nadu, India
e-mail: sundarvtu@gmail.com

S. Pakiriswamy
School of Electrical and Computing, Vel Tech University, Avadi,
Chennai 600062, Tamil Nadu, India
e-mail: sarasujivat@gmail.com

S. Vaidyanathan and C. Volos (eds.), *Advances and Applications in Chaotic Systems*, Studies in Computational Intelligence 636,
DOI 10.1007/978-3-319-30279-9_18

1 Introduction

Chaotic systems are defined as nonlinear dynamical systems which are sensitive to initial conditions, topologically mixing and with dense periodic orbits. Sensitivity to initial conditions of chaotic systems is popularly known as the *butterfly effect*. Small changes in an initial state will make a very large difference in the behavior of the system at future states.

Some classical paradigms of 3-D chaotic systems in the literature are Lorenz system [15], Rössler system [29], ACT system [2], Sprott systems [37], Chen system [9], Lü system [16], Cai system [7], Tigan system [48], etc.

Many new chaotic systems have been discovered in the recent years such as Zhou system [132], Zhu system [133], Li system [14], Wei–Yang system [126], Sundarapandian systems [40, 45], Vaidyanathan systems [59, 60, 62–65, 68, 79, 80, 94, 97, 101, 112, 114, 116, 118, 119, 121], Pehlivan system [21], Sampath–Vaidyanathan system [30], etc.

The synchronization of chaotic systems is a phenomenon that occurs when two or more chaotic systems are coupled or when a chaotic system drives another chaotic system. Because of the butterfly effect which causes exponential divergence of the trajectories of two identical chaotic systems started with nearly the same initial conditions, the synchronization of chaotic systems is a challenging research problem in the chaos literature [3, 4].

Chaos theory has applications in several fields of science and engineering such as chemical reactors [69, 73, 75, 77, 81, 85–87], biological systems [67, 70–72, 74, 76, 78, 82–84, 88–92], memristors [1, 22, 125], lasers [6], oscillations [49], robotics [11, 124], electrical circuits [17, 123], cryptosystems [28, 50], secure communications [127, 128], etc.

Major works on synchronization of chaotic systems deal with the complete synchronization (CS) of a pair of chaotic systems called the *master* and *slave* systems. The design goal of the complete synchronization is to apply the output of the master system to control the slave system so that the output of the slave system tracks the output of the master system asymptotically with time.

Pecora and Carroll pioneered the research on synchronization of chaotic systems with their seminal papers [8, 20]. The active control method [38, 52, 57, 103, 107] is commonly used when the system parameters are available for measurement. Adaptive control method [39, 53, 61, 93, 104, 106, 113, 117, 120] is commonly used when some or all the system parameters are not available for measurement and estimates for the uncertain parameters of the systems.

Backstepping control method [23–27, 47, 109, 115, 122] is also used for the synchronization of chaotic systems, which is a recursive method for stabilizing the origin of a control system in strict-feedback form. Sliding mode control method [46, 54, 56, 58, 66, 95, 96, 110, 111] is also a popular method for the synchronization of chaotic systems.

In the chaos literature, many types of synchronization schemes have been proposed such as complete synchronization [38, 52, 57, 103, 107], anti-synchronization

[41–43, 55, 102], hybrid synchronization [12, 44, 51, 105, 108], generalized synchronization [5, 10, 130], projective synchronization [36, 129, 131], generalized projective synchronization [18, 19, 31–35, 98–100], etc.

In generalized projective synchronization (GPS), the responses of the synchronized dynamical states synchronize up to a constant scaling matrix α. The complete synchronization (CS) and anti-synchronization (AS) are special cases of the generalized projective synchronization where the scaling matrix $\alpha = I$ and $\alpha = -I$, respectively. Hybrid, projective and generalized synchronization are special cases of the generalized projective synchronization of chaotic systems.

In this work, we first announce a 3-D six-term novel chaotic system with a quartic nonlinearity. Next, the qualitative properties of the novel chaotic system are discussed in detail. This work derives new results for adaptive controller design for the generalized projective synchronization (GPS) of the identical novel chaotic systems.

This work is organized as follows. Section 2 discusses the dynamic equations of the novel chaotic system. Section 3 discusses the qualitative properties of the novel chaotic system. In this section, we show that the novel chaotic system has three unstable equilibrium points. The Lyapunov exponents of the novel chaotic system are obtained as $L_1 = 0.1507$, $L_2 = 0$ and $L_3 = -0.9521$, while the Kaplan–Yorke dimension of the novel chaotic system is obtained as $D_{KY} = 2.1583$. The maximal Lyapunov exponent (MLE) of the novel chaotic system is obtained as $L_1 = 0.1507$.

In Sect. 4, we derive new GPS results for the adaptive controller design for identical novel chaotic systems, when the system parameters are unknown. In Sect. 5, we summarize the main results obtained in this work.

2 A Six-Term 3-D Novel Chaotic System

In this work, we announce a six-term novel chaotic system described by the 3-D dynamics

$$\begin{aligned} \dot{x}_1 &= x_2 \\ \dot{x}_2 &= x_1 - ax_2 - x_1 x_3 \\ \dot{x}_3 &= -bx_3 + x_1^4 \end{aligned} \tag{1}$$

where x_1, x_2, x_3 are the states and a, b are constant, positive, parameters.

The system (1) exhibits a *strange chaotic attractor* when the parameter values are taken as

$$a = 0.5, \quad b = 0.3 \tag{2}$$

For numerical simulations, we take the initial values of the Vaidyanathan chaotic system (1) as

$$x_1(0) = 0.1, \quad x_2(0) = 0.2, \quad x_3(0) = 0.1 \tag{3}$$

Figure 1 shows the 3-D phase portrait of the novel chaotic system (1). Figures 2, 3 and 4 show the 2-D projections of the novel chaotic system (1) on the (x_1, x_2), (x_2, x_3) and (x_1, x_3) coordinate planes respectively.

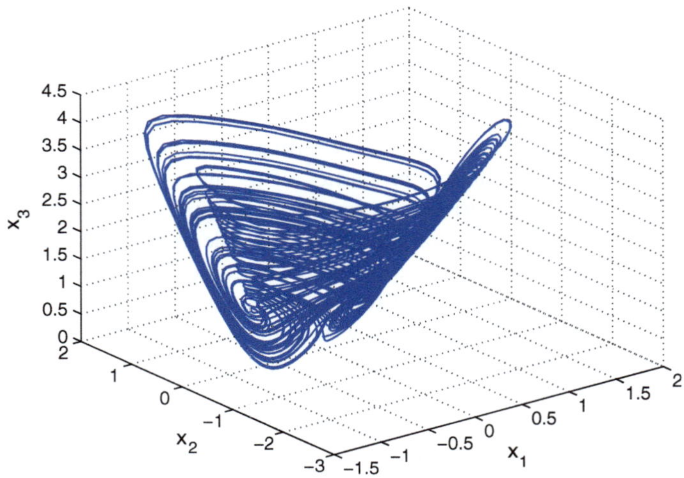

Fig. 1 3-D phase portrait of the Vaidyanathan chaotic system

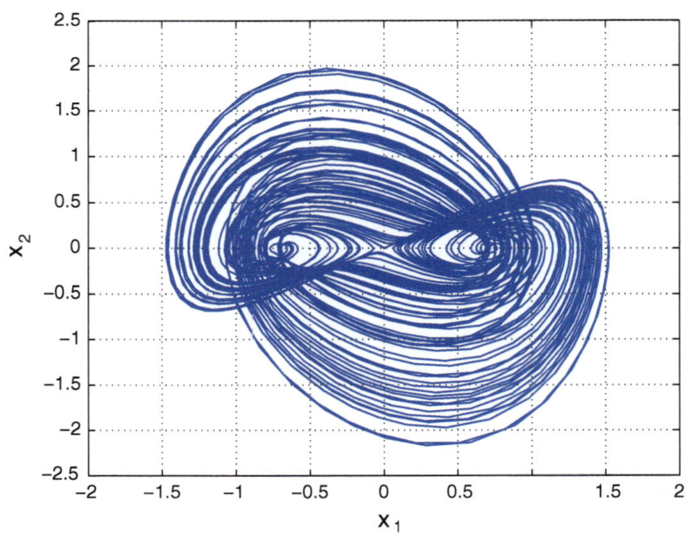

Fig. 2 2-D projection of the Vaidyanathan chaotic system on the (x_1, x_2) plane

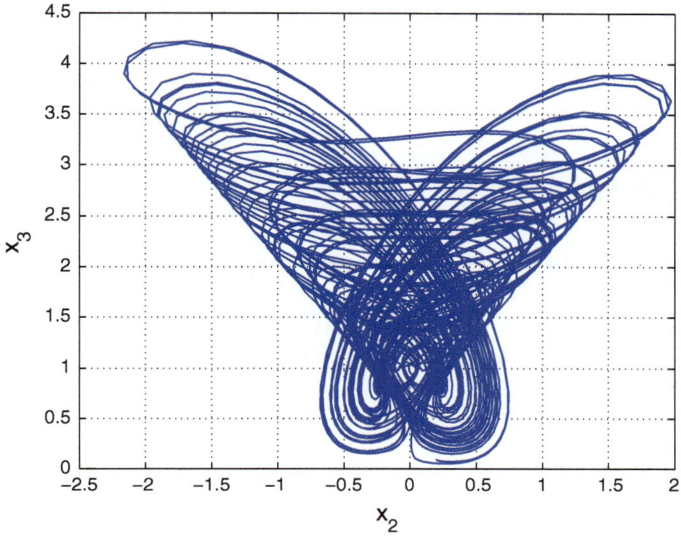

Fig. 3 2-D projection of the Vaidyanathan chaotic system on the (x_2, x_3) plane

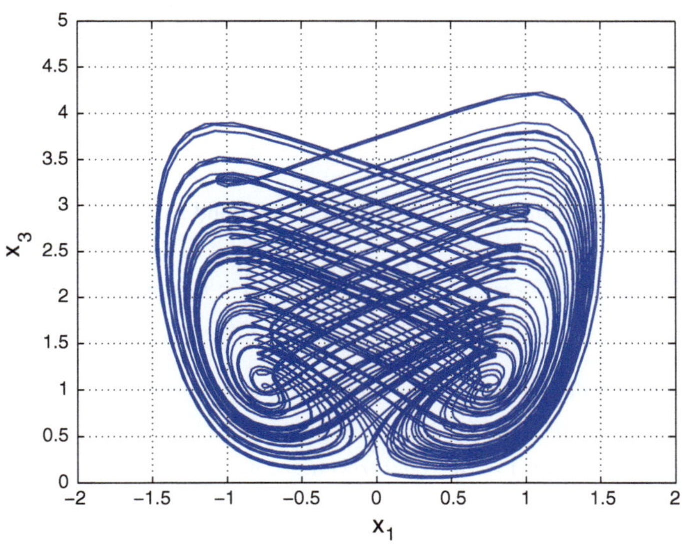

Fig. 4 2-D projection of the Vaidyanathan chaotic system on the (x_1, x_3) plane

3 Qualitative Properties of the 3-D Novel Chaotic System

In this section, we give a dynamic analysis of the 3-D novel jerk chaotic system (1).

3.1 Dissipativity

In vector notation, the novel system (1) can be expressed as

$$\dot{\mathbf{x}} = f(\mathbf{x}) = \begin{bmatrix} f_1(x_1, x_2, x_3) \\ f_2(x_1, x_2, x_3) \\ f_3(x_1, x_2, x_3) \end{bmatrix}, \tag{4}$$

where

$$\begin{cases} f_1(x_1, x_2, x_3) = x_2 \\ f_2(x_1, x_2, x_3) = x_1 - ax_2 - x_1x_3 \\ f_3(x_1, x_2, x_3) = -bx_3 + x_1^4 \end{cases} \tag{5}$$

Let Ω be any region in $\mathbf{R}^3$ with a smooth boundary and also, $\Omega(t) = \Phi_t(\Omega)$, where Φ_t is the flow of f. Furthermore, let $V(t)$ denote the volume of $\Omega(t)$.

By Liouville's theorem, we know that

$$\dot{V}(t) = \int\limits_{\Omega(t)} (\nabla \cdot f)\, dx_1\, dx_2\, dx_3 \tag{6}$$

The divergence of the novel jerk system (4) is found as:

$$\nabla \cdot f = \frac{\partial f_1}{\partial x_1} + \frac{\partial f_2}{\partial x_2} + \frac{\partial f_3}{\partial x_3} = -a - b = -\mu < 0 \tag{7}$$

where $\mu = a + b > 0$.

Inserting the value of $\nabla \cdot f$ from (7) into (6), we get

$$\dot{V}(t) = \int\limits_{\Omega(t)} (-\mu)\, dx_1\, dx_2\, dx_3 = -\mu V(t) \tag{8}$$

Integrating the first order linear differential equation (8), we get

$$V(t) = \exp(-\mu t) V(0) \tag{9}$$

Since $\mu > 0$, it follows from Eq. (9) that $V(t) \to 0$ exponentially as $t \to \infty$. This shows that the novel 3-D chaotic system (1) is dissipative. Hence, the system limit sets are ultimately confined into a specific limit set of zero volume, and the asymptotic motion of the novel chaotic system (1) settles onto a strange attractor of the system.

3.2 Equilibrium Points

The equilibrium points of the novel chaotic system (1) are obtained by solving the equations

$$\begin{cases} f_1(x_1, x_2, x_3) = x_2 & = 0 \\ f_2(x_1, x_2, x_3) = x_1 - ax_2 - x_1x_3 & = 0 \\ f_3(x_1, x_2, x_3) = -bx_3 + x_1^4 & = 0 \end{cases} \tag{10}$$

We take the parameter values as in the chaotic case, *viz.* $a = 0.5$ and $b = 0.3$.

Solving the Eq. (10), we get three equilibrium points of the novel chaotic system (1), *viz.*

$$E_0 = \begin{bmatrix} 0 \\ 0 \\ 0 \end{bmatrix}, \quad E_1 = \begin{bmatrix} 0.7401 \\ 0 \\ 1 \end{bmatrix} \text{ and } E_2 = \begin{bmatrix} -0.7401 \\ 0 \\ 1 \end{bmatrix} \tag{11}$$

To test the stability type of the equilibrium point E_1, we calculate the Jacobian matrix of the novel chaotic system (1) at any point x:

$$J(x) = \begin{bmatrix} 0 & 1 & 0 \\ 1 - x_3 & -a & -x_1 \\ 4x_1^3 & 0 & -b \end{bmatrix} = \begin{bmatrix} 0 & 1 & 0 \\ 1 - x_3 & -0.5 & -x_1 \\ 4x_1^3 & 0 & -0.3 \end{bmatrix} \tag{12}$$

We find that

$$J_0 \triangleq J(E_0) = \begin{bmatrix} 0 & 1 & 0 \\ 1 & -0.5 & 0 \\ 0 & 0 & -0.3 \end{bmatrix} \tag{13}$$

The matrix J_0 has the eigenvalues

$$\lambda_1 = 0.7808, \quad \lambda_2 = -0.3, \quad \lambda_3 = -1.2808 \tag{14}$$

This shows that the equilibrium point E_0 is a saddle-point, which is unstable.

Next, we find that

$$J_1 \triangleq J(E_1) = \begin{bmatrix} 0 & 1 & 0 \\ 0 & -0.5 & -0.7401 \\ 1.6216 & 0 & -0.3 \end{bmatrix} \tag{15}$$

The matrix J_1 has the eigenvalues

$$\lambda_1 = -1.3486, \quad \lambda_{2,3} = 0.2743 \pm 0.9026i \tag{16}$$

This shows that the equilibrium point E_1 is a saddle-focus, which is unstable.

A simple calculation shows that $J_2 = J(E_2)$ has the same eigenvalues as J_1. Thus, the equilibrium point E_2 is also a saddle-focus, which is unstable.

Hence, the novel chaotic system (1) has three unstable equilibrium points, *viz.* E_0, E_1 and E_2.

3.3 Lyapunov Exponents and Kaplan–Yorke Dimension

We take the parameter values of the novel system as (1) $a = 0.5$ and $b = 0.3$. We take the initial state of the system (1) as given in (3).

Then the Lyapunov exponents of the novel system (1) are numerically obtained using MATLAB as

$$L_1 = 0.1507, \quad L_2 = 0, \quad L_3 = -0.9521 \tag{17}$$

Thus, the maximal Lyapunov exponent (MLE) of the novel system (1) is positive, which means that the system has a chaotic behavior.

Since $L_1 + L_2 + L_3 = -0.8014 < 0$, it follows that the novel chaotic system (1) is dissipative.

Also, the Kaplan–Yorke dimension of the novel chaotic system (1) is obtained as

$$D_{KY} = 2 + \frac{L_1 + L_2}{|L_3|} = 2.1583 \tag{18}$$

which is fractional.

4 Adaptive Controller Design for the GPS of Novel Chaotic Systems

In this section, we design an adaptive controller for the generalized projective synchronization (GPS) of the identical novel chaotic systems, when the system parameters are unknown.

As the master system, we consider the novel chaotic system

$$\begin{aligned} \dot{x}_1 &= x_2 \\ \dot{x}_2 &= x_1 - ax_2 - x_1x_3 \\ \dot{x}_3 &= -bx_3 + x_1^4 \end{aligned} \tag{19}$$

where x_1, x_2, x_3 are the states and a, b are unknown parameters.

As the slave system, we consider the controlled novel chaotic system

$$\begin{aligned} \dot{y}_1 &= y_2 + u_1 \\ \dot{y}_2 &= y_1 - ay_2 - y_1 y_3 + u_2 \\ \dot{y}_3 &= -by_3 + y_1^4 + u_3 \end{aligned} \tag{20}$$

where y_1, y_2, y_3 are the states and u_1, u_2, u_3 are adaptive controls to be designed using estimates for unknown parameters.

For the GPS of the identical novel chaotic systems (19) and (20), we define the GPS synchronization error as

$$\begin{aligned} e_1 &= y_1 - m_1 x_1 \\ e_2 &= y_2 - m_2 x_2 \\ e_3 &= y_3 - m_3 x_3 \end{aligned} \tag{21}$$

where m_1, m_2, m_3 are real scaling constants.

The error dynamics is obtained by differentiating (21) as

$$\begin{aligned} \dot{e}_1 &= y_2 - m_1 x_2 + u_1 \\ \dot{e}_2 &= y_1 - m_2 x_1 - ae_2 - y_1 y_3 + m_2 x_1 x_3 + u_2 \\ \dot{e}_3 &= -be_3 + y_1^4 - m_3 x_1^4 + u_3 \end{aligned} \tag{22}$$

We consider the adaptive controller defined by

$$\begin{aligned} u_1 &= -y_2 + m_1 x_2 - k_1 e_1 \\ u_2 &= -y_1 + m_2 x_1 + \hat{a}(t) e_2 + y_1 y_3 - m_2 x_1 x_3 - k_2 e_2 \\ u_3 &= \hat{b}(t) e_3 - y_1^4 + m_3 x_1^4 - k_3 e_3 \end{aligned} \tag{23}$$

where k_1, k_2, k_3 are positive gain constants and $\hat{a}(t), \hat{b}(t)$ are estimates of the unknown parameters a, b, respectively.

Substituting (23) into (22), we get the closed-loop control system

$$\begin{aligned} \dot{e}_1 &= -k_1 e_1 \\ \dot{e}_2 &= -\left[a - \hat{a}(t)\right] e_2 - k_2 e_2 \\ \dot{e}_3 &= -\left[b - \hat{b}(t)\right] e_3 - k_3 e_3 \end{aligned} \tag{24}$$

We define the parameter estimation errors as

$$\begin{aligned} e_a(t) &= a - \hat{a}(t) \\ e_b(t) &= b - \hat{b}(t) \end{aligned} \tag{25}$$

Using (25), we can simplify the error dynamics as

$$\begin{aligned} \dot{e}_1 &= -k_1 e_1 \\ \dot{e}_2 &= -e_a e_2 - k_2 e_2 \\ \dot{e}_3 &= -e_b e_3 - k_3 e_3 \end{aligned} \tag{26}$$

Differentiating (25) with respect to t, we obtain

$$\begin{aligned} \dot{e}_a &= -\dot{\hat{a}}(t) \\ \dot{e}_b &= -\dot{\hat{b}}(t) \end{aligned} \tag{27}$$

We use adaptive control theory to find an update law for the parameter estimates. We consider the quadratic candidate Lyapunov function defined by

$$V(e_1, e_2, e_3, e_a, e_b) = \frac{1}{2}\left(e_1^2 + e_2^2 + e_3^2 + e_a^2 + e_b^2\right) \tag{28}$$

Differentiating V along the trajectories of (26) and (27), we obtain

$$\dot{V} = -k_1 e_1^2 - k_2 e_2^2 - k_3 e_3^2 + e_a\left[-e_2^2 - \dot{\hat{a}}\right] + e_b\left[-e_3^2 - \dot{\hat{b}}\right] \tag{29}$$

In view of Eq. (29), we take the parameter update law as

$$\begin{aligned} \dot{\hat{a}} &= -e_2^2 \\ \dot{\hat{b}} &= -e_3^2 \end{aligned} \tag{30}$$

Next, we state and prove the main result of this section.

Theorem 1 *The adaptive control law defined by (23) and the parameter update law (30) achieve global and exponential generalized projective synchronization (GPS) between the identical Vaidyanathan systems (19) and (20) with unknown parameters, where k_1, k_2, k_3 are positive gain constants.*

Proof We consider the quadratic Lyapunov function defined by (28), which is clearly a positive definite function on $\mathbf{R}^7$.

By substituting the parameter update law (30) into (29), we obtain the time-derivative of V as

$$\dot{V} = -k_1 e_1^2 - k_2 e_2^2 - k_3 e_3^2 \tag{31}$$

From (31), it is clear that $\dot{V}$ is a negative semi-definite function on $\mathbf{R}^7$.

Thus, we conclude that the GPS error vector $\mathbf{e}(t)$ and the parameter estimation error are globally bounded, *i.e.*

$$\left[\, e_1(t)\; e_2(t)\; e_3(t)\; e_a(t)\; e_b(t) \,\right]^T \in \mathbf{L}_\infty.$$

We define $k = \min\{k_1, k_2, k_3\}$.

Then it follows from (31) that

$$\dot{V} \leq -k\|\mathbf{e}(t)\|^2 \tag{32}$$

Thus, we have

$$k\|\mathbf{e}(t)\|^2 \leq -\dot{V} \tag{33}$$

Integrating the inequality (33) from 0 to t, we get

$$k\int_0^t \|\mathbf{e}(\tau)\|^2 d\tau \;\leq\; V(0) - V(t) \tag{34}$$

From (34), it follows that $\mathbf{e} \in \mathbf{L}_2$.

Using (26), we conclude that $\dot{\mathbf{e}} \in \mathbf{L}_\infty$.

Using Barbalat's lemma [13], we conclude that $\mathbf{e}(t) \to \mathbf{0}$ exponentially as $t \to \infty$ for all initial conditions $\mathbf{e}(0) \in \mathbf{R}^3$.

This completes the proof. □

For numerical simulations using MATLAB, we use the classical fourth order Runge–Kutta method with $h = 10^{-8}$ for solving systems of differential equations.

The parameter values of the Vaidyanathan chaotic systems are taken as in the chaotic case (2), *i.e.*

$$a = 0.5, \quad b = 0.3 \tag{35}$$

We take the gains as

$$k_1 = 8, \quad k_2 = 8, \quad k_3 = 8 \tag{36}$$

We the GPS scales as

$$m_1 = 2.3, \quad m_2 = 1.4, \quad m_3 = 1.7 \tag{37}$$

As initial values of the master system (19), we take

$$x_1(0) = 3.2, \quad x_2(0) = -2.9, \quad x_3(0) = 1.5 \tag{38}$$

As initial values of the slave system (20), we take

$$y_1(0) = 1.7, \quad y_2(0) = 1.2, \quad y_3(0) = 4.3 \tag{39}$$

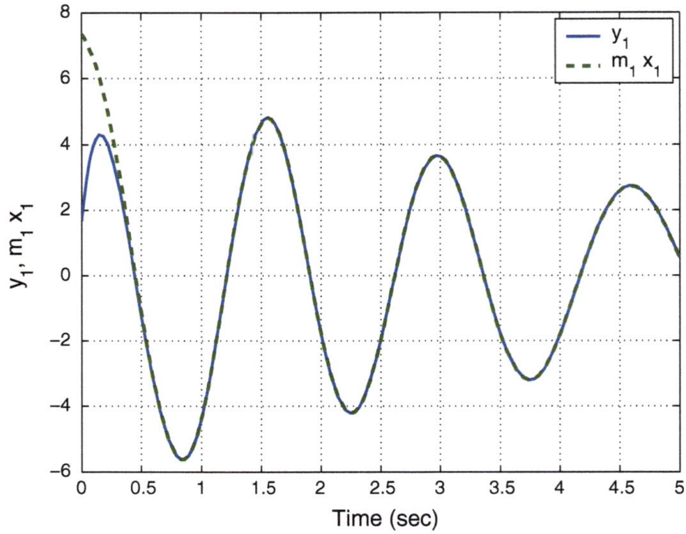

Fig. 5 Synchronization of the states y_1 and m_1x_1

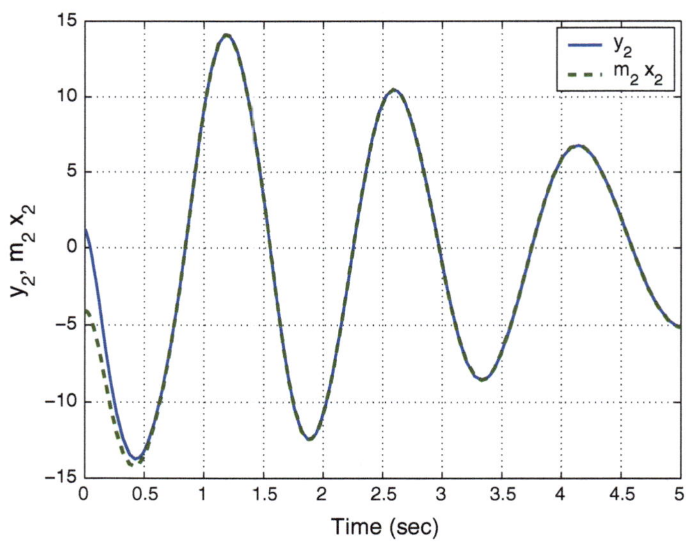

Fig. 6 Synchronization of the states y_2 and m_2x_2

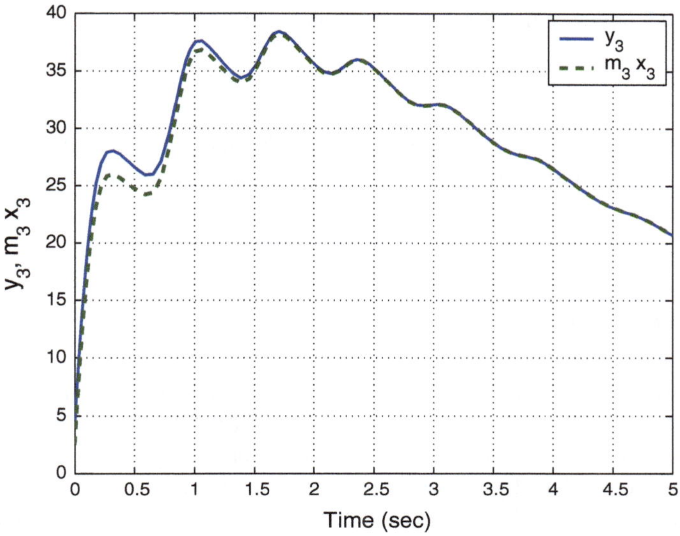

Fig. 7 Synchronization of the states y_3 and m_3x_3

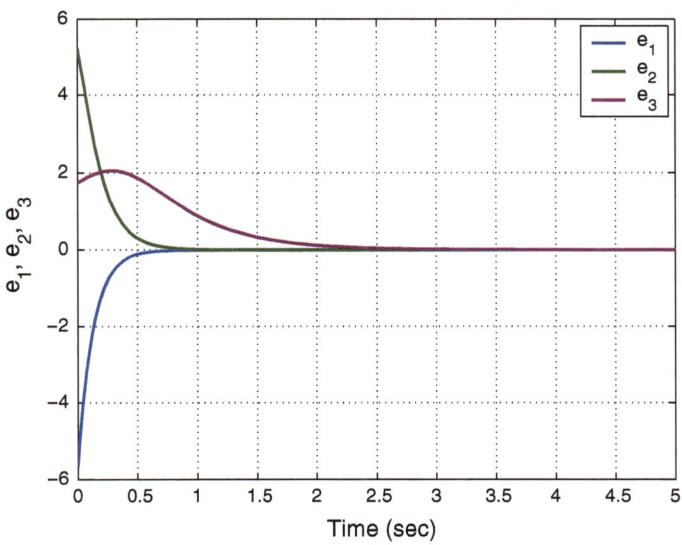

Fig. 8 Time-history of the GPS synchronization errors

As initial values of the parameter estimates, we take

$$\hat{a}(0) = 5.2, \quad \hat{b}(0) = 9.4 \tag{40}$$

Figures 5, 6 and 7 depict the GPS of the identical novel chaotic systems (19) and (20). Figure 8 depicts the time-history of the GPS synchronization errors e_1, e_2, e_3.

5 Conclusions

In this work, we described a 3-D six-term novel chaotic system with a quartic nonlinearity. We discussed the qualitative properties of the novel chaotic system are discussed in detail. We showed that the novel chaotic system has three unstable equilibrium points. The Lyapunov exponents of the novel chaotic system have been obtained as $L_1 = 0.1507$, $L_2 = 0$ and $L_3 = -0.9521$, while the Kaplan–Yorke dimension of the novel chaotic system has been obtained as $D_{KY} = 2.1583$. Using Lyapunov stability theory, we also derived an adaptive controller for the generalized projective synchronization (GPS) of the identical novel chaotic systems with unknown parameters. MATLAB plots have been depicted to illustrate the phase portraits of the novel chaotic system and also the GPS results for the novel chaotic systems using adaptive control.

References

1. Abdurrahman A, Jiang H, Teng Z (2015) Finite-time synchronization for memristor-based neural networks with time-varying delays. Neural Netw 69:20–28
2. Arneodo A, Coullet P, Tresser C (1981) Possible new strange attractors with spiral structure. Commun Math Phys 79(4):573–576
3. Azar AT, Vaidyanathan S (2015a) Chaos modeling and control systems design, Studies in computational intelligence, vol 581. Springer, Germany
4. Azar AT, Vaidyanathan S (2015b) Computational intelligence applications in modeling and control. Studies in computational intelligence, vol 575. Springer, Germany
5. Basnarkov L, Duane GS, Kocarev L (2014) Generalized synchronization and coherent structures in spatially extended systems. Chaos Solitons Fractals 59:35–41
6. Behnia S, Afrang S, Akhshani A, Mabhouti K (2013) A novel method for controlling chaos in external cavity semiconductor laser. Optik 124(8):757–764
7. Cai G, Tan Z (2007) Chaos synchronization of a new chaotic system via nonlinear control. J Uncertain Syst 1(3):235–240
8. Carroll TL, Pecora LM (1991) Synchronizing chaotic circuits. IEEE Trans Circuits Syst 38(4):453–456
9. Chen G, Ueta T (1999) Yet another chaotic attractor. Int J Bifurc Chaos 9(7):1465–1466
10. Hu A, Xu Z (2011) Multi-stable chaotic attractors in generalized synchronization. Commun Nonlinear Sci Numer Simul 16(8):3237–3244
11. Islam MM, Murase K (2005) Chaotic dynamics of a behaviour-based miniature mobile robot: effects of environment and control structure. Neural Netw 18(2):123–144

12. Karthikeyan R, Sundarapandian V (2014) Hybrid synchronization of four-scroll systems via active control. J Electr Eng 65(2):97–103
13. Khalil HK (2001) Nonlinear Syst, 3rd edn. Prentice Hall, New Jersey
14. Li D (2008) A three-scroll chaotic attractor. Phys Lett A 372(4):387–393
15. Lorenz EN (1963) Deterministic periodic flow. J Atmosph Sci 20(2):130–141
16. Lü J, Chen G (2002) A new chaotic attractor coined. Int J Bifurc Chaos 12(3):659–661
17. Matouk AE (2011) Chaos, feedback control and synchronization of a fractional-order modified autonomous Van der Pol-Duffing circuit. Commun Nonlinear Sci Numer Simul 16(2): 975–986
18. Pakiriswamy S, Vaidyanathan S (2012a) Generalized projective synchronization of hyperchaotic Lü and hyperchaotic Cai systems via active control. Lect Notes Inst Comput Sci, Soc-Inf Telecommun Eng 84:53–62
19. Pakiriswamy S, Vaidyanathan S (2012b) Generalized projective synchronization of three-scroll chaotic systems via active control. Lect Notes Inst Comput Sci, Soc-Inf Telecommun Eng 85:146–155
20. Pecora LM, Carroll TL (1990) Synchronization in chaotic systems. Phys Rev Lett 64(8):821–824
21. Pehlivan I, Moroz IM, Vaidyanathan S (2014) Analysis, synchronization and circuit design of a novel butterfly attractor. J Sound Vib 333(20):5077–5096
22. Pham VT, Volos CK, Vaidyanathan S, Le TP, Vu VY (2015) A memristor-based hyperchaotic system with hidden attractors: Dynamics, synchronization and circuital emulating. J Eng Sci Technol Rev 8(2):205–214
23. Rasappan S, Vaidyanathan S (2012a) Global chaos synchronization of WINDMI and Coullet chaotic systems by backstepping control. Far East J Math Sci 67(2):265–287
24. Rasappan S, Vaidyanathan S (2012b) Hybrid synchronization of n-scroll Chua and Lur'e chaotic systems via backstepping control with novel feedback. Arch Control Sci 22(3): 343–365
25. Rasappan S, Vaidyanathan S (2012c) Synchronization of hyperchaotic Liu system via backstepping control with recursive feedback. Commun Comput Inf Sci 305:212–221
26. Rasappan S, Vaidyanathan S (2013) Hybrid synchronization of n-scroll chaotic Chua circuits using adaptive backstepping control design with recursive feedback. Malays J Math Sci 7(2):219–246
27. Rasappan S, Vaidyanathan S (2014) Global chaos synchronization of WINDMI and Coullet chaotic systems using adaptive backstepping control design. Kyungpook Math J 54(1):293–320
28. Rhouma R, Belghith S (2011) Cryptanalysis of a chaos-based cryptosystem. Commun Nonlinear Sci Numer Simul 16(2):876–884
29. Rössler OE (1976) An equation for continuous chaos. Phys Lett A 57(5):397–398
30. Sampath S, Vaidyanathan S, Volos CK, Pham VT (2015) An eight-term novel four-scroll chaotic System with cubic nonlinearity and its circuit simulation. J Eng Sci Technol Rev 8(2):1–6
31. Sarasu P, Sundarapandian V (2011a) Active controller design for generalized projective synchronization of four-scroll chaotic systems. Int J Syst Signal Control Eng Appl 4(2):26–33
32. Sarasu P, Sundarapandian V (2011b) The generalized projective synchronization of hyperchaotic Lorenz and hyperchaotic Qi systems via active control. Int J Soft Comput 6(5):216–223
33. Sarasu P, Sundarapandian V (2012a) Adaptive controller design for the generalized projective synchronization of 4-scroll systems. Int J Syst Signal Control Eng Appl 5(2):21–30
34. Sarasu P, Sundarapandian V (2012b) Generalized projective synchronization of three-scroll chaotic systems via adaptive control. Eur J Sci Res 72(4):504–522
35. Sarasu P, Sundarapandian V (2012c) Generalized projective synchronization of two-scroll systems via adaptive control. Int J Soft Comput 7(4):146–156
36. Shi Y, Zhu P, Qin K (2014) Projective synchronization of different chaotic neural networks with mixed time delays based on an integral sliding mode controller. Neurocomputing 123:443–449
37. Sprott JC (1994) Some simple chaotic flows. Phys Rev E 50(2):647–650

38. Sundarapandian V (2010) Output regulation of the Lorenz attractor. Far East J Math Sci 42(2):289–299
39. Sundarapandian V (2013a) Adaptive control and synchronization design for the Lu-Xiao chaotic system. Lect Notes Electr Eng 131:319–327
40. Sundarapandian V (2013b) Analysis and anti-synchronization of a novel chaotic system via active and adaptive controllers. J Eng Sci Technol Rev 6(4):45–52
41. Sundarapandian V, Karthikeyan R (2011a) Anti-synchronization of hyperchaotic Lorenz and hyperchaotic Chen systems by adaptive control. Int J Syst Signal Control Eng Appl 4(2):18–25
42. Sundarapandian V, Karthikeyan R (2011b) Anti-synchronization of Lü and Pan chaotic systems by adaptive nonlinear control. Eur J Sci Res 64(1):94–106
43. Sundarapandian V, Karthikeyan R (2012a) Adaptive anti-synchronization of uncertain Tigan and Li systems. J Eng Appl Sci 7(1):45–52
44. Sundarapandian V, Karthikeyan R (2012b) Hybrid synchronization of hyperchaotic Lorenz and hyperchaotic Chen systems via active control. J Eng Appl Sci 7(3):254–264
45. Sundarapandian V, Pehlivan I (2012) Analysis, control, synchronization, and circuit design of a novel chaotic system. Math Comput Model 55(7–8):1904–1915
46. Sundarapandian V, Sivaperumal S (2011) Sliding controller design of hybrid synchronization of four-wing chaotic systems. Int J Soft Comput 6(5):224–231
47. Suresh R, Sundarapandian V (2013) Global chaos synchronization of a family of n-scroll hyperchaotic chua circuits using backstepping control with recursive feedback. Far East J Math Sci 73(1):73–95
48. Tigan G, Opris D (2008) Analysis of a 3D chaotic system. Chaos Solitons Fractals 36: 1315–1319
49. Tuwankotta JM (2006) Chaos in a coupled oscillators system with widely spaced frequencies and energy-preserving non-linearity. Int J Non-Linear Mech 41(2):180–191
50. Usama M, Khan MK, Alghathbar K, Lee C (2010) Chaos-based secure satellite imagery cryptosystem. Comput Math Appl 60(2):326–337
51. Vaidyanathan S (2011a) Hybrid chaos synchronization of Liu and Lü systems by active nonlinear control. Commun Comput Inf Sci 204:1–10
52. Vaidyanathan S (2011b) Output regulatoin of Arneodo-Coullet chaotic system. Commun Comput Inf Sci 133:98–107
53. Vaidyanathan S (2012a) Adaptive controller and synchronizer design for the Qi-Chen chaotic system. Lect Notes Inst Comput Sci, Soc-Inf Telecommun Eng 85:124–133
54. Vaidyanathan S (2012b) Analysis and synchronization of the hyperchaotic Yujun systems via sliding mode control. Adv Intell Syst Comput 176:329–337
55. Vaidyanathan S (2012c) Anti-synchronization of Sprott-L and Sprott-M chaotic systems via adaptive control. Int J Control Theory Appl 5(1):41–59
56. Vaidyanathan S (2012d) Global chaos control of hyperchaotic Liu system via sliding control method. Int J Control Theory Appl 5(2):117–123
57. Vaidyanathan S (2012e) Output regulatoin of the Liu chaotic system. Appl Mech Mater 110–116:3982–3989
58. Vaidyanathan S (2012f) Sliding mode control based global chaos control of Liu-Liu-Liu-Su chaotic system. Int J Control Theory Appl 5(1):15–20
59. Vaidyanathan S (2013a) A new six-term 3-D chaotic system with an exponential nonlinearity. Far East J Math Sci 79(1):135–143
60. Vaidyanathan S (2013b) Analysis and adaptive synchronization of two novel chaotic systems with hyperbolic sinusoidal and cosinusoidal nonlinearity and unknown parameters. J Eng Sci Technol Rev 6(4):53–65
61. Vaidyanathan S (2013c) Analysis, control and synchronization of hyperchaotic Zhou system via adaptive control. Adv Intell Syst Comput 177:1–10
62. Vaidyanathan S (2014a) A new eight-term 3-D polynomial chaotic system with three quadratic nonlinearities. Far East J Math Sci 84(2):219–226
63. Vaidyanathan S (2014b) Analysis and adaptive synchronization of eight-term 3-D polynomial chaotic systems with three quadratic nonlinearities. Eur Phys J: Spec Top 223(8):1519–1529

64. Vaidyanathan S (2014c) Analysis, control and synchronisation of a six-term novel chaotic system with three quadratic nonlinearities. Int J Model Identif Control 22(1):41–53
65. Vaidyanathan S (2014d) Generalized projective synchronisation of novel 3-D chaotic systems with an exponential non-linearity via active and adaptive control. Int J Model Identif Control 22(3):207–217
66. Vaidyanathan S (2014e) Global chaos synchronization of identical Li-Wu chaotic systems via sliding mode control. Int J Model Identif Control 22(2):170–177
67. Vaidyanathan S (2015a) 3-cells cellular neural network (CNN) attractor and its adaptive biological control. Int J PharmTech Res 8(4):632–640
68. Vaidyanathan S (2015b) A 3-D novel highly chaotic system with four quadratic nonlinearities, its adaptive control and anti-synchronization with unknown parameters. J Eng Sci Technol Rev 8(2):106–115
69. Vaidyanathan S (2015c) A novel chemical chaotic reactor system and its adaptive control. Int J ChemTech Res 8(7):146–158
70. Vaidyanathan S (2015d) Adaptive backstepping control of enzymes-substrates system with ferroelectric behaviour in brain waves. Int J PharmTech Res 8(2):256–261
71. Vaidyanathan S (2015e) Adaptive biological control of generalized Lotka-Volterra three-species biological system. Int J PharmTech Res 8(4):622–631
72. Vaidyanathan S (2015f) Adaptive chaotic synchronization of enzymes-substrates system with ferroelectric behaviour in brain waves. Int J PharmTech Res 8(5):964–973
73. Vaidyanathan S (2015g) Adaptive control of a chemical chaotic reactor. Int J PharmTech Res 8(3):377–382
74. Vaidyanathan S (2015h) Adaptive control of the FitzHugh-Nagumo chaotic neuron model. Int J PharmTech Res 8(6):117–127
75. Vaidyanathan S (2015i) Adaptive synchronization of chemical chaotic reactors. Int J ChemTech Res 8(2):612–621
76. Vaidyanathan S (2015j) Adaptive synchronization of generalized Lotka-Volterra three-species biological systems. Int J PharmTech Res 8(5):928–937
77. Vaidyanathan S (2015k) Adaptive synchronization of novel 3-D chemical chaotic reactor systems. Int J ChemTech Res 8(7):159–171
78. Vaidyanathan S (2015l) Adaptive synchronization of the identical FitzHugh-Nagumo chaotic neuron models. Int J PharmTech Res 8(6):167–177
79. Vaidyanathan S (2015m) Analysis, control and synchronization of a 3-D novel jerk chaotic system with two quadratic nonlinearities. Kyungpook Math J 55:563–586
80. Vaidyanathan S (2015n) Analysis, properties and control of an eight-term 3-D chaotic system with an exponential nonlinearity. Int J Model Identifn Control 23(2):164–172
81. Vaidyanathan S (2015o) Anti-synchronization of brusselator chemical reaction systems via adaptive control. Int J ChemTech Res 8(6):759–768
82. Vaidyanathan S (2015p) Chaos in neurons and adaptive control of Birkhoff-Shaw strange chaotic attractor. Int J PharmTech Res 8(5):956–963
83. Vaidyanathan S (2015q) Chaos in neurons and synchronization of Birkhoff-Shaw strange chaotic attractors via adaptive control. Int J PharmTech Res 8(6):1–11
84. Vaidyanathan S (2015r) Coleman-Gomatam logarithmic competitive biology models and their ecological monitoring. Int J PharmTech Res 8(6):94–105
85. Vaidyanathan S (2015s) Dynamics and control of brusselator chemical reaction. Int J ChemTech Res 8(6):740–749
86. Vaidyanathan S (2015t) Dynamics and control of tokamak system with symmetric and magnetically confined plasma. Int J ChemTech Res 8(6):795–803
87. Vaidyanathan S (2015u) Global chaos synchronization of chemical chaotic reactors via novel sliding mode control method. Int J ChemTech Res 8(7):209–221
88. Vaidyanathan S (2015v) Global chaos synchronization of the forced Van der Pol chaotic oscillators via adaptive control method. Int J PharmTech Res 8(6):156–166
89. Vaidyanathan S (2015w) Global chaos synchronization of the Lotka-Volterra biological systems with four competitive species via active control. Int J PharmTech Res 8(6):206–217

90. Vaidyanathan S (2015x) Lotka-Volterra population biology models with negative feedback and their ecological monitoring. Int J PharmTech Res 8(5):974–981
91. Vaidyanathan S (2015y) Lotka-Volterra two species competitive biology models and their ecological monitoring. Int J PharmTech Res 8(6):32–44
92. Vaidyanathan S (2015z) Output regulation of the forced Van der Pol chaotic oscillator via adaptive control method. Int J PharmTech Res 8(6):106–116
93. Vaidyanathan S, Azar AT (2015a) Analysis and control of a 4-D novel hyperchaotic system, vol. 581, pp. 3–17
94. Vaidyanathan S, Azar AT (2015b) Analysis, control and synchronization of a nine-term 3-D novel chaotic system. In: Azar AT, Vaidyanathan S (eds) Chaos modelling and control systems design. Studies in computational intelligence, vol 581. Springer, Germany, pp 19–38
95. Vaidyanathan S, Azar AT (2015c) Anti-synchronization of identical chaotic systems using sliding mode control and an application to Vaidhyanathan-Madhavan chaotic systems. Stud Comput Intell 576:527–547
96. Vaidyanathan S, Azar AT (2015d) Hybrid synchronization of identical chaotic systems using sliding mode control and an application to Vaidhyanathan chaotic systems. Stud Comput Intell 576:549–569
97. Vaidyanathan S, Madhavan K (2013) Analysis, adaptive control and synchronization of a seven-term novel 3-D chaotic system. Int J Control Theory Appl 6(2):121–137
98. Vaidyanathan S, Pakiriswamy S (2011) The design of active feedback controllers for the generalized projective synchronization of hyperchaotic Qi and hyperchaotic Lorenz systems. Commun Comput Inf Sci 245:231–238
99. Vaidyanathan S, Pakiriswamy S (2012) Generalized projective synchronization of double-scroll chaotic systems using active feedback control. Lect Notes Inst Comput Sci Soc-Inf Telecommun Eng 84:111–118
100. Vaidyanathan S, Pakiriswamy S (2013) Generalized projective synchronization of six-term Sundarapandian chaotic systems by adaptive control. Int J Control Theory Appl 6(2):153–163
101. Vaidyanathan S, Pakiriswamy S (2015) A 3-D novel conservative chaotic System and its generalized projective synchronization via adaptive control. J Eng Sci Technol Rev 8(2):52–60
102. Vaidyanathan S, Rajagopal K (2011a) Anti-synchronization of Li and T chaotic systems by active nonlinear control. Commun Comput Inf Sci 198:175–184
103. Vaidyanathan S, Rajagopal K (2011b) Global chaos synchronization of hyperchaotic Pang and Wang systems by active nonlinear control. Commun Comput Inf Sci 204:84–93
104. Vaidyanathan S, Rajagopal K (2011c) Global chaos synchronization of Lü and Pan systems by adaptive nonlinear control. Commun Comput Inf Sci 205:193–202
105. Vaidyanathan S, Rajagopal K (2011d) Hybrid synchronization of hyperchaotic Wang-Chen and hyperchaotic Lorenz systems by active non-linear control. Int J Syst Signal Control Eng Appl 4(3):55–61
106. Vaidyanathan S, Rajagopal K (2012) Global chaos synchronization of hyperchaotic Pang and hyperchaotic Wang systems via adaptive control. Int J Soft Comput 7(1):28–37
107. Vaidyanathan S, Rasappan S (2011a) Global chaos synchronization of hyperchaotic Bao and Xu systems by active nonlinear control. Commun Comput Inf Sci 198:10–17
108. Vaidyanathan S, Rasappan S (2011b) Hybrid synchronization of hyperchaotic Qi and Lü systems by nonlinear control. Commun Comput Inf Sci 131:585–593
109. Vaidyanathan S, Rasappan S (2014) Global chaos synchronization of n-scroll Chua circuit and Lur'e system using backstepping control design with recursive feedback. Arab J Sci Eng 39(4):3351–3364
110. Vaidyanathan S, Sampath S (2011) Global chaos synchronization of hyperchaotic Lorenz systems by sliding mode control. Commun Comput Inf Sci 205:156–164
111. Vaidyanathan S, Sampath S (2012) Anti-synchronization of four-wing chaotic systems via sliding mode control. Int J Autom Comput 9(3):274–279
112. Vaidyanathan S, Volos C (2015) Analysis and adaptive control of a novel 3-D conservative no-equilibrium chaotic system. Arch Control Sci 25(3):333–353

113. Vaidyanathan S, Volos C, Pham VT (2014a) Hyperchaos, adaptive control and synchronization of a novel 5-D hyperchaotic system with three positive Lyapunov exponents and its SPICE implementation. Arch Control Sci 24(4):409–446
114. Vaidyanathan S, Volos C, Pham VT, Madhavan K, Idowu BA (2014b) Adaptive backstepping control, synchronization and circuit simulation of a 3-D novel jerk chaotic system with two hyperbolic sinusoidal nonlinearities. Arch Control Sci 24(3):375–403
115. Vaidyanathan S, Idowu BA, Azar AT (2015a) Backstepping controller design for the global chaos synchronization of Sprott's jerk systems. Stud Comput Intell 581:39–58
116. Vaidyanathan S, Rajagopal K, Volos CK, Kyprianidis IM, Stouboulos IN (2015b) Analysis, adaptive control and synchronization of a seven-term novel 3-D chaotic system with three quadratic nonlinearities and its digital implementation in LabVIEW. J Eng Sci Technol Rev 8(2):130–141
117. Vaidyanathan S, Volos C, Pham VT, Madhavan K (2015c) Analysis, adaptive control and synchronization of a novel 4-D hyperchaotic hyperjerk system and its SPICE implementation. Arch Control Sci 25(1):135–158
118. Vaidyanathan S, Volos CK, Kyprianidis IM, Stouboulos IN, Pham VT (2015d) Analysis, adaptive control and anti-synchronization of a six-term novel jerk chaotic system with two exponential nonlinearities and its circuit simulation. J Eng Sci Technol Rev 8(2):24–36
119. Vaidyanathan S, Volos CK, Pham VT (2015e) Analysis, adaptive control and adaptive synchronization of a nine-term novel 3-D chaotic system with four quadratic nonlinearities and its circuit simulation. J Eng Sci Technol Rev 8(2):181–191
120. Vaidyanathan S, Volos CK, Pham VT (2015f) Analysis, control, synchronization and SPICE implementation of a novel 4-D hyperchaotic Rikitake dynamo system without equilibrium. J Eng Sci Technol Rev 8(2):232–244
121. Vaidyanathan S, Volos CK, Pham VT (2015g) Global chaos control of a novel nine-term chaotic system via sliding mode control. In: Azar AT, Zhu Q (eds) Advances and applications in sliding mode control systems, studies in computational intelligence, vol 576. Springer, Germany, pp 571–590
122. Vaidyanathan S, Volos CK, Rajagopal K, Kyprianidis IM, Stouboulos IN (2015h) Adaptive backstepping controller design for the anti-synchronization of identical WINDMI chaotic systems with unknown parameters and its SPICE implementation. J Eng Sci Technol Rev 8(2):74–82
123. Volos CK, Kyprianidis IM, Stouboulos IN, Anagnostopoulos AN (2009) Experimental study of the dynamic behavior of a double scroll circuit. J Appl Funct Anal 4:703–711
124. Volos CK, Kyprianidis IM, Stouboulos IN (2013) Experimental investigation on coverage performance of a chaotic autonomous mobile robot. Robot Auton Syst 61(12):1314–1322
125. Volos CK, Kyprianidis IM, Stouboulos IN, Tlelo-Cuautle E, Vaidyanathan S (2015) Memristor: A new concept in synchronization of coupled neuromorphic circuits. J Eng Sci Technol Rev 8(2):157–173
126. Wei Z, Yang Q (2010) Anti-control of Hopf bifurcation in the new chaotic system with two stable node-foci. Appl Math Comput 217(1):422–429
127. Yang J, Zhu F (2013) Synchronization for chaotic systems and chaos-based secure communications via both reduced-order and step-by-step sliding mode observers. Commun Nonlinear Sci Numer Simul 18(4):926–937
128. Yang J, Chen Y, Zhu F (2014) Singular reduced-order observer-based synchronization for uncertain chaotic systems subject to channel disturbance and chaos-based secure communication. Appl Math Comput 229:227–238
129. Yu J, Hu C, Jiang H, Fan X (2014) Projective synchronization for fractional neural networks. Neural Netw 49:87–95
130. Yuan Z, Xu Z, Guo L (2012) Generalized synchronization of two bidirectionally coupled discrete dynamical systems. Commun Nonlinear Sci Numer Simul 17(2):992–1002
131. Zhang D, Xu J (2010) Projective synchronization of different chaotic time-delayed neural networks based on integral sliding mode controller. Appl Math Comput 217(1):164–174

132. Zhou W, Xu Y, Lu H, Pan L (2008) On dynamics analysis of a new chaotic attractor. Phys Lett A 372(36):5773–5777
133. Zhu C, Liu Y, Guo Y (2010) Theoretic and numerical study of a new chaotic system. Intell Inf Manag 2:104–109

A Novel 4-D Hyperchaotic Chemical Reactor System and Its Adaptive Control

Sundarapandian Vaidyanathan and Abdesselem Boulkroune

Abstract Chaos in nonlinear dynamics occurs widely in physics, chemistry, biology, ecology, secure communications, cryptosystems and many scientific branches. Chaotic systems have important applications in science and engineering. In this work, we derive a twelve-term novel 4-D hyperchaotic system by introducing a state feedback control to the 3-D chemical chaotic reactor obtained by Huang, Yang, J Math Chem 38(1):107–117, 2015, [11]. The phase portraits of the twelve-term novel hyperchaotic chemical reactor system are depicted and the qualitative properties of the novel hyperchaotic system are discussed. The Lyapunov exponents of the novel hyperchaotic chemical reactor system are obtained as $L_1 = 0.2263$, $L_2 = 0.0365$, $L_3 = 0$ and $L_4 = -10.8396$. Also, the Kaplan–Yorke dimension of the novel hyperchaotic chemical reactor system is obtained as $D_{KY} = 3.0240$. Since the sum of the Lyapunov exponents is negative, the novel hyperchaotic system is dissipative. Next, an adaptive controller is designed to globally stabilize the novel hyperchaotic system with unknown parameters. Finally, an adaptive controller is also designed to achieve global chaos synchronization of the identical hyperchaotic systems with unknown parameters. MATLAB simulations are depicted to illustrate all the main results derived in this work.

Keywords Chaos · Chaotic systems · Chemical reactor · Hyperchaos · Hyperchaotic systems · Adaptive control · Synchronization

---

S. Vaidyanathan (✉)
Research and Development Centre, Vel Tech University, Avadi, Chennai 600062, Tamil Nadu, India
e-mail: sundarvtu@gmail.com

A. Boulkroune
LAJ, University of Jijel, BP. 98, 18000 Ouled-aissa, Jijel, Algeria
e-mail: boulkroune2002@yahoo.fr

S. Vaidyanathan and C. Volos (eds.), *Advances and Applications in Chaotic Systems*, Studies in Computational Intelligence 636,
DOI 10.1007/978-3-319-30279-9_19

1 Introduction

Chaos theory describes the quantitative study of unstable aperiodic dynamic behaviour in deterministic nonlinear dynamical systems. For the motion of a dynamical system to be chaotic, the system variables should contain some nonlinear terms and the system must satisfy three properties: boundedness, infinite recurrence and sensitive dependence on initial conditions [3].

Some classical paradigms of 3-D chaotic systems in the literature are Lorenz system [18], Rössler system [26], ACT system [2], Sprott systems [33], Chen system [6], Lü system [19], Cai system [4], Tigan system [43], etc.

Many new chaotic systems have been discovered in the recent years such as Zhou system [114], Zhu system [116], Li system [16], Wei–Yang system [110], Sundarapandian systems [36, 40], Vaidyanathan systems [50, 52, 54–57, 61, 72, 73, 87, 88, 90, 96, 98, 100, 103, 104, 106], Pehlivan system [20], Sampath system [28], etc.

Chaos theory has applications in several fields of science and engineering such as chemical reactors [62, 66, 68, 70, 74, 78–80], biological systems [60, 63–65, 67, 69, 71, 75–77, 81–85], memristors [1, 22, 107], etc.

A hyperchaotic system is defined as a chaotic system with at least two positive Lyapunov exponents [3]. Thus, the dynamics of a hyperchaotic system can expand in several different directions simultaneously. Thus, the hyperchaotic systems have more complex dynamical behaviour and they have miscellaneous applications in engineering such as secure communications [7, 15, 111], cryptosystems [10, 25, 115], fuzzy logic [32, 113], electrical circuits [109, 112], etc.

The first 4-D hyperchaotic system was found by Rössler [27]. Many hyperchaotic systems have been reported in the chaos literature such as hyperchaotic Lorenz system [12], hyperchaotic Lü system [5], hyperchaotic Chen system [17], hyperchaotic Wang system [108], hyperchaotic Vaidyanathan systems [51, 59, 86, 97, 102, 105], hyperchaotic Pham system [21], etc.

The control of a chaotic or hyperchaotic system aims to stabilize or regulate the system with the help of a feedback control. There are many methods available for controlling a chaotic system such as active control [34, 44, 45], adaptive control [35, 46, 53], sliding mode control [48, 49], backstepping control [99], etc.

The synchronization of chaotic systems aims to synchronize the states of master and slave systems asymptotically with time. There are many methods available for chaos synchronization such as active control [13, 29, 30, 91, 93], adaptive control [31, 37–39, 47, 89, 92], sliding mode control [41, 58, 95, 101], backstepping control [23, 24, 42, 94], etc.

Recently, Huang and Yang derived a chemical reactor model by considering reactor dynamics with five steps [11]. For the non-dimensionalized dynamical evolutions of the Huang–Yang chaotic reactor, the Lyapunov exponents were obtained as $L_1 = 0.4001$, $L_2 = 0$ and $L_3 = -11.8762$.

In this research work, we derive a novel 4-D hyperchaotic chemical reactor system by adding a state feedback control to the 3-D Huang–Yang chemical reactor.

This work is organized as follows. Section 2 describes the dynamics of the twelve-term novel 4-D hyperchaotic chemical reactor system. Section 3 details the qualitative properties of the novel hyperchaotic chemical system. The Lyapunov exponents of the novel hyperchaotic chemical system are obtained as $L_1 = 0.4119$, $L_2 = 0.0434$, $L_3 = 0$ and $L_4 = -11.9003$, while the Kaplan–Yorke dimension of the novel hyperchaotic system is obtained as $D_{KY} = 3.0383$.

In Sect. 4, we design an adaptive controller to globally stabilize the novel hyperchaotic chemical reactor system with unknown parameters. In Sect. 5, an adaptive controller is designed to achieve global chaos synchronization of the identical novel hyperchaotic chemical reactor systems with unknown parameters. Section 6 contains the conclusions of this work.

2 A Novel 4-D Hyperchaotic Chemical Reactor System

The well-stirred chemical reactor dynamics model of Huang–Yang [11] consist of the following five steps given below.

$$A_1 + X \underset{k_1}{\overset{k_{-1}}{\rightleftharpoons}} 2X \tag{1a}$$

$$X + Y \overset{k_2}{\rightarrow} 2Y \tag{1b}$$

$$A_5 + Y \overset{k_3}{\rightarrow} A_2 \tag{1c}$$

$$X + Z \overset{k_4}{\rightarrow} A_3 \tag{1d}$$

$$A_5 + Z \underset{k_5}{\overset{k_{-5}}{\rightleftharpoons}} 2Z \tag{1e}$$

Equations (1a) and (1e) indicate reversible steps, while Eqs. (1c)–(1e) indicate non-reversible steps of the Huang–Yang chemical reactor [11].

In (1), A_1, A_4, A_5 are initiators and A_2, A_3 are products. The intermediates whose dynamics are followed are X, Y and Z.

Assuming an ideal mixture and a well-stirred reactor, the macroscopic rate equations for the Huang–Yang chemical reactor can be written in non-dimensionalized form as

$$\begin{cases} \dot{x} = a_1 x - k_{-1} x^2 - xy - xz \\ \dot{y} = xy - a_5 y \\ \dot{z} = a_4 z - xz - k_{-5} z^2 \end{cases} \tag{2}$$

In (2), x, y and z are the mole fractions of X, Y and Z. Also, the rate constants k_1, k_3 and k_5 are incorporated in the parameters a_1, a_5 and a_4.

To simplify the notations, we rename the constants and express the chemical reactor system (2) as

$$\begin{cases} \dot{x} = ax - px^2 - xy - xz \\ \dot{y} = xy - cy \\ \dot{z} = bz - xz - qz^2 \end{cases} \tag{3}$$

Huang and Yang [11] showed that the chemical reactor system (3) is *chaotic,* when the system parameters are chosen as

$$a = 30, \quad b = 16.5, \quad c = 10, \quad p = 0.415, \quad q = 0.5 \tag{4}$$

The Lyapunov exponents of the Huang–Yang chemical reactor system (3) for the parameter values (4) are numerically obtained as

$$L_1 = 0.2138, \quad L_2 = 0, \quad L_3 = -10.8292 \tag{5}$$

In this section, we derive a twelve-term novel hyperchaotic system by introducing a state feedback control to the Huang–Yang chemical reactor system (3).

Our novel hyperchaotic chemical chaotic reactor system is described by the 4-D dynamics

$$\begin{cases} \dot{x} = ax - px^2 - xy - xz \\ \dot{y} = xy - cy + sw \\ \dot{z} = bz - xz - qz^2 \\ \dot{w} = -r(x + z) \end{cases} \tag{6}$$

where x, y, z, w are the states and a, b, c, p, q, r, s are constant positive parameters.

The 4-D system (6) consists of twelve terms on the right hand side with six quadratic nonlinearities.

The system (6) exhibits a *strange hyperchaotic attractor* for the parameter values

$$a = 30, \quad b = 16.5, \quad c = 10, \quad p = 0.415, \quad q = 0.5, \quad r = 0.07, \quad s = 0.0001 \tag{7}$$

For numerical simulations, we take the initial conditions as

$$x(0) = 3, \quad y(0) = 0, \quad z(0) = 0.1, \quad w(0) = 0.1 \tag{8}$$

Figures 1, 2, 3 and 4 the 3-D projection of the novel hyperchaotic system (6) on the (x, y, z), (x, y, w), (x, z, w) and (y, z, w) spaces, respectively.

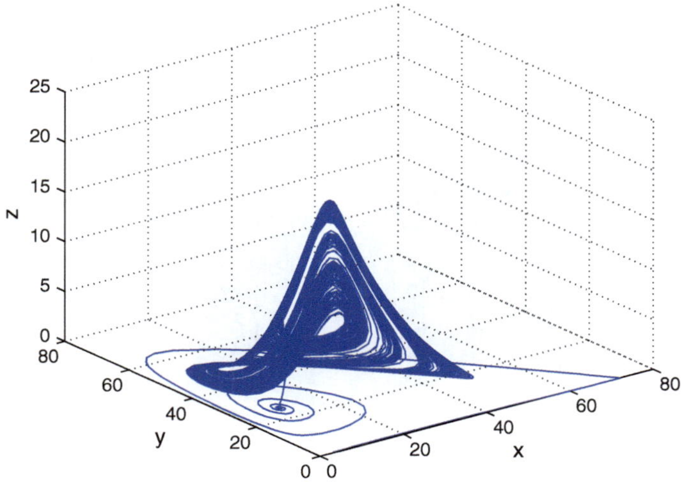

Fig. 1 3-D projection of the novel hyperchaotic system on the (x, y, z) space

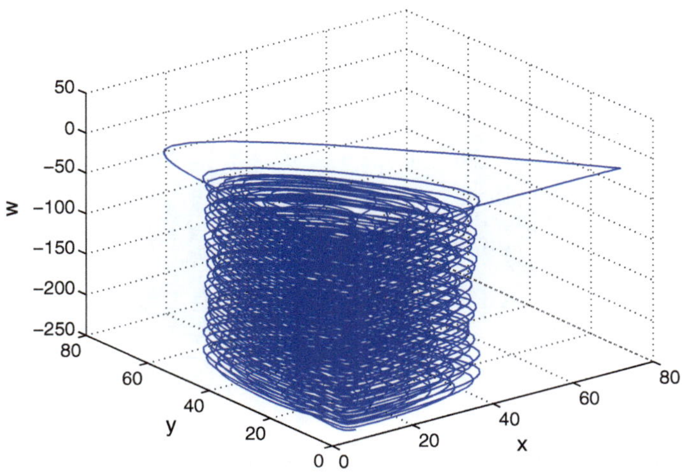

Fig. 2 3-D projection of the novel hyperchaotic system on the (x, y, w) space

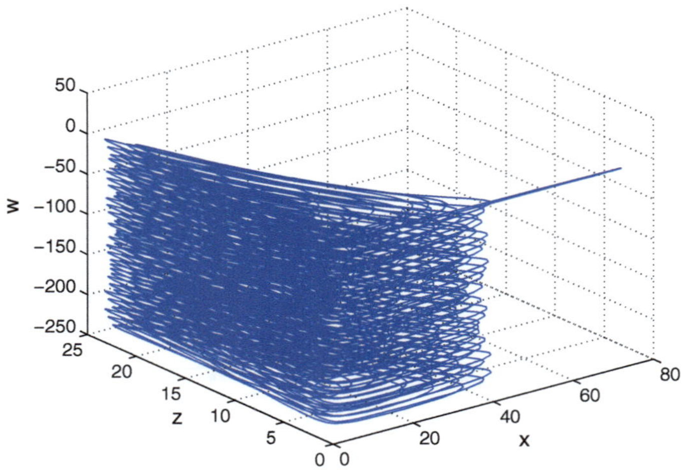

Fig. 3 3-D projection of the novel hyperchaotic system on the (x, z, w) space

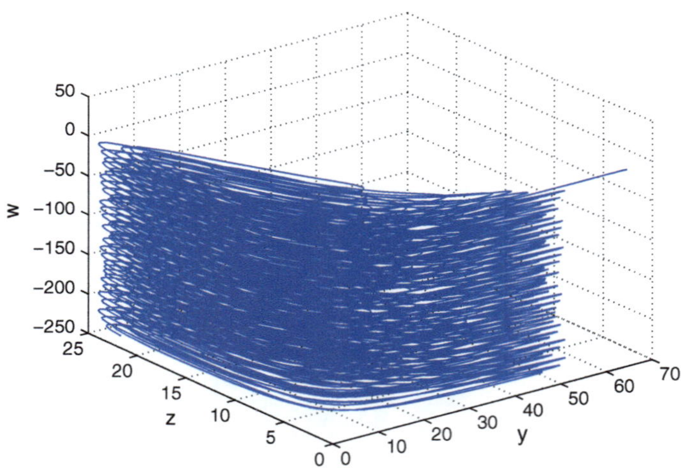

Fig. 4 3-D projection of the novel hyperchaotic system on the (y, z, w) space

3 Analysis of the Novel 4-D Hyperchaotic Chemical Reactor System

In this section, we give a dynamic analysis of the novel 4-D hyperchaotic chemical reactor system (6). We take the parameter values as in the hyperchaotic case (7).

3.1 Equilibrium Points

It is easy to see that the hyperchaotic chemical reactor system (6) has an equilibrium at the origin.

We take the parameter values as in the hyperchaotic case (7).

To test the stability type of the equilibrium point $E_0 = \mathbf{0}$, we calculate the Jacobian matrix of the novel hyperchaotic system (6) at $E_0 = \mathbf{0}$:

We find that

$$J \triangleq J(E_0) = \begin{bmatrix} a & 0 & 0 & 0 \\ 0 & -c & 0 & s \\ 0 & 0 & b & 0 \\ -r & 0 & -r & 0 \end{bmatrix} = \begin{bmatrix} 30 & 0 & 0 & 0 \\ 0 & -10 & 0 & 0.0001 \\ 0 & 0 & 16.5 & 0 \\ -0.07 & 0 & -0.07 & 0 \end{bmatrix} \tag{9}$$

The matrix J has the eigenvalues

$$\lambda_1 = -10, \quad \lambda_2 = 0, \quad \lambda_3 = 16.5, \quad \lambda_4 = 30 \tag{10}$$

This shows that the equilibrium point $E_0 = \mathbf{0}$ is a saddle-point, which is unstable.

3.2 Invariance

It is easy to see that the (z, w) plane is invariant under the flow of the novel 4-D hyperchaotic system (6). The invariant motion along the (z, w)-plane is characterized by the planar dynamics

$$\begin{cases} \dot{z} = bz - qz^2 \\ \dot{w} = -rz \end{cases} \tag{11}$$

which is unstable because b is a positive constant.

3.3 Lyapunov Exponents and Kaplan–Yorke Dimension

We take the parameter values of the 4-D chemical reactor system (6) as in the hyperchaotic case (7).

We take the initial state of the novel system (6) as given in (8).

Then the Lyapunov exponents of the system (6) are numerically obtained using MATLAB as

$$L_1 = 0.2263, \quad L_2 = 0.0365, \quad L_3 = 0, \quad L_4 = -10.8396 \tag{12}$$

Since there are two positive Lyapunov exponents in (12), the novel system (6) exhibits *hyperchaotic* behavior.

Since $L_1 + L_2 + L_3 + L_4 = -10.6768 < 0$, it follows that the novel hyperchaotic system (6) is dissipative.

The Kaplan–Yorke dimension [8, 9] of a chaotic system of order n is defined as

$$D_{KY} = j + \frac{L_1 + \cdots + L_j}{|L_{j+1}|} \tag{13}$$

where $L_1 \geq L_2 \geq \cdots \geq L_n$ are the Lyapunov exponents of the chaotic system and j is the largest integer for which $L_1 + L_2 + \cdots + L_j \geq 0$. (Kaplan–Yorke conjecture states that for typical chaotic systems, $D_{KY} \approx D_L$, the information dimension of the system.)

Thus, the Kaplan–Yorke dimension of the novel hyperchaotic chemical reactor system (6) is calculated as

$$D_{KY} = 3 + \frac{L_1 + L_2 + L_3}{|L_4|} = 3.0240, \tag{14}$$

which is fractional.

4 Adaptive Control of the Novel 4-D Hyperchaotic Chemical Reactor System

In this section, we use adaptive control method to derive an adaptive feedback control law for globally stabilizing the novel 4-D hyperchaotic chemical reactor system with unknown parameters.

Thus, we consider the novel 4-D hyperchaotic system given by

$$\begin{cases} \dot{x} = ax - px^2 - xy - xz + u_x \\ \dot{y} = xy - cy + sw + u_y \\ \dot{z} = bz - xz - qz^2 + u_z \\ \dot{w} = -r(x+z) + u_w \end{cases} \tag{15}$$

In (15), x, y, z, w are the states and u_x, u_y, u_z, u_w are the adaptive controls to be determined using estimates of the unknown system parameters.

We consider the adaptive feedback control law

$$\begin{cases} u_x = -\hat{a}(t)x + \hat{p}(t)x^2 + xy + xz - k_x x \\ u_y = -xy + \hat{c}(t)y - \hat{s}(t)w - k_y y \\ u_z = -\hat{b}(t)z + xz + \hat{q}(t)z^2 - k_z z \\ u_w = \hat{r}(t)(x+z) - k_w w \end{cases} \tag{16}$$

where k_x, k_y, k_z, k_w are positive gain constants.

Substituting (16) into (15), we get the closed-loop plant dynamics as

$$\begin{cases} \dot{x} = [a - \hat{a}(t)]x - [p - \hat{p}(t)]x^2 - k_x x \\ \dot{y} = -[c - \hat{c}(t)]y + [s - \hat{s}(t)]w - k_y y \\ \dot{z} = [b - \hat{b}(t)]z - [q - \hat{q}(t)]z^2 - k_z z \\ \dot{w} = -[r - \hat{r}(t)](x + z) - k_w w \end{cases} \tag{17}$$

The parameter estimation errors are defined as

$$\begin{cases} e_a(t) = a - \hat{a}(t) \\ e_b(t) = b - \hat{b}(t) \\ e_c(t) = c - \hat{c}(t) \\ e_p(t) = p - \hat{p}(t) \\ e_q(t) = q - \hat{q}(t) \\ e_r(t) = r - \hat{r}(t) \\ e_s(t) = s - \hat{s}(t) \end{cases} \tag{18}$$

In view of (18), we can simplify the plant dynamics (17) as

$$\begin{cases} \dot{x} = e_a x - e_p x^2 - k_x x \\ \dot{y} = -e_c y + e_s w - k_y y \\ \dot{z} = e_b z - e_q z^2 - k_z z \\ \dot{w} = -e_r(x + z) - k_w w \end{cases} \tag{19}$$

Differentiating (18) with respect to t, we obtain

$$\begin{cases} \dot{e}_a = -\dot{\hat{a}} \\ \dot{e}_b = -\dot{\hat{b}} \\ \dot{e}_c = -\dot{\hat{c}} \\ \dot{e}_p = -\dot{\hat{p}} \\ \dot{e}_q = -\dot{\hat{q}} \\ \dot{e}_r = -\dot{\hat{r}} \\ \dot{e}_s = -\dot{\hat{s}} \end{cases} \tag{20}$$

We consider the quadratic candidate Lyapunov function defined by

$$V = \frac{1}{2}\left(x^2 + y^2 + z^2 + w^2\right) + \frac{1}{2}\left(e_a^2 + e_b^2 + e_c^2 + e_p^2 + e_q^2 + e_r^2 + e_s^2\right) \tag{21}$$

Differentiating V along the trajectories of (19) and (20), we obtain

$$\begin{aligned} \dot{V} = & -k_x x^2 - k_y y^2 - k_z z^2 - k_w w^2 + e_a\left[x^2 - \dot{\hat{a}}\right] + e_b\left[z^2 - \dot{\hat{b}}\right] + e_c\left[-y^2 - \dot{\hat{c}}\right] \\ & + e_p\left[-x^3 - \dot{\hat{p}}\right] + e_q\left[-z^3 - \dot{\hat{q}}\right] + e_r\left[-(x+z)w - \dot{\hat{r}}\right] + e_s\left[yw - \dot{\hat{s}}\right] \end{aligned} \tag{22}$$

In view of (22), we take the parameter update law as

$$\begin{cases} \dot{\hat{a}}(t) = x^2 \\ \dot{\hat{b}}(t) = z^2 \\ \dot{\hat{c}}(t) = -y^2 \\ \dot{\hat{p}}(t) = -x^3 \\ \dot{\hat{q}}(t) = -z^3 \\ \dot{\hat{r}}(t) = -(x+z)w \\ \dot{\hat{s}}(t) = yw \end{cases} \tag{23}$$

Next, we state and prove the main result of this section.

Theorem 1 *The novel 4-D hyperchaotic chemical reactor system (15) with unknown system parameters is globally and exponentially stabilized for all initial conditions by the adaptive control law (16) and the parameter update law (23), where k_x, k_y, k_z, k_w are positive gain constants.*

Proof We prove this result by applying Lyapunov stability theory [14].

We consider the quadratic Lyapunov function defined by (21), which is clearly a positive definite function on $\mathbf{R}^{11}$.

By substituting the parameter update law (23) into (22), we obtain the time-derivative of V as

$$\dot{V} = -k_x x^2 - k_y y^2 - k_z z^2 - k_w w^2 \tag{24}$$

From (24), it is clear that $\dot{V}$ is a negative semi-definite function on $\mathbf{R}^{11}$.

Thus, we can conclude that the state vector $X(t) = (x(t), y(t), z(t), w(t))$ and the parameter estimation error are globally bounded.

We define $k = \min\{k_x, k_y, k_z, k_w\}$.

Then it follows from (24) that

$$\dot{V} \leq -k\|X(t)\|^2 \tag{25}$$

Thus, we have

$$k\|X(t)\|^2 \leq -\dot{V} \tag{26}$$

Integrating the inequality (26) from 0 to t, we get

$$k \int_0^t \|X(\tau)\|^2 \, d\tau \leq V(0) - V(t) \tag{27}$$

From (27), it follows that $X \in \mathbf{L}_2$.

Using (19), we can conclude that $\dot{X} \in \mathbf{L}_\infty$.

Using Barbalat's lemma [14], we conclude that $X(t) \to 0$ exponentially as $t \to \infty$ for all initial conditions $X(0) \in \mathbf{R}^4$.

This completes the proof. ■

For the numerical simulations, the classical fourth-order Runge–Kutta method with step size $h = 10^{-8}$ is used to solve the systems (15) and (23), when the adaptive control law (16) is applied.

The parameter values of the novel 4-D hyperchaotic system (15) are taken as in the hyperchaotic case (7), i.e.

$$a = 30, \;\; b = 16.5, \;\; c = 10, \;\; p = 0.415, \;\; q = 0.5, \;\; r = 0.07, \;\; s = 0.0001 \tag{28}$$

We take the positive gain constants as $k_x = 20$, $k_y = 10$, $k_z = 10$ and $k_w = 10$.

Furthermore, as initial conditions of the novel 4-D hyperchaotic chemical reactor system (15), we take

$$x(0) = 5.1, \;\; y(0) = 2.3, \;\; z(0) = 1.8, \;\; w(0) = 2.7 \tag{29}$$

Also, as initial conditions of the parameter estimates, we take

$$\hat{a}(0) = 2, \; \hat{b}(0) = 5, \; \hat{c}(0) = 3, \; \hat{p}(0) = 12, \; \hat{q}(0) = 24, \; \hat{r}(0) = 8, \; \hat{s}(0) = 10 \tag{30}$$

In Fig. 5, the exponential convergence of the controlled states of the novel 4-D hyperchaotic chemical reactor system (15) is depicted.

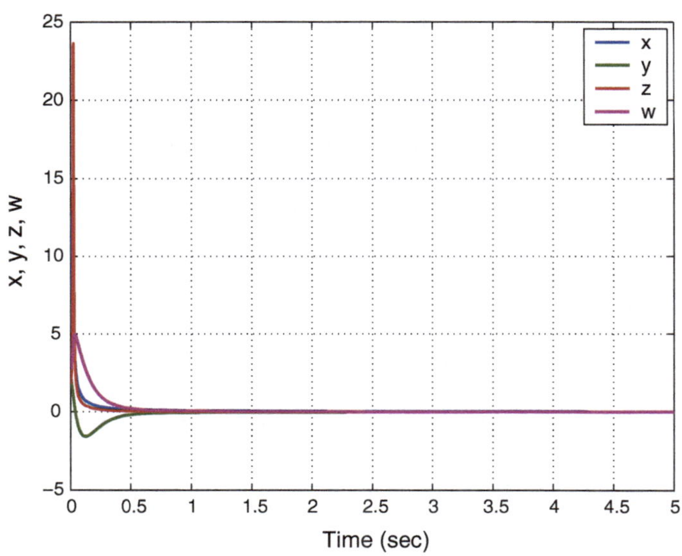

Fig. 5 Time-history of the controlled states x, y, z, w

5 Adaptive Synchronization of the Identical Novel Hyperchaotic Chemical Reactor Systems

In this section, we use adaptive control method to derive an adaptive feedback control law for globally synchronizing identical novel 4-D hyperchaotic chemical chaotic reactor systems with unknown parameters.

As the master system, we consider the novel 4-D hyperchaotic system given by

$$\begin{cases} \dot{x}_1 &= ax_1 - px_1^2 - x_1y_1 - x_1z_1 \\ \dot{y}_1 &= x_1y_1 - cy_1 + sw_1 \\ \dot{z}_1 &= bz_1 - x_1z_1 - qz_1^2 \\ \dot{w}_1 &= -r(x_1 + z_1) \end{cases} \tag{31}$$

In (31), x_1, y_1, z_1, w_1 are the states and a, b, c, p, q, r, s are unknown parameters.

As the slave system, we consider the novel 4-D hyperchaotic system given by

$$\begin{cases} \dot{x}_2 &= ax_2 - px_2^2 - x_2y_2 - x_2z_2 + u_x \\ \dot{y}_2 &= x_2y_2 - cy_2 + sw_2 + u_y \\ \dot{z}_2 &= bz_2 - x_2z_2 - qz_2^2 + u_z \\ \dot{w}_2 &= -r(x_2 + z_2) + u_w \end{cases} \tag{32}$$

In (32), y_1, y_2, y_3, y_4 are the states and u_x, u_y, u_z, u_w are the adaptive controls.

The synchronization error is defined by

$$\begin{cases} e_x &= x_2 - x_1 \\ e_y &= y_2 - y_1 \\ e_z &= z_2 - z_1 \\ e_w &= w_2 - w_1 \end{cases} \tag{33}$$

Then the synchronization error dynamics is obtained as

$$\begin{cases} \dot{e}_x &= ae_x - p(x_2^2 - x_1^2) - x_2y_2 + x_1y_1 - x_2z_2 + x_1z_1 + u_x \\ \dot{e}_y &= -ce_y + x_2y_2 - x_1y_1 + se_w + u_y \\ \dot{e}_z &= be_z - x_2z_2 + x_1z_1 - q(z_2^2 - z_1^2) + u_z \\ \dot{e}_w &= -r(e_x + e_z) + u_w \end{cases} \tag{34}$$

We consider the adaptive feedback control law

$$\begin{cases} u_x &= -\hat{a}(t)e_x + \hat{p}(t)(x_2^2 - x_1^2) + x_2y_2 - x_1y_1 + x_2z_2 - x_1z_1 - k_xe_x \\ u_y &= \hat{c}(t)e_y - x_2y_2 + x_1y_1 - \hat{s}(t)e_w - k_ye_y \\ u_z &= -\hat{b}(t)e_z + x_2z_2 - x_1z_1 + \hat{q}(t)(z_2^2 - z_1^2) - k_ze_z \\ u_w &= \hat{r}(t)(e_x + e_z) - k_we_w \end{cases} \tag{35}$$

In (35), k_x, k_y, k_z, k_w are positive gain constants.

Substituting (35) into (34), we get the closed-loop error dynamics as

$$\begin{cases} \dot{e}_x = [a - \hat{a}(t)]e_x - [p - \hat{p}(t)](x_2^2 - x_1^2) - k_x e_x \\ \dot{e}_y = -[c - \hat{c}(t)]e_y + [s - \hat{s}(t)]e_w - k_y e_y \\ \dot{e}_z = [b - \hat{b}(t)]e_z - [q - \hat{q}(t)](z_2^2 - z_1^2) - k_z e_z \\ \dot{e}_w = -[r - \hat{r}(t)](e_x + e_z) - k_w e_w \end{cases} \tag{36}$$

The parameter estimation errors are defined as

$$\begin{cases} e_a(t) = a - \hat{a}(t) \\ e_b(t) = b - \hat{b}(t) \\ e_c(t) = c - \hat{c}(t) \\ e_p(t) = p - \hat{p}(t) \\ e_q(t) = q - \hat{q}(t) \\ e_r(t) = r - \hat{r}(t) \\ e_s(t) = s - \hat{s}(t) \end{cases} \tag{37}$$

In view of (37), we can simplify the error dynamics (36) as

$$\begin{cases} \dot{e}_x = e_a e_x - e_p(x_2^2 - x_1^2) - k_x e_x \\ \dot{e}_y = -e_c e_y + e_s e_w - k_y e_y \\ \dot{e}_z = e_b e_z - e_q(z_2^2 - z_1^2) - k_z e_z \\ \dot{e}_w = -e_r(e_x + e_z) - k_w e_w \end{cases} \tag{38}$$

Differentiating (37) with respect to t, we obtain

$$\begin{cases} \dot{e}_a = -\dot{\hat{a}} \\ \dot{e}_b = -\dot{\hat{b}} \\ \dot{e}_c = -\dot{\hat{c}} \\ \dot{e}_p = -\dot{\hat{p}} \\ \dot{e}_q = -\dot{\hat{q}} \\ \dot{e}_r = -\dot{\hat{r}} \\ \dot{e}_s = -\dot{\hat{s}} \end{cases} \tag{39}$$

We consider the quadratic candidate Lyapunov function defined by

$$V = \frac{1}{2}\left(e_x^2 + e_y^2 + e_z^2 + e_w^2\right) + \frac{1}{2}\left(e_a^2 + e_b^2 + e_c^2 + e_p^2 + e_q^2 + e_r^2 + e_s^2\right) \tag{40}$$

Differentiating V along the trajectories of (38) and (39), we obtain

$$\begin{aligned}\dot{V} &= -k_x e_x^2 - k_y e_y^2 - k_z e_z^2 - k_w e_w^2 + e_a\left[e_x^2 - \dot{\hat{a}}\right] + e_b\left[e_z^2 - \dot{\hat{b}}\right] \\ &+ e_c\left[-e_y^2 - \dot{\hat{c}}\right] + e_p\left[-e_x(x_2^2 - x_1^2) - \dot{\hat{p}}\right] + e_q\left[-e_z(z_2^2 - z_1^2) - \dot{\hat{q}}\right] \\ &+ e_r\left[-(e_x + e_z)e_w - \dot{\hat{r}}\right] + e_s\left[e_y e_w - \dot{\hat{s}}\right]\end{aligned} \tag{41}$$

In view of (41), we take the parameter update law as

$$\begin{cases} \dot{\hat{a}}(t) = e_x^2 \\ \dot{\hat{b}}(t) = e_z^2 \\ \dot{\hat{c}}(t) = -e_y^2 \\ \dot{\hat{p}}(t) = -e_x(x_2^2 - x_1^2) \\ \dot{\hat{q}}(t) = -e_z(z_2^2 - z_1^2) \\ \dot{\hat{r}}(t) = -(e_x + e_z)e_w \\ \dot{\hat{s}}(t) = e_y e_w \end{cases} \tag{42}$$

Theorem 2 *The novel hyperchaotic chemical reactor systems (31) and (32) with unknown system parameters are globally and exponentially synchronized for all initial conditions by the adaptive control law (35) and the parameter update law (42), where k_x, k_y, k_z, k_w are positive gain constants.*

Proof We consider the quadratic Lyapunov function defined by (40), which is clearly a positive definite function on $\mathbf{R}^{11}$.

By substituting the parameter update law (42) into (41), we obtain

$$\dot{V} = -k_x e_x^2 - k_y e_y^2 - k_z e_z^2 - k_w e_w^2 \tag{43}$$

From (43), it is clear that $\dot{V}$ is a negative semi-definite function on $\mathbf{R}^{11}$.

Thus, we can conclude that the error vector $\mathbf{e}(t) = (e_x(t), e_y(t), e_z(t), e_w(t))$ and the parameter estimation error are globally bounded.

We define $k = \min\{k_1, k_2, k_3, k_4\}$.

Then it follows from (43) that

$$\dot{V} \leq -k\|\mathbf{e}(t)\|^2 \tag{44}$$

Thus, we have

$$k\|\mathbf{e}(t)\|^2 \leq -\dot{V} \tag{45}$$

Integrating the inequality (45) from 0 to t, we get

$$k\int_0^t \|\mathbf{e}(\tau)\|^2 \, d\tau \leq V(0) - V(t) \tag{46}$$

From (46), it follows that $\mathbf{e} \in \mathbf{L}_2$. Using (38), we can conclude that $\dot{\mathbf{e}} \in \mathbf{L}_\infty$.

Using Barbalat's lemma [14], we conclude that $\mathbf{e}(t) \to 0$ exponentially as $t \to \infty$ for all initial conditions $\mathbf{e}(0) \in \mathbf{R}^4$. ■

For the numerical simulations, the classical fourth-order Runge–Kutta method with step size $h = 10^{-8}$ is used to solve the systems (31), (32) and (42), when the adaptive control law (35) is applied.

The parameter values of the novel hyperchaotic systems are taken as in the hyperchaotic case (7).

We take the positive gain constants as $k_x = 20$, $k_y = 10$, $k_z = 10$ and $k_w = 10$.

Furthermore, as initial conditions of the master system (31), we take

$$x_1(0) = 5.2, \quad x_2(0) = 2.8, \quad x_3(0) = 4.3, \quad x_4(0) = 7.6 \tag{47}$$

As initial conditions of the slave system (32), we take

$$y_1(0) = 6.4, \quad y_2(0) = 3.5, \quad y_3(0) = 7.2, \quad y_4(0) = 5.4 \tag{48}$$

Also, as initial conditions of the parameter estimates, we take

$$\hat{a}(0) = 12, \ \hat{b}(0) = 4, \ \hat{c}(0) = 1, \ \hat{p}(0) = 8, \ \hat{q}(0) = 6, \ \hat{r}(0) = 5, \ \hat{s}(0) = 7 \tag{49}$$

Figures 6, 7, 8 and 9 describe the complete synchronization of the novel hyperchaotic systems (31) and (32), while Fig. 10 describes the time-history of the synchronization errors e_x, e_y, e_z, e_w.

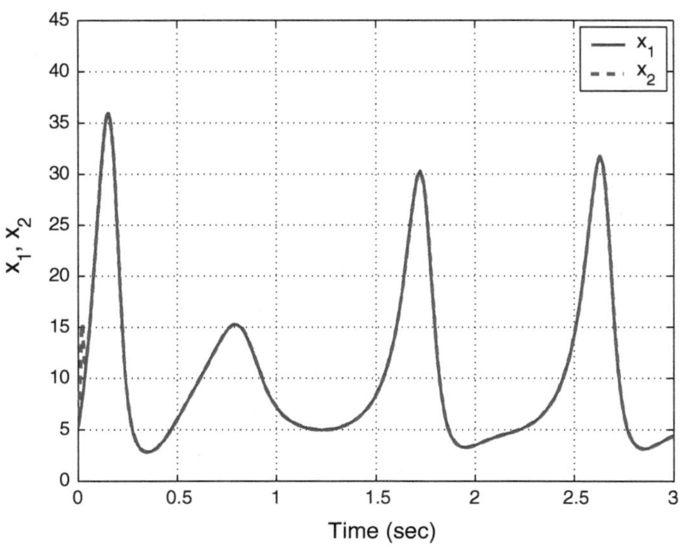

Fig. 6 Synchronization of the states x_1 and x_2

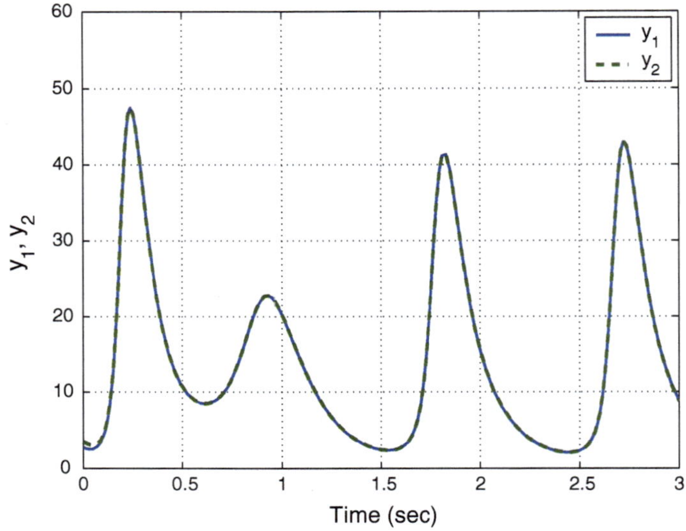

Fig. 7 Synchronization of the states y_1 and y_2

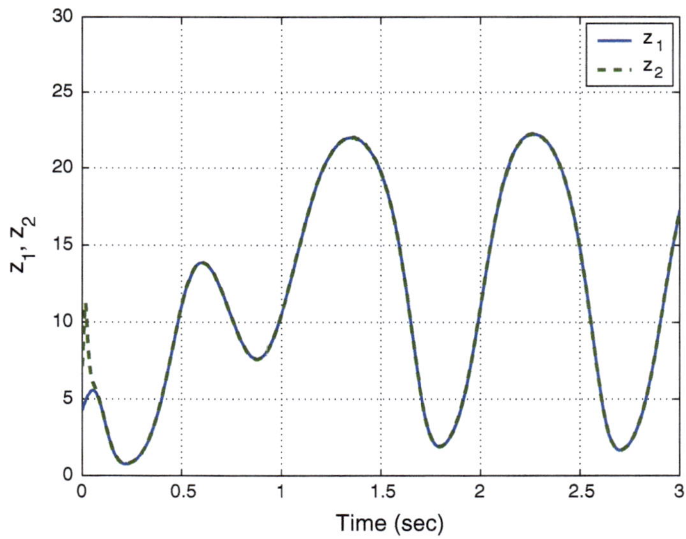

Fig. 8 Synchronization of the states z_1 and z_2

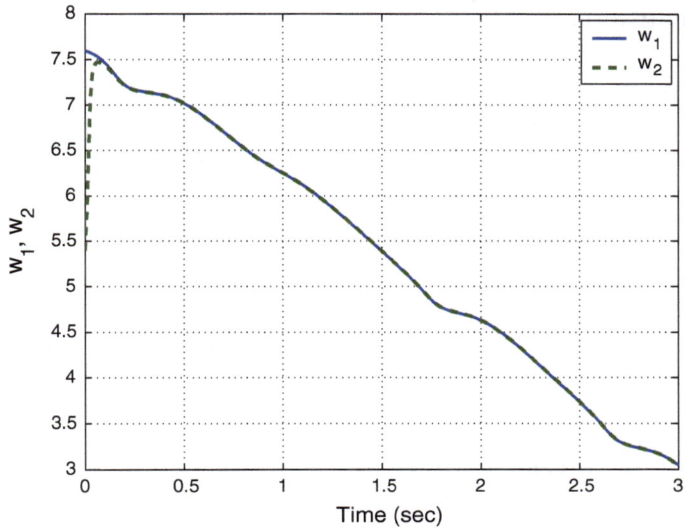

Fig. 9 Synchronization of the states w_1 and w_2

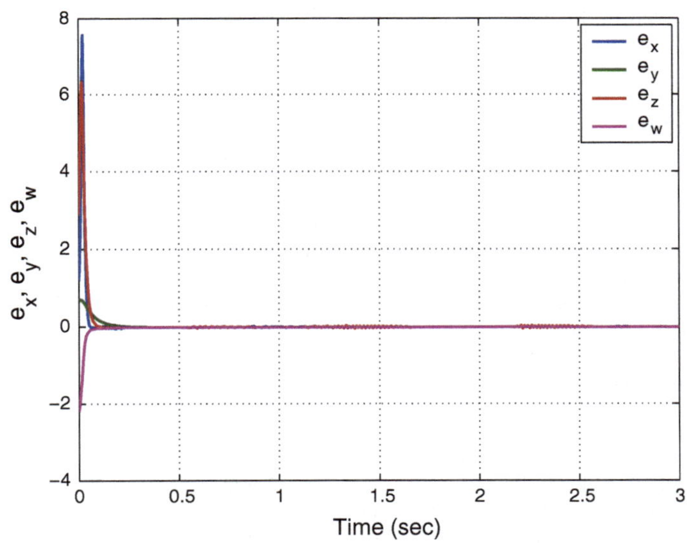

Fig. 10 Time-history of the synchronization errors e_x, e_y, e_z, e_w

6 Conclusions

Chaos and hyperchaos have important applications in science and engineering. Since a hyperchaotic system has at least two positive Lyapunov exponents, the dynamics of a hyperchaotic system can expand in several different directions simultaneously. Thus, hyperchaotic systems have more complex behaviour than chaotic systems and they have miscellaneous applications in areas like secure communications, cryptosystems, etc. In this work, we derived a twelve-term novel 4-D hyperchaotic system by introducing a state feedback control to the 3-D chemical chaotic reactor obtained by Huang and Yang [11]. The qualitative properties of the novel hyperchaotic system were discussed in detail. The Lyapunov exponents of the novel hyperchaotic chemical reactor system were obtained as $L_1 = 0.2263$, $L_2 = 0.0365$, $L_3 = 0$ and $L_4 = -10.8396$, while the Kaplan–Yorke dimension of the novel hyperchaotic chemical reactor system was derived as $D_{KY} = 3.0240$. Since the sum of the Lyapunov exponents is negative, the novel hyperchaotic system is dissipative. Next, an adaptive controller was designed to globally stabilize the novel hyperchaotic system with unknown parameters. Finally, an adaptive controller was also designed to achieve global chaos synchronization of the identical hyperchaotic systems with unknown parameters. MATLAB simulations were shown to validate and illustrate the main results derived in this work.

References

1. Abdurrahman A, Jiang H, Teng Z (2015) Finite-time synchronization for memristor-based neural networks with time-varying delays. Neural Netw 69:20–28
2. Arneodo A, Coullet P, Tresser C (1981) Possible new strange attractors with spiral structure. Commun Math Phys 79(4):573–576
3. Azar AT, Vaidyanathan S (2015) Chaos modeling and control systems design, vol 581. Springer, Germany
4. Cai G, Tan Z (2007) Chaos synchronization of a new chaotic system via nonlinear control. J Uncertain Syst 1(3):235–240
5. Chen A, Lu J, Lü J, Yu S (2006) Generating hyperchaotic Lü attractor via state feedback control. Phys A 364:103–110
6. Chen G, Ueta T (1999) Yet another chaotic attractor. Int J Bifurc Chaos 9(7):1465–1466
7. Filali RL, Benrejeb M, Borne P (2014) On observer-based secure communication design using discrete-time hyperchaotic systems. Commun Nonlinear Sci Numer Simul 19(5):1424–1432
8. Grassberger P, Procaccia I (1983) Characterization of strange attractors. Phys Rev Lett 50:346–349
9. Grassberger P, Procaccia I (1983) Measuring the strangeness of strange attractors. Phys D 9:189–208
10. Hammami S (2015) State feedback-based secure image cryptosystem using hyperchaotic synchronization. ISA Trans 54:52–59
11. Huang Y, Yang XS (2005) Chaoticity of some chemical attractors: a computer assisted proof. J Math Chem 38(1):107–117
12. Jia Q (2007) Hyperchaos generated from the Lorenz chaotic system and its control. Phys Lett A 366:217–222

13. Karthikeyan R, Sundarapandian V (2014) Hybrid chaos synchronization of four-scroll systems via active control. J Electr Eng 65(2):97–103
14. Khalil HK (2001) Nonlinear systems, 3rd edn. Prentice Hall, New Jersey
15. Li C, Liao X, Wong KW (2005) Lag synchronization of hyperchaos with application to secure communications. Chaos Solitons Fractals 23(1):183–193
16. Li D (2008) A three-scroll chaotic attractor. Phys Lett A 372(4):387–393
17. Li X (2009) Modified projective synchronization of a new hyperchaotic system via nonlinear control. Commun Theor Phys 52:274–278
18. Lorenz EN (1963) Deterministic periodic flow. J Atmos Sci 20(2):130–141
19. Lü J, Chen G (2002) A new chaotic attractor coined. Int J Bifurc Chaos 12(3):659–661
20. Pehlivan I, Moroz IM, Vaidyanathan S (2014) Analysis, synchronization and circuit design of a novel butterfly attractor. J Sound Vib 333(20):5077–5096
21. Pham VT, Volos C, Jafari S, Wang X, Vaidyanathan S (2014) Hidden hyperchaotic attractor in a novel simple memristive neural network. Optoelectron Adv Mater Rapid Commun 8(11–12):1157–1163
22. Pham VT, Volos CK, Vaidyanathan S, Le TP, Vu VY (2015) A memristor-based hyperchaotic system with hidden attractors: Dynamics, synchronization and circuital emulating. J Eng Sci Technol Rev 8(2):205–214
23. Rasappan S, Vaidyanathan S (2013) Hybrid synchronization of n-scroll Chua circuits using adaptive backstepping control design with recursive feedback. Malays J Math Sci 73(1):73–95
24. Rasappan S, Vaidyanathan S (2014) Global chaos synchronization of WINDMI and Coullet chaotic systems using adaptive backstepping control design. Kyungpook Math J 54(1):293–320
25. Rhouma R, Belghith S (2008) Cryptanalysis of a new image encryption algorithm based on hyper-chaos. Phys Lett A 372(38):5973–5978
26. Rössler OE (1976) An equation for continuous chaos. Phys Lett A 57(5):397–398
27. Rössler OE (1979) An equation for hyperchaos. Phys Lett A 71:155–157
28. Sampath S, Vaidyanathan S, Volos CK, Pham VT (2015) An eight-term novel four-scroll chaotic System with cubic nonlinearity and its circuit simulation. J Eng Sci Technol Rev 8(2):1–6
29. Sarasu P, Sundarapandian V (2011) Active controller design for generalized projective synchronization of four-scroll chaotic systems. Int J Syst Signal Control Eng Appl 4(2):26–33
30. Sarasu P, Sundarapandian V (2011) The generalized projective synchronization of hyperchaotic Lorenz and hyperchaotic Qi systems via active control. Int J Soft Comput 6(5):216–223
31. Sarasu P, Sundarapandian V (2012) Generalized projective synchronization of two-scroll systems via adaptive control. Int J Soft Comput 7(4):146–156
32. Senouci A, Boukabou A (2014) Predictive control and synchronization of chaotic and hyperchaotic systems based on a $T-S$ fuzzy model. Math Comput Simul 105:62–78
33. Sprott JC (1994) Some simple chaotic flows. Phys Rev E 50(2):647–650
34. Sundarapandian V (2010) Output regulation of the Lorenz attractor. Far East J Math Sci 42(2):289–299
35. Sundarapandian V (2013) Adaptive control and synchronization design for the Lu-Xiao chaotic system. Lect Notes Electr Eng 131:319–327
36. Sundarapandian V (2013) Analysis and anti-synchronization of a novel chaotic system via active and adaptive controllers. J Eng Sci Technol Rev 6(4):45–52
37. Sundarapandian V, Karthikeyan R (2011) Anti-synchronization of hyperchaotic Lorenz and hyperchaotic Chen systems by adaptive control. Int J Syst Signal Control Eng Appl 4(2):18–25
38. Sundarapandian V, Karthikeyan R (2011) Anti-synchronization of Lü and Pan chaotic systems by adaptive nonlinear control. Eur J Sci Res 64(1):94–106
39. Sundarapandian V, Karthikeyan R (2012) Adaptive anti-synchronization of uncertain Tigan and Li systems. J Eng Appl Sci 7(1):45–52
40. Sundarapandian V, Pehlivan I (2012) Analysis, control, synchronization, and circuit design of a novel chaotic system. Math Comput Model 55(7–8):1904–1915

41. Sundarapandian V, Sivaperumal S (2011) Sliding controller design of hybrid synchronization of four-wing chaotic systems. Int J Soft Comput 6(5):224–231
42. Suresh R, Sundarapandian V (2013) Global chaos synchronization of a family of n-scroll hyperchaotic Chua circuits using backstepping control with recursive feedback. Far East J Math Sci 7(2):219–246
43. Tigan G, Opris D (2008) Analysis of a 3D chaotic system. Chaos Solitons Fractals 36: 1315–1319
44. Vaidyanathan S (2011) Output regulation of Arneodo-Coullet chaotic system. Commun Comput Inf Sci 133:98–107
45. Vaidyanathan S (2011) Output regulation of the unified chaotic system. Commun Comput Inf Sci 198:1–9
46. Vaidyanathan S (2012) Adaptive controller and syncrhonizer design for the Qi-Chen chaotic system. Lect Notes Inst Comput Sci Soc-Inf Telecommun Eng 84:73–82
47. Vaidyanathan S (2012) Anti-synchronization of Sprott-L and Sprott-M chaotic systems via adaptive control. Int J Control Theory Appl 5(1):41–59
48. Vaidyanathan S (2012) Global chaos control of hyperchaotic Liu system via sliding control method. Int J Control Theory Appl 5(2):117–123
49. Vaidyanathan S (2012) Sliding mode control based global chaos control of Liu-Liu-Liu-Su chaotic system. Int J Control Theory Appl 5(1):15–20
50. Vaidyanathan S (2013) A new six-term 3-D chaotic system with an exponential nonlinearity. Far East J Math Sci 79(1):135–143
51. Vaidyanathan S (2013) A ten-term novel 4-D hyperchaotic system with three quadratic nonlinearities and its control. Int J Control Theory Appl 6(2):97–109
52. Vaidyanathan S (2013) Analysis and adaptive synchronization of two novel chaotic systems with hyperbolic sinusoidal and cosinusoidal nonlinearity and unknown parameters. J Eng Sci Technol Rev 6(4):53–65
53. Vaidyanathan S (2013) Analysis, control and synchronization of hyperchaotic Zhou system via adaptive control. Adv Intell Syst Comput 177:1–10
54. Vaidyanathan S (2014) A new eight-term 3-D polynomial chaotic system with three quadratic nonlinearities. Far East J Math Sci 84(2):219–226
55. Vaidyanathan S (2014) Analysis and adaptive synchronization of eight-term 3-D polynomial chaotic systems with three quadratic nonlinearities. Eur Phys J: Spec Top 223(8):1519–1529
56. Vaidyanathan S (2014) Analysis, control and synchronisation of a six-term novel chaotic system with three quadratic nonlinearities. Int J Model Ident Control 22(1):41–53
57. Vaidyanathan S (2014) Generalized projective synchronisation of novel 3-D chaotic systems with an exponential non-linearity via active and adaptive control. Int J Model Ident Control 22(3):207–217
58. Vaidyanathan S (2014) Global chaos synchronisation of identical Li-Wu chaotic systems via sliding mode control. Int J Model Ident Control 22(2):170–177
59. Vaidyanathan S (2014) Qualitative analysis and control of an eleven-term novel 4-D hyperchaotic system with two quadratic nonlinearities. Int J Control Theory Appl 7:35–47
60. Vaidyanathan S (2015) 3-cells Cellular Neural Network (CNN) attractor and its adaptive biological control. Int J PharmTech Res 8(4):632–640
61. Vaidyanathan S (2015) A 3-D novel highly chaotic system with four quadratic nonlinearities, its adaptive control and anti-synchronization with unknown parameters. J Eng Sci Technol Rev 8(2):106–115
62. Vaidyanathan S (2015) A novel chemical chaotic reactor system and its adaptive control. Int J ChemTech Res 8(7):146–158
63. Vaidyanathan S (2015) Adaptive backstepping control of enzymes-substrates system with ferroelectric behaviour in brain waves. Int J PharmTech Res 8(2):256–261
64. Vaidyanathan S (2015) Adaptive biological control of generalized Lotka-Volterra three-species biological system. Int J PharmTech Res 8(4):622–631
65. Vaidyanathan S (2015) Adaptive chaotic synchronization of enzymes-substrates system with ferroelectric behaviour in brain waves. Int J PharmTech Res 8(5):964–973

66. Vaidyanathan S (2015) Adaptive control of a chemical chaotic reactor. Int J PharmTech Res 8(3):377–382
67. Vaidyanathan S (2015) Adaptive control of the FitzHugh-Nagumo chaotic neuron model. Int J PharmTech Res 8(6):117–127
68. Vaidyanathan S (2015) Adaptive synchronization of chemical chaotic reactors. Int J ChemTech Res 8(2):612–621
69. Vaidyanathan S (2015) Adaptive synchronization of generalized Lotka-Volterra three-species biological systems. Int J PharmTech Res 8(5):928–937
70. Vaidyanathan S (2015) Adaptive synchronization of novel 3-D chemical chaotic reactor systems. Int J ChemTech Res 8(7):159–171
71. Vaidyanathan S (2015) Adaptive synchronization of the identical FitzHugh-Nagumo chaotic neuron models. Int J PharmTech Res 8(6):167–177
72. Vaidyanathan S (2015) Analysis, control and synchronization of a 3-D novel jerk chaotic system with two quadratic nonlinearities. Kyungpook Math J 55:563–586
73. Vaidyanathan S (2015) Analysis, properties and control of an eight-term 3-D chaotic system with an exponential nonlinearity. Int J Model Ident Control 23(2):164–172
74. Vaidyanathan S (2015) Anti-synchronization of brusselator chemical reaction systems via adaptive control. Int J ChemTech Res 8(6):759–768
75. Vaidyanathan S (2015) Chaos in neurons and adaptive control of Birkhoff-Shaw strange chaotic attractor. Int J PharmTech Res 8(5):956–963
76. Vaidyanathan S (2015) Chaos in neurons and synchronization of Birkhoff-Shaw strange chaotic attractors via adaptive control. Int J PharmTech Res 8(6):1–11
77. Vaidyanathan S (2015) Coleman-Gomatam logarithmic competitive biology models and their ecological monitoring. Int J PharmTech Res 8(6):94–105
78. Vaidyanathan S (2015) Dynamics and control of brusselator chemical reaction. Int J ChemTech Res 8(6):740–749
79. Vaidyanathan S (2015) Dynamics and control of tokamak system with symmetric and magnetically confined plasma. Int J ChemTech Res 8(6):795–803
80. Vaidyanathan S (2015) Global chaos synchronization of chemical chaotic reactors via novel sliding mode control method. Int J ChemTech Res 8(7):209–221
81. Vaidyanathan S (2015) Global chaos synchronization of the forced Van der Pol chaotic oscillators via adaptive control method. Int J PharmTech Res 8(6):156–166
82. Vaidyanathan S (2015) Global chaos synchronization of the Lotka-Volterra biological systems with four competitive species via active control. Int J PharmTech Res 8(6):206–217
83. Vaidyanathan S (2015) Lotka-Volterra population biology models with negative feedback and their ecological monitoring. Int J PharmTech Res 8(5):974–981
84. Vaidyanathan S (2015) Lotka-Volterra two species competitive biology models and their ecological monitoring. Int J PharmTech Res 8(6):32–44
85. Vaidyanathan S (2015) Output regulation of the forced Van der Pol chaotic oscillator via adaptive control method. Int J PharmTech Res 8(6):106–116
86. Vaidyanathan S, Azar AT (2015) Analysis and control of a 4-D novel hyperchaotic system. Stud Comput Intell 581:3–17
87. Vaidyanathan S, Azar AT (2015) Analysis, control and synchronization of a nine-term 3-D novel chaotic system. In: Azar AT, Vaidyanathan S (eds) Chaos modelling and control systems design. Studies in computational intelligence, vol 581. Springer, Germany, pp 19–38
88. Vaidyanathan S, Madhavan K (2013) Analysis, adaptive control and synchronization of a seven-term novel 3-D chaotic system. Int J Control Theory Appl 6(2):121–137
89. Vaidyanathan S, Pakiriswamy S (2013) Generalized projective synchronization of six-term Sundarapandian chaotic systems by adaptive control. Int J Control Theory Appl 6(2):153–163
90. Vaidyanathan S, Pakiriswamy S (2015) A 3-D novel conservative chaotic system and its generalized projective synchronization via adaptive control. J Eng Sci Technol Rev 8(2):52–60
91. Vaidyanathan S, Rajagopal K (2011) Hybrid synchronization of hyperchaotic Wang-Chen and hyperchaotic Lorenz systems by active non-linear control. Int J Syst Signal Control Eng Appl 4(3):55–61

92. Vaidyanathan S, Rajagopal K (2012) Global chaos synchronization of hyperchaotic Pang and hyperchaotic Wang systems via adaptive control. Int J Soft Comput 7(1):28–37
93. Vaidyanathan S, Rasappan S (2011) Global chaos synchronization of hyperchaotic Bao and Xu systems by active nonlinear control. Commun Comput Inf Sci 198:10–17
94. Vaidyanathan S, Rasappan S (2014) Global chaos synchronization of n-scroll Chua circuit and Lur'e system using backstepping control design with recursive feedback. Arab J Sci Eng 39(4):3351–3364
95. Vaidyanathan S, Sampath S (2012) Anti-synchronization of four-wing chaotic systems via sliding mode control. Int J Autom Comput 9(3):274–279
96. Vaidyanathan S, Volos C (2015) Analysis and adaptive control of a novel 3-D conservative no-equilibrium chaotic system. Arch Control Sci 25(3):333–353
97. Vaidyanathan S, Volos C, Pham VT (2014) Hyperchaos, adaptive control and synchronization of a novel 5-D hyperchaotic system with three positive Lyapunov exponents and its SPICE implementation. Arch Control Sci 24(4):409–446
98. Vaidyanathan S, Volos C, Pham VT, Madhavan K, Idowu BA (2014) Adaptive backstepping control, synchronization and circuit simulation of a 3-D novel jerk chaotic system with two hyperbolic sinusoidal nonlinearities. Arch Control Sci 24(3):375–403
99. Vaidyanathan S, Idowu BA, Azar AT (2015) Backstepping controller design for the global chaos synchronization of Sprott's jerk systems. Stud Comput Intell 581:39–58
100. Vaidyanathan S, Rajagopal K, Volos CK, Kyprianidis IM, Stouboulos IN (2015) Analysis, adaptive control and synchronization of a seven-term novel 3-D chaotic system with three quadratic nonlinearities and its digital implementation in LabVIEW. J Eng Sci Technol Rev 8(2):130–141
101. Vaidyanathan S, Sampath S, Azar AT (2015) Global chaos synchronisation of identical chaotic systems via novel sliding mode control method and its application to Zhu system. Int J Model Ident Control 23(1):92–100
102. Vaidyanathan S, Volos C, Pham VT, Madhavan K (2015) Analysis, adaptive control and synchronization of a novel 4-D hyperchaotic hyperjerk system and its SPICE implementation. Nonlinear Dyn 25(1):135–158
103. Vaidyanathan S, Volos CK, Kyprianidis IM, Stouboulos IN, Pham VT (2015) Analysis, adaptive control and anti-synchronization of a six-term novel jerk chaotic system with two exponential nonlinearities and its circuit simulation. J Eng Sci Technol Rev 8(2):24–36
104. Vaidyanathan S, Volos CK, Pham VT (2015) Analysis, adaptive control and adaptive synchronization of a nine-term novel 3-D chaotic system with four quadratic nonlinearities and its circuit simulation. J Eng Sci Technol Rev 8(2):181–191
105. Vaidyanathan S, Volos CK, Pham VT (2015) Analysis, control, synchronization and SPICE implementation of a novel 4-D hyperchaotic Rikitake dynamo system without equilibrium. J Eng Sci Technol Rev 8(2):232–244
106. Vaidyanathan S, Volos CK, Pham VT (2015) Global chaos control of a novel nine-term chaotic system via sliding mode control. In: Azar AT, Zhu Q (eds) Advances and applications in sliding mode control systems. Studies in computational intelligence, vol 576. Springer, Germany, pp 571–590
107. Volos CK, Kyprianidis IM, Stouboulos IN, Tlelo-Cuautle E, Vaidyanathan S (2015) Memristor: A new concept in synchronization of coupled neuromorphic circuits. J Eng Sci Technol Rev 8(2):157–173
108. Wang J, Chen Z (2008) A novel hyperchaotic system and its complex dynamics. Int J Bifurc Chaos 18:3309–3324
109. Wei X, Yunfei F, Qiang L (2012) A novel four-wing hyper-chaotic system and its circuit implementation. Procedia Eng 29:1264–1269
110. Wei Z, Yang Q (2010) Anti-control of Hopf bifurcation in the new chaotic system with two stable node-foci. Appl Math Comput 217(1):422–429
111. Wu X, Zhu C, Kan H (2015) An improved secure communication scheme based passive synchronization of hyperchaotic complex nonlinear system. Appl Math Comput 252: 201–214

112. Yujun N, Xingyuan W, Mingjun W, Huaguang Z (2010) A new hyperchaotic system and its circuit implementation. Commun Nonlinear Sci Numer Simul 15(11):3518–3524
113. Zhang H, Liao X, Yu J (2005) Fuzzy modeling and synchronization of hyperchaotic systems. Chaos Solitons Fractals 26(3):835–843
114. Zhou W, Xu Y, Lu H, Pan L (2008) On dynamics analysis of a new chaotic attractor. Phys Lett A 372(36):5773–5777
115. Zhu C (2012) A novel image encryption scheme based on improved hyperchaotic sequences. Optics Commun 285(1):29–37
116. Zhu C, Liu Y, Guo Y (2010) Theoretic and numerical study of a new chaotic system. Intell Inf Manag 2:104–109

A Novel 5-D Hyperchaotic System with a Line of Equilibrium Points and Its Adaptive Control

Sundarapandian Vaidyanathan

Abstract Chaos theory describes the qualitative study of unstable aperiodic behavior in deterministic nonlinear dynamical systems. Chaos theory has applications in several fields of science and engineering. A hyperchaotic system is generally defined as a chaotic system with at least two positive Lyapunov exponents. In this research work, we announce a novel 5-D hyperchaotic system with an infinite line of equilibrium points. The novel 5-D hyperchaotic system has fifteen terms on the right hand side with two quadratic nonlinearities. The phase portraits of the 5-D novel hyperchaotic system are depicted and the qualitative properties of the novel hyperchaotic system are discussed. All the equilibrium points of the novel 5-D hyperchaotic system are unstable. The Lyapunov exponents of the 5-D novel hyperchaotic system are obtained as $L_1 = 1.2995, L_2 = 0.2505, L_3 = 0.0615, L_4 = 0$ and $L_5 = -17.5932$. The maximal Lyapunov exponent of the novel hyperchaotic system is $L_1 = 1.2995$. Also, the Kaplan–Yorke dimension of the 5-D novel hyperchaotic system is obtained as $D_{KY} = 4.0916$. Since the sum of the Lyapunov exponents is negative, the 5-D novel hyperchaotic system is dissipative. Next, an adaptive controller is designed to globally stabilize the novel hyperchaotic system with unknown parameters. Finally, an adaptive controller is also designed to achieve global chaos synchronization of the identical hyperchaotic systems with unknown parameters. MATLAB simulations are depicted to illustrate all the main results derived in this work for the 5-D novel hyperchaotic system.

Keywords Chaos · Chaotic systems · Hyperchaos · Hyperchaotic systems · Line equilibrium · Adaptive control · Synchronization

---

S. Vaidyanathan (✉)
Research and Development Centre, Vel Tech University,
Avadi, Chennai 600062, Tamil Nadu, India
e-mail: sundarvtu@gmail.com

S. Vaidyanathan and C. Volos (eds.), *Advances and Applications in Chaotic Systems*, Studies in Computational Intelligence 636,
DOI 10.1007/978-3-319-30279-9_20

1 Introduction

In the last few decades, Chaos theory has become a very important and active research field, employing many applications in different disciplines like physics, chemistry, biology, ecology, engineering and economics, among others [3]. Some classical paradigms of 3-D chaotic systems in the literature are Lorenz system [17], Rössler system [25], ACT system [2], Sprott systems [32], Chen system [6], Lü system [18], Cai system [4], Tigan system [42], etc.

Many new chaotic systems have been discovered in the recent years such as Zhou system [113], Zhu system [115], Li system [15], Wei-Yang system [109], Sundarapandian systems [35, 39], Vaidyanathan systems [49, 51, 53–56, 60, 71, 72, 86, 87, 89, 95, 97, 99, 102, 103, 105], Pehlivan system [19], Sampath system [27], etc.

Chaos theory has applications in several fields of science and engineering such as chemical reactors [61, 65, 67, 69, 73, 77–79], biological systems [59, 62–64, 66, 68, 70, 74–76, 80–84], memristors [1, 21, 106], etc.

A hyperchaotic system is defined as a chaotic system with at least two positive Lyapunov exponents [3]. Thus, the dynamics of a hyperchaotic system can expand in several different directions simultaneously. Hyperchaotic systems possess complex behaviour and they have miscellaneous applications in engineering such as secure communications [7, 13, 110], cryptosystems [8, 24, 114], fuzzy logic [31, 112], electrical circuits [108, 111], etc.

The minimal dimension of an autonomous hyperchaotic system is four. The first 4-D hyperchaotic system was found by Rössler [26]. Many hyperchaotic systems have been reported in the chaos literature such as hyperchaotic Lorenz system [10], hyperchaotic Lü system [5], hyperchaotic Chen system [16], hyperchaotic Wang system [107], hyperchaotic Vaidyanathan systems [50, 58, 85, 96, 101, 104], hyperchaotic Pham system [20], etc.

The control of a chaotic or hyperchaotic system aims to stabilize or regulate the system with the help of a feedback control. There are many methods available for controlling a chaotic system such as active control [33, 43, 44], adaptive control [34, 45, 52], sliding mode control [47, 48], backstepping control [98], etc.

The synchronization of chaotic systems aims to synchronize the states of master and slave systems asymptotically with time. There are many methods available for chaos synchronization such as active control [11, 28, 29, 90, 92], adaptive control [30, 36–38, 46, 88, 91], sliding mode control [40, 57, 94, 100], backstepping control [22, 23, 41, 93], etc.

Recently, there is good interest in the chaos literature in the finding of chaotic or hyperchaotic systems with line equilibrium [9, 14].

In this research work, we announce a novel 5-D hyperchaotic system with a line of equilibrium points given by

$$E_k = \left[0,\ 0,\ 0,\ k,\ -k\right]^T, \quad (k \in \mathbf{R}) \tag{1}$$

All the equilibrium points (1) of the novel 5-D hyperchaotic system are unstable. The novel 5-D hyperchaotic system has fifteen terms on the right hand side with two quadratic nonlinearities. We have also designed adaptive controllers for stabilization and synchronization of the novel hyperchaotic systems when the system parameters are unknown.

This work is organized as follows. Section 2 describes the dynamic equations and phase portraits of the fifteen-term novel 5-D hyperchaotic system. Section 3 details the qualitative properties of the novel hyperchaotic system. The Lyapunov exponents of the 5-D novel hyperchaotic system are obtained as $L_1 = 1.2995$, $L_2 = 0.2505$, $L_3 = 0.0615$, $L_4 = 0$ and $L_5 = -17.5932$. The maximal Lyapunov exponent of the novel hyperchaotic system is $L_1 = 1.2995$. Also, the Kaplan–Yorke dimension of the 5-D novel hyperchaotic system is obtained as $D_{KY} = 4.0916$. Since the sum of the Lyapunov exponents is negative, the 5-D novel hyperchaotic system is dissipative.

In Sect. 4, we design an adaptive controller to globally stabilize the novel 5-D hyperchaotic system with unknown parameters. In Sect. 5, an adaptive controller is designed to achieve global chaos synchronization of the identical novel 5-D hyperchaotic systems with unknown parameters. Section 6 summarizes the main results derived in this work.

2 A Novel 5-D Hyperchaotic System

In [58], Vaidyanathan derived an eleven-term novel 4-D hyperchaotic system with only two quadratic nonlinearities, which is described by the dynamics

$$\begin{cases} \dot{x}_1 = a(x_2 - x_1) + x_3 + x_4 \\ \dot{x}_2 = cx_1 - x_1x_3 + x_4 \\ \dot{x}_3 = -bx_3 + x_1x_2 \\ \dot{x}_4 = -d(x_1 + x_2) \end{cases} \tag{2}$$

where x_1, x_2, x_3, x_4 are the states and a, b, c, d are constant positive parameters.

In [58], Vaidyanathan showed that the system (2) exhibits a *strange hyperchaotic attractor*, when the parameters take the values

$$a = 12, \quad b = 4, \quad c = 100, \quad d = 5 \tag{3}$$

In [58], Vaidyanathan showed that the system (2) has a unique equilibrium at the origin, which is a saddle-point. Hence, $\mathbf{x} = \mathbf{0}$ is an unstable equilibrium of the hyperchaotic system.

In [58], the Lyapunov exponents of the Vaidyanathan hyperchaotic system (2) for the parameter values (3) have been numerically obtained using MATLAB as

$$L_1 = 1.3981, \; L_2 = 0.2393, \; L_3 = 0, \; L_4 = -17.6509 \tag{4}$$

Also, the Kaplan–Yorke dimension of the Vaidyanathan hyperchaotic system (2) has been calculated as

$$D_{KY} = 3 + \frac{L_1 + L_2 + L_3}{|L_4|} = 3.0928 \tag{5}$$

In this research work, we obtain a novel 5-D hyperchaotic system by adding a state feedback control to the 4-D Vaidyanathan hyperchaotic system as follows:

$$\begin{cases} \dot{x}_1 = a(x_2 - x_1) + x_3 + x_4 + x_5 \\ \dot{x}_2 = cx_1 - x_1x_3 + x_4 + x_5 \\ \dot{x}_3 = -bx_3 + x_1x_2 \\ \dot{x}_4 = -d(x_1 + x_2) \\ \dot{x}_5 = -p(x_1 + x_2) \end{cases} \tag{6}$$

where x_1, x_2, x_3, x_4, x_5 are the states and a, b, c, d, p are constant positive parameters.

The 5-D system (6) has fifteen terms on the right hand side with only two quadratic nonlinearities.

The 5-D system (6) exhibits a *strange hyperchaotic attractor* when the parameter values are taken as

$$a = 12, \quad b = 4, \quad c = 100, \quad d = 5, \quad p = 1 \tag{7}$$

For numerical simulations, we take the initial conditions of the 5-D system (6) as

$$x_1(0) = 0.5, \quad x_2(0) = 1.1, \quad x_3(0) = 0.8, \quad x_4(0) = 0.1, \quad x_5(0) = 1.2 \tag{8}$$

The Lyapunov exponents of the 5-D system (6) for the parameter values (7) and for the initial conditions (8) are numerically determined using MATLAB as

$$\begin{cases} L_1 = 1.2995 \\ L_2 = 0.2505 \\ L_3 = 0.0615 \\ L_4 = 0 \\ L_5 = -17.5932 \end{cases} \tag{9}$$

Thus, the 5-D system (6) is hyperchaotic with three positive Lyapunov exponents.

Figures 1, 2, 3 and 4 the 3-D projection of the 5-D hyperchaotic system (6) on the (x_1, x_2, x_3), (x_1, x_3, x_5), (x_2, x_3, x_4) and (x_3, x_4, x_5) spaces, respectively.

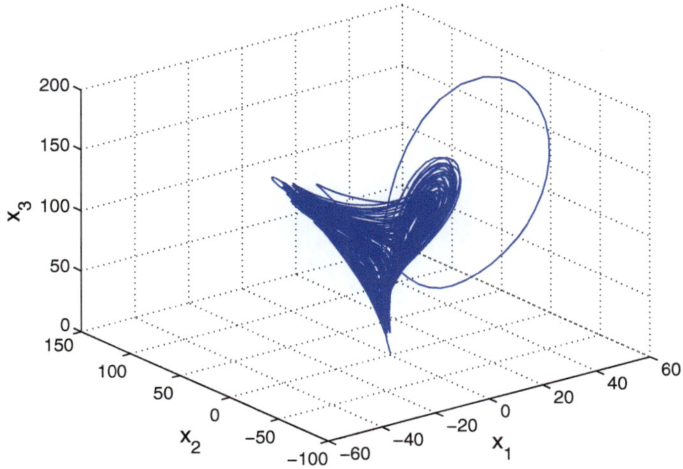

Fig. 1 3-D projection of the novel hyperchaotic system on the (x_1, x_2, x_3) space

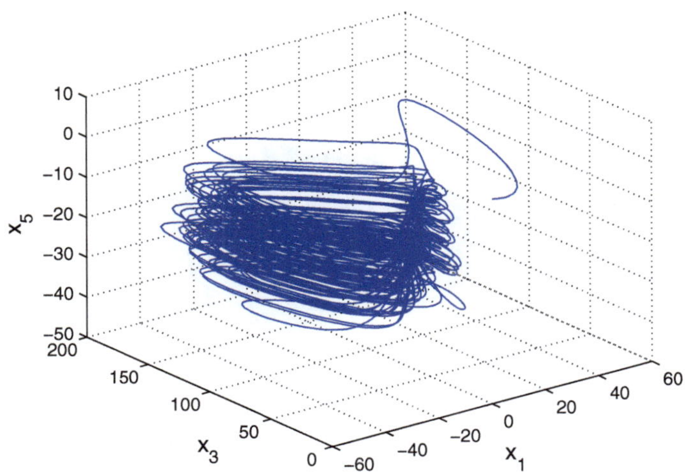

Fig. 2 3-D projection of the novel hyperchaotic system on the (x_1, x_3, x_5) space

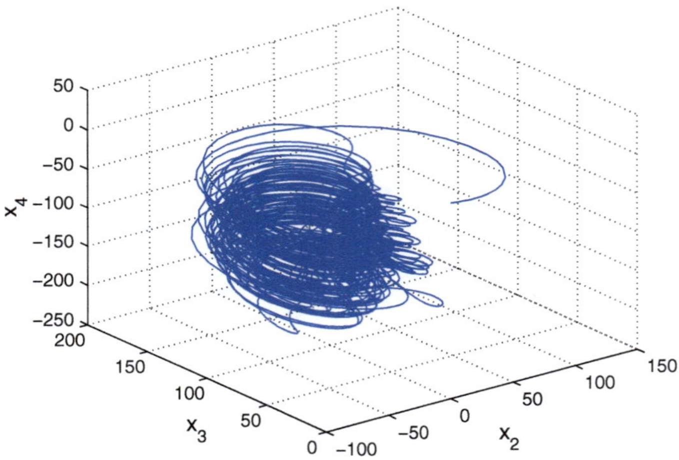

Fig. 3 3-D projection of the novel hyperchaotic system on the (x_2, x_3, x_4) space

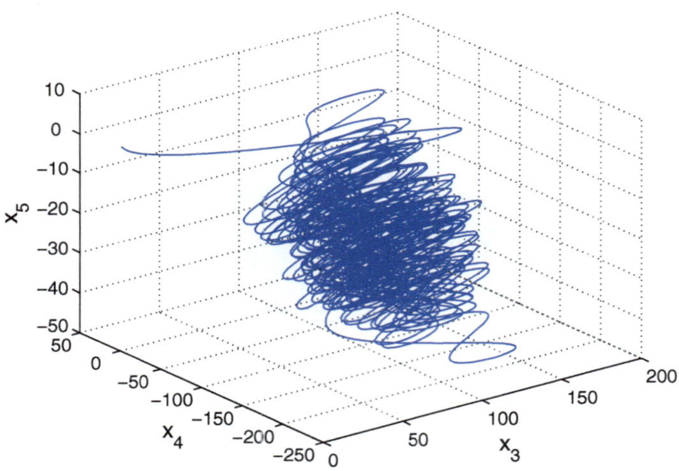

Fig. 4 3-D projection of the novel hyperchaotic system on the (x_3, x_4, x_5) space

3 Analysis of the Novel 5-D Hyperchaotic System

3.1 Dissipativity

In vector notation, the novel 5-D hyperchaotic system (6) can be expressed as

$$\dot{\mathbf{x}} = f(\mathbf{x}) = \begin{bmatrix} f_1(x_1, x_2, x_3, x_4, x_5) \\ f_2(x_1, x_2, x_3, x_4, x_5) \\ f_3(x_1, x_2, x_3, x_4, x_5) \\ f_4(x_1, x_2, x_3, x_4, x_5) \\ f_5(x_1, x_2, x_3, x_4, x_5) \end{bmatrix}, \tag{10}$$

where

$$\begin{cases} f_1(x_1, x_2, x_3, x_4, x_5) = a(x_2 - x_1) + x_3 + x_4 + x_5 \\ f_2(x_1, x_2, x_3, x_4, x_5) = cx_1 - x_1x_3 + x_4 + x_5 \\ f_3(x_1, x_2, x_3, x_4, x_5) = -bx_3 + x_1x_2 \\ f_4(x_1, x_2, x_3, x_4, x_5) = -d(x_1 + x_2) \\ f_5(x_1, x_2, x_3, x_4, x_5) = -p(x_1 + x_2) \end{cases} \tag{11}$$

Let Ω be any region in $\mathbf{R}^5$ with a smooth boundary and also, $\Omega(t) = \Phi_t(\Omega)$, where Φ_t is the flow of f. Furthermore, let $V(t)$ denote the volume of $\Omega(t)$.

By Liouville's theorem, we know that

$$\dot{V}(t) = \int\limits_{\Omega(t)} (\nabla \cdot f)\, dx_1\, dx_2\, dx_3\, dx_4\, dx_5 \tag{12}$$

The divergence of the novel 5-D system (6) is found as:

$$\nabla \cdot f = \frac{\partial f_1}{\partial x_1} + \frac{\partial f_2}{\partial x_2} + \frac{\partial f_3}{\partial x_3} + \frac{\partial f_1}{\partial x_4} + \frac{\partial f_1}{\partial x_5} = -a - b = -\mu \tag{13}$$

where $\mu = a + b$.

For the choice of parameter values given in (7), we find that $\mu = 16 > 0$.

Inserting the value of $\nabla \cdot f$ from (13) into (12), we get

$$\dot{V}(t) = \int\limits_{\Omega(t)} (-\mu)\, dx_1\, dx_2\, dx_3\, dx_4\, dx_5 = -\mu V(t) \tag{14}$$

Integrating the first order linear differential equation (14), we get

$$V(t) = \exp(-\mu t) V(0) \tag{15}$$

Since $\mu > 0$, it follows from (15) that $V(t) \to 0$ exponentially as $t \to \infty$. This shows that the novel 5-D hyperchaotic system (6) is dissipative. Hence, the system limit sets are ultimately confined into a specific limit set of zero volume, and the asymptotic motion of the novel 5-D hyperchaotic system (6) settles onto a strange attractor of the system.

3.2 Equilibrium Points

The equilibrium points of the 5-D novel hyperchaotic system (6) are obtained by solving the equations

$$a(x_2 - x_1) + x_3 + x_4 + x_5 = 0 \tag{16a}$$

$$cx_1 - x_1 x_3 + x_4 + x_5 = 0 \tag{16b}$$

$$-bx_3 + x_1 x_2 = 0 \tag{16c}$$

$$-d(x_1 + x_2) = 0 \tag{16d}$$

$$-p(x_1 + x_2) = 0 \tag{16e}$$

We take the parameter values as in the Eq. (7).

It is easy to see that the 5-D novel hyperchaotic system (6) has a line of equilibrium points given by

$$E_k = \begin{bmatrix} 0 \\ 0 \\ 0 \\ k \\ -k \end{bmatrix}, \quad (k \in \mathbf{R}) \tag{17}$$

The Jacobian matrix of the 5-D hyperchaotic system (6) at any equilibrium point E_k is obtained as the constant matrix

$$J = J(E_k) = \begin{bmatrix} -a & a & 1 & 1 & 1 \\ c & 0 & 0 & 1 & 1 \\ 0 & 0 & -b & 0 & 0 \\ -d & -d & 0 & 0 & 0 \\ -p & -p & 0 & 0 & 0 \end{bmatrix} = \begin{bmatrix} -12 & 12 & 1 & 1 & 1 \\ 100 & 0 & 0 & 1 & 1 \\ 0 & 0 & -4 & 0 & 0 \\ -5 & -5 & 0 & 0 & 0 \\ -1 & -1 & 0 & 0 & 0 \end{bmatrix} \tag{18}$$

The eigenvalues of the matrix $J(\mathbf{x}^\star)$ are numerically obtained as

$$\lambda_1 = -41.2426, \ \lambda_2 = -4, \ \lambda_3 = 0, \ \lambda_4 = 0.6305, \ \lambda_5 = 28.6121 \tag{19}$$

Thus, all the line equilibrium points E_k are unstable.

3.3 Lyapunov Exponents and Kaplan–Yorke Dimension

For the parameter values given in the Eq. (7) and for the initial conditions (8), the Lyapunov exponents of the 5-D novel hyperchaotic system (6) are numerically calculated using MATLAB as

$$L_1 = 1.2995, \quad L_2 = 0.2505, \quad L_3 = 0.0615, \quad L_4 = 0, \quad L_5 = -17.5932 \tag{20}$$

Thus, the 5-D novel hyperchaotic Lorenz system (6) has three positive Lyapunov exponents. Also, the maximal Lyapunov exponent (MLE) of the system (6) is obtained as $L_1 = 1.2995$.

Since the sum of the Lyapunov exponents is negative, the novel hyperchaotic system (6) is dissipative.

Also, the Kaplan–Yorke dimension of the novel hyperchaotic system (6) is obtained as

$$D_{KY} = 4 + \frac{L_1 + L_2 + L_3 + L_4}{|L_5|} = 4.0916 \tag{21}$$

which is fractional.

Since the 5-D hyperchaotic Lorenz system (6) has three positive Lyapunov exponents, it has a very complex dynamics and the system trajectories can be expended in three different directions.

4 Adaptive Control of the Novel 5-D Hyperchaotic System

In this section, we use adaptive control method to derive an adaptive feedback control law for globally stabilizing the novel 5-D hyperchaotic system with unknown parameters.

Thus, we consider the novel 5-D hyperchaotic system given by

$$\begin{cases} \dot{x}_1 = a(x_2 - x_1) + x_3 + x_4 + x_5 + u_1 \\ \dot{x}_2 = cx_1 - x_1x_3 + x_4 + x_5 + u_2 \\ \dot{x}_3 = -bx_3 + x_1x_2 + u_3 \\ \dot{x}_4 = -d(x_1 + x_2) + u_4 \\ \dot{x}_5 = -p(x_1 + x_2) + u_5 \end{cases} \tag{22}$$

In (22), x_1, x_2, x_3, x_4, x_5 are the states and u_1, u_2, u_3, u_4, u_5 are the adaptive controls to be determined using estimates $\hat{a}(t)$, $\hat{b}(t)$, $\hat{c}(t)$, $\hat{d}(t)$, $\hat{p}(t)$ for the unknown parameters a, b, c, d, p, respectively.

We consider the adaptive feedback control law

$$\begin{cases} u_1 = -\hat{a}(t)(x_2 - x_1) - x_3 - x_4 - x_5 - k_1x_1 \\ u_2 = -\hat{c}(t)x_1 + x_1x_3 - x_4 - x_5 - k_2x_2 \\ u_3 = \hat{b}(t)x_3 - x_1x_2 - k_3x_3 \\ u_4 = \hat{d}(t)(x_1 + x_2) - k_4x_4 \\ u_5 = \hat{p}(t)(x_1 + x_2) - k_5x_5 \end{cases} \tag{23}$$

where k_1, k_2, k_3, k_4, k_5 are positive gain constants.

Substituting (23) into (22), we get the closed-loop plant dynamics as

$$\begin{cases} \dot{x}_1 = [a - \hat{a}(t)](x_2 - x_1) - k_1x_1 \\ \dot{x}_2 = [c - \hat{c}(t)]x_1 - k_2x_2 \\ \dot{x}_3 = -[b - \hat{b}(t)]x_3 - k_3x_3 \\ \dot{x}_4 = -[d - \hat{d}(t)](x_1 + x_2) - k_4x_4 \\ \dot{x}_5 = -[p - \hat{p}(t)](x_1 + x_2) - k_5x_5 \end{cases} \tag{24}$$

The parameter estimation errors are defined as

$$\begin{cases} e_a(t) = a - \hat{a}(t) \\ e_b(t) = b - \hat{b}(t) \\ e_c(t) = c - \hat{c}(t) \\ e_d(t) = d - \hat{d}(t) \\ e_p(t) = p - \hat{p}(t) \end{cases} \tag{25}$$

In view of (25), we can simplify the plant dynamics (24) as

$$\begin{cases} \dot{x}_1 = e_a(x_2 - x_1) - k_1x_1 \\ \dot{x}_2 = e_cx_1 - k_2x_2 \\ \dot{x}_3 = -e_bx_3 - k_3x_3 \\ \dot{x}_4 = -e_d(x_1 + x_2) - k_4x_4 \\ \dot{x}_5 = -e_p(x_1 + x_2) - k_5x_5 \end{cases} \tag{26}$$

Differentiating (25) with respect to t, we obtain

$$\begin{cases} \dot{e}_a(t) = -\dot{\hat{a}}(t) \\ \dot{e}_b(t) = -\dot{\hat{b}}(t) \\ \dot{e}_c(t) = -\dot{\hat{c}}(t) \\ \dot{e}_d(t) = -\dot{\hat{d}}(t) \\ \dot{e}_p(t) = -\dot{\hat{p}}(t) \end{cases} \tag{27}$$

We consider the quadratic candidate Lyapunov function defined by

$$V(\mathbf{x}, e_a, e_b, e_c, e_d, e_p) = \frac{1}{2}\sum_{i=1}^{5} x_i^2 + \frac{1}{2}\left(e_a^2 + e_b^2 + e_c^2 + e_d^2 + e_p^2\right) \tag{28}$$

Differentiating V along the trajectories of (26) and (27), we obtain

$$\begin{aligned}\dot{V} &= -k_1x_1^2 - k_2x_2^2 - k_3x_3^2 - k_4x_4^2 - k_5x_5^2 \\ &\quad + e_a\left[x_1(x_2 - x_1) - \dot{\hat{a}}\right] + e_b\left[-x_3^2 - \dot{\hat{b}}\right] + e_c\left[-x_3^2 - \dot{\hat{c}}\right] \\ &\quad + e_d\left[-x_4(x_1 + x_2) - \dot{\hat{d}}\right] + e_p\left[-x_5(x_1 + x_2) - \dot{\hat{p}}\right]\end{aligned} \tag{29}$$

In view of (29), we take the parameter update law as

$$\begin{cases} \dot{\hat{a}}(t) = x_1(x_2 - x_1) \\ \dot{\hat{b}}(t) = -x_3^2 \\ \dot{\hat{c}}(t) = x_1x_2 \\ \dot{\hat{d}}(t) = -x_4(x_1 + x_2) \\ \dot{\hat{p}}(t) = -x_5(x_1 + x_2) \end{cases} \tag{30}$$

Next, we state and prove the main result of this section.

Theorem 1 *The novel 5-D hyperchaotic system with unknown system (22) parameters is globally and exponentially stabilized for all initial conditions by the adaptive control law (23) and the parameter update law (30), where k_i, $(i = 1, \ldots, 5)$ are positive gain constants.*

Proof We prove this result by applying Lyapunov stability theory [12].

We consider the quadratic Lyapunov function defined by (28), which is clearly a positive definite function on $\mathbf{R}^{10}$.

By substituting the parameter update law (30) into (29), we obtain the time-derivative of V as

$$\dot{V} = -k_1x_1^2 - k_2x_2^2 - k_3x_3^2 - k_4x_4^2 - k_5x_5^2 \tag{31}$$

From (31), it is clear that $\dot{V}$ is a negative semi-definite function on $\mathbf{R}^{10}$.

Thus, we can conclude that the state vector $\mathbf{x}(t)$ and the parameter estimation error are globally bounded, i.e.

$$\left[x_1(t)\ x_2(t)\ x_3(t)\ x_4(t)\ x_5(t)\ e_a(t)\ e_b(t)\ e_c(t)\ e_d(t)\ e_p(t)\right]^T \in \mathbf{L}_\infty.$$

We define $k = \min\{k_1, k_2, k_3, k_4, k_5\}$.

Then it follows from (31) that

$$\dot{V} \le -k\|\mathbf{x}(t)\|^2 \tag{32}$$

Thus, we have

$$k\|\mathbf{x}(t)\|^2 \le -\dot{V} \tag{33}$$

Integrating the inequality (33) from 0 to t, we get

$$k\int_0^t \|\mathbf{x}(\tau)\|^2 \, d\tau \;\le\; V(0) - V(t) \tag{34}$$

From (34), it follows that $\mathbf{x} \in \mathbf{L}_2$.

Using (26), we can conclude that $\dot{\mathbf{x}} \in \mathbf{L}_\infty$.

Using Barbalat's lemma [12], we conclude that $\mathbf{x}(t) \to 0$ exponentially as $t \to \infty$ for all initial conditions $\mathbf{x}(0) \in \mathbf{R}^5$.

This completes the proof. ■

For the numerical simulations, the classical fourth-order Runge–Kutta method with step size $h = 10^{-8}$ is used to solve the systems (22) and (30), when the adaptive control law (23) is applied.

The parameter values of the novel 4-D hyperchaotic system (22) are taken as in the hyperchaotic case (7), i.e.

$$a = 12, \quad b = 4, \quad c = 100, \quad d = 5, \quad p = 1 \tag{35}$$

We take the positive gain constants as

$$k_1 = 5, \quad k_2 = 5, \quad k_3 = 5, \quad k_4 = 5, \quad k_5 = 5 \tag{36}$$

Furthermore, as initial conditions of the novel 5-D hyperchaotic system (22), we take

$$x_1(0) = 12.5, \quad x_2(0) = -16.2, \quad x_3(0) = 19.4, \quad x_4(0) = 20.8, \quad x_5(0) = 14.3 \tag{37}$$

Also, as initial conditions of the parameter estimates, we take

$$\hat{a}(0) = 1.4, \quad \hat{b}(0) = 20.3, \quad \hat{c}(0) = 10.5, \quad \hat{d}(0) = 15.7, \quad \hat{p}(0) = 22.1 \tag{38}$$

In Fig. 5, the exponential convergence of the controlled states of the novel 5-D hyperchaotic system (22) is shown.

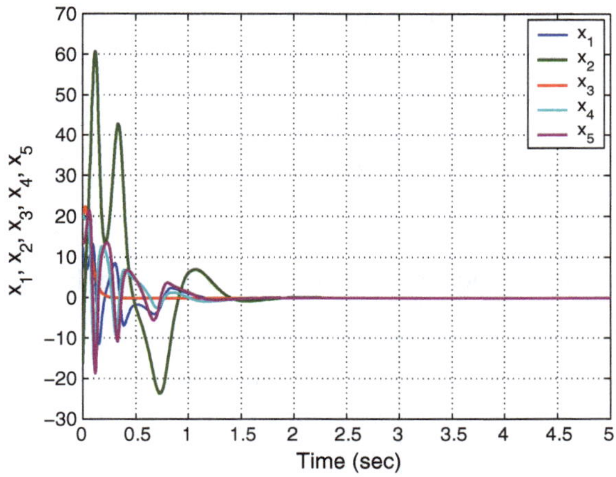

Fig. 5 Time-history of the controlled states x_1, x_2, x_3, x_4, x_5

5 Adaptive Synchronization of the Identical Novel Hyperchaotic Systems

In this section, we use adaptive control method to derive an adaptive feedback control law for globally synchronizing identical novel 5-D hyperchaotic systems with unknown parameters. We use Lyapunov stability theory to prove the main adaptive control result.

As the master system, we consider the novel 5-D hyperchaotic system given by

$$\begin{cases} \dot{x}_1 = a(x_2 - x_1) + x_3 + x_4 + x_5 \\ \dot{x}_2 = cx_1 - x_1x_3 + x_4 + x_5 \\ \dot{x}_3 = -bx_3 + x_1x_2 \\ \dot{x}_4 = -d(x_1 + x_2) \\ \dot{x}_5 = -p(x_1 + x_2) \end{cases} \tag{39}$$

In (39), x_1, x_2, x_3, x_4, x_5 are the states and a, b, c, d, p are unknown system parameters.

As the slave system, we consider the novel 5-D hyperchaotic system given by

$$\begin{cases} \dot{y}_1 = a(y_2 - y_1) + y_3 + y_4 + y_5 + u_1 \\ \dot{y}_2 = cy_1 - y_1y_3 + y_4 + y_5 + u_2 \\ \dot{y}_3 = -by_3 + y_1y_2 + u_3 \\ \dot{y}_4 = -d(y_1 + y_2) + u_4 \\ \dot{y}_5 = -p(y_1 + y_2) + u_5 \end{cases} \tag{40}$$

The synchronization error between the novel hyperchaotic systems (39) and (40) is defined by

$$e_i = y_i - x_i, \quad (i = 1, 2, \ldots, 5) \tag{41}$$

Then the synchronization error dynamics is obtained as

$$\begin{cases} \dot{e}_1 = a(e_2 - e_1) + e_3 + e_4 + e_5 + u_1 \\ \dot{e}_2 = ce_1 + e_4 + e_5 - y_1y_3 + x_1x_3 + u_2 \\ \dot{e}_3 = -be_3 + y_1y_2 - x_1x_2 + u_3 \\ \dot{e}_4 = -d(e_1 + e_2) + u_4 \\ \dot{e}_5 = -p(e_1 + e_2) + u_5 \end{cases} \tag{42}$$

We consider the adaptive feedback control law

$$\begin{cases} u_1 = -\hat{a}(t)(e_2 - e_1) - e_3 - e_4 - e_5 - k_1e_1 \\ u_2 = -\hat{c}(t)e_1 - e_4 - e_5 + y_1y_3 - x_1x_3 - k_2e_2 \\ u_3 = \hat{b}(t)e_3 - y_1y_2 + x_1x_2 - k_3e_3 \\ u_4 = \hat{d}(t)(e_1 + e_2) - k_4e_4 \\ u_5 = \hat{p}(t)(e_1 + e_2) - k_5e_5 \end{cases} \tag{43}$$

where k_1, k_2, k_3, k_4, k_5 are positive gain constants.

Substituting (43) into (42), we get the closed-loop error dynamics as

$$\begin{cases} \dot{e}_1 = \left[a - \hat{a}(t)\right](e_2 - e_1) - k_1e_1 \\ \dot{e}_2 = \left[c - \hat{c}(t)\right]e_1 - k_2e_2 \\ \dot{e}_3 = -\left[b - \hat{b}(t)\right]e_3 - k_3e_3 \\ \dot{e}_4 = -\left[d - \hat{d}(t)\right](e_1 + e_2) - k_4e_4 \\ \dot{e}_5 = -\left[p - \hat{p}(t)\right](e_1 + e_2) - k_5e_5 \end{cases} \tag{44}$$

The parameter estimation errors are defined as

$$\begin{cases} e_a(t) = a - \hat{a}(t) \\ e_b(t) = b - \hat{b}(t) \\ e_c(t) = c - \hat{c}(t) \\ e_d(t) = d - \hat{d}(t) \\ e_p(t) = p - \hat{p}(t) \end{cases} \tag{45}$$

In view of (45), we can simplify the error dynamics (44) as

$$\begin{cases} \dot{e}_1 = e_a(e_2 - e_1) - k_1e_1 \\ \dot{e}_2 = e_ce_1 - k_2e_2 \\ \dot{e}_3 = -e_be_3 - k_3e_3 \\ \dot{e}_4 = -e_d(e_1 + e_2) - k_4e_4 \\ \dot{e}_5 = -e_p(e_1 + e_2) - k_5e_5 \end{cases} \tag{46}$$

Differentiating (45) with respect to t, we obtain

$$\begin{cases} \dot{e}_a(t) = -\dot{\hat{a}}(t) \\ \dot{e}_b(t) = -\dot{\hat{b}}(t) \\ \dot{e}_c(t) = -\dot{\hat{c}}(t) \\ \dot{e}_d(t) = -\dot{\hat{d}}(t) \\ \dot{e}_p(t) = -\dot{\hat{p}}(t) \end{cases} \tag{47}$$

We use adaptive control theory to find an update law for the parameter estimates. We consider the quadratic candidate Lyapunov function defined by

$$V(\mathbf{e}, e_a, e_b, e_c, e_d, e_p) = \frac{1}{2}\sum_{i=1}^{5} e_i^2 + \frac{1}{2}\left(e_a^2 + e_b^2 + e_c^2 + e_d^2 + e_p^2\right) \tag{48}$$

Differentiating V along the trajectories of (46) and (47), we obtain

$$\begin{aligned} \dot{V} &= -k_1 e_1^2 - k_2 e_2^2 - k_3 e_3^2 - k_4 e_4^2 - k_5 e_5^2 \\ &\quad + e_a\left[e_1(e_2 - e_1) - \dot{\hat{a}}\right] + e_b\left[-e_3^2 - \dot{\hat{b}}\right] + e_c\left[e_1 e_2 - \dot{\hat{c}}\right] \\ &\quad + e_d\left[-e_4(e_1 + e_2) - \dot{\hat{d}}\right] + e_p\left[-e_5(e_1 + e_2) - \dot{\hat{p}}\right] \end{aligned} \tag{49}$$

In view of (49), we take the parameter update law as

$$\begin{cases} \dot{\hat{a}}(t) = e_1(e_2 - e_1) \\ \dot{\hat{b}}(t) = -e_3^2 \\ \dot{\hat{c}}(t) = e_1 e_2 \\ \dot{\hat{d}}(t) = -e_4(e_1 + e_2) \\ \dot{\hat{p}}(t) = -e_5(e_1 + e_2) \end{cases} \tag{50}$$

Next, we state and prove the main result of this section.

Theorem 2 *The novel 5-D hyperchaotic systems (39) and (40) with unknown system parameters are globally and exponentially synchronized for all initial conditions by the adaptive control law (43) and the parameter update law (50), where k_i, $(i = 1, \ldots, 5)$ are positive gain constants.*

Proof We prove this result by applying Lyapunov stability theory [12].

We consider the quadratic Lyapunov function defined by (48), which is clearly a positive definite function on $\mathbf{R}^{10}$.

By substituting the parameter update law (50) into (49), we obtain

$$\dot{V} = -k_1 e_1^2 - k_2 e_2^2 - k_3 e_3^2 - k_4 e_4^2 - k_5 e_5^2 \tag{51}$$

From (51), it is clear that $\dot{V}$ is a negative semi-definite function on $\mathbf{R}^{10}$.

Thus, we can conclude that the error vector $\mathbf{e}(t)$ and the parameter estimation error are globally bounded, i.e.

$$\left[e_1(t)\ e_2(t)\ e_3(t)\ e_4(t)\ e_5(t)\ e_a(t)\ e_b(t)\ e_c(t)\ e_d(t)\ e_p(t)\right]^T \in \mathbf{L}_\infty. \tag{52}$$

We define $k = \min\{k_1, k_2, k_3, k_4, k_5\}$.

Then it follows from (51) that

$$\dot{V} \le -k\|\mathbf{e}(t)\|^2 \tag{53}$$

Thus, we have

$$k\|\mathbf{e}(t)\|^2 \le -\dot{V} \tag{54}$$

Integrating the inequality (54) from 0 to t, we get

$$k\int_0^t \|\mathbf{e}(\tau)\|^2\,d\tau \ \le\ V(0) - V(t) \tag{55}$$

From (55), it follows that $\mathbf{e} \in \mathbf{L}_2$.

Using (46), we can conclude that $\dot{\mathbf{e}} \in \mathbf{L}_\infty$.

Using Barbalat's lemma [12], we conclude that $\mathbf{e}(t) \to 0$ exponentially as $t \to \infty$ for all initial conditions $\mathbf{e}(0) \in \mathbf{R}^5$.

This completes the proof. ∎

For the numerical simulations, the classical fourth-order Runge–Kutta method with step size $h = 10^{-8}$ is used to solve the 5-D systems (39), (40) and (50), when the adaptive control law (43) is applied.

The parameter values of the novel hyperchaotic systems are taken as in the hyperchaotic case (7), i.e.

$$a = 12,\quad b = 4,\quad c = 100,\quad d = 5,\quad p = 1 \tag{56}$$

We take the positive gain constants as $k_i = 5$ for $i = 1, 2, \ldots, 5$.

Furthermore, as initial conditions of the master system (39), we take

$$x_1(0) = 6.3,\quad x_2(0) = 5.8,\quad x_3(0) = -2.7,\quad x_4(0) = 14.5,\quad x_5(0) = -16.7 \tag{57}$$

As initial conditions of the slave system (40), we take

$$y_1(0) = 15.4,\quad y_2(0) = -13.5,\quad y_3(0) = 14.2,\quad y_4(0) = -29.4,\quad y_5(0) = 22.8 \tag{58}$$

Also, as initial conditions of the parameter estimates, we take

$$\hat{a}(0) = 2.1, \quad \hat{b}(0) = 4.7, \quad \hat{c}(0) = 3.2, \quad \hat{d}(0) = 12.4, \quad \hat{p}(0) = 25.8 \tag{59}$$

Figures 6, 7, 8, 9 and 10 describe the complete synchronization of the novel 5-D hyperchaotic systems (39) and (40), while Fig. 11 describes the time-history of the synchronization errors e_1, e_2, e_3, e_4, e_5.

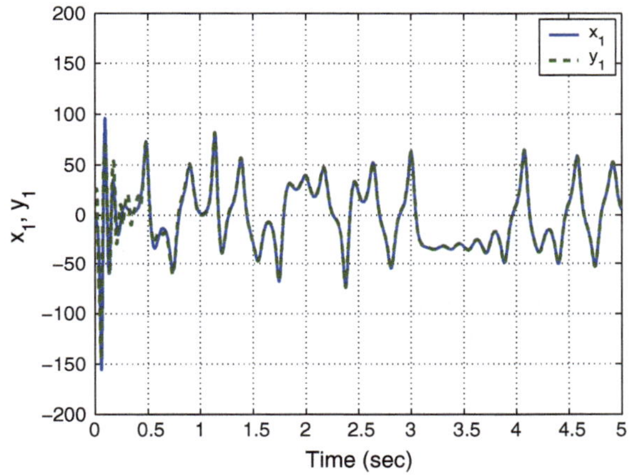

Fig. 6 Synchronization of the states x_1 and y_1

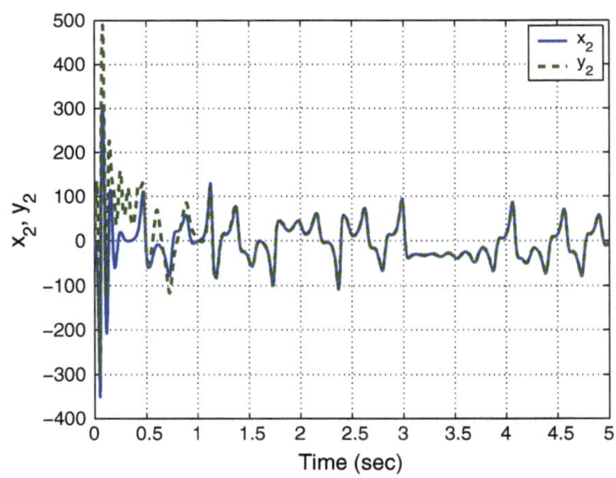

Fig. 7 Synchronization of the states x_2 and y_2

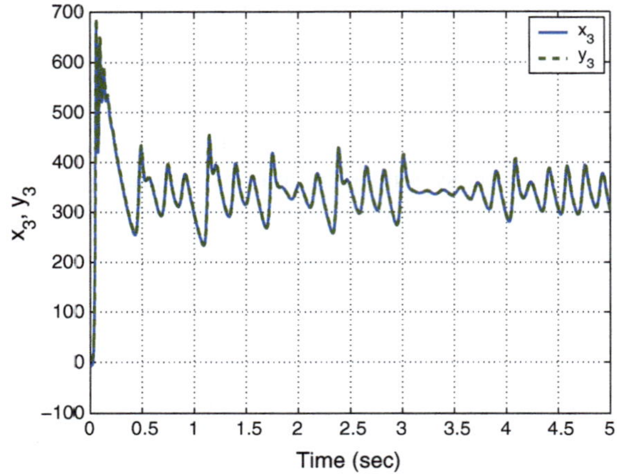

Fig. 8 Synchronization of the states x_3 and y_3

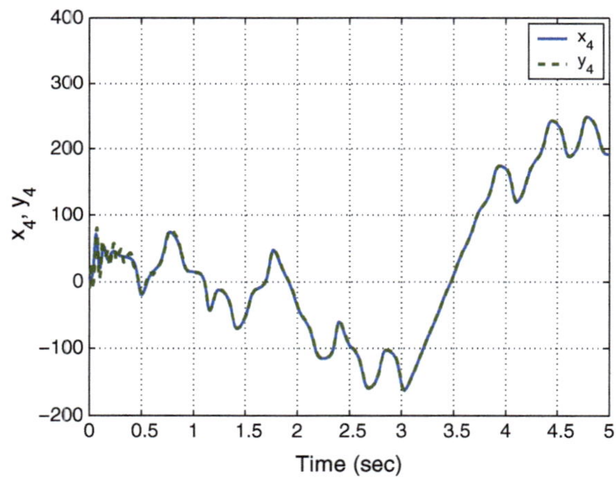

Fig. 9 Synchronization of the states x_4 and y_4

Fig. 10 Synchronization of the states x_5 and y_5

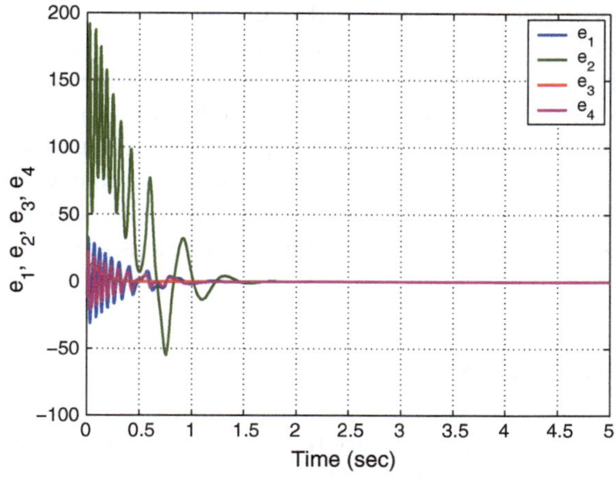

Fig. 11 Time-history of the synchronization errors e_1, e_2, e_3, e_4, e_5

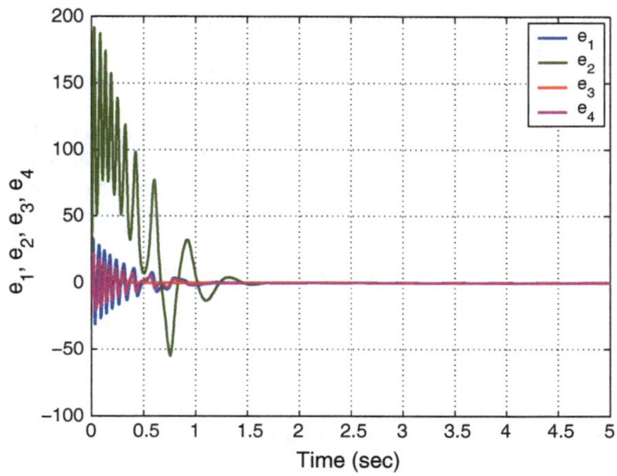

6 Conclusions

In this research work, we described a novel 5-D hyperchaotic system with an infinite line of equilibrium points. All the equilibrium points of the novel 5-D hyperchaotic system are unstable. The novel 5-D hyperchaotic system has fifteen terms on the right hand side with two quadratic nonlinearities. The phase portraits of the 5-D novel hyperchaotic system are depicted and the qualitative properties of the novel hyperchaotic system are discussed. The Lyapunov exponents of the 5-D novel hyperchaotic system have been obtained as $L_1 = 1.2995$, $L_2 = 0.2505$, $L_3 = 0.0615$, $L_4 = 0$ and $L_5 = -17.5932$, The maximal Lyapunov exponent of the novel hyperchaotic system is $L_1 = 1.2995$. while the Kaplan–Yorke dimension of the 5-D novel hyperchaotic system is obtained as $D_{KY} = 4.0916$. Next, an adaptive controller was designed to globally stabilize the novel hyperchaotic system with unknown parameters. Finally, an adaptive controller was also designed to achieve global chaos synchronization of the identical hyperchaotic systems with unknown parameters. MATLAB simulations were shown to illustrate all the main results derived in this work.

References

1. Abdurrahman A, Jiang H, Teng Z (2015) Finite-time synchronization for memristor-based neural networks with time-varying delays. Neural Netw 69:20–28
2. Arneodo A, Coullet P, Tresser C (1981) Possible new strange attractors with spiral structure. Commun Math Phys 79(4):573–576
3. Azar AT, Vaidyanathan S (2015) Chaos modeling and control systems design, vol 581. Springer, Germany
4. Cai G, Tan Z (2007) Chaos synchronization of a new chaotic system via nonlinear control. J. Uncertain Syst 1(3):235–240
5. Chen A, Lu J, Lü J, Yu S (2006) Generating hyperchaotic Lü attractor via state feedback control. Physica A 364:103–110
6. Chen G, Ueta T (1999) Yet another chaotic attractor. Int J Bifurc Chaos 9(7):1465–1466
7. Filali RL, Benrejeb M, Borne P (2014) On observer-based secure communication design using discrete-time hyperchaotic systems. Commun Nonlinear Sci Numer Simul 19(5):1424–1432
8. Hammami S (2015) State feedback-based secure image cryptosystem using hyperchaotic synchronization. ISA Trans 54:52–59
9. Jafari S, Sprott JC (2013) Simple chaotic flows with a line equilibrium. Chaos Solitons Fractals 57:79–84
10. Jia Q (2007) Hyperchaos generated from the Lorenz chaotic system and its control. Phys Lett A 366:217–222
11. Karthikeyan R, Sundarapandian V (2014) Hybrid chaos synchronization of four-scroll systems via active control. J Electr Eng 65(2):97–103
12. Khalil HK (2001) Nonlinear systems, 3rd edn. Prentice Hall, New Jersey
13. Li C, Liao X, Wong KW (2005) Lag synchronization of hyperchaos with application to secure communications. Chaos Solitons Fractals 23(1):183–193
14. Li C, Sprott JC, Thio W (2014) Bistability in a hyperchaotic system with a line equilibrium. J Exp Theor Phys 118(3):494–500
15. Li D (2008) A three-scroll chaotic attractor. Phys Lett A 372(4):387–393
16. Li X (2009) Modified projective synchronization of a new hyperchaotic system via nonlinear control. Commun Theor Phys 52:274–278
17. Lorenz EN (1963) Deterministic periodic flow. J Atmos Sci 20(2):130–141
18. Lü J, Chen G (2002) A new chaotic attractor coined. Int J Bifurc Chaos 12(3):659–661
19. Pehlivan I, Moroz IM, Vaidyanathan S (2014) Analysis, synchronization and circuit design of a novel butterfly attractor. J Sound Vib 333(20):5077–5096
20. Pham VT, Volos C, Jafari S, Wang X, Vaidyanathan S (2014) Hidden hyperchaotic attractor in a novel simple memristive neural network. Optoelectron Adv Mater Rapid Commun 8(11–12):1157–1163
21. Pham VT, Volos CK, Vaidyanathan S, Le TP, Vu VY (2015) A memristor-based hyperchaotic system with hidden attractors: dynamics, synchronization and circuital emulating. J Eng Sci Technol Rev 8(2):205–214

22. Rasappan S, Vaidyanathan S (2013) Hybrid synchronization of n-scroll Chua circuits using adaptive backstepping control design with recursive feedback. Malays J Math Sci 73(1):73–95
23. Rasappan S, Vaidyanathan S (2014) Global chaos synchronization of WINDMI and Coullet chaotic systems using adaptive backstepping control design. Kyungpook Math J 54(1):293–320
24. Rhouma R, Belghith S (2008) Cryptanalysis of a new image encryption algorithm based on hyper-chaos. Phys Lett A 372(38):5973–5978
25. Rössler OE (1976) An equation for continuous chaos. Phys Lett A 57(5):397–398
26. Rössler OE (1979) An equation for hyperchaos. Phys Lett A 71:155–157
27. Sampath S, Vaidyanathan S, Volos CK, Pham VT (2015) An eight-term novel four-scroll chaotic system with cubic nonlinearity and its circuit simulation. J Eng Sci Technol Rev 8(2):1–6
28. Sarasu P, Sundarapandian V (2011) Active controller design for generalized projective synchronization of four-scroll chaotic systems. Int J Syst Signal Control Eng Appl 4(2):26–33
29. Sarasu P, Sundarapandian V (2011) The generalized projective synchronization of hyperchaotic Lorenz and hyperchaotic Qi systems via active control. Int J Soft Comput 6(5):216–223
30. Sarasu P, Sundarapandian V (2012) Generalized projective synchronization of two-scroll systems via adaptive control. Int J Soft Comput 7(4):146–156
31. Senouci A, Boukabou A (2014) Predictive control and synchronization of chaotic and hyperchaotic systems based on a T–S fuzzy model. Math Comput Simul 105:62–78
32. Sprott JC (1994) Some simple chaotic flows. Phys Rev E 50(2):647–650
33. Sundarapandian V (2010) Output regulation of the Lorenz attractor. Far East J Math Sci 42(2):289–299
34. Sundarapandian V (2013) Adaptive control and synchronization design for the Lu-Xiao chaotic system. Lect Notes Electr Eng 131:319–327
35. Sundarapandian V (2013) Analysis and anti-synchronization of a novel chaotic system via active and adaptive controllers. J Eng Sci Technol Rev 6(4):45–52
36. Sundarapandian V, Karthikeyan R (2011) Anti-synchronization of hyperchaotic Lorenz and hyperchaotic Chen systems by adaptive control. Int J Syst Signal Control Eng Appl 4(2):18–25
37. Sundarapandian V, Karthikeyan R (2011) Anti-synchronization of Lü and Pan chaotic systems by adaptive nonlinear control. Eur J Sci Res 64(1):94–106
38. Sundarapandian V, Karthikeyan R (2012) Adaptive anti-synchronization of uncertain Tigan and Li systems. J Eng Appl Sci 7(1):45–52
39. Sundarapandian V, Pehlivan I (2012) Analysis, control, synchronization, and circuit design of a novel chaotic system. Math Comput Model 55(7–8):1904–1915
40. Sundarapandian V, Sivaperumal S (2011) Sliding controller design of hybrid synchronization of four-wing chaotic systems. Int J Soft Comput 6(5):224–231
41. Suresh R, Sundarapandian V (2013) Global chaos synchronization of a family of n-scroll hyperchaotic Chua circuits using backstepping control with recursive feedback. Far East J Math Sci 7(2):219–246
42. Tigan G, Opris D (2008) Analysis of a 3D chaotic system. Chaos Solitons Fractals 36: 1315–1319
43. Vaidyanathan S (2011) Output regulation of Arneodo-Coullet chaotic system. Commun Comput Inf Sci 133:98–107
44. Vaidyanathan S (2011) Output regulation of the unified chaotic system. Commun Comput Inf Sci 198:1–9
45. Vaidyanathan S (2012) Adaptive controller and syncrhonizer design for the Qi-Chen chaotic system. Lect Notes Inst Comput Sci Soc-Inf Telecommun Eng 84:73–82
46. Vaidyanathan S (2012) Anti-synchronization of Sprott-L and Sprott-M chaotic systems via adaptive control. Int J Control Theory Appl 5(1):41–59
47. Vaidyanathan S (2012) Global chaos control of hyperchaotic Liu system via sliding control method. Int J Control Theory Appl 5(2):117–123

48. Vaidyanathan S (2012) Sliding mode control based global chaos control of Liu-Liu-Liu-Su chaotic system. Int J Control Theory Appl 5(1):15–20
49. Vaidyanathan S (2013) A new six-term 3-D chaotic system with an exponential nonlinearity. Far East J Math Sci 79(1):135–143
50. Vaidyanathan S (2013) A ten-term novel 4-D hyperchaotic system with three quadratic nonlinearities and its control. Int J Control Theory Appl 6(2):97–109
51. Vaidyanathan S (2013) Analysis and adaptive synchronization of two novel chaotic systems with hyperbolic sinusoidal and cosinusoidal nonlinearity and unknown parameters. J Eng Sci Technol Rev 6(4):53–65
52. Vaidyanathan S (2013) Analysis, control and synchronization of hyperchaotic Zhou system via adaptive control. Adv Intell Syst Comput 177:1–10
53. Vaidyanathan S (2014) A new eight-term 3-D polynomial chaotic system with three quadratic nonlinearities. Far East J Math Sci 84(2):219–226
54. Vaidyanathan S (2014) Analysis and adaptive synchronization of eight-term 3-D polynomial chaotic systems with three quadratic nonlinearities. Eur Phys J Spec Top 223(8):1519–1529
55. Vaidyanathan S (2014) Analysis, control and synchronisation of a six-term novel chaotic system with three quadratic nonlinearities. Int J Model Identif Control 22(1):41–53
56. Vaidyanathan S (2014) Generalized projective synchronisation of novel 3-D chaotic systems with an exponential non-linearity via active and adaptive control. Int J Model Identif Control 22(3):207–217
57. Vaidyanathan S (2014) Global chaos synchronisation of identical Li-Wu chaotic systems via sliding mode control. Int J Model Identif Control 22(2):170–177
58. Vaidyanathan S (2014) Qualitative analysis and control of an eleven-term novel 4-D hyperchaotic system with two quadratic nonlinearities. Int J Control Theory Appl 7:35–47
59. Vaidyanathan S (2015) 3-cells Cellular Neural Network (CNN) attractor and its adaptive biological control. Int J PharmTech Res 8(4):632–640
60. Vaidyanathan S (2015) A 3-D novel highly chaotic system with four quadratic nonlinearities, its adaptive control and anti-synchronization with unknown parameters. J Eng Sci Technol Rev 8(2):106–115
61. Vaidyanathan S (2015) A novel chemical chaotic reactor system and its adaptive control. Int J ChemTech Res 8(7):146–158
62. Vaidyanathan S (2015) Adaptive backstepping control of enzymes-substrates system with ferroelectric behaviour in brain waves. Int J PharmTech Res 8(2):256–261
63. Vaidyanathan S (2015) Adaptive biological control of generalized Lotka-Volterra three-species biological system. Int J PharmTech Res 8(4):622–631
64. Vaidyanathan S (2015) Adaptive chaotic synchronization of enzymes-substrates system with ferroelectric behaviour in brain waves. Int J PharmTech Res 8(5):964–973
65. Vaidyanathan S (2015) Adaptive control of a chemical chaotic reactor. Int J PharmTech Res 8(3):377–382
66. Vaidyanathan S (2015) Adaptive control of the FitzHugh-Nagumo chaotic neuron model. Int J PharmTech Res 8(6):117–127
67. Vaidyanathan S (2015) Adaptive synchronization of chemical chaotic reactors. Int J ChemTech Res 8(2):612–621
68. Vaidyanathan S (2015) Adaptive synchronization of generalized Lotka-Volterra three-species biological systems. Int J PharmTech Res 8(5):928–937
69. Vaidyanathan S (2015) Adaptive synchronization of novel 3-D chemical chaotic reactor systems. Int J ChemTech Res 8(7):159–171
70. Vaidyanathan S (2015) Adaptive synchronization of the identical FitzHugh-Nagumo chaotic neuron models. Int J PharmTech Res 8(6):167–177
71. Vaidyanathan S (2015) Analysis, control and synchronization of a 3-D novel jerk chaotic system with two quadratic nonlinearities. Kyungpook Math J 55:563–586
72. Vaidyanathan S (2015) Analysis, properties and control of an eight-term 3-D chaotic system with an exponential nonlinearity. Int J Model Identif Control 23(2):164–172

73. Vaidyanathan S (2015) Anti-synchronization of brusselator chemical reaction systems via adaptive control. Int J ChemTech Res 8(6):759–768
74. Vaidyanathan S (2015) Chaos in neurons and adaptive control of Birkhoff-Shaw strange chaotic attractor. Int J PharmTech Res 8(5):956–963
75. Vaidyanathan S (2015) Chaos in neurons and synchronization of Birkhoff-Shaw strange chaotic attractors via adaptive control. Int J PharmTech Res 8(6):1–11
76. Vaidyanathan S (2015) Coleman-Gomatam logarithmic competitive biology models and their ecological monitoring. Int J PharmTech Res 8(6):94–105
77. Vaidyanathan S (2015) Dynamics and control of brusselator chemical reaction. Int J ChemTech Res 8(6):740–749
78. Vaidyanathan S (2015) Dynamics and control of tokamak system with symmetric and magnetically confined plasma. Int J ChemTech Res 8(6):795–803
79. Vaidyanathan S (2015) Global chaos synchronization of chemical chaotic reactors via novel sliding mode control method. Int J ChemTech Res 8(7):209–221
80. Vaidyanathan S (2015) Global chaos synchronization of the forced Van der Pol chaotic oscillators via adaptive control method. Int J PharmTech Res 8(6):156–166
81. Vaidyanathan S (2015) Global chaos synchronization of the Lotka-Volterra biological systems with four competitive species via active control. Int J PharmTech Res 8(6):206–217
82. Vaidyanathan S (2015) Lotka-Volterra population biology models with negative feedback and their ecological monitoring. Int J PharmTech Res 8(5):974–981
83. Vaidyanathan S (2015) Lotka-Volterra two species competitive biology models and their ecological monitoring. Int J PharmTech Res 8(6):32–44
84. Vaidyanathan S (2015) Output regulation of the forced Van der Pol chaotic oscillator via adaptive control method. Int J PharmTech Res 8(6):106–116
85. Vaidyanathan S, Azar AT (2015) Analysis and control of a 4-D novel hyperchaotic system. Stud Comput Intell 581:3–17
86. Vaidyanathan S, Azar AT (2015) Analysis, control and synchronization of a nine-term 3-D novel chaotic system. In: Azar AT, Vaidyanathan S (eds) Chaos modelling and control systems design, studies in computational intelligence, vol 581. Springer, Germany, pp 19–38
87. Vaidyanathan S, Madhavan K (2013) Analysis, adaptive control and synchronization of a seven-term novel 3-D chaotic system. Int J Control Theory Appl 6(2):121–137
88. Vaidyanathan S, Pakiriswamy S (2013) Generalized projective synchronization of six-term Sundarapandian chaotic systems by adaptive control. Int J Control Theory Appl 6(2):153–163
89. Vaidyanathan S, Pakiriswamy S (2015) A 3-D novel conservative chaotic System and its generalized projective synchronization via adaptive control. J Eng Sci Technol Rev 8(2):52–60
90. Vaidyanathan S, Rajagopal K (2011) Hybrid synchronization of hyperchaotic Wang-Chen and hyperchaotic Lorenz systems by active non-linear control. Int J Syst Signal Control Eng Appl 4(3):55–61
91. Vaidyanathan S, Rajagopal K (2012) Global chaos synchronization of hyperchaotic Pang and hyperchaotic Wang systems via adaptive control. Int J Soft Comput 7(1):28–37
92. Vaidyanathan S, Rasappan S (2011) Global chaos synchronization of hyperchaotic Bao and Xu systems by active nonlinear control. Commun Comput Inf Sci 198:10–17
93. Vaidyanathan S, Rasappan S (2014) Global chaos synchronization of n-scroll Chua circuit and Lur'e system using backstepping control design with recursive feedback. Arab J Sci Eng 39(4):3351–3364
94. Vaidyanathan S, Sampath S (2012) Anti-synchronization of four-wing chaotic systems via sliding mode control. Int J Autom Comput 9(3):274–279
95. Vaidyanathan S, Volos C (2015) Analysis and adaptive control of a novel 3-D conservative no-equilibrium chaotic system. Arch Control Sci 25(3):333–353
96. Vaidyanathan S, Volos C, Pham VT (2014) Hyperchaos, adaptive control and synchronization of a novel 5-D hyperchaotic system with three positive Lyapunov exponents and its SPICE implementation. Arch Control Sci 24(4):409–446

97. Vaidyanathan S, Volos C, Pham VT, Madhavan K, Idowu BA (2014) Adaptive backstepping control, synchronization and circuit simulation of a 3-D novel jerk chaotic system with two hyperbolic sinusoidal nonlinearities. Arch Control Sci 24(3):375–403
98. Vaidyanathan S, Idowu BA, Azar AT (2015) Backstepping controller design for the global chaos synchronization of Sprott's jerk systems. Stud Comput Intell 581:39–58
99. Vaidyanathan S, Rajagopal K, Volos CK, Kyprianidis IM, Stouboulos IN (2015) Analysis, adaptive control and synchronization of a seven-term novel 3-D chaotic system with three quadratic nonlinearities and its digital implementation in LabVIEW. J Eng Sci Technol Rev 8(2):130–141
100. Vaidyanathan S, Sampath S, Azar AT (2015) Global chaos synchronisation of identical chaotic systems via novel sliding mode control method and its application to Zhu system. Int J Model Identif Control 23(1):92–100
101. Vaidyanathan S, Volos C, Pham VT, Madhavan K (2015) Analysis, adaptive control and synchronization of a novel 4-D hyperchaotic hyperjerk system and its SPICE implementation. Nonlinear Dyn 25(1):135–158
102. Vaidyanathan S, Volos CK, Kyprianidis IM, Stouboulos IN, Pham VT (2015) Analysis, adaptive control and anti-synchronization of a six-term novel jerk chaotic system with two exponential nonlinearities and its circuit simulation. J Eng Sci Technol Rev 8(2):24–36
103. Vaidyanathan S, Volos CK, Pham VT (2015) Analysis, adaptive control and adaptive synchronization of a nine-term novel 3-D chaotic system with four quadratic nonlinearities and its circuit simulation. J Eng Sci Technol Rev 8(2):181–191
104. Vaidyanathan S, Volos CK, Pham VT (2015) Analysis, control, synchronization and SPICE implementation of a novel 4-D hyperchaotic Rikitake dynamo system without equilibrium. J Eng Sci Technol Rev 8(2):232–244
105. Vaidyanathan S, Volos CK, Pham VT (2015) Global chaos control of a novel nine-term chaotic system via sliding mode control. In: Azar AT, Zhu Q (eds) Advances and applications in sliding mode control systems, studies in computational intelligence, vol 576. Springer, Germany, pp 571–590
106. Volos CK, Kyprianidis IM, Stouboulos IN, Tlelo-Cuautle E, Vaidyanathan S (2015) Memristor: A new concept in synchronization of coupled neuromorphic circuits. J Eng Sci Technol Rev 8(2):157–173
107. Wang J, Chen Z (2008) A novel hyperchaotic system and its complex dynamics. Int J Bifurc Chaos 18:3309–3324
108. Wei X, Yunfei F, Qiang L (2012) A novel four-wing hyper-chaotic system and its circuit implementation. Procedia Eng 29:1264–1269
109. Wei Z, Yang Q (2010) Anti-control of Hopf bifurcation in the new chaotic system with two stable node-foci. Appl Math Comput 217(1):422–429
110. Wu X, Zhu C, Kan H (2015) An improved secure communication scheme based passive synchronization of hyperchaotic complex nonlinear system. Appl Math Comput 252:201–214
111. Yujun N, Xingyuan W, Mingjun W, Huaguang Z (2010) A new hyperchaotic system and its circuit implementation. Commun Nonlinear Sci Numer Simul 15(11):3518–3524
112. Zhang H, Liao X, Yu J (2005) Fuzzy modeling and synchronization of hyperchaotic systems. Chaos Solitons Fractals 26(3):835–843
113. Zhou W, Xu Y, Lu H, Pan L (2008) On dynamics analysis of a new chaotic attractor. Phys Lett A 372(36):5773–5777
114. Zhu C (2012) A novel image encryption scheme based on improved hyperchaotic sequences. Opt Commun 285(1):29–37
115. Zhu C, Liu Y, Guo Y (2010) Theoretic and numerical study of a new chaotic system. Intell Inf Manag 2:104–109

Analysis, Control and Circuit Simulation of a Novel 3-D Finance Chaotic System

S. Vaidyanathan, Ch.K. Volos, O.I. Tacha, I.M. Kyprianidis, I.N. Stouboulos and V.-T. Pham

Abstract There is a growing interest in developing nonlinear dynamical systems for economic models displaying chaotic behaviour. In this work, we describe an eight-term novel 3-D finance chaotic system consisting of two nonlinearities (one quadratic and one quartic). The phase portraits of the novel 3-D finance chaotic system are depicted using MATLAB. We give a dynamic analysis of the novel 3-D finance chaotic system. The novel chaotic system has three equilibrium points of which one equilibrium point on the x_2 axis is a saddle point, while the other two equilibrium points are saddle-foci. The novel finance chaotic system has rotation symmetry about the x_2 axis. The Lyapunov exponents of the novel finance chaotic system are obtained as $L_1 = 0.1209$, $L_2 = 0$ and $L_3 = -0.4321$, while the Kaplan–Yorke dimension of the novel finance chaotic system is obtained as $D_{KY} = 2.2798$. Since the sum of the Lyapunov exponents is negative, the novel chaotic system is dissipative. Next, we derive new results for the global chaos control of the novel finance chaotic system with unknown parameters using adaptive control method. The chaos control problem aims to regulate the states of the novel finance chaotic system to desired constant values. The main adaptive control result for the novel finance chaotic system is established using Lyapunov stability theory. Finally, an electronic circuit realization of the novel finance chaotic system using Spice is presented in detail to confirm the feasibility of the theoretical model.

Keywords Chaos · Chaotic systems · Finance system · Chaos control · Adaptive control

S. Vaidyanathan (✉)
Research and Development Centre, Vel Tech University,
Avadi, Chennai 600062, Tamil Nadu, India
e-mail: sundarvtu@gmail.com

Ch.K. Volos · O.I. Tacha · I.M. Kyprianidis · I.N. Stouboulos
Department of Physics, Aristotle University of Thessaloniki,
54124 Thessaloniki, Greece

V.-T. Pham
School of Electronics and Telecommunications,
Hanoi University of Science and Technology, Hanoi, Vietnam
email: pvt3010@gmail.com

S. Vaidyanathan and C. Volos (eds.), *Advances and Applications in Chaotic Systems*, Studies in Computational Intelligence 636,
DOI 10.1007/978-3-319-30279-9_21

1 Introduction

In the last few decades, Chaos theory has become a very important and active research field, employing many applications in different disciplines like physics, chemistry, biology, ecology, engineering and economics, among others [3, 13, 15, 25]. Some classical paradigms of 3-D chaotic systems in literature are the Lorenz system [17], Rössler system [22], ACT system [2], Sprott systems [24], Chen system [6], Lü system [18], Cai system [5], Tigan system [30], etc.

Many new chaotic systems have been discovered in the recent years such as Zhou system [90], Zhu system [91], Li system [16], Sundarapandian systems [28, 29], Vaidyanathan systems [36, 38, 40–43, 46, 57, 58, 72–75, 77, 79, 81, 82, 84], Pehlivan system [20], Sampath system [23], etc.

Chaos theory has applications in several fields of science and engineering such as chemical reactors [47, 51, 53, 55, 59, 63–65], biological systems [45, 48–50, 52, 54, 56, 60–62, 66–70], memristors [1, 21, 89], etc.

The control of a chaotic system aims to stabilize or regulate the system with the help of a feedback control. There are many methods available for controlling a chaotic system such as active control [26, 31, 32], adaptive control [27, 33, 37, 39, 44, 71, 76, 80, 83], sliding mode control [34, 35], backstepping control [19, 78, 85], etc.

In recent years, there is significant interest in applying nonlinear dynamical systems to model finance systems displaying chaotic behaviour [4, 7–10, 86–88]. The study of complexity of economy and finance systems has important theoretical and practical meaning and it is the developmental direction of complex nonlinear economic systems.

In this work, we describe an eight-term novel 3-D finance chaotic system consisting of two nonlinearities (one quadratic and one quartic). Our novel 3-D finance chaotic system is obtained by modifying the dynamics of the finance chaotic system described in [10].

This work is organized as follows. Section 2 describes the dynamic equations and phase portraits of the novel 3-D finance chaotic system. Section 3 details the qualitative analysis and properties of the novel finance chaotic system.

In Sect. 3, the Lyapunov exponents of the novel finance chaotic system are obtained as $L_1 = 0.1209$, $L_2 = 0$ and $L_3 = -0.4321$, while the Kaplan–Yorke dimension of the novel chaotic system is obtained as $D_{KY} = 2.2798$. The maximal Lyapunov novel exponent of the novel finance chaotic system is $L_1 = 0.1209$. Since the sum of the Lyapunov exponents is negative, the novel finance chaotic system is dissipative.

In Sect. 4, we derive new results for the global chaos control of the novel finance chaotic system with unknown parameters.

In Sect. 5, an electronic circuit realization of the novel finance chaotic system using Spice is presented to confirm the feasibility of the theoretical model. Section 6 contains the conclusions of this work.

2 A Novel 3-D Finance Chaotic System

In [10], Gao and Ma studied the finance chaotic system described by the dynamics

$$\begin{cases} \dot{x}_1 = x_3 + (x_2 - a)x_1 \\ \dot{x}_2 = 1 - bx_2 - x_1^2 \\ \dot{x}_3 = -x_1 - cx_3 \end{cases} \tag{1}$$

where x_1, x_2, x_3 are the states and a, b, c are constant, positive, parameters.

In the finance model (1), x_1 denotes the interest rate, x_2 denotes the investment demand and x_3 denotes the price index. The parameter a denotes the savings, b denotes the investment cost and c denotes the commodities demand elasticity.

In [10], it was shown that the system (1) exhibits a *strange chaotic attractor* for the parameter values

$$a = 6, \quad b = 0.1, \quad c = 1 \tag{2}$$

For the initial values

$$x_1(0) = 0.6, \quad x_2(0) = 0.2, \quad x_3(0) = 0.8 \tag{3}$$

and for the parameter values (2), the Lyapunov exponents of the finance chaotic system (1) can be numerically obtained as

$$L_1 = 0.0833, \quad L_2 = 0, \quad L_3 = -0.4101 \tag{4}$$

The Kaplan–Yorke dimension [11, 12] of a chaotic system of order n is defined as

$$D_{KY} = j + \frac{L_1 + \cdots + L_j}{|L_{j+1}|} \tag{5}$$

where $L_1 \geq L_2 \geq \cdots \geq L_n$ are the Lyapunov exponents of the chaotic system and j is the largest integer for which $L_1 + L_2 + \cdots + L_j \geq 0$. (Kaplan–Yorke conjecture states that for typical chaotic systems, $D_{KY} \approx D_L$, the information dimension of the system.)

Thus, the Kaplan–Yorke dimension of the finance chaotic system (1) can be numerically obtained as

$$D_{KY} = 2 + \frac{L_1 + L_2}{|L_3|} = 2.2031 \tag{6}$$

In this section, we describe an eight-term novel finance chaotic system, which is obtained by modifying the second equation in the dynamics of (1) as

$$\begin{cases} \dot{x}_1 = x_3 + (x_2 - a)x_1 \\ \dot{x}_2 = 1 - bx_2 - x_1^4 \\ \dot{x}_3 = -x_1 - cx_3 \end{cases} \tag{7}$$

The novel finance system (7) is an eight-term polynomial system with two nonlinearities (one quadratic and one quartic).

The novel finance system (7) exhibits a *strange chaotic attractor* for the parameter values

$$a = 7.5, \quad b = 0.1, \quad c = 1 \tag{8}$$

For numerical simulations, we take the initial conditions as

$$x_1(0) = 0.6, \quad x_2(0) = 0.2, \quad x_3(0) = 0.8 \tag{9}$$

In this paper, we shall show that for the parameter values (8) and for the initial conditions (9), the Lyapunov exponents of the novel finance chaotic system (7) are given by

$$L_1 = 0.1209, \quad L_2 = 0, \quad L_3 = -0.4321 \tag{10}$$

Thus, the Kaplan–Yorke dimension of the novel finance chaotic system (7) can be numerically obtained as

$$D_{KY} = 2 + \frac{L_1 + L_2}{|L_3|} = 2.2798 \tag{11}$$

From Eqs. (4) and (10), we note that the maximal Lyapunov exponent of the Gao–Ma finance chaotic system (1) is $L_1 = 0.0833$, while the maximal Lyapunov exponent of the novel finance chaotic system (7) is $L_1 = 0.1209$.

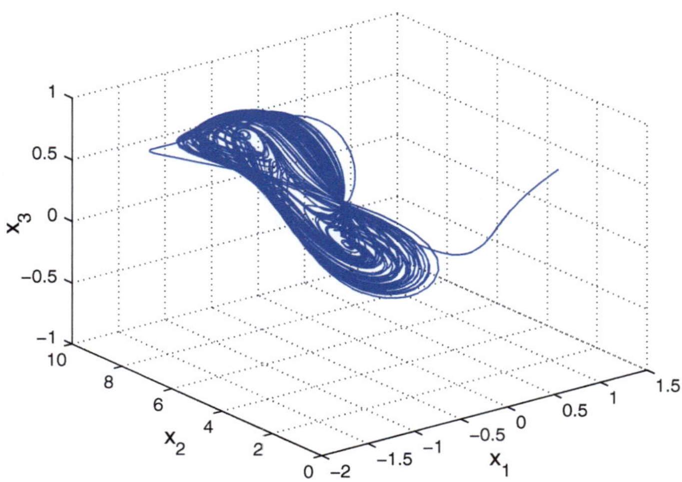

Fig. 1 3-D phase portrait of the novel finance chaotic system

From Eqs. (6) and (11), we note that the Kaplan–Yorke dimension of the Gao–Ma finance chaotic system (1) is $D_{KY} = 2.2031$, while the Kaplan–Yorke dimension of the novel finance chaotic system (7) is $D_{KY} = 2.2798$.

The above observations show that the novel finance chaotic system (7) displays more complexity than the Gao–Ma finance chaotic system (1).

Figure 1 depicts the 3-D phase portrait of the novel finance chaotic system (7), while Figs. 2, 3 and 4 depict the 2-D projection of the novel finance chaotic system (7) on the (x_1, x_2), (x_2, x_3) and (x_1, x_3) planes, respectively.

From Figs. 1, 2, 3 and 4, we deduce that the strange chaotic attractor of the novel finance chaotic system (7) can be classified as a *two-scroll* chaotic attractor.

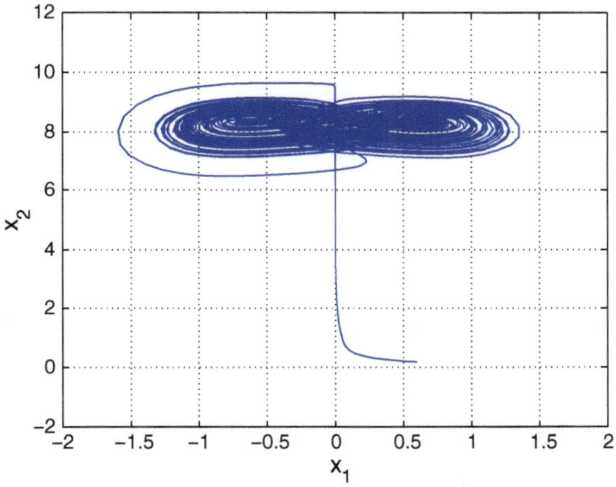

Fig. 2 2-D projection of the novel finance chaotic system on the (x_1, x_2) plane

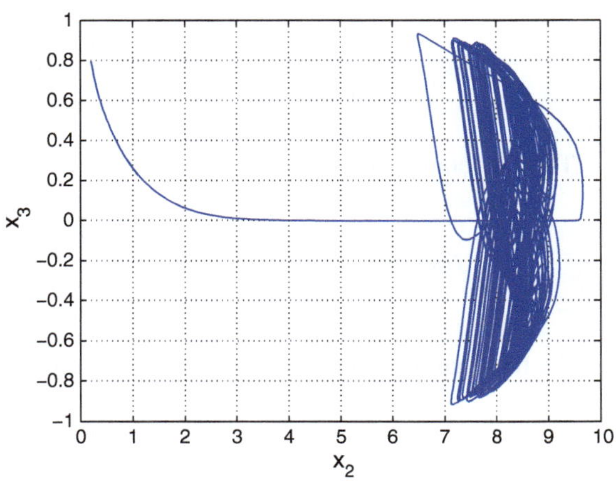

Fig. 3 2-D projection of the novel finance chaotic system on the (x_2, x_3) plane

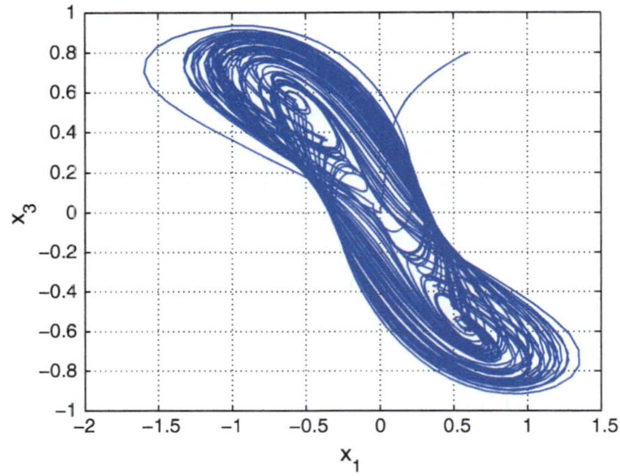

Fig. 4 2-D projection of the novel finance chaotic system on the (x_1, x_3) plane

3 Analysis of the Novel 3-D Finance Chaotic System

In this section, we give a dynamic analysis of the 3-D novel chaotic system (7). We take the parameter values as in the chaotic case (8), viz. $a = 7.5$, $b = 0.1$ and $c = 1$.

3.1 Equilibrium Points

It is easy to see that the system (7) has three equilibrium points, viz.

$$E_1 = \begin{bmatrix} 0 \\ 10 \\ 0 \end{bmatrix}, E_2 = \begin{bmatrix} 0.6223 \\ 8.5000 \\ -0.6223 \end{bmatrix} \text{ and } E_3 = \begin{bmatrix} -0.6223 \\ 8.5000 \\ 0.6223 \end{bmatrix} \tag{12}$$

The Jacobian of the system (7) at any point $\mathbf{x} \in \mathbf{R}^3$ is calculated as

$$J(\mathbf{x}) = \begin{bmatrix} x_2 - a & x_1 & 1 \\ -4x_1^3 & -b & 0 \\ -1 & 0 & -c \end{bmatrix} = \begin{bmatrix} x_2 - 7.5 & x_1 & 1 \\ -4x_1^3 & -0.1 & 0 \\ -1 & 0 & -1 \end{bmatrix} \tag{13}$$

We find that the matrix $J_1 = J(E_1)$ has the eigenvalues

$$\lambda_1 = -0.6861, \quad \lambda_2 = -0.1, \quad \lambda_3 = 2.1861 \tag{14}$$

This shows that the equilibrium point E_1 is a saddle-point, which is unstable.

Next, we find that the matrix $J_2 = J(E_2)$ has the eigenvalues

$$\lambda_1 = -0.6369, \quad \lambda_{2,3} = 0.2685 \pm 0.9326i \tag{15}$$

This shows that the equilibrium point E_2 is a saddle-focus, which is unstable. We also find that the matrix $J_3 = J(E_3)$ has the eigenvalues

$$\lambda_1 = -0.6369, \quad \lambda_{2,3} = 0.2685 \pm 0.9326i \tag{16}$$

This shows that the equilibrium point E_3 is also a saddle-focus, which is unstable.

3.2 Rotation Symmetry About the x_2-axis

It is easy to see that the system (7) is invariant under the change of coordinates

$$(x_1, x_2, x_3) \mapsto (-x_1, x_2, -x_3) \tag{17}$$

Thus, it follows that the 3-D novel finance chaotic system (7) has rotation symmetry about the x_2-axis and that any non-trivial trajectory must have a twin trajectory.

3.3 Invariance

It is easy to see that the x_2-axis is invariant under the flow of the 3-D novel finance chaotic system (7). The invariant motion along the x_2-axis is characterized by

$$\dot{x}_2 = 1 - bx_2, \quad (b > 0) \tag{18}$$

which is stable, but not asymptotically stable.

3.4 Lyapunov Exponents and Kaplan–Yorke Dimension

We take the parameter values of the novel system (7) as in the chaotic case (8). We take the initial state of the novel system (7) as given in (9).

Then the Lyapunov exponents of the system (7) are numerically obtained as

$$L_1 = 0.1209, \quad L_2 = 0, \quad L_3 = -0.4321 \tag{19}$$

Thus, the maximal Lyapunov exponent of the novel finance chaotic system (7) is found as $L_1 = 0.1209$.

Also, the Kaplan–Yorke dimension of the novel chaotic system (7) is found as

$$D_{KY} = 2 + \frac{L_1 + L_2}{|L_3|} = 2.2798 \tag{20}$$

4 Adaptive Control of the Novel 3-D Finance Chaotic System

In this section, we consider the controlled novel 3-D finance chaotic system given by

$$\begin{cases} \dot{x}_1 = x_3 + (x_2 - a)x_1 + u_1 \\ \dot{x}_2 = 1 - bx_2 - x_1^4 + u_2 \\ \dot{x}_3 = -x_1 - cx_3 + u_3 \end{cases} \tag{21}$$

where x_1, x_2, x_3 are the states and a, b, c are unknown system parameters.

We consider the research problem of finding adaptive controls u_1, u_2, u_3 so as to regulate the states x_1, x_2, and x_3 of the system (21) to desired constant values α, β and γ respectively.

Thus, we define the *control error* as

$$\begin{cases} e_1(t) = x_1(t) - \alpha \\ e_2(t) = x_2(t) - \beta \\ e_3(t) = x_3(t) - \gamma \end{cases} \tag{22}$$

Then the error dynamics is determined as

$$\begin{cases} \dot{e}_1 = (e_3 + \gamma) + (e_2 + \beta)(e_1 + \alpha) - a(e_1 + \alpha) + u_1 \\ \dot{e}_2 = 1 - b(e_2 + \beta) - (e_1 + \alpha)^4 + u_2 \\ \dot{e}_3 = -(e_1 + \alpha) - c(e_3 + \gamma) + u_3 \end{cases} \tag{23}$$

We consider the adaptive feedback control law

$$\begin{cases} u_1 = -(e_3 + \gamma) - (e_2 + \beta)(e_1 + \alpha) + \hat{a}(t)(e_1 + \alpha) - k_1 e_1 \\ u_2 = -1 + \hat{b}(t)(e_2 + \beta) + (e_1 + \alpha)^4 - k_2 e_2 \\ u_3 = (e_1 + \alpha) + \hat{c}(t)(e_3 + \gamma) - k_3 e_3 \end{cases} \tag{24}$$

where k_1, k_2, k_3 are positive gain constants.

Substituting (24) into (23), we get the closed-loop plant dynamics as

$$\begin{cases} \dot{e}_1 = -[a - \hat{a}(t)](e_1 + \alpha) - k_1 e_1 \\ \dot{e}_2 = -[b - \hat{b}(t)](e_2 + \beta) - k_2 e_2 \\ \dot{e}_3 = -[c - \hat{c}(t)](e_3 + \gamma) - k_3 e_3 \end{cases} \tag{25}$$

The parameter estimation errors are defined as

$$\begin{cases} e_a(t) = a - \hat{a}(t) \\ e_b(t) = b - \hat{b}(t) \\ e_c(t) = c - \hat{c}(t) \end{cases} \tag{26}$$

In view of (26), we can simplify the plant dynamics (25) as

$$\begin{cases} \dot{e}_1 = -e_a(e_1 + \alpha) - k_1 e_1 \\ \dot{e}_2 = -e_b(e_2 + \beta) - k_2 e_2 \\ \dot{e}_3 = -e_c(e_3 + \gamma) - k_3 e_3 \end{cases} \tag{27}$$

Differentiating (26) with respect to t, we obtain

$$\begin{cases} \dot{e}_a(t) = -\dot{\hat{a}}(t) \\ \dot{e}_b(t) = -\dot{\hat{b}}(t) \\ \dot{e}_c(t) = -\dot{\hat{c}}(t) \end{cases} \tag{28}$$

We consider the quadratic candidate Lyapunov function defined by

$$V(\mathbf{e}, e_a, e_b, e_c) = \frac{1}{2}\left(e_1^2 + e_2^2 + e_3^2\right) + \frac{1}{2}\left(e_a^2 + e_b^2 + e_c^2\right) \tag{29}$$

Differentiating V along the trajectories of (27) and (28), we obtain

$$\begin{cases} \dot{V} = -k_1 e_1^2 - k_2 e_2^2 - k_3 e_3^2 + e_a\left[-e_1(e_1 + \alpha) - \dot{\hat{a}}\right] \\ \qquad + e_b\left[-e_2(e_2 + \beta) - \dot{\hat{b}}\right] + e_c\left[-e_3(e_3 + \gamma) - \dot{\hat{c}}\right] \end{cases} \tag{30}$$

In view of (30), we take the parameter update law as

$$\begin{cases} \dot{\hat{a}}(t) = -e_1(e_1 + \alpha) \\ \dot{\hat{b}}(t) = -e_2(e_2 + \beta) \\ \dot{\hat{c}}(t) = -e_3(e_3 + \gamma) \end{cases} \tag{31}$$

Next, we state and prove the main result of this section.

Theorem 1 *The states x_1, x_2 and x_3 of the novel 3-D finance chaotic system (21) with unknown system parameters are globally and exponentially regulated for all initial conditions to the desired constant values α, β and γ, respectively, by the adaptive control law (24) and the parameter update law (31), where k_1, k_2, k_3 are positive gain constants.*

Proof We prove this result by applying Lyapunov stability theory [14].

We consider the quadratic Lyapunov function defined by (29), which is clearly a positive definite function on $\mathbf{R}^6$.

By substituting the parameter update law (31) into (30), we obtain the time-derivative of V as

$$\dot{V} = -k_1 e_1^2 - k_2 e_2^2 - k_3 e_3^2 \tag{32}$$

From (32), it is clear that $\dot{V}$ is a negative semi-definite function on $\mathbf{R}^6$.

Thus, we can conclude that the state vector $\mathbf{x}(t)$ and the parameter estimation error are globally bounded, i.e.

$$\left[\, e_1(t)\ e_2(t)\ e_3(t)\ e_a(t)\ e_b(t)\ e_c(t) \,\right]^T \in \mathbf{L}_\infty.$$

We define $k = \min\{k_1, k_2, k_3\}$.

Then it follows from (32) that

$$\dot{V} \le -k \|\mathbf{e}(t)\|^2 \tag{33}$$

Thus, we have

$$k \|\mathbf{e}(t)\|^2 \le -\dot{V} \tag{34}$$

Integrating the inequality (34) from 0 to t, we get

$$k \int_0^t \|\mathbf{e}(\tau)\|^2 \, d\tau \le V(0) - V(t) \tag{35}$$

From (35), it follows that $\mathbf{e} \in \mathbf{L}_2$.

Using (27), we can conclude that $\dot{\mathbf{e}} \in \mathbf{L}_\infty$.

Using Barbalat's lemma [14], we conclude that $\mathbf{e}(t) \to 0$ exponentially as $t \to \infty$ for all initial conditions $\mathbf{e}(0) \in \mathbf{R}^3$.

Hence, it follows that the states x_1, x_2 and x_3 of the novel 3-D finance chaotic system (21) with unknown system parameters a, b and c are globally and exponentially regulated for all initial conditions to the desired constant values α, β and γ, respectively, by the adaptive control law (24) and the parameter update law (31).

This completes the proof. ■

For the numerical simulations, the classical fourth-order Runge–Kutta method with step size $h = 10^{-8}$ is used to solve the systems (21) and (31), when the adaptive control law (24) is applied.

The parameter values of the novel 3-D finance chaotic system (21) are taken as in the chaotic case (8), i.e.

$$a = 7.5, \quad b = 0.1, \quad c = 1 \tag{36}$$

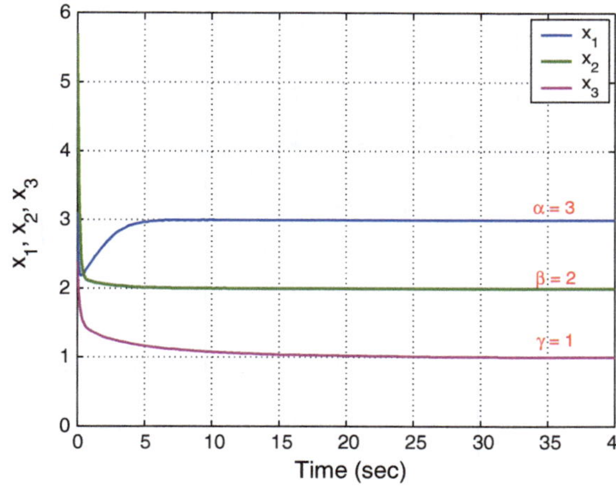

Fig. 5 Time-history of the regulated states x_1, x_2, x_3

We take the positive gain constants as

$$k_1 = 10, \quad k_2 = 10, \quad k_3 = 10 \tag{37}$$

Furthermore, as initial conditions of the novel 3-D finance chaotic system (21), we take

$$x_1(0) = 3.1, \quad x_2(0) = 5.7, \quad x_3(0) = 2.4 \tag{38}$$

As initial conditions of the parameter estimates, we take

$$\hat{a}(0) = 3.4, \quad \hat{b}(0) = 2.8, \quad \hat{c}(0) = 4.7 \tag{39}$$

As the desired values of the states, we take

$$\alpha = 3, \quad \beta = 2, \quad \gamma = 1 \tag{40}$$

In Fig. 5, the exponential convergence of the regulated states of the 3-D novel finance chaotic system (21) is depicted.

It is seen that the states x_1, x_2, x_3 of the novel finance chaotic system (21) converge to the desired steady-state values in 30 s.

5 Circuit Simulation of the Novel Finance Chaotic System

In this section, circuit implementation of the novel finance chaotic system is studied. The electronic circuit realizing the new system (7) is designed using operational amplifier approach [74, 77, 83] and shown in Fig. 6. Each state variable of system

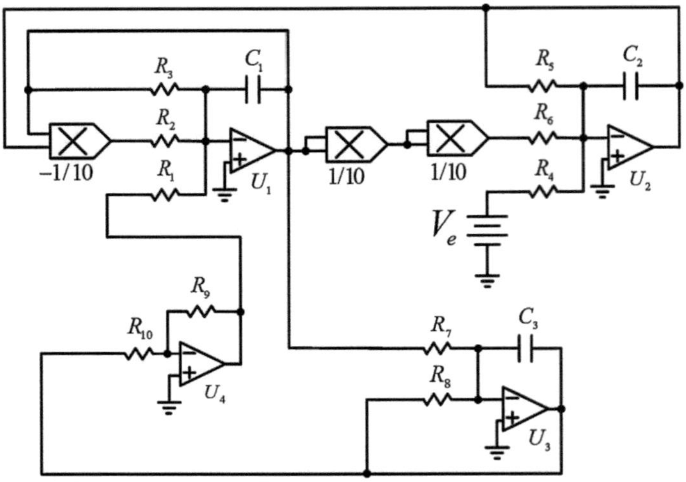

Fig. 6 The designed electronic circuit schematic of the novel finance chaotic system

(7), i.e. x_1, x_2, x_3 is implemented as the voltage across the corresponding capacitors C_1, C_2, and C_3, respectively. The circuital equations of the designed circuit are

$$\begin{cases} \frac{dv_{C_1}}{dt} = \frac{1}{R_1C_1}v_{C_3} + \frac{1}{10R_2C_1}v_{C_1}v_{C_2} - \frac{1}{R_3C_1}v_{C_1} \\ \frac{dv_{C_2}}{dt} = -\frac{1}{R_4C_2}V_e - \frac{1}{R_5C_2}v_{C_2} - \frac{1}{1000R_6C_2}v_{C_1}^4 \\ \frac{dv_{C_3}}{dt} = -\frac{1}{R_7C_3}v_{C_1} - \frac{1}{R_8C_3}v_{C_3} \end{cases} \tag{41}$$

where v_{C_1}, v_{C_2}, and v_{C_3} denote the voltages across the capacitors C_1, C_2, and C_3, respectively.

The TL084 operational amplifiers are used in this work. The power supplies of the operational amplifiers are $\pm 15\,V_{DC}$. The values of components in Fig. 6 are chosen to match the parameters of system (7) as follows: $R_1 = R_4 = R_7 = R_8 = R_9 = R_{10} = 400\,\text{k}\Omega, R_2 = 40\,\text{k}\Omega, R_3 = 53.333\,\text{k}\Omega, R_5 = 4\,\text{M}\Omega, R_6 = 0.4\,\text{k}\Omega, V_e = -1\,V_{DC}$, and $C_1 = C_2 = C_3 = 1\,\text{nF}$.

The designed circuit is implemented by using the electronic simulation package Cadence OrCAD. The obtained results are presented in Figs. 7, 8 and 9, which show the chaotic attractors in $v_{C_1} - v_{C_2}$, $v_{C_2} - v_{C_3}$, and $v_{C_1} - v_{C_3}$ planes, respectively. It is easy to see a good agreement between the circuital attractors and theoretical ones.

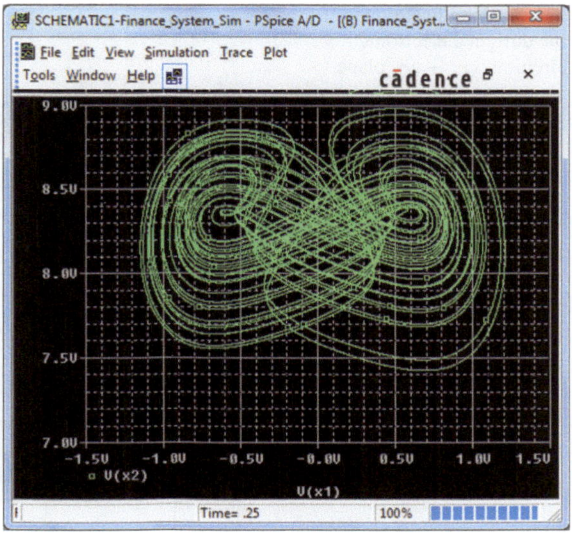

Fig. 7 Chaotic attractor of the designed electronic circuit in $v_{C_1} - v_{C_2}$ plane

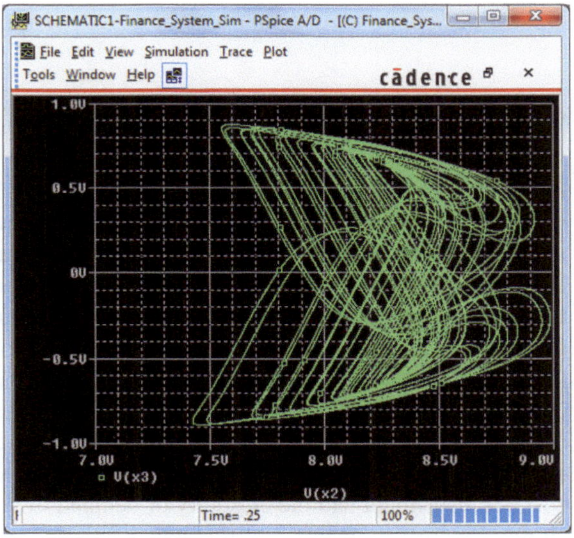

Fig. 8 Chaotic attractor of the designed electronic circuit in $v_{C_2} - v_{C_3}$ plane

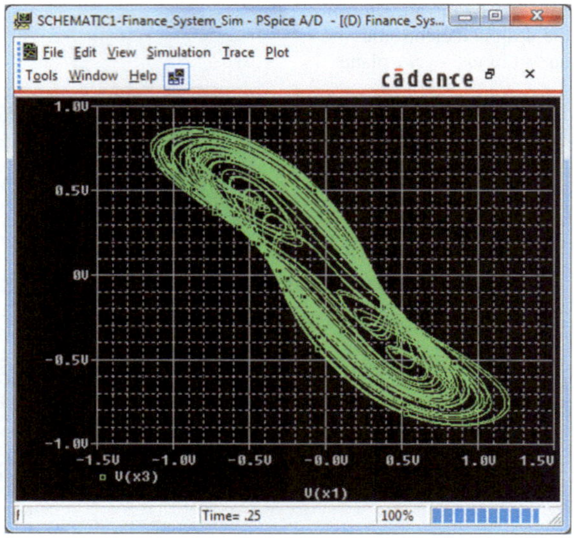

Fig. 9 Chaotic attractor of the designed electronic circuit in $v_{C_1} - v_{C_3}$ plane

6 Conclusions

In this chapter, the dynamics of a eight-term novel 3-D finance chaotic system consisting of two nonlinearities (one quadratic and one quartic), were investigated. The model describes the time variation of three state variables: the interest rate (x_1), the investment demand (x_2) and the price index (x_3). From the economical point of view, the factors that influence changes in the variable x_1 mainly come from the investment market and the structural adjustment from the prices. The changing rate of x_2 is in proportion to the rate of investment, and in proportion to an inversion with the cost of investment and interest rates. Furthermore, changes in x_3 are controlled by a contradiction between supply and demand in commercial markets and are also influenced by inflation rates. So, as parameters in the proposed system the saving amount, cost per investment and the elasticity of demand of commercial markets, have been chosen. Interesting features of this novel finance system, such as a chaotic behavior and a rotation symmetry about the x_2 axis were investigated. Also, new results for the global chaos control of the proposed finance chaotic system with unknown parameters using adaptive control method was presented. This, approach is a very interesting research subject due to the fact that the chaos control problem aims to regulate the states of the novel finance chaotic system to desired constant values. In economy, this result has a significant interest, especially in cases in which finance systems need to be regulated. Finally, an electronic circuit realization of the novel finance chaotic system using Spice was presented in detail to confirm the feasibility of the theoretical model.

References

1. Abdurrahman A, Jiang H, Teng Z (2015) Finite-time synchronization for memristor-based neural networks with time-varying delays. Neural Netw 69:20–28
2. Arneodo A, Coullet P, Tresser C (1981) Possible new strange attractors with spiral structure. Commun Math Phys 79(4):573–576
3. Azar AT, Vaidyanathan S (2015) Chaos modeling and control systems design, vol 581. Springer, Berlin
4. Bouali S, Buscarino A, Fortuna L, Frasca M, Gambuzza LV (2012) Emulating complex business cycles by using an electronic analogue. Nonlinear Anal: Real World Appl 13(6):2459–2465
5. Cai G, Tan Z (2007) Chaos synchronization of a new chaotic system via nonlinear control. J Uncertain Syst 1(3):235–240
6. Chen G, Ueta T (1999) Yet another chaotic attractor. Int J Bifur Chaos 9(7):1465–1466
7. Chen WC (2008) Dynamics and control of a financial system with time-delayed feedbacks. Chaos, Solitons Fractals 37(4):1198–1207
8. Chen WC (2008) Nonlinear dynamics and chaos in a fractional-order financial system. Chaos, Solitons Fractals 36(5):1305–1314
9. Fanti L, Manfredi P (2007) Chaotic business cycles and fiscal policy: an IS-LM model with distributed tax collection lags. Chaos, Solitons Fractals 32(2):736–744
10. Gao Q, Ma J (2009) Chaos and Hopf bifurcation of a finance system. Nonlinear Dyn 58(1–2):209–216
11. Grassberger P, Procaccia I (1983) Characterization of strange attractors. Phys Rev Lett 50:346–349
12. Grassberger P, Procaccia I (1983) Measuring the strangeness of strange attractors. Phys D: Nonlinear Phenom 9:189–208
13. Hilborn RC (2001) Chaos and nonlinear dynamics: an introduction for scientists and engineers, 2nd edn. Oxford University Press, Oxford, UK
14. Khalil HK (2001) Nonlinear Systems, 3rd edn. Prentice Hall, New Jersey
15. Lakshmanan M, Rajasekhar S (2003) Nonlinear dynamics: integrability, chaos, and patterns. Springer, Berlin
16. Li D (2008) A three-scroll chaotic attractor. Phys Lett A 372(4):387–393
17. Lorenz EN (1963) Deterministic periodic flow. J Atmos Sci 20(2):130–141
18. Lü J, Chen G (2002) A new chaotic attractor coined. Int J Bifurc Chaos 12(3):659–661
19. Njah AN, Sunday OD (2009) Generalization on the chaos control of 4-D chaotic systems using recursive backstepping nonlinear controller. Chaos, Solitons Fractals 41(5):2371–2376
20. Pehlivan I, Moroz IM, Vaidyanathan S (2014) Analysis, synchronization and circuit design of a novel butterfly attractor. J Sound Vib 333(20):5077–5096
21. Pham VT, Volos CK, Vaidyanathan S, Le TP, Vu VY (2015) A memristor-based hyperchaotic system with hidden attractors: dynamics, synchronization and circuital emulating. J Eng Sci Technol Rev 8(2):205–214
22. Rössler OE (1976) An equation for continuous chaos. Phys Lett A 57(5):397–398
23. Sampath S, Vaidyanathan S, Volos CK, Pham VT (2015) An eight-term novel four-scroll chaotic system with cubic nonlinearity and its circuit simulation. J Eng Sci Technol Rev 8(2):1–6
24. Sprott JC (1994) Some simple chaotic flows. Phys Rev E 50(2):647–650
25. Strogatz SH (1994) Nonlinear dynamics and chaos: with applications to physics, biology, chemistry, and engineering. Perseus Books, Massachussetts
26. Sundarapandian V (2010) Output regulation of the Lorenz attractor. Far East J Math Sci 42(2):289–299
27. Sundarapandian V (2013) Adaptive control and synchronization design for the Lu-Xiao chaotic system. Lect Notes Electr Eng 131:319–327
28. Sundarapandian V (2013) Analysis and anti-synchronization of a novel chaotic system via active and adaptive controllers. J Eng Sci Technol Rev 6(4):45–52
29. Sundarapandian V, Pehlivan I (2012) Analysis, control, synchronization, and circuit design of a novel chaotic system. Mathe Comput Modell 55(7–8):1904–1915

30. Tigan G, Opris D (2008) Analysis of a 3D chaotic system. Chaos, Solitons Fractals 36:1315–1319
31. Vaidyanathan S (2011) Output regulation of Arneodo-Coullet chaotic system. Commun Comput Inf Sci 133:98–107
32. Vaidyanathan S (2011) Output regulation of the unified chaotic system. Commun Compute Inf Sci 198:1–9
33. Vaidyanathan S (2012) Adaptive controller and syncrhonizer design for the Qi-Chen chaotic system. Advances in computer science and information technology. Computer science and engineering, vol 84., Lecture notes of the institute for computer sciences, social-informatics and telecommunications engineeringSpringer, Berlin, pp 73–82
34. Vaidyanathan S (2012) Global chaos control of hyperchaotic Liu system via sliding control method. Int J Control Theory Appl 5(2):117–123
35. Vaidyanathan S (2012) Sliding mode control based global chaos control of Liu-Liu-Liu-Su chaotic system. Int J Control Theory Appl 5(1):15–20
36. Vaidyanathan S (2013) A new six-term 3-D chaotic system with an exponential nonlinearity. Far East J Math Sci 79(1):135–143
37. Vaidyanathan S (2013) A ten-term novel 4-D hyperchaotic system with three quadratic nonlinearities and its control. Int J Control Theory Appl 6(2):97–109
38. Vaidyanathan S (2013) Analysis and adaptive synchronization of two novel chaotic systems with hyperbolic sinusoidal and cosinusoidal nonlinearity and unknown parameters. J Eng Sci Technol Rev 6(4):53–65
39. Vaidyanathan S (2013) Analysis, control and synchronization of hyperchaotic Zhou system via adaptive control. Adv Intell Syst Comput 177:1–10
40. Vaidyanathan S (2014) A new eight-term 3-D polynomial chaotic system with three quadratic nonlinearities. Far East J Math Sci 84(2):219–226
41. Vaidyanathan S (2014) Analysis and adaptive synchronization of eight-term 3-D polynomial chaotic systems with three quadratic nonlinearities. Eur Phys J: Special Top 223(8):1519–1529
42. Vaidyanathan S (2014) Analysis, control and synchronisation of a six-term novel chaotic system with three quadratic nonlinearities. Int J Model, Identif Control 22(1):41–53
43. Vaidyanathan S (2014) Generalized projective synchronisation of novel 3-D chaotic systems with an exponential non-linearity via active and adaptive control. Int J Model, Identif Control 22(3):207–217
44. Vaidyanathan S (2014) Qualitative analysis and control of an eleven-term novel 4-D hyperchaotic system with two quadratic nonlinearities. Int J Control Theory Appl 7:35–47
45. Vaidyanathan S (2015) 3-cells cellular neural network (CNN) attractor and its adaptive biological control. Int J PharmTech Res 8(4):632–640
46. Vaidyanathan S (2015) A 3-D novel highly chaotic system with four quadratic nonlinearities, its adaptive control and anti-synchronization with unknown parameters. J Eng Sci Technol Rev 8(2):106–115
47. Vaidyanathan S (2015) A novel chemical chaotic reactor system and its adaptive control. Int J ChemTech Res 8(7):146–158
48. Vaidyanathan S (2015) Adaptive backstepping control of enzymes-substrates system with ferroelectric behaviour in brain waves. Int J PharmTech Res 8(2):256–261
49. Vaidyanathan S (2015) Adaptive biological control of generalized Lotka-Volterra three-species biological system. Int J PharmTech Res 8(4):622–631
50. Vaidyanathan S (2015) Adaptive chaotic synchronization of enzymes-substrates system with ferroelectric behaviour in brain waves. Int Jo PharmTech Res 8(5):964–973
51. Vaidyanathan S (2015) Adaptive control of a chemical chaotic reactor. Int J PharmTech Res 8(3):377–382
52. Vaidyanathan S (2015) Adaptive control of the FitzHugh-Nagumo chaotic neuron model. Int J PharmTech Res 8(6):117–127
53. Vaidyanathan S (2015) Adaptive synchronization of chemical chaotic reactors. Int J ChemTech Res 8(2):612–621

54. Vaidyanathan S (2015) Adaptive synchronization of generalized Lotka-Volterra three-species biological systems. Int J PharmTech Res 8(5):928–937
55. Vaidyanathan S (2015) Adaptive synchronization of novel 3-D chemical chaotic reactor systems. Int J ChemTech Res 8(7):159–171
56. Vaidyanathan S (2015) Adaptive synchronization of the identical FitzHugh-Nagumo chaotic neuron models. Int J PharmTech Res 8(6):167–177
57. Vaidyanathan S (2015) Analysis, control and synchronization of a 3-D novel jerk chaotic system with two quadratic nonlinearities. Kyungpook Math J 55:563–586
58. Vaidyanathan S (2015) Analysis, properties and control of an eight-term 3-D chaotic system with an exponential nonlinearity. Int J Model, Identif Control 23(2):164–172
59. Vaidyanathan S (2015) Anti-synchronization of brusselator chemical reaction systems via adaptive control. Int J ChemTech Res 8(6):759–768
60. Vaidyanathan S (2015) Chaos in neurons and adaptive control of Birkhoff-Shaw strange chaotic attractor. Int J PharmTech Res 8(5):956–963
61. Vaidyanathan S (2015) Chaos in neurons and synchronization of Birkhoff-Shaw strange chaotic attractors via adaptive control. Int J PharmTech Res 8(6):1–11
62. Vaidyanathan S (2015) Coleman-Gomatam logarithmic competitive biology models and their ecological monitoring. Int J PharmTech Res 8(6):94–105
63. Vaidyanathan S (2015) Dynamics and control of brusselator chemical reaction. Int J PharmTech Res 8(6):740–749
64. Vaidyanathan S (2015) Dynamics and control of tokamak system with symmetric and magnetically confined plasma. Int J PharmTech Res 8(6):795–803
65. Vaidyanathan S (2015) Global chaos synchronization of chemical chaotic reactors via novel sliding mode control method. Int J PharmTech Res 8(7):209–221
66. Vaidyanathan S (2015) Global chaos synchronization of the forced Van der Pol chaotic oscillators via adaptive control method. Int J PharmTech Res 8(6):156–166
67. Vaidyanathan S (2015) Global chaos synchronization of the Lotka-Volterra biological systems with four competitive species via active control. Int J PharmTech Res 8(6):206–217
68. Vaidyanathan S (2015) Lotka-Volterra population biology models with negative feedback and their ecological monitoring. Int J PharmTech Res 8(5):974–981
69. Vaidyanathan S (2015) Lotka-Volterra two species competitive biology models and their ecological monitoring. Int J PharmTech Res 8(6):32–44
70. Vaidyanathan S (2015) Output regulation of the forced Van der Pol chaotic oscillator via adaptive control method. Int J PharmTech Res 8(6):106–116
71. Vaidyanathan S, Azar AT (2015) Analysis and control of a 4-D novel hyperchaotic system. Stud Comput Intell 581:3–17
72. Vaidyanathan S, Azar AT (2015) Analysis, control and synchronization of a nine-term 3-D novel chaotic system. In: Azar AT, Vaidyanathan S (eds) Chaos modelling and control systems design, studies in computational intelligence, vol 581. Springer, Berlin, pp 19–38
73. Vaidyanathan S, Madhavan K (2013) Analysis, adaptive control and synchronization of a seven-term novel 3-D chaotic system. Int J Control Theory Appl 6(2):121–137
74. Vaidyanathan S, Pakiriswamy S (2015) A 3-D novel conservative chaotic System and its generalized projective synchronization via adaptive control. J Eng Sci Techn Rev 8(2):52–60
75. Vaidyanathan S, Volos C (2015) Analysis and adaptive control of a novel 3-D conservative no-equilibrium chaotic system. Arch Control Sci 25(3):333–353
76. Vaidyanathan S, Volos C, Pham VT (2014) Hyperchaos, adaptive control and synchronization of a novel 5-D hyperchaotic system with three positive Lyapunov exponents and its SPICE implementation. Arch Control Sci 24(4):409–446
77. Vaidyanathan S, Volos C, Pham VT, Madhavan K, Idowu BA (2014) Adaptive backstepping control, synchronization and circuit simulation of a 3-D novel jerk chaotic system with two hyperbolic sinusoidal nonlinearities. Arch Control Sci 24(3):375–403
78. Vaidyanathan S, Idowu BA, Azar AT (2015) Backstepping controller design for the global chaos synchronization of Sprott's jerk systems. Stud Comput Intell 581:39–58

79. Vaidyanathan S, Rajagopal K, Volos CK, Kyprianidis IM, Stouboulos IN (2015) Analysis, adaptive control and synchronization of a seven-term novel 3-D chaotic system with three quadratic nonlinearities and its digital implementation in LabVIEW. J Eng Sci Technol Rev 8(2):130–141
80. Vaidyanathan S, Volos C, Pham VT, Madhavan K (2015) Analysis, adaptive control and synchronization of a novel 4-D hyperchaotic hyperjerk system and its SPICE implementation. Arch Control Sci 25(1):135–158
81. Vaidyanathan S, Volos CK, Kyprianidis IM, Stouboulos IN, Pham VT (2015) Analysis, adaptive control and anti-synchronization of a six-term novel jerk chaotic system with two exponential nonlinearities and its circuit simulation. J Eng Sci Technol Rev 8(2):24–36
82. Vaidyanathan S, Volos CK, Pham VT (2015) Analysis, adaptive control and adaptive synchronization of a nine-term novel 3-D chaotic system with four quadratic nonlinearities and its circuit simulation. J Eng Sci Technol Rev 8(2):181–191
83. Vaidyanathan S, Volos CK, Pham VT (2015) Analysis, control, synchronization and SPICE implementation of a novel 4-D hyperchaotic Rikitake dynamo system without equilibrium. J Eng Sci Technol Rev 8(2):232–244
84. Vaidyanathan S, Volos CK, Pham VT (2015) Global chaos control of a novel nine-term chaotic system via sliding mode control. In: Azar AT, Zhu Q (eds) Advances and applications in sliding mode control systems, studies in computational intelligence, vol 576. Springer, Berlin, pp 571–590
85. Vincent UE, Njah AN, Laoye JA (2007) Controlling chaos and deterministic directed transport in inertia ratchets using backstepping control. Phys D 231(2):130–136
86. Volos CK, Stavrinides SG, Kyprianidis IM, Stouboulos IN, Magafas I, Anagnostopoulos AN (2011) Nonlinear financial dynamics from an engineer's point of view. J Eng Sci Technol Rev 4(3):281–285
87. Volos CK, Kyprianidis IM, Stouboulos IN (2012) Synchronization phenomena in coupled nonlinear systems applied in economic cycles. WSEAS Trans Syst 11(12):681–690
88. Volos CK, Kyprianidis IM, Stouboulos IN (2015) The effect of foreign direct investment in economic growth from the perspective of nonlinear dynamics. J Eng Sci Technol Rev 8(1):1–7
89. Volos CK, Kyprianidis IM, Stouboulos IN, Tlelo-Cuautle E, Vaidyanathan S (2015) Memristor: a new concept in synchronization of coupled neuromorphic circuits. J Eng Sci Technol Rev 8(2):157–173
90. Zhou W, Xu Y, Lu H, Pan L (2008) On dynamics analysis of a new chaotic attractor. Phys Lett A 372(36):5773–5777
91. Zhu C, Liu Y, Guo Y (2010) Theoretic and numerical study of a new chaotic system. Intell Inf Manage 2:104–109

A Novel Highly Hyperchaotic System and Its Adaptive Control

Sundarapandian Vaidyanathan

Abstract In this work, we describe a twelve-term novel highly hyperchaotic system with four quadratic nonlinearities and an exponential nonlinearity. The phase portraits of the twelve-term novel hyperchaotic system are depicted and the qualitative properties of the novel hyperchaotic system are discussed. The novel hyperchaotic system has two unstable equilibrium points. The Lyapunov exponents of the novel hyperchaotic system are obtained as $L_1 = 14.5577, L_2 = 0.1225, L_3 = 0$ and $L_4 = -36.3884$. The maximal Lyapunov exponent of the novel hyperchaotic system has a high value, viz. $L_1 = 14.5577$. Thus, the novel 4-D system shows highly hyperchaotic behavior. Also, the Kaplan–Yorke dimension of the novel hyperchaotic system is obtained as $D_{KY} = 3.4045$, which is a high value for a 4-D hyperchaotic system. Since the sum of the Lyapunov exponents is negative, the novel hyperchaotic system is dissipative. Next, an adaptive controller is designed to globally stabilize the highly hyperchaotic system with unknown parameters. Finally, an adaptive controller is also designed to achieve global chaos synchronization of the identical highly hyperchaotic systems with unknown parameters. MATLAB simulations are presented to depict the phase portraits of the novel highly hyperchaotic system and illustrate all the main adaptive control results derived in this work.

Keywords Chaos · Chaotic systems · Hyperchaos · Hyperchaotic systems · Adaptive control · Synchronization

1 Introduction

In the last few decades, Chaos theory has become a very important and active research field, employing many applications in different disciplines like physics, chemistry, biology, ecology, engineering and economics, among others [3].

S. Vaidyanathan (✉)
Research and Development Centre, Vel Tech University, Avadi, Chennai 600062, Tamil Nadu, India
e-mail: sundarvtu@gmail.com

S. Vaidyanathan and C. Volos (eds.), *Advances and Applications in Chaotic Systems*, Studies in Computational Intelligence 636,
DOI 10.1007/978-3-319-30279-9_22

Some classical paradigms of 3-D chaotic systems in the literature are Lorenz system [15], Rössler system [23], ACT system [2], Sprott systems [30], Chen system [6], Lü system [16], Cai system [4], Tigan system [40], etc.

Many new chaotic systems have been discovered in the recent years such as Zhou system [112], Zhu system [114], Li system [13], Wei-Yang system [108], Sundarapandian systems [33, 37], Vaidyanathan systems [47, 49, 51–54, 58, 69, 70, 84, 85, 87, 93, 95, 98, 101, 102, 104], Pehlivan system [17], Sampath system [25], etc.

Chaos theory has applications in several fields of science and engineering such as chemical reactors [59, 63, 65, 67, 71, 75–77], biological systems [57, 60–62, 64, 66, 68, 72–74, 78–82], memristors [1, 19, 105], etc.

A hyperchaotic system is defined as a chaotic system with at least two positive Lyapunov exponents [3]. Thus, the dynamics of a hyperchaotic system can expand in several different directions simultaneously. Thus, the hyperchaotic systems have more complex dynamical behaviour and they have miscellaneous applications in engineering such as secure communications [7, 12, 109], cryptosystems [8, 22, 113], fuzzy logic [29, 111], electrical circuits [107, 110], etc.

The minimum dimension of an autonomous, continuous-time, hyperchaotic system is four. The first 4-D hyperchaotic system was found by Rössler [24]. Many hyperchaotic systems have been reported in the chaos literature such as hyperchaotic Lorenz system [9], hyperchaotic Lü system [5], hyperchaotic Chen system [14], hyperchaotic Wang system [106], hyperchaotic Vaidyanathan systems [48, 56, 83, 94, 96, 100, 103], hyperchaotic Pham system [18], etc.

The control of a chaotic or hyperchaotic system aims to stabilize or regulate the system with the help of a feedback control. There are many control methods such as active control [31, 41, 42], adaptive control [32, 43, 50], sliding mode control [45, 46], backstepping control [97], etc.

The synchronization of chaotic systems aims to synchronize the states of master and slave systems asymptotically with time. There are many control methods for synchronization such as active control [10, 26, 27, 88, 90], adaptive control [28, 34–36, 44, 86, 89], sliding mode control [38, 55, 92, 99], backstepping control [20, 21, 39, 91], etc.

In this research work, we announce a twelve-term novel 4-D hyperchaotic system with four quadratic nonlinearities and an exponential nonlinearity. Section 2 describes the dynamic equations and phase portraits of the twelve-term novel 4-D hyperchaotic system. Section 3 details the qualitative properties of the novel highly hyperchaotic system. The Lyapunov exponents of the novel hyperchaotic system are obtained as $L_1 = 14.5577$, $L_2 = 0.1225$, $L_3 = 0$ and $L_4 = -36.2884$. Since the maximal Lyapunov exponent of the novel hyperchaotic system is very high, viz. $L_1 = 14.5577$, it follows that the novel hyperchaotic system is highly hyperchaotic.

2 A Novel 4-D Hyperchaotic System

In this section, we describe a twelve-term novel hyperchaotic system described by

$$\begin{cases} \dot{x}_1 = a(x_2 - x_1) + x_2x_3 + qx_4 \\ \dot{x}_2 = bx_1 - cx_1x_3 + qx_4 \\ \dot{x}_3 = \exp(x_1x_2) - px_3 + x_1^2 + x_2^2 \\ \dot{x}_4 = -rx_2 \end{cases} \tag{1}$$

where x_1, x_2, x_3, x_4 are the states and a, b, c, p, q, r are constant positive parameters.

The system (1) exhibits a *strange hyperchaotic attractor* for the parameter values

$$a = 13.8, \quad b = 41, \quad c = 0.4, \quad p = 10.6, \quad q = 2.8, \quad r = 3.5 \tag{2}$$

For numerical simulations, we take the initial conditions as

$$x_1(0) = 0.4, \quad x_2(0) = 0.2, \quad x_3(0) = 0.2, \quad x_4(0) = 0.4 \tag{3}$$

Figures 1, 2, 3 and 4 show the 3-D projection of the novel hyperchaotic system (1) on the (x_1, x_2, x_3), (x_1, x_2, x_4), (x_1, x_3, x_4) and (x_2, x_3, x_4) spaces, respectively.

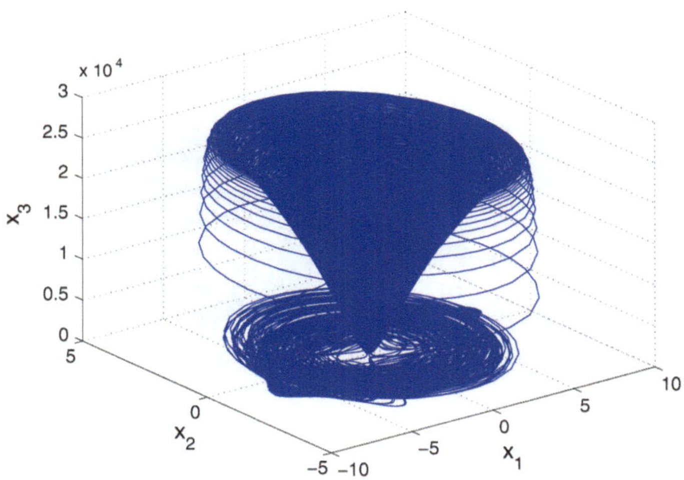

Fig. 1 3-D projection of the novel hyperchaotic system on the (x_1, x_2, x_3) space

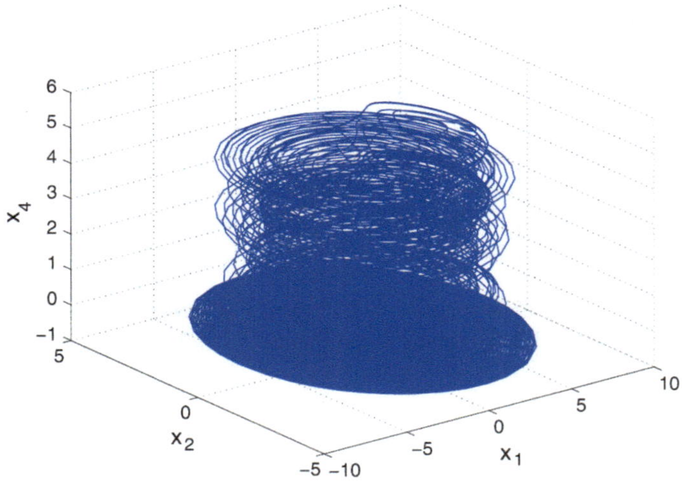

Fig. 2 3-D projection of the novel hyperchaotic system on the (x_1, x_2, x_4) space

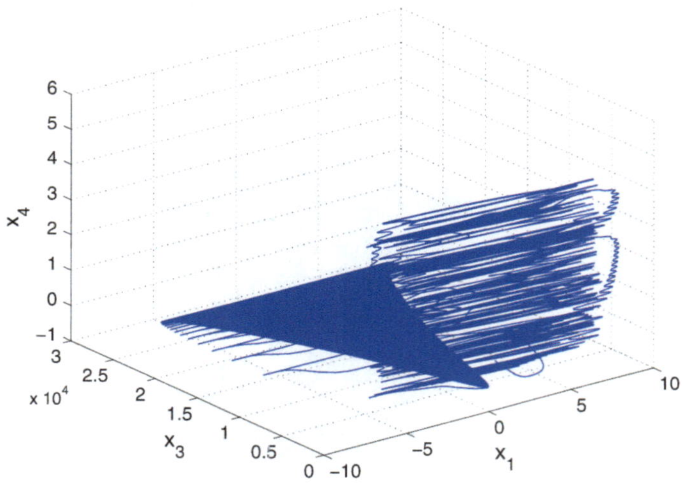

Fig. 3 3-D projection of the novel hyperchaotic system on the (x_1, x_3, x_4) space

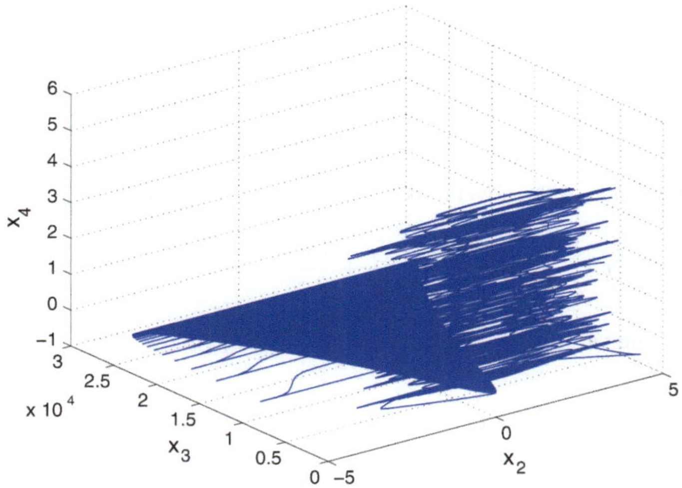

Fig. 4 3-D projection of the novel hyperchaotic system on the (x_2, x_3, x_4) space

3 Analysis of the Novel 4-D Hyperchaotic System

In this section, we give a dynamic analysis of the 4-D novel hyperchaotic system (1). We take the parameter values as in the hyperchaotic case (2).

3.1 Dissipativity

In vector notation, the novel hyperchaotic system (1) can be expressed as

$$\dot{\mathbf{x}} = f(\mathbf{x}) = \begin{bmatrix} f_1(x_1, x_2, x_3, x_4) \\ f_2(x_1, x_2, x_3, x_4) \\ f_3(x_1, x_2, x_3, x_4) \\ f_4(x_1, x_2, x_3, x_4) \end{bmatrix}, \tag{4}$$

where

$$\begin{cases} f_1(x_1, x_2, x_3, x_4) = a(x_2 - x_1) + x_2 x_3 + q x_4 \\ f_2(x_1, x_2, x_3, x_4) = b x_1 - c x_1 x_3 + q x_4 \\ f_3(x_1, x_2, x_3, x_4) = \exp(x_1 x_2) - p x_3 + x_1^2 + x_2^2 \\ f_4(x_1, x_2, x_3, x_4) = -r x_2 \end{cases} \tag{5}$$

Let Ω be any region in $\mathbf{R}^4$ with a smooth boundary and also, $\Omega(t) = \Phi_t(\Omega)$, where Φ_t is the flow of f. Furthermore, let $V(t)$ denote the hypervolume of $\Omega(t)$.

By Liouville's theorem, we know that

$$\dot{V}(t) = \int\limits_{\Omega(t)} (\nabla \cdot f)\, dx_1\, dx_2\, dx_3\, dx_4 \tag{6}$$

The divergence of the novel hyperchaotic system (4) is found as:

$$\nabla \cdot f = \frac{\partial f_1}{\partial x_1} + \frac{\partial f_2}{\partial x_2} + \frac{\partial f_3}{\partial x_3} + \frac{\partial f_4}{\partial x_4} = -(a + p) = -\mu < 0 \tag{7}$$

Inserting the value of $\nabla \cdot f$ from (7) into (6), we get

$$\dot{V}(t) = \int\limits_{\Omega(t)} (-\mu)\, dx_1\, dx_2\, dx_3\, dx_4 = -\mu V(t) \tag{8}$$

Integrating the first order linear differential equation (8), we get

$$V(t) = \exp(-\mu t) V(0) \tag{9}$$

Since $\mu > 0$, it follows from Eq. (9) that $V(t) \to 0$ exponentially as $t \to \infty$. This shows that the novel hyperchaotic system (1) is dissipative. Hence, the system limit sets are ultimately confined into a specific limit set of zero hypervolume, and the asymptotic motion of the novel hyperchaotic system (1) settles onto a strange attractor of the system.

3.2 Equilibrium Points

We take the parameter values as in the hyperchaotic case (2).

It is easy to see that the system (1) has two equilibrium points given by

$$E_1 = \begin{bmatrix} 0 \\ 0 \\ 0.0943 \\ 0 \end{bmatrix} \text{ and } E_2 = \begin{bmatrix} -38.0946 \\ 0 \\ -137 \\ -187.7521 \end{bmatrix} \tag{10}$$

The Jacobian matrix of the system (1) at any point $x \in \mathbf{R}^4$ is given by

$$J(x) = \begin{bmatrix} -a & a + x_3 & x_2 & q \\ b - cx_1 & 0 & -cx_1 & q \\ x_2 \exp(x_1 x_2) + 2x_1 & x_1 \exp(x_1 x_2) + 2x_2 & -p & 0 \\ 0 & -r & 0 & 0 \end{bmatrix} \tag{11}$$

Thus, the Jacobian matrix at E_1 is obtained as

$$J_1 = J(E_1) = \begin{bmatrix} -13.8 & 13.8943 & 0 & 2.8 \\ 41 & 0 & 0 & 2.8 \\ 0 & 0 & -10.6 & 0 \\ 0 & -3.5 & 0 & 0 \end{bmatrix} \tag{12}$$

which has the eigenvalues

$$\lambda_1 = -31.8864, \quad \lambda_2 = -10.6, \quad \lambda_3 = 0.9848, \quad \lambda_4 = 17.1015 \tag{13}$$

This shows that E_1 is a saddle-point, which is unstable.
Next, the Jacobian matrix at E_1 is obtained as

$$J_2 = J(E_2) = \begin{bmatrix} -13.8 & -123.2 & 0 & 2.8 \\ 56.2378 & 0 & 15.2378 & 2.8 \\ -76.1892 & -38.0946 & -10.6 & 0 \\ 0 & -3.5 & 0 & 0 \end{bmatrix} \tag{14}$$

which has the eigenvalues

$$\lambda_1 = -0.0669, \quad \lambda_2 = 7.7477, \quad \lambda_{3,4} = -16.0404 \pm 87.4911\, i \tag{15}$$

This shows that E_2 is a saddle-focus, which is unstable.

3.3 Rotation Symmetry About the x_3-Axis

It is easy to see that the novel 4-D hyperchaotic system (1) is invariant under the change of coordinates

$$(x_1, x_2, x_3, x_4) \mapsto (-x_1, -x_2, x_3, -x_4) \tag{16}$$

Since the transformation (16) persists for all values of the system parameters, it follows that the novel 4-D hyperchaotic system (1) has rotation symmetry about the x_3-axis and that any non-trivial trajectory must have a twin trajectory.

3.4 Invariance

It is easy to see that the x_3-axis is invariant under the flow of the 4-D novel hyperchaotic system (1).

The invariant motion along the x_3-axis is characterized by the scalar dynamics

$$\dot{x}_3 = 1 - px_3, \quad (c > 0) \tag{17}$$

which is stable, but not asymptotically stable.

3.5 Lyapunov Exponents and Kaplan–Yorke Dimension

We take the parameter values of the novel system (1) as in the hyperchaotic case (2), i.e.

$$a = 13.8, \ \ b = 41, \ \ c = 0.4, \ \ p = 10.6, \ \ q = 2.8, \ \ r = 3.5 \tag{18}$$

We take the initial state of the novel system (1) as given in (3).

Then the Lyapunov exponents of the system (1) are numerically obtained using MATLAB as

$$L_1 = 14.5577, \ \ L_2 = 0.1225, \ \ L_3 = 0, \ \ L_4 = -36.3884 \tag{19}$$

Since there are two positive Lyapunov exponents in (19), the novel system (1) exhibits *hyperchaotic* behavior.

Since the maximal Lyapunov exponent of the system (1) has a high value, viz. $L_1 = 14.5577$, the system is highly hyperchaotic.

Since $L_1 + L_2 + L_3 + L_4 = -21.6082 < 0$, it follows that the novel hyperchaotic system (1) is dissipative.

Also, the Kaplan–Yorke dimension of the novel hyperchaotic system (1) is calculated as

$$D_{KY} = 3 + \frac{L_1 + L_2 + L_3}{|L_4|} = 3.4045 \tag{20}$$

The high value of D_{KY} shows the complex behaviour of the novel hyperchaotic system (1).

4 Adaptive Control of the Novel Highly Hyperchaotic System

In this section, we apply adaptive control method to derive an adaptive feedback control law for globally stabilizing the novel 4-D highly hyperchaotic system with unknown parameters.

Thus, we consider the controlled novel 4-D hyperchaotic system given by

$$\begin{cases} \dot{x}_1 = a(x_2 - x_1) + x_2x_3 + qx_4 + u_1 \\ \dot{x}_2 = bx_1 - cx_1x_3 + qx_4 + u_2 \\ \dot{x}_3 = \exp(x_1x_2) - px_3 + x_1^2 + x_2^2 + u_3 \\ \dot{x}_4 = -rx_2 + u_4 \end{cases} \tag{21}$$

In (21), x_1, x_2, x_3, x_4 are the states and u_1, u_2, u_3, u_4 are the adaptive controls to be determined using estimates of the unknown system parameters.

We consider the adaptive feedback control law

$$\begin{cases} u_1 = -\hat{a}(t)(x_2 - x_1) - x_2x_3 - \hat{q}(t)x_4 - k_1x_1 \\ u_2 = -\hat{b}(t)x_1 + \hat{c}(t)x_1x_3 - \hat{q}(t)x_4 - k_2x_2 \\ u_3 = -\exp(x_1x_2) + \hat{p}(t)x_3 - x_1^2 - x_2^2 - k_3x_3 \\ u_4 = \hat{r}(t)x_2 - k_4x_4 \end{cases} \tag{22}$$

where k_1, k_2, k_3, k_4 are positive gain constants.

Substituting (22) into (21), we get the closed-loop plant dynamics as

$$\begin{cases} \dot{x}_1 = [a - \hat{a}(t)](x_2 - x_1) + [q - \hat{q}(t)]x_4 - k_1x_1 \\ \dot{x}_2 = [b - \hat{b}(t)]x_1 - [c - \hat{c}(t)]x_1x_3 + [q - \hat{q}(t)]x_4 - k_2x_2 \\ \dot{x}_3 = -[p - \hat{p}(t)]x_3 - k_3x_3 \\ \dot{x}_4 = -[r - \hat{r}(t)]x_2 - k_4x_4 \end{cases} \tag{23}$$

The parameter estimation errors are defined as

$$\begin{cases} e_a(t) = a - \hat{a}(t) \\ e_b(t) = b - \hat{b}(t) \\ e_c(t) = c - \hat{c}(t) \\ e_p(t) = p - \hat{p}(t) \\ e_q(t) = q - \hat{q}(t) \\ e_r(t) = r - \hat{r}(t) \end{cases} \tag{24}$$

In view of (24), we can simplify the plant dynamics (23) as

$$\begin{cases} \dot{x}_1 = e_a(x_2 - x_1) + e_qx_4 - k_1x_1 \\ \dot{x}_2 = e_bx_1 - e_cx_1x_3 + e_qx_4 - k_2x_2 \\ \dot{x}_3 = -e_px_3 - k_3x_3 \\ \dot{x}_4 = -e_rx_2 - k_4x_4 \end{cases} \tag{25}$$

Differentiating (24) with respect to t, we obtain

$$\begin{cases} \dot{e}_a(t) = -\dot{\hat{a}}(t) \\ \dot{e}_b(t) = -\dot{\hat{b}}(t) \\ \dot{e}_c(t) = -\dot{\hat{c}}(t) \\ \dot{e}_p(t) = -\dot{\hat{p}}(t) \\ \dot{e}_q(t) = -\dot{\hat{q}}(t) \\ \dot{e}_r(t) = -\dot{\hat{r}}(t) \end{cases} \tag{26}$$

We consider the quadratic candidate Lyapunov function defined by

$$V(\mathbf{x}, e_a, e_b, e_c, e_p, e_q, e_r) = \frac{1}{2}\sum_{i=1}^{4} x_i^2 + \frac{1}{2}\left(e_a^2 + e_b^2 + e_c^2 + e_p^2 + e_q^2 + e_r^2\right) \tag{27}$$

Differentiating V along the trajectories of (25) and (26), we obtain

$$\begin{cases} \dot{V} = -k_1x_1^2 - k_2x_2^2 - k_3x_3^2 - k_4x_4^2 + e_a\left[x_1(x_2 - x_1) - \dot{\hat{a}}\right] \\ \quad + e_b\left[x_1x_2 - \dot{\hat{b}}\right] + e_c\left[-x_1x_2x_3 - \dot{\hat{c}}\right] + e_p\left[-x_3^2 - \dot{\hat{p}}\right] \\ \quad + e_q\left[x_1x_4 + x_2x_4 - \dot{\hat{q}}\right] + e_r\left[-x_2x_4 - \dot{\hat{r}}\right] \end{cases} \tag{28}$$

In view of (28), we take the parameter update law as

$$\begin{cases} \dot{\hat{a}}(t) = x_1(x_2 - x_1) \\ \dot{\hat{b}}(t) = x_1x_2 \\ \dot{\hat{c}}(t) = -x_1x_2x_3 \\ \dot{\hat{p}}(t) = -x_3^2 \\ \dot{\hat{q}}(t) = (x_1 + x_2)x_4 \\ \dot{\hat{r}}(t) = -x_2x_4 \end{cases} \tag{29}$$

Next, we state and prove the main result of this section.

Theorem 1 *The novel 4-D hyperchaotic system (21) with unknown system parameters is globally and exponentially stabilized for all initial conditions by the adaptive control law (22) and the parameter update law (29), where k_1, k_2, k_3, k_4 are positive gain constants.*

Proof We prove this result by applying Lyapunov stability theory [11].

We consider the quadratic Lyapunov function defined by (27), which is clearly a positive definite function on $\mathbf{R}^{10}$.

By substituting the parameter update law (29) into (28), we obtain the time-derivative of V as

$$\dot{V} = -k_1x_1^2 - k_2x_2^2 - k_3x_3^2 - k_4x_4^2 \tag{30}$$

From (30), it is clear that $\dot{V}$ is a negative semi-definite function on $\mathbf{R}^{10}$.

Thus, we can conclude that the state vector $\mathbf{x}(t)$ and the parameter estimation error are globally bounded, i.e.

$$\left[x_1(t)\ x_2(t)\ x_3(t)\ x_4(t)\ e_a(t)\ e_b(t)\ e_c(t)\ e_p(t)\ e_q(t)\ e_r(t)\right]^T \in \mathbf{L}_\infty.$$

We define $k = \min\{k_1, k_2, k_3, k_4\}$.

Then it follows from (30) that

$$\dot{V} \le -k\|\mathbf{x}(t)\|^2 \tag{31}$$

Thus, we have

$$k\|\mathbf{x}(t)\|^2 \le -\dot{V} \tag{32}$$

Integrating the inequality (32) from 0 to t, we get

$$k\int_0^t \|\mathbf{x}(\tau)\|^2\, d\tau \;\le\; V(0) - V(t) \tag{33}$$

From (33), it follows that $\mathbf{x} \in \mathbf{L}_2$.

Using (25), we can conclude that $\dot{\mathbf{x}} \in \mathbf{L}_\infty$.

Using Barbalat's lemma [11], we conclude that $\mathbf{x}(t) \to 0$ exponentially as $t \to \infty$ for all initial conditions $\mathbf{x}(0) \in \mathbf{R}^4$.

Thus, the novel 4-D hyperchaotic system (21) with unknown system parameters is globally and exponentially stabilized for all initial conditions by the adaptive control law (22) and the parameter update law (29).

This completes the proof. ■

For the numerical simulations, the classical fourth-order Runge–Kutta method with step size $h = 10^{-8}$ is used to solve the systems (21) and (29), when the adaptive control law (22) is applied.

The parameter values of the novel highly hyperchaotic system (21) are taken as in the hyperchaotic case (2), i.e.

$$a = 13.8,\quad b = 41,\quad c = 0.4,\quad p = 10.6,\quad q = 2.8,\quad r = 3.5 \tag{34}$$

We take the positive gain constants as $k_i = 6$ for $i = 1, 2, 3, 4$.

Furthermore, as initial conditions of the novel 4-D hyperchaotic system (21), we take

$$x_1(0) = 3.2,\quad x_2(0) = 6.8,\quad x_3(0) = 4.7,\quad x_4(0) = 10.3 \tag{35}$$

Also, as initial conditions of the parameter estimates, we take

$$\hat{a}(0) = 3,\ \hat{b}(0) = 11,\ \hat{c}(0) = 14,\ \hat{p}(0) = 6,\ \hat{q}(0) = 1,\ \hat{r}(0) = 7 \tag{36}$$

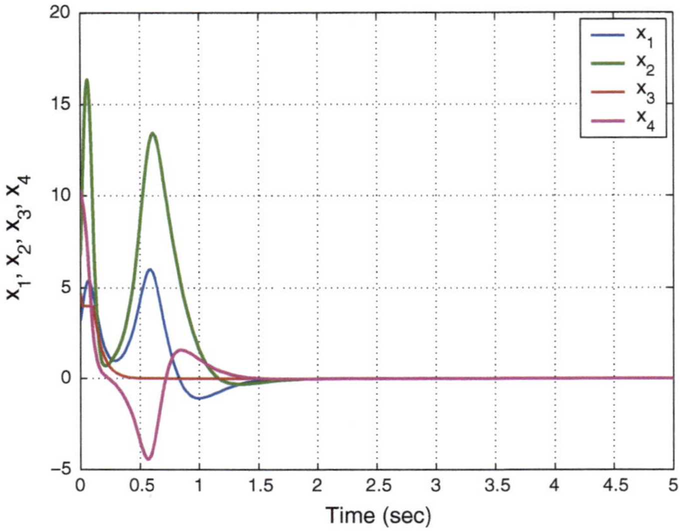

Fig. 5 Time-history of the controlled states x_1, x_2, x_3, x_4

In Fig. 5, the exponential convergence of the controlled states of the novel 4-D hyperchaotic system (21) is shown.

5 Adaptive Synchronization of the Identical Novel Hyperchaotic Systems

In this section, we apply adaptive control method to derive an adaptive feedback control law for globally synchronizing identical novel 4-D highly hyperchaotic systems with unknown parameters.

As the master system, we consider the novel 4-D hyperchaotic system given by

$$\begin{cases} \dot{x}_1 = a(x_2 - x_1) + x_2 x_3 + q x_4 \\ \dot{x}_2 = b x_1 - c x_1 x_3 + q x_4 \\ \dot{x}_3 = \exp(x_1 x_2) - p x_3 + x_1^2 + x_2^2 \\ \dot{x}_4 = -r x_2 \end{cases} \tag{37}$$

In (37), x_1, x_2, x_3, x_4 are the states and a, b, c, p are unknown system parameters.

As the slave system, we consider the 4-D novel hyperchaotic system given by

$$\begin{cases} \dot{y}_1 = a(y_2 - y_1) + y_2 y_3 + q y_4 + u_1 \\ \dot{y}_2 = b y_1 - c y_1 y_3 + q y_4 + u_2 \\ \dot{y}_3 = \exp(y_1 y_2) - p y_3 + y_1^2 + y_2^2 + u_3 \\ \dot{y}_4 = -r y_2 + u_4 \end{cases} \tag{38}$$

The synchronization error between the novel hyperchaotic systems (37) and (38) is defined by

$$\begin{cases} e_1 = y_1 - x_1 \\ e_2 = y_2 - x_2 \\ e_3 = y_3 - x_3 \\ e_4 = y_4 - x_4 \end{cases} \tag{39}$$

Then the synchronization error dynamics is obtained as

$$\begin{cases} \dot{e}_1 = a(e_2 - e_1) + qe_4 + y_2 y_3 - x_2 x_3 + u_1 \\ \dot{e}_2 = be_1 + qe_4 - c(y_1 y_3 - x_1 x_3) + u_2 \\ \dot{e}_3 = -pe_3 + \exp(y_1 y_2) - \exp(x_1 x_2) + y_1^2 + y_2^2 - x_1^2 - x_2^2 + u_3 \\ \dot{e}_4 = -re_2 + u_4 \end{cases} \tag{40}$$

We consider the adaptive feedback control law

$$\begin{cases} u_1 = -\hat{a}(t)(e_2 - e_1) - \hat{q}(t)e_4 - y_2 y_3 + x_2 x_3 - k_1 e_1 \\ u_2 = -\hat{b}(t)e_1 - \hat{q}(t)e_4 + \hat{c}(t)(y_1 y_3 - x_1 x_3) - k_2 e_2 \\ u_3 = \hat{p}(t)e_3 - \exp(y_1 y_2) + \exp(x_1 x_2) - y_1^2 - y_2^2 + x_1^2 + x_2^2 - k_3 e_3 \\ u_4 = \hat{r}(t)e_2 - k_4 e_4 \end{cases} \tag{41}$$

where k_1, k_2, k_3, k_4 are positive gain constants.

Substituting (41) into (40), we get the closed-loop error dynamics as

$$\begin{cases} \dot{e}_1 = \left[a - \hat{a}(t)\right](e_2 - e_1) + \left[q - \hat{q}(t)\right]e_4 - k_1 e_1 \\ \dot{e}_2 = \left[b - \hat{b}(t)\right]e_1 + \left[q - \hat{q}(t)\right]e_4 - \left[c - \hat{c}(t)\right](y_1 y_3 - x_1 x_3) - k_2 e_2 \\ \dot{e}_3 = -\left[p - \hat{p}(t)\right]e_3 - k_3 e_3 \\ \dot{e}_4 = -\left[r - \hat{r}(t)\right]e_2 - k_4 e_4 \end{cases} \tag{42}$$

The parameter estimation errors are defined as

$$\begin{cases} e_a(t) = a - \hat{a}(t) \\ e_b(t) = b - \hat{b}(t) \\ e_c(t) = c - \hat{c}(t) \\ e_p(t) = p - \hat{p}(t) \\ e_q(t) = q - \hat{q}(t) \\ e_r(t) = r - \hat{r}(t) \end{cases} \tag{43}$$

In view of (43), we can simplify the error dynamics (42) as

$$\begin{cases} \dot{e}_1 = e_a(e_2 - e_1) + e_q e_4 - k_1 e_1 \\ \dot{e}_2 = e_b e_1 + e_q e_4 - e_c(y_1 y_3 - x_1 x_3) - k_2 e_2 \\ \dot{e}_3 = -e_p e_3 - k_3 e_3 \\ \dot{e}_4 = -e_r e_2 - k_4 e_4 \end{cases} \tag{44}$$

Differentiating (43) with respect to t, we obtain

$$\begin{cases} \dot{e}_a(t) = -\dot{\hat{a}}(t) \\ \dot{e}_b(t) = -\dot{\hat{b}}(t) \\ \dot{e}_c(t) = -\dot{\hat{c}}(t) \\ \dot{e}_p(t) = -\dot{\hat{p}}(t) \\ \dot{e}_q(t) = -\dot{\hat{q}}(t) \\ \dot{e}_r(t) = -\dot{\hat{r}}(t) \end{cases} \tag{45}$$

We use adaptive control theory to find an update law for the parameter estimates. We consider the quadratic candidate Lyapunov function defined by

$$V(\mathbf{e}, e_a, e_b, e_c, e_p, e_q, e_r) = \frac{1}{2}\sum_{i=1}^{4} e_i^2 + \frac{1}{2}\left(e_a^2 + e_b^2 + e_c^2 + e_p^2 + e_q^2 + e_r^2\right) \tag{46}$$

Differentiating V along the trajectories of (44) and (45), we obtain

$$\begin{aligned} \dot{V} = &-k_1 e_1^2 - k_2 e_2^2 - k_3 e_3^2 - k_4 e_4^2 + e_a\left[e_1(e_2 - e_1) - \dot{\hat{a}}\right] \\ &+ e_b\left[e_1 e_2 - \dot{\hat{b}}\right] + e_c\left[-e_2(y_1 y_3 - x_1 x_3) - \dot{\hat{c}}\right] + e_p\left[-e_3^2 - \dot{\hat{p}}\right] \\ &+ e_q\left[e_1 e_4 + e_2 e_4 - \dot{\hat{q}}\right] + e_r\left[-e_2 e_4 - \dot{\hat{r}}\right] \end{aligned} \tag{47}$$

In view of (47), we take the parameter update law as

$$\begin{cases} \dot{\hat{a}}(t) = e_1(e_2 - e_1) \\ \dot{\hat{b}}(t) = e_1 e_2 \\ \dot{\hat{c}}(t) = -e_2(y_1 y_3 - x_1 x_3) \\ \dot{\hat{p}}(t) = -e_3^2 \\ \dot{\hat{q}}(t) = (e_1 + e_2)e_4 \\ \dot{\hat{r}}(t) = -e_2 e_4 \end{cases} \tag{48}$$

Theorem 2 *The novel highly hyperchaotic systems (37) and (38) with unknown system parameters are globally and exponentially synchronized for all initial conditions by the adaptive control law (41) and the parameter update law (48), where* k_1, k_2, k_3, k_4 *are positive gain constants.*

Proof We prove this result by applying Lyapunov stability theory [11].

We consider the quadratic Lyapunov function defined by (46), which is clearly a positive definite function on $\mathbf{R}^{10}$.

By substituting the parameter update law (48) into (47), we obtain

$$\dot{V} = -k_1 e_1^2 - k_2 e_2^2 - k_3 e_3^2 - k_4 e_4^2 \tag{49}$$

From (49), it is clear that $\dot{V}$ is a negative semi-definite function on $\mathbf{R}^{10}$.

Thus, we can conclude that the error vector $\mathbf{e}(t)$ and the parameter estimation error are globally bounded. We define $k = \min\{k_1, k_2, k_3, k_4\}$.

Then it follows from (49) that

$$\dot{V} \le -k\|\mathbf{e}(t)\|^2 \tag{50}$$

Thus, we have

$$k\|\mathbf{e}(t)\|^2 \le -\dot{V} \tag{51}$$

Integrating the inequality (51) from 0 to t, we get

$$k \int_0^t \|\mathbf{e}(\tau)\|^2 \, d\tau \le V(0) - V(t) \tag{52}$$

From (52), it follows that $\mathbf{e} \in \mathbf{L}_2$. Using (44), we can conclude that $\dot{\mathbf{e}} \in \mathbf{L}_\infty$.

Using Barbalat's lemma [11], we conclude that $\mathbf{e}(t) \to 0$ exponentially as $t \to \infty$ for all initial conditions $\mathbf{e}(0) \in \mathbf{R}^4$. ■

For the numerical simulations, the classical fourth-order Runge–Kutta method with step size $h = 10^{-8}$ is used to solve the systems (37), (38) and (48), when the adaptive control law (41) is applied.

The parameter values of the novel hyperchaotic systems are taken as in the hyperchaotic case (2).

We take the positive gain constants as $k_i = 6$ for $i = 1, \ldots, 4$.

Furthermore, as initial conditions of the master system (37), we take

$$x_1(0) = -3.9, \quad x_2(0) = 2.8, \quad x_3(0) = 1.7, \quad x_4(0) = -8.3 \tag{53}$$

As initial conditions of the slave system (38), we take

$$y_1(0) = 2.4, \quad y_2(0) = 3.5, \quad y_3(0) = -4.2, \quad y_4(0) = 9.2 \tag{54}$$

Also, as initial conditions of the parameter estimates, we take

$$\hat{a}(0) = 2, \quad \hat{b}(0) = 6, \quad \hat{c}(0) = 4, \quad \hat{p}(0) = 7, \quad \hat{q}(0) = 12, \quad \hat{r}(0) = 5 \tag{55}$$

Figures 6, 7, 8 and 9 describe the complete synchronization of the novel hyperchaotic systems (37) and (38), while Fig. 10 describes the time-history of the synchronization errors e_1, e_2, e_3, e_4.

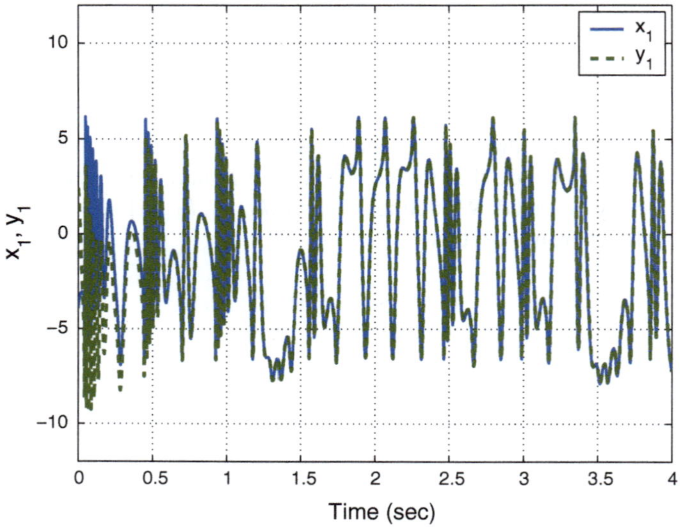

Fig. 6 Synchronization of the states x_1 and y_1

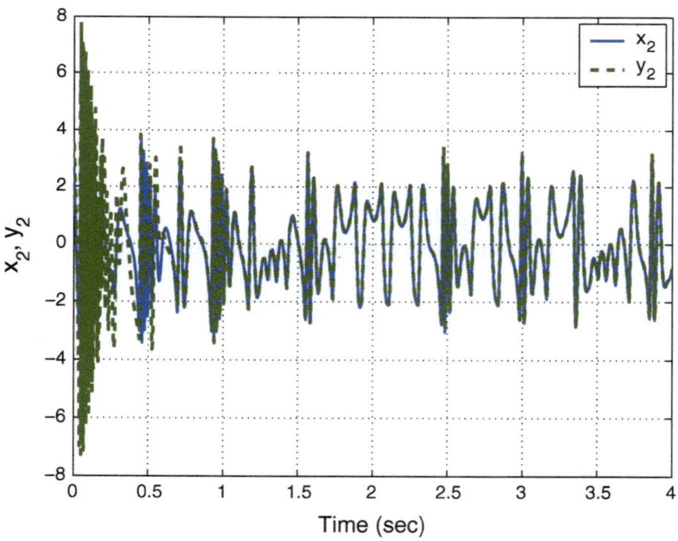

Fig. 7 Synchronization of the states x_2 and y_2

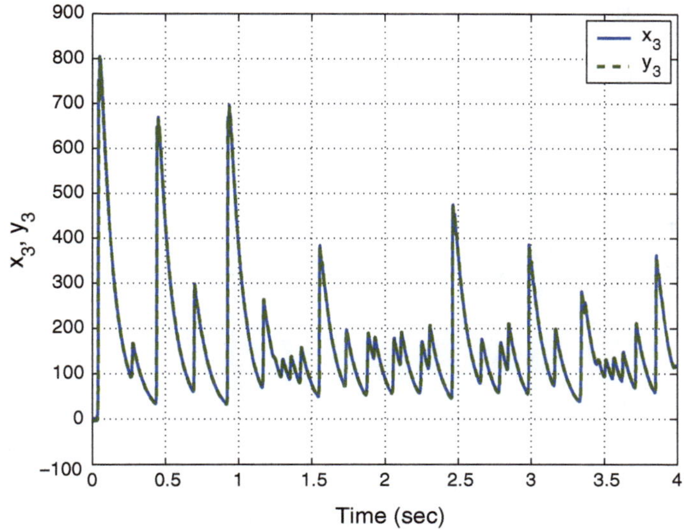

Fig. 8 Synchronization of the states x_3 and y_3

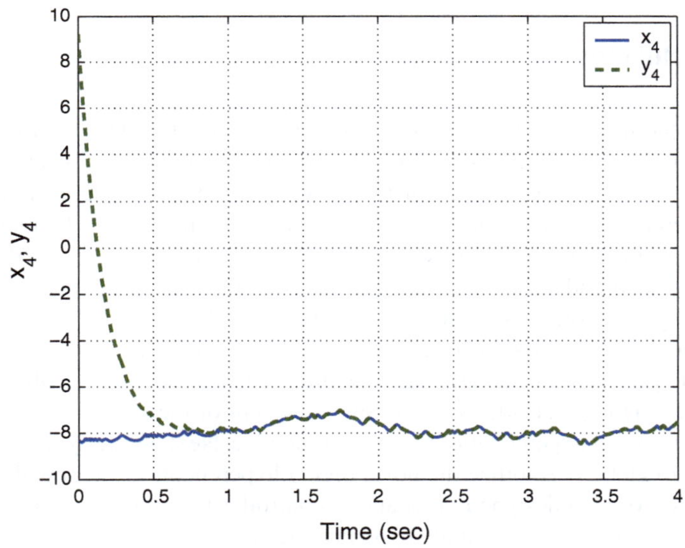

Fig. 9 Synchronization of the states x_4 and y_4

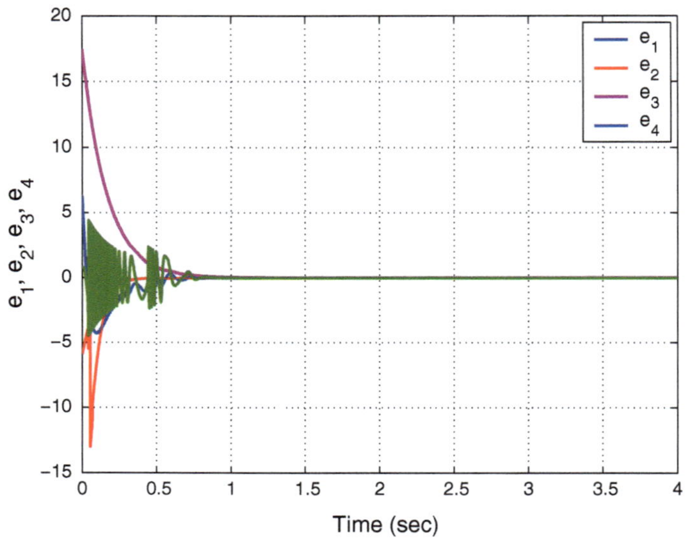

Fig. 10 Time-history of the synchronization errors e_1, e_2, e_3, e_4

6 Conclusions

In this research work, we announced a twelve-term novel 4-D highly hyperchaotic system with four quadratic nonlinearities and an exponential nonlinearity. We discussed the qualitative properties of the novel hyperchaotic system in detail. The novel hyperchaotic system has two unstable equilibrium points. The Lyapunov exponents of the novel hyperchaotic system have been found as $L_1 = 14.5577$, $L_2 = 0.1225$, $L_3 = 0$ and $L_4 = -36.3884$. Since the maximal Lyapunov exponent of the novel hyperchaotic system has a high value, viz. $L_1 = 14.5577$, the novel 4-D system shows highly hyperchaotic behavior. Also, the Kaplan–Yorke dimension of the novel hyperchaotic system has been calculated as $D_{KY} = 3.4045$, which is a high value for a 4-D hyperchaotic system. Since the sum of the Lyapunov exponents is negative, the novel hyperchaotic system is dissipative. Next, we designed an adaptive controller to globally stabilize the novel highly hyperchaotic system with unknown parameters. We also designed an adaptive controller to achieve global chaos synchronization of the identical highly hyperchaotic systems with unknown parameters. MATLAB simulations were shown to depict the phase portraits of the novel highly hyperchaotic system and illustrate all the main control results derived in this work.

References

1. Abdurrahman A, Jiang H, Teng Z (2015) Finite-time synchronization for memristor-based neural networks with time-varying delays. Neural Netw 69:20–28
2. Arneodo A, Coullet P, Tresser C (1981) Possible new strange attractors with spiral structure. Commun Math Phys 79(4):573–576
3. Azar AT, Vaidyanathan S (2015) Chaos modeling and control systems design, vol 581. Springer, Germany
4. Cai G, Tan Z (2007) Chaos synchronization of a new chaotic system via nonlinear control. J Uncertain Syst 1(3):235–240
5. Chen A, Lu J, Lü J, Yu S (2006) Generating hyperchaotic Lü attractor via state feedback control. Phys A 364:103–110
6. Chen G, Ueta T (1999) Yet another chaotic attractor. Int J Bifurc Chaos 9(7):1465–1466
7. Filali RL, Benrejeb M, Borne P (2014) On observer-based secure communication design using discrete-time hyperchaotic systems. Commun Nonlinear Sci Numer Simul 19(5):1424–1432
8. Hammami S (2015) State feedback-based secure image cryptosystem using hyperchaotic synchronization. ISA Trans 54:52–59
9. Jia Q (2007) Hyperchaos generated from the Lorenz chaotic system and its control. Phys Lett A 366:217–222
10. Karthikeyan R, Sundarapandian V (2014) Hybrid chaos synchronization of four-scroll systems via active control. J Electr Eng 65(2):97–103
11. Khalil HK (2001) Nonlinear systems, 3rd edn. Prentice Hall, New Jersey
12. Li C, Liao X, Wong KW (2005) Lag synchronization of hyperchaos with application to secure communications. Chaos Solitons Fractals 23(1):183–193
13. Li D (2008) A three-scroll chaotic attractor. Phys Lett A 372(4):387–393
14. Li X (2009) Modified projective synchronization of a new hyperchaotic system via nonlinear control. Commun Theor Phys 52:274–278
15. Lorenz EN (1963) Deterministic periodic flow. J Atmos Sci 20(2):130–141
16. Lü J, Chen G (2002) A new chaotic attractor coined. Int J Bifurc Chaos 12(3):659–661
17. Pehlivan I, Moroz IM, Vaidyanathan S (2014) Analysis, synchronization and circuit design of a novel butterfly attractor. J Sound Vib 333(20):5077–5096
18. Pham VT, Volos C, Jafari S, Wang X, Vaidyanathan S (2014) Hidden hyperchaotic attractor in a novel simple memristive neural network. Optoelectron Adv Mater Rapid Commun 8(11–12):1157–1163
19. Pham VT, Volos CK, Vaidyanathan S, Le TP, Vu VY (2015) A memristor-based hyperchaotic system with hidden attractors: dynamics, synchronization and circuital emulating. J Eng Sci Technol Rev 8(2):205–214
20. Rasappan S, Vaidyanathan S (2013) Hybrid synchronization of n-scroll Chua circuits using adaptive backstepping control design with recursive feedback. Malays J Math Sci 73(1):73–95
21. Rasappan S, Vaidyanathan S (2014) Global chaos synchronization of WINDMI and Coullet chaotic systems using adaptive backstepping control design. Kyungpook Math J 54(1): 293–320
22. Rhouma R, Belghith S (2008) Cryptanalysis of a new image encryption algorithm based on hyper-chaos. Phys Lett A 372(38):5973–5978
23. Rössler OE (1976) An equation for continuous chaos. Phys Lett A 57(5):397–398
24. Rössler OE (1979) An equation for hyperchaos. Phys Lett A 71:155–157
25. Sampath S, Vaidyanathan S, Volos CK, Pham VT (2015) An eight-term novel four-scroll chaotic system with cubic nonlinearity and its circuit simulation. J Eng Sci Technol Rev 8(2):1–6
26. Sarasu P, Sundarapandian V (2011) Active controller design for generalized projective synchronization of four-scroll chaotic systems. Int J Syst Signal Control Eng Appl 4(2):26–33
27. Sarasu P, Sundarapandian V (2011) The generalized projective synchronization of hyperchaotic Lorenz and hyperchaotic Qi systems via active control. Int J Soft Comput 6(5):216–223

28. Sarasu P, Sundarapandian V (2012) Generalized projective synchronization of two-scroll systems via adaptive control. Int J Soft Comput 7(4):146–156
29. Senouci A, Boukabou A (2014) Predictive control and synchronization of chaotic and hyperchaotic systems based on a $T - S$ fuzzy model. Math Comput Simul 105:62–78
30. Sprott JC (1994) Some simple chaotic flows. Phys Rev E 50(2):647–650
31. Sundarapandian V (2010) Output regulation of the Lorenz attractor. Far East J Math Sci 42(2):289–299
32. Sundarapandian V (2013) Adaptive control and synchronization design for the Lu-Xiao chaotic system. Lect Notes Electr Eng 131:319–327
33. Sundarapandian V (2013) Analysis and anti-synchronization of a novel chaotic system via active and adaptive controllers. J Eng Sci Technol Rev 6(4):45–52
34. Sundarapandian V, Karthikeyan R (2011) Anti-synchronization of hyperchaotic Lorenz and hyperchaotic Chen systems by adaptive control. Int J Syst Signal Control Eng Appl 4(2):18–25
35. Sundarapandian V, Karthikeyan R (2011) Anti-synchronization of Lü and Pan chaotic systems by adaptive nonlinear control. Eur J Sci Res 64(1):94–106
36. Sundarapandian V, Karthikeyan R (2012) Adaptive anti-synchronization of uncertain Tigan and Li systems. J Eng Appl Sci 7(1):45–52
37. Sundarapandian V, Pehlivan I (2012) Analysis, control, synchronization, and circuit design of a novel chaotic system. Math Comput Model 55(7–8):1904–1915
38. Sundarapandian V, Sivaperumal S (2011) Sliding controller design of hybrid synchronization of four-wing chaotic systems. Int J Soft Comput 6(5):224–231
39. Suresh R, Sundarapandian V (2013) Global chaos synchronization of a family of n-scroll hyperchaotic Chua circuits using backstepping control with recursive feedback. Far East J Math Sci 7(2):219–246
40. Tigan G, Opris D (2008) Analysis of a 3D chaotic system. Chaos Solitons Fractals 36: 1315–1319
41. Vaidyanathan S (2011) Output regulation of Arneodo-Coullet chaotic system. Commun Comput Inf Sci 133:98–107
42. Vaidyanathan S (2011) Output regulation of the unified chaotic system. Commun Comput Inf Sci 198:1–9
43. Vaidyanathan S (2012) Adaptive controller and syncrhonizer design for the Qi-Chen chaotic system. Lect Notes Inst Comput Sci Soc-Inf Telecommun Eng 84:73–82
44. Vaidyanathan S (2012) Anti-synchronization of Sprott-L and Sprott-M chaotic systems via adaptive control. Int J Control Theory Appl 5(1):41–59
45. Vaidyanathan S (2012) Global chaos control of hyperchaotic Liu system via sliding control method. Int J Control Theory Appl 5(2):117–123
46. Vaidyanathan S (2012) Sliding mode control based global chaos control of Liu-Liu-Liu-Su chaotic system. Int J Control Theory Appl 5(1):15–20
47. Vaidyanathan S (2013) A new six-term 3-D chaotic system with an exponential nonlinearity. Far East J Math Sci 79(1):135–143
48. Vaidyanathan S (2013) A ten-term novel 4-D hyperchaotic system with three quadratic nonlinearities and its control. Int J Control Theory Appl 6(2):97–109
49. Vaidyanathan S (2013) Analysis and adaptive synchronization of two novel chaotic systems with hyperbolic sinusoidal and cosinusoidal nonlinearity and unknown parameters. J Eng Sci Technol Rev 6(4):53–65
50. Vaidyanathan S (2013) Analysis, control and synchronization of hyperchaotic Zhou system via adaptive control. Adv Intell Syst Comput 177:1–10
51. Vaidyanathan S (2014) A new eight-term 3-D polynomial chaotic system with three quadratic nonlinearities. Far East J Math Sci 84(2):219–226
52. Vaidyanathan S (2014) Analysis and adaptive synchronization of eight-term 3-D polynomial chaotic systems with three quadratic nonlinearities. Eur Phys J: Spec Top 223(8):1519–1529
53. Vaidyanathan S (2014) Analysis, control and synchronisation of a six-term novel chaotic system with three quadratic nonlinearities. Int J Modell Identif Control 22(1):41–53

54. Vaidyanathan S (2014) Generalized projective synchronisation of novel 3-D chaotic systems with an exponential non-linearity via active and adaptive control. Int J Modell Identif Control 22(3):207–217
55. Vaidyanathan S (2014) Global chaos synchronisation of identical Li-Wu chaotic systems via sliding mode control. Int J Modell Identif Control 22(2):170–177
56. Vaidyanathan S (2014) Qualitative analysis and control of an eleven-term novel 4-D hyperchaotic system with two quadratic nonlinearities. Int J Control Theory Appl 7:35–47
57. Vaidyanathan S (2015) 3-cells cellular neural network (CNN) attractor and its adaptive biological control. Int J Pharm Tech Res 8(4):632–640
58. Vaidyanathan S (2015) A 3-D novel highly chaotic system with four quadratic nonlinearities, its adaptive control and anti-synchronization with unknown parameters. J Eng Sci Technol Rev 8(2):106–115
59. Vaidyanathan S (2015) A novel chemical chaotic reactor system and its adaptive control. Int J Chem Tech Res 8(7):146–158
60. Vaidyanathan S (2015) Adaptive backstepping control of enzymes-substrates system with ferroelectric behaviour in brain waves. Int J Pharm Tech Res 8(2):256–261
61. Vaidyanathan S (2015) Adaptive biological control of generalized Lotka-Volterra three-species biological system. Int J Pharm Tech Res 8(4):622–631
62. Vaidyanathan S (2015) Adaptive chaotic synchronization of enzymes-substrates system with ferroelectric behaviour in brain waves. Int J Pharm Tech Res 8(5):964–973
63. Vaidyanathan S (2015) Adaptive control of a chemical chaotic reactor. Int J Pharm Tech Res 8(3):377–382
64. Vaidyanathan S (2015) Adaptive control of the FitzHugh-Nagumo chaotic neuron model. Int J Pharm Tech Res 8(6):117–127
65. Vaidyanathan S (2015) Adaptive synchronization of chemical chaotic reactors. Int J Chem Tech Res 8(2):612–621
66. Vaidyanathan S (2015) Adaptive synchronization of generalized Lotka-Volterra three-species biological systems. Int J Pharm Tech Res 8(5):928–937
67. Vaidyanathan S (2015) Adaptive synchronization of novel 3-D chemical chaotic reactor systems. Int J Chem Tech Res 8(7):159–171
68. Vaidyanathan S (2015) Adaptive synchronization of the identical FitzHugh-Nagumo chaotic neuron models. Int J Pharm Tech Res 8(6):167–177
69. Vaidyanathan S (2015) Analysis, control and synchronization of a 3-D novel jerk chaotic system with two quadratic nonlinearities. Kyungpook Math J 55:563–586
70. Vaidyanathan S (2015) Analysis, properties and control of an eight-term 3-D chaotic system with an exponential nonlinearity. Int J Modell Identif Control 23(2):164–172
71. Vaidyanathan S (2015) Anti-synchronization of Brusselator chemical reaction systems via adaptive control. Int J Chem Tech Res 8(6):759–768
72. Vaidyanathan S (2015) Chaos in neurons and adaptive control of Birkhoff-Shaw strange chaotic attractor. Int J Pharm Tech Res 8(5):956–963
73. Vaidyanathan S (2015) Chaos in neurons and synchronization of Birkhoff-Shaw strange chaotic attractors via adaptive control. Int J Pharm Tech Res 8(6):1–11
74. Vaidyanathan S (2015) Coleman-Gomatam logarithmic competitive biology models and their ecological monitoring. Int J Pharm Tech Res 8(6):94–105
75. Vaidyanathan S (2015) Dynamics and control of Brusselator chemical reaction. Int J Chem Tech Res 8(6):740–749
76. Vaidyanathan S (2015) Dynamics and control of Tokamak system with symmetric and magnetically confined plasma. Int J Chem Tech Res 8(6):795–803
77. Vaidyanathan S (2015) Global chaos synchronization of chemical chaotic reactors via novel sliding mode control method. Int J Chem Tech Res 8(7):209–221
78. Vaidyanathan S (2015) Global chaos synchronization of the forced Van der Pol chaotic oscillators via adaptive control method. Int J Pharm Tech Res 8(6):156–166
79. Vaidyanathan S (2015) Global chaos synchronization of the Lotka-Volterra biological systems with four competitive species via active control. Int J Pharm Tech Res 8(6):206–217

80. Vaidyanathan S (2015) Lotka-Volterra population biology models with negative feedback and their ecological monitoring. Int J Pharm Tech Res 8(5):974–981
81. Vaidyanathan S (2015) Lotka-Volterra two species competitive biology models and their ecological monitoring. Int J Pharm Tech Res 8(6):32–44
82. Vaidyanathan S (2015) Output regulation of the forced Van der Pol chaotic oscillator via adaptive control method. Int J Pharm Tech Res 8(6):106–116
83. Vaidyanathan S, Azar AT (2015) Analysis and control of a 4-D novel hyperchaotic system. Stud Comput Intell 581:3–17
84. Vaidyanathan S, Azar AT (2015) Analysis, control and synchronization of a nine-term 3-D novel chaotic system. In: Azar AT, Vaidyanathan S (eds) Chaos modelling and control systems design, vol 581. Studies in computational intelligence. Springer, Germany, pp 19–38
85. Vaidyanathan S, Madhavan K (2013) Analysis, adaptive control and synchronization of a seven-term novel 3-D chaotic system. Int J Control Theory Appl 6(2):121–137
86. Vaidyanathan S, Pakiriswamy S (2013) Generalized projective synchronization of six-term Sundarapandian chaotic systems by adaptive control. Int J Control Theory Appl 6(2):153–163
87. Vaidyanathan S, Pakiriswamy S (2015) A 3-D novel conservative chaotic system and its generalized projective synchronization via adaptive control. J Eng Sci Technol Rev 8(2): 52–60
88. Vaidyanathan S, Rajagopal K (2011) Hybrid synchronization of hyperchaotic Wang-Chen and hyperchaotic Lorenz systems by active non-linear control. Int J Syst Signal Control Eng Appl 4(3):55–61
89. Vaidyanathan S, Rajagopal K (2012) Global chaos synchronization of hyperchaotic Pang and hyperchaotic Wang systems via adaptive control. Int J Soft Comput 7(1):28–37
90. Vaidyanathan S, Rasappan S (2011) Global chaos synchronization of hyperchaotic Bao and Xu systems by active nonlinear control. Commun Comput Inf Sci 198:10–17
91. Vaidyanathan S, Rasappan S (2014) Global chaos synchronization of n-scroll Chua circuit and Lur'e system using backstepping control design with recursive feedback. Arab J Sci Eng 39(4):3351–3364
92. Vaidyanathan S, Sampath S (2012) Anti-synchronization of four-wing chaotic systems via sliding mode control. Int J Autom Comput 9(3):274–279
93. Vaidyanathan S, Volos C (2015) Analysis and adaptive control of a novel 3-D conservative no-equilibrium chaotic system. Arch Control Sci 25(3):333–353
94. Vaidyanathan S, Volos C, Pham VT (2014) Hyperchaos, adaptive control and synchronization of a novel 5-D hyperchaotic system with three positive Lyapunov exponents and its SPICE implementation. Arch Control Sci 24(4):409–446
95. Vaidyanathan S, Volos C, Pham VT, Madhavan K, Idowu BA (2014) Adaptive backstepping control, synchronization and circuit simulation of a 3-D novel jerk chaotic system with two hyperbolic sinusoidal nonlinearities. Arch Control Sci 24(3):375–403
96. Vaidyanathan S, Azar AT, Rajagopal K, Alexander P (2015) Design and SPICE implementation of a 12-term novel hyperchaotic system and its synchronisation via active control. Int J Modell Identif Control 23(3):267–277
97. Vaidyanathan S, Idowu BA, Azar AT (2015) Backstepping controller design for the global chaos synchronization of Sprott's jerk systems. Stud Comput Intell 581:39–58
98. Vaidyanathan S, Rajagopal K, Volos CK, Kyprianidis IM, Stouboulos IN (2015) Analysis, adaptive control and synchronization of a seven-term novel 3-D chaotic system with three quadratic nonlinearities and its digital implementation in LabVIEW. J Eng Sci Technol Rev 8(2):130–141
99. Vaidyanathan S, Sampath S, Azar AT (2015) Global chaos synchronisation of identical chaotic systems via novel sliding mode control method and its application to Zhu system. Int J Modell Identif Control 23(1):92–100
100. Vaidyanathan S, Volos C, Pham VT, Madhavan K (2015) Analysis, adaptive control and synchronization of a novel 4-D hyperchaotic hyperjerk system and its SPICE implementation. Nonlinear Dyn 25(1):135–158

101. Vaidyanathan S, Volos CK, Kyprianidis IM, Stouboulos IN, Pham VT (2015) Analysis, adaptive control and anti-synchronization of a six-term novel jerk chaotic system with two exponential nonlinearities and its circuit simulation. J Eng Sci Technol Rev 8(2):24–36
102. Vaidyanathan S, Volos CK, Pham VT (2015) Analysis, adaptive control and adaptive synchronization of a nine-term novel 3-D chaotic system with four quadratic nonlinearities and its circuit simulation. J Eng Sci Technol Rev 8(2):181–191
103. Vaidyanathan S, Volos CK, Pham VT (2015) Analysis, control, synchronization and SPICE implementation of a novel 4-D hyperchaotic Rikitake dynamo system without equilibrium. J Eng Sci Technol Rev 8(2):232–244
104. Vaidyanathan S, Volos CK, Pham VT (2015) Global chaos control of a novel nine-term chaotic system via sliding mode control. In: Azar AT, Zhu Q (eds) Advances and applications in sliding mode control systems, vol 576. Studies in computational intelligence. Springer, Germany, pp 571–590
105. Volos CK, Kyprianidis IM, Stouboulos IN, Tlelo-Cuautle E, Vaidyanathan S (2015) Memristor: a new concept in synchronization of coupled neuromorphic circuits. J Eng Sci Technol Rev 8(2):157–173
106. Wang J, Chen Z (2008) A novel hyperchaotic system and its complex dynamics. Int J Bifurc Chaos 18:3309–3324
107. Wei X, Yunfei F, Qiang L (2012) A novel four-wing hyper-chaotic system and its circuit implementation. Procedia Eng 29:1264–1269
108. Wei Z, Yang Q (2010) Anti-control of Hopf bifurcation in the new chaotic system with two stable node-foci. Appl Math Comput 217(1):422–429
109. Wu X, Zhu C, Kan H (2015) An improved secure communication scheme based passive synchronization of hyperchaotic complex nonlinear system. Appl Math Comput 252: 201–214
110. Yujun N, Xingyuan W, Mingjun W, Huaguang Z (2010) A new hyperchaotic system and its circuit implementation. Commun Nonlinear Sci Numer Simul 15(11):3518–3524
111. Zhang H, Liao X, Yu J (2005) Fuzzy modeling and synchronization of hyperchaotic systems. Chaos Solitons Fractals 26(3):835–843
112. Zhou W, Xu Y, Lu H, Pan L (2008) On dynamics analysis of a new chaotic attractor. Phys Lett A 372(36):5773–5777
113. Zhu C (2012) A novel image encryption scheme based on improved hyperchaotic sequences. Opt Commun 285(1):29–37
114. Zhu C, Liu Y, Guo Y (2010) Theoretic and numerical study of a new chaotic system. Intell Inf Manag 2:104–109

Sliding Mode Controller Design for the Global Stabilization of Chaotic Systems and Its Application to Vaidyanathan Jerk System

Sundarapandian Vaidyanathan

Abstract Chaos in nonlinear dynamics occurs widely in physics, chemistry, biology, ecology, secure communications, cryptosystems and many scientific branches. Control of chaotic systems is an important research problem in chaos theory. Sliding mode control is an important method used to solve various problems in control systems engineering. In robust control systems, the sliding mode control is often adopted due to its inherent advantages of easy realization, fast response and good transient performance as well as insensitivity to parameter uncertainties and disturbance. In this work, we derive a novel sliding mode control method for the global stabilization of chaotic systems. The general control result derived using novel sliding mode control method is proved using Lyapunov stability theory. As an application of the general result, the problem of global stabilization of the Vaidyanathan jerk chaotic system (2015) is studied and a new sliding mode controller is derived. The Lyapunov exponents of the Vaidyanathan jerk system are obtained as $L_1 = 0.12476, L_2 = 0$ and $L_3 = -1.12451$. Since the Vaidyanathan jerk system has a positive Lyapunov exponent, it is chaotic. The Maximal Lyapunov Exponent (MLE) of the Vaidyanathan jerk system is given by $L_1 = 0.12476$. Also, the Kaplan–Yorke dimension of the Vaidyanathan jerk system is obtained as $D_{KY} = 2.11095$. Numerical simulations using MATLAB have been shown to depict the phase portraits of the Vaidyanathan jerk system and the sliding mode controller design for the global stabilization of the Vaidyanathan jerk system.

Keywords Chaos · Chaotic systems · Jerk systems · Sliding mode control · Stabilization

---

S. Vaidyanathan (✉)
Research and Development Centre, Vel Tech University, Avadi 600062, Tamil Nadu, India
e-mail: sundarvtu@gmail.com

S. Vaidyanathan and C. Volos (eds.), *Advances and Applications in Chaotic Systems*, Studies in Computational Intelligence 636,
DOI 10.1007/978-3-319-30279-9_23

1 Introduction

Chaos theory describes the quantitative study of unstable aperiodic dynamic behaviour in deterministic nonlinear dynamical systems. For the motion of a dynamical system to be chaotic, the system variables should contain some nonlinear terms and the system must satisfy three properties: boundedness, infinite recurrence and sensitive dependence on initial conditions [9].

The Lyapunov exponent is a measure of the divergence of phase points that are initially very close and can be used to quantify chaotic systems. It is common to refer to the largest Lyapunov exponent as the *Maximal Lyapunov Exponent* (MLE). A positive maximal Lyapunov exponent and phase space compactness are usually taken as defining conditions for a chaotic system.

Some classical paradigms of 3-D chaotic systems in the literature are Lorenz system [30], Rössler system [40], ACT system [1], Sprott systems [43], Chen system [15], Lü system [31], Liu system [29], Cai system [14], Chen–Lee system [16], Tigan system [51], etc.

Many new chaotic systems have been discovered in the recent years such as Zhou system [116], Zhu system [117], Li system [25], Wei–Yang system [111], Sundarapandian systems [48, 49], Vaidyanathan systems [60, 63, 65–68, 70, 81, 95–97, 99, 101, 104, 106–108], Pehlivan system [35], Sampath system [41], Pham system [36],etc.

Chaos theory and control systems have many important applications in science and engineering [2, 10–13, 118]. Some commonly known applications are oscillators [22, 42], lasers [26, 113], chemical reactors [71, 75, 77, 79, 82, 86–88], biological systems [69, 72–74, 76, 78, 83–85, 89, 91–93], ecology [18, 45], encryption [24, 115], cryptosystems [39, 52], mechanical systems [4–8], secure communications [17, 33, 114], robotics [32, 34, 109], cardiology [38, 112], intelligent control [3, 27], neural networks [20, 21, 28], finance [19, 44], memristors [37, 110], etc.

Control or regulation of a chaotic system deals with the design of a state feedback control law so as to stabilize or regulate the trajectories of the chaotic system. Many techniques have been devised for the global control of chaotic systems such as the active control method [46, 47, 55, 56, 102], adaptive control method [61, 62, 64, 90, 94, 100, 103, 105], sliding mode control method [50, 57–59, 98], etc.

In the sliding mode control theory, the control dynamics will have two sequential modes, viz. the reaching mode and the sliding mode. Basically, a sliding mode controller (SMC) design consists of two parts: hyperplane design and controller design. A hyperplane is first designed via the pole-placement approach in the modern control theory and a controller is then designed based on the sliding condition. The stability of the overall system is guaranteed by the sliding condition and by a stable hyperplane.

The sliding mode control method is an effective control tool which has the advantages of low sensitivity to parameter variations in the plant and disturbances affecting the plant.

In this work, we use a novel sliding mode control method for deriving a general result for the global stabilization of chaotic systems using sliding mode control (SMC) theory.

The general control result derived using novel sliding mode control method is proved using Lyapunov stability theory. As an application of the general result, the problem of global stabilization of the Vaidyanathan jerk chaotic system [80] is studied and a new sliding mode controller is derived.

This work is organized as follows. Section 2 contains the problem statement of global chaos control of chaotic systems. Section 3 describes the novel sliding mode controller design for globally stabilizing chaotic systems. The general control result derived using novel sliding mode control method is proved using Lyapunov stability theory.

Section 4 describes the Vaidyanathan jerk chaotic system [80] and its properties. The Lyapunov exponents of the Vaidyanathan jerk system are obtained as $L_1 = 0.12476$, $L_2 = 0$ and $L_3 = -1.12451$. Since the Vaidyanathan jerk system has a positive Lyapunov exponent, it is chaotic. The Maximal Lyapunov Exponent (MLE) of the Vaidyanathan jerk system is given by $L_1 = 0.12476$. Also, the Kaplan–Yorke dimension of the Vaidyanathan jerk system is obtained as $D_{KY} = 2.11095$.

Section 5 describes the application of the general result derived in Sect. 3 for the global chaos control of the Vaidyanathan jerk chaotic system [80]. Numerical simulations using MATLAB have been shown to depict the phase portraits of the Vaidyanathan jerk system and the sliding mode controller design for the global stabilization of the Vaidyanathan jerk system. Section 6 contains the conclusions of this work.

2 Problem Statement

This section gives a problem statement for the global chaos control of a given chaotic system.

To start with, we consider a chaotic system given by

$$\dot{\mathbf{x}} = A\mathbf{x} + f(\mathbf{x}) + \mathbf{u} \tag{1}$$

where $\mathbf{x} \in \mathbf{R}^n$ denotes the state of the system, $A \in \mathbf{R}^{n\times n}$ denotes the matrix of system parameters, $f(\mathbf{x}) \in \mathbf{R}^n$ contains the nonlinear parts of the system and $\mathbf{u}$ is the control.

Thus, the global chaos control for the given chaotic system (1) can be stated as follows: Find a feedback controller $\mathbf{u}(\mathbf{x})$ so as to render the state $\mathbf{x}(t)$ of the corresponding closed-loop system to be globally asymptotically stable for all values of $\mathbf{x}(0) \in \mathbf{R}^n$, i.e.

$$\lim_{t\to\infty} \|\mathbf{x}(t)\| = 0 \quad \text{for all} \;\; \mathbf{x}(0) \in \mathbf{R}^n \tag{2}$$

3 A Novel Sliding Mode Control Method for Global Stabilization of Chaotic Systems

This section details the main results of this work, viz. novel sliding mode controller design for achieving global asymptotic stabilization of a given chaotic system.

First, we start the design by setting the control as

$$\mathbf{u}(t) = -f(\mathbf{x}) + Bv(t) \tag{3}$$

In Eq. (3), $B \in \mathbf{R}^n$ is chosen such that (A, B) is completely controllable.

By substituting (3) into (1), we get the closed-loop plant dynamics

$$\dot{\mathbf{x}} = A\mathbf{x} + Bv \tag{4}$$

The system (4) is a linear time-invariant control system with single input v.

Next, we start the sliding controller design by defining the sliding variable as

$$s(\mathbf{x}) = C\mathbf{x} = c_1 x_1 + c_2 x_2 + \cdots + c_n x_n, \tag{5}$$

where $C \in \mathbf{R}^{1\times n}$ is a constant vector to be determined.

The sliding manifold S is defined as the hyperplane

$$S = \{\mathbf{x} \in \mathbf{R}^n : \; s(\mathbf{x}) = C\mathbf{x} = 0\} \tag{6}$$

We shall assume that a sliding motion occurs on the hyperplane S.

In sliding mode, the following equations must be satisfied:

$$s = 0 \tag{7a}$$

$$\dot{s} = CA\mathbf{x} + CBv = 0 \tag{7b}$$

We assume that

$$CB \neq 0 \tag{8}$$

The sliding motion is influenced by the equivalent control derived from (7b) as

$$v_{\text{eq}}(t) = -(CB)^{-1}\, CA\mathbf{x}(t) \tag{9}$$

By substituting (9) into (4), we obtain the equivalent system dynamics in the sliding phase as

$$\dot{\mathbf{x}} = A\mathbf{x} - (CB)^{-1}CA\mathbf{x} = E\mathbf{x}, \tag{10}$$

where

$$E = \left[I - B(CB)^{-1}C\right]A \tag{11}$$

We note that E is independent of the control and has at most $(n-1)$ non-zero eigenvalues, depending on the chosen switching surface, while the associated eigenvectors belong to $\ker(C)$.

Since (A, B) is controllable, we can use sliding control theory [53, 54] to choose B and C so that E has any desired $(n-1)$ stable eigenvalues.

This shows that the dynamics (10) is globally asympotically stable.

Finally, for the sliding controller design, we apply a novel sliding control law, viz.

$$\dot{s} = -ks - qs^2 \operatorname{sgn}(s) \tag{12}$$

In (12), $\operatorname{sgn}(\cdot)$ denotes the *sign* function and the SMC constants $k > 0$, $q > 0$ are found in such a way that the sliding condition is satisfied and that the sliding motion will occur.

By combining Eqs. (7b), (9) and (12), we finally obtain the sliding mode controller $v(t)$ as

$$v(t) = -(CB)^{-1}\left[C(kI + A)\mathbf{x} + qs^2 \operatorname{sgn}(s)\right] \tag{13}$$

Next, we establish the main result of this section.

Theorem 1 *The chaotic system (1) is globally asymptotically stabilized for all initial conditions* $\mathbf{x}(0)$ *in* $\mathbf{R}^n$, *where* v *is defined by the novel sliding mode control law (13),* $B \in \mathbf{R}^{n\times 1}$ *is such that* (A, B) *is controllable,* $C \in \mathbf{R}^{1\times n}$ *is such that* $CB \neq 0$ *and the matrix* E *defined by (11) has* $(n-1)$ *stable eigenvalues.*

Proof Upon substitution of the control laws (3) and (13) into the system dynamics (1), we obtain the closed-loop system as

$$\dot{\mathbf{x}} = A\mathbf{x} - B(CB)^{-1}\left[C(kI + A)\mathbf{x} + qs^2 \operatorname{sgn}(s)\right] \tag{14}$$

We shall show that the error dynamics (14) is globally asymptotically stable by considering the quadratic Lyapunov function

$$V(\mathbf{x}) = \frac{1}{2}\, s^2(\mathbf{x}) \tag{15}$$

The sliding mode motion is characterized by the equations

$$s(\mathbf{x}) = 0 \quad \text{and} \quad \dot{s}(\mathbf{x}) = 0 \tag{16}$$

By the choice of E, the dynamics in the sliding mode given by Eq. (10) is globally asymptotically stable.

When $s(\mathbf{x}) \neq 0$, $V(\mathbf{x}) > 0$.

Also, when $s(\mathbf{x}) \neq 0$, differentiating V along the error dynamics (14) or the equivalent dynamics (12), we get

$$\dot{V}(\mathbf{x}) = s\dot{s} = -ks^2 - qs^3 \operatorname{sgn}(s) < 0 \tag{17}$$

Hence, by Lyapunov stability theory [23], the error dynamics (14) is globally asymptotically stable for all $\mathbf{x}(0) \in \mathbf{R}^n$.

This completes the proof. ■

4 Vaidyanathan Jerk Chaotic System and Its Properties

In this section, we describe the Vaidyanathan jerk chaotic system [80] and discuss its dynamic properties.

The Vaidyanathan jerk chaotic system [80] is described by the 3-D dynamics

$$\begin{aligned} \dot{x}_1 &= x_2 \\ \dot{x}_2 &= x_3 \\ \dot{x}_3 &= ax_1 - bx_2 - x_3 - x_1^2 - x_2^2 \end{aligned} \tag{18}$$

where x_1, x_2, x_3 are the states and a, b are constant, positive, parameters.

In [80], it was shown that the system (18) exhibits a *strange chaotic attractor*, when the parameters take the values

$$a = 7.5 \quad b = 4 \tag{19}$$

For numerical simulations, we take the initial values of the Vaidyanathan jerk chaotic system (18) as

$$x_1(0) = 0.2, \quad x_2(0) = 0.6, \quad x_3(0) = 0.4 \tag{20}$$

For the parameter values in (19) and the initial values in (20), the Lyapunov exponents of the Vaidyanathan jerk system (18) are numerically obtained as

$$L_1 = 0.12476, \quad L_2 = 0, \quad L_3 = -1.12451 \tag{21}$$

Since the sum of the Lyapunov exponents in (21) is negative, the Vaidyanathan jerk system (18) is dissipative.

The Kaplan–Yorke dimension of the Vaidyanathan jerk system (18) is calculated as

$$D_{KY} = 2 + \frac{L_1 + L_2}{|L_3|} = 2.11095, \tag{22}$$

which is fractional.

It is easy to show that the Vaidyanathan hyperjerk system (18) has two equilibrium points given by

$$E_0 = \begin{bmatrix} 0 \\ 0 \\ 0 \end{bmatrix} \quad \text{and} \quad E_1 = \begin{bmatrix} 7.5 \\ 0 \\ 0 \end{bmatrix} \tag{23}$$

In [80], it was shown that both E_0 and E_1 are saddle-focus points, and hence they are unstable.

For the initial conditions given in (20), phase portraits of the Vaidyanathan jerk system (18) are plotted using MATLAB.

Figures 1 shows the strange chaotic attractor of the Vaidyanathan jerk system (18). Figures 2, 3 and 4 show the 2-D projection of the Vaidyanathan jerk system (18) on the (x_1, x_2), (x_2, x_3) and (x_1, x_3) planes, respectively.

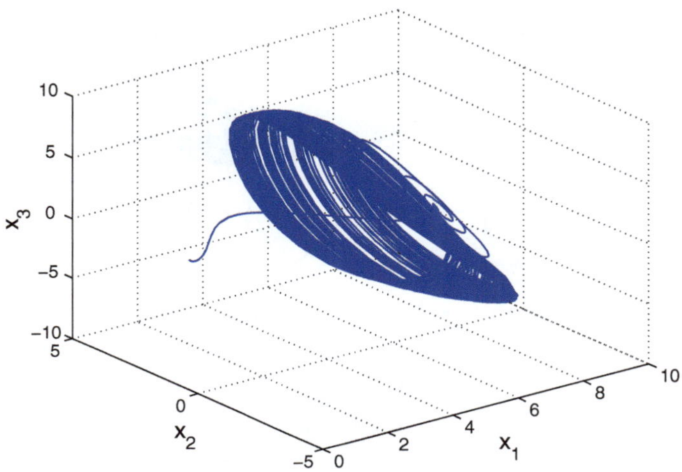

Fig. 1 Strange chaotic attractor of the Vaidyanathan jerk system

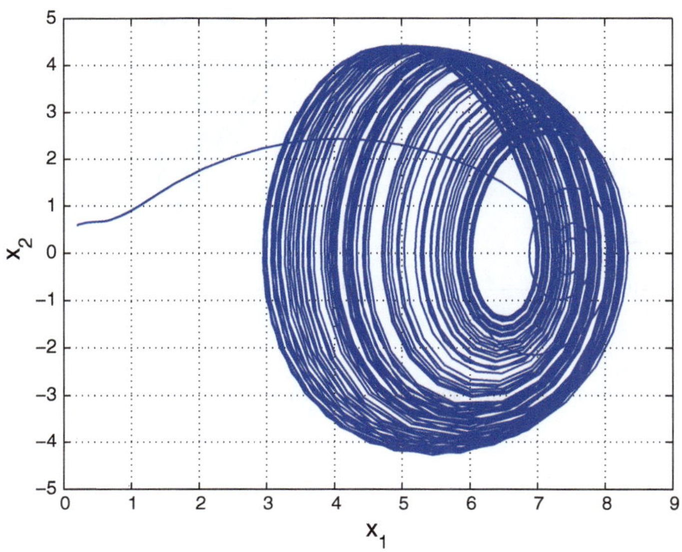

Fig. 2 2-D projection of the Vaidyanathan jerk system on the (x_1, x_2) plane

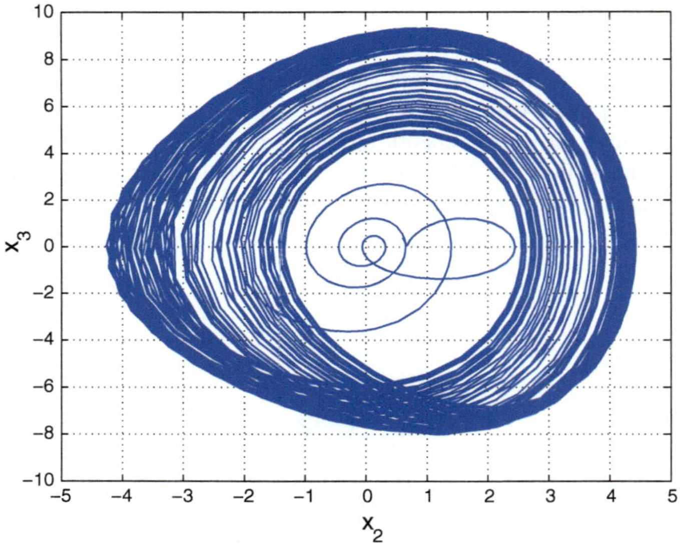

Fig. 3 2-D projection of the Vaidyanathan jerk system on the (x_2, x_3) plane

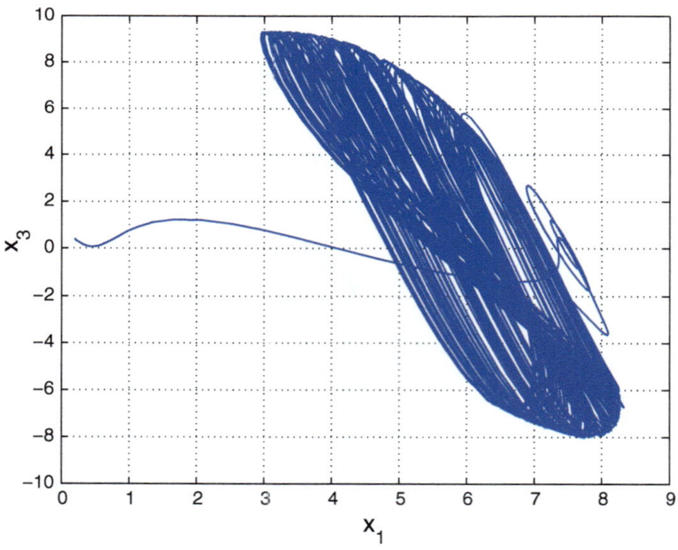

Fig. 4 2-D projection of the Vaidyanathan jerk system on the (x_1, x_3) plane

5 Sliding Mode Controller Design for the Global Stabilization of Vaidyanathan Jerk System

In this section, we describe the sliding mode controller design for the global stabilization of Vaidyanathan jerk system [80] by applying the novel method described by Theorem 1 in Sect. 3.

Thus, we consider the controlled Vaidyanathan jerk system given by

$$\begin{aligned} \dot{x}_1 &= x_2 + u_1 \\ \dot{x}_2 &= x_3 + u_2 \\ \dot{x}_3 &= ax_1 - bx_2 - x_3 - x_1^2 - x_2^2 + u_3 \end{aligned} \tag{24}$$

In matrix form, we can write the error dynamics (24) as

$$\dot{\mathbf{x}} = A\mathbf{x} + \psi(\mathbf{x}) + \mathbf{u} \tag{25}$$

The matrices A and ψ in (25) are given by

$$A = \begin{bmatrix} 0 & 1 & 0 \\ 0 & 0 & 1 \\ a & -b & -1 \end{bmatrix}, \quad \psi(\mathbf{x}) = \begin{bmatrix} 0 \\ 0 \\ -x_1^2 - x_2^2 \end{bmatrix} \tag{26}$$

We follow the procedure given in Sect. 3 for the construction of the novel sliding controller to achieve global stabilization of the Vaidyanathan jerk system (24).

We take the parameter values of a and b as in the chaotic case, i.e.

$$a = 7.5, \quad b = 4 \tag{27}$$

First, we set $\mathbf{u}$ as

$$\mathbf{u}(t) = -\psi(\mathbf{x}) + Bv(t) \tag{28}$$

where B is selected such that (A, B) is completely controllable.

A simple choice of B is

$$B = \begin{bmatrix} 1 \\ 1 \\ 1 \end{bmatrix} \tag{29}$$

It can be easily checked that (A, B) is completely controllable.

Next, we take the sliding variable as

$$s(\mathbf{x}) = C\mathbf{x} = \begin{bmatrix} 1 & -1 & -20 \end{bmatrix} \mathbf{e} = x_1 - x_2 - 20x_3 \tag{30}$$

If we define $E = [I - B(CB)^{-1}C]A$, then the matrix E has the eigenvalues

$$\text{eig}(E) = \{0, -1.7500 \pm 2.1243i\} \tag{31}$$

which shows that the motion along the sliding manifold is globally asymptotically stable.

Next, we take the sliding mode gains as

$$k = 5, \quad q = 0.2 \tag{32}$$

From Eq. (13) in Sect. 3, we obtain the novel sliding control v as

$$v(t) = -7.25x_1 + 3.8x_2 - 4.05x_3 + 0.01s^2 \,\text{sgn}(s) \tag{33}$$

As an application of Theorem 1 to the identical Vaidyanathan jerk chaotic system, we obtain the following main result of this section.

Theorem 2 *Vaidyanathan jerk chaotic system (24) is globally and asymptotically stabilized for all initial conditions* $\mathbf{x}(0) \in \mathbf{R}^3$ *with the sliding controller* $\mathbf{u}$ *defined by (28), where* $\psi(\mathbf{x})$ *is defined by (26), B is defined by (29) and v is defined by (33).* ■

For numerical simulations, we use MATLAB for solving the systems of differential equations using the classical fourth-order Runge–Kutta method with step size $h = 10^{-8}$.

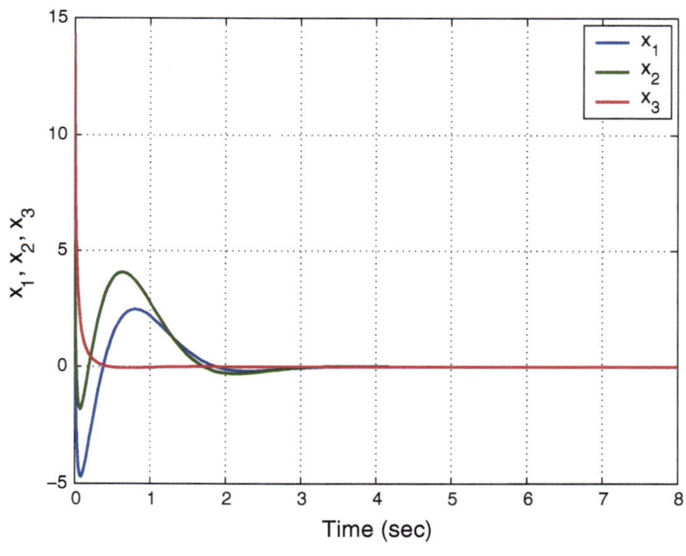

Fig. 5 Time-history of the controlled states x_1, x_2, x_3

The parameter values of the Vaidyanathan jerk system are taken as in the chaotic case, viz. $a = 7.5$ and $b = 4$.

The sliding mode gains are taken as $k = 5$ and $q = 0.2$.

As an initial condition for the Vaidyanathan jerk system (24), we take

$$x_1(0) = 5.7, \quad x_2(0) = 8.2, \quad x_3(0) = 14.3 \tag{34}$$

Figure 5 shows the time-history of the controlled states x_1, x_2, x_3.

6 Conclusions

In this work, we derived a novel sliding mode control method for the global stabilization of chaotic systems. The general control result derived using novel sliding mode control method was proved using Lyapunov stability theory. As an application of the general result, the problem of global stabilization of the Vaidyanathan jerk chaotic system (2015) was studied and a new sliding mode controller has been derived. The Lyapunov exponents of the Vaidyanathan jerk system have been obtained as $L_1 = 0.12476$, $L_2 = 0$ and $L_3 = -1.12451$. Since the Vaidyanathan jerk system has a positive Lyapunov exponent, it is chaotic. The Maximal Lyapunov Exponent (MLE) of the Vaidyanathan jerk system is seen as $L_1 = 0.12476$. Also, the Kaplan–Yorke dimension of the Vaidyanathan jerk system has been derived as $D_{KY} = 2.11095$. Numerical simulations using MATLAB were shown to depict the phase portraits of the Vaidyanathan jerk system and the sliding mode controller design for the global stabilization of the Vaidyanathan jerk system.

References

1. Arneodo A, Coullet P, Tresser C (1981) Possible new strange attractors with spiral structure. Commun Math Phys 79(4):573–576
2. Azar AT (2010) Fuzzy systems. IN-TECH, Vienna
3. Azar AT (2012) Overview of type-2 fuzzy logic systems. Int J Fuzzy Syst Appl 2(4):1–28
4. Azar AT, Serrano FE (2014) Robust IMC-PID tuning for cascade control systems with gain and phase margin specifications. Neural Comput Appl 25(5):983–995
5. Azar AT, Serrano FE (2015) Adaptive sliding mode control of the Furuta pendulum. In: Azar AT, Zhu Q (eds) Advances and applications in sliding mode control systems, studies in computational intelligence, vol 576. Springer, Berlin, pp 1–42
6. Azar AT, Serrano FE (2015) Deadbeat control for multivariable systems with time varying delays. In: Azar AT, Vaidyanathan S (eds) Chaos modeling and control systems design, studies in computational intelligence, vol 581. Springer, Berlin, pp 97–132
7. Azar AT, Serrano FE (2015) Design and modeling of anti wind up PID controllers. In: Zhu Q, Azar AT (eds) Complex system modelling and control through intelligent soft computations, studies in fuzziness and soft computing, vol 319. Springer, Cham, pp 1–44
8. Azar AT, Serrano FE (2015) Stabilizatoin and control of mechanical systems with backlash. In: Vaidyanathan S, Azar AT (eds) Handbook of research on advanced intelligent control

engineering and automation, Advances in Computational Intelligence and Robotics (ACIR). IGI-Global, Hershey, pp 1–60
9. Azar AT, Vaidyanathan S (2015) Chaos modeling and control systems design, vol 581. Springer, Berlin
10. Azar AT, Vaidyanathan S (2015) Chaos modeling and control systems design, studies in computational intelligence, vol 581. Springer, Berlin
11. Azar AT, Vaidyanathan S (2015) Computational intelligence applications in modeling and control, studies in computational intelligence, vol 575. Springer, Cham
12. Azar AT, Vaidyanathan S (2015) Handbook of research on advanced intelligent control engineering and automation., Advances in computational intelligence and robotics (ACIR)IGI-Global, Pennsylvania
13. Azar AT, Zhu Q (2015) Advances and applications in sliding mode control systems, studies in computational intelligence, vol 576. Springer, Cham
14. Cai G, Tan Z (2007) Chaos synchronization of a new chaotic system via nonlinear control. J Uncertain Syst 1(3):235–240
15. Chen G, Ueta T (1999) Yet another chaotic attractor. Int J Bifurc Chaos 9(7):1465–1466
16. Chen HK, Lee CI (2004) Anti-control of chaos in rigid body motion. Chaos Solitons Fractals 21(4):957–965
17. Feki M (2003) An adaptive chaos synchronization scheme applied to secure communication. Chaos Solitons Fractals 18(1):141–148
18. Gibson WT, Wilson WG (2013) Individual-based chaos: Extensions of the discrete logistic model. J Theor Biol 339:84–92
19. Guégan D (2009) Chaos in economics and finance. Annu Rev Control 33(1):89–93
20. Huang X, Zhao Z, Wang Z, Li Y (2012) Chaos and hyperchaos in fractional-order cellular neural networks. Neurocomputing 94:13–21
21. Kaslik E, Sivasundaram S (2012) Nonlinear dynamics and chaos in fractional-order neural networks. Neural Netw 32:245–256
22. Kengne J, Chedjou JC, Kenne G, Kyamakya K (2012) Dynamical properties and chaos synchronization of improved Colpitts oscillators. Commun Nonlinear Sci Numer Simul 17(7):2914–2923
23. Khalil HK (2001) Nonlinear systems, 3rd edn. Prentice Hall, Upper Saddle River
24. Lang J (2015) Color image encryption based on color blend and chaos permutation in the reality-preserving multiple-parameter fractional Fourier transform domain. Opt Commun 338:181–192
25. Li D (2008) A three-scroll chaotic attractor. Phys Lett A 372(4):387–393
26. Li N, Pan W, Yan L, Luo B, Zou X (2014) Enhanced chaos synchronization and communication in cascade-coupled semiconductor ring lasers. Commun Nonlinear Sci Numer Simul 19(6):1874–1883
27. Li Z, Chen G (2006) Integration of fuzzy logic and chaos theory, studies in fuzziness and soft computing, vol 187. Springer, Berlin
28. Lian S, Chen X (2011) Traceable content protection based on chaos and neural networks. Appl Soft Comput 11(7):4293–4301
29. Liu C, Liu T, Liu L, Liu K (2004) A new chaotic attractor. Chaos Solitions Fractals 22(5):1031–1038
30. Lorenz EN (1963) Deterministic periodic flow. J Atmos Sci 20(2):130–141
31. Lü J, Chen G (2002) A new chaotic attractor coined. Int J Bifurc Chaos 12(3):659–661
32. Mondal S, Mahanta C (2014) Adaptive second order terminal sliding mode controller for robotic manipulators. J Frankl Inst 351(4):2356–2377
33. Murali K, Lakshmanan M (1998) Secure communication using a compound signal from generalized chaotic systems. Phys Lett A 241(6):303–310
34. Nehmzow U, Walker K (2005) Quantitative description of robot-environment interaction using chaos theory. Robot Auton Syst 53(3–4):177–193
35. Pehlivan I, Moroz IM, Vaidyanathan S (2014) Analysis, synchronization and circuit design of a novel butterfly attractor. J Sound Vib 333(20):5077–5096

36. Pham VT, Vaidyanathan S, Volos CK, Jafari S (2015a) Hidden attractors in a chaotic system with an exponential nonlinear term. Eur Phys J Spec Top 224(8):1507–1517
37. Pham VT, Volos CK, Vaidyanathan S, Le TP, Vu VY (2015b) A memristor-based hyperchaotic system with hidden attractors: dynamics, synchronization and circuital emulating. J Eng Sci Technol Rev 8(2):205–214
38. Qu Z (2011) Chaos in the genesis and maintenance of cardiac arrhythmias. Prog Biophysi Mol Biol 105(3):247–257
39. Rhouma R, Belghith S (2011) Cryptoanalysis of a chaos based cryptosystem on DSP. Commun Nonlinear Sci Numer Simul 16(2):876–884
40. Rössler OE (1976) An equation for continuous chaos. Phys Lett A 57(5):397–398
41. Sampath S, Vaidyanathan S, Volos CK, Pham VT (2015) An eight-term novel four-scroll chaotic system with cubic nonlinearity and its circuit simulation. J Eng Sci Technol Rev 8(2):1–6
42. Sharma A, Patidar V, Purohit G, Sud KK (2012) Effects on the bifurcation and chaos in forced Duffing oscillator due to nonlinear damping. Commun Nonlinear Sci Numer Simul 17(6):2254–2269
43. Sprott JC (1994) Some simple chaotic flows. Phys Rev E 50(2):647–650
44. Sprott JC (2004) Competition with evolution in ecology and finance. Phys Lett A 325(5–6):329–333
45. Suérez I (1999) Mastering chaos in ecology. Ecol Model 117(2–3):305–314
46. Sundarapandian V (2003) A relation between the output regulation and the observer design for nonlinear systems. Appl Math Lett 16:235–242
47. Sundarapandian V (2010) Output regulation of the Lorenz attractor. Far East J Math Sci 42:289–299
48. Sundarapandian V (2013) Analysis and anti-synchronization of a novel chaotic system via active and adaptive controllers. J Eng Sci Technol Rev 6(4):45–52
49. Sundarapandian V, Pehlivan I (2012) Analysis, control, synchronization, and circuit design of a novel chaotic system. Math Comput Model 55(7–8):1904–1915
50. Sundarapandian V, Sivaperumal S (2012) Sliding controller design of hybrid synchronization of four-wing chaotic systems. Int J Soft Comput 6(5):224–231
51. Tigan G, Opris D (2008) Analysis of a 3D chaotic system. Chaos Solitons Fractals 36: 1315–1319
52. Usama M, Khan MK, Alghatbar K, Lee C (2010) Chaos-based secure satellite imagery cryptosystem. Comput Math Appl 60(2):326–337
53. Utkin VI (1977) Variable structure systems with sliding modes. IEEE Trans Autom Control 22(2):212–222
54. Utkin VI (1993) Sliding mode control design principles and applications to electric drives. IEEE Trans Ind Electron 40(1):23–36
55. Vaidyanathan S (2011) Output regulation of Arneodo–Coullet chaotic system. Commun Comput Inf Sci 133:98–107
56. Vaidyanathan S (2011) Output regulation of the unified chaotic system. Commun Comput Inf Sci 198:1–9
57. Vaidyanathan S (2012) Analysis and synchronization of the hyperchaotic Yujun systems via sliding mode control. Adv Intell Syst Comput 176:329–337
58. Vaidyanathan S (2012) Global chaos control of hyperchaotic Liu system via sliding control method. Int J Control Theory Appl 5(2):117–123
59. Vaidyanathan S (2012) Sliding control based global chaos control of Liu-Liu-Liu-Su chaotic system. Int J Control Theory Appl 5(1):15–20
60. Vaidyanathan S (2013) A new six-term 3-D chaotic system with an exponential nonlinearity. Far East J Math Sci 79(1):135–143
61. Vaidyanathan S (2013) A ten-term novel 4-D hyperchaotic system with three quadratic nonlinearities and its control. Int J Control Theory Appl 6:97–109
62. Vaidyanathan S (2013) Adaptive control and synchronization design for the Lu-Xiao chaotic system. Lect Notes Electr Eng 131:319–327

63. Vaidyanathan S (2013) Analysis and adaptive synchronization of two novel chaotic systems with hyperbolic sinusoidal and cosinusoidal nonlinearity and unknown parameters. J Eng Sci Technol Rev 6(4):53–65
64. Vaidyanathan S (2013) Analysis, control and synchronization of hyperchaotic Zhou system via adaptive control. Adv Intell Syst Comput 177:1–10
65. Vaidyanathan S (2014) A new eight-term 3-D polynomial chaotic system with three quadratic nonlinearities. Far East J Math Sci 84(2):219–226
66. Vaidyanathan S (2014) Analysis and adaptive synchronization of eight-term 3-D polynomial chaotic systems with three quadratic nonlinearities. Eur Phys J Spec Top 223(8):1519–1529
67. Vaidyanathan S (2014) Analysis, control and synchronisation of a six-term novel chaotic system with three quadratic nonlinearities. Int J Model Identif Control 22(1):41–53
68. Vaidyanathan S (2014) Generalized projective synchronisation of novel 3-D chaotic systems with an exponential non-linearity via active and adaptive control. Int J Model Identif Control 22(3):207–217
69. Vaidyanathan S (2015) 3-cells Cellular Neural Network (CNN) attractor and its adaptive biological control. Int J PharmTech Res 8(4):632–640
70. Vaidyanathan S (2015) A 3-D novel highly chaotic system with four quadratic nonlinearities, its adaptive control and anti-synchronization with unknown parameters. J Eng Sci Technol Rev 8(2):106–115
71. Vaidyanathan S (2015) A novel chemical chaotic reactor system and its adaptive control. Int J ChemTech Res 8(7):146–158
72. Vaidyanathan S (2015) Adaptive backstepping control of enzymes-substrates system with ferroelectric behaviour in brain waves. Int J PharmTech Res 8(2):256–261
73. Vaidyanathan S (2015) Adaptive biological control of generalized Lotka–Volterra three-species biological system. Int J PharmTech Res 8(4):622–631
74. Vaidyanathan S (2015) Adaptive chaotic synchronization of enzymes-substrates system with ferroelectric behaviour in brain waves. Int J PharmTech Res 8(5):964–973
75. Vaidyanathan S (2015) Adaptive control of a chemical chaotic reactor. Int J PharmTech Res 8(3):377–382
76. Vaidyanathan S (2015) Adaptive control of the FitzHugh–Nagumo chaotic neuron model. Int J PharmTech Res 8(6):117–127
77. Vaidyanathan S (2015) Adaptive synchronization of chemical chaotic reactors. Int J ChemTech Res 8(2):612–621
78. Vaidyanathan S (2015) Adaptive synchronization of generalized Lotka–Volterra three-species biological systems. Int J PharmTech Res 8(5):928–937
79. Vaidyanathan S (2015) Adaptive synchronization of novel 3-D chemical chaotic reactor systems. Int J ChemTech Res 8(7):159–171
80. Vaidyanathan S (2015) Analysis, control, and synchronization of a 3-D novel jerk chaotic system with two quadratic nonlinearities. Kyungpook Math J 55:563–586
81. Vaidyanathan S (2015) Analysis, properties and control of an eight-term 3-D chaotic system with an exponential nonlinearity. Int J Model Identif Control 23(2):164–172
82. Vaidyanathan S (2015) Anti-synchronization of brusselator chemical reaction systems via adaptive control. Int J ChemTech Res 8(6):759–768
83. Vaidyanathan S (2015) Chaos in neurons and adaptive control of Birkhoff–Shaw strange chaotic attractor. Int J PharmTech Res 8(5):956–963
84. Vaidyanathan S (2015) Chaos in neurons and synchronization of Birkhoff–Shaw strange chaotic attractors via adaptive control. Int J PharmTech Res 8(6):1–11
85. Vaidyanathan S (2015) Coleman–Gomatam logarithmic competitive biology models and their ecological monitoring. Int J PharmTech Res 8(6):94–105
86. Vaidyanathan S (2015) Dynamics and control of brusselator chemical reaction. Int J ChemTech Res 8(6):740–749
87. Vaidyanathan S (2015) Dynamics and control of tokamak system with symmetric and magnetically confined plasma. Int J ChemTech Res 8(6):795–803

88. Vaidyanathan S (2015) Global chaos synchronization of chemical chaotic reactors via novel sliding mode control method. Int J ChemTech Res 8(7):209–221
89. Vaidyanathan S (2015) Global chaos synchronization of the forced Van der Pol chaotic oscillators via adaptive control method. Int J PharmTech Res 8(6):156–166
90. Vaidyanathan S (2015) Hyperchaos, qualitative analysis, control and synchronisation of a ten-term 4-D hyperchaotic system with an exponential nonlinearity and three quadratic nonlinearities. Int J Model Identif Control 23:380–392
91. Vaidyanathan S (2015) Lotka–Volterra population biology models with negative feedback and their ecological monitoring. Int J PharmTech Res 8(5):974–981
92. Vaidyanathan S (2015) Lotka–Volterra two species competitive biology models and their ecological monitoring. Int J PharmTech Res 8(6):32–44
93. Vaidyanathan S (2015) Output regulation of the forced Van der Pol chaotic oscillator via adaptive control method. Int J PharmTech Res 8(6):106–116
94. Vaidyanathan S, Azar AT (2015) Analysis and control of a 4-D novel hyperchaotic system. Stud Comput Intell 581:3–17
95. Vaidyanathan S, Azar AT (2015) Analysis, control and synchronization of a nine-term 3-D novel chaotic system. In: Azar AT, Vaidyanathan S (eds) Chaos Model Control Syst Des, Stud Comput Intell, vol 581. Springer, Cham, pp 19–38
96. Vaidyanathan S, Madhavan K (2013) Analysis, adaptive control and synchronization of a seven-term novel 3-D chaotic system. Int J Control Theory Appl 6(2):121–137
97. Vaidyanathan S, Pakiriswamy S (2015) A 3-D novel conservative chaotic system and its generalized projective synchronization via adaptive control. J Eng Sci Technol Rev 8(2): 52–60
98. Vaidyanathan S, Sampath S (2012) Anti-synchronization of four-wing chaotic systems via sliding mode control. Int J Autom Comput 9(3):274–279
99. Vaidyanathan S, Volos C (2015) Analysis and adaptive control of a novel 3-D conservative no-equilibrium chaotic system. Arch Control Sci 25(3):333–353
100. Vaidyanathan S, Volos C, Pham VT (2014a) Hyperchaos, adaptive control and synchronization of a novel 5-D hyperchaotic system with three positive Lyapunov exponents and its SPICE implementation. Arch Control Sci 24:409–446
101. Vaidyanathan S, Volos C, Pham VT, Madhavan K, Idowu BA (2014b) Adaptive backstepping control, synchronization and circuit simulation of a 3-D novel jerk chaotic system with two hyperbolic sinusoidal nonlinearities. Arch Control Sci 24(3):375–403
102. Vaidyanathan S, Azar AT, Rajagopal K, Alexander P (2015) Design and SPICE implementation of a 12-term novel hyperchaotic system and its synchronisation via active control. Int J Model Identif Control 23:267–277
103. Vaidyanathan S, Pham VT, Volos CK (2015) A 5-D hyperchaotic Rikitake dynamo system with hidden attractors. Eur Phys J Spec Top 224:1575–1592
104. Vaidyanathan S, Rajagopal K, Volos CK, Kyprianidis IM, Stouboulos IN (2015) Analysis, adaptive control and synchronization of a seven-term novel 3-D chaotic system with three quadratic nonlinearities and its digital implementation in LabVIEW. J Eng Sci Technol Rev 8(2):130–141
105. Vaidyanathan S, Volos C, Pham VT (2015d) Analysis, control, synchronization and SPICE implementation of a novel 4-D hyperchaotic Rikitake dynamo system without equilibrium. J Eng Sci Technol Rev 8:232–244
106. Vaidyanathan S, Volos CK, Kyprianidis IM, Stouboulos IN, Pham VT (2015e) Analysis, adaptive control and anti-synchronization of a six-term novel jerk chaotic system with two exponential nonlinearities and its circuit simulation. J Eng Sci Technol Rev 8(2):24–36
107. Vaidyanathan S, Volos CK, Pham VT (2015f) Analysis, adaptive control and adaptive synchronization of a nine-term novel 3-D chaotic system with four quadratic nonlinearities and its circuit simulation. J Eng Sci Technol Rev 8(2):174–184
108. Vaidyanathan S, Volos CK, Pham VT (2015g) Global chaos control of a novel nine-term chaotic system via sliding mode control. In: Azar AT, Zhu Q (eds) Advances and Applications in Sliding Mode Control Systems, Studies in Computational Intelligence, vol 576. Springer, Cham, pp 571–590

109. Volos CK, Kyprianidis IM, Stouboulos IN (2013) Experimental investigation on coverage performance of a chaotic autonomous mobile robot. Robot Auton Syst 61(12):1314–1322
110. Volos CK, Kyprianidis IM, Stouboulos IN, Tlelo-Cuautle E, Vaidyanathan S (2015) Memristor: a new concept in synchronization of coupled neuromorphic circuits. J Eng Sci Technol Rev 8(2):157–173
111. Wei Z, Yang Q (2010) Anti-control of Hopf bifurcation in the new chaotic system with two stable node-foci. Appl Math Comput 217(1):422–429
112. Witte CL, Witte MH (1991) Chaos and predicting varix hemorrhage. Med Hypotheses 36(4):312–317
113. Yuan G, Zhang X, Wang Z (2014) Generation and synchronization of feedback-induced chaos in semiconductor ring lasers by injection-locking. Optik- Int J Light Electron Opt 125(8):1950–1953
114. Zaher AA, Abu-Rezq A (2011) On the design of chaos-based secure communication systems. Commun Nonlinear Syst Numer Simul 16(9):3721–3727
115. Zhang X, Zhao Z, Wang J (2014) Chaotic image encryption based on circular substitution box and key stream buffer. Signal Process: Image Commun 29(8):902–913
116. Zhou W, Xu Y, Lu H, Pan L (2008) On dynamics analysis of a new chaotic attractor. Phys Lett A 372(36):5773–5777
117. Zhu C, Liu Y, Guo Y (2010) Theoretic and numerical study of a new chaotic system. Intell Inf Manag 2:104–109
118. Zhu Q, Azar AT (2015) Complex system modelling and control through intelligent soft computations, studies in fuzzines and soft computing, vol 319. Springer, Berlin

Adaptive Control and Synchronization of a Rod-Type Plasma Torch Chaotic System via Backstepping Control Method

Sundarapandian Vaidyanathan

Abstract In this work, we first describe the Ghorui jerk chaotic system (2000) describing a strange attractor of a thermal arc plasma system based on triple convection theory. The phase portraits of the rod-type plasma torch chaotic system are displayed and the dynamic properties of the rod-type plasma torch chaotic system are discussed. We show that the rod-type plasma torch chaotic system has three unstable equilibrium points on the x_1-axis. The Lyapunov exponents of the rod-type plasma torch chaotic system are obtained as $L_1 = 0.3451$, $L_2 = 0$ and $L_3 = -1.3509$. Clearly, the Maximal Lyapunov Exponent (MLE) of the rod-type plasma torch chaotic system is given by $L_1 = 0.3451$. Since the sum of the Lyapunov exponents of the rod-type plasma torch chaotic system is negative, the chaotic system is dissipative. Also, the Kaplan–Yorke dimension of the rod-type plasma torch chaotic system is obtained as $D_{KY} = 2.2555$. Next, an adaptive backstepping controller is designed to globally stabilize the rod-type plasma torch chaotic system with unknown parameters. Moreover, an adaptive backstepping controller is also designed to achieve global chaos synchronization of the identical rod-type plasma torch chaotic systems with unknown parameters. The backstepping control method is a recursive procedure that links the choice of a Lyapunov function with the design of a controller and guarantees global asymptotic stability of strict feedback systems. MATLAB simulations have been shown to illustrate all the main results derived in this work.

Keywords Chaos · Chaotic systems · Jerk systems · Triple convection · Plasma torch system · Backstepping control · Adaptive control · Synchronization

S. Vaidyanathan (✉)
Research and Development Centre, Vel Tech University, Avadi, Chennai 600062, Tamil Nadu, India
e-mail: sundarvtu@gmail.com

S. Vaidyanathan and C. Volos (eds.), *Advances and Applications in Chaotic Systems*, Studies in Computational Intelligence 636,
DOI 10.1007/978-3-319-30279-9_24

1 Introduction

Chaos theory deals with the qualitative study of chaotic dynamical systems and their applications in science and engineering. A dynamical system is called *chaotic* if it satisfies the three properties: boundedness, infinite recurrence and sensitive dependence on initial conditions [3].

Some classical paradigms of 3-D chaotic systems in the literature are Lorenz system [18], Rössler system [30], ACT system [2], Sprott systems [38], Chen system [7], Lü system [19], Cai system [5], Tigan system [49], etc.

Many new chaotic systems have been discovered in the recent years such as Zhou system [130], Zhu system [131], Li system [15], Wei-Yang system [125], Sundarapandian systems [41, 46], Vaidyanathan systems [59, 60, 62–65, 68, 79, 80, 94, 97, 99, 108, 111, 113, 115, 117, 118], Pehlivan system [22], Sampath system [31], etc.

Chaos theory has applications in several fields of science and engineering such as chemical reactors [69, 73, 75, 77, 81, 85–87], biological systems [67, 70–72, 74, 76, 78, 82–84, 88–92], memristors [1, 23, 123], lasers [4], oscillations [50], robotics [11, 122], electrical circuits [20, 121], cryptosystems [29, 51], secure communications [127, 128], etc.

Many methods have been designed for control and regulation of chaotic systems such as active control [39, 40, 53], adaptive control [109, 116, 119], backstepping control [16, 124], sliding mode control [56, 58], etc.

Synchronization of chaotic systems is a phenomenon that occurs when two or more chaotic systems are coupled or when a chaotic system drives another chaotic system. Because of the butterfly effect which causes exponential divergence of the trajectories of two identical chaotic systems started with nearly the same initial conditions, the synchronization of chaotic systems is a challenging research problem in the chaos literature [3].

Pecora and Carroll pioneered the research on synchronization of chaotic systems with their seminal papers [6, 21]. The active control method [13, 32, 33, 45, 52, 57, 100, 101, 104] is typically used when the system parameters are available for measurement. Adaptive control method [34–36, 42–44, 55, 61, 93, 98, 102, 103, 110, 114] is typically used when some or all the system parameters are not available for measurement and estimates for the uncertain parameters of the systems.

Sampled-data feedback control method [9, 17, 126, 129] and time-delay feedback control method [8, 12, 37] are also used for synchronization of chaotic systems. Backstepping control method [24–28, 48, 105, 112, 120] is also used for the synchronization of chaotic systems. Backstepping control is a recursive method for stabilizing the origin of a control system in strict-feedback form [14]. Another popular method for the synchronization of chaotic systems is the sliding mode control method [47, 54, 66, 95, 96, 106, 107], which is a nonlinear control method that alters the dynamics of a nonlinear system by application of a discontinuous control signal that forces the system to “slide” along a cross-section of the system’s normal behavior.

In the recent decades, there is some good interest in finding jerk chaotic systems, which are described by the third-order ODE

$$\dddot{x} = j(x, \dot{x}, \ddot{x}) \tag{1}$$

The differential equation (1) is called "jerk system" because the third order time derivative in mechanical systems is called *jerk*.

By defining phase variables $x_1 = x$, $x_2 = \dot{x}$ and $x_3 = \ddot{x}$, the jerk differential equation (1) can be expressed as a 3-D system given by

$$\begin{cases} \dot{x}_1 = x_2 \\ \dot{x}_2 = x_3 \\ \dot{x}_3 = j(x_1, x_2, x_3) \end{cases} \tag{2}$$

Thermal plasma technology is of great importance in industry where it is applied in the manufacture of novel materials, eliminating poisonous waste, and in enabling secure and effective production. The efficiency of plasma technology in modern industry is affected mainly by the instruments used to produce plasma such as plasma fluctuations. In the recent decades, a diagrammatic plasma torch has been proposed for studying fluctuations in practical tests. Especially, it was shown that the inherent variations in plasma instruments can exhibit chaotic dynamical behaviour [10].

In this research work, we study an important jerk 3-D system, viz. Ghorui's rod-type plasma torch chaotic system [10] which is based on triple convection theory.

Also, we use backstepping control method for the global stabilization and synchronization of the Ghorui's rod-type plasma torch chaotic system.

In control theory, backstepping is a control technique used to derive control laws associated with an appropriate Lyapunov function, which guarantees the stability of nonlinear systems. The idea of backstepping control method is to recursively select some appropriate functions of state variables in pseudo-control inputs for lower dimension subsystems of the overall system. Each backstepping stage yields a new pseudo-control design, which is expressed in terms of the pseudo-control designs obtained from the previous design stages. Finally, a backstepping feedback controller is obtained when the backstepping design procedure is terminated. This backstepping feedback controller achieves the original design objective due to the final Lyapunov function and it is produced by summing up the Lyapunov functions associated with each individual design stage.

This research work is organized as follows. Section 2 describes the dynamic equations and phase portraits of the Ghorui's rod-type plasma torch chaotic system. Section 3 details the qualitative properties of the rod-type plasma torch chaotic system. We show that the rod-type plasma torch chaotic system has three unstable equilibrium points on the x_1-axis. The Lyapunov exponents of the rod-type plasma torch chaotic system are obtained as $L_1 = 0.3451$, $L_2 = 0$ and $L_3 = -1.3509$, while the Kaplan–Yorke dimension of the rod-type plasma torch chaotic system is obtained as $D_{KY} = 2.2555$.

In Sect. 4, we design an adaptive backstepping controller to globally stabilize the rod-type plasma torch chaotic system with unknown parameters. In Sect. 5, an adaptive backstepping controller is designed to achieve global chaos synchronization of the identical rod-type plasma torch chaotic systems with unknown parameters. Section 6 contains the conclusions of this work.

2 Rod-Plasma Torch 3-D Chaotic System

The following differential equation is considered for the modelling of a thermal arc plasma based on triple convection theory [10]:

$$\dddot{F} + \Omega_2 \ddot{F} + \Omega_1 \dot{F} + \Omega_0 F = \pm F^3 \tag{3}$$

Thermo-physical parameters such as the plasma torch tool, flow speed of plasma gas, and arc current determine the parameters of Eq. (3). To study the dynamical behaviour of the plasma torch, the coefficients in Eq. (3) are considered as in [10].

Thus, we rewrite the Ghorui's thermal arc plasma equation (3) as

$$\dddot{F} + \ddot{F} + b\dot{F} - aF = -F^3 \tag{4}$$

where a and b are constant, positive, parameters.

The state-space model of the rod-plasma torch equation (4) can be described as follows.

$$\begin{cases} \dot{x}_1 = x_2 \\ \dot{x}_2 = x_3 \\ \dot{x}_3 = ax_1 - bx_2 - x_3 - x_1^3 \end{cases} \tag{5}$$

where x_1, x_2, x_3 are the states and a, b are constant, positive parameters.

In [10], it was shown that the system (5) exhibits a *strange chaotic attractor* for the parameter values

$$a = 130, \quad b = 50 \tag{6}$$

For numerical simulations, we take the initial conditions as

$$x_1(0) = 0.1, \quad x_2(0) = 0.1, \quad x_3(0) = 0.1 \tag{7}$$

Figure 1 shows the 3-D phase portrait of the rod-type plasma torch chaotic system (5). Figures 2, 3 and 4 show the 2-D projection of the rod-type plasma torch chaotic system (5) on the (x_1, x_2), (x_2, x_3) and (x_1, x_3) planes, respectively.

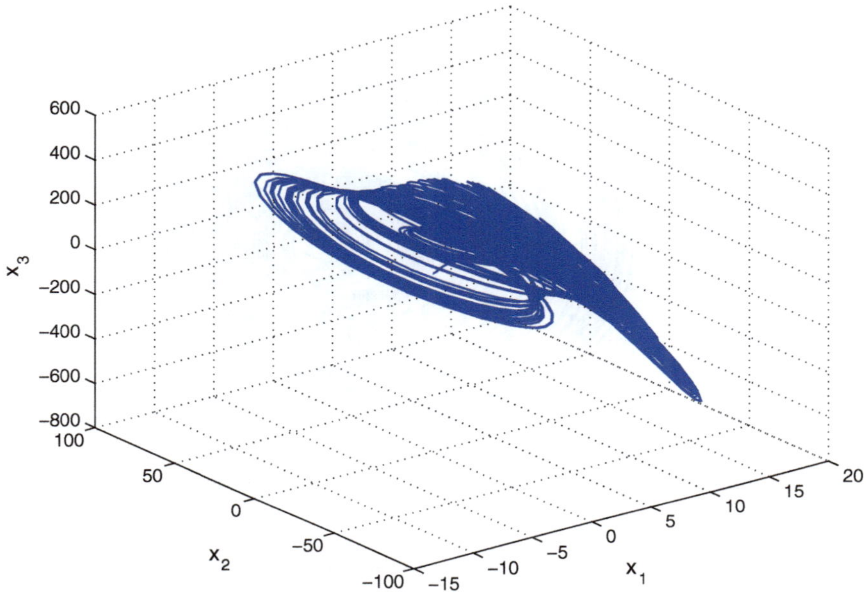

Fig. 1 3-D phase portrait of the rod-type plasma torch chaotic system

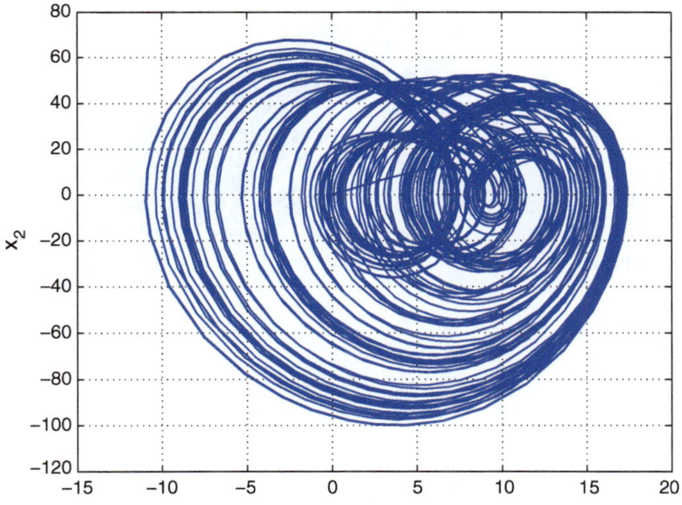

Fig. 2 2-D projection of the rod-type plasma torch chaotic system on the (x_1, x_2) plane

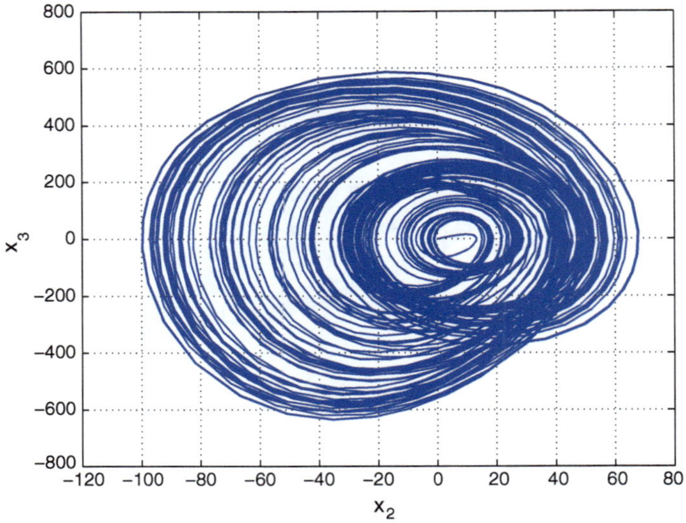

Fig. 3 2-D projection of the rod-type plasma torch chaotic system on the (x_2, x_3) plane

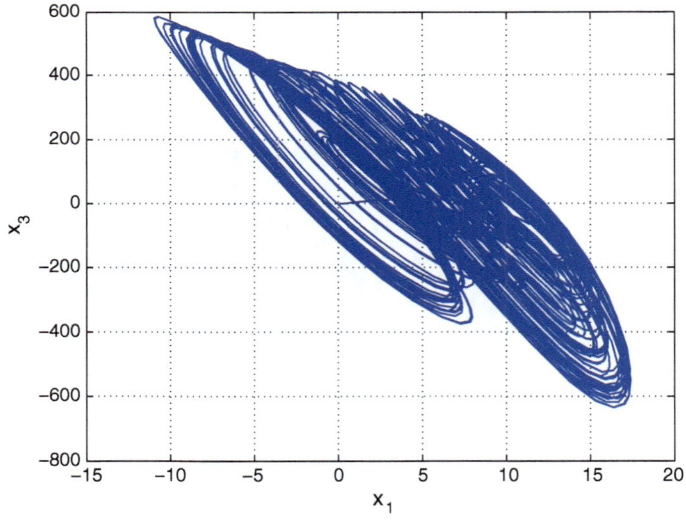

Fig. 4 2-D projection of the rod-type plasma torch chaotic system on the (x_1, x_3) plane

3 Analysis of the 3-D Rod-Type Plasma Torch Chaotic System

3.1 Dissipativity

In vector notation, the rod-type plasma torch system (5) can be expressed as

$$\dot{\mathbf{x}} = f(\mathbf{x}) = \begin{bmatrix} f_1(x_1, x_2, x_3) \\ f_2(x_1, x_2, x_3) \\ f_3(x_1, x_2, x_3) \end{bmatrix}, \tag{8}$$

where

$$\begin{cases} f_1(x_1, x_2, x_3) = x_2 \\ f_2(x_1, x_2, x_3) = x_3 \\ f_3(x_1, x_2, x_3) = ax_1 - bx_2 - x_3 - x_1^3 \end{cases} \tag{9}$$

Let Ω be any region in $\mathbf{R}^3$ with a smooth boundary and also, $\Omega(t) = \Phi_t(\Omega)$, where Φ_t is the flow of f. Furthermore, let $V(t)$ denote the volume of $\Omega(t)$.

By Liouville's theorem, we know that

$$\dot{V}(t) = \int\limits_{\Omega(t)} (\nabla \cdot f)\, dx_1\, dx_2\, dx_3 \tag{10}$$

The divergence of the novel jerk system (8) is found as:

$$\nabla \cdot f = \frac{\partial f_1}{\partial x_1} + \frac{\partial f_2}{\partial x_2} + \frac{\partial f_3}{\partial x_3} = -1 < 0 \tag{11}$$

since b is a positive parameter.

Inserting the value of $\nabla \cdot f$ from (11) into (10), we get

$$\dot{V}(t) = \int\limits_{\Omega(t)} (-b)\, dx_1\, dx_2\, dx_3 = -V(t) \tag{12}$$

Integrating the first order linear differential equation (12), we get

$$V(t) = \exp(-t)V(0) \tag{13}$$

Since $b > 0$, it follows from Eq. (13) that $V(t) \to 0$ exponentially as $t \to \infty$. This shows that the rod-type plasma torch chaotic system (5) is dissipative. Hence, the system limit sets are ultimately confined into a specific limit set of zero volume, and the asymptotic motion of the rod-type plasma torch chaotic system (5) settles onto a strange attractor of the system.

3.2 Equilibrium Points

The equilibrium points of the rod-type plasma torch chaotic system (5) are obtained by solving the equations

$$\begin{cases} f_1(x_1, x_2, x_3) = x_2 & = 0 \\ f_2(x_1, x_2, x_3) = x_3 & = 0 \\ f_3(x_1, x_2, x_3) = ax_1 - bx_2 - x_3 - x_1^3 & = 0 \end{cases} \tag{14}$$

We take the parameter values as in the chaotic case, viz. $a = 130$ and $b = 50$.

Solving the Eq. (14), we get three equilibrium points of the rod-type plasma torch chaotic system (5) as

$$E_0 = \begin{bmatrix} 0 \\ 0 \\ 0 \end{bmatrix}, \quad E_1 = \begin{bmatrix} \sqrt{130} \\ 0 \\ 0 \end{bmatrix}, \quad E_2 = \begin{bmatrix} -\sqrt{130} \\ 0 \\ 0 \end{bmatrix} \tag{15}$$

To test the stability type of the equilibrium points, we calculate the Jacobian matrix of the rod-type plasma torch chaotic system (5) at any point x:

$$J(x) = \begin{bmatrix} 0 & 1 & 0 \\ 0 & 0 & 1 \\ 130 - 3x_1^2 & -50 & -1 \end{bmatrix} \tag{16}$$

We find that

$$J_0 \triangleq J(E_0) = \begin{bmatrix} 0 & 1 & 0 \\ 0 & 0 & 1 \\ 130 & -50 & -1 \end{bmatrix} \tag{17}$$

The matrix J_0 has the eigenvalues

$$\lambda_1 = 2.2650, \quad \lambda_{2,3} = -1.6325 \pm 7.3980\, i \tag{18}$$

This shows that the equilibrium point E_0 is a saddle-focus, which is unstable.

Next, we find that

$$J_1 \triangleq J(E_0) = \begin{bmatrix} 0 & 1 & 0 \\ 0 & 0 & 1 \\ -260 & -50 & -1 \end{bmatrix} \tag{19}$$

The matrix J_1 has the eigenvalues

$$\lambda_1 = -4.1312, \quad \lambda_{2,3} = 1.5656 \pm 7.7772\, i \tag{20}$$

This shows that the equilibrium point E_1 is a saddle-focus, which is unstable.

We also find that

$$J_2 \triangleq J(E_0) = \begin{bmatrix} 0 & 1 & 0 \\ 0 & 0 & 1 \\ -260 & -50 & -1 \end{bmatrix} \tag{21}$$

The matrix J_2 has the eigenvalues

$$\lambda_1 = -4.1312, \quad \lambda_{2,3} = 1.5656 \pm 7.7772\, i \tag{22}$$

This shows that the equilibrium point E_2 is a saddle-focus, which is unstable.

Thus, the rod-type plasma torch chaotic system (5) has three unstable equilibrium points on the x_1-axis.

3.3 *Invariance*

The rod-type plasma torch chaotic system (5) is invariant under the coordinates transformation

$$(x_1, x_2, x_3) \mapsto (-x_1, -x_2, -x_3) \tag{23}$$

This shows that the rod-type plasma torch chaotic system (5) has point-reflection symmetry about the origin and every non-trivial trajectory of the system (5) must have a twin trajectory.

3.4 *Lyapunov Exponents and Kaplan–Yorke Dimension*

We take the parameter values of the rod-type plasma torch system (5) as $a = 130$ and $b = 50$. We take the initial state of the rod-type plasma torch system (5) as given in (7).

Then the Lyapunov exponents of the rod-type plasma torch system (5) are numerically obtained using MATLAB as

$$L_1 = 0.3451, \quad L_2 = 0, \quad L_3 = -1.3509 \tag{24}$$

Thus, the maximal Lyapunov exponent (MLE) of the rod-type plasma torch system (5) is positive, which means that the system has a chaotic behavior.

Since $L_1 + L_2 + L_3 = -1.0058 < 0$, it follows that the rod-type plasma torch chaotic system (5) is dissipative.

Also, the Kaplan–Yorke dimension of the rod-type plasma torch chaotic system (5) is obtained as

$$D_{KY} = 2 + \frac{L_1 + L_2}{|L_3|} = 2.2555 \tag{25}$$

4 Adaptive Control of the Rod-Type Plasma Torch Chaotic System

In this section, we use backstepping control method to derive an adaptive feedback control law for globally stabilizing the rod-type plasma torch chaotic system with unknown parameters.

Thus, we consider the rod-type plasma torch chaotic system given by

$$\begin{cases} \dot{x}_1 = x_2 \\ \dot{x}_2 = x_3 \\ \dot{x}_3 = ax_1 - bx_2 - x_3 - x_1^3 + u \end{cases} \tag{26}$$

In (26), x_1, x_2, x_3 are the states, a, b are unknown constant parameters, and u is a backstepping control law to be determined using estimates $\hat{a}(t)$ and $\hat{b}(t)$ for the unknown parameters a and b, respectively.

The parameter estimation errors are defined as:

$$\begin{cases} e_a(t) = a - \hat{a}(t) \\ e_b(t) = b - \hat{b}(t) \end{cases} \tag{27}$$

Differentiating (27) with respect to t, we obtain the following equations:

$$\begin{cases} \dot{e}_a(t) = -\dot{\hat{a}}(t) \\ \dot{e}_b(t) = -\dot{\hat{b}}(t) \end{cases} \tag{28}$$

Next, we shall state and prove the main result of this section.

Theorem 1 *The 3-D rod-type plasma torch chaotic system (26), with unknown parameters a and b, is globally and exponentially stabilized by the adaptive feedback control law,*

$$u(t) = -\left[3 + \hat{a}(t)\right] x_1 - \left[5 - \hat{b}(t)\right] x_2 - 2x_3 + x_1^3 - kz_3 \tag{29}$$

where $k > 0$ is a gain constant,

$$z_3 = 2x_1 + 2x_2 + x_3, \tag{30}$$

and the update law for the parameter estimates $\hat{a}(t), \hat{b}(t))$ is given by

$$\begin{cases} \dot{\hat{a}}(t) = x_1 z_3 \\ \dot{\hat{b}}(t) = -x_2 z_3 \end{cases} \tag{31}$$

Proof We prove this result via Lyapunov stability theory [14].

First, we define a quadratic Lyapunov function

$$V_1(z_1) = \frac{1}{2}\, z_1^2 \tag{32}$$

where

$$z_1 = x_1 \tag{33}$$

Differentiating V_1 along the dynamics (26), we get

$$\dot{V}_1 = z_1 \dot{z}_1 = x_1 x_2 = -z_1^2 + z_1(x_1 + x_2) \tag{34}$$

Now, we define

$$z_2 = x_1 + x_2 \tag{35}$$

Using (35), we can simplify the Eq. (34) as

$$\dot{V}_1 = -z_1^2 + z_1 z_2 \tag{36}$$

Secondly, we define a quadratic Lyapunov function

$$V_2(z_1, z_2) = V_1(z_1) + \frac{1}{2}\, z_2^2 = \frac{1}{2}\left(z_1^2 + z_2^2\right) \tag{37}$$

Differentiating V_2 along the dynamics (26), we get

$$\dot{V}_2 = -z_1^2 - z_2^2 + z_2(2x_1 + 2x_2 + x_3) \tag{38}$$

Now, we define

$$z_3 = 2x_1 + 2x_2 + x_3 \tag{39}$$

Using (39), we can simplify the Eq. (38) as

$$\dot{V}_2 = -z_1^2 - z_2^2 + z_2 z_3 \tag{40}$$

Finally, we define a quadratic Lyapunov function

$$V(z_1, z_2, z_3, e_a, e_b) = V_2(z_1, z_2) + \frac{1}{2} z_3^2 + \frac{1}{2} e_a^2 + \frac{1}{2} e_b^2 \tag{41}$$

Differentiating V along the dynamics (26), we get

$$\dot{V} = -z_1^2 - z_2^2 - z_3^2 + z_3(z_3 + z_2 + \dot{z}_3) - e_a\dot{\hat{a}} - e_b\dot{\hat{b}} \tag{42}$$

Equation (42) can be written compactly as

$$\dot{V} = -z_1^2 - z_2^2 - z_3^2 + z_3 S - e_a\dot{\hat{a}} - e_b\dot{\hat{b}} \tag{43}$$

where

$$S = z_3 + z_2 + \dot{z}_3 = z_3 + z_2 + 2\dot{x}_1 + 2\dot{x}_2 + \dot{x}_3 \tag{44}$$

A simple calculation gives

$$S = (3 + a)x_1 + (5 - b)x_2 + 2x_3 - x_1^3 + u \tag{45}$$

Substituting the adaptive control law (29) into (45), we obtain

$$S = \left[a - \hat{a}(t)\right] x_1 - \left[b - \hat{b}(t)\right] x_2 - kz_3 \tag{46}$$

Using the definitions (28), we can simplify (46) as

$$S = e_a x_1 - e_b x_2 - kz_3 \tag{47}$$

Substituting the value of S from (47) into (43), we obtain

$$\dot{V} = -z_1^2 - z_2^2 - (1 + k)z_3^2 + e_a\left(x_1 z_3 - \dot{\hat{a}}\right) + e_b\left(-x_2 z_3 - \dot{\hat{b}}\right) \tag{48}$$

Substituting the update law (31) into (48), we get

$$\dot{V} = -z_1^2 - z_2^2 - (1 + k)z_3^2, \tag{49}$$

which is a negative semi-definite function on $\mathbf{R}^5$.

From (49), it follows that the vector $\mathbf{z}(t) = (z_1(t), z_2(t), z_3(t))$ and the parameter estimation error $(e_a(t), e_b(t))$ are globally bounded, i.e.

$$\left[z_1(t)\ z_2(t)\ z_3(t)\ e_a(t)\ e_b(t)\right] \in \mathbf{L}_\infty \tag{50}$$

Also, it follows from (49) that

$$\dot{V} \le -z_1^2 - z_2^2 - z_3^2 = -\|\mathbf{z}\|^2 \tag{51}$$

That is,

$$\|\mathbf{z}\|^2 \le -\dot{V} \tag{52}$$

Integrating the inequality (52) from 0 to t, we get

$$\int_0^t |\mathbf{z}(\tau)|^2 \, d\tau \le V(0) - V(t) \tag{53}$$

From (53), it follows that $\mathbf{z}(t) \in \mathbf{L}_2$.

From Eq. (26), it can be deduced that $\dot{\mathbf{z}}(t) \in \mathbf{L}_\infty$.

Thus, using Barbalat's lemma [14], we conclude that $\mathbf{z}(t) \to \mathbf{0}$ exponentially as $t \to \infty$ for all initial conditions $\mathbf{z}(0) \in \mathbf{R}^3$.

Hence, it is immediate that $\mathbf{x}(t) \to \mathbf{0}$ exponentially as $t \to \infty$ for all initial conditions $\mathbf{x}(0) \in \mathbf{R}^3$.

This completes the proof. ■

For the numerical simulations, the classical fourth-order Runge–Kutta method with step size $h = 10^{-8}$ is used to solve the system of differential equations (26) and (31), when the adaptive control law (29) is applied.

The parameter values of the novel jerk chaotic system (26) are taken as

$$a = 130, \quad b = 50 \tag{54}$$

We take the positive gain constant as $k = 10$.

Furthermore, as initial conditions of the novel rod-type plasma torch chaotic system (26), we take

$$x_1(0) = 12.6, \quad x_2(0) = 8.2, \quad x_3(0) = 6.3 \tag{55}$$

Also, as initial conditions of the parameter estimates $\hat{a}(t)$ and $\hat{b}(t)$, we take

$$\hat{a}(0) = 15.9, \quad \hat{b}(0) = 26.7 \tag{56}$$

In Fig. 5, the exponential convergence of the controlled states $x_1(t), x_2(t), x_3(t)$ is depicted, when the adaptive control law (29) and (31) are implemented.

5 Adaptive Synchronization of the Identical Rod-Type Plasma Torch Chaotic Systems

In this section, we use backstepping control method to derive an adaptive control law for globally and exponentially synchronizing the identical 3-D novel rod-type plasma torch chaotic systems with unknown parameters.

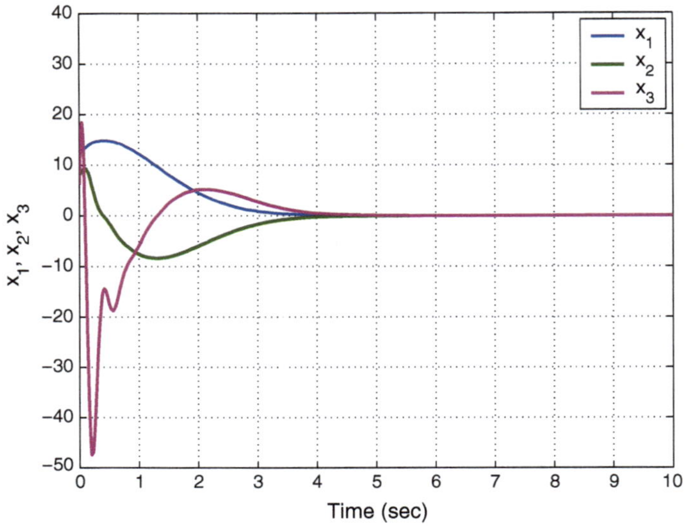

Fig. 5 Time-history of the controlled states x_1, x_2, x_3

As the master system, we consider the rod-type plasma torch chaotic system given by

$$\begin{cases} \dot{x}_1 = x_2 \\ \dot{x}_2 = x_3 \\ \dot{x}_3 = ax_1 - bx_2 - x_3 - x_1^3 \end{cases} \tag{57}$$

where x_1, x_2, x_3 are the states of the system, and a, b are unknown, constant parameters.

As the slave system, we consider the rod-type plasma torch chaotic system given by

$$\begin{cases} \dot{y}_1 = y_2 \\ \dot{y}_2 = y_3 \\ \dot{y}_3 = ay_1 - by_2 - y_3 - y_1^3 + u \end{cases} \tag{58}$$

where y_1, y_2, y_3 are the states of the system, and u is a backstepping control to be determined using estimates $\hat{a}(t)$ and $\hat{b}(t)$ for the unknown parameters a and b, respectively.

We define the synchronization error between the states of the master system (57) and the slave system (58) as

$$\begin{cases} e_1 = y_1 - x_1 \\ e_2 = y_2 - x_2 \\ e_3 = y_3 - x_3 \end{cases} \tag{59}$$

Then the error dynamics is easily obtained as

$$\begin{cases} \dot{e}_1 = e_2 \\ \dot{e}_2 = e_3 \\ \dot{e}_3 = ae_1 - be_2 - e_3 - y_1^3 + x_1^3 + u \end{cases} \tag{60}$$

The parameter estimation errors are defined as:

$$\begin{cases} e_a(t) = a - \hat{a}(t) \\ e_b(t) = b - \hat{b}(t) \end{cases} \tag{61}$$

Differentiating (61) with respect to t, we obtain

$$\begin{cases} \dot{e}_a(t) = -\dot{\hat{a}}(t) \\ \dot{e}_b(t) = -\dot{\hat{b}}(t) \end{cases} \tag{62}$$

Theorem 2 *The identical 3-D novel rod-type plasma torch chaotic systems (57) and (58) with unknown parameters a and b are globally and exponentially synchronized by the adaptive control law*

$$u = -\left[3 + \hat{a}(t)\right] e_1 - \left[5 - \hat{b}(t)\right] e_2 - 2e_3 + y_1^3 - x_1^3 - kz_3 \tag{63}$$

where $k > 0$ is a gain constant,

$$z_3 = 2e_1 + 2e_2 + e_3, \tag{64}$$

and the update law for the parameter estimates $\hat{a}(t)$, $\hat{b}(t)$ is given by

$$\begin{cases} \dot{\hat{a}}(t) = e_1 z_3 \\ \dot{\hat{b}}(t) = -e_2 z_3 \end{cases} \tag{65}$$

Proof First, we define a quadratic Lyapunov function

$$V_1(z_1) = \frac{1}{2} z_1^2 \tag{66}$$

where

$$z_1 = e_1 \tag{67}$$

Differentiating V_1 along the error dynamics (60), we get

$$\dot{V}_1 = z_1 \dot{z}_1 = e_1 e_2 = -z_1^2 + z_1(e_1 + e_2) \tag{68}$$

Now, we define

$$z_2 = e_1 + e_2 \tag{69}$$

Using (69), we can simplify the Eq. (68) as

$$\dot{V}_1 = -z_1^2 + z_1 z_2 \tag{70}$$

Secondly, we define a quadratic Lyapunov function

$$V_2(z_1, z_2) = V_1(z_1) + \frac{1}{2} z_2^2 = \frac{1}{2} \left(z_1^2 + z_2^2 \right) \tag{71}$$

Differentiating V_2 along the error dynamics (60), we get

$$\dot{V}_2 = -z_1^2 - z_2^2 + z_2(2e_1 + 2e_2 + e_3) \tag{72}$$

Now, we define

$$z_3 = 2e_1 + 2e_2 + e_3 \tag{73}$$

Using (73), we can simplify the Eq. (72) as

$$\dot{V}_2 = -z_1^2 - z_2^2 + z_2 z_3 \tag{74}$$

Finally, we define a quadratic Lyapunov function

$$V(z_1, z_2, z_3, e_a, e_b) = V_2(z_1, z_2) + \frac{1}{2} z_3^2 + \frac{1}{2} e_a^2 + \frac{1}{2} e_b^2 \tag{75}$$

Differentiating V along the error dynamics (60), we get

$$\dot{V} = -z_1^2 - z_2^2 - z_3^2 + z_3(z_3 + z_2 + \dot{z}_3) - e_a \dot{\hat{a}} - e_b \dot{\hat{b}} \tag{76}$$

Equation (76) can be written compactly as

$$\dot{V} = -z_1^2 - z_2^2 - z_3^2 + z_3 S - e_a \dot{\hat{a}} - e_b \dot{\hat{b}} \tag{77}$$

where

$$S = z_3 + z_2 + \dot{z}_3 = z_3 + z_2 + 2\dot{e}_1 + 2\dot{e}_2 + \dot{e}_3 \tag{78}$$

A simple calculation gives

$$S = (3 + a)e_1 + (5 - b)e_2 + 2e_3 - y_1^3 + x_1^3 + u \tag{79}$$

Substituting the adaptive control law (63) into (79), we obtain

$$S = \left[a - \hat{a}(t)\right] e_1 - \left[b - \hat{b}(t)\right] e_2 - kz_3 \tag{80}$$

Using the definitions (62), we can simplify (80) as

$$S = e_a e_1 - e_b e_2 - kz_3 \tag{81}$$

Substituting the value of S from (81) into (77), we obtain

$$\dot{V} = -z_1^2 - z_2^2 - (1+k)z_3^2 + e_a \left[e_1 z_3 - \dot{\hat{a}}\right] + e_b \left[-e_2 z_3 - \dot{\hat{b}}\right] \tag{82}$$

Substituting the update law (65) into (82), we get

$$\dot{V} = -z_1^2 - z_2^2 - (1+k)z_3^2, \tag{83}$$

which is a negative semi-definite function on $\mathbf{R}^5$.

From (83), it follows that the vector $\mathbf{z}(t) = (z_1(t), z_2(t), z_3(t))$ and the parameter estimation error $(e_a(t), e_b(t))$ are globally bounded, i.e.

$$\left[z_1(t)\ z_2(t)\ z_3(t)\ e_a(t)\ e_b(t) \right] \in \mathbf{L}_\infty \tag{84}$$

Also, it follows from (83) that

$$\dot{V} \leq -z_1^2 - z_2^2 - z_3^2 = -\|\mathbf{z}\|^2 \tag{85}$$

That is,

$$\|\mathbf{z}\|^2 \leq -\dot{V} \tag{86}$$

Integrating the inequality (86) from 0 to t, we get

$$\int_0^t |\mathbf{z}(\tau)|^2 \, d\tau \leq V(0) - V(t) \tag{87}$$

From (87), it follows that $\mathbf{z}(t) \in \mathbf{L}_2$.

From Eq. (60), it can be deduced that $\dot{\mathbf{z}}(t) \in \mathbf{L}_\infty$.

Thus, using Barbalat's lemma [14], we conclude that $\mathbf{z}(t) \to \mathbf{0}$ exponentially as $t \to \infty$ for all initial conditions $\mathbf{z}(0) \in \mathbf{R}^3$.

Hence, it is immediate that $\mathbf{e}(t) \to \mathbf{0}$ exponentially as $t \to \infty$ for all initial conditions $\mathbf{e}(0) \in \mathbf{R}^3$.

This completes the proof. ■

For the numerical simulations, the classical fourth-order Runge–Kutta method with step size $h = 10^{-8}$ is used to solve the system of rod-type plasma torch chaotic systems, which are taken as the master and slave systems.

The parameter values of the rod-type plasma torch chaotic systems are taken as in the chaotic case, i.e. $a = 130$ and $b = 50$.

We take the positive gain constant as $k = 10$.

Furthermore, as initial conditions of the master system (57), we take

$$x_1(0) = 7.2, \quad x_2(0) = 4.9, \quad x_3(0) = -10.1 \tag{88}$$

As initial conditions of the slave system (58), we take

$$y_1(0) = -14.8, \quad y_2(0) = -16.5, \quad y_3(0) = 8.3 \tag{89}$$

Also, as initial conditions of the parameter estimates $\hat{a}(t)$ and $\hat{b}(t)$, we take

$$\hat{a}(0) = 12.7, \quad \hat{b}(0) = 28.5 \tag{90}$$

In Figs. 6, 7 and 8, the complete synchronization of the identical 3-D rod-type plasma torch chaotic systems (57) and (58) is depicted.

Also, in Fig. 9, the time-history of the synchronization errors $e_1(t)$, $e_2(t)$, $e_3(t)$, is depicted.

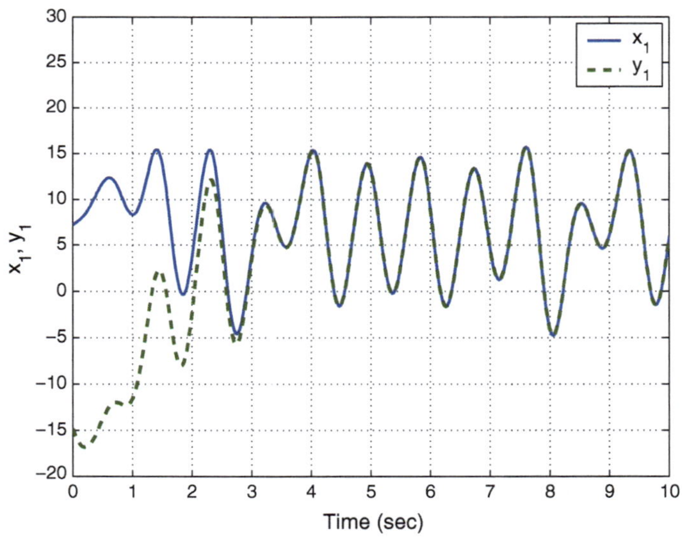

Fig. 6 Synchronization of the states x_1 and y_1

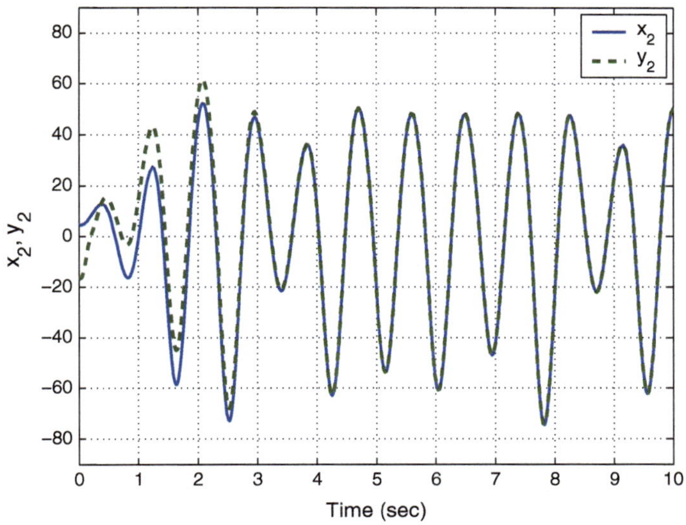

Fig. 7 Synchronization of the states x_2 and y_2

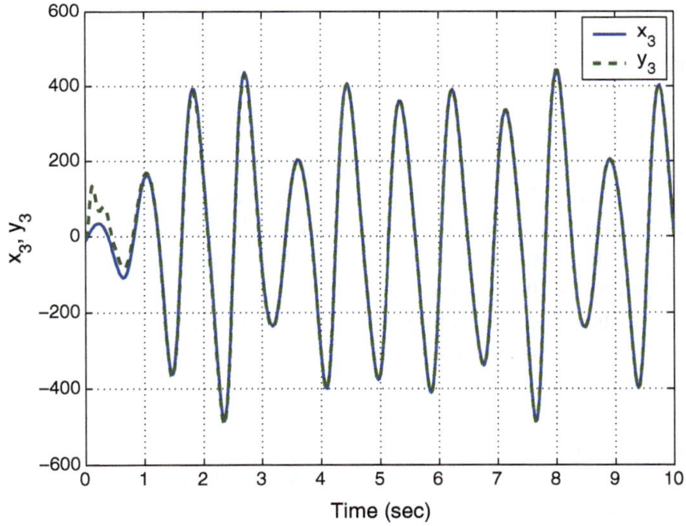

Fig. 8 Synchronization of the states x_3 and y_3

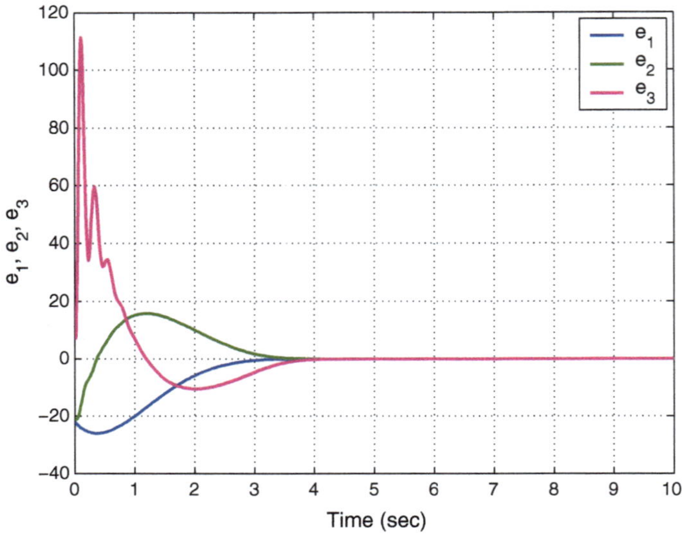

Fig. 9 Time-history of the synchronization errors e_1, e_2, e_3

6 Conclusions

In this work, we first described the Ghorui jerk chaotic system (2000) describing a strange attractor of a thermal arc plasma system based on triple convection theory. The phase portraits of the rod-type plasma torch chaotic system were displayed and the dynamic properties of the rod-type plasma torch chaotic system were discussed. We established that the rod-type plasma torch chaotic system has three unstable equilibrium points on the x_1-axis. The Lyapunov exponents of the rod-type plasma torch chaotic system have been obtained as $L_1 = 0.3451, L_2 = 0$ and $L_3 = -1.3509$. Since the sum of the Lyapunov exponents of the rod-type plasma torch chaotic system is negative, the chaotic system is dissipative. Also, the Kaplan–Yorke dimension of the rod-type plasma torch chaotic system has been derived as $D_{KY} = 2.2555$. Next, an adaptive backstepping controller was designed to globally stabilize the rod-type plasma torch chaotic system with unknown parameters. Moreover, an adaptive backstepping controller was also designed to achieve global chaos synchronization of the identical rod-type plasma torch chaotic systems with unknown parameters. MATLAB simulations were shown to illustrate all the main results derived in this work.

References

1. Abdurrahman A, Jiang H, Teng Z (2015) Finite-time synchronization for memristor-based neural networks with time-varying delays. Neural Netw 69:20–28
2. Arneodo A, Coullet P, Tresser C (1981) Possible new strange attractors with spiral structure. Commun Math Phys 79(4):573–576
3. Azar AT, Vaidyanathan S (2015) Chaos modeling and control systems design, vol 581. Springer, Germany
4. Behnia S, Afrang S, Akhshani A, Mabhouti K (2013) A novel method for controlling chaos in external cavity semiconductor laser. Optik 124(8):757–764
5. Cai G, Tan Z (2007) Chaos synchronization of a new chaotic system via nonlinear control. J Uncertain Syst 1(3):235–240
6. Carroll TL, Pecora LM (1991) Synchronizing chaotic circuits. IEEE Trans Circuits Syst 38(4):453–456
7. Chen G, Ueta T (1999) Yet another chaotic attractor. Int J Bifurc Chaos 9(7):1465–1466
8. Chen WH, Wei D, Lu X (2014) Global exponential synchronization of nonlinear time-delay Lur'e systems via delayed impulsive control. Commun Nonlinear Sci Numer Simul 19(9):3298–3312
9. Gan Q, Liang Y (2012) Synchronization of chaotic neural networks with time delay in the leakage term and parametric uncertainties based on sampled-data control. J Frankl Inst 349(6):1955–1971
10. Ghorui S, Sahasrabudhe SN, Muryt PSS, Das AK, Venkatramani N (2000) Experimental evidence of chaotic behavior in atmosphere pressure arc discharge. IEEE Trans Plasma Sci 28(1):253–260
11. Islam MM, Murase K (2005) Chaotic dynamics of a behaviour-based miniature mobile robot: effects of environment and control structure. Neural Netw 18(2):123–144
12. Jiang GP, Zheng WX, Chen G (2004) Global chaos synchronization with channel time-delay. Chaos Solitons Fractals 20(2):267–275
13. Karthikeyan R, Sundarapandian V (2014) Hybrid chaos synchronization of four-scroll systems via active control. J Electr Eng 65(2):97–103
14. Khalil HK (2001) Nonlinear systems, 3rd edn. Prentice Hall, New Jersey
15. Li D (2008) A three-scroll chaotic attractor. Phys Lett A 372(4):387–393
16. Li GH, Zhou SP, Yang K (2007) Controlling chaos in Colpitts oscillator. Chaos Solitons Fractals 33:582–587
17. Li N, Zhang Y, Nie Z (2011) Synchronization for general complex dynamical networks with sampled-data. Neurocomputing 74(5):805–811
18. Lorenz EN (1963) Deterministic periodic flow. J Atmos Sci 20(2):130–141
19. Lü J, Chen G (2002) A new chaotic attractor coined. Int J Bifurc Chaos 12(3):659–661
20. Matouk AE (2011) Chaos, feedback control and synchronization of a fractional-order modified autonomous Van der Pol-Duffing circuit. Commun Nonlinear Sci Numer Simul 16(2):975–986
21. Pecora LM, Carroll TL (1990) Synchronization in chaotic systems. Phys Rev Lett 64(8):821–824
22. Pehlivan I, Moroz IM, Vaidyanathan S (2014) Analysis, synchronization and circuit design of a novel butterfly attractor. J Sound Vib 333(20):5077–5096
23. Pham VT, Volos CK, Vaidyanathan S, Le TP, Vu VY (2015) A memristor-based hyperchaotic system with hidden attractors: dynamics, synchronization and circuital emulating. J Eng Sci Technol Rev 8(2):205–214
24. Rasappan S, Vaidyanathan S (2012) Global chaos synchronization of WINDMI and Coullet chaotic systems by backstepping control. Far East J Math Sci 67(2):265–287
25. Rasappan S, Vaidyanathan S (2012) Hybrid synchronization of n-scroll Chua and Lur'e chaotic systems via backstepping control with novel feedback. Arch Control Sci 22(3):343–365
26. Rasappan S, Vaidyanathan S (2012) Synchronization of hyperchaotic Liu system via backstepping control with recursive feedback. Commun Comput Inf Sci 305:212–221

27. Rasappan S, Vaidyanathan S (2013) Hybrid synchronization of n-scroll chaotic Chua circuits using adaptive backstepping control design with recursive feedback. Malays J Math Sci 7(2):219–246
28. Rasappan S, Vaidyanathan S (2014) Global chaos synchronization of WINDMI and Coullet chaotic systems using adaptive backstepping control design. Kyungpook Math J 54(1):293–320
29. Rhouma R, Belghith S (2011) Cryptanalysis of a chaos-based cryptosystem. Commun Nonlinear Sci Numer Simul 16(2):876–884
30. Rössler OE (1976) An equation for continuous chaos. Phys Lett A 57(5):397–398
31. Sampath S, Vaidyanathan S, Volos CK, Pham VT (2015) An eight-term novel four-scroll chaotic system with cubic nonlinearity and its circuit simulation. J Eng Sci Technol Rev 8(2):1–6
32. Sarasu P, Sundarapandian V (2011) Active controller design for the generalized projective synchronization of four-scroll chaotic systems. Int J Syst Signal Control Eng Appl 4(2):26–33
33. Sarasu P, Sundarapandian V (2011) The generalized projective synchronization of hyperchaotic Lorenz and hyperchaotic Qi systems via active control. Int J Soft Comput 6(5):216–223
34. Sarasu P, Sundarapandian V (2012) Adaptive controller design for the generalized projective synchronization of 4-scroll systems. Int J Syst Signal Control Eng Appl 5(2):21–30
35. Sarasu P, Sundarapandian V (2012) Generalized projective synchronization of three-scroll chaotic systems via adaptive control. Eur J Sci Res 72(4):504–522
36. Sarasu P, Sundarapandian V (2012) Generalized projective synchronization of two-scroll systems via adaptive control. Int J Soft Comput 7(4):146–156
37. Shahverdiev EM, Shore KA (2009) Impact of modulated multiple optical feedback time delays on laser diode chaos synchronization. Opt Commun 282(17):3568–3572
38. Sprott JC (1994) Some simple chaotic flows. Phys Rev E 50(2):647–650
39. Sundarapandian V (2010) Output regulation of the Lorenz attractor. Far East J Math Sci 42(2):289–299
40. Sundarapandian V (2011) Output regulation of the Arneodo-Coullet chaotic system. Commun Comput Inf Sci 133:98–107
41. Sundarapandian V (2013) Analysis and anti-synchronization of a novel chaotic system via active and adaptive controllers. J Eng Sci Technol Rev 6(4):45–52
42. Sundarapandian V, Karthikeyan R (2011) Anti-synchronization of hyperchaotic Lorenz and hyperchaotic Chen systems by adaptive control. Int J Syst Signal Control Eng Appl 4(2):18–25
43. Sundarapandian V, Karthikeyan R (2011) Anti-synchronization of Lü and Pan chaotic systems by adaptive nonlinear control. Eur J Sci Res 64(1):94–106
44. Sundarapandian V, Karthikeyan R (2012) Adaptive anti-synchronization of uncertain Tigan and Li systems. J Eng Appl Sci 7(1):45–52
45. Sundarapandian V, Karthikeyan R (2012) Hybrid synchronization of hyperchaotic Lorenz and hyperchaotic Chen systems via active control. J Eng Appl Sci 7(3):254–264
46. Sundarapandian V, Pehlivan I (2012) Analysis, control, synchronization, and circuit design of a novel chaotic system. Math Comput Modell 55(7–8):1904–1915
47. Sundarapandian V, Sivaperumal S (2011) Sliding controller design of hybrid synchronization of four-wing chaotic systems. Int J Soft Comput 6(5):224–231
48. Suresh R, Sundarapandian V (2013) Global chaos synchronization of a family of n-scroll hyperchaotic Chua circuits using backstepping control with recursive feedback. Far East J Math Sci 73(1):73–95
49. Tigan G, Opris D (2008) Analysis of a 3D chaotic system. Chaos Solitons Fractals 36:1315–1319
50. Tuwankotta JM (2006) Chaos in a coupled oscillators system with widely spaced frequencies and energy-preserving non-linearity. Int J Non-Linear Mech 41(2):180–191
51. Usama M, Khan MK, Alghathbar K, Lee C (2010) Chaos-based secure satellite imagery cryptosystem. Comput Math Appl 60(2):326–337

52. Vaidyanathan S (2011) Hybrid chaos synchronization of Liu and Lü systems by active nonlinear control. Commun Comput Inf Sci 204:1–10
53. Vaidyanathan S (2011) Output regulation of the unified chaotic system. Commun Comput Inf Sci 204:84–93
54. Vaidyanathan S (2012) Analysis and synchronization of the hyperchaotic Yujun systems via sliding mode control. Adv Intell Syst Comput 176:329–337
55. Vaidyanathan S (2012) Anti-synchronization of Sprott-L and Sprott-M chaotic systems via adaptive control. Int J Control Theory Appl 5(1):41–59
56. Vaidyanathan S (2012) Global chaos control of hyperchaotic Liu system via sliding control method. Int J Control Theory Appl 5(2):117–123
57. Vaidyanathan S (2012) Output regulation of the Liu chaotic system. Appl Mech Mater 110–116:3982–3989
58. Vaidyanathan S (2012) Sliding mode control based global chaos control of Liu-Liu-Liu-Su chaotic system. Int J Control Theory Appl 5(1):15–20
59. Vaidyanathan S (2013) A new six-term 3-D chaotic system with an exponential nonlinearity. Far East J Math Sci 79(1):135–143
60. Vaidyanathan S (2013) Analysis and adaptive synchronization of two novel chaotic systems with hyperbolic sinusoidal and cosinusoidal nonlinearity and unknown parameters. J Eng Sci Technol Rev 6(4):53–65
61. Vaidyanathan S (2013) Analysis, control and synchronization of hyperchaotic Zhou system via adaptive control. Adv Intell Syst Comput 177:1–10
62. Vaidyanathan S (2014) A new eight-term 3-D polynomial chaotic system with three quadratic nonlinearities. Far East J Math Sci 84(2):219–226
63. Vaidyanathan S (2014) Analysis and adaptive synchronization of eight-term 3-D polynomial chaotic systems with three quadratic nonlinearities. Eur Phys J: Spec Top 223(8):1519–1529
64. Vaidyanathan S (2014) Analysis, control and synchronisation of a six-term novel chaotic system with three quadratic nonlinearities. Int J Modell Identif Control 22(1):41–53
65. Vaidyanathan S (2014) Generalized projective synchronisation of novel 3-D chaotic systems with an exponential non-linearity via active and adaptive control. Int J Modell Identif Control 22(3):207–217
66. Vaidyanathan S (2014) Global chaos synchronization of identical Li-Wu chaotic systems via sliding mode control. Int J Modell Identif Control 22(2):170–177
67. Vaidyanathan S (2015) 3-cells cellular neural network (CNN) attractor and its adaptive biological control. Int J Pharm Tech Res 8(4):632–640
68. Vaidyanathan S (2015) A 3-D novel highly chaotic system with four quadratic nonlinearities, its adaptive control and anti-synchronization with unknown parameters. J Eng Sci Technol Rev 8(2):106–115
69. Vaidyanathan S (2015) A novel chemical chaotic reactor system and its adaptive control. Int J Chem Tech Res 8(7):146–158
70. Vaidyanathan S (2015) Adaptive backstepping control of enzymes-substrates system with ferroelectric behaviour in brain waves. Int J Pharm Tech Res 8(2):256–261
71. Vaidyanathan S (2015) Adaptive biological control of generalized Lotka-Volterra three-species biological system. Int J Pharm Tech Res 8(4):622–631
72. Vaidyanathan S (2015) Adaptive chaotic synchronization of enzymes-substrates system with ferroelectric behaviour in brain waves. Int J Pharm Tech Res 8(5):964–973
73. Vaidyanathan S (2015) Adaptive control of a chemical chaotic reactor. Int J Pharm Tech Res 8(3):377–382
74. Vaidyanathan S (2015) Adaptive control of the FitzHugh-Nagumo chaotic neuron model. Int J Pharm Tech Res 8(6):117–127
75. Vaidyanathan S (2015) Adaptive synchronization of chemical chaotic reactors. Int J Chem Tech Res 8(2):612–621
76. Vaidyanathan S (2015) Adaptive synchronization of generalized Lotka-Volterra three-species biological systems. Int J Pharm Tech Res 8(5):928–937

77. Vaidyanathan S (2015) Adaptive synchronization of novel 3-D chemical chaotic reactor systems. Int J Chem Tech Res 8(7):159–171
78. Vaidyanathan S (2015) Adaptive synchronization of the identical FitzHugh-Nagumo chaotic neuron models. Int J Pharm Tech Res 8(6):167–177
79. Vaidyanathan S (2015) Analysis, control, and synchronization of a 3-D novel jerk chaotic system with two quadratic nonlinearities. Kyungpook Math J 55:563–586
80. Vaidyanathan S (2015) Analysis, properties and control of an eight-term 3-D chaotic system with an exponential nonlinearity. Int J Modell Identif Control 23(2):164–172
81. Vaidyanathan S (2015) Anti-synchronization of Brusselator chemical reaction systems via adaptive control. Int J Chem Tech Res 8(6):759–768
82. Vaidyanathan S (2015) Chaos in neurons and adaptive control of Birkhoff-Shaw strange chaotic attractor. Int J Pharm Tech Res 8(5):956–963
83. Vaidyanathan S (2015) Chaos in neurons and synchronization of Birkhoff-Shaw strange chaotic attractors via adaptive control. Int J Pharm Tech Res 8(6):1–11
84. Vaidyanathan S (2015) Coleman-Gomatam logarithmic competitive biology models and their ecological monitoring. Int J Pharm Tech Res 8(6):94–105
85. Vaidyanathan S (2015) Dynamics and control of Brusselator chemical reaction. Int J Chem Tech Res 8(6):740–749
86. Vaidyanathan S (2015) Dynamics and control of Tokamak system with symmetric and magnetically confined plasma. Int J Chem Tech Res 8(6):795–803
87. Vaidyanathan S (2015) Global chaos synchronization of chemical chaotic reactors via novel sliding mode control method. Int J Chem Tech Res 8(7):209–221
88. Vaidyanathan S (2015) Global chaos synchronization of the forced Van der Pol chaotic oscillators via adaptive control method. Int J Pharm Tech Res 8(6):156–166
89. Vaidyanathan S (2015) Global chaos synchronization of the Lotka-Volterra biological systems with four competitive species via active control. Int J Pharm Tech Res 8(6):206–217
90. Vaidyanathan S (2015) Lotka-Volterra population biology models with negative feedback and their ecological monitoring. Int J Pharm Tech Res 8(5):974–981
91. Vaidyanathan S (2015) Lotka-Volterra two species competitive biology models and their ecological monitoring. Int J Pharm Tech Res 8(6):32–44
92. Vaidyanathan S (2015) Output regulation of the forced Van der Pol chaotic oscillator via adaptive control method. Int J Pharm Tech Res 8(6):106–116
93. Vaidyanathan S, Azar AT (2015) Analysis and control of a 4-D novel hyperchaotic system. In: Azar AT, Vaidyanathan S (eds) Chaos modeling and control systems design, vol 581. Studies in computational intelligence. Springer, Germany, pp 19–38
94. Vaidyanathan S, Azar AT (2015) Analysis, control and synchronization of a nine-term 3-D novel chaotic system. In: Azar AT, Vaidyanathan S (eds) Chaos modelling and control systems design. Studies in computational intelligence, vol 581. Springer, Germany, pp 19–38
95. Vaidyanathan S, Azar AT (2015) Anti-synchronization of identical chaotic systems using sliding mode control and an application to Vaidhyanathan-Madhavan chaotic systems. Stud Comput Intell 576:527–547
96. Vaidyanathan S, Azar AT (2015) Hybrid synchronization of identical chaotic systems using sliding mode control and an application to Vaidhyanathan chaotic systems. Stud Comput Intell 576:549–569
97. Vaidyanathan S, Madhavan K (2013) Analysis, adaptive control and synchronization of a seven-term novel 3-D chaotic system. Int J Control Theory Appl 6(2):121–137
98. Vaidyanathan S, Pakiriswamy S (2013) Generalized projective synchronization of six-term Sundarapandian chaotic systems by adaptive control. Int J Control Theory Appl 6(2):153–163
99. Vaidyanathan S, Pakiriswamy S (2015) A 3-D novel conservative chaotic system and its generalized projective synchronization via adaptive control. J Eng Sci Technol Rev 8(2):52–60
100. Vaidyanathan S, Rajagopal K (2011) Anti-synchronization of Li and T chaotic systems by active nonlinear control. Commun Comput Inf Sci 198:175–184

101. Vaidyanathan S, Rajagopal K (2011) Global chaos synchronization of hyperchaotic Pang and Wang systems by active nonlinear control. Commun Comput Inf Sci 204:84–93
102. Vaidyanathan S, Rajagopal K (2011) Global chaos synchronization of Lü and Pan systems by adaptive nonlinear control. Commun Comput Inf Sci 205:193–202
103. Vaidyanathan S, Rajagopal K (2012) Global chaos synchronization of hyperchaotic Pang and hyperchaotic Wang systems via adaptive control. Int J Soft Comput 7(1):28–37
104. Vaidyanathan S, Rasappan S (2011) Global chaos synchronization of hyperchaotic Bao and Xu systems by active nonlinear control. Commun Comput Inf Sci 198:10–17
105. Vaidyanathan S, Rasappan S (2014) Global chaos synchronization of n-scroll Chua circuit and Lur'e system using backstepping control design with recursive feedback. Arab J Sci Eng 39(4):3351–3364
106. Vaidyanathan S, Sampath S (2011) Global chaos synchronization of hyperchaotic Lorenz systems by sliding mode control. Commun Comput Inf Sci 205:156–164
107. Vaidyanathan S, Sampath S (2012) Anti-synchronization of four-wing chaotic systems via sliding mode control. Int J Autom Comput 9(3):274–279
108. Vaidyanathan S, Volos C (2015) Analysis and adaptive control of a novel 3-D conservative no-equilibrium chaotic system. Arch Control Sci 25(3):333–353
109. Vaidyanathan S, Volos C, Pham VT (2014) Hyperchaos, adaptive control and synchronization of a novel 5-D hyperchaotic system with three positive Lyapunov exponents and its SPICE implementation. Arch Control Sci 24(4):409–446
110. Vaidyanathan S, Volos C, Pham VT (2014) Hyperchaos, adaptive control and synchronization of a novel 5-D hyperchaotic system with three positive Lyapunov exponents and its SPICE implementation. Arch Control Sci 24(4):409–446
111. Vaidyanathan S, Volos C, Pham VT, Madhavan K, Idowu BA (2014) Adaptive backstepping control, synchronization and circuit simulation of a 3-D novel jerk chaotic system with two hyperbolic sinusoidal nonlinearities. Arch Control Sci 24(3):375–403
112. Vaidyanathan S, Idowu BA, Azar AT (2015) Backstepping controller design for the global chaos synchronization of Sprott's jerk systems. Stud Comput Intell 581:39–58
113. Vaidyanathan S, Rajagopal K, Volos CK, Kyprianidis IM, Stouboulos IN (2015) Analysis, adaptive control and synchronization of a seven-term novel 3-D chaotic system with three quadratic nonlinearities and its digital implementation in LabVIEW. J Eng Sci Technol Rev 8(2):130–141
114. Vaidyanathan S, Volos C, Pham VT, Madhavan K (2015) Analysis, adaptive control and synchronization of a novel 4-D hyperchaotic hyperjerk system and its SPICE implementation. Arch Control Sci 25(1):5–28
115. Vaidyanathan S, Volos CK, Kyprianidis IM, Stouboulos IN, Pham VT (2015) Analysis, adaptive control and anti-synchronization of a six-term novel jerk chaotic system with two exponential nonlinearities and its circuit simulation. J Eng Sci Technol Rev 8(2):24–36
116. Vaidyanathan S, Volos CK, Madhavan K (2015) Analysis, control, synchronization and SPICE implementation of a novel 4-D hyperchaotic Rikitake dynamo system without equilibrium. J Eng Sci Technol Rev 8(2):232–244
117. Vaidyanathan S, Volos CK, Pham VT (2015) Analysis, adaptive control and adaptive synchronization of a nine-term novel 3-D chaotic system with four quadratic nonlinearities and its circuit simulation. J Eng Sci Technol Rev 8(2):181–191
118. Vaidyanathan S, Volos CK, Pham VT (2015) Global chaos control of a novel nine-term chaotic system via sliding mode control. In: Azar AT, Zhu Q (eds) Advances and applications in sliding mode control systems, vol 576. Studies in computational intelligence. Springer, Germany, pp 571–590
119. Vaidyanathan S, Volos CK, Pham VT, Madhavan K (2015) Analysis, adaptive control and synchronization of a novel 4-D hyperchaotic hyperjerk system and its SPICE implementation. Arch Control Sci 25(1):135–158
120. Vaidyanathan S, Volos CK, Rajagopal K, Kyprianidis IM, Stouboulos IN (2015) Adaptive backstepping controller design for the anti-synchronization of identical WINDMI chaotic systems with unknown parameters and its SPICE implementation. J Eng Sci Technol Rev 8(2):74–82

121. Volos CK, Kyprianidis IM, Stouboulos IN, Anagnostopoulos AN (2009) Experimental study of the dynamic behavior of a double scroll circuit. J Appl Funct Anal 4:703–711
122. Volos CK, Kyprianidis IM, Stouboulos IN (2013) Experimental investigation on coverage performance of a chaotic autonomous mobile robot. Robot Auton Syst 61(12):1314–1322
123. Volos CK, Kyprianidis IM, Stouboulos IN, Tlelo-Cuautle E, Vaidyanathan S (2015) Memristor: a new concept in synchronization of coupled neuromorphic circuits. J Eng Sci Technol Rev 8(2):157–173
124. Wang X, Ge C (2008) Controlling and tracking of Newton-Leipnik system via backstepping design. Int J Nonlinear Sci 5(2):133–139
125. Wei Z, Yang Q (2010) Anti-control of Hopf bifurcation in the new chaotic system with two stable node-foci. Appl Math Comput 217(1):422–429
126. Xiao X, Zhou L, Zhang Z (2014) Synchronization of chaotic Lur'e systems with quantized sampled-data controller. Commun Nonlinear Sci Numer Simul 19(6):2039–2047
127. Yang J, Zhu F (2013) Synchronization for chaotic systems and chaos-based secure communications via both reduced-order and step-by-step sliding mode observers. Commun Nonlinear Sci Numer Simul 18(4):926–937
128. Yang J, Chen Y, Zhu F (2014) Singular reduced-order observer-based synchronization for uncertain chaotic systems subject to channel disturbance and chaos-based secure communication. Appl Math Comput 229:227–238
129. Zhang H, Zhou J (2012) Synchronization of sampled-data coupled harmonic oscillators with control inputs missing. Syst Control Lett 61(12):1277–1285
130. Zhou W, Xu Y, Lu H, Pan L (2008) On dynamics analysis of a new chaotic attractor. Phys Lett A 372(36):5773–5777
131. Zhu C, Liu Y, Guo Y (2010) Theoretic and numerical study of a new chaotic system. Intell Inf Manag 2:104–109

Analysis, Adaptive Control and Synchronization of a Novel 3-D Chaotic System with a Quartic Nonlinearity

Sundarapandian Vaidyanathan

Abstract In this work, we describe a seven-term novel chaotic system with a quartic nonlinearity and two quadratic nonlinearities. The phase portraits of the novel chaotic system are illustrated and the dynamic properties of the chaotic system are discussed. The novel 3-D chaotic system has three unstable equilibrium points. We show that the equilibrium point at the origin is a saddle point, while the other two equilibrium points are saddle-foci. We show that novel 3-D chaotic system has rotation symmetry about the x_3-axis. We also show that the x_3-axis is invariant under the flow of the 3-D novel chaotic system. The Lyapunov exponents of the novel 3-D chaotic system are obtained as $L_1 = 0.75364$, $L_2 = 0$ and $L_3 = -2.50392$, while the Kaplan–Yorke dimension of the novel chaotic system is obtained as $D_{KY} = 2.3010$. Since the sum of the Lyapunov exponents is negative, the novel chaotic system is dissipative. Next, we derive new results for the global chaos control of the novel 3-D chaotic system with unknown parameters using adaptive control method. We also derive new results for the global chaos synchronization of the identical novel 3-D chaotic systems with unknown parameters using adaptive control method. The main control results are established using Lyapunov stability theory. MATLAB simulations are depicted to illustrate the phase portraits of the novel 3-D chaotic system and also the adaptive control results derived in this work.

Keywords Chaos · Chaotic systems · Chaos control · Adaptive control · Synchronization

---

S. Vaidyanathan (✉)
Research and Development Centre, Vel Tech University, Avadi,
Chennai 600062, Tamil Nadu, India
e-mail: sundarvtu@gmail.com

S. Vaidyanathan and C. Volos (eds.), *Advances and Applications in Chaotic Systems*, Studies in Computational Intelligence 636,
DOI 10.1007/978-3-319-30279-9_25

1 Introduction

In the last few decades, Chaos theory has become a very important and active research field, employing many applications in different disciplines like physics, chemistry, biology, ecology, engineering and economics, among others [3].

Some classical paradigms of 3-D chaotic systems in the literature are Lorenz system [11], Rössler system [20], ACT system [2], Sprott systems [25], Chen system [6], Lü system [12], Cai system [5], Tigan system [35], etc.

Many new chaotic systems have been discovered in the recent years such as Zhou system [108], Zhu system [109], Li system [10], Sundarapandian systems [28, 32], Vaidyanathan systems [44, 46, 48–51, 55, 66, 67, 81, 82, 84, 90, 92, 95, 98, 99, 101], Pehlivan system [15], Sampath system [21], etc.

Chaos theory has applications in several fields of science and engineering such as chemical reactors [56, 60, 62, 64, 68, 72–74], biological systems [54, 57–59, 61, 63, 65, 69–71, 75–79], memristors [1, 16, 105], lasers [4], oscillations [36], robotics [7, 104], electrical circuits [13, 103], cryptosystems [19, 37], secure communications [106, 107], etc.

The control of a chaotic system aims to stabilize or regulate the system with the help of a feedback control. There are many methods available for controlling a chaotic system such as active control [26, 38, 39], adaptive control [27, 40, 45, 47, 53, 80, 91, 97, 100], sliding mode control [42, 43], backstepping control [14, 94, 102], etc.

There are many methods available for chaos synchronization such as active control [8, 22, 23, 85, 87, 93], adaptive control [24, 29–31, 41, 83, 86], sliding mode control [33, 52, 89, 96], backstepping control [17, 18, 34, 88], etc.

In this research work, we announce a seven-term novel 3-D chaotic system with a quartic nonlinearity and two quadratic nonlinearities. Using adaptive control method, we have also derived new results for the global chaos control of the novel 3-D chaotic system and global chaos synchronization of the identical novel 3-D chaotic systems when the system parameters are unknown.

This work is organized as follows. Section 2 describes the dynamic equations and phase portraits of the seven-term novel 3-D chaotic system. Section 3 details the dynamic analysis and properties of the novel 3-D chaotic system. The Lyapunov exponents of the novel chaotic system are obtained as $L_1 = 0.75364$, $L_2 = 0$ and $L_3 = -2.50392$, while the Kaplan–Yorke dimension of the novel chaotic system is obtained as $D_{KY} = 2.3010$. We show that novel 3-D chaotic system has rotation symmetry about the x_3-axis. We also that the x_3-axis is invariant under the flow of the 3-D novel chaotic system.

In Sect. 4, we derive new results for the global chaos control of the novel 3-D chaotic system with unknown parameters. In Sect. 5, we derive new results for the global chaos synchronization of the identical novel 3-D chaotic systems with unknown parameters. Section 6 contains a summary of the main results derived in this work.

2 A Novel 3-D Chaotic System

In this section, we describe a seven-term novel chaotic system, which is given by the 3-D dynamics

$$\begin{cases} \dot{x}_1 = x_2 + x_2 x_3 \\ \dot{x}_2 = a x_1 - x_2 - x_1 x_3 \\ \dot{x}_3 = -b x_3 + x_1^4 \end{cases} \tag{1}$$

where x_1, x_2, x_3 are the states and a, b are constant, positive parameters.

The novel 3-D system (1) is a seven-term polynomial system with a quartic nonlinearity and two quadratic nonlinearities.

The system (1) exhibits a *strange chaotic attractor* for the parameter values

$$a = 4, \quad b = 0.75 \tag{2}$$

For numerical simulations, we take the initial conditions as

$$x_1(0) = 1, \quad x_2(0) = 1, \quad x_3(0) = 1 \tag{3}$$

Figure 1 depicts the 3-D phase portrait of the novel chaotic system (1), while Figs. 2, 3 and 4 depict the 2-D projection of the novel chaotic system (1) on the (x_1, x_2), (x_2, x_3) and (x_1, x_3) planes, respectively.

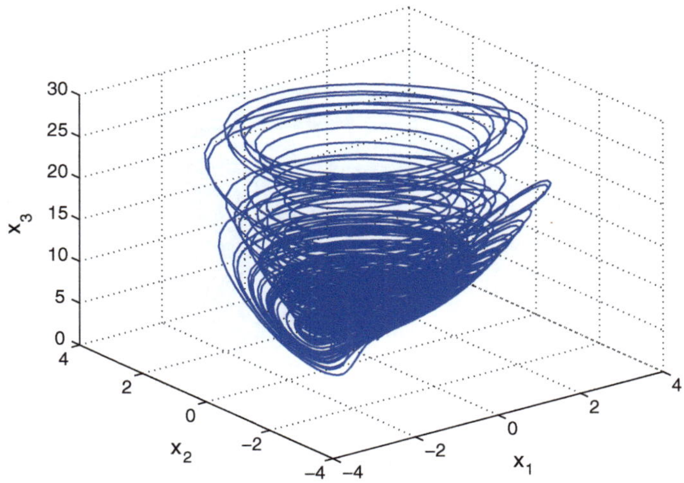

Fig. 1 3-D phase portrait of the novel highly chaotic system

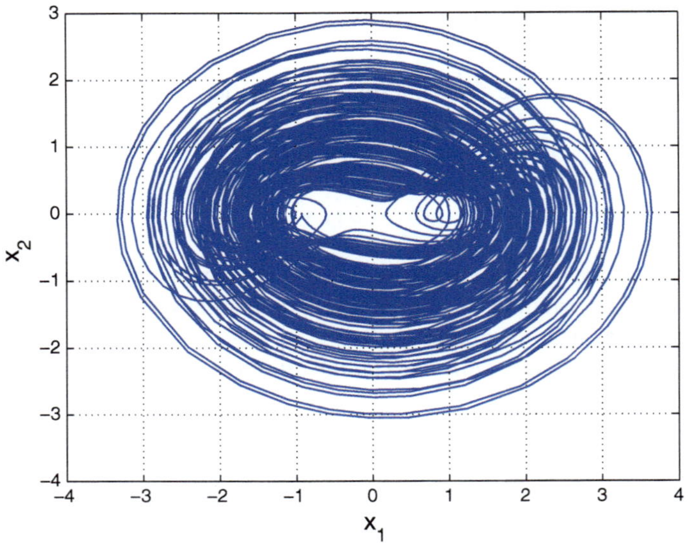

Fig. 2 2-D projection of the novel highly chaotic system on the (x_1, x_2) plane

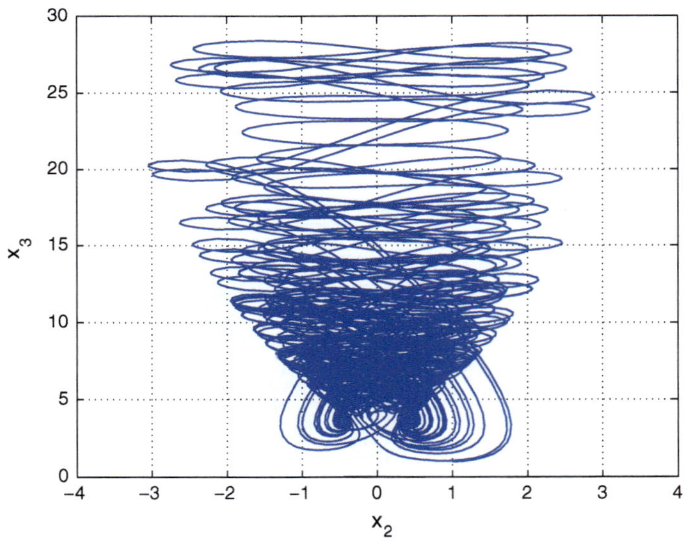

Fig. 3 2-D projection of the novel highly chaotic system on the (x_2, x_3) plane

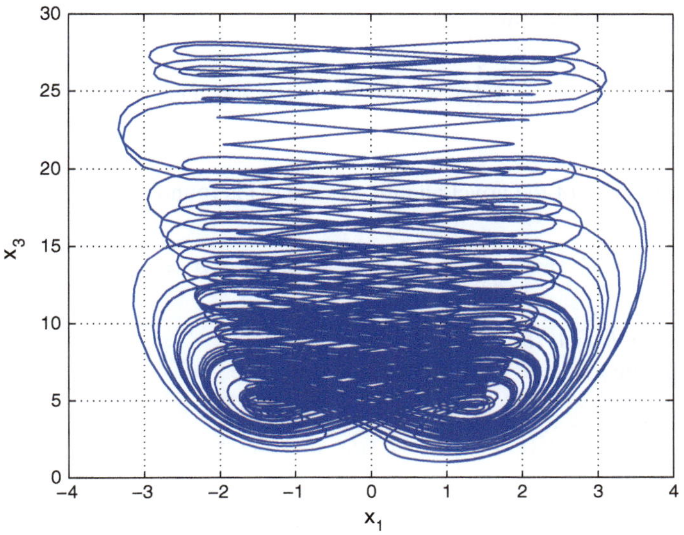

Fig. 4 2-D projection of the novel highly chaotic system on the (x_1, x_3) plane

3 Analysis of the Novel 3-D Highly Chaotic System

In this section, we give a dynamic analysis of the 3-D novel highly chaotic system (1). We take the parameter values as in the chaotic case (2), viz. $a = 4$ and $b = 0.75$.

3.1 Dissipativity

In vector notation, the novel chaotic system (1) can be expressed as

$$\dot{\mathbf{x}} = f(\mathbf{x}) = \begin{bmatrix} f_1(x_1, x_2, x_3) \\ f_2(x_1, x_2, x_3) \\ f_3(x_1, x_2, x_3) \end{bmatrix}, \tag{4}$$

where

$$\begin{cases} f_1(x_1, x_2, x_3) = x_2 + x_2 x_3 \\ f_2(x_1, x_2, x_3) = a x_1 - x_2 - x_1 x_3 \\ f_3(x_1, x_2, x_3) = -b x_3 + x_1^4 \end{cases} \tag{5}$$

Let Ω be any region in $\mathbf{R}^3$ with a smooth boundary and also, $\Omega(t) = \Phi_t(\Omega)$, where Φ_t is the flow of f. Furthermore, let $V(t)$ denote the volume of $\Omega(t)$.

By Liouville's theorem, we know that

$$\dot{V}(t) = \int\limits_{\Omega(t)} (\nabla \cdot f)\, dx_1\, dx_2\, dx_3 \tag{6}$$

The divergence of the novel chaotic system (4) is found as

$$\nabla \cdot f = \frac{\partial f_1}{\partial x_1} + \frac{\partial f_2}{\partial x_2} + \frac{\partial f_3}{\partial x_3} = -1 - b = -\mu < 0 \tag{7}$$

since $\mu = 1 + b = 1.75 > 0$.

Inserting the value of $\nabla \cdot f$ from (7) into (6), we get

$$\dot{V}(t) = \int\limits_{\Omega(t)} (-\mu)\, dx_1\, dx_2\, dx_3 = -\mu V(t) \tag{8}$$

Integrating the first order linear differential equation (8), we get

$$V(t) = \exp(-\mu t) V(0) \tag{9}$$

Since $\mu > 0$, it follows from Eq. (9) that $V(t) \to 0$ exponentially as $t \to \infty$. This shows that the novel chaotic system (1) is dissipative.

Hence, the system limit sets are ultimately confined into a specific limit set of zero volume, and the asymptotic motion of the novel chaotic system (1) settles onto a strange attractor of the system.

3.2 Equilibrium Points

We take the parameter values as in the chaotic case (2), viz. $a = 4$ and $b = 0.75$.

It is easy to see that the system (1) has three equilibrium points, viz.

$$E_0 = \begin{bmatrix} 0 \\ 0 \\ 0 \end{bmatrix}, \quad E_1 = \begin{bmatrix} 3^{\frac{1}{4}} \\ 0 \\ 4 \end{bmatrix}, \quad E_2 = \begin{bmatrix} -3^{\frac{1}{4}} \\ 0 \\ 4 \end{bmatrix} \tag{10}$$

The Jacobian of the system (1) at any point $\mathbf{x} \in \mathbf{R}^3$ is calculated as

$$J(\mathbf{x}) = \begin{bmatrix} 0 & 1 + x_3 & x_2 \\ 4 - x_3 & -1 & -x_1 \\ 4x_1^3 & 0 & -0.75 \end{bmatrix} \tag{11}$$

The Jacobian of the system (1) at E_0 is obtained as

$$J_0 = J(E_0) = \begin{bmatrix} 0 & 1 & 0 \\ 4 & -1 & 0 \\ 0 & 0 & -0.75 \end{bmatrix} \tag{12}$$

We find that the matrix J_0 has the eigenvalues

$$\lambda_1 = -0.75, \;\; \lambda_2 = -2.5616, \;\; \lambda_3 = 1.5616 \tag{13}$$

This shows that the equilibrium point E_0 is a saddle-point, which is unstable.
The Jacobian of the system (1) at E_1 is obtained as

$$J_1 = J(E_1) = \begin{bmatrix} 0 & 5 & 0 \\ 0 & -1 & -1.3161 \\ 9.1180 & 0 & -0.75 \end{bmatrix} \tag{14}$$

We find that the matrix J_1 has the eigenvalues

$$\lambda_1 = -4.5204, \;\; \lambda_{2,3} = 1.3852 \pm 3.3696i \tag{15}$$

This shows that the equilibrium point E_1 is a saddle-focus, which is unstable.
The Jacobian of the system (1) at E_2 is obtained as

$$J_2 = J(E_2) = \begin{bmatrix} 0 & 5 & 0 \\ 0 & -1 & -1.3161 \\ 9.1180 & 0 & -0.75 \end{bmatrix} \tag{16}$$

We find that the matrix J_2 has the eigenvalues

$$\lambda_1 = -4.5204, \;\; \lambda_{2,3} = 1.3852 \pm 3.3696i \tag{17}$$

This shows that the equilibrium point E_2 is a saddle-focus, which is unstable.

3.3 Symmetry and Invariance

It is easy to see that the system (1) is invariant under the change of coordinates

$$(x_1, x_2, x_3) \mapsto (-x_1, -x_2, x_3) \tag{18}$$

Thus, it follows that the 3-D novel chaotic system (1) has rotation symmetry about the x_3-axis and that any non-trivial trajectory must have a twin trajectory.

Next, it is easy to see that the x_3-axis is invariant under the flow of the 3-D novel chaotic system (1). The invariant motion along the x_3-axis is characterized by

$$\dot{x}_3 = -bx_3, \quad (b > 0) \tag{19}$$

which is globally exponentially stable.

3.4 Lyapunov Exponents and Kaplan–Yorke Dimension

We take the parameter values of the novel system (1) as in the chaotic case (2), i.e.

$$a = 4, \quad b = 0.75 \tag{20}$$

We take the initial state of the novel system (1) as given in (3), i.e.

$$x_1(0) = 1, \quad x_2(0) = 1, \quad x_3(0) = 1 \tag{21}$$

Then the Lyapunov exponents of the system (1) are numerically obtained as

$$L_1 = 0.75364, \quad L_2 = 0, \quad L_3 = -2.50392 \tag{22}$$

Since the sum of the Lyapunov exponents is negative, we conclude that the novel chaotic system (1) is dissipative.

Also, the Kaplan–Yorke dimension of the novel chaotic system (1) is found as

$$D_{KY} = 2 + \frac{L_1 + L_2}{|L_3|} = 2.3010 \tag{23}$$

4 Adaptive Control of the Novel 3-D Chaotic System

In this section, we use adaptive control method to derive an adaptive feedback control law for globally stabilizing the novel 3-D chaotic system with unknown system parameters.

Thus, we consider the novel 3-D chaotic system with controls given by

$$\begin{cases} \dot{x}_1 = x_2 + x_2 x_3 + u_1 \\ \dot{x}_2 = a x_1 - x_2 - x_1 x_3 + u_2 \\ \dot{x}_3 = -b x_3 + x_1^4 + u_3 \end{cases} \tag{24}$$

In (24), x_1, x_2, x_3 are the states and u_1, u_2, u_3 are the adaptive controls to be determined using estimates $\hat{a}(t)$, $\hat{b}(t)$ for the unknown parameters a, b, respectively.

We consider the adaptive feedback control law

$$\begin{cases} u_1 = -x_2 - x_2 x_3 - k_1 x_1 \\ u_2 = -\hat{a}(t) x_1 + x_2 + x_1 x_3 - k_2 x_2 \\ u_3 = \hat{b}(t) x_3 - x_1^4 - k_3 x_3 \end{cases} \tag{25}$$

where k_1, k_2, k_3 are positive gain constants.

Substituting (25) into (24), we get the closed-loop plant dynamics as

$$\begin{cases} \dot{x}_1 = -k_1 x_1 \\ \dot{x}_2 = [a - \hat{a}(t)] x_1 - k_2 x_2 \\ \dot{x}_3 = -[b - \hat{b}(t)] x_3 - k_3 x_3 \end{cases} \tag{26}$$

The parameter estimation errors are defined as

$$\begin{cases} e_a(t) = a - \hat{a}(t) \\ e_b(t) = b - \hat{b}(t) \end{cases} \tag{27}$$

In view of (27), we can simplify the plant dynamics (26) as

$$\begin{cases} \dot{x}_1 = -k_1 x_1 \\ \dot{x}_2 = e_a x_1 - k_2 x_2 \\ \dot{x}_3 = -e_b x_3 - k_3 x_3 \end{cases} \tag{28}$$

Differentiating (27) with respect to t, we obtain

$$\begin{cases} \dot{e}_a(t) = -\dot{\hat{a}}(t) \\ \dot{e}_b(t) = -\dot{\hat{b}}(t) \end{cases} \tag{29}$$

We consider the quadratic candidate Lyapunov function defined by

$$V(\mathbf{x}, e_a, e_b) = \frac{1}{2}\left(x_1^2 + x_2^2 + x_3^2\right) + \frac{1}{2}\left(e_a^2 + e_b^2\right) \tag{30}$$

Differentiating V along the trajectories of (28) and (29), we obtain

$$\dot{V} = -k_1 x_1^2 - k_2 x_2^2 - k_3 x_3^2 + e_a\left[x_1 x_2 - \dot{\hat{a}}\right] + e_b\left[-x_3^2 - \dot{\hat{b}}\right] \tag{31}$$

In view of (31), we take the parameter update law as

$$\begin{cases} \dot{\hat{a}}(t) = x_1 x_2 \\ \dot{\hat{b}}(t) = -x_3^2 \end{cases} \tag{32}$$

Next, we state and prove the main result of this section.

Theorem 1 *The novel 3-D chaotic system (24) with unknown system parameters is globally and exponentially stabilized for all initial conditions by the adaptive control law (25) and the parameter update law (32), where k_1, k_2, k_3 are positive gain constants.*

Proof We prove this result by applying Lyapunov stability theory [9].

We consider the quadratic Lyapunov function defined by (30), which is clearly a positive definite function on $\mathbf{R}^5$.

By substituting the parameter update law (32) into (31), we obtain the time-derivative of V as

$$\dot{V} = -k_1 x_1^2 - k_2 x_2^2 - k_3 x_3^2 \tag{33}$$

From (33), it is clear that $\dot{V}$ is a negative semi-definite function on $\mathbf{R}^5$.

Thus, we can conclude that the state vector $\mathbf{x}(t)$ and the parameter estimation error are globally bounded i.e.

$$\left[\, x_1(t) \;\, x_2(t) \;\, x_3(t) \;\, e_a(t) \;\, e_b(t) \,\right]^T \in \mathbf{L}_\infty.$$

We define $k = \min\{k_1, k_2, k_3\}$.

Then it follows from (33) that

$$\dot{V} \leq -k\|\mathbf{x}(t)\|^2 \tag{34}$$

Thus, we have

$$k\|\mathbf{x}(t)\|^2 \leq -\dot{V} \tag{35}$$

Integrating the inequality (35) from 0 to t, we get

$$k\int_0^t \|\mathbf{x}(\tau)\|^2 \, d\tau \;\leq\; V(0) - V(t) \tag{36}$$

From (36), it follows that $\mathbf{x} \in \mathbf{L}_2$.

Using (28), we can conclude that $\dot{\mathbf{x}} \in \mathbf{L}_\infty$.

Using Barbalat's lemma [9], we conclude that $\mathbf{x}(t) \to 0$ exponentially as $t \to \infty$ for all initial conditions $\mathbf{x}(0) \in \mathbf{R}^3$.

Hence, the novel highly chaotic system (24) with unknown system parameters is globally and exponentially stabilized for all initial conditions by the adaptive control law (25) and the parameter update law (32).

This completes the proof. □

For the numerical simulations, the classical fourth-order Runge–Kutta method with step size $h = 10^{-8}$ is used to solve the systems (24) and (32), when the adaptive control law (25) is applied.

The parameter values of the novel 3-D chaotic system (24) are taken as in the chaotic case (2), i.e.

$$a = 4, \quad b = 0.75 \tag{37}$$

We take the positive gain constants as

$$k_1 = 6, \quad k_2 = 6, \quad k_3 = 6 \tag{38}$$

Furthermore, as initial conditions of the novel highly chaotic system (24), we take

$$x_1(0) = 12.7, \quad x_2(0) = 24.8, \quad x_3(0) = -16.9 \tag{39}$$

Also, as initial conditions of the parameter estimates, we take

$$\hat{a}(0) = 15.7, \quad \hat{b}(0) = 14.3 \tag{40}$$

In Fig. 5, the exponential convergence of the controlled states of the 3-D novel chaotic system (24) is depicted.

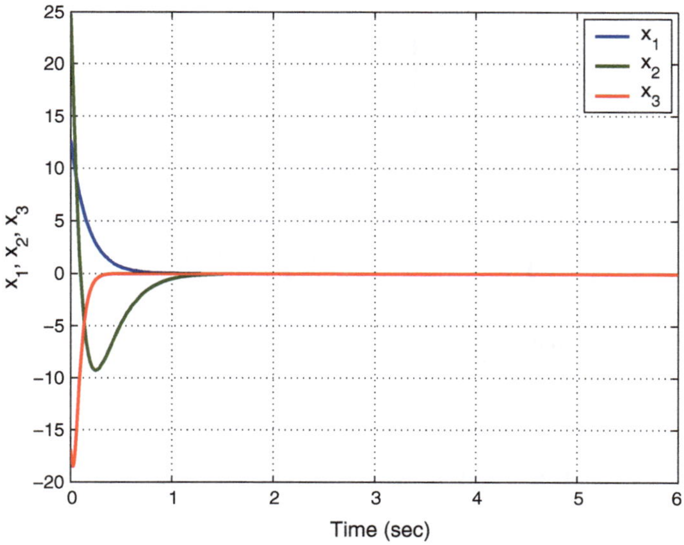

Fig. 5 Time-history of the controlled states x_1, x_2, x_3

5 Adaptive Synchronization of the Identical Novel Chaotic Systems

In this section, we use adaptive control method to derive an adaptive feedback control law for globally synchronizing identical novel 3-D chaotic systems with unknown parameters. The main result is established using Lyapunov stability theory.

As the master system, we consider the novel 3-D chaotic system given by

$$\begin{cases} \dot{x}_1 = x_2 + x_2 x_3 \\ \dot{x}_2 = a x_1 - x_2 - x_1 x_3 \\ \dot{x}_3 = -b x_3 + x_1^4 \end{cases} \tag{41}$$

In (41), x_1, x_2, x_3 are the states and a, b, c, p are unknown system parameters.

As the slave system, we consider the novel 3-D chaotic system given by

$$\begin{cases} \dot{y}_1 = y_2 + y_2 y_3 + u_1 \\ \dot{y}_2 = a y_1 - y_2 - y_1 y_3 + u_2 \\ \dot{y}_3 = -b y_3 + y_1^4 + u_3 \end{cases} \tag{42}$$

In (42), y_1, y_2, y_3 are the states and u_1, u_2, u_3 are the adaptive controls to be determined using estimates $\hat{a}(t), \hat{b}(t)$ for the unknown parameters a, b, respectively.

The synchronization error between the novel chaotic systems is defined by

$$\begin{cases} e_1 = y_1 - x_1 \\ e_2 = y_2 - x_2 \\ e_3 = y_3 - x_3 \end{cases} \tag{43}$$

Then the error dynamics is obtained as

$$\begin{cases} \dot{e}_1 = e_2 + y_2 y_3 - x_2 x_3 + u_1 \\ \dot{e}_2 = a e_1 - e_2 - y_1 y_3 + x_1 x_3 + u_2 \\ \dot{e}_3 = -b e_3 + y_1^4 - x_1^4 + u_3 \end{cases} \tag{44}$$

We consider the adaptive feedback control law

$$\begin{cases} u_1 = -e_2 - y_2 y_3 + x_2 x_3 - k_1 e_1 \\ u_2 = -\hat{a}(t) e_1 + e_2 + y_1 y_3 - x_1 x_3 - k_2 e_2 \\ u_3 = \hat{b}(t) e_3 - y_1^4 + x_1^4 - k_3 e_3 \end{cases} \tag{45}$$

where k_1, k_2, k_3 are positive gain constants.

Substituting (45) into (44), we get the closed-loop error dynamics as

$$\begin{cases} \dot{e}_1 = -k_1 e_1 \\ \dot{e}_2 = [a - \hat{a}(t)] e_1 - k_2 e_2 \\ \dot{e}_3 = -[b - \hat{b}(t)] e_3 - k_3 e_3 \end{cases} \tag{46}$$

The parameter estimation errors are defined as

$$\begin{cases} e_a(t) = a - \hat{a}(t) \\ e_b(t) = b - \hat{b}(t) \end{cases} \tag{47}$$

In view of (47), we can simplify the error dynamics (46) as

$$\begin{cases} \dot{e}_1 = -k_1 e_1 \\ \dot{e}_2 = e_a e_1 - k_2 e_2 \\ \dot{e}_3 = -e_b e_3 - k_3 e_3 \end{cases} \tag{48}$$

Differentiating (47) with respect to t, we obtain

$$\begin{cases} \dot{e}_a(t) = -\dot{\hat{a}}(t) \\ \dot{e}_b(t) = -\dot{\hat{b}}(t) \end{cases} \tag{49}$$

We consider the quadratic candidate Lyapunov function defined by

$$V(\mathbf{e}, e_a, e_b) = \frac{1}{2}\left(e_1^2 + e_2^2 + e_3^2\right) + \frac{1}{2}\left(e_a^2 + e_b^2\right) \tag{50}$$

Differentiating V along the trajectories of (48) and (49), we obtain

$$\dot{V} = -k_1 e_1^2 - k_2 e_2^2 - k_3 e_3^2 + e_a\left[e_1 e_2 - \dot{\hat{a}}\right] + e_b\left[-e_3^2 - \dot{\hat{b}}\right] \tag{51}$$

In view of (51), we take the parameter update law as

$$\begin{cases} \dot{\hat{a}}(t) = e_1 e_2 \\ \dot{\hat{b}}(t) = -e_3^2 \end{cases} \tag{52}$$

Next, we state and prove the main result of this section.

Theorem 2 *The novel 3-D chaotic systems (41) and (42) with unknown system parameters are globally and exponentially synchronized for all initial conditions by the adaptive control law (45) and the parameter update law (52), where k_1, k_2, k_3 are positive gain constants.*

Proof We prove this result by applying Lyapunov stability theory [9].

We consider the quadratic Lyapunov function defined by (50), which is clearly a positive definite function on $\mathbf{R}^5$.

By substituting the parameter update law (52) into (51), we obtain

$$\dot{V} = -k_1 e_1^2 - k_2 e_2^2 - k_3 e_3^2 \tag{53}$$

From (53), it is clear that $\dot{V}$ is a negative semi-definite function on $\mathbf{R}^5$.

Thus, we can conclude that the error vector $\mathbf{e}(t)$ and the parameter estimation error are globally bounded, i.e.

$$\left[\, e_1(t)\; e_2(t)\; e_3(t)\; e_a(t)\; e_b(t) \,\right]^T \in \mathbf{L}_\infty. \tag{54}$$

We define $k = \min\{k_1, k_2, k_3\}$.

Then it follows from (53) that

$$\dot{V} \leq -k\|\mathbf{e}(t)\|^2 \tag{55}$$

Thus, we have

$$k\|\mathbf{e}(t)\|^2 \leq -\dot{V} \tag{56}$$

Integrating the inequality (56) from 0 to t, we get

$$k \int_0^t \|\mathbf{e}(\tau)\|^2 \, d\tau \;\leq\; V(0) - V(t) \tag{57}$$

From (57), it follows that $\mathbf{e} \in \mathbf{L}_2$.

Using (48), we can conclude that $\dot{\mathbf{e}} \in \mathbf{L}_\infty$.

Using Barbalat's lemma [9], we conclude that $\mathbf{e}(t) \to 0$ exponentially as $t \to \infty$ for all initial conditions $\mathbf{e}(0) \in \mathbf{R}^3$.

This completes the proof. □

For the numerical simulations, the classical fourth-order Runge–Kutta method with step size $h = 10^{-8}$ is used to solve the systems (41), (42) and (52), when the adaptive control law (45) is applied.

The parameter values of the novel chaotic systems are taken as in the chaotic case (2), i.e. $a = 4$ and $b = 0.75$. We take the positive gain constants as $k_i = 6$ for $i = 1, 2, 3$.

As initial conditions of the master system (41), we take $x_1(0) = 2.4, x_2(0) = -1.7$ and $x_3(0) = -2.3$.

As initial conditions of the slave system (42), we take $y_1(0) = 3.6$, $y_2(0) = 2.2$ and $y_3(0) = 5.1$.

Also, as initial conditions of the parameter estimates, we take $\hat{a}(0) = 8.1$ and $\hat{b}(0) = 7.4$.

Figures 6, 7 and 8 describe the complete synchronization of the novel highly chaotic systems (41) and (42), while Fig. 9 describes the time-history of the synchronization errors e_1, e_2, e_3.

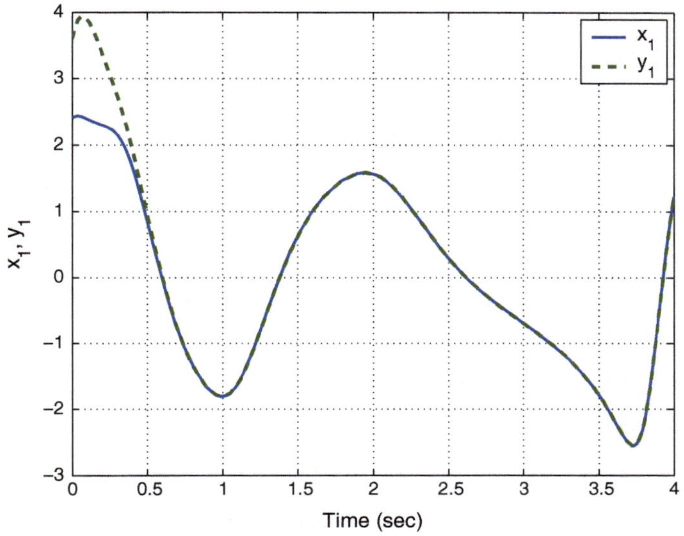

Fig. 6 Synchronization of the states x_1 and y_1

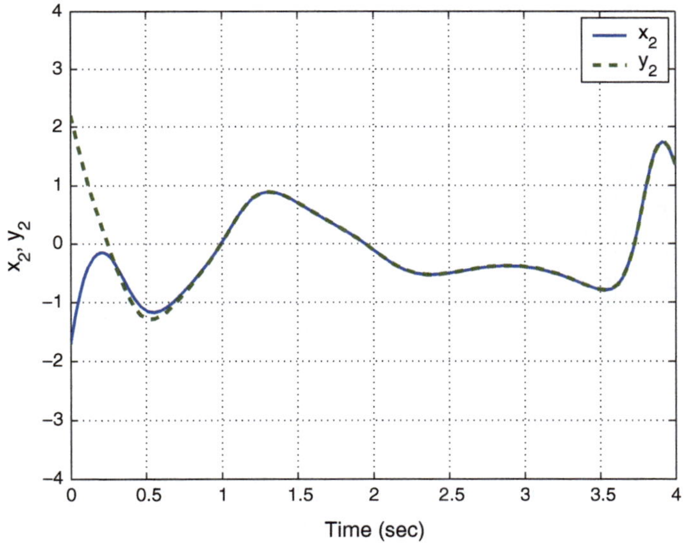

Fig. 7 Synchronization of the states x_2 and y_2

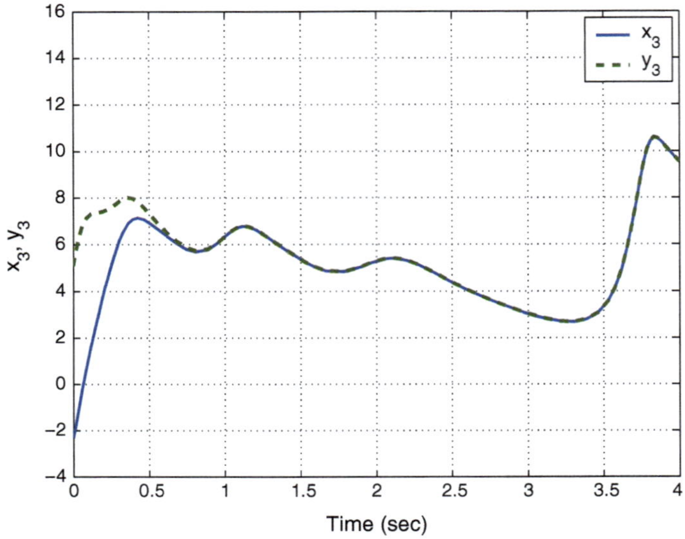

Fig. 8 Synchronization of the states x_3 and y_3

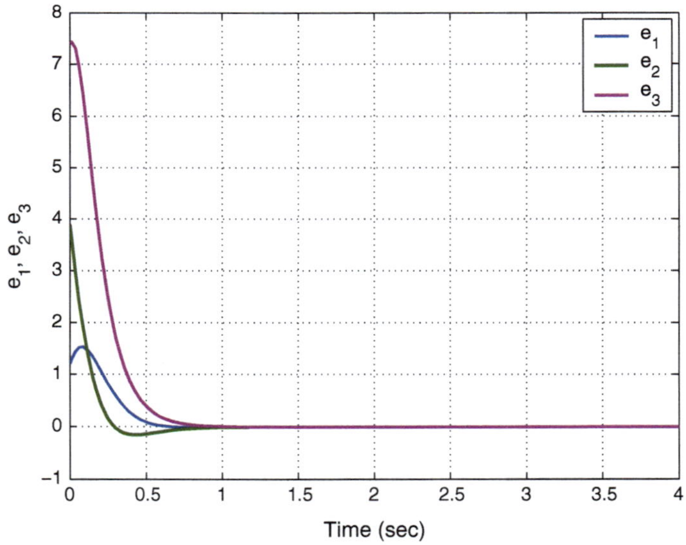

Fig. 9 Time-history of the synchronization errors e_1, e_2, e_3

6 Conclusions

In this work, we announced a seven-term novel chaotic system with a quartic nonlinearity and two quadratic nonlinearities. The phase portraits of the novel chaotic system were illustrated and the dynamic properties of the chaotic system were discussed. We showed that the novel 3-D chaotic system has three unstable equilibrium points. Explicitly, we established that the equilibrium point at the origin is a saddle point, while the other two equilibrium points are saddle-foci. We showed that novel 3-D chaotic system has rotation symmetry about the x_3-axis. We also showed that the x_3-axis is invariant under the flow of the 3-D novel chaotic system. The Lyapunov exponents of the novel 3-D chaotic system have been obtained as $L_1 = 0.75364$, $L_2 = 0$ and $L_3 = -2.50392$, while the Kaplan–Yorke dimension of the novel chaotic system has been derived as $D_{KY} = 2.3010$. Since the sum of the Lyapunov exponents is negative, the novel chaotic system is dissipative. Next, we established new results for the global chaos control of the novel 3-D chaotic system with unknown parameters using adaptive control method. We also established new results for the global chaos synchronization of the identical novel 3-D chaotic systems with unknown parameters using adaptive control method. The main control results were proved using Lyapunov stability theory. MATLAB simulations have been shown to illustrate the phase portraits of the novel 3-D chaotic system and also the adaptive control results derived in this work.

References

1. Abdurrahman A, Jiang H, Teng Z (2015) Finite-time synchronization for memristor-based neural networks with time-varying delays. Neural Netw 69:20–28
2. Arneodo A, Coullet P, Tresser C (1981) Possible new strange attractors with spiral structure. Commun Math Phys 79(4):573–576
3. Azar AT, Vaidyanathan S (2015) Chaos modeling and control systems design, vol 581. Springer, Cham
4. Behnia S, Afrang S, Akhshani A, Mabhouti K (2013) A novel method for controlling chaos in external cavity semiconductor laser. Opt 124(8):757–764
5. Cai G, Tan Z (2007) Chaos synchronization of a new chaotic system via nonlinear control. J Uncertain Syst 1(3):235–240
6. Chen G, Ueta T (1999) Yet another chaotic attractor. Int J Bifurc Chaos 9(7):1465–1466
7. Islam MM, Murase K (2005) Chaotic dynamics of a behaviour-based miniature mobile robot: effects of environment and control structure. Neural Netw 18(2):123–144
8. Karthikeyan R, Sundarapandian V (2014) Hybrid chaos synchronization of four-scroll systems via active control. J Electr Eng 65(2):97–103
9. Khalil HK (2001) Nonlinear systems, 3rd edn. Prentice Hall, New Jersey
10. Li D (2008) A three-scroll chaotic attractor. Phys Lett A 372(4):387–393
11. Lorenz EN (1963) Deterministic periodic flow. J Atmos Sci 20(2):130–141
12. Lü J, Chen G (2002) A new chaotic attractor coined. Int J Bifurc Chaos 12(3):659–661
13. Matouk AE (2011) Chaos, feedback control and synchronization of a fractional-order modified autonomous Van der Pol-Duffing circuit. Commun Nonlinear Sci Numer Simul 16(2):975–986

14. Njah AN, Sunday OD (2009) Generalization on the chaos control of 4-D chaotic systems using recursive backstepping nonlinear controller. Chaos Solitons Fractals 41(5):2371–2376
15. Pehlivan I, Moroz IM, Vaidyanathan S (2014) Analysis, synchronization and circuit design of a novel butterfly attractor. J Sound Vib 333(20):5077–5096
16. Pham VT, Volos CK, Vaidyanathan S, Le TP, Vu VY (2015) A memristor-based hyperchaotic system with hidden attractors: Dynamics, synchronization and circuital emulating. J Eng Sci Technol Rev 8(2):205–214
17. Rasappan S, Vaidyanathan S (2013) Hybrid synchronization of n-scroll Chua circuits using adaptive backstepping control design with recursive feedback. Malays J Math Sci 73(1):73–95
18. Rasappan S, Vaidyanathan S (2014) Global chaos synchronization of WINDMI and coullet chaotic systems using adaptive backstepping control design. Kyungpook Math J 54(1): 293–320
19. Rhouma R, Belghith S (2011) Cryptanalysis of a chaos-based cryptosystem. Commun Nonlinear Sci Numer Simul 16(2):876–884
20. Rössler OE (1976) An equation for continuous chaos. Phys Lett A 57(5):397–398
21. Sampath S, Vaidyanathan S, Volos CK, Pham VT (2015) An eight-term novel four-scroll chaotic system with cubic nonlinearity and its circuit simulation. J Eng Sci Technol Rev 8(2):1–6
22. Sarasu P, Sundarapandian V (2011) Active controller design for generalized projective synchronization of four-scroll chaotic systems. Int J Syst Signal Control Eng Appl 4(2):26–33
23. Sarasu P, Sundarapandian V (2011) The generalized projective synchronization of hyperchaotic Lorenz and hyperchaotic Qi systems via active control. Int J Soft Comput 6(5): 216–223
24. Sarasu P, Sundarapandian V (2012) Generalized projective synchronization of two-scroll systems via adaptive control. Int J Soft Comput 7(4):146–156
25. Sprott JC (1994) Some simple chaotic flows. Phys Rev E 50(2):647–650
26. Sundarapandian V (2010) Output regulation of the Lorenz attractor. Far East J Math Sci 42(2):289–299
27. Sundarapandian V (2013) Adaptive control and synchronization design for the Lu-Xiao chaotic system. Lect Notes Electr Eng 131:319–327
28. Sundarapandian V (2013) Analysis and anti-synchronization of a novel chaotic system via active and adaptive controllers. J Eng Sci Technol Rev 6(4):45–52
29. Sundarapandian V, Karthikeyan R (2011) Anti-synchronization of hyperchaotic Lorenz and hyperchaotic Chen systems by adaptive control. Int J Syst Signal Control Eng Appl 4(2):18–25
30. Sundarapandian V, Karthikeyan R (2011) Anti-synchronization of Lü and Pan chaotic systems by adaptive nonlinear control. Eur J Sci Res 64(1):94–106
31. Sundarapandian V, Karthikeyan R (2012) Adaptive anti-synchronization of uncertain Tigan and Li systems. J Eng Appl Sci 7(1):45–52
32. Sundarapandian V, Pehlivan I (2012) Analysis, control, synchronization, and circuit design of a novel chaotic system. Math Comput Model 55(7–8):1904–1915
33. Sundarapandian V, Sivaperumal S (2011) Sliding controller design of hybrid synchronization of four-wing chaotic systems. Int J Soft Comput 6(5):224–231
34. Suresh R, Sundarapandian V (2013) Global chaos synchronization of a family of n-scroll hyperchaotic Chua circuits using backstepping control with recursive feedback. Far East J Math Sci 7(2):219–246
35. Tigan G, Opris D (2008) Analysis of a 3D chaotic system. Chaos, Solitons Fractals 36: 1315–1319
36. Tuwankotta JM (2006) Chaos in a coupled oscillators system with widely spaced frequencies and energy-preserving non-linearity. Int J Non-Linear Mech 41(2):180–191
37. Usama M, Khan MK, Alghathbar K, Lee C (2010) Chaos-based secure satellite imagery cryptosystem. Comput Math Appl 60(2):326–337
38. Vaidyanathan S (2011) Output regulation of Arneodo-Coullet chaotic system. Commun Comput Inf Sci 133:98–107

39. Vaidyanathan S (2011) Output regulation of the unified chaotic system. Commun Comput Inf Sci 198:1–9
40. Vaidyanathan S (2012) Adaptive controller and syncrhonizer design for the Qi-Chen chaotic system. Lect Notes Insti Compu Sci, Soc-Inf Telecommun Eng 84:73–82
41. Vaidyanathan S (2012) Anti-synchronization of Sprott-L and Sprott-M chaotic systems via adaptive control. Int J Control Theory Appl 5(1):41–59
42. Vaidyanathan S (2012) Global chaos control of hyperchaotic Liu system via sliding control method. Int J Control Theory Appl 5(2):117–123
43. Vaidyanathan S (2012) Sliding mode control based global chaos control of Liu-Liu-Liu-Su chaotic system. Int J Control Theory Appl 5(1):15–20
44. Vaidyanathan S (2013) A new six-term 3-D chaotic system with an exponential nonlinearity. Far East J Math Sci 79(1):135–143
45. Vaidyanathan S (2013) A ten-term novel 4-D hyperchaotic system with three quadratic nonlinearities and its control. Int J Control Theory Appl 6(2):97–109
46. Vaidyanathan S (2013) Analysis and adaptive synchronization of two novel chaotic systems with hyperbolic sinusoidal and cosinusoidal nonlinearity and unknown parameters. J Eng Sci Technol Rev 6(4):53–65
47. Vaidyanathan S (2013) Analysis, control and synchronization of hyperchaotic Zhou system via adaptive control. Adv Intell Syst Comput 177:1–10
48. Vaidyanathan S (2014) A new eight-term 3-D polynomial chaotic system with three quadratic nonlinearities. Far East J Math Sci 84(2):219–226
49. Vaidyanathan S (2014) Analysis and adaptive synchronization of eight-term 3-D polynomial chaotic systems with three quadratic nonlinearities. Eur Phys J: Spec Top 223(8):1519–1529
50. Vaidyanathan S (2014) Analysis, control and synchronisation of a six-term novel chaotic system with three quadratic nonlinearities. Int J Model, Identif Control 22(1):41–53
51. Vaidyanathan S (2014) Generalised projective synchronisation of novel 3-D chaotic systems with an exponential non-linearity via active and adaptive control. Int J Model, Identif Control 22(3):207–217
52. Vaidyanathan S (2014) Global chaos synchronisation of identical Li-Wu chaotic systems via sliding mode control. Int J Model, Identif Control 22(2):170–177
53. Vaidyanathan S (2014) Qualitative analysis and control of an eleven-term novel 4-D hyperchaotic system with two quadratic nonlinearities. Int J Control Theory Appl 7:35–47
54. Vaidyanathan S (2015) 3-cells Cellular Neural Network (CNN) attractor and its adaptive biological control. Int J PharmTech Res 8(4):632–640
55. Vaidyanathan S (2015) A 3-D novel highly chaotic system with four quadratic nonlinearities, its adaptive control and anti-synchronization with unknown parameters. J Eng Sci Technol Rev 8(2):106–115
56. Vaidyanathan S (2015) A novel chemical chaotic reactor system and its adaptive control. Int J ChemTech Res 8(7):146–158
57. Vaidyanathan S (2015) Adaptive backstepping control of enzymes-substrates system with ferroelectric behaviour in brain waves. Int J PharmTech Res 8(2):256–261
58. Vaidyanathan S (2015) Adaptive biological control of generalized Lotka-Volterra three-species biological system. Int J PharmTech Res 8(4):622–631
59. Vaidyanathan S (2015) Adaptive chaotic synchronization of enzymes-substrates system with ferroelectric behaviour in brain waves. Int J PharmTech Res 8(5):964–973
60. Vaidyanathan S (2015) Adaptive control of a chemical chaotic reactor. Int J PharmTech Res 8(3):377–382
61. Vaidyanathan S (2015) Adaptive control of the FitzHugh-Nagumo chaotic neuron model. Int J PharmTech Res 8(6):117–127
62. Vaidyanathan S (2015) Adaptive synchronization of chemical chaotic reactors. Int J ChemTech Res 8(2):612–621
63. Vaidyanathan S (2015) Adaptive synchronization of generalized Lotka-Volterra three-species biological systems. Int J PharmTech Res 8(5):928–937

64. Vaidyanathan S (2015) Adaptive synchronization of novel 3-D chemical chaotic reactor systems. Int J ChemTech Res 8(7):159–171
65. Vaidyanathan S (2015) Adaptive synchronization of the identical FitzHugh-Nagumo chaotic neuron models. Int J PharmTech Res 8(6):167–177
66. Vaidyanathan S (2015) Analysis, control and synchronization of a 3-D novel jerk chaotic system with two quadratic nonlinearities. Kyungpook Math J 55:563–586
67. Vaidyanathan S (2015) Analysis, properties and control of an eight-term 3-D chaotic system with an exponential nonlinearity. Int J Model, Identif Control 23(2):164–172
68. Vaidyanathan S (2015) Anti-synchronization of brusselator chemical reaction systems via adaptive control. Int J ChemTech Res 8(6):759–768
69. Vaidyanathan S (2015) Chaos in neurons and adaptive control of Birkhoff-Shaw strange chaotic attractor. Int J PharmTech Res 8(5):956–963
70. Vaidyanathan S (2015) Chaos in neurons and synchronization of Birkhoff-Shaw strange chaotic attractors via adaptive control. Int J PharmTech Res 8(6):1–11
71. Vaidyanathan S (2015) Coleman-Gomatam logarithmic competitive biology models and their ecological monitoring. Int J PharmTech Res 8(6):94–105
72. Vaidyanathan S (2015) Dynamics and control of brusselator chemical reaction. Int J ChemTech Res 8(6):740–749
73. Vaidyanathan S (2015) Dynamics and control of tokamak system with symmetric and magnetically confined plasma. Int J ChemTech Res 8(6):795–803
74. Vaidyanathan S (2015) Global chaos synchronization of chemical chaotic reactors via novel sliding mode control method. Int J ChemTech Res 8(7):209–221
75. Vaidyanathan S (2015) Global chaos synchronization of the forced Van der Pol chaotic oscillators via adaptive control method. Int J PharmTech Res 8(6):156–166
76. Vaidyanathan S (2015) Global chaos synchronization of the Lotka-Volterra biological systems with four competitive species via active control. Int J PharmTech Res 8(6):206–217
77. Vaidyanathan S (2015) Lotka-Volterra population biology models with negative feedback and their ecological monitoring. Int J PharmTech Res 8(5):974–981
78. Vaidyanathan S (2015) Lotka-Volterra two species competitive biology models and their ecological monitoring. Int J PharmTech Res 8(6):32–44
79. Vaidyanathan S (2015) Output regulation of the forced Van der Pol chaotic oscillator via adaptive control method. Int J PharmTech Res 8(6):106–116
80. Vaidyanathan S, Azar AT (2015) Analysis and control of a 4-D novel hyperchaotic system. Stud Comput Intell 581:3–17
81. Vaidyanathan S, Azar AT (2015) Analysis, control and synchronization of a nine-term 3-D novel chaotic system. In: Azar AT, Vaidyanathan S (eds) Chaos modelling and control systems design, studies in computational intelligence, vol 581. Springer, Berlin, pp 19–38
82. Vaidyanathan S, Madhavan K (2013) Analysis, adaptive control and synchronization of a seven-term novel 3-D chaotic system. Int J Control Theory Appl 6(2):121–137
83. Vaidyanathan S, Pakiriswamy S (2013) Generalized projective synchronization of six-term Sundarapandian chaotic systems by adaptive control. Int J Control Theory Appl 6(2):153–163
84. Vaidyanathan S, Pakiriswamy S (2015) A 3-D novel conservative chaotic system and its generalized projective synchronization via adaptive control. J Eng Sci Technol Rev 8(2): 52–60
85. Vaidyanathan S, Rajagopal K (2011) Hybrid synchronization of hyperchaotic Wang-Chen and hyperchaotic Lorenz systems by active non-linear control. Int J Syst Signal Control Eng Appl 4(3):55–61
86. Vaidyanathan S, Rajagopal K (2012) Global chaos synchronization of hyperchaotic Pang and hyperchaotic Wang systems via adaptive control. Int J Soft Comput 7(1):28–37
87. Vaidyanathan S, Rasappan S (2011) Global chaos synchronization of hyperchaotic Bao and Xu systems by active nonlinear control. Commun Comput Inf Sci 198:10–17
88. Vaidyanathan S, Rasappan S (2014) Global chaos synchronization of n-scroll Chua circuit and Lur'e system using backstepping control design with recursive feedback. Arabian J Sci Eng 39(4):3351–3364

89. Vaidyanathan S, Sampath S (2012) Anti-synchronization of four-wing chaotic systems via sliding mode control. Int J Autom Comput 9(3):274–279
90. Vaidyanathan S, Volos C (2015) Analysis and adaptive control of a novel 3-D conservative no-equilibrium chaotic system. Arch Control Sci 25(3):333–353
91. Vaidyanathan S, Volos C, Pham VT (2014) Hyperchaos, adaptive control and synchronization of a novel 5-D hyperchaotic system with three positive Lyapunov exponents and its SPICE implementation. Arch Control Sci 24(4):409–446
92. Vaidyanathan S, Volos C, Pham VT, Madhavan K, Idowu BA (2014) Adaptive backstepping control, synchronization and circuit simulation of a 3-D novel jerk chaotic system with two hyperbolic sinusoidal nonlinearities. Arch Control Sci 24(3):375–403
93. Vaidyanathan S, Azar AT, Rajagopal K, Alexander P (2015) Design and SPICE implementation of a 12-term novel hyperchaotic system and its synchronisation via active control. Int J Model, Identif Control 23(3):267–277
94. Vaidyanathan S, Idowu BA, Azar AT (2015) Backstepping controller design for the global chaos synchronization of Sprott's jerk systems. Stud Comput Intell 581:39–58
95. Vaidyanathan S, Rajagopal K, Volos CK, Kyprianidis IM, Stouboulos IN (2015) Analysis, adaptive control and synchronization of a seven-term novel 3-D chaotic system with three quadratic nonlinearities and its digital implementation in LabVIEW. J Eng Sci Technol Rev 8(2):130–141
96. Vaidyanathan S, Sampath S, Azar AT (2015) Global chaos synchronisation of identical chaotic systems via novel sliding mode control method and its application to Zhu system. Int J Model, Identif Control 23(1):92–100
97. Vaidyanathan S, Volos C, Pham VT, Madhavan K (2015) Analysis, adaptive control and synchronization of a novel 4-D hyperchaotic hyperjerk system and its SPICE implementation. Nonlinear Dyn 25(1):135–158
98. Vaidyanathan S, Volos CK, Kyprianidis IM, Stouboulos IN, Pham VT (2015) Analysis, adaptive control and anti-synchronization of a six-term novel jerk chaotic system with two exponential nonlinearities and its circuit simulation. J Eng Sci Technol Rev 8(2):24–36
99. Vaidyanathan S, Volos CK, Pham VT (2015) Analysis, adaptive control and adaptive synchronization of a nine-term novel 3-D chaotic system with four quadratic nonlinearities and its circuit simulation. J Eng Sci Technol Rev 8(2):181–191
100. Vaidyanathan S, Volos CK, Pham VT (2015) Analysis, control, synchronization and SPICE implementation of a novel 4-D hyperchaotic Rikitake dynamo system without equilibrium. J Eng Sci Technol Rev 8(2):232–244
101. Vaidyanathan S, Volos CK, Pham VT (2015) Global chaos control of a novel nine-term chaotic system via sliding mode control. In: Azar AT, Zhu Q (eds) Advances and applications in sliding mode control systems, studies in computational intelligence, vol 576. Springer, Berlin, pp 571–590
102. Vincent UE, Njah AN, Laoye JA (2007) Controlling chaos and deterministic directed transport in inertia ratchets using backstepping control. Phys D 231(2):130–136
103. Volos CK, Kyprianidis IM, Stouboulos IN, Anagnostopoulos AN (2009) Experimental study of the dynamic behavior of a double scroll circuit. J Appl Func Anal 4:703–711
104. Volos CK, Kyprianidis IM, Stouboulos IN (2013) Experimental investigation on coverage performance of a chaotic autonomous mobile robot. Robot Auton Syst 61(12):1314–1322
105. Volos CK, Kyprianidis IM, Stouboulos IN, Tlelo-Cuautle E, Vaidyanathan S (2015) Memristor: A new concept in synchronization of coupled neuromorphic circuits. J Eng Sci Technol Rev 8(2):157–173
106. Yang J, Zhu F (2013) Synchronization for chaotic systems and chaos-based secure communications via both reduced-order and step-by-step sliding mode observers. Commun Nonlinear Sci Numer Simul 18(4):926–937
107. Yang J, Chen Y, Zhu F (2014) Singular reduced-order observer-based synchronization for uncertain chaotic systems subject to channel disturbance and chaos-based secure communication. Appl Math Comput 229:227–238

108. Zhou W, Xu Y, Lu H, Pan L (2008) On dynamics analysis of a new chaotic attractor. Phys Lett A 372(36):5773–5777
109. Zhu C, Liu Y, Guo Y (2010) Theoretic and numerical study of a new chaotic system. Intell Inf Manag 2:104–109

MIX
Papier aus verantwortungsvollen Quellen
Paper from responsible sources
FSC® C105338

If you have any concerns about our products,
you can contact us on
ProductSafety@springernature.com

In case Publisher is established outside the EU,
the EU authorized representative is:
Springer Nature Customer Service Center GmbH
Europaplatz 3, 69115 Heidelberg, Germany

Printed by Libri Plureos GmbH
in Hamburg, Germany